PROBABILITY AND STATISTICS: THEORY AND APPLICATIONS

PROBABILITY AND STATISTICS: THEORY AND APPLICATIONS

D. A. S. FRASER
University of Toronto

DUXBURY PRESS
North Scituate, Massachusetts

Duxbury Press
A DIVISION OF WADSWORTH PUBLISHING COMPANY, INC.

Probability and Statistics: Theory and Applications was edited, designed, and prepared for composition by Service to Publishers, Inc. The cover was designed by Elwood Blankenship.

The cover design shows a radar display of air-traffic patterns in a metropolitan area.

L. C. Catalog Card No.: 75-11276
ISBN: 0-87872-099-5

PRINTED IN THE UNITED STATES OF AMERICA

1 2 3 4 5 6 7 8 9 10—80 79 78 77 76

PREFACE

WHY

In recent years important developments and trends have taken place in probability and statistics. Greatly increased attention is being given to the *analysis of data,* particularly to the development of inference methods that can be applied easily to data without restrictive models. The availability of desk and large-scale *computers* has opened up large areas of applications—the simple and direct completion of rather complicated analyses, on the one hand, and the numerical derivation of needed distributions and percentage points, on the other hand. The *likelihood function* is emerging as a basic means of assessing data and making inferences, greatly extending the range of standard methods. And in a related way the likelihood function *as it depends on the sample space* has strengthened theoretical methods and largely eclipsed the one-time prominence of sufficiency. The increased attention to the analysis of data has also highlighted the essential *unity of statistical inference,* with estimation, tests, and confidence regions appearing as closely related components. The availability of computers has also led to the development of *exact inference methods* for distribution forms other than the familiar normal, thus allowing proper attention to longer-tailed distributions. On the *instructional* side there is an increased need to organize introductory probability in a way that is adequate and appropriate to both the study of statistics and the continued study of probability. *Conditional probability* is receiving new attention and wider application; and on a larger scale there is an increasing need for the *clear organization* of basic probability concepts. Paralleling these important positive developments and trends there has been an unfortunate deemphasis—almost neglect—of basic *principles of experimentation,* no doubt caused in part by the broader attention that has been given to design and analysis details.

These developments and trends are important to all of probability and statistics, and particularly important to an introductory course in probability and statistics; for clearly they bear on almost all areas of theory and application. Existing texts do not reflect the important positive trends just described. For example, likelihood function analysis is almost entirely absent from existing texts; indeed, the likelihood function itself is incorrectly defined in most texts. And quite generally the texts omit or neglect reasonable contact with the very fundamental principles of experimentation.

This text was designed for a full-year course on probability and statistics, blending theory and application. The text was developed to bring together in an organized way the important positive trends just described and to focus on the basic principles of experimentation. The material was assembled to provide a broad coverage course in itself or a basic course for continued study in probability or in statistics.

The text presupposes a one-year course in calculus and assumes a concurrent second course in calculus introducing multiple integration and the elements of linear algebra.

WHAT

The likelihood function is used prominently. Its correct definition leads to current likelihood analysis methods of inference and it also provides by far the shortest route to the classical concepts of sufficiency and minimal sufficiency.

The normal distribution is treated as just one of many distribution forms that can be proposed for applications. Recent theoretical results and computer methods are used to develop inference results for other distribution forms.

Conditional probability is given special and detailed attention. Recent examinations of conditional probability affirm it as the most basic concept of probability and statistics; it is *probability* itself, in applications.

The three methods of inference—estimation, significance tests, and confidence intervals—are introduced together as unified components of statistical inference. The introduction is made in an applied inference context using *symmetry* and *large samples,* two basic tools from probability theory; the usual formal theories are developed later.

The organization is in terms of concepts—concepts of probability and concepts of statistics. The notation is flexible, suitable for the informality of data analysis and for the formality of probability theory. An underlying theme is the basic goal of science—experimentation and the determination of cause–effect relations.

A large number of illustrations and geometrical pictures are used to clarify, illustrate, and give insight. In particular, plots are included for all the common distributions, both regular distributions and derived central and noncentral distributions. Thus some feeling for the form of a distribution accompanies the functional expression of the distribution.

Special attention is given to a broad selection of examples and problems covering theory, applications, data analysis, and computer usage. In addition, most chapters have a supplemental section that elaborates on theoretical points of interest, or provides the background material for references within chapters, or extends some interesting point or development within a chapter. These supplements are assembled at the back of the book; they can be used as supplemental reading for the particularly concerned student or as part of the basic material for a much more detailed and intensive course.

HOW

The material in this text is organized around the basic concepts of probability and statistics. These concepts are introduced section by section, chapter by chapter, building a solid, integrated framework for probability and for statistics. For example, the binomial distribution for independent Bernoulli trials is derived in Chapter 3, after the discussion of *marginal distributions* and after the treatment of *independence.* Probability and statistics have a logical framework and the concepts are introduced in a logical progression.

For ***probability theory,*** the most fundamental concepts are those of *event* and *probability.* Thus in Chapter 1 probabilities are introduced on a class of events; this introduction is, of course, immediately paralleled in terms of sample spaces. The initial introduction involving just probabilities and events has the great advantage of allowing events to be treated later as primary entities. Various sample spaces are then easily examined, sample spaces suggested by convenience perhaps or by utility; and many basic results are seen trivially to be independent of the supporting sample space.

Chapter 2 is concerned with probability on real spaces. Again the essential concepts are those of event and probability. At this stage there is no need for *functions*—or *random variables*—for they deflect attention from the basic ingredients, the probabilities and the events.

Chapter 3 is concerned with marginal probability, with the use of functions to map into a simpler space or to restrict attention to a smaller class of events. Here the capital-letter notation X is first used—where functions are really involved. Earlier sections use the quite flexible lowercase letter notation x, where of course the essential use of x is to form events $\{x : x \leq 3\}$ on a sample space. The dual notation is consistent and it emphasizes that the function notation serves to refer back to an earlier sample space; for example, $\{X \leq 3\}$ is an abbreviation for $\{s : X(s) \leq 3\}$, where the lowercase letter s is used for events on the earlier sample space. Indeed, the capital-letter notation should be used with extreme caution. For with conditional probability involving, say, normal(μ, 1) "random variables" $X_1, \ldots, X_n$, it *does* matter what the functions $X_1, \ldots, X_n$ are—different results with different functions. Accordingly, we avoid such "random variable" functions unless the functions are given explicitly.

Conditional probability in theory and in applications is a primary but often neglected concept. Chapter 4 focuses on the definition of conditional probability and its importance in applications. A selection model is used in Section 4.1; it gives a precise concrete interpretation for conditional probabilities.

Various aspects of means and variances are presented in Chapter 5. The development is organized so that most formulas are immediately available in multivariate form—no change in essential form, just the substitution of vectors for reals. Section 5.4 includes a wide range of problems recording means and variances for all the common distributions. An important section, 5.5, is included covering conditional means and variances.

Some results on limiting distributions and limiting functions are surveyed in Chapter 6.

For ***statistics*** the central theme is statistical inference leading to experimental design and the determination of cause–effect relations. In Chapter 7 statistical inference is introduced in the context of applications. This introduction uses the *symmetry* and *large-sample* results from the chapters on probability theory. And the applied context allows an easy and informal first contact with estimation, tests of significance, and confidence intervals—three closely related aspects of statistical inference; examples are taken from applications.

The more formal development of statistics in Chapter 8 is organized around the concepts of *likelihood.* On the applied side we obtain the *likelihood function* as a means of directly assessing data; the direct use of likelihood for statistical inference is a recent trend in statistics and it has great flexibility for applications. The likelihood function is introduced in Section 8.1B for data on the lifetime of crucibles used in an electrolytic reduction process. On the theoretical side we obtain the *mapping* that produces likelihood functions, called the *likelihood statistic;* this concept replaces the classical concept of a minimal sufficient statistic and has greater justifications for its use. Normal distribution theory and exponential models are introduced at this stage;

and as part of this we obtain the exact distributions for inference based on the normal model.

In Chapter 9 estimation theory is introduced, initially in an applied context of combining available estimates; a surveyor wishing to adjust data is used as an illustration in Section 9.1. This combining of estimates is generalized progressively and leads directly to the Gauss–Markov theorem itself. Then a full statistical model is introduced and the standard estimation results are obtained involving maximum likelihood, the information inequality, and the use of sufficiency and completeness.

In Chapter 10 hypothesis testing theory is developed; this includes the fundamental lemma, uniformly most powerful tests, and chi-square tests for qualitative data. Supplemental sections examine UMP unbiased and invariant tests and some formalities of confidence region theory.

Simple and general linear models are examined in Chapter 11, not just for normal error but for arbitrary error forms. For the location and scale model a computer program is available for the distribution of the t-statistic and the distribution of the standard deviation. The common distribution theory for quadratic forms is special to the normal error model; the development in this chapter gives this distribution theory as a special case but also provides inference methods more generally.

The concluding Chapter 12 surveys the basic principles of experimentation and introduces factorial designs; randomized blocks and covariance analysis are included in the supplemental sections. An adequate introduction to statistics must examine the elements of experimental design, as these elements comprise the fundamentals of science itself.

Answers to selected exercises and problems are recorded at the back of the book. A solutions manual for instructors, prepared by Gordon Fick with my participation, is available to instructors through the publisher.

A computer program for the inference methods in Chapter 11 has been developed by Gordon Fick. For copies write to: Statistics Section, Att.: Gordon Fick, Mathematics Department, University of Toronto, Toronto, Canada, M5S 1A1 (a card deck is available for a small service charge).

ACKNOWLEDGMENTS

I wish to thank the many students who have participated in the development of the present material through several revisions over a six-year period. In particular, Lane Bishop, David Brenner, Michael Evans, Gordon Fick, Georges Monette, Kai W. Ng, and Allan Wilks have worked very closely on the final revision and assisted immensely with the intricacies inherent in the proof stage.

My deep appreciation goes to Gordon Fick for his help and advice during the development of the book, for many fruitful suggestions during proofing, and for his central involvement with the solutions manual and the computer program for the nonnormal inference methods. My deep appreciation also goes to Frances Mitchell for her very careful, patient, and concerned typing of the manuscript and the solutions manual.

D. A. S. FRASER

CONTENTS

PREFACE v

CHAPTER 1 PROBABILITY 1

1.1 Introduction / 1
1.2 Events / 9
1.3 Probabilities / 19
1.4 Calculating Probabilities / 28

CHAPTER 2 PROBABILITY ON THE LINE AND PLANE 40

2.1 Probability on the Real Line: Discrete / 40
2.2 Probability on the Real Line: Absolutely Continuous / 51
2.3 Probability on the Plane / 69
2.4 Change of Variables / 80

CHAPTER 3 MARGINAL PROBABILITY 98

3.1 Data Reduction / 98
3.2 Marginal Probability on the Line and Plane / 105
3.3 Statistical Independence / 119
3.4 Counting Sample Points / 128
3.5 Marginal Distributions for Frequencies / 136

CHAPTER 4 CONDITIONAL PROBABILITY 152

4.1 Definition and Interpretation / 152
4.2 Conditional Probability on the Plane / 163

CHAPTER 5 MEAN VALUE FOR REAL AND VECTOR DISTRIBUTIONS 180

5.1 Definition of the Mean / 180
5.2 The Location of a Distribution / 186
5.3 The Scaling of a Distribution / 193
5.4 The Location and Scaling of a Sample / 207
5.5 Conditional Location and Scaling / 217
5.6 Moments of a Distribution / 224
5.7 Moment-Generating and Characteristic Functions / 230

CHAPTER 6 LIMITING DISTRIBUTIONS AND LIMITING FUNCTIONS 244

6.1 Limiting Distributions / 244
6.2 The Weak Law of Large Numbers and The Central Limit Theorem / 254

CHAPTER 7 STATISTICAL INFERENCE 268

7.1 The Statistical Model and the Data / 268
7.2 Inference from Large Samples / 275
7.3 Inference for Finite Populations / 286
7.4 Inference from Symmetry / 292

CHAPTER 8 THE LIKELIHOOD FUNCTION IN STATISTICAL INFERENCE 308

8.1 The Observed Likelihood Function / 308
8.2 Distribution Theory for the Normal Model / 319
8.3 Statistical Inference for the Normal Model / 326
8.4 An Important Statistic / 332
8.5 Models That Have Exponential Form / 341
8.6 Inference from Large Samples Using Likelihood / 348

CHAPTER 9 ESTIMATION 362

9.1 Some Traditional Methods / 362
9.2 Linear Least Squares / 370
9.3 Combining Unbiased Estimators / 379
9.4 The Information Inequality for Unbiased Estimators / 389
9.5 Likelihood and Completeness for Unbiased Estimators / 398

CHAPTER 10 TESTING STATISTICAL HYPOTHESES 410

10.1 The Fundamental Lemma / 410
10.2 Composite Hypotheses / 421
10.3 Some Other Methods in Hypothesis Testing / 430
10.4 Testing a Statistical Model / 442

CHAPTER 11 LINEAR MODELS 456

11.1 The Location-Scale Models / 456
11.2 Calculating Projections and Residuals / 468
11.3 The Regression Model / 476
11.4 Inference with the Regression Model / 482

CHAPTER 12 THE DESIGN OF EXPERIMENTS 490

12.1 One-Factor Design / 490
12.2 Two-Factor Design / 501

APPENDIX I SUPPLEMENTARY MATERIAL **511**

1.5 Probability / 511
2.5 Probability on the Line and Plane / 520
3.6 Marginal Probability / 526
4.3 Conditional Probability / 537
5.8 Mean Value / 544
6.3 Limiting Functions / 554
7.5 Testing Randomness / 563
9.6 Bayes Estimation / 572
10.5 Confidence Intervals and Confidence Regions / 579
12.3 Three-Factor Design / 584
12.4 Randomized Blocks / 590
12.5 Analysis of Covariance / 594

APPENDIX II TABLES **599**

ANSWERS **605**

INDEX **615**

1

PROBABILITY

1.1

INTRODUCTION

In the past, most scientific investigations, particularly quantitative investigations, have been concerned with *deterministic relationships:* for example, with the extension of a spring as determined by the force applied, or the pressure in a flask containing a gas as determined by the temperature, or the rate of a chemical process as determined by the amount of a catalyst, or the height reached by a projectile as determined by the force applied, or *a response variable as a function of an input variable*.

A very close investigation of such systems, however, shows that results do not, in fact, reproduce exactly. That is, for any prescribed initial conditions, the results are found to vary from repetition to repetition by small amounts. This *variation* can derive from many sources of variation—from error in the measurement of the response variable, from variation in the initial materials and conditions (although kept constant as nearly as possible), and from variation in the internal operation of the system being investigated.

More recently, quantitative investigations have become prominent in the biological and social sciences. In these areas the relationships are no longer deterministic in quite the same way. The almost negligible variation found in the traditional physical system is replaced by a very prominent variation that can nearly conceal response effects that come from changes in the input variables. And the moderate need in the ''deterministic'' context for an adequate description of variation becomes an overwhelming need in these new areas.

Probability theory provides *the description for variation in systems operated under essentially constant conditions*. In addition, it can be combined with deterministic theory to provide the description for variation in systems operated under conditions that are changed by design. Probability theory is a *descriptive* part of science, describing variation; it is thus applicable in essentially all areas of scientific endeavor. Probability theory is developed and discussed in Chapters 1 through 6.

Variation is found in the full range of physical, biological, and social systems. In one direction we have the system of tossing a coin and recording a specified characteristic of the result. This is an example of *games of chance,* which involve coins, dice, cards, and roulette wheels. Such games of chance represent the historical origins of probability theory when prominent mathematicians assisted French gamblers in the seventeenth century. In the other direction, that of scientific areas, we have, for example, the measurement of a reaction time of a particular kind of animal in relation to treatment by a drug. In probability theory the games of chance provide a wide spectrum of examples in which input conditions are relatively easily controlled and kept constant; these examples are used freely in the chapters on probability theory. The more scientific examples are involved with statistical theory and are largely withheld until later chapters on that theory. Sometimes illustrations are chosen from everyday activities (e.g., meeting buses at a terminal, selecting from a menu at a restaurant, or wrapping Christmas packages). We try to avoid such illustrations, as they seem to promote casual application without attention to conditions needed for validity.

Statistical theory is developed and discussed in Chapters 7 through 12. This theory builds on the use of probability to describe variation and is concerned with operating the system, obtaining observations on the system, and drawing conclusions concerning unknowns of the system; most important, it is concerned with designing input conditions so as to obtain conclusions concerning input–response and cause–effect relationships. Statistical theory is a *prescriptive* part of science, prescribing the scientist's method in many areas of scientific endeavor. There will be more on this beginning in Chapter 7.

For the study of variation we must control and hold constant the various input variables or factors affecting the system under investigation. We must then decide what is to be recorded concerning the result of a performance of the system. For the coin-tossing example we could record whether the event "heads" H occurs; or we could record whether the event "tails" T occurs. And for the reaction-time example we could record whether the reaction time lies in the interval $[2.5, 3.5)$. Then for any chosen event or characteristic, we can examine a large number of independent repetitions of the system and record the proportion of occurrences of the event in those repetitions; for the event H the proportion might be 0.53. Of course, with another large number of repetitions, the proportion may be slightly different. But it is a physical phenomenon that the proportion of occurrences for the event tends to a limit as the number of repetitions increases indefinitely; the limit is called the *probability of the event*. A system with this property is called a stable system or a *random system*. In cases where the proportion behaves systematically or does not tend to a limit, experience indicates that the input variables have not been controlled and kept constant; in other words, the system is not random.

The probability of an event is viewed as a property of a performance in the future or of a concealed performance in the past. The interpretation for such a probability is in terms of the long-run proportion of occurrences.

Consider this in more detail.

A EVENTS

Events are the primary elements of probability, statistics—indeed of science itself; as an example, consider the event "recovery" in the pharmaceutical investigation of a new antibiotic. As a simple example involving a random system, we will examine the rolling of two dice. An investigator may be interested in the event "double 1"; let D designate this event. Or he may be interested in the event "the dice are less than 3 inches from each other"; let L designate this event. Or he may be interested in the event "a particular die is within 10° of being lined up with the rectangular table"; let W designate this event.

We can think of an event as a *set of descriptions* for the response or outcome on a performance of the system. Or we can think of an event as a *proposition* concerning the outcome. Or we can think of an event in the everyday sense of a *kind of event*.

Thus an event can *occur* or *not occur* on a performance of the system; and, correspondingly, the set of descriptions can hold or not hold, and the proposition can be true or false. We use capital letters to designate events, for example A, B, C, ...

Consider some *operations* on events:

1. The event "A intersect B," designated $A \wedge B$ or AB, is the event "both A and B." For example, $D \wedge L$ is the event "two 1s less than 3 inches from each other." The operation intersection can also be called *meet* or *conjunction*.
2. The event "A union B," designated $A \vee B$, is the event "either A or B (or both)." For example, $D \vee L$ is the event "double 1 or the dice are less than 3 inches from each other." The operation union can also be called *join* or *disjunction*.
3. The event "A complement," designated A^c, is the event "not A." For example, L^c is the event "the dice are at least 3 inches from each other." The operation *complementation* can also be called *negation*.

As a special event, let $\varnothing$ designate the *null event;* this handles, for example, the event "$A \wedge A^c$," which *cannot occur*. And let $\mathcal{U}$ designate the *universal event;* this handles, for example, the event "$A \vee A^c$," which *necessarily occurs*.

Consider some *definitions* for events:

1. The events A and B are *disjoint* if $A \wedge B = \varnothing$, that is, if they cannot occur simultaneously. Disjoint events can also be called *mutually exclusive* events. For example, A and A^c are disjoint.
2. The event A is *contained in* the event B, designated $A \subset B$, if $A \wedge B = A$. This can be interpreted as: the occurrence of A implies the occurrence of B.
3. The class of events $\{B_1, \ldots, B_k\}$ is a *partition* of the event A if $A = B_1 \vee B_2 \vee \cdots \vee B_k$ and if events B_i and B_j with $i \neq j$ are disjoint. For example, $\{DL, DL^c\}$ is a partition of the event D, "double 1."

B EXAMPLE

Consider the tossing of a certain first coin; we shall consider a second coin later. We might be interested in a variety of events: a certain face is upward, the coin is within a certain distance of the edge of the table, the orientation of the coin as determined by the upward face is within a certain angle of the directions established by the edge of the table. Suppose that we are interested in the event H_1, "heads for this coin."

Consider the operations applied to the event H_1. If we ignore the possibility that the coin stands on edge, then H_1^c is the event, say T_1, "tails for this coin." The event $H_1 \wedge T_1 = \varnothing$ is the null event, and $H_1 \vee T_1 = \mathcal{U}$ is the universal event. Thus, accepting the operations, we become interested effectively in the class $\mathcal{B} = \{\varnothing, H_1, T_1, \mathcal{U}\}$ of events. Note from an elementary point of view that we have produced a partition of the universal event:

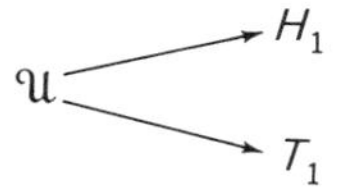

The coin was tossed 100 times and for each event of interest a record was kept of whether or not it occurred on each repetition. For this it is convenient to let 1 designate

TABLE 1.1
Recording the event A

Event A	Indicator $I_A(R_1)$	$I_A(R_2)$	$\cdots$	$I_A(R_{100})$	Frequency $\sum_1^{100} I_A(R_i)$	Proportion $\hat{P}(A) = \sum_1^{100} I_A(R_i)/100$
$\varnothing$	0	0		0	0	0
H_1	1	0		1	53	0.53
T_1	0	1		0	47	0.47
$\mathfrak{U}$	1	1		1	100	1.00

occurrence and 0 designate nonoccurrence. The record for an event A is then the succession of values of an *indicator function* I_A: thus $I_A(R_i)$ is equal to 1 or 0 according as the event A occurs or does not occur on the ith repetition R_i. For the succession of 100 performances we obtained heads, tails, tails, . . ., heads; the tabulation is as given in Table 1.1.

For any event A the total $\Sigma\, I_A(R_i)$ is the number of occurrences or the *frequency* of A in the 100 repetitions; it is a measure of the event A, call it the *weight of occurrence* of A. The average $\hat{P}(A) = \Sigma\, I_A(R_i)/100$ is the *proportion of occurrences* of A in the 100 repetitions; it is a measure of the event A, call it the *estimated probability* of A.

If we were to examine another 100 repetitions of the system, we would find typically that the estimated probabilities $\hat{P}(H_1)$ and $\hat{P}(T_1)$ were slightly different. Necessarily, of course, $\hat{P}(\varnothing) = 0$ and $\hat{P}(\mathfrak{U}) = 1$.

For a larger batch of repetitions, say 1000, we could calculate estimated probabilities in a similar way. And for different batches of 1000 we would find typically that the estimated probabilities tend to vary less than for batches of 100.

As the number of repetitions is increased indefinitely, we would observe the physical phenomenon that the estimated probability for an event A approaches a limiting value, say $P(A)$; this is a measure of the event A, the long-run measure of A, call it the *probability* or *true probability* of A.

Of course, beyond a certain point the coin may show appreciable wear, the initial conditions for the tossing of the coin would be different, and we would expect the probability of the event to be slightly changed.

Now suppose that the coin is symmetric, one face to the other, except for very delicate labeling that indicates heads for one face and tails for the other face. And suppose that the tossing of the coin is done carefully to ensure dynamic symmetry between the two faces. Then we do not need the large number of repetitions. For the symmetry between the two faces implies symmetry between the true measures for the two faces: $P(H_1) = P(T_1)$. Of course, the sum of the proportions for the two faces is always 1. Thus $P(H_1) + P(T_1) = 1$, and it follows that

$$\bar{P}(\varnothing) = 0, \qquad \bar{P}(H_1) = 1/2 = \bar{P}(T_1), \qquad \bar{P}(\mathfrak{U}) = 1;$$

note that we are using a bar over P to indicate a probability evaluated from symmetry. This gives us an approximation to the true probabilities, to the degree that symmetry represents the coin tossing.

Now consider the model that we have constructed for the variation in the coin-tossing system. For the events of interest we have a class

$$\mathcal{B} = \{\varnothing, H_1, T_1, \mathcal{U}\},$$

which is closed under the standard operations $\wedge$, $\vee$, and ^{c} on events. And for the long-run evaluation of the events in $\mathcal{B}$ we have a measure or function P defined on $\mathcal{B}$. From the 100 performances, we have the probability measure P given by

$\mathcal{B}$	$\varnothing$	H_1	T_1	$\mathcal{U}$
P	0	0.53	0.47	1.00

as an approximation to the true measure; and from symmetry we have the probability measure P given by

$\mathcal{B}$	$\varnothing$	H_1	T_1	$\mathcal{U}$
P	0	0.5	0.5	1.00

as an approximation to the true measure.

In summary: we have constructed a model $(\mathcal{B}, P)$, where $\mathcal{B}$ is an appropriate class of events and P is the measure that gives the probability for each of the events in $\mathcal{B}$. In an application one or other of the two approximations might be used for P—with appropriate cautions.

C EXAMPLE (*continuation*)

Consider the tossing of the preceding coin together with a second coin. And suppose that we are interested in the event H_1, "heads for the first coin," and the event H_2, "heads for the second coin." We are now *observing in greater depth*.

The standard operations applied to the events H_1 and H_2 generate 16 events altogether. For example, $H_1^c = T_1$, $H_2^c = T_2$, say; and, for example, $(H_1H_2)^c$ is the event "at least one tail." The various events can be labeled in a simple manner and assembled as the class $\mathcal{B}_2$:

$$\mathcal{B}_2 = \{\underline{\varnothing}, H_1H_2, H_1T_2, T_1H_2, T_1T_2, \underline{H_1}, \underline{T_1}, H_2, T_2,$$
$$H_1H_2 \vee T_1T_2, H_1T_2 \vee T_1H_2, (H_1H_2)^c, (H_1T_2)^c, (T_1H_2)^c, (T_1T_2)^c, \underline{\mathcal{U}}\}$$

The events in the original class $\mathcal{B}$ have been underlined. Thus note that the new class, $\mathcal{B}_2$, is an *enlargement* of $\mathcal{B}$; that is, $\mathcal{B} \subset \mathcal{B}_2$. Also note from the point of view of partitions that we have a finer partition of the universal event:

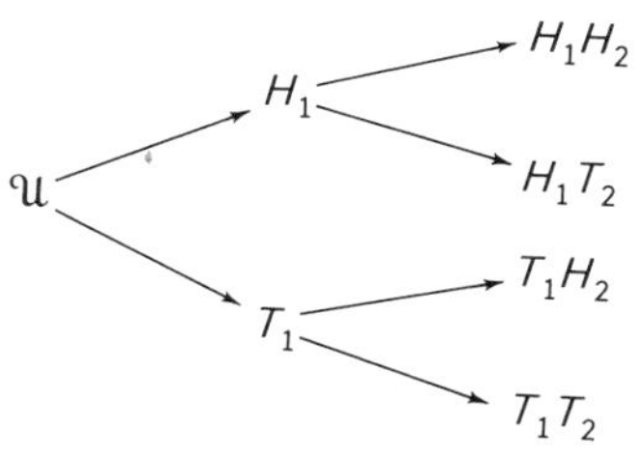

For some later considerations, three of the events in $\mathcal{B}_2$ are of somewhat special interest and we give them alternative designations:

$$E_0 = T_1T_2, \qquad E_1 = H_1T_2 \vee T_1H_2, \qquad E_2 = H_1H_2.$$

Note that E_j is the event "exactly j heads for the two coins."

The pair of coins was tossed 100 times, in fact, the 100 tosses discussed earlier when we mentioned only the first coin. The outcomes, with the increased description, were heads–heads, tails–heads, . . ., heads–tails, giving the tabulation shown in Table 1.2. Note that it would suffice to tabulate the four elementary events H_1H_2, H_1T_2, T_1H_2, T_1T_2. For there is *additivity* such that for any other event we can obtain the indicator, the frequency, or the proportion by adding the corresponding values for component elementary events. Thus $H_1H_2 \vee T_1T_2$ has disjoint components H_1H_2 and T_1T_2 and $\widehat{P}(H_1H_2 \vee T_1T_2) = \widehat{P}(H_1H_2) + \widehat{P}(T_1T_2)$. Also note that $\widehat{P}$ here is an *extension* of $\widehat{P}$ in the earlier example; the earlier values of $\widehat{P}$ are underlined.

We have now constructed the model $(\mathcal{B}_2, P)$ for the coin-tossing system. The class $\mathcal{B}_2$ is an appropriate class of events, an enlargement of the original class $\mathcal{B}$. And P is the measure that gives the probability for each of the events in $\mathcal{B}_2$; it is an extension of the original P to the enlarged class of events. From the 100 performances we have the measure P given in Table 1.3.

Now suppose that each coin is symmetric, one face to the other, except for delicate labeling that indicates heads for one face and tails for the other face. And suppose that the tossing is done carefully to ensure dynamic symmetry. Then we do not need the large number of repetitions. For the symmetry among the four events H_1H_2, H_1T_2, T_1H_2, T_1T_2 gives $P(H_1H_2) = P(H_1T_2) = P(T_1H_2) = P(T_1T_2)$; and the sum of the proportions for the four events is 1. Accordingly,

TABLE 1.2
Recording the event A

Event A	Indicator $I_A(R_1)$	$I_A(R_2)$	$\cdots$	$I_A(R_{100})$	Frequency $\sum_1^{100} I_A(R_i)$	Proportion $\widehat{P}(A) = \sum_1^{100} I_A(R_i)/100$
$\underline{\varnothing}$	0	0		0	0	$\underline{0}$
$H_1H_2 = E_2$	1	0		0	25	0.25
H_1T_2	0	0		1	28	0.28
T_1H_2	0	1		0	23	0.23
$T_1T_2 = E_0$	0	0		0	24	0.24
$\underline{H_1}$	1	0		1	53	$\underline{0.53}$
$\underline{T_1}$	0	1		0	47	$\underline{0.47}$
H_2	1	1		0	48	0.48
T_2	0	0		1	52	0.52
$H_1H_2 \vee T_1T_2$	1	0		0	49	0.49
$H_1T_2 \vee T_1H_2 = E_1$	0	1		1	51	0.51
$(H_1H_2)^c$	0	1		1	75	0.75
$(H_1T_2)^c$	1	1		0	72	0.72
$(T_1H_2)^c$	1	0		1	77	0.77
$(T_1T_2)^c$	1	1		1	76	0.76
$\underline{\mathcal{U}}$	1	1		1	100	$\underline{1.00}$

TABLE 1.3
Measure P obtained from 100 performances

$\mathcal{B}_2$	P	$\mathcal{B}_2$	P	$\mathcal{B}_2$	P
$\varnothing$	0	H_1	0.53	$(H_1H_2)^c$	0.75
H_1H_2	0.25	T_1	0.47	$(H_1T_2)^c$	0.72
H_1T_2	0.28	H_2	0.48	$(T_1H_2)^c$	0.77
T_1H_2	0.23	T_2	0.52	$(T_1T_2)^c$	0.76
T_1T_2	0.24	$H_1H_2 \vee T_1T_2$	0.49	$\mathcal{U}$	1.00
		$H_1T_2 \vee T_1H_2$	0.51		

$$\bar{P}(H_1H_2) = \bar{P}(H_1T_2) = \bar{P}(T_1H_2) = \bar{P}(T_1T_2) = 0.25;$$

and the additivity mentioned above gives the probabilities for the other events. Thus from symmetry we obtain the measure P given in Table 1.4.

TABLE 1.4
Measure P obtained from symmetry

$\mathcal{B}_2$	P	$\mathcal{B}_2$	P	$\mathcal{B}_2$	P
$\varnothing$	0	H_1	1/2	$(H_1H_2)^c$	3/4
H_1H_2	1/4	T_1	1/2	$(H_1T_2)^c$	3/4
H_1T_2	1/4	H_2	1/2	$(T_1H_2)^c$	3/4
T_1H_2	1/4	T_2	1/2	$(T_1T_2)^c$	3/4
T_1T_2	1/4	$H_1H_2 \vee T_1T_2$	1/2	$\mathcal{U}$	1
		$H_1T_2 \vee T_1H_2$	1/2		

D COMMENTS

For our probability model we have chosen events as the basic entities; events are examined formally in Section 1.2. Alternatively, we could have chosen a labeling for the various possible outcomes at some degree or depth of observation. However, we would then have had to change the outcome description in going from the example in Section B to that in Section C.

The choice of events as the basic entities is slightly more complicated, but it does pay some rich dividends. For example, many results in later chapters are available immediately in an invariant form, invariant of the degree or depth of labeling of the outcomes of performances; we thus avoid the need for direct proofs of the invariance. Labeling, however, is a *convenience* and we shall introduce it in Section 1.2 as an easy device for presenting events.

E EXERCISES

1 Consider the tossing of a coin as described in Section B but now allow for the rare event E, "the coin stands on edge."

(a) Record the class $\mathcal{B}'$ of events generated from H, T, and E by the standard operations; it has eight elements and the operations can be used to form convenient designations.
(b) Tabulate the estimated probability measure $\widehat{P}$ for the events in $\mathcal{B}'$; use the data in Section B.

2 A water sample can be described as pure, P (no coliform bacteria); tolerable, T (from 1 to 10 coliform bacteria); or unsafe, U (more than 10 coliform bacteria). Present the 8 possible events obtained by the standard operations; give a brief description of each.

3 A lot containing 1000 trailer wheel bearings can be described as premium, P (no defective bearings); tolerable, T (from 1 to 5 defective bearings); or unsatisfactory, U (more than 5 defective bearings). Present the 8 possible events obtained by the standard operations; give a brief description of each.

4 The offspring of a certain mating of mice can be described as male, M; female, F; black fur, B; brown fur, R, where $\{B, R\}$ is a partition of the universal event. Record the class $\mathcal{B}$ of 16 events generated from the given events.

F PROBLEMS

5 A water sample can be described as pure, P (no coliform bacteria); safe, S (from 0 to 10 coliform bacteria); or unsafe, U (more than 10 coliform bacteria); compare with Exercise 2. For each of the 8 events as determined in Exercise 2, record a simple designation in terms of the events P, S, and U.

6 The binary numbers 0 and 1 have been used for the indicator functions in the tabulations of results in Sections B and C. The binary numbers can also be used to form alternative labels for events in relation to the "smallest," or *elementary,* events. Consider Exercise 1. The label $(1, 0, 0)$ can be used for H, $(0, 1, 0)$ for T, and $(0, 0, 1)$ for E; and then, for example, $(1, 1, 0)$ for $H \vee T$, and so on, where the three positions correspond to the three elementary events and 1(0) corresponds to the presence (absence) of the elementary event as a component.

(a) In terms of this labeling, show that

$$\mathcal{B}' = \{(i_1, i_2, i_3) : i_j = 0, 1, j = 1, 2, 3\}.$$

(b) Verify that

$$(i_1, i_2, i_3) \wedge (i'_1, i'_2, i'_3) = (\min(i_1, i'_1), \min(i_2, i'_2), \min(i_3, i'_3)),$$

$$(i_1, i_2, i_2) \vee (i'_1, i'_2, i'_3) = (\max(i_1, i'_1), \max(i_2, i'_2), \max(i_3, i'_3)),$$

$$(i_1, i_2, i_3)^c = (1 - i_1, 1 - i_2, 1 - i_3).$$

7 Let $\mathcal{A} = \{A_1, \ldots, A_k\}$ be a partition of the universal event $\mathcal{U}$ into k different nonnull events. Let $\mathcal{B}$ be the set of events generated from $\mathcal{A}$ by the standard operations; then in the manner used in Problem 6, $\mathcal{B}$ can be presented in the following form:

$$\mathcal{B} = \{(i_1, \ldots, i_k) : i_j = 0, 1, j = 1, \ldots, k\}.$$

Determine the operations $\wedge$, $\vee$, and ^{c} in terms of this representation.

8 Let $\mathcal{A} = \{A_1, A_2\}$ be a class of events for the outcome of a certain system. And let $\mathcal{B}$ be the set of events generated from $\mathcal{A}$ by the standard operations. Then it follows easily that the number of elements in $\mathcal{B}$ (the *cardinality* of $\mathcal{B}$) is either 2, 4, 8, or 16. For each of these values give necessary and sufficient conditions in terms of the events A_1A_2, $A_1A_2^c$, $A_1^cA_2$, and $A_1^cA_2^c$.

1.2

EVENTS

We now consider *events* in more detail. There is, of course, an easy way of doing this—by means of sets on a sample space—and we examine this easy way in Sections B and C. The easy way, however, is at a price—complications arise whenever we change our depth of observation of the real-world system under investigation. To rise above these complications we first examine, in Section A, events as basic entities of the mathematical model. And, indeed, this is worthwhile. For, later, we shall have the substantial advantage of knowing that various results for events and probabilities are independent of the sample space used.

We examine *probabilities* in Section 1.3, and some rules for calculating probabilities in Section 1.4.

A ALGEBRA OF EVENTS

The standard operations $\wedge$, $\vee$, and ^{c} are almost trivial for the events in the examples of Sections 1.1B and 1.1C. We now need these operations for more general situations and we must examine the basic rules for them: for example, $A \vee B = B \vee A$.

Consider a random system, and let $\mathcal{B}$ be the class of events of interest. We now examine properties that go naturally with such a class $\mathcal{B}$.

We have interpreted an event as a set of descriptions or as a proposition concerning the outcome of the system being investigated. And we have presented the operations $\wedge$, $\vee$, and ^{c} as logical operations on these propositions. Some of the basic rules for these logical operations are now assembled. They do form a somewhat unwieldy collection, but fortunately our subsequent analysis will not need to work directly with them. Rather, they are assembled here to crystallize the formal framework that we are working in.

The rules of logic are best presented in an abstract setting. For this, let $\mathcal{B} = \{A, B, \ldots\}$ be a class of undefined elements $A, B, \ldots$; in an application the elements will designate events for the outcome of the system. And let $\wedge$ and $\vee$ be operations that combine two elements in $\mathcal{B}$ to form a single element of $\mathcal{B}$, and ^{c} be an operation that takes a single element of $\mathcal{B}$ into a single element of $\mathcal{B}$; in an application the operations will correspond to the logical operations discussed in Section 1.1. A class $\mathcal{B}$ with operations that satisfy the natural properties for events is called a Boolean algebra:

DEFINITION 1

A **Boolean algebra** is a class $\mathcal{B} = \{A, B, \ldots\}$ with operations $\wedge$, $\vee$, and ^{c} such that

(a) Reflexive: $AA = A$ $\quad A \vee A = A$

(b) Commutative: $AB = BA$ $\quad A \vee B = B \vee A$

(c) Associative: $A(BC) = (AB)C$ $\quad A \vee (B \vee C) = (A \vee B) \vee C$

(d) Distributive: $A(B \vee C) = AB \vee AC$ $\quad A \vee (BC) = (A \vee B)(A \vee C)$

(e) Distributive: $(AB)^c = A^c \vee B^c$ $\quad (A \vee B)^c = A^cB^c$

(f) Reflexive: $(A^c)^c = A$

(g) There are distinguished elements $\varnothing$ and $\mathcal{U} = \varnothing^c$ such that

$$AA^c = \varnothing, \qquad A \vee A^c = \mathcal{U},$$

$$A\varnothing = \varnothing, \quad A\mathcal{U} = A, \qquad A \vee \varnothing = A, \quad A \vee \mathcal{U} = \mathcal{U}.$$

As examples consider the classes $\mathcal{B}$ and $\mathcal{B}_2$ for the coin tossing discussed in Sections 1.1B and 1.1C. In most cases we will want our class of events to be *more* than a Boolean algebra. Consider a physicist who counts α-particles in a 7.5-second interval from a radioactive source. Let E_i be the event "i particles in the 7.5-second interval." The physicist might be interested in the event "more than five particles"; in symbols this would be $E_6 \vee E_7 \cdots$. Thus we must be able to apply our operations a *countable* number of times. For this we define a Boolean σ-algebra; the σ denotes *countable*. In the following definition we use $A \subset B$; recall that this is formally defined by the condition $A \wedge B = A$.

DEFINITION 2

A **Boolean σ-algebra** $\mathcal{B}$ is a Boolean algebra such that

(a) For any sequence of elements $A_1, A_2, \ldots$ in $\mathcal{B}$, there is an **intersection** element C (denoted $\wedge_1^\infty A_i$) in $\mathcal{B}$: C is contained in ($\subset$) each A_i; and any other element in $\mathcal{B}$ that is contained in each A_i is also contained in C.

(b) For any sequence of elements $A_1, A_2, \ldots$ in $\mathcal{B}$, there is a **union** element D (denoted $\vee_1^\infty A_i$) in $\mathcal{B}$: D contains ($\supset$) each A_i; and any other element in $\mathcal{B}$ that contains each A_i also contains D.

In terms of our logical framework, Boolean σ-algebras have applications throughout science and technology. For consider the pharmaceutical assessment of the antibiotic mentioned in Section A; the investigator would be interested in "recovery" or its complement for each patient in a clinical trial, as well as for combinations of these.

We could easily construct an example that has the physicist count α-particles. But we wait, as we will soon have a wealth of examples based on a sample space, the concept to be developed next.

In conclusion, we assume that the class $\mathcal{B}$ of events for our random system is a Boolean σ-algebra.

B THE SAMPLE SPACE

Consider a random system and let $\mathcal{B}$ be the class of events of interest. As part of this we have, of course, implicitly decided on the depth of observation, the degree to which we are going to recognize characteristics of the result of a performance. For example, with the coin-tossing system we could recognize only the face showing for the first coin, or only the faces showing for the first and the second coins, or perhaps only certain characteristics of location and orientation of the coins.

Let s designate a possible outcome of the system, given an agreed depth of observation; we call such an outcome a *sample point*. We now define: The *sample space* is the set $\mathcal{S} = \{s\}$ of possible outcomes, the set of sample points. As alternative notation we might use t and $\mathfrak{T}$ or, say, x and $\mathfrak{X}$. As an application, consider the following pharmaceutical illustration.

EXAMPLE 1

Life or death: Consider the administration of a drug to a rabbit obtained from a certain genetic strain. And suppose that we have decided to observe only whether the animal lives $x_1 = 0$ or dies $x_1 = 1$ in a test period. Then the sample space is

$$\mathcal{S} = \{x_1\} = \{0, 1\}.$$

Now suppose that we have decided to observe whether the animal lives $x_1 = 0$ or dies $x_1 = 1$ and whether a second animal tested lives $x_2 = 0$ or dies $x_2 = 1$. Then the sample space becomes

$$\begin{aligned}\mathcal{S}_2 &= \{(x_1, x_2) : x_i = 0, 1\} = \{(0, 0), (0, 1), (1, 0), (1, 1)\} \\ &= \{0, 1\} \times \{0, 1\} = \{0, 1\}^2;\end{aligned}$$

this is the Cartesian product of the set $\{0, 1\}$ with itself. The sample spaces $\mathcal{S}$ and $\mathcal{S}_2$ are presented in Figure 1.1.

EXAMPLE 2

Performance rating: Consider a psychology experiment in which a subject's performance is rated from 1 to 6. For two subjects let x_i be the rating for the ith subject. Then the sample space is

$$\begin{aligned}\mathcal{S} &= \{(x_1, x_2) : x_i = 1, 2, \ldots, 6\} \\ &= \{1, 2, 3, 4, 5, 6\} \times \{1, 2, 3, 4, 5, 6\} = \{1, 2, 3, 4, 5, 6\}^2;\end{aligned}$$

this is the Cartesian product of the set $\{1, 2, 3, 4, 5, 6\}$ with itself.

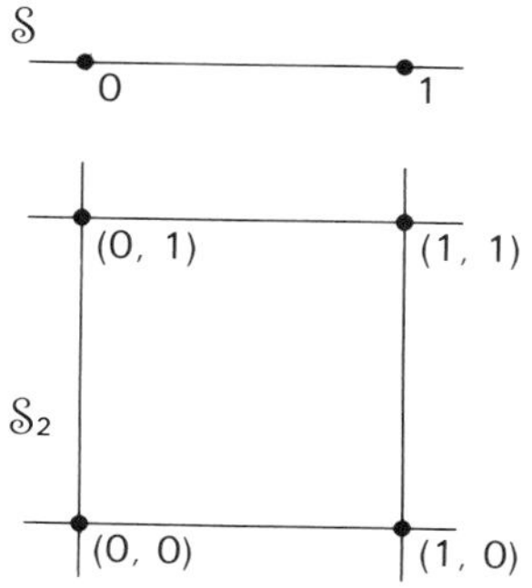

FIGURE 1.1
Sample spaces $\mathcal{S}$ for a single animal and $\mathcal{S}_2$ for two animals. Also, sample spaces $\mathcal{S}$ and $\mathcal{S}_2$ in Example 3.

Sometimes we shall use a lowercase letter, say s, to refer to a particular outcome in $\mathcal{S}$. And sometimes we shall use it to refer to an arbitrary outcome in $\mathcal{S}$, as, for example, in defining a subset of $\mathcal{S}$ such as $\{s: s > 5\} = \{s > 5\}$; recall the example on α-particles. There are some real advantages to keeping the dual usage and not making a formal distinction in the notation. Of course, the context will always make clear the distinction as needed.

Sometimes it is convenient to let the sample space be larger than needed, to be a convenient and simple mathematical space. For example, in the rating of a single subject's performance, the sample space is $\{1, 2, 3, 4, 5, 6\}$, but it may be convenient to use the space of natural numbers $N = \{1, 2, 3, \ldots\}$, or the space of integers $Z = \{\ldots, -1, 0, 1, \ldots\}$, or even the space R of real numbers. Of course, we shall have ample opportunity later to indicate that certain sample points in effect are not possible outcomes.

C EVENTS REPRESENTED BY SUBSETS

Consider a random system. Let $\mathcal{B}$ be the class of events of interest, and let $\mathcal{S}$ be the sample space corresponding to a chosen depth of observation. Now consider an event B in $\mathcal{B}$. The event can be viewed as a proposition concerning the outcome. Accordingly, we can examine each outcome in $\mathcal{S}$, collect those outcomes for which the proposition is true, and obtain a subset, say A, of the space $\mathcal{S}$. The event B is thus represented as a subset A of $\mathcal{S}$, the truth set for the event.

Let $\mathcal{A} = \{A\}$ consist of the subsets of $\mathcal{S}$ that correspond to events in $\mathcal{B} = \{B\}$. Often it will be convenient to refer to $\mathcal{A}$ itself as the class of events for the system. Clearly the operation intersection $\wedge$ for events becomes the operation intersection $\cap$ for sets, union $\vee$ becomes union $\cup$, and complement ^{c} becomes ordinary complement ^{c}; and, of course, the null event $\varnothing$ becomes the empty set $\varnothing$, and disjoint events B_1 and B_2 $(B_1 \wedge B_2 = \varnothing)$ become disjoint sets A_1 and A_2 $(A_1 \cap A_2 = \varnothing)$.

Sometimes $\mathcal{A}$ may consist of all the subsets of the sample space, as in the following examples.

EXAMPLE 3

Tossing two coins: Consider the tossing of two coins as discussed in Sections 1.1B and 1.1C. And suppose that we recognize only the face of the first coin: let $x_1 (= 0, 1)$ be the number of heads showing for that coin. Then $\mathcal{S} = \{x_1\} = \{0, 1\}$. The class of all subsets of $\mathcal{S}$ is

$$\mathcal{A} = \{\varnothing, \{1\}, \{0\}, \{0, 1\}\},$$

where $\varnothing$ now designates the empty set. The class $\mathcal{A}$ has four elements, and they can be put in one-to-one correspondence with the elements of $\mathcal{B}$ in Section 1.1B. Now suppose that we observe the face on the first coin and the face on the second coin: let x_i be the number of heads for ith coin. Then $\mathcal{S}_2 = \{(x_1, x_2)\} = \{0, 1\}^2$. The class of all subsets of $\mathcal{S}_2$ is

$$\mathcal{A}_2 = \{\varnothing, \{(0, 0)\}, \ldots, \{(0, 0), (0, 1)\}, \ldots, \{(0, 0), (0, 1), (1, 0)\}, \ldots, \mathcal{S}_2\}.$$

The class $\mathcal{A}_2$ has 16 elements and they can be put in one-to-one correspondence with the elements of $\mathcal{B}_2$ in Section 1.1C. Note that the event H_1 is represented by the set $\{1\}$ relative to $\mathcal{S}$ and by the set $\{(1, 0), (1, 1)\}$ relative to $\mathcal{S}_2$. See Figure 1.1 and compare with Example 1.

EXAMPLE 4

Tossing two dice: Suppose that we recognize only the points on the first die and the points on the second die: let x_i be the number of points on the ith die. The sample space is $\mathcal{S} = \{(x_1, x_2) : x_i = 1, 2, \ldots, 6\}$; see Figure 1.2. The event L, "at least one 6," can be represented by the subset

$$L = \{(1, 6), (2, 6), \ldots, (6, 6), (6, 5), \ldots, (6, 2), (6, 1)\}.$$

The event T, "total points is 7," can be represented by the subset

$$T = \{(1, 6), (2, 5), (3, 4), (4, 3), (5, 2), (6, 1)\}.$$

Note, for example, that $L \cap T = \{(1, 6), (6, 1)\}$ is the event "a 1 and a 6." Compare with Example 2.

EXAMPLE 5

One card from a deck: A card is dealt from a well-shuffled deck of 52 playing cards and its designation is recorded. The sample space is

$$\mathcal{S} = \{\text{AS, KS}, \ldots, \text{2S, AH}, \ldots, \text{2H, AD}, \ldots, \text{2D, AC}, \ldots, \text{2C}\},$$

where, for example, AS designates "ace of spades" and 2C designates "2 of clubs"; $\mathcal{S}$ contains 52 points. Let $\mathcal{A}$ be the class of all subsets of $\mathcal{S}$; then

$$\mathcal{A} = \{\varnothing, \{\text{AS}\}, \ldots, \{\text{2C}\}, \{\text{AS, KS}\}, \ldots, \{\text{3C, 2C}\}, \ldots, \mathcal{S}\}.$$

A little calculation shows that $\mathcal{A}$ has 2^{52} elements (corresponding to the presence or absence of each sample point).

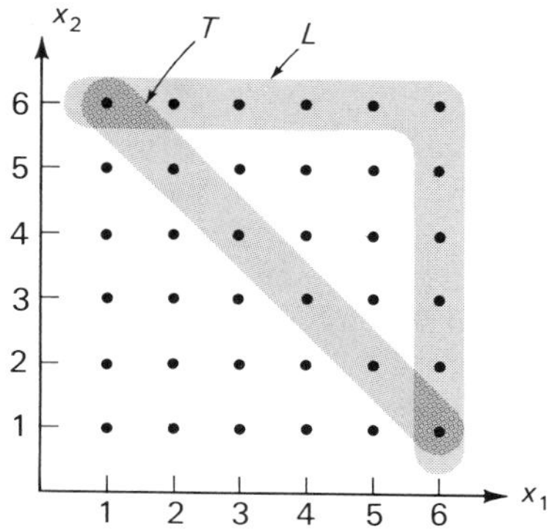

FIGURE 1.2
Sample space $\mathcal{S}$ for two dice. Also, sample space $\mathcal{S}$ in Example 2.

With a finite or countable sample space $\mathcal{S}$ it is quite natural to take $\mathcal{A}$ to be the class of all subsets of $\mathcal{S}$. The class of all subsets of a space $\mathcal{S}$ is called the *power set* of $\mathcal{S}$ and designated $\mathcal{P}(\mathcal{S})$.

In Example 5 the sample space has 52 points and we noted that the class $\mathcal{A}$ has 2^{52} points, a *very* large number. Perhaps this should be a warning. Indeed, if we consider a sample space as large as the real line R, as, for example, with a single measurement of a physical quantity, then it can be shown that the power set of R is too big, too complicated, and in fact impossible for some quite reasonable problems involving variation. Often we must settle for a class $\mathcal{A}$ smaller than the power set; we now examine properties for such a class.

D ALGEBRA OF SUBSETS

Consider a sample space $\mathcal{S}$ and let $\mathcal{A}$ be a class of subsets of $\mathcal{S}$. If $\mathcal{A}$ is to work as a class of events, we must be able to apply the standard operations $\cap$, $\cup$, ^{c}, $\cap_1^\infty$, and $\cup_1^\infty$ and remain with the elements of $\mathcal{A}$. Such a class of subsets is called a σ-algebra:

DEFINITION 3

A **σ-algebra** is a nonempty class $\mathcal{A} = \{A\}$ of subsets of a space $\mathcal{S}$ such that

(a) Closed under σ-union: If $A_1, A_2, \ldots \in \mathcal{A}$, then $\cup_1^\infty A_i \in \mathcal{A}$.

(b) Closed under complement: If $A \in \mathcal{A}$, then $A^c = \mathcal{S} - A \in \mathcal{A}$.

Closure under σ-intersection can be proved from (a) and (b); see Problem 20.

The combination $(\mathcal{S}, \mathcal{A})$ of a sample space $\mathcal{S}$ and a σ-algebra $\mathcal{A}$ over $\mathcal{S}$ is called a *measurable space*.

In Section A we assumed that our class of events satisfied the logic axioms of a Boolean σ-algebra. It is natural, then, to inquire whether our present class $\mathcal{A}$, given by Definition 3, satisfies the axioms of a Boolean σ-algebra. Fortunately, the answer is "yes": a σ-algebra $\mathcal{A}$ over a sample space $\mathcal{S}$ is automatically a Boolean σ-algebra. Thus quite freely, from now on, *we shall express our algebra of events as a σ-algebra over a suitable sample space*. Of course, we should remember, as indicated by Example 3, that many different sample spaces can be used to present a given Boolean σ-algebra of events; the advantage of the Boolean approach is that it gives us an invariant framework for building the probability model.

The verification that the axioms of a Boolean σ-algebra hold for subsets of a space is straightforward. For example, consider $(A_1A_2)^c = A_1^c \cup A_2^c$. A point s belongs to the left side, if and only if s does not belong to A_1A_2; s does not belong to A_1A_2 if and only if s does not belong to A_i for some i; s does not belong to A_i for some i if and only if s belongs to A_i^c for some i; s belongs to A_i^c for some i if and only if s belongs to the right side. This argument can be abbreviated as

$$s \in (A_1A_2)^c \quad \leftrightarrow \quad s \notin A_1A_2 \quad \leftrightarrow \quad s \notin A_i \text{ for some } i$$
$$\leftrightarrow \quad s \in A_i^c \text{ for some } i \quad \leftrightarrow \quad s \in A_1^c \cup A_2^c.$$

Consider now some examples in which the class $\mathcal{A}$ is smaller than the power set.

EXAMPLE 6

One measurement by a physicist: Consider a physicist measuring a current flow, and suppose that he makes a single measurement y. The sample space is

$$\mathcal{S} = \{y : y \in R\} = R,$$

the real line; note that we are allowing any positive or negative value for y. The physicist might be interested in the event that y falls in an interval, say $(1, 2)$; or in a finite union of intervals, say $(-2, -1) \cup (1, 2)$. Certainly we would want the class $\mathcal{A}$ to include all the intervals (a, b). Accordingly, we take $\mathcal{A}$ to be the smallest σ-algebra that includes all the intervals; of course, any finite or countable union of intervals is an element of $\mathcal{A}$. Problem 25 shows that it is meaningful to talk about a smallest σ-algebra. An element of $\mathcal{A}$ is called a *Borel set* on the real line; and the class $\mathcal{A}$ is called the *Borel class* on the real line and designated $\mathcal{B}^1$. It is of some interest that the class of countable unions of intervals is only a very small step toward the Borel class; and also that the Borel class is only a very small step toward the power set $\mathcal{P}(R)$.

EXAMPLE 7

***k* measurements by a physicist:** Consider the physicist in Example 6 and suppose that he makes k measurements $y_1, \ldots, y_k$ on the current flow. The sample space then is

$$\mathcal{S} = \{(y_1, \ldots, y_k) : y_i \in R\} = R^k,$$

the k-dimensional space over the reals. The physicist might be interested in the event that $\mathbf{y}' = (y_1, \ldots, y_k)$ falls in a rectangle, say, $(a_1, b_1) \times \cdots \times (a_k, b_k)$; or in some finite union of rectangles; see Figure 1.3. Certainly we would want the class $\mathcal{A}$ to include all the rectangles $(a_1, b_1) \times \cdots \times (a_k, b_k)$; and we would want to include rectangles partially closed in various ways. Accordingly, we take $\mathcal{A}$ to be the smallest σ-algebra that includes all such rectangles; this is called the *Borel class* on R^k and is designated $\mathcal{B}^k$. It is of interest that the class of countable

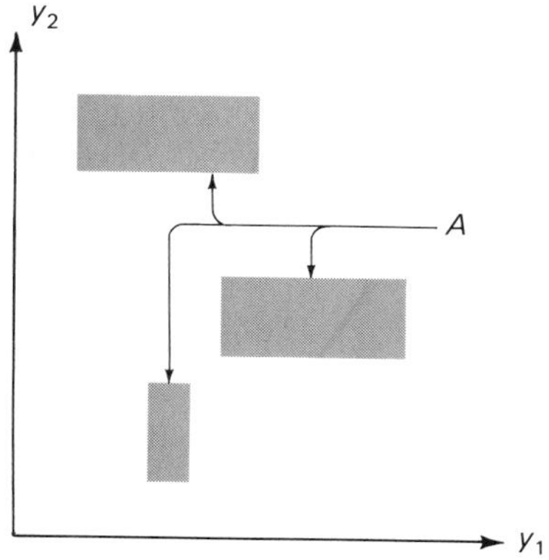

FIGURE 1.3
Set A, a finite union of rectangles.

unions of rectangles is only a very small step toward the Borel class; and also that the Borel class is only a very small step toward the power set $\mathcal{P}(R^k)$.

In general, if the sample space $\mathcal{S}$ is R^1 or say R^k, we shall take the algebra of events to be the Borel class over that space. In a sense the Borel class represents the logical consequences of considering intervals and assuming closure under σ-union and σ-intersection.

E INDICATOR FUNCTION FOR AN EVENT

We now introduce the *indicator function* I_A as an alternative way of representing a subset A. We shall see later that indicator functions provide a simple mechanical route to otherwise complicated results. Recall the brief use of indicators in Sections 1.1B and 1.1C.

DEFINITION 4

The **indicator function** I_A for a subset $A \subset \mathcal{S}$ is the real-valued function on $\mathcal{S}$ given by

$$(1) \qquad I_A(s) = 1 \quad \text{if } s \in A$$
$$\qquad = 0 \quad \text{otherwise.}$$

For an example, see Figure 1.4.

Now, instead of thinking of a subset A of a space $\mathcal{S}$, we can think of a special kind of real-valued function I_A over $\mathcal{S}$. Of course, an indicator function takes only the values 0 and 1, but perhaps we can think ahead to more general functions that take, say, "part" of a sample point or take a sample point with a positive or negative "weight."

Consider some of the basic properties that make indicator functions very useful.

1. *For any set A,* $I_A = 1 - I_{A^c}$.

Proof The proof can be given in the abbreviated form used in Section D:

$$I_A(s) = 1 \quad \leftrightarrow \quad s \in A \quad \leftrightarrow \quad s \notin A^c \quad \leftrightarrow \quad I_{A^c}(s) = 0 \quad \leftrightarrow \quad 1 - I_{A^c}(s) = 1.$$

2. *For any sets* $A_1, A_2, \ldots,$

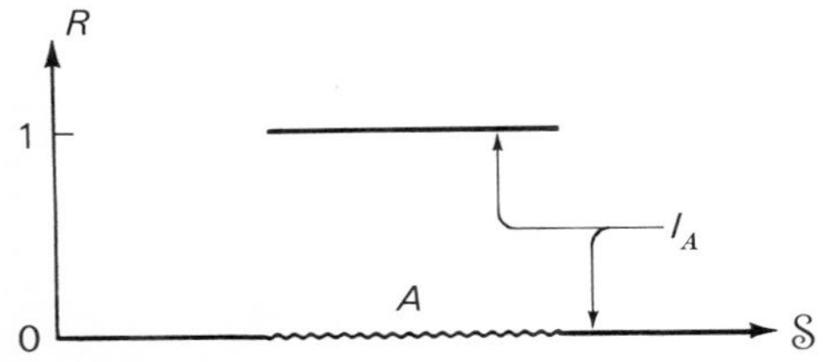

FIGURE 1.4
Indicator function for set A.

$$I_{\cap_1^\infty A_i} = \prod_1^\infty I_{A_i}.$$

3. *For any disjoint sets $A_1, A_2, \ldots,$*

$$I_{\cup_1^\infty A_i} = \sum_1^\infty I_{A_i}.$$

The proofs for these follow the pattern indicated for property 1.

As a small indication of the usefulness of indicator functions, consider the following calculation: for any sets A and B,

$$(2)\qquad I_{A\cup B} = 1 - I_{(A\cup B)^c} = 1 - I_{A^cB^c} = 1 - (1 - I_A)(1 - I_B)$$
$$= I_A + I_B - I_{AB}.$$

F EXERCISES: GAMES OF CHANCE

1 Consider tossing a die and recording y_1, the number of points showing.
(a) Record the sample space $\mathcal{S}$ for y_1.
(b) Present the following events as subsets of $\mathcal{S}$: A, "the number is even"; B, "the number is divisible by 3"; and C, "the number is less than 4"; $A \wedge B$, $A \wedge C$, and $A^c \vee B^c$.

2 Consider tossing a die and a coin and recording (y_1, y_2), where y_1 is the number of points on the die and y_2 is the number of heads.
(a) Record the sample space $\mathcal{S}$ for (y_1, y_2).
(b) Present the following events as subsets of $\mathcal{S}$: "y_1 is even"; "y_1 is divisible by 3"; $y_1 + y_2 = 1$; $y_1 + y_2 = 2$; and $y_1 + y_2 \leq 2$.

3 Two dice are tossed and (y_1, y_2) is recorded, where y_i is the number of points on the ith die.
(a) Record the sample $\mathcal{S}$ for (y_1, y_2).
(b) Present the following events as subsets of $\mathcal{S}$: A, "the total points is 6"; B, "the total points is 7"; $|y_2 - y_1| = 3$; and $|y_2 - y_1| = 4$. Record these events on a diagram.

4 Two dice are tossed and $\{y_{(1)}, y_{(2)}\}$ is recorded, where $y_{(1)}$ is the smaller number showing and $y_{(2)}$ is the larger.
(a) Record the sample space $\mathcal{S}$ for $\{y_{(1)}, y_{(2)}\}$.
(b) Present the following events as subsets of $\mathcal{S}$: A, "the total points is 6"; $y_{(2)} - y_{(1)} = 3$; and $A \wedge \{y_{(2)} - y_{(1)} = 3\}$. Record these events on a diagram.

5 Three coins are tossed; let $y_i = 0(1)$ according as the ith coin is tails (heads).
(a) Record in full the sample space $\mathcal{S}$ for (y_1, y_2, y_3).
(b) Let H_i be the event "heads on the ith toss." Present H_1, H_1H_2, and $H_1H_2H_3$ as subsets of $\mathcal{S}$.
(c) Let E_j be the event "exactly j of the events H_1, H_2, and H_3 occur." Present E_0, E_1, E_2, and E_3 as subsets of $\mathcal{S}$.

6 Consider an urn containing 95 green balls and 5 black balls. The balls are mixed and two balls are drawn in succession, giving (y_1, y_2), where $y_i = g(b)$ if the ith ball is green (black). Record the sample space $\mathcal{S}$ for (y_1, y_2). Present the following events as subsets of $\mathcal{S}$: E_0, E_1, and E_2, where E_j is the event "exactly j black balls."

7 An urn contains 3 green balls, marked g_1, g_2, and g_3, and one black ball, marked b. The balls are thoroughly mixed and two balls are drawn in succession, yielding (y_1, y_2), where y_i is the mark on the ith ball. Record the sample space $\mathcal{S}$. Present the following events as subsets of $\mathcal{S}$: E_0, E_1, and E_2, where E_j is the event "exactly j black balls."

8 A deck of 52 playing cards is shuffled and two cards are dealt in succession face up. Describe the sample space $\mathcal{S}$ for (y_1, y_2), where y_i is the designation on the ith card dealt.

How many points does $\mathcal{S}$ contain? Describe the subset A for the event "both are spades." How many points does A contain?

9 A deck of 52 playing cards is shuffled and a card is dealt; the card is returned to the deck, the deck is shuffled, and a second card is dealt. Describe the sample space for (y_1, y_2), where y_i is the designation on the ith card dealt. How many points does $\mathcal{S}$ contain? Describe the subset A for the event "both are spades." How many points does A contain?

10 Four cards are numbered 1, 2, 3, and 4. The cards are shuffled face down and dealt in succession, giving (y_1, y_2, y_3, y_4); as the cards are dealt, the dealer calls the numbers 1, 2, 3, and 4 in succession.

(a) Present the sample space $\mathcal{S}$ for (y_1, y_2, y_3, y_4).

(b) Let A_i be the event "the ith card is called correctly." Present A_1 as a subset of $\mathcal{S}$.

(c) Let E_j be the event "exactly j of the events A_1, A_2, A_3, and A_4." Present E_2, E_3, and E_4 as subsets of $\mathcal{S}$.

G EXERCISES: INDUSTRY AND SCIENCE

11 Transformers are mass-produced and packaged in lots of 100. A particular lot is subjected to a sampling plan: two transformers are sampled in succession, tested, and (y_1, y_2) is recorded, where $y_i = g(b)$ if the ith transformer is good (bad). Record the sample space $\mathcal{S}$ for (y_1, y_2) for a typical lot; certain exclusions would occur with exceptional lots. Present the following events as subsets of $\mathcal{S}$: E_0, E_1, and E_2, where E_j is the event "exactly j bad items."

12 A physicist records (y_1, y_2), where y_1 (y_2) is the number of α-particles in a first (second) 7.5-second interval.

(a) Present the sample space $\mathcal{S}$ for (y_1, y_2) and make a diagram.

(b) Present the following events as subsets of $\mathcal{S}$ and mark them on the diagram: the average $\bar{y} = (y_1 + y_2)/2 \leq 3.5$; $\max(y_1, y_2) \leq 4$; and $\min(y_1, y_2) \leq 2$.

13 A psychologist records $(y_1, \ldots, y_n)$, where y_i is the logarithm of the reaction intensity for the ith subject. For simplicity take $n = 2$. Present the following events as subsets and mark them on a diagram: $\bar{y} = (y_1 + y_2)/2 \leq 5$; $\max(y_1, y_2) \leq 5$; and $\min(y_1, y_2) \leq 3$.

14 Details as in Exercise 13 with $n = 2$. Let $y_{(1)}$ be the smallest, $y_{(2)}$ the second smallest, . . ., $y_{(n)}$ the largest of $y_1, \ldots, y_n$; $y_{(1)}, \ldots, y_{(n)}$ are called the *order statistics*. Present the following events as subsets and mark them on a diagram: $y_{(2)} \leq 4$; $y_{(1)} \leq 2$; and $y_{(2)} - y_{(1)} \leq 2$.

15 A chemist makes n determinations $(y_1, \ldots, y_n)$ on the rate of a chemical reaction; positive and negative values are possible. The sample space $\mathcal{S} = R^n$. For simplicity take $n = 2$; thus $\mathcal{S} = R^2$. Present the following events as subsets and mark them on a diagram: $y_1 \leq 5$; $y_2 \leq 5$; $y_1 + y_2 \leq 5$; and $|y_2 - y_1| \leq 5$.

16 Details as in Exercise 15 with $n = 2$. Present the following events as subsets and mark them on a diagram: $y_{(1)} \leq 5$; $y_{(2)} \leq 5$; and $y_{(2)} - y_{(1)} \leq 5$. For notation, see Exercise 14.

17 Details as in Exercise 15 with $n = 2$. The sample average is $\bar{y} = (y_1 + \cdots + y_n)/n$; the sample variance is $s_y^2 = \Sigma_1^n (y_i - \bar{y})^2/(n - 1)$. Present the following events as subsets and mark them on a diagram: $\bar{y} - 4.1 \leq 3/\sqrt{2}$; and $s_y^2 \leq 4$.

18 Details as in Exercise 15 with $n = 2$. The t-statistic relative to the location 4.1 is

$$t(y_1, \ldots, y_n) = \frac{\bar{y} - 4.1}{s_y/\sqrt{n}}.$$

Present the following event as a subset and mark it on a diagram: $t(y_1, y_2) \leq 3$.

H PROBLEMS

19 Consider subsets A_i of a sample space $\mathcal{S}$. The relation $(A_1A_2)^c = A_1^c \cup A_2^c$ discussed in Section C can be generalized; prove De Morgan's laws:

$$(\cap_1^\infty A_i)^c = \cup_1^\infty A_i^c; \qquad (\cup_1^\infty A_i)^c = \cap_1^\infty A_i^c.$$

Thus *complement of intersection* is *union of complements; complement of union* is *intersection of complements*.

20 Let $\mathcal{A}$ be a σ-algebra of subsets of a sample space $\mathcal{S}$. Prove that $\mathcal{A}$ is closed under σ-intersection.

21 Consider a sample space $\mathcal{S}$. An indicator function is a map from $\mathcal{S}$ into $\{0, 1\}$. The class of all indicator functions can be designated $2^{\mathcal{S}}$, referring to binary-valued functions on $\mathcal{S}$.
(a) Show that the power set $\mathcal{P}(\mathcal{S})$ is equivalent to $2^{\mathcal{S}}$. Compare with Problems 1.1.6 and 1.1.7.
(b) Prove that

$$I_{\cup_1^\infty A_i} = \max_1^\infty I_{A_i}, \qquad I_{\cap_1^\infty A_i} = \min_1^\infty I_{A_i}.$$

22 Let $A_1, \ldots, A_m$ be subsets of the sample space $\mathcal{S}$; let I_i be the indicator function for the set A_i; and let E_j be the event "exactly j of the events $A_1, \ldots, A_m$." Prove that

$$\begin{aligned} I_{E_0} &= (1 - I_1) \cdots (1 - I_m) \\ &= 1 - \sum_{i_1} I_{i_1} + \sum_{i_1 < i_2} I_{i_1} I_{i_2} - + \cdots + (-1)^m I_1 \cdots I_m \\ &= 1 - T_1 + T_2 - + \cdots + (-1)^m T_m, \end{aligned}$$

where $T_r = \sum_{i_1 < \cdots < i_r} I_{i_1} \cdots I_{i_r}$.

23 Let $\mathcal{A}_0 = \{A_i\}$ be a class of subsets A on a sample space $\mathcal{S}$. Show that there exists a σ-algebra on $\mathcal{S}$ that contains $\mathcal{A}_0$.

24 (*continuation*) Let $\{\mathcal{A}_\alpha\}$ be the set of all σ-algebras $\mathcal{A}_\alpha$ such that $\mathcal{A}_0 \subset \mathcal{A}_\alpha$; α ranges over some index set. Show that $\mathcal{A} = \cap \mathcal{A}_\alpha$ is a σ-algebra over $\mathcal{S}$.

25 (*continuation*) If $\mathcal{B}_\alpha \subset \mathcal{B}_\beta$, then we say that $\mathcal{B}_\alpha$ is smaller than $\mathcal{B}_\beta$. Show that it is meaningful to speak of the smallest σ-algebra containing a class $\mathcal{A}_0$ of subsets of $\mathcal{S}$.

1.3

PROBABILITIES

Probabilities are what the probability model is all about. In this section we examine the simple mathematical framework for probabilities. Then in the next section we examine various computational rules for handling probabilities.

A MEASURING LENGTH, AREA, AND PROBABILITY

Consider the coin tossing in Sections 1.1C and 1.2C with sample space

$$\mathcal{S}_2 = \{(x_1, x_2)\} = \{(0, 0), (0, 1), (1, 0), (1, 1)\},$$

where x_i is the number of heads for the ith coin. The symmetry discussed in Section 1.1C gives a probability 1/4 at each of the four points in $\mathcal{S}_2$. And the additivity mentioned there shows that the probability for any set of points is the sum of the values at component points.

From a slightly different point of view we can think of the four points in $\mathcal{S}_2$ as four points on the plane R^2; and we can think of attaching weights of value 1/4 at each of the four points. We could then picture breaking off some rectangle or even some more complicated set and measuring its weight; we would obtain, of course, the value given by the symmetric $\bar{P}$ discussed in Section 1.1C.

And in a more general way, we could think of the plane R^2 as being a plate of varying density. We could then break off some rectangle or even some more complicated set and measure its weight; for any set A we would obtain its weight measure, say $\mu(A)$. If the plate were of uniform density and with unit weight for unit area, the measure μ gives the area for any set on the plane.

And in a simpler way we could think of the line R as being a rod of varying density. We could then break off some interval or even some more complicated set and measure its weight; for any set A, we would obtain its weight measure, say $\mu(A)$. If the rod were of uniform density with unit weight for unit length, the measure μ gives the length of any set on the line. We shall see that measuring probability is a simple special case of such examples of measuring weight.

B FORMALITIES OF MEASURING

Consider a sample space $\mathcal{S}$ with events given by a class $\mathcal{A}$ of subsets of $\mathcal{S}$.

We will find the notion of a partition very convenient. For the present notation, we have:

DEFINITION 1

A class $\{A_1, A_2, \ldots\}$ of subsets of $\mathcal{S}$ is a countable partition of a subset A if

$$A = A_1 \cup A_2 \cup \cdots = \cup_1^\infty A_i$$

and if A_i and A_j are disjoint for $i \neq j$.

In a sense we have taken the set A and broken it into a countable number of subsets $A_1, A_2, \ldots$; see Figure 1.5. And if we think of the plane as a plate of varying density, we have taken a piece A of the plane and broken it into a countable number of pieces $A_1, A_2, \ldots$. Note that the definition covers the case of a finite partition by having most of the $A_i = \varnothing$.

Now consider the formal definition of a measure μ. For this we allow the value $+\infty$, as, for example, the weight of the whole plane viewed, say, as a plate of uniform thickness.

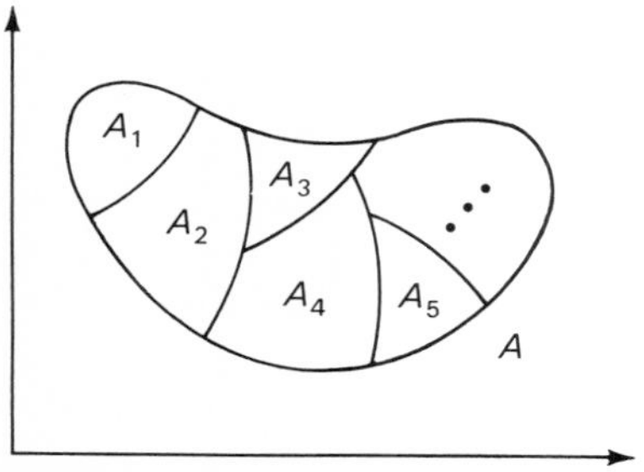

FIGURE 1.5
Set A partitioned into $A_1, A_2, \ldots$.

DEFINITION 2

A **measure** μ is a function from $\mathcal{A}$ into $[0, \infty]$ such that
(a) Nontrivial: $\mu(A) < \infty$ for some A in $\mathcal{A}$.
(b) σ-additive: if $\{A_1, A_2, \ldots\}$ is a countable partition of A with A_i in $\mathcal{A}$, then

$$\mu(A) = \sum_1^{\infty} \mu(A_i).$$

It is straightforward to prove that $\mu(\phi) = 0$ and to prove *finite additivity:*

$$\mu(\cup_1^n A_i) = \sum_1^n \mu(A_i),$$

where the A_i are disjoint; see Problems 11 and 12.

EXAMPLE 1

Coin tossing from Sections 1.1B and 1.1C: For observing only the first coin we have $\mathcal{S} = \{0, 1\}$. Suppose that we consider the points of $\mathcal{S}$ as points on the real line and then consider subsets of the real line. The measure P based on the 100 repetitions can be presented in terms of sets rather than the events in Section 1.1B:

$$\begin{aligned} P(A) &= 0 && \text{if neither } 0, 1 \in A \\ &= 0.53 && \text{only } 1 \in A \\ &= 0.47 && \text{only } 0 \in A \\ &= 1.00 && \text{both } 0, 1 \in A. \end{aligned}$$

For observing both coins we have $\mathcal{S}_2 = \{0, 1\}^2$. Suppose that we consider the points of $\mathcal{S}$ as points on the plane, and then consider subsets of the plane. The measure P based on the 100 repetitions can be presented in terms of sets on the plane; we define $P(A)$ to be the earlier $P(B)$ of Section 1.1C, where B is the event corresponding to the subset A; see Figure 1.6.

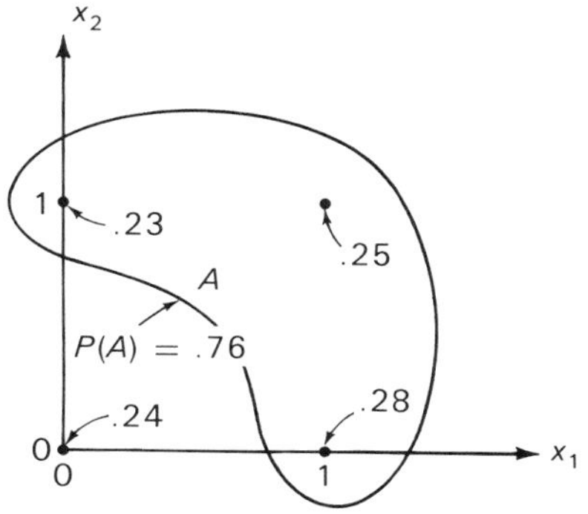

FIGURE 1.6
Probability of at least one head based on 100 repetitions.

EXAMPLE 2

Cardinality: The number of elements in a set A is a measure of the set. For notation, let

$c(A)$ = number of elements in A = cardinality of A.

Then for a σ-algebra $\mathcal{A}$ over a sample space $\mathcal{S}$, the function c is a measure, called the *cardinality measure*.

For a sample space such as R^1 or $R^2 \ldots$, we have that every nonfinite set has cardinality measure $+\infty$. We obtain a more useful measure if we count only those points that fall in some designated subset $\mathcal{S}_0$ of interest; accordingly, we define a measure μ_1 by

$$\mu_1(A) = c(A \cap \mathcal{S}_0).$$

For example, consider the plane R^2 and let $\mathcal{S}_0$ be the set $\mathcal{S}_2$ described in the preceding example. Then for the set A in Figure 1.6, we have $\mu_1(A) = 3$.

EXAMPLE 3

If we think of unit weights attached to the points of $\mathcal{S}_0$ in the preceding example, the measure μ_1 gives the weight for a set. It is easy to think more generally of attaching different weights at various points in a designated subset $\mathcal{S}_0$. Consider a finite or countable set $\mathcal{S}_0 = \{a_1, a_2, \ldots\}$ and let $m(a_i) \geq 0$ be the weight attached to the point a_i. Then we define a measure μ_2 by

$$\mu_2(A) = \sum_{a_i \in A} m(a_i),$$

where the summation is over all points a_i in A. If we define $m(s) = 0$ for points s not in $\mathcal{S}_0$, we can write

$$\text{(1)} \qquad \mu_2(A) = \sum_{s \in A} m(s),$$

where only a finite or countable number of terms have $m(s)$ different from zero and the summation is effectively over such terms. It is trivial to verify that μ_2 satisfies the σ-additivity and thus satisfies the formal Definition 2 for a measure. The measure μ_2 is called *symmetric* if the weights $m(a_i) = m$ are all equal. Note then that

$$\text{(2)} \qquad \mu_2(A) = m\mu_1(A).$$

C THREE KINDS OF MEASURE

We have been considering measures that range from one extreme with value $+\infty$ for every set except $\varnothing$ to near simplicity in the other direction as in Example 1. Consider some simple kinds of measures:

DEFINITION 3

A measure μ_1 is a **σ-finite measure** if $\mathcal{S}$ can be expressed as a countable union $\cup_1^\infty A_i$ of sets A_i each having finite measure $\mu(A_i) < \infty$.

Consider the plane R^2 as a plate of uniform density and unit weight per unit area. Then the weight measure gives the area of sets on the plane. It is σ-finite; for we can break the plane into a countable number of unit squares each of area 1. The cardinality function on the line or plane is not σ-finite. If we do not allow the value $+\infty$ for a measure, we obtain a finite measure:

DEFINITION 4

A measure μ is a **finite measure** if $\mu(\mathcal{S}) < \infty$.

The measure μ_2 defined in Example 3 is a finite measure in the case that $\Sigma_1^\infty m(a_i) < \infty$. For the coin-tossing example (Example 1) the measure for the whole space is 1:

DEFINITION 5

A measure μ is a **probability measure** if $\mu(\mathcal{S}) = 1$.

For a probability measure we shall usually use the letter P in place of μ.

D MEASURING PROBABILITY

From the simple coin-tossing example we can picture probability as a total weight of 1 distributed around on the points of the sample space. The additivity mentioned in Section 1.1 becomes the σ-additivity of a measure, and the total weight of 1 is the special property for probabilities. We summarize this as:

DEFINITION 6

A probability measure P is a nonnegative function on a σ-algebra $\mathcal{A}$ such that
(a) σ-additive: if $\{A_1, A_2, \ldots\}$ is a countable partition of A, then

$$P(A) = \sum_1^\infty P(A_i).$$

(b) Normed: $P(\mathcal{S}) = 1$.

Of course for a probability measure P, we have $P(\phi) = 0$; and we have finite additivity: $P(\cup_1^n A_i) = \Sigma_1^n P(A_i)$, provided that the A_i are disjoint (Problems 11 and 12). Note that we can think of the measure P as being defined on the corresponding Boolean σ-algebra.

We can obtain a fairly general illustration of a probability measure by specializing the measure μ_2 in Example 3. We use the notation $p(a_i)$ for the weight at a_i in $\mathcal{S}_0$, and we require that the sum $\Sigma_1^\infty p(A_i) = 1$ is equal to unity. This gives the measure P_2:

(3) $$P_2(A) = \sum_{a_i \in A} p(a_i).$$

This is a probability measure; for as a special case of μ_2 it is σ-additive; and of course $P_2(\mathcal{S}) = \Sigma\ p(a_i) = 1$. The measure P_2 is called a *discrete measure* on the space $\mathcal{S}$.

EXAMPLE 4

Coin tossing from Sections 1.1B and 1.1C: For observing only the first coin, we take $\mathcal{S}_0 = \{0, 1\}$ and $\mathcal{S}$ to be the real line R. The measure based on the 100 repetitions is obtained by defining p as follows:

$p(0) = 0.47, \qquad p(1) = 0.53.$

The measure based on symmetry is obtained by defining p as follows:

$p(0) = 1/2, \qquad p(1) = 1/2.$

For observing both coins we take $\mathcal{S}_0 = \{0, 1\}^2$ and $\mathcal{S}$ to be the plane R^2. The measure based on the 100 repetitions is obtained by defining p as follows:

$p(0, 0) = 0.24, \qquad p(1, 0) = 0.28,$

$p(0, 1) = 0.23, \qquad p(1, 1) = 0.25;$

see Figure 1.6. The measure based on symmetry is obtained by defining p as follows:

$p(0, 0) = 1/4, \qquad p(1, 0) = 1/4,$

$p(0, 1) = 1/4, \qquad p(1, 1) = 1/4.$

EXAMPLE 5

Binomial frequencies: Consider some distributions of probability on the real line $\mathcal{S} = R$. For a simple example take $\mathcal{S}_0 = \{0, 1, 2\}$ and define p by

$p(0) = 1/4, \qquad p(1) = 1/2, \qquad p(2) = 1/4;$

see Figure 1.7. This arises with a pair of symmetric coins if we recognize only the events E_0, E_1, and E_2, that is, recognize only the number of heads showing; this is easily checked from the concluding results in Example 4. For a more general example, take $\mathcal{S}_0 = \{0, 1, \ldots, n\}$ and define p by

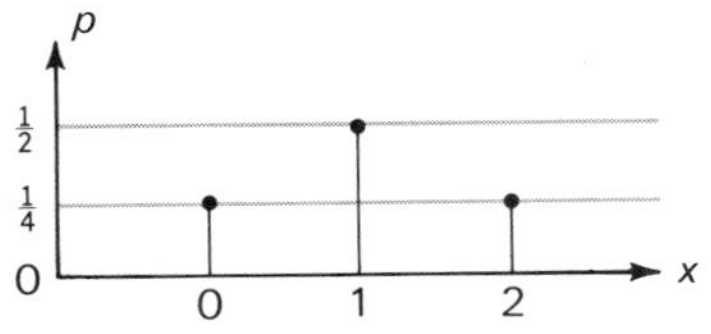

FIGURE 1.7
Binomial(2, 1/2) probabilities.

(4) $$p(s) = \frac{n!}{s!(n-s)!}(1/2)^n;$$

note that *n factorial* is defined by $n! = n(n-1)\cdots 2\cdot 1$ for positive integers n and $n! = 1$ for $n = 0$. These probabilities on $\mathcal{S}_0$ add up to 1 by the binomial theorem. We will see later that this distribution occurs with n symmetric coins if we recognize only the number of heads showing. This distribution is called the *binomial(n, 1/2) distribution*. The same distribution arises in the context of Example 1.2.1 (drug administered to rabbits: observe life or death in a test period) if we recognize only the number of deaths occurring and if the probability of a death is 1/2. More on this in Chapter 3, when we have developed the necessary concepts.

EXAMPLE 6

Poisson frequencies: For a slightly different example on the real line, take $\mathcal{S}_0 = \{0, 1, 2, \ldots\}$ and define p by

(5) $$p(s) = \frac{\lambda^s}{s!}e^{-\lambda} \qquad \text{for } s \in \mathcal{S}_0,$$

where λ is a nonnegative number. We check that the total is 1:

$$\Sigma\, p(s) = e^{-\lambda}\sum_0^\infty \frac{\lambda^s}{s!} = e^{-\lambda}e^{\lambda} = 1.$$

This distribution arises for the physicist counting the number of α-particles emitted by a radioactive source in a 7.5-second interval (mentioned in Section 1.2A); the value for λ would be the long-run average number of particles in a 7.5-second interval; in Section 2.1D we shall return to this illustration in more detail. This distribution is called the *Poisson(λ) distribution* and arises in many applications where we count the number of occurrences of some event in an interval of space or time: red blood cells in a square on a blood slide; accidents of a particular type in a week. See Figure 1.8 for the Poisson(λ) distribution with $\lambda = 2$.

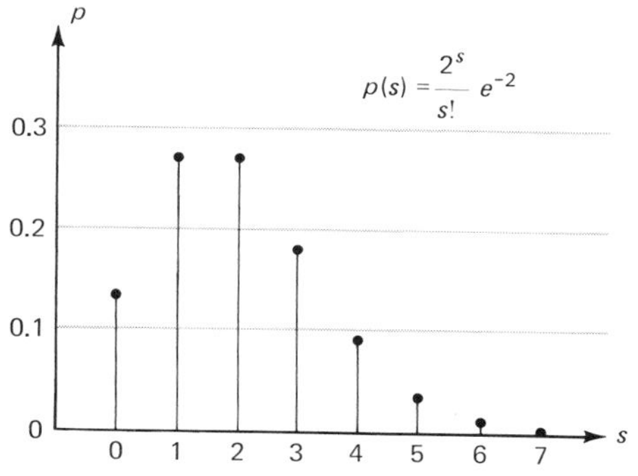

FIGURE 1.8
Poisson(λ) probabilities with $\lambda = 2$.

E SYMMETRIC PROBABILITY DISTRIBUTIONS

For the coin-tossing example in Section 1.1C, we saw how symmetry among the faces for each coin gave symmetry among the four elementary events H_1H_2, H_1T_2, T_1H_2, and T_1T_2. We will see many examples with such symmetry; in fact, almost all the traditional games of chance are based on symmetry. For an appropriate general model we now define a symmetric probability measure: a probability measure P as given by (3) is symmetric with respect to $\mathcal{S}_0 = \{a_1, \ldots, a_N\}$ if the probability weights $p(a_i) = p$ are all equal. Since $\Sigma_1^N p(a_i) = 1$, it follows that $p(a_i) = 1/N$; thus

$$(6) \qquad \bar{P}_2(A) = \sum_{a_i \in A} \frac{1}{N} = \frac{1}{N} c(A \cap \mathcal{S}_0) = \frac{c(A \cap \mathcal{S}_0)}{c(\mathcal{S}_0)}.$$

Note that the probability is the number of $\mathcal{S}_0$-points in A divided by the total number of $\mathcal{S}_0$-points. This is sometimes called the classical definition of probability, owing to its origins in the traditional games of chance.

EXAMPLE 7

Tossing two symmetric dice (Example 1.2.4): Consider two symmetric dice and suppose that they are tossed symmetrically. Let y_i be the number of points showing on the ith die. The space of possible outcomes is

$$\mathcal{S}_0 = \{(y_1, y_2) : y_i = 1, \ldots, 6\} = \{1, 2, 3, 4, 5, 6\}^2.$$

There are 36 points in $\mathcal{S}_0$ and the symmetry of the dice and tossing gives symmetry among the 36 points. This implies that the probability measure is the symmetric measure (6) with $N = 36$.

$$\bar{P}(A) = \frac{\text{number of possible points in } A}{\text{total number of possible points}} = \frac{c(A \cap \mathcal{S}_0)}{36}.$$

Thus we have the probability 1/36 at each of the marked points in Figure 1.2. And for the events discussed in Example 1.2.4, we have

$$\bar{P}(L \text{ ``at least one 6''}) = 11/36,$$

$$\bar{P}(T \text{ ``total points is 7''}) = 6/36,$$

$$\bar{P}(T \cap L) = 2/36.$$

F THE PROBABILITY MODEL

Consider a random system. We now have the mathematical structure that is needed to describe the variation observed from performance to performance of the system.

In Section 1.2 we considered the class $\mathcal{B}$ of events that we are interested in. The natural properties for such a class have been summarized in the requirement that $\mathcal{B}$ be a Boolean σ-algebra. In practice we shall often use a sample space $\mathcal{S}$ as a convenience; the events can then be expressed easily as subsets A of $\mathcal{S}$, subsets that belong to a σ-algebra $\mathcal{A}$ over $\mathcal{S}$. Of course, this convenience comes at a price: if we decide to observe more about the system, then we shall require a new sample space and a new σ-algebra; and we will then have a different representation for each of the original events in $\mathcal{B}$.

In the present section we have considered the function P that gives the long-run proportion of occurrences for any event. The natural properties for such a function have been summarized in the requirement that P be a probability measure. Our model, then, is $(\mathcal{B}, P)$, where $\mathcal{B}$ is the Boolean σ-algebra and P is the probability measure. $(\mathcal{B}, P)$ is called a *Boolean probability space*. If we include the sample space $\mathcal{S}$, then the model becomes $(\mathcal{S}, \mathcal{A}, P)$ and is called a *probability space*.

G EXERCISES: GAMES OF CHANCE

1 Three symmetric coins are tossed; let $x_i = 0(1)$ according as the ith coin is tails (heads).
(a) Record the effective sample space $\mathcal{S}_0$ for the outcomes (x_1, x_2, x_3).
(b) Determine the symmetric probability measure $\bar{P}$.
(c) Calculate $\bar{P}(E_0)$, $\bar{P}(E_1)$, $\bar{P}(E_2)$, and $\bar{P}(E_3)$, where E_j is the event "exactly j heads."

2 *Tossing two symmetric dice* (Example 7). Calculate the probabilities $P(T_2)$, $P(T_3)$, . . ., $P(T_{12})$, where T_j is the event "total of the points is j."

3 For tossing two coins and counting the heads on each, the effective sample space is $\mathcal{S}_0 = \{(x_1, x_2) : x_i = 0, 1\}$. Suppose that the coins are similar but possibly biased. Then in Section 3.5A we shall see that the measure P_2 in (3) is appropriate with

$$p(x_1, x_2) = p^{x_1}q^{1-x_1}p^{x_2}q^{1-x_2}, \qquad x_i = 0, 1,$$

where $p, q \geq 0$ and $p + q = 1$.
(a) Tabulate the values of the function p for the four sample points.
(b) Calculate the probabilities $P(H_1)$ and $P(T_1)$, where $H_1(T_1)$ are the events "heads (tails) for the first coin."

4 *Balls in an urn* (Exercise 1.2.7). An urn contains 3 green balls marked g_1, g_2, and g_3 and 1 black ball marked b. The balls are thoroughly mixed and two balls are drawn in succession, yielding (y_1, y_2), where y_i is the mark on the ith ball drawn.
(a) Determine the symmetric probability measure.
(b) Calculate $P(E_0)$, $P(E_1)$, and $P(E_2)$, where E_j is the occurrence of exactly j black balls.

5 *Cards from a deck* (Exercises 1.2.8 and 1.2.9). Calculate the symmetric probability measure for (a) two cards without replacement (Exercise 1.2.8); (b) two cards with replacement (Exercise 1.2.9).

6 *Calling cards* (Exercise 1.2.10). Four cards are numbered 1, 2, 3, and 4. The cards are thoroughly shuffled and dealt face up in succession as the dealer calls successively 1, 2, 3, and 4.
(a) Record the symmetric probability measure on the sample space $\mathcal{S}$ described earlier.
(b) Calculate $P(E_0)$, $P(E_1)$, $P(E_2)$, $P(E_3)$, and $P(E_4)$, where E_j is the event "exactly j cards called correctly." Another version has a secretary shuffling four letters and placing them in the four addressed envelopes.

H EXERCISES: INDUSTRY AND SCIENCE

7 *Transformers in a lot:* A lot contains 4 transformers serial marked 1, 2, 3, and 4; the first 3 are good (g) and the last defective (b). The transformers are thoroughly mixed and 2 are sampled in succession (a random sample of 2).
(a) Define an appropriate sample space having a symmetric probability measure; record the symmetric probability measure.
(b) Calculate $P(E_0)$, $P(E_1)$, and $P(E_2)$, where E_j is the occurrence of j defective transformers.

8 *Lifetime of transistors:* The lifetime t of a certain transistor is such that $u = e^{-t/5}$ with values

in (0, 1) has a very simple probability measure describing its variation: $P((a, b)) = b - a$ for any interval (a, b) contained in (0, 1).

(a) Calculate $P(0.2 < u < 0.8)$, $P(0.4 < u < 0.6)$.

(b) Calculate the probability that the first decimal place of u is even.

9 *Round-off error:* The values of a certain function are rounded off to the nearest integer. The error e is found to behave very much as the outcome of a random system that produces values in $(-0.5, 0.5)$ and has a simple probability measure: $P((a, b)) = b - a$ for any interval (a, b) contained in $(-0.5, 0.5)$.

(a) Calculate $P(|e| < 0.3)$, $P(|e| < 0.1)$.

(b) Calculate the probability that the first decimal place is even.

(c) Calculate the probability that the error is positive *and* the first decimal place is even.

10 *Round-off error (continuation):* The values for a pair of functions are rounded off to the nearest integers. The error (e_1, e_2) is found to behave very much as the outcome of a random system that produces values in $(-0.5, 0.5)^2$ and has a simple probability measure: $P(A)$ = area of A for any simple set contained in $(-0.5, 0.5)^2$.

(a) Calculate $P(|e_1| < 0.1)$ and $P(|e_2| < 0.1)$.

(b) Calculate $P(\max |e_i| < 0.1)$.

(c) Calculate $P((e_1^2 + e_2^2)^{1/2} < 0.1)$.

PROBLEMS

11 The definition of a measure μ required that there be a set A that has $\mu(A) < \infty$. Use the definition of a measure to prove the "obvious," that $\mu(\phi) = 0$.

12 (*continuation*) The definition of a measure is in terms of σ-additivity. Prove finite additivity,

$$\mu(\cup_1^k A_i) = \sum_1^k \mu(A_i),$$

where the A_i are disjoint.

13 (*continuation*) Prove monotonicity: If $A \subset B$, then $\mu(A) \leq \mu(B)$.

14 The Poisson(λ) probability measure in Example 6 has

$$P((-\infty, b]) = \sum_{s=0}^{b} \frac{\lambda^s}{s!} e^{-\lambda} \qquad \text{for } b = 0, 1, 2, \ldots.$$

Show that $P((-\infty, b])$ is a monotone-decreasing function of λ ($\lambda > 0$). (*Hint:* Differentiate with respect to λ and simplify.)

15 The binomial(n, p) probability measure has

$$P((-\infty, b]) = \sum_0^b \frac{n!}{s!(n-s)!} p^s(1-p)^{n-s} \qquad \text{for } b = 0, 1, \ldots, n,$$

where $p \in [0, 1]$. Show that $P((-\infty, b])$ is a monotone-decreasing function of p.

1.4

CALCULATING PROBABILITIES

We have been examining the mathematical framework for probabilities. We now examine some convenient rules that are useful for calculating probabilities. Our starting point is Definition 1.3.6 and we derive the formulas deductively. The properties in the definition can be summarized as (a) σ-additivity, and (b) total probability is 1.

A ADDITIVITY

We have noted in Section 1.3 that a probability measure P has the *null property* $P(\varnothing) = 0$ and the *additivity property*

(1) $$P(\cup_1^k A_i) = \sum_1^k P(A_i)$$

provided that the A_i are disjoint. The additivity property is often used in its simplest form with two component sets. For convenience we record this as a proposition:

PROPOSITION 1

If A and B are disjoint, then

$$P(A \cup B) = P(A) + P(B).$$

A special form of this additivity can be very useful:

PROPOSITION 2

The probability of an event A is 1 minus the probability of the complementary event:

(2) $$P(A) = 1 - P(A^c).$$

Proof The sample space $\mathcal{S}$ can be partitioned into the two sets A and A^c; thus

$$1 = P(\mathcal{S}) = P(A) + P(A^c),$$

and the proposition follows by rearrangement.

EXAMPLE 1

Symmetric coin: A symmetric coin is tossed 4 times and (x_1, x_2, x_3, x_4) is observed, where $x_i = 0(1)$ according as the ith coin is tails (heads). We have symmetry among the $2^4 = 16$ sample points; thus we use the symmetric measure in Section 1.3E. The event "at least one head" has many sample points, but the complementary event is very simple:

$$\begin{aligned} P(\text{at least one head}) &= 1 - P(\text{all tails}) \\ &= 1 - 1/16 = 15/16. \end{aligned}$$

B PARTIAL ADDITIVITY

If two events A and B are not disjoint, then the formula for $P(A \cup B)$ must reflect the overlap:

PROPOSITION 3

For arbitrary events A and B,

(3) $P(A \cup B) = P(A) + P(B) - P(AB).$

Proof The event $A \cup B$ can be partitioned into three events,

$A \cup B = AB^c \cup AB \cup A^cB;$

see Figure 1.9. Additivity then gives

$$\begin{aligned} P(A \cup B) &= P(AB^c) + P(AB) + P(A^cB) \\ &= [P(AB^c) + P(AB)] + [P(A^cB) + P(AB)] - P(AB) \\ &= P(A) + P(B) - P(AB). \end{aligned}$$

Sometimes the probabilities $P(A)$, $P(B)$, and $P(AB)$ are available in an immediate and direct way:

EXAMPLE 2

Symmetric dice: Two symmetric dice are tossed and the number of points on each is observed:

$$\begin{aligned} P(\text{at least one six}) &= P(\text{first die is a 6}) + P(\text{second die is a 6}) \\ &\quad - P(\text{both dice show 6}) \\ &= 1/6 + 1/6 - 1/36 = 11/36. \end{aligned}$$

The preceding proposition can be generalized to cover three events, A, B, and C:

PROPOSITION 4

For arbitrary events A, B, and C:

(4) $$\begin{aligned} P(A \cup B \cup C) = {} & P(A) + P(B) + P(C) \\ & - P(AB) - P(BC) - P(AC) + P(ABC). \end{aligned}$$

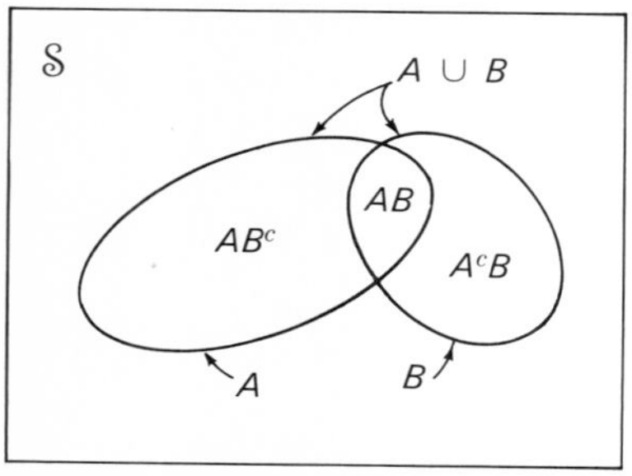

FIGURE 1.9
Event $A \cup B$ can be partitioned as $AB^c \cup AB \cup A^cB$.

This can be proved from Proposition 3, or by the method used with Proposition 3, or by a simple procedure with indicator functions to be presented in Supplement Section 1.5A.

C MONOTONICITY

Probabilities have a very simple, almost trivial, monotone property:

PROPOSITION 5

Probability is a monotone function of events: if $A \subset B$, then $P(A) \leq P(B)$.

Proof If $A \subset B$, then B can be partitioned as

$$B = A \cup A^c B,$$

which produces

$$P(B) = P(A) + P(A^c B) \geq P(A).$$

Note that we used the nonnegativity of $P(A^c B)$.

EXAMPLE 3

Serial system: A certain control system is composed of m component systems. Suppose that the full system functions if and only if each of its m components functions; such a system is called a *serial system*. Let A be the event "the system functions," and let A_j be the event "the jth component system functions." The serial property can then be expressed as

$$A = \cap_1^m A_j.$$

Thus $A \subset A_j$; and from the proposition it follows that $P(A) \leq P(A_j)$. This gives the rather obvious result,

$$P(\text{system functions}) \leq P(j\text{th component functions})$$

for arbitrary j.

D MONOTONE CONTINUITY

A sequence of events $A_1, A_2, \ldots$ is called a *monotone sequence* if

$$A_1 \subset A_2 \subset A_3 \subset \cdots \qquad \text{(monotone increasing)}$$

or if

$$A_1 \supset A_2 \supset A_3 \supset \cdots \qquad \text{(monotone decreasing)}.$$

For a monotone sequence we can define the *limit set* in a straightforward manner:

(5) $$\lim_{i\to\infty} A_i = \cup_1^\infty A_i \qquad \text{if monotone increasing}$$

$= \cap_1^\infty A_i$ if monotone decreasing;

see Figure 1.10. For monotone sequences we have a *continuity property:*

PROPOSITION 6

If $A_1, A_2, \ldots$ is a monotone sequence, then

$$P(\lim_{i\to\infty} A_i) = \lim_{i\to\infty} P(A_i).$$

Proof This proposition is just σ-additivity slightly disguised. First consider the increasing case. Let $B_i = A_i - A_{i-1}$ be the difference set $A_i A_{i-1}^c$:

$$B_1 = A_1$$

$$B_2 = A_2 - A_1$$

$$B_3 = A_3 - A_2.$$

Note that in going from A_{i-1} to A_i the set of "new" points is given by B_i; it then follows by induction that

$$\cup_1^n B_i = A_n = \cup_1^n A_i; \qquad \cup_1^\infty B_i = \cup_1^\infty A_i = \lim_{i\to\infty} A_i.$$

The sets $B_1, B_2, \ldots$, however, form a partition of $\cup_1^\infty B_i = \cup_1^\infty A_i$; thus

$$P(\cup_1^\infty A_i) = P(\cup_1^\infty B_i) = \sum_1^\infty P(B_i) = \lim_{n\to\infty} \sum_1^n P(B_i)$$

$$= \lim_{n\to\infty} P(\cup_1^n B_i) = \lim_{n\to\infty} P(A_n),$$

which proves the proposition for the increasing case. The proof for the decreasing case follows by complementation: if $A_1, A_2, \ldots$ is monotone decreasing, then A_1^c, $A_2^c, \ldots$ is monotone increasing.

EXAMPLE 4

Endless tossing: A symmetric coin is tossed until the first head appears; what is the probability that the first head appears on an even-numbered toss? Let C_i be the

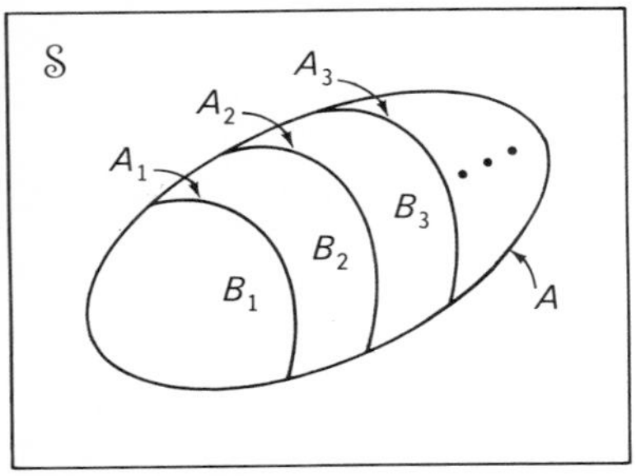

FIGURE 1.10
Monotone-increasing sequence with limit set A.

occurrence of the first head on the ith toss; let $A_{2n} = C_2 \cup C_4 \cup \cdots \cup C_{2n}$ be the occurrence of the first head on an even-numbered toss in the first $2n$ tosses; then the event of interest is $\lim_{n\to\infty} A_{2n}$ and we can use Proposition 6.

To cover all possible outcomes we must consider a countable number of tosses, and to have symmetry we must consider each possibility for each toss in the sequence. For this, let $i_j = 0(1)$ if the jth toss is tails (heads); then the sample space is

$$\mathcal{S} = \{(i_1, i_2, \ldots) : i_j = 0, 1\};$$

note that i_j is effectively the indicator function for heads on the jth toss. This is a complicated sample space, and we do not yet have the background for describing all the subsets or being sure there is a probability for each subset.

To see how big the sample space is, consider the following. A sample point $(i_1, i_2, \ldots)$ can be represented as a binary expansion, $0.i_1i_2\ldots$. This maps $\mathcal{S}$ into $[0, 1]$ in a way that is almost one to one; the exceptions are pairs such as $0.101\dot{1}$ and $0.11000\ldots$, which describe the same point just as $0.69\dot{9}$ and $0.7000\ldots$ in decimal expansion describe the same point.

In spite of not yet being able to determine the full probability model, we can, however, calculate the probability for the event C_2 using symmetry for the first two tosses, for C_4 using symmetry for the first four tosses, and so on:

$$P(A_{2n}) = P(C_2) + P(C_4) + \cdots + P(C_{2n})$$

$$= \frac{1}{2^2} + \frac{1}{2^4} + \cdots + \frac{1}{2^{2n}},$$

$$\lim_{n\to\infty} P(A_{2n}) = \frac{1}{4}\frac{1}{1 - 1/4} = \frac{1}{3}.$$

There are some interesting aspects to the mapping of $\mathcal{S}$ into $[0, 1]$. The probability is $1/2$ for each of the values 0, 1 in the first position; thus the probability is $1/2$ for each of the intervals $[0, 1/2)$, $[1/2, 1]$. The probability is $1/4$ for each of 00, 01, 10, 11 in the first two positions; thus the probability is $1/4$ for each of the intervals $[0, 1/4)$, $[1/4, 1/2)$, $[1/2, 3/4)$, $[3/4, 1]$. And so on. We thus obtain a uniform distribution of probability on the interval $[0, 1]$!

E ALTERNATIVE DEFINITION OF PROBABILITY

The continuity proposition (Proposition 6) has an important special case: if $A_1, A_2, \ldots$ is monotone decreasing with $\lim_{i\to\infty} A_i = \varnothing$, then

(6) $$\lim_{i\to\infty} P(A_i) = 0.$$

A function P with the preceding property is called *continuous at* $\varnothing$.

Definition 1.3.6 for a probability measure can now be presented in an alternative form that is often convenient:

DEFINITION 7

A probability measure P is a nonnegative function on a σ-algebra $\mathcal{A}$ such that
(a) Additive: if A_1 and A_2 are disjoint, then

$$P(A_1 \cup A_2) = P(A_1) + P(A_2).$$

(b) Continuous at $\varnothing$: if $A_1, A_2, \ldots$ is monotone decreasing to $\varnothing$, then

$$\lim_{i\to\infty} P(A_i) = 0.$$

(c) Normed: $P(\mathcal{S}) = 1$.

The results in Section D establish Definition 7 from Definition 1.3.6; Problem 17 is to establish the reverse.

Consider a monotone-decreasing sequence $A_1, A_2, \ldots$ with $\lim A_i = A$. Sometimes we will find that

$$\lim_{i\to\infty} P(A_i) = 0,$$

and yet A is not empty. Of course, by (6) we have that $P(A) = 0$. Such a set A may present the points at which some condition or proposition does not hold. We shall then speak of the condition holding *except on a set of probability zero*. Or we shall speak of the condition holding *with probability 1;* or holding *almost surely,* abbreviated *a.s.*

F EXERCISES: GAMES OF CHANCE

1 Consider the tossing of two symmetric dice. Calculate the probability that at least one die shows an even number; use the expression for $P(A \cup B)$.

2 Four cards are numbered 1, 2, 3, and 4. The cards are shuffled and dealt in succession while the dealer calls 1, 2, 3, and 4 in correspondence. Calculate the probability that at least one card is called incorrectly; see Exercises 1.2.10 and 1.3.6.

3 (*continuation*) Calculate the probability that at least 2 cards are called incorrectly.

4 A deck of 52 playing cards is thoroughly shuffled and 2 cards are dealt in succession. Let A_j be the occurrence of an ace on the jth draw; and let E_j be the occurrence of exactly j aces. Express E_j in terms of the A_j and calculate $P(E_0)$, $P(E_1)$, and $P(E_2)$; see Exercise 1.2.8.

5 A deck of 52 playing cards is shuffled; a card is dealt, observed, and returned to the deck; the deck is shuffled and a card is dealt. Let A_j be the occurrence of an ace on the jth draw; let E_j be the occurrence of exactly j aces. Express E_j in terms of the A_j; and calculate $P(E_0)$, $P(E_1)$, and $P(E_2)$; see Exercise 1.2.9.

6 A deck of 52 playing cards is thoroughly shuffled and dealt in succession while the dealer calls AS, KS, . . ., 2C in succession. Calculate the probability that at least one card is called incorrectly; that exactly one card is called incorrectly.

7 Each of two players, A and B, tosses a symmetric coin; a player wins if he gets heads and his opponent gets tails; if neither wins the double toss is called a draw. Two players toss repeatedly until there is a win. Calculate the probability that player A wins:
(a) Use the method in Example 4.
(b) Define a very elementary probability space and use symmetry.

8 A player tosses a symmetric coin until he obtains the first head. Calculate the probability that the first head occurs on a toss with number divisible by 3; see Example 4.

9 Each of two players tosses a symmetric coin; they are interested in a *match:* both show heads or both show tails. For a single trial, calculate the probability for a match. The players decide to toss repetitively until they obtain a match. Calculate the probability that a match occurs on an even-numbered trial; assume the symmetry and the results of Example 4.

G EXERCISES: INDUSTRY AND SCIENCE

10 Suppose that a shipment of 4 calculators has one defective. Suppose the calculators are thoroughly shuffled. The buyer tests the calculators one at a time. Calculate the probability that

(a) The last one tested is the defective calculator.

(b) The defective calculator is found before the last test. Compare with Exercise 1.3.7.

11 The error (e_1, e_2) in rounding off a pair of functions is found to behave as the outcome of a random system that produces values in $(-0.5, 0.5)^2$ and has a simple probability measure: $P(A)$ = area of A for any simple set $A \subset (-0.5, 0.5)^2$. Use the results calculated in Exercise 1.3.10 to obtain the probability that

(a) At least one of $|e_1|$, $|e_2|$ is ≥ 0.1.

(b) At least one of $|e_1|$, $|e_2|$ is < 0.1.

(c) $e_1^2 + e_2^2$ is ≥ 0.01.

12 A certain blood smear is obtained on a grid $(-5, +5) \times (-5, +5)$ on the plane. One particular cell is known to have a different color under ultraviolet light. Available theory says that the probability that the cell is in the set A is given by (area A)/100 for a set $A \subset (-5, 5)^2$. With the appropriate lighting the microscope makes a sweep covering one unit on either side of the first axis; calculate the probability of detection. Alternatively, the microscope makes a sweep within one unit of the second axis; calculate the probability of detection. The microscope makes a sweep within one unit of each of the axes; calculate the probability of detection.

H PROBLEMS

13 Use Proposition 3 to prove Proposition 4:

$$P(A_1 \cup A_2 \cup A_3) = \sum_i P(A_i) - \sum_{i<j} P(A_iA_j) + P(A_1A_2A_3). \tag{7}$$

14 (*continuation*) Let A_1, A_2, and A_3 be subsets of the sample space $\mathcal{S}$ and let E_r be "the occurrence of exactly r of A_1, A_2, and A_3." Prove the following formulas in the order $P(E_0)$, $P(E_3)$, $P(E_2)$, and $P(E_1)$:

$$P(E_0) = 1 - \sum P(A_i) + \sum_{i<j} P(A_iA_j) - P(A_1A_2A_3),$$

$$P(E_1) = \sum P(A_i) - 2 \sum_{i<j} P(A_iA_j) + 3P(A_1A_2A_3),$$

$$P(E_2) = \sum_{i<j} P(A_iA_j) - 3P(A_1A_2A_3),$$

$$P(E_3) = P(A_1A_2A_3).$$

15 Let $E_{[r]}$ be the event "at least r of $A_1, \ldots, A_3$." Use Problem 14 to prove the following:

$$P(E_{[0]}) = 1,$$

$$P(E_{[1]}) = \sum P(A_i) - \sum_{i<j} P(A_iA_j) + P(A_1A_2A_3),$$

$$P(E_{[2]}) = \sum_{i<j} P(A_iA_j) - 2P(A_1A_2A_3),$$

$$P(E_{[3]}) = P(A_1A_2A_3).$$

16 Three cards are numbered 1, 2, and 3. The cards are shuffled and dealt in succession while the dealer calls 1, 2, and 3 in correspondence. Use the results in Problem 14 to calculate $P(E_0)$, $P(E_1)$, $P(E_2)$, $P(E_3)$, where E_j is the event "exactly j cards called correctly." Of course, these probabilities are easily calculated by counting. But the formulas do illustrate the method needed for n cards; see Problems 1.5.3 and 1.5.4 in Supplement Section 1.5.

17 Show that Definition 7 implies Definition 1.3.6. (*Hint:* To show that *finite additivity* plus *continuity* implies *σ-additivity,* consider a finite union together with a "tail sum"; i.e., $A_1 \cup A_2 \cup \cdots \cup A_k \cup (\cup_{k+1}^{\infty} A_j)$ and let $k \to \infty$.)

18 *Generalization of Problem* 13: Let $A_1, \ldots, A_m$ be subsets of the sample space $\mathcal{S}$. Prove by induction that

$$P(\cup_1^m A_i) = S_1 - S_2 + S_3 - \cdots (-1)^{m-1}S_m,$$

where $S_r = \Sigma_{i_1<\cdots<i_r} P(A_{i_1} \cdots A_{i_r})$. A routine mechanical proof can be obtained from the results in Supplement Section 1.5.

19 For a monotone sequence $A_1, A_2, \ldots$ of subsets of a sample space $\mathcal{S}$, show that

$$I_{\lim_{i\to\infty} A_i} = \lim_{i\to\infty} I_i;$$

that is, show that

$$I_{\lim_{i\to\infty} A_i}(s) = \lim_{i\to\infty} I_i(s) \qquad \text{for each } s \in \mathcal{S}.$$

Recall the abbreviated method of proof in Section 1.2D.

20 (a) Prove: if $A_1, A_2, \ldots$ is a monotone-increasing sequence, then

$$\mu(\lim_{i\to\infty} A_i) = \lim_{i\to\infty} \mu(A_i).$$

(b) Prove: if $A_1, A_2, \ldots$ is a monotone-decreasing sequence and if $\mu(A_1)$, say, is finite, then

$$\mu(\lim_{i\to\infty} A_i) = \lim_{i\to\infty} \mu(A_i).$$

21 (*continuation*) Prove: if $A_1, A_2, \ldots$ is a monotone-decreasing sequence of sets having $\mu(A_1) < \infty$ and if $\lim A_i = \varnothing$, then

$$\lim_{i\to\infty} \mu(A_i) = 0.$$

Note: Consider a monotone-decreasing sequence $A_1, A_2, \ldots$ with $\lim A_i = A$. Sometimes we shall find that

$$\lim_{i\to\infty} \mu(A_i) = 0,$$

and yet A is not empty. Of course, $\mu(A) = 0$. Such a set A may present the points at which some condition or proposition does not hold. We shall then speak of the condition holding except on *a set of measure*(μ) *zero;* or holding *almost everywhere*(μ), abbreviated *a.e.*(μ).

SUPPLEMENTARY MATERIAL

Supplement Section 1.5 is on pages 511–519; it further develops some ideas in the preceding sections and helps toward making this text self-contained.

An indicator function takes the values 0 and 1; a simple generalization, a *simple function,* takes a finite number of values. The notion of *mean value* is introduced for such simple functions; this leads to easy proofs for *some probability formulas* recorded in the problems.

The *extension theorem* for measures is formally recorded. Among other things, it takes the notion of length for an interval and extends it to a *length measure* on the line, and similarly takes the notion of area for a rectangle and extends it to an *area measure* on the plane.

NOTES AND REFERENCES

The following are some textbooks that develop introductory probability theory.

Feller, W. (1968). *An Introduction to Probability Theory and Its Applications,* Vol. 1, 3rd ed. New York: John Wiley & Sons, Inc.

Fraser, D. A. S. (1958). *Statistics, An Introduction.* Huntington, N.Y.: Krieger Publishing Co.

Freund, J. E. (1971). *Mathematical Statistics.* Englewood Cliffs, N.J.: Prentice-Hall, Inc.

Hoel, P. G., S. C. Port, and C. J. Stone (1971). *Introduction to Probability Theory.* Boston: Houghton Mifflin Company.

Hogg, R. V., and A. T. Craig (1971). *Introduction to Mathematical Statistics,* 3rd ed. New York: Macmillan Publishing Co., Inc.

Kalbfleisch, J. G. (1971). *Probability and Statistical Inference.* Waterloo, Ontario: J. G. Kalbfleisch, University of Waterloo.

Mendenhall, W., and R. L. Scheaffer (1972). *Mathematical Statistics with Applications.* North Scituate, Mass.: Duxbury Press.

Parzen, E. (1960). *Modern Probability Theory and Its Applications.* New York: John Wiley & Sons, Inc.

Probability theory as a branch of mathematics has been developing since the seventeenth century. The current axiomatic formulation in terms of a probability measure is due to Kolmogorov (1933); this formulation brought the needed precision to probability theory. Some theory of Boolean algebras may be found in Halmos (1963). The extension theorem for measures may be found in advanced books on measure theory, for example, Ash (1972) and Halmos (1950).

Ash, R. B. (1972). *Real Analysis and Probability.* New York: Academic Press.

Halmos, P. R. (1950). *Measure Theory.* New York: Van Nostrand Reinhold Company.

Halmos, P. R. (1963). *Lectures on Boolean Algebras.* Van Nostrand Reinhold Company.

Kolmogorov, A. N. (1933). *Foundations of the Theory of Probability* (as translated, 1950). New York: Chelsea Publishing Company.

In Chapter 1 we examined the probability model and then investigated some routine methods for calculating probabilities.

A central interest of probability theory is of course the calculation of probabilities for events of interest in the real world. To approach this in detail we need to examine rather closely the standard models encountered in science, engineering, and other applications.

Most data in science, engineering, and business come in the form of real numbers. Sometimes a particular piece of data will be a measurement or determination of a physical variable with possible values on an interval of the real line. As examples of such quantitative determinations, consider the weight increase of an animal under a diet treatment, the temperature at which a certain transducer fails, and the thickness of cold-rolled steel plate. At other times a particular piece of data will be a count of how many objects in a sample of objects under investigation have some particular quality or characteristic. As examples of such qualitative data, consider the number of rabbits surviving a certain drug treatment, the number of bearings passing a simple "go, no go" test of diameter, and the number of α particles received in a 7.5-second time interval with a radioactive source.

In Chapter 2 we discuss various ways of describing a response that involves one or several real numbers. In the first two sections we examine ***real-valued responses:*** *first discrete responses, such as qualitative counts, and then continuous responses, such as quantitative determinations. The third section examines* ***vector-valued responses*** *involving several counts or determinations. The concluding section discusses* ***changes of variable,*** *the process of relabeling the response variable of a system under investigation.*

2

PROBABILITY ON THE LINE AND PLANE

2.1

PROBABILITY ON THE REAL LINE: DISCRETE

Consider a real-valued response from a random system, such as the measurement of a failure temperature for a transducer from a certain production run. By our discussions in Chapter 1 we know that we are interested in events such as the intervals on R. Closure under the standard operations then leads to the Borel class $\mathcal{B}^1$ of sets on R. For the full probability description we then need the function P that gives the probability for *each* set in the class $\mathcal{B}^1$. Actually, however, we are interested only in intervals and perhaps unions of intervals; indeed, we have no way in general of presenting an arbitrary Borel set. We first examine an alternative and simple way of presenting probability, by a distribution function that gives probabilities only for very special intervals.

A DISTRIBUTION FUNCTION ON THE REAL LINE

Consider a distribution of probability on the real line R as given by a measure; for example, this could be the distribution describing breaking load for a new design of concrete beam. We now define a distribution function that presents probability only for the special subsets $(-\infty, y]$, that is, from $-\infty$ up to and including a point y:

DEFINITION 1

The distribution function F is a **real-valued function** on R such that

(1) $F(y) = P((-\infty, y])$

at each point y.

The distribution function is sometimes called the *cumulative distribution function*.

If we have the distribution function F, we can calculate the probability for any interval $(a, b]$:

$$\begin{aligned} P((a, b]) &= P((-\infty, b]) - P((-\infty, a]) \\ &= F(b) - F(a). \end{aligned}$$

For notation it is convenient to use $(a, b]$ as an operator that says "take a difference" in the obvious way; thus

(2) $$P((a, b]) = (a, b]F = F(b) - F(a).$$

We can calculate the probability at a point b by first calculating it for an interval $(b - \delta, b]$ and then taking the limit as δ goes to zero:

$$P((b - \delta, b]) = F(b) - F(b - \delta),$$

$$P(\{b\}) = \lim_{\delta \to 0} P((b - \delta, b]) = F(b) - F(b - 0),$$

where we designate the limit from the left as

$$F(b - 0) = \lim_{\delta \to 0} F(b - \delta) \qquad \text{with } \delta > 0,$$

and we have used the continuity of Proposition 1.4.6. Again for notation it is convenient to use $(b - 0, b]$ as an operator; thus

(3) $$P(\{b\}) = (b - 0, b]F = F(b) - F(b - 0).$$

EXAMPLE 1

Heads with two symmetric coins: Consider tossing two symmetric coins and suppose that only the number y of heads is recorded: see Example 1.3.5. The possible outcomes are 0, 1, and 2 with corresponding probabilities 1/4, 1/2, and 1/4. The distribution function can be calculated easily:

$$\begin{aligned} F(y) &= 0 && \text{if } -\infty < y < 0, \\ &= 1/4 && 0 \le y < 1, \\ &= 3/4 && 1 \le y < 2, \\ &= 1 && 2 \le y < \infty; \end{aligned}$$

see Figure 2.1. The probability on the interval (0.5, 1.3] is

$$P((0.5, 1.3]) = F(1.3) - F(0.5) = 3/4 - 1/4 = 1/2 = P(\{1\}).$$

This example has probability at the discrete points 0, 1, and 2. A natural

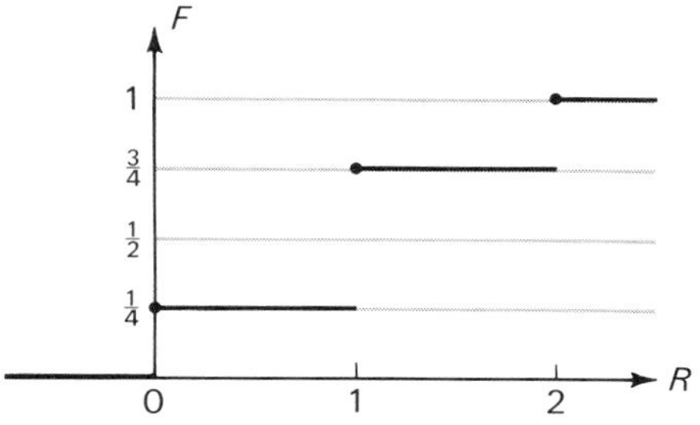

FIGURE 2.1
Distribution function for the number of heads when two symmetric coins are tossed.

generalization has probability $p(y)$ at points y in a finite or countable set $\mathcal{S}_0$; see the measure P_2 in Section 1.3D.

EXAMPLE 2

Orientation with two symmetric coins: Consider tossing two symmetric coins and recording the angle that the line joining them makes (measured positively) with the front edge of the table; any angle from 0° to 180° is possible. Symmetry gives, for example,

$$P((0, 10]) = P((10, 20]) = \cdots = P((170, 180]) = 1/18,$$

$$P((0, 5]) = P((5, 10]) = \cdots = P((175, 180]) = 1/36;$$

thus

$$\begin{aligned} F(y) &= P((-\infty, y]) \\ &= 0 && \text{if } -\infty < y < 0, \\ &= y/180 && 0 \le y < 180, \\ &= 1 && 180 \le y < \infty; \end{aligned}$$

see Figure 2.2. The probability on the interval $(-10, 80]$ is

$$\begin{aligned} P((-10, 80]) &= F(80) - F(-10) = 80/180 - 0 = 4/9 \\ &= P((0, 80]). \end{aligned}$$

This example has probability spread uniformly over the interval (0, 180] at the rate or density 1/180 per unit length. A natural generalization has probability on R spread at the rate $f(y)$ per unit length at the point y; this gives probability $f(y)\,dy$ for a small interval of length dy at the point y, gives probability $\int_c^d f(y)\,dy$, *the area under f between c and d,* for the interval (c, d), and, accordingly, gives probability 0 at a mathematical point.

B PROPERTIES OF A DISTRIBUTION FUNCTION

Consider a distribution of probability on the real line R as given by the measure P. Let F be the corresponding distribution function. We shall prove the following *necessary* properties of a distribution function F:

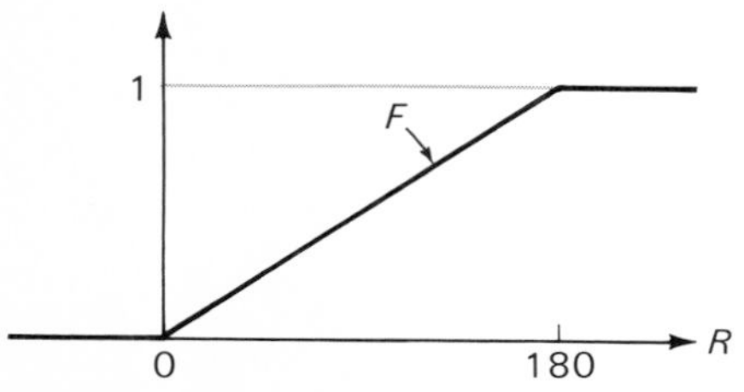

FIGURE 2.2
Distribution function F for the angle when two symmetric coins are tossed.

1. Monotone: $F(b) - F(a) = (a, b]F \geq 0$ for $b > a$.
2. Right continuous: $F(y + 0) = F(y)$ for all y.
3. Normed: $F(-\infty) = 0$, $F(+\infty) = 1$.

The proofs follow from the various properties that we have established for the probability measure P. For property 1 we note that $F(b) - F(a)$ is the probability for the interval $(a, b]$ and is thus ≥ 0. For property 2 we consider the interval $(y, y + \delta]$ with $\delta > 0$ and note that $\lim_{\delta \to 0} (y, y + \delta] = \varnothing$; thus $F(y + 0) - F(y)$ is equal to $P(\varnothing) = 0$ by the continuity Proposition 1.4.6. And for property 3 we note that $(-\infty, y]$ has limit $\varnothing$ as $y \to -\infty$ and limit R as $y \to +\infty$; thus $F(-\infty) = 0$, $F(+\infty) = 1$ by the same continuity property.

We now argue that properties 1–3 are also *sufficient* for a real-valued function F on R to be the distribution function of some measure P. We can define a function P for any interval $(a, b]$ by

$$P((a, b]) = (a, b]F = F(b) - F(a),$$

and property 1 ensures that this is ≥ 0. And for a finite union of disjoint intervals, the measure would be the sum of such values. The Extension Theorem recorded in Supplement Section 1.5D then gives a unique extension of P to all the sets in $\mathcal{B}^1$. Thus the distribution function F uniquely determines the probability measure P on the Borel sets. This is the same kind of argument as is used to extend the notion of length from intervals to the Borel sets on R; see Supplement Section 1.5E.

We now have two ways of presenting a distribution of probability on the sample space R: by means of the measure P or by means of the distribution function F. It is convenient, then, to have the general term *distribution* and to speak of a distribution with measure P or a distribution with distribution function F.

For notation we shall usually use capital letters such as F, G, H, . . . to designate distribution functions. Sometimes subscripts will be added: F_1, G_3, ... And sometimes, if the probability on R has come from somewhere else by means of a real-valued function, say Y, we will use that function as a subscript: F_Y.

C DISCRETE COMPONENT OF A DISTRIBUTION

In Example 1 we saw that the points 0, 1, and 2 received chunks of probability 1/4, 1/2, and 1/4. In Example 2 we saw that the probability was spread out continuously so that the probability at any mathematical point was zero. We now examine more formally the occurrence of chunks of probability.

Consider a distribution on the real line with measure P and distribution function F. Let p be the function that gives probability at a point:

DEFINITION 2

The **probability function** p is a real-valued function on R such that

(4) $\quad p(y) = P(\{y\}) = F(y) - F(y - 0)$

at each point y.

The probability function p is, in effect, a restriction of P to certain special sets, the sets that contain a single point. In Example 1 the function p has the values 1/4, 1/2, and 1/4 at the points 0, 1, and 2 and the value 0 elsewhere. In Example 2, p has the value zero everywhere.

A point at which $p(y) > 0$ is called a discrete point or, more exactly, a point having discrete probability. Consider the set of points having discrete probability:

$$\mathcal{S}_d = \{y : p(y) > 0\}. \tag{5}$$

It is straightforward to show that $\mathcal{S}_d$ has a finite or at most countable number of points; see Problem 16.

Now let F_d be a "distribution function" for the discrete probabilities; specifically, let F_d record the total discrete probability for any interval $(-\infty, y]$:

$$F_d(y) = P((-\infty, y] \cap \mathcal{S}_d) = \sum_{t \le y} p(t). \tag{6}$$

If all the probability is at discrete points, then $P(\mathcal{S}_d) = 1$ and $F_d(+\infty) = 1$; and if all the probability is spread continuously, then $P(\mathcal{S}_d) = 0$ and $F_d(+\infty) = 0$.

D DISCRETE DISTRIBUTIONS

A distribution is called "discrete" if all the probability is at discrete points in the manner just described. This is the kind of distribution that is useful for the qualitative counts mentioned in the preface to this section; for example, the distribution describing the number of α particles in a 7.5-second interval.

DEFINITION 3

A distribution is **discrete** if its measure and distribution function have the forms

$$P(A) = \sum_{y \in A} p(y), \qquad F(y) = \sum_{t \le y} p(t), \tag{7}$$

where $p(y)$ is given by Definition 2.

This is the measure P_2 in Section 1.3D, specialized to the real line R.

With a discrete distribution it is straightforward to calculate the probability for a set—sum the probabilities at component points.

Consider a simple example.

EXAMPLE 3

The Bernoulli(p) distribution: For the tossing of a possibly biased coin, let x be the number of heads observed. The effective sample space for x is $\{0, 1\}$, and all the probability must occur at the points 0 and 1 in the enlarged space $\mathcal{S} = R$. Let

$$p(0) = P(\{0\}) = q, \qquad p(1) = P(\{1\}) = p,$$

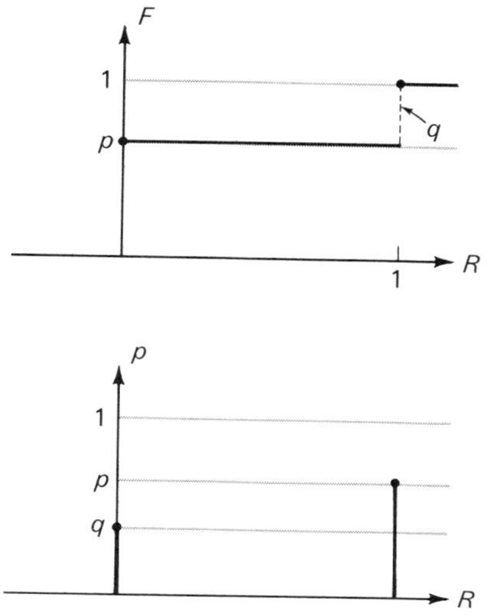

FIGURE 2.3
Distribution and probability functions for the Bernoulli(p) distribution.

where p, $q \geq 0$ and $p + q = 1$, and let $p(x) = 0$ for $x \neq 0, 1$. This gives a very simple discrete distribution; see Figure 2.3.

The example presents the Bernoulli distribution in the context of a classical game of chance, coin tossing. The applications for this distribution are, however, wide-ranging. Consider the administration of a drug to a certain kind of animal and the observation of, say, survival for a specified time interval; the mass production of an electronic component and the observation for a single component of, say, good (as opposed to defective) quality; the production of α-particles by a radioactive source and the observation of, say, ''no particles'' in a 7.5-second interval of time.

Now consider a discrete distribution with probability function p. The following are *necessary* properties of a probability function p; they follow from Definition 3.

1. Nonnegative: $p(y) \geq 0$ for all y.
2. Discrete: $p(y) = 0$ except at a finite or countable number of points.
3. Normed: $\Sigma\, p(y) = 1$.

Property 2 follows in a sense from 3 because the summation is meaningful only if there are at most a countable number of points with $p(y) > 0$; see Problem 16.

We now note that the properties 1–3 are also *sufficient* for a real-valued function p on R to be the probability function of a discrete distribution on R. This follows by noting that the properties give a function p as used in defining the discrete probability measure P_2 in Section 1.3D.

EXAMPLE 4

The Poisson(λ) distribution: Consider the Poisson distribution as described briefly in Example 1.3.6:

$$p(y) = \frac{\lambda^y}{y!} e^{-\lambda} \qquad \text{for } y = 0, 1, 2, \ldots,$$

$$= 0 \qquad \text{otherwise;}$$

$$P(A) = P(y \in A) = \sum_{y \in A} p(y) = \sum_{y \in A} \frac{\lambda^y}{y!} e^{-\lambda} \qquad \text{for nonnegative integers } y.$$

The Poisson probability function with $\lambda = 2$ is plotted in Figure 1.8; the values are best obtained by desk calculator or computer: $p(0) = 0.1353$, $p(1) = 0.2707$, $p(2) = 0.2707$, $p(3) = 0.1804, \ldots$. For example,

$$P(-\infty < y \leq 1.5) = 0.1353 + 0.2707 = 0.4060,$$

$$P(0.5 < y \leq 2) = 0.2707 + 0.2707 = 0.5414,$$

$$P(0.5 < y < 2) = 0.2707.$$

Recall from Example 1.3.6 that the Poisson distribution is useful for applications involving frequency counts of the occurrence of some event in an interval of space or time: blood cells on a slide, accidents in a week, and so on.

In the preceding example, note that we are using y to form events for a particular Poisson distribution; for example, the event $0.5 < y \leq 2$ in the general sense of Section 1.1A or the event $\{y : 0.5 < y \leq 2\}$ in the subset sense of Section 1.2C; the corresponding probability is

$$P(0.5 < y \leq 2) = P(\{y : 0.5 < y \leq 2\}) = 0.5414.$$

This is notationally very convenient—to use a particular letter, typically lowercase, for constructing sets on a designated space with a designated distribution. Note that when the letter *is* used, it is used only to present an event or to construct a subset as part of a probability calculation.

E LARGE-SAMPLE DATA FROM A DISCRETE DISTRIBUTION

Rutherford and Geiger took 2608 independent observations on the number y of α-particles in a 7.5-second interval; see Table 2.1. As scientists they were interested in many things but certainly in the pattern of variation for the response y. There are some theoretical reasons for thinking that the Poisson distribution would properly describe this variation. Accordingly, they would be interested in the appropriate Poisson distribution, that is, in estimating the true value for the parameter λ.

The column f_i records the frequency for each value y_i of the number of α-particles in a 7.5-second interval. The column $\hat{p}_i$ records the corresponding proportion or estimated probability. Thus, without any assumption concerning a pattern among the probabilities, we would estimate the probability for, say, $y = 2$ as 0.1469.

Now suppose that we accept the Poisson distribution for the number y; the theoretical grounds for this will be examined in Supplement Section 4.3. We are then faced with choosing a value for λ that makes the Poisson probabilities close in some

TABLE 2.1
Observations of α-particles in 7.5 second interval

Sample point y_i	Frequency f_i	Estimated probability $\hat{p}_i$	Estimated Poisson probability $\hat{\hat{p}}_i$	Expected Poisson frequency $e_i = 2608\hat{\hat{p}}_i$	Standardized residual $d_i = 2(\sqrt{f_i} - \sqrt{e_i})$
0	57	0.0219	0.0209	54.4	0.35
1	203	0.0778	0.0807	210.5	−0.52
2	383	0.1469	0.1562	407.3	−1.23
3	525	0.2013	0.2015	525.4	−0.02
4	532	0.2040	0.1950	508.4	1.03
5	408	0.1564	0.1509	393.5	0.72
6	273	0.1047	0.0973	253.9	1.18
7	139	0.0533	0.0538	140.4	−0.11
8	45	0.0172	0.0260	67.9	−3.06
9	27	0.0104	0.0112	29.2	0.42
10	10	0.0038	0.0043	11.3	−0.40
11	4	0.0015	0.0015	4.0	
12	2	0.0008	0.0005	1.3	
13	0	0.0000	0.0002	0.4	
14	0	0.0000	0.0000	0.1	
11–14[a]	(6)	(0.0023)	(0.0022)	(5.8)	(0.08)
	2608	1.0000	1.0000	2608.0	

[a] For comparing frequencies with expected frequencies it is conventional to group together cells with "expected" frequencies less than 5, thus the combined range 11–14.

sense to the observed proportions; this is essentially a statistical problem and the theoretical aspects will be examined in Chapter 9. For the present, however, we can argue rather directly to an appropriate value.

The mean or long-run average value of y for the Poisson(λ) distribution is just λ; see Problem 20 in Supplement Section 1.5. The Rutherford and Geiger data have a total of

$$0 \times 57 + 1 \times 203 + 2 \times 383 + \cdots = 10{,}094$$

α-particles for a total of 2608 time intervals. The average number of α-particles for a 7.5-second interval is thus

$$\frac{10{,}094}{2{,}608} = 3.87.$$

This is an average value for 2608 repetitions; as such it should provide a reasonable approximation for the mean or long-run average value λ. Accordingly, we will use $\hat{\lambda} = 3.87$ as an estimate of the true value λ. The estimated Poisson probabilities based on $\hat{\lambda} = 3.87$ are recorded in the column $\hat{\hat{p}}_i$ in Table 2.1. Thus on the basis of the Poisson pattern we would estimate the probability for, say, $y = 2$ as 0.1562. Note, of course, that this estimate is based not just on the frequency at the value 2 but on all the frequencies in the table.

Parenthetically, we can note that the average can be calculated for other time intervals. For example, the total count 10,094 occurred with an overall time interval of $2608 \times 7.5 = 19{,}560.0$ seconds. The average value for a 1-second interval is thus

$$\frac{10{,}094}{19{,}560} = \frac{3.87}{7.5} = 0.516,$$

and for a 2-second time interval is 1.032.

Now suppose that we want to check whether the Poisson pattern is appropriate. In part, this is a question of fit and is examined in detail in Section 10.4. For the present we record the calculations and defer the justifications until Chapter 10. We can compare the observed frequencies f_i with "expected" frequencies

$$e_i = 2608\hat{\hat{p}}_i$$

obtained by apportioning 2608 according to the proportions $\hat{\hat{p}}_i$. The comparison with expectation is made in suitable units by calculating the standardized residual

$$d_i = 2(\sqrt{f_i} - \sqrt{e_i});$$

some justifications for this will be examined in Chapter 6. The values of d_i are recorded in the right-hand column. If a Poisson distribution describes the counting process, then, subject to minor qualifications, the deviations will fall in the intervals $(-1, +1)$, $(-2, +2)$, and $(-3, +3)$ with probabilities $68\frac{1}{4}$, $95\frac{1}{2}$, and $99\frac{3}{4}$ percent approximately. The observed values seem moderately in accord with this distribution, thus lending some reassurance to the use of the Poisson pattern.

An overall test of the residuals is also available from Section 10.4; the details would go as follows. The test is made by calculating

$$\chi^2 = \sum d_i^2 = 14.60$$

and comparing with a chi-square distribution (Appendix II) with 10 degrees of freedom (12 residuals, less 1 because $\Sigma e_i = \Sigma f_i$, less 1 because the parameter λ has been fitted to the data). The observed value 14.60 is a reasonable value for this chi-square distribution, thus giving some overall assurance to the use of the Poisson pattern.

F EXERCISES

1 A symmetric die is tossed and the number y of points is recorded. Let F be the distribution function for y.
   (a) Present the distribution function F and make a sketch of it.
   (b) Use the distribution function to calculate the probabilities for the following events: $3 < y \le 5$; and $4 < y \le 7$.

2 (*continuation*) For a suitable price a gambling house will pay $x = y^2$ on a roll of the die. Let G be the distribution function for x.
   (a) Present the distribution function G and make a sketch of it.
   (b) Calculate the probabilities for the following events: $9 < x \le 25$; and $16 < x \le 49$.

3 For the Poisson(1) distribution, calculate the values of $p(y)$ successively; use $e^{-1} = 0.3679$. Calculate $P(y < 1)$; $P(y = 1)$; and $P(y > 1)$.

4 At a certain time of day the number of incoming calls for a 1-minute interval is approximately Poisson(2). Calculate the probability that less than 2 calls occur in a 1-minute interval; that exactly 2 occur in a 1-minute interval; and that more than 2 occur in a 1-minute interval.

5 In a certain city the number of fatal accidents is approximately Poisson with an average of 1 per week. For a 3-week period, calculate the probability for zero fatal accidents; for less than 3; and for 3 or more.

6 The number of fish caught by a fisherman at a certain location is approximately Poisson, with an average of 1 per hour. Calculate the probability that a 3-hour fishing period produces zero fish.

7 The number of defects of a certain kind in steel cable is approximately Poisson, with an average of 2 per 1000 feet. Calculate the probability of zero defects in a 250-foot coil; and 2 or more defects in a 250-foot coil.

G EXERCISES

Some common discrete distributions are recorded in these exercises together with an indication of various applications. The reader should become familiar with general characteristics of each. The grounds for using some of these in applications will be examined in Section 3.5.

8 The *uniform*$\{1, 2, \ldots, N\}$ *distribution* has probability function

$$p(y) = 1/N \qquad \text{if } y = 1, 2, \ldots, N$$

and zero otherwise; see Figure 2.4.

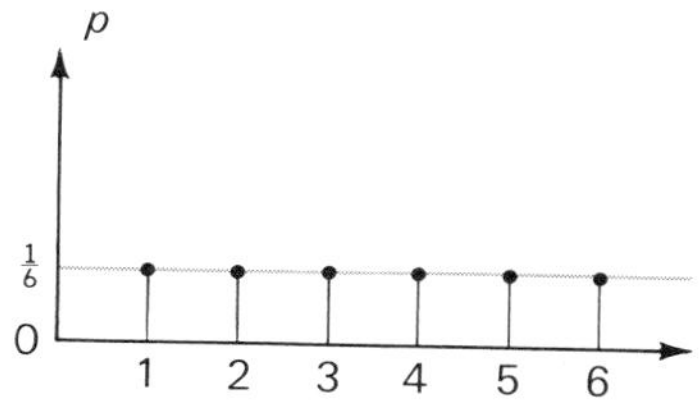

FIGURE 2.4
Uniform distribution on $\{1, 2, \ldots, 6\}$.

Applications: A symmetric die produces a score y; $N = 6$. A computer can generate *random bits,* 0, 1, with equal probability 1/2; $N = 2$, and the value 2 is identified with 0. A computer can generate *random digits,* 0, . . ., 9, with equal probability 1/10; $N = 10$, and the value 10 is identified with 0. Computer generation uses arithmetic operations and cyclic effects can occur, thus voiding the constant conditions presupposed for a random system.
(a) Present the distribution function F.
(b) Calculate the probability for the following intervals: $(-\infty, N/2]$ and $(N/4, 3N/4]$.

9 The *geometric*(p) *distribution* has probability function

$$p(y) = pq^y \qquad \text{if } y = 0, 1, 2, \ldots$$

and zero otherwise, where $p > 0$, $q \geq 0$ and $p + q = 1$; see Figure 2.5.

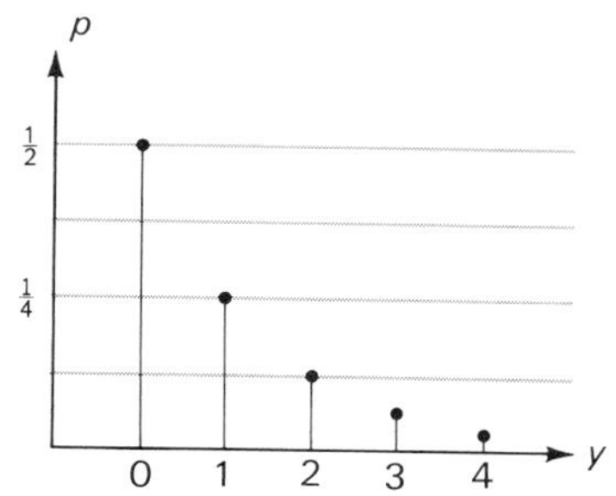

FIGURE 2.5
Geometric(1/2) distribution: number of tails until the first head.

Applications: A coin with probability p for heads is tossed until a head occurs; y is the number of tails obtained. A subject tries a test having probability p of success; without learning, y is the number of failures until the first success.

(a) Show that $p(\cdot)$ is a probability function for a discrete distribution.

(b) Calculate the distribution function and sketch it for $p = 2/3$.

10 The *binomial(n, p) distribution* has probability function

$$p(y) = \frac{n!}{y!(n-y)!} p^y q^{n-y} \qquad \text{if } y = 0, 1, \ldots, n$$

and zero otherwise, where $p, q \geq 0$ and $p + q = 1$; note that $p^0 = 1$ for all values of p; see Figure 2.6.

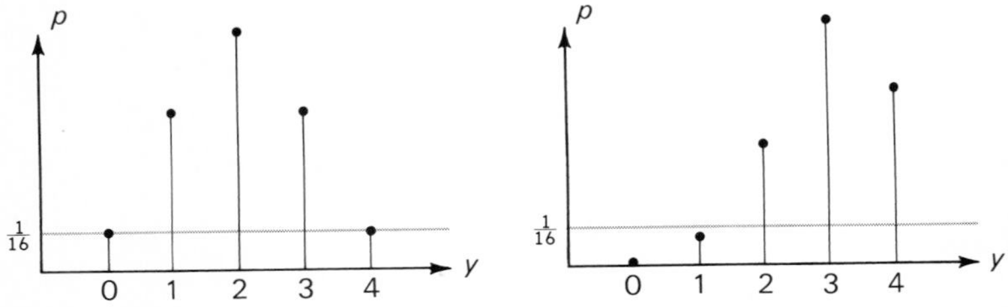

FIGURE 2.6
Binomial(4, 1/2) and (4, 3/4) distributions.

Applications: A symmetric coin is tossed n times and the number y of heads is counted; $p = 1/2$. Each of n rabbits is separately subjected to a drug treatment having probability p of success; y is the number of successes. Each of n electronic systems has a probability p of failure; y is the number of failures.

(a) Tabulate $p(\cdot)$ for the case $n = 4$, $p = 1/2$; calculate the probability for the event $1 \leq y \leq 3$.

(b) Tabulate $p(\cdot)$ for the case $n = 4$, $p = 3/4$; calculate the probability for the event $1 \leq y \leq 3$.

11 The *negative binomial(r, p) distribution* has probability function

$$p(y) = \frac{(y + r - 1)!}{y!(r-1)!} p^r q^y \qquad \text{if } y = 0, 1, \ldots$$

and zero otherwise, where r is a positive integer and $p > 0$, $q \geq 0$, $p + q = 1$; see Figure 2.7.

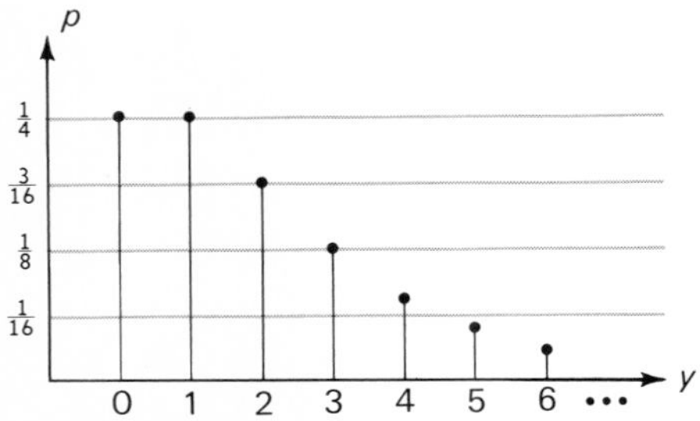

FIGURE 2.7
Negative binomial(2, 1/2) distribution.

Applications: A die is tossed until the r aces occur; y is the number of nonaces obtained ($p = 1/6$). Separate components have a probability p of being good; y is the number of defectives until r good components have been obtained.

(a) Show that $r = 1$ gives the geometric(p) distribution.
(b) Tabulate the probability function for the case $r = 2$, $p = 1/2$.

12 The *hypergeometric*(N, n, D/N) *distribution* has probability function

$$p(y) = \frac{n!}{y!(n-y)!}\,\frac{D!}{(D-y)!}\,\frac{(N-D)!}{(N-D-n+y)!}\,\frac{(N-n)!}{N!}$$

for integer values such that $\max(0, n - N + D) \leq y \leq \min(n, D)$ and zero otherwise; see Figure 2.8.

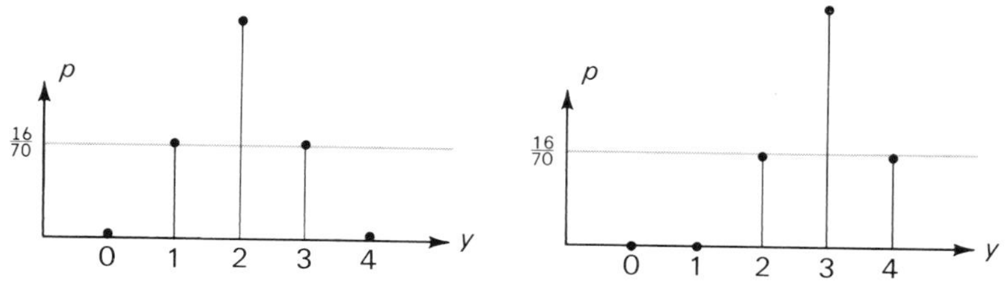

FIGURE 2.8
Hypergeometric(8, 4, 1/2) and (8, 4, 3/4) distributions.

Applications: An urn contains D black balls and $N - D$ white balls; the balls are mixed and n balls are drawn; y is the number of black balls. A lot contains D defectives and $N - D$ good items; the items are mixed and n items are drawn; y is the number of defectives.
(a) Tabulate $p(\cdot)$ for $N = 8$, $n = 4$, and $D/N = 4/8$; calculate the probability that $1 \leq y \leq 3$.
(b) Tabulate $p(\cdot)$ for $N = 8$, $n = 4$, and $D/N = 6/8$; calculate the probability that $1 \leq y \leq 3$.

H PROBLEMS

13 Define G_d by $G_d(y) = F_d(y)/p_d$ provided that $p_d = F_d(+\infty) \neq 0$. Show that G_d is a distribution function for a discrete distribution.

14 Let $\mathcal{S}_c = \{y : F \text{ is continuous at } y\}$; and define F_c by $F_c(y) = P(\mathcal{S}_c \cap (-\infty, y])$. Show that F_c is a continuous function.

15 (*continuation*) Define $G_c(y) = F_c(y)/p_c$ provided that $p_c = F_c(\infty) \neq 0$. Show that G_c is a distribution function. Note that $F(y) = p_d G_d(y) + p_c G_c(y)$; thus the distribution function has been decomposed into a mixture of discrete and continuous components.

16 Show that the set $\mathcal{S}_d$ in Section C is finite or countable. For this, let $A_n = \{y : p(y) > 1/n\}$ and prove that $\mathcal{S}_d = \cup_1^\infty A_n$.

2.2

PROBABILITY ON THE REAL LINE: ABSOLUTELY CONTINUOUS

We now discuss the distribution for a response that can take any value over a continuous range. Recall the quantitative determinations mentioned in the preface to this chapter; as an example, consider the distribution that describes the weight increase of an animal under a diet treatment.

A ABSOLUTELY CONTINUOUS COMPONENT OF A DISTRIBUTION

First, consider in general a distribution on the real line with measure P and distribution function F. Let f be a function that gives probability density at a point:

DEFINITION 1

The **probability density function** f is a real-valued function on R defined by

$$f(y) = dF(y)/dy = F'(y) \qquad \text{if } F'(y) \text{ exists}$$

and is equal to zero otherwise.

For the number of heads when two coins are tossed (Example 2.1.1), we have $f(y) = 0$ everywhere. And for the orientation when two coins are tossed (Example 2.1.2), we have $f(y) = 1/180$ on $(0, 180)$ and $= 0$ elsewhere.

A point at which $F'(y)$ exists is called a *point having probability density*.

Consider the set of all points having probability density:

(1) $$\mathcal{S}_a = \{y : F'(y) \text{ exists}\},$$

where the subscript a denotes *absolutely continuous*. The set $\mathcal{S}_a$ is *most* of the real line R. The complementary set $R - \mathcal{S}_a$, where $F'(y)$ does not exist, contains, of course, the set $\mathcal{S}_a$ of discrete points and, indeed, it may contain other points, but it is *not* large; advanced analysis shows that it has length measure zero (for some details, see Supplement Sections 1.5E and 1.5F).

Now let F_a be a ''distribution function'' for the absolutely continuous probability; specifically, let F_a record the total absolutely continuous probability for any interval $(-\infty, y]$:

(2) $$F_a(y) = P((-\infty, y] \cap \mathcal{S}_a) = \int_{-\infty}^{y} f(t)\, dt.$$

We can justify the expression in terms of an integral by using ordinary calculus provided that $F'(y)$ is continuous. In general, however, we would need the concept of the general integral and some related analysis. Supplement Section 5.8 contains a brief introduction to the general integral. For all our examples, however, we will be able to use the ordinary integral.

If all the probability is at points having probability density, then $P(\mathcal{S}_a) = 1$ and $F_a(+\infty) = 1$; in this case the absolutely continuous component accounts for the full distribution.

B ABSOLUTELY CONTINUOUS DISTRIBUTIONS

A distribution is called ''absolutely continuous'' if all the probability is at points having probability density. This is the kind of distribution that is useful for the quantitative determinations discussed in the preface to this chapter.

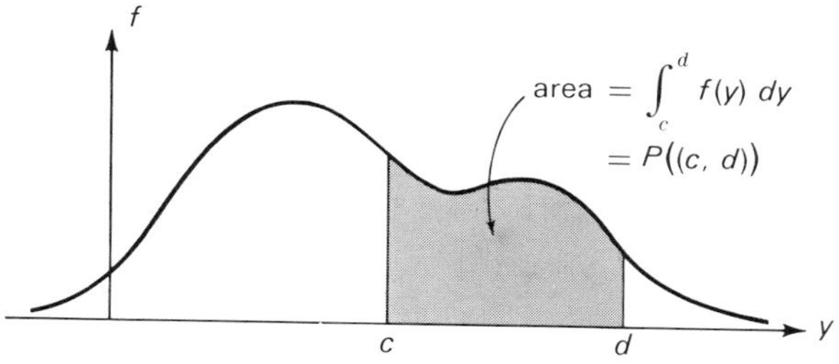

FIGURE 2.9
Probability density f. The probability for the interval (c, d) is given by the area over (c, d) and under f.

DEFINITION 2

A distribution is **absolutely continuous** if its measure and distribution function have the forms

$$P(A) = \int_A f(y)\,dy, \qquad F(y) = \int_{-\infty}^{y} f(t)\,dt,$$

where $f(y)$ is given by Definition 1.

With an absolutely continuous distribution it is straightforward to calculate the probability for a set—integrate over the set. For example, to calculate the probability for an interval (c, d), we determine the *area* $\int_c^d f(y)\,dy$ over the interval (c, d) and under the density f; see Figure 2.9. And we would obtain the same value for each of the intervals $[c, d)$, $(c, d]$, and $[c, d]$, for clearly the area over a *mathematical* point such as c or d is zero.

EXAMPLE 1

Uniform(a, b) distribution: The uniform distribution on the interval (a, b) is defined by

$$f(y) = \frac{1}{b-a} \qquad \text{if } a < y < b$$

and zero otherwise; see Figure 2.10. Note that f is a multiple, $1/(b-a)$, of the indicator function $I_{(a,b)}$ of the interval (a, b). The distribution function F is

$$\begin{aligned} F(y) = \int_{-\infty}^{y} f(t)\,dt &= 0 && \text{if } -\infty < y \le a, \\ &= \frac{y-a}{b-a} && a < y < b, \\ &= 1 && b \le y < \infty. \end{aligned}$$

Probabilities are easily calculated; for example, with an interval (c, d) contained in (a, b),

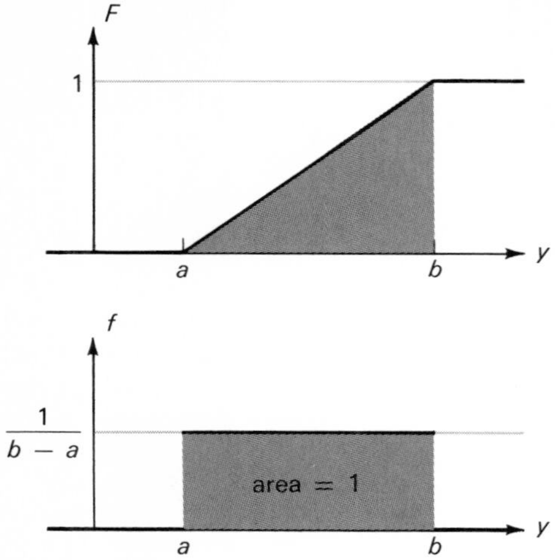

FIGURE 2.10
Distribution function and probability density function for the uniform (a, b) distribution.

$$P((c, d)) = \int_c^d f(y)\, dy = \int_c^d \frac{1}{b - a}\, dy = \frac{d - c}{b - a},$$

which is, of course, just the length of the interval (c, d) as a proportion of the length of the full interval; see Exercises 1.3.8 and 1.3.9.

For applications of this distribution, consider the following. The orientation when two coins are tossed [Example 2.1.2 with $(a, b) = (0, 180)$]. The round-off error with certain functions (Exercise 1.3.9); with round-off to the nearest integer, $(a, b) = (-0.5, 0.5)$. Certain exponential expressions for lifetime in the no-aging case; Exercise 1.3.8 with $u = e^{-t/5}$ and $(a, b) = (0, 1)$. The binary expansion $0.i_1i_2i_3 \cdots = \Sigma_1^\infty i_j 2^{-j}$, where $i_1, i_2, \ldots$ are the indicators for heads in a succession of tosses with a symmetric coin (Example 1.4.4); or where $i_1, i_2, \ldots$ are random bits (Exercise 2.1.8); here $(a, b) = (0, 1)$; the distribution is approximate for a finite expansion such as $0.i_1i_2i_3i_4i_51$. The decimal expansion $0.i_1i_2i_3 \cdots = \Sigma_1^\infty i_j \cdot 10^{-j}$, where $i_1, i_2, \ldots$ are random digits (Exercise 2.1.8). The theoretical basis for some of these applications will be available from Section 3.3.

Now consider an absolutely continuous distribution with probability density function f. The following are necessary properties of a density function f; they follow from Definitions 1 and 2.

1. Nonnegative: $f(y) \geq 0$ for all y.
2. Measurable: $\{y : f(y) \leq c\} \in \mathcal{B}^1$ for all c.
3. Normed: $\int_{-\infty}^{\infty} f(y)\, dy = 1$.

Property 1 follows from Definition 1. Property 3 says that $F_a(+\infty) = 1$. These are the basic conditions for an absolutely continuous distribution. Property 2 says that an interval condition on f must be an event in the Borel class $\mathcal{B}^1$; the proof is beyond our present scope; throughout the text and problems we will *assume* that this property holds.

Properties 1–3 are also *sufficient* for a function f to define an absolutely continuous distribution. This follows easily by noting that

$$\int_{-\infty}^{y} f(t)\,dt$$

satisfies the conditions for a distribution function as recorded in Section 2.1B. The expressions for $P(A)$ and $F(y)$ in Definition 2 are not affected if we change f at a point, or on a countable set of points, or even on a more general set of length measure zero; formally this needs the theory of the general integral. For our purposes we assume that f is a function that *does* give probability density, as discussed at the beginning of this section.

C NORMAL DISTRIBUTION

Many quantitative responses produce values that concentrate around a general level for the variable and drop off rather sharply in a bell-shaped curve such as in Figure 2.11. The normal distribution has traditionally been used as an approximation for such a pattern of variation. For example, it seems to approximate very closely the controlled laboratory measurement of a physical constant, say the gravitational constant. If coordinates are taken relative to the general location and are appropriately scaled, then the normal distribution has the density function

$$g(z) = \frac{1}{\sqrt{2\pi}} e^{-z^2/2};$$

this is the *standard normal distribution,* and its density function g is plotted in Figure 2.11. Property 1 for a density is immediate; property 3 is somewhat complicated but follows easily from calculations in Section 2.3E.

Probabilities for the standard normal distribution can be obtained from the standard normal distribution function G, which is tabulated in Appendix II:

$$P((-1,\ 1)) = \int_{-1}^{+1} g(z)\,dz = 2(G(1) - G(0)) = 68.26\%,$$

$$P((-2,\ 2)) = \int_{-2}^{+2} g(z)\,dz = 2(G(2) - G(0)) = 95.45\%,$$

$$P((-3,\ 3)) = \int_{-3}^{+3} g(z)\,dz = 2(G(3) - G(0)) = 99.73\%;$$

these calculations use the symmetry of g about the origin.

Now suppose that we relabel the variation z by the transformation

$$y = \mu + \sigma z,$$

which scales it by the factor σ (>0) and locates it at the general level μ. The standard normal then becomes the *general normal,* which can describe a response with values located at a general level μ and scaled about μ by the factor σ; see Figure 2.11.

Let f be the density function and F the distribution function for this relabeled response variable y. The event that the original variable is less than or equal to a value

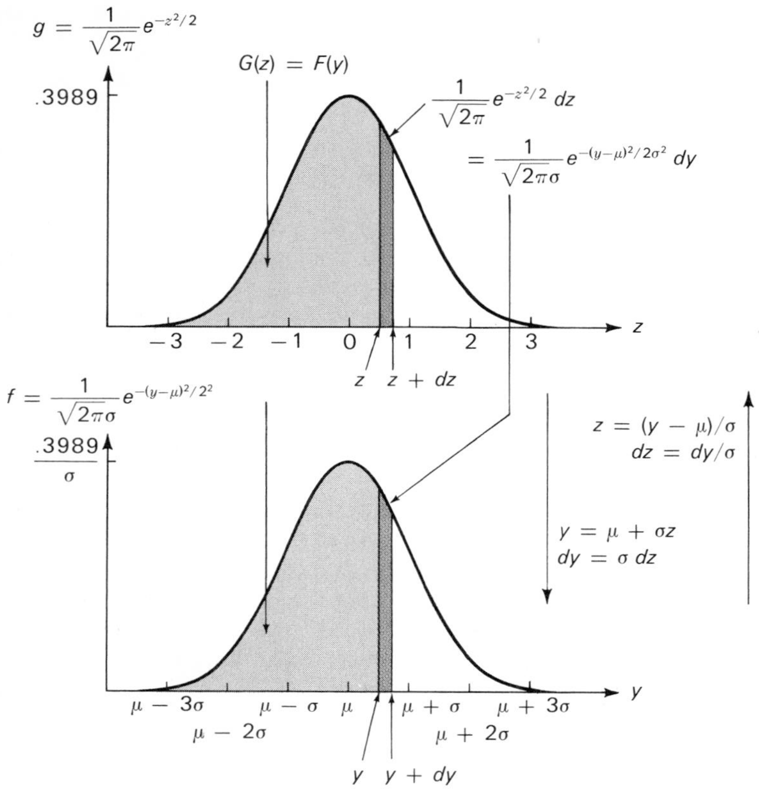

FIGURE 2.11
Standard normal density function g and density function f for the general normal located at μ and scaled by σ.

z is the same as the event that the new variable is less than or equal to the value $y = \mu + \sigma z$. Solving for z gives $z = (y - \mu)/\sigma$. Thus we obtain the distribution function

$$F(y) = G\left(\frac{y - \mu}{\sigma}\right) = \int_{-\infty}^{(y-\mu)/\sigma} \frac{1}{\sqrt{2\pi}} e^{-t^2/2}\, dt$$

$$= \int_{-\infty}^{y} \frac{1}{\sqrt{2\pi}\sigma} e^{-(s-\mu)^2/2\sigma^2}\, ds,$$

where $z = (y - \mu)/\sigma$ becomes $t = (s - \mu)/\sigma$ for the variables in the integration. Then by differentiation we obtain the density function

$$f(y) = \frac{1}{\sqrt{2\pi}\sigma} e^{-(y-\mu)^2/2\sigma^2} = \frac{1}{\sqrt{2\pi}\sigma} \exp\left\{-\frac{(y - \mu)^2}{2\sigma^2}\right\}.$$

This is called the *normal*(μ, σ) *distribution,* the normal located at μ and scaled by σ or, as we shall see later, the normal with *mean* μ and *standard deviation* σ.

Perhaps the most convenient way to remember an absolutely continuous distribution is in terms of the *probability differential,*

$$\frac{1}{\sqrt{2\pi}} e^{-z^2/2}\, dz = \frac{1}{\sqrt{2\pi}\sigma} e^{-(y-\mu)^2/2\sigma^2}\, dy;$$

this represents the probability in an increment dz for the original variable z and in the corresponding increment dy for the relabeled variable y; see Figure 2.11. Note that the density changes by the factor $1/\sigma$ and length changes by the factor σ; these compensate each other, giving the same area above corresponding increments. The differential can be interpreted as a linear form; or it can be viewed as a formal piece of an integral used for calculating probabilities.

Probabilities for the general normal are easily obtained from the tables for the standard normal; consider an example.

EXAMPLE 2

A psychologist's measuring instrument produces a response y that is normal(500, 100) for a certain population of students. Find the probability that a measurement y is less than or equal to 600:

$$P(y \le 600) = P\left(\frac{y - 500}{100} \le \frac{600 - 500}{100}\right)$$

$$= P(z \le 1) = G(1) = 0.8413.$$

Note that y is used to form events for the measurement variable and z is used to form events for the standardized variable describing the variation.

D LARGE-SAMPLE DATA FROM AN ABSOLUTELY CONTINUOUS DISTRIBUTION

In an application we can always think of taking a large number of repetitions, recording the frequency for some event of interest, and then dividing by the number of repetitions to obtain the proportion or estimated probability for the event. However, if we are interested in the density function at some point y, we cannot fruitfully follow the preceding procedure because the probability *at* a point is zero. But we can take small intervals and use the procedure to estimate the corresponding probabilities.

In practice, values will occur to three decimal places, say; or to five decimal places; or within intervals if the real line has been partitioned into short intervals or *cells*. Of course, then the distribution is technically a discrete distribution. But it is usually more convenient to remain with the continuous distribution as a reasonable approximation and to acknowledge that there may be effects due to round-off to the cell midpoints.

Grumell and Dunningham have examined the percent ash in 250 samples of coal from a particular source. The values are recorded in Table 2.2 in terms of frequencies for various cells on the range of the variable; some cells with low frequencies are combined and recorded at the beginning and at the end of the table.

The column $\hat{p}$ records the proportion or estimated probability for each of the given cells. These estimated probabilities are recorded in Figure 2.12 as rectangular areas over the cells; the resulting figure is called a *histogram*. Thus, without any assumption concerning a pattern among the probabilities, we would estimate the probability for the interval (16.00, 16.99) as 0.156. Note that this is not an estimate of the density

TABLE 2.2
Cell frequencies for coal samples

Interval or cell	Frequency f	Proportion or estimated probability $\hat{p}$	Estimated normal probability $\hat{\hat{p}}$	Expected normal frequency $e = 250\hat{\hat{p}}$	Standardized residual $2(\sqrt{f} - \sqrt{e})$
(9.00–12.99)	(16)	0.064	0.064	16.01	−0.00
9.00– 9.99	1	0.004			
10.00–10.99	3	0.012			
11.00–11.99	3	0.012			
12.00–12.99	9	0.036			
13.00–13.99	13	0.052	0.062	15.64	−0.70
14.00–14.99	27	0.108	0.096	24.02	0.59
15.00–15.99	28	0.112	0.128	31.88	−0.71
16.00–16.99	39	0.156	0.147	36.85	0.35
17.00–17.99	42	0.168	0.148	37.02	0.79
18.00–18.99	34	0.136	0.128	32.08	0.33
19.00–19.99	19	0.076	0.097	24.28	−1.14
20.00–20.99	14	0.056	0.064	15.88	−0.49
21.00–21.99	10	0.040			
22.00–22.99	4	0.016			
23.00–23.99	3	0.012			
24.00–24.99	0	0			
25.00–25.99	1	0.004			
(21.00–25.99)	(18)	0.072	0.065	16.34	0.40
	250	1.00		250.00	

function itself but rather of the integral of the density over the particular interval. Typically, with a larger number of repetitions it would be possible to slightly reduce the cell width. Continuing in this way it is possible in a limiting sense to estimate the density function itself, assuming continuity.

Now suppose that there is some background information to support a normal distribution for the response y. We are then faced with choosing values for μ and σ that make the normal probabilities close to the observed proportions; this is essentially a statistical problem, and the theoretical aspects will be examined in Chapter 10. For the present, however, we record some informal discussion in support of the appropriate calculations.

First we remark that μ can be interpreted as the long-run average for the normal variable y and σ^2 as the long-run average of $(y - \mu)^2$; details will be given in Chapter 5. For the $n = 250$ repetitions, the average value (based on cell midpoints) is $\bar{y} = \Sigma_1^n y_i/n = 17.015$ and the average squared deviation is

$$s_y^2 = \frac{1}{n-1} \sum (y_i - \bar{y})^2 = 6.97 = (2.64)^2;$$

the use of $n - 1$ rather than n compensates for the use of $\bar{y}$ rather than the true (unknown) μ; details will be given in Chapters 5 and 7. The normal probabilities based on $\mu = 17.015$ and $\sigma = 2.64$ are recorded as $\hat{\hat{p}}$ in the table. Thus, on the basis of the

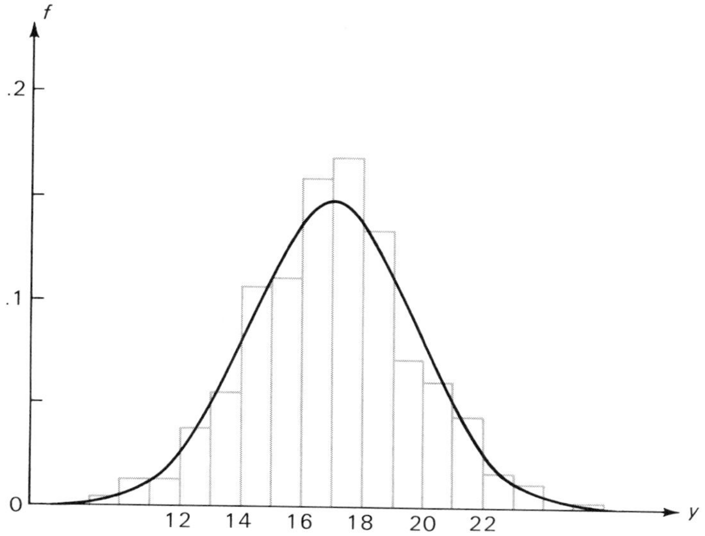

FIGURE 2.12
Histogram; the fitted normal density.

normal pattern of variation, we would estimate the probability for the interval (16.00, 16.99) as 0.147. Note that this estimate is based not just on the frequency for the interval but on all the frequencies in the table.

Now suppose that we want to check whether the normal pattern is appropriate. This is a question of fit and the details are examined in Section 10.4. For the present we record the calculations and defer the justifications until Chapter 10. We can compare the observed frequencies f with the expected frequencies

$$e = 250\widehat{\hat{p}}$$

by calculating the standardized residuals

$$d = 2(\sqrt{f} - \sqrt{e}).$$

The values of d are recorded in the right-hand column of Table 2.2. If we have fitted the true distribution for variation, the residuals will behave approximately as values for the standard normal distribution. The observed values seem quite reasonable for the standard normal and we thus have some reassurance in the use of the normal distribution for the response y.

An overall test of the residuals is also available from Section 10.4; the details would go as follows. The test is made by calculating

$$\chi^2 = \sum d_i^2 = 3.89,$$

and comparing with a chi-square distribution (Appendix II) with 7 degrees of freedom (10 deviations, less 1 because $\Sigma e_i = \Sigma f_i$, less 2 because two parameters μ and σ^2 have been fitted to the data). The observed value 3.89 is a reasonable value for this chi-square distribution, thus giving some overall assurance to the use of the normal pattern.

E EXERCISES

1 Consider the function G given by

$$G(y) = 0 \quad \text{if } -\infty < y < 0,$$
$$= y^2 \quad 0 \leq y < 1,$$
$$= 1 \quad 1 \leq y < \infty.$$

(a) Show that G is a distribution function and sketch it.
(b) Calculate the probability density function g; it is sometimes called the *triangular distribution*.
(c) Calculate the *median* ζ, the value that has 1/2 the probability on each side of it.

2 A psychologist's measuring procedure gives a response that is normal (500, 100).
(a) Calculate the probabilities: $P(350 < y < 650)$; $P(y < 500)$; and $P(y < 400)$.
(b) Use the differential to calculate $P(599\frac{1}{2} < y < 600\frac{1}{2})$.

3 A manufacturer produces incandescent bulbs that have a normal (1100, 100) lifetime. He guarantees that his bulbs have a lifetime of 1000 hours. Calculate the proportion of bulbs that will fail the guarantee.

4 A biologist has found that a certain diet treatment gives a weight increase in pounds that is normal (5, 2.2). Calculate the probability that the increase is more than 6 pounds; that the increase is negative.

5 The brain weight in grams for a certain racial group is approximately normal (1400, 125). Calculate the probability that an individual chosen at random has a brain weight between 1300 and 1500; and between 1200 and 1600.

F EXERCISES

Some of the common distributions are recorded in these exercises and the subsequent problems. The reader should become familiar with general characteristics of each. Some interrelations will be examined later. In each case assume the measurability property 2 for a density function.

6 *Standard exponential distribution* has density function

$$g(z) = e^{-z} \quad \text{if } 0 < z < \infty$$

and zero otherwise; see Figure 2.13 with $\theta = 1$.
(a) Verify that g is a density function.
(b) Determine the distribution function G.
(c) Calculate $\tilde{z}$ such that $G(\tilde{z}) = 1/2$. This value $G^{-1}(1/2)$ is called the median or half-life.

7 The *exponential*(θ) *distribution* has density function

$$f(y) = \theta^{-1}e^{-y/\theta} \quad \text{if } 0 < y < \infty$$

and zero otherwise; see Figure 2.13.

Applications: Some electronic and industrial components show no signs of aging yet have a constant failure rate; the lifetime is exponential(θ), where θ is the average lifetime. The time interval between successive α-particles in a radioactive disintegration is exponential(θ), where θ is the average interval.
(a) Verify that f is a density function.
(b) Determine the distribution function F.
(c) Calculate the half-life $\tilde{y}$; see Exercise 6.
(d) Show that the exponential (θ) distribution is obtained if the standard exponential z is relabeled as $y = \theta z$; use the approach involving events as in Section C.

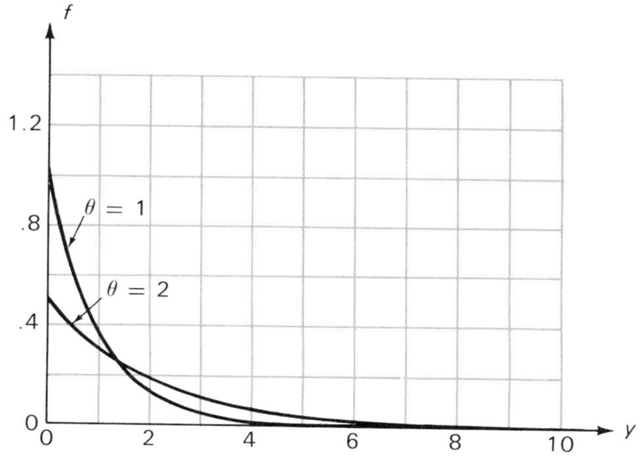

FIGURE 2.13
Exponential(θ) distribution: $\theta = 1, 2$.

8 The *standard Cauchy distribution* (*the Witch of Agnesi*) has density function

$$g(z) = \frac{1}{\pi(1 + z^2)};$$

see Figure 2.14.

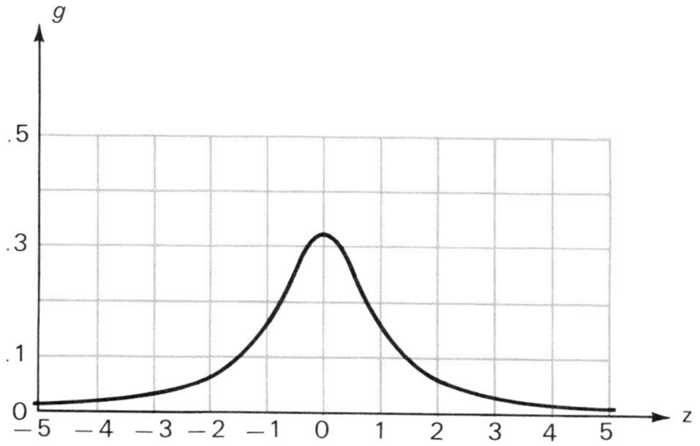

FIGURE 2.14
Standard Cauchy distribution.

Applications: The Cauchy distribution for variation has a bell-shaped curve like the normal but has more probability in the tails of the distribution. Some response variables produce extreme values much more frequently than the normal; the Cauchy provides an extreme pattern for such variation.

(a) Verify that g is a density function.
(b) Calculate the distribution function G.
(c) The general Cauchy (μ, σ) distribution is obtained by the relabeling $y = \mu + \sigma z$. Determine the density function f for y; use the approach involving events as in Section C.

9 The *Laplace* or *double exponential distribution* has density function

$$g(z) = (1/2) \exp \{-|z|\};$$

see Figure 2.15.

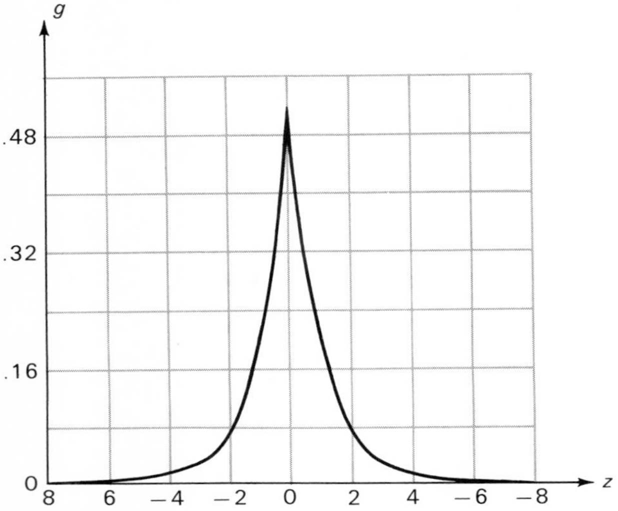

FIGURE 2.15
Laplace distribution.

Applications: This distribution is primarily of mathematical interest. However, it can provide a tentative distribution for variation with more probability in the tails than the normal, but the cusp at the origin makes it somewhat unnatural.

(a) Verify that g is a density function.

(b) Calculate the distribution function G.

10 The *Pareto*(α) *distribution* has the density function

$$g(z) = \alpha(1 + z)^{-(\alpha+1)} \qquad \text{if } 0 < z < \infty$$

and zero otherwise; see Figure 2.16.

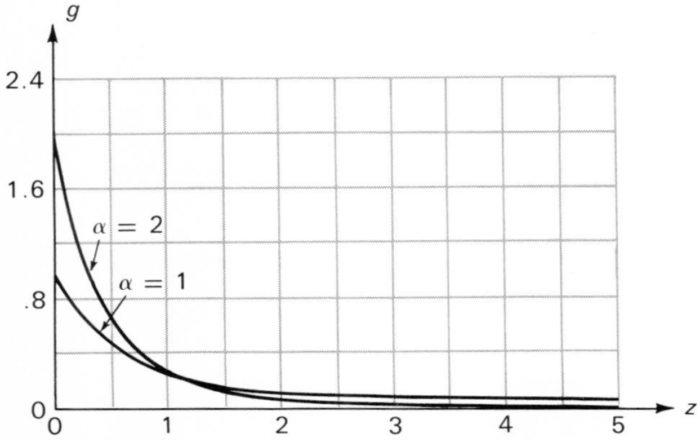

FIGURE 2.16
Pareto(α) distribution: $\alpha = 1, 2$.

Applications: In a population the distribution of incomes above a threshhold value. The distribution of the amounts of insurance claims above a threshhold value.

(a) Verify that g is a density function; $\alpha > 0$.
(b) Calculate the distribution function G.
(c) Determine the median $\tilde{z}$.

11 The *standard logistic distribution* has density function

$$g(z) = e^{-z}(1 + e^{-z})^{-2} = (1/4)\operatorname{sech}^2(z/2);$$

see Figures 2.17 and 2.25.

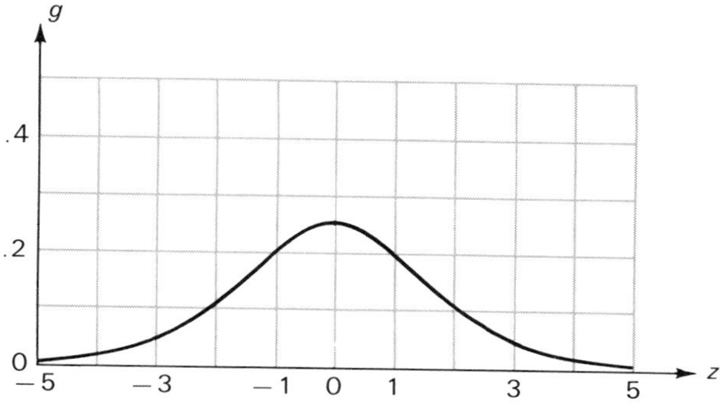

FIGURE 2.17
Standard logistic distribution.

Applications: The logistic has a bell-shaped curve like the normal but has slightly more probability in the tails; often used in place of the normal if a simple distribution function is needed. The logistic distribution function is often used to describe growth: z is time in appropriate units and $G(z)$ is attained height. Or to describe chemical reactions: z is time and $G(z)$ is the amount of a product, a product that catalyzes the reaction.

(a) Verify that g is a density function.
(b) Calculate the distribution function G.
(c) Determine the median $\tilde{z}$.

12 The *standard lognormal*(τ) *distribution* has density function

$$h(z) = \frac{1}{\sqrt{2\pi}\tau} \exp\left\{\frac{-(\ln z)^2}{2\tau^2}\right\} z^{-1} \qquad \text{if } 0 < z < \infty$$

and zero otherwise; see Figure 2.18.

Applications: The particle size in naturally occurring mixtures. In certain cases, the critical dose of a drug just causing a reaction.

(a) Show that h is a density function; $\tau > 0$.
(b) Express the distribution function in terms of the distribution function G for the standard normal.
(c) Determine the median $\tilde{z}$.

13 The *standard Weibull*(β) *distribution* has density function

$$g(z) = \beta z^{\beta-1} \exp\{-z^{\beta}\} \qquad \text{if } 0 < z < \infty$$

and zero otherwise; see Figure 2.19.

Applications: The breaking strength of certain materials. The lifetime of certain components when the conditions for the exponential are not fulfilled.

(a) Show that g is a probability density function; $\beta > 0$.
(b) Calculate the distribution function G.
(c) Determine the median $\tilde{z}$.

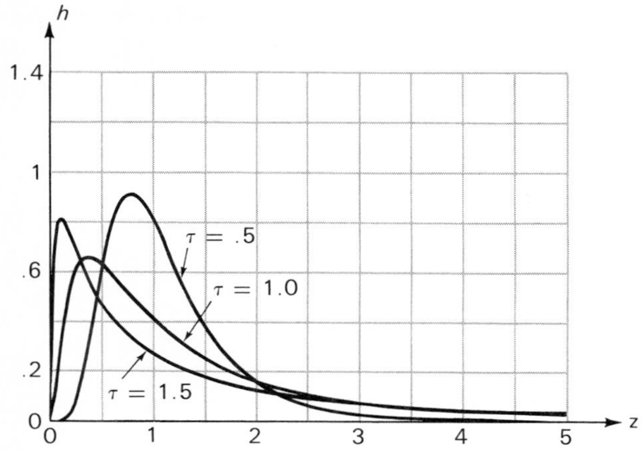

FIGURE 2.18
Lognormal(τ) distribution with $\tau = 0.5,\ 1.0,\ 1.5$.

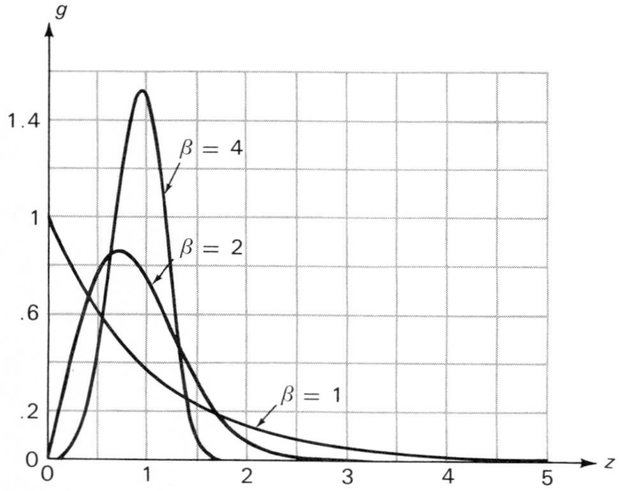

FIGURE 2.19
Weibull(β) distribution with $\beta = 1,\ 2,\ 4$.

14 The *standard extreme value distribution* has density function

$$g(z) = e^{-z} \exp\{-e^{-z}\}, \qquad -\infty < z < \infty;$$

see Figure 2.20.

Applications: The largest value of a response over a period of time: rainfall, earthquakes, flood flows.

(a) Show that g is a probability density function.

(b) Calculate the distribution function.

(c) Determine the median $\tilde{z}$.

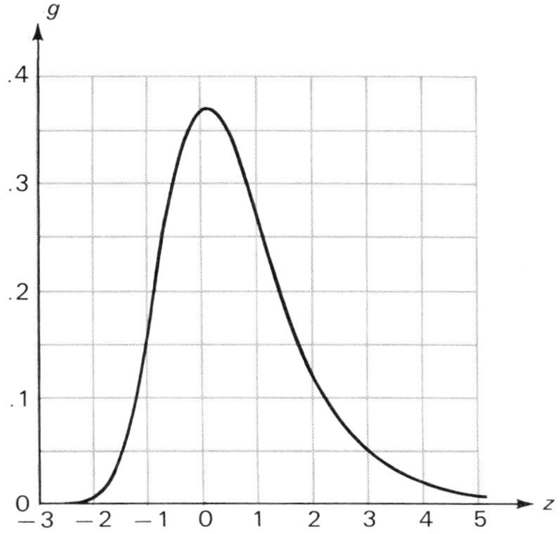

FIGURE 2.20
Standard extreme-value distribution.

G PROBLEMS: DISTRIBUTIONS RELATED TO THE NORMAL

15 The *gamma(p) distribution* has density function

$$h(u) = \Gamma^{-1}(p)e^{-u}u^{p-1} \qquad \text{if } 0 < u < \infty$$

and zero otherwise. The gamma function $\Gamma(\cdot)$ is defined in Problem 21; see Figure 2.21.

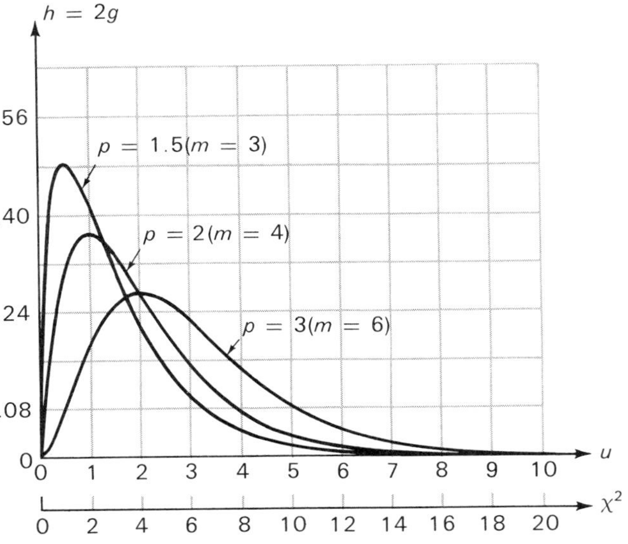

FIGURE 2.21
Gamma(p) distribution for u with $p = 1.5, 2, 3$ and chi-square(m) distribution for χ^2 with $m = 3, 4, 6$.

Applications: The lifetime of certain components when the conditions for the exponential are not fulfilled. With repetitions $z_1, \ldots, z_m$ from the standard normal we will see that $u = \chi^2/2 = (z_1^2 + \cdots + z_m^2)/2$ is gamma$(m/2)$. Use the properties of the gamma function (Problem 21) to show that h is a probability density function; $p > 0$.

16 The *chi-square(m) distribution* has density function

$$g(\chi^2) = \Gamma^{-1}\left(\frac{m}{2}\right) e^{-\chi^2/2}\left(\frac{\chi^2}{2}\right)^{(m/2)-1} \cdot \frac{1}{2} \qquad \text{if } 0 < \chi^2 < \infty$$

and zero otherwise, where χ^2 designates the variable and m is usually a positive integer; see Figure 2.21. Percentage points are available from Appendix II.

Applications: With repetitions $z_1, \ldots, z_m$ from the standard normal, we will see that $\chi^2 = z_1^2 + \cdots + z_m^2$ is chi-square(m).

(a) Use properties of the gamma function (Problem 21) to show that g is a probability density function; $m > 0$.

(b) Use the approach involving events in Section C to show that $u = \chi^2/2$ is gamma$(m/2)$.

17 The *beta(p, q) distribution* has density function

$$g(u) = \frac{\Gamma(p + q)}{\Gamma(p)\Gamma(q)} u^{p-1}(1 - u)^{q-1} \qquad \text{if } 0 < u < 1$$

and zero otherwise; see Figure 2.22.

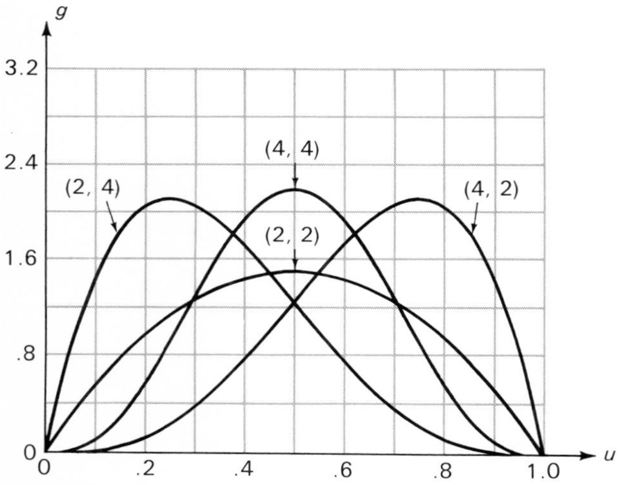

FIGURE 2.22
Beta(p, q) distribution: $(p, q) = (2, 2), (2, 4), (4, 2), (4, 4)$.

Applications: With repetitions $z_1, z_2, \ldots$ from the standard normal, we will see that $u = \Sigma_1^m z_i^2 / \Sigma_1^{m+n} z_i^2$ is beta$(m/2, n/2)$. Use properties of the beta function (Problem 22) to show that g is a probability density function; $p, q > 0$.

18 The *canonical F(m, n) distribution* has density function

$$h(G) = \frac{\Gamma\left(\frac{m + n}{2}\right)}{\Gamma\left(\frac{m}{2}\right)\Gamma\left(\frac{n}{2}\right)} \frac{G^{(m/2)-1}}{(1 + G)^{(m+n)/2}} \qquad \text{if } 0 < G < \infty$$

and zero otherwise; see Figure 2.23.

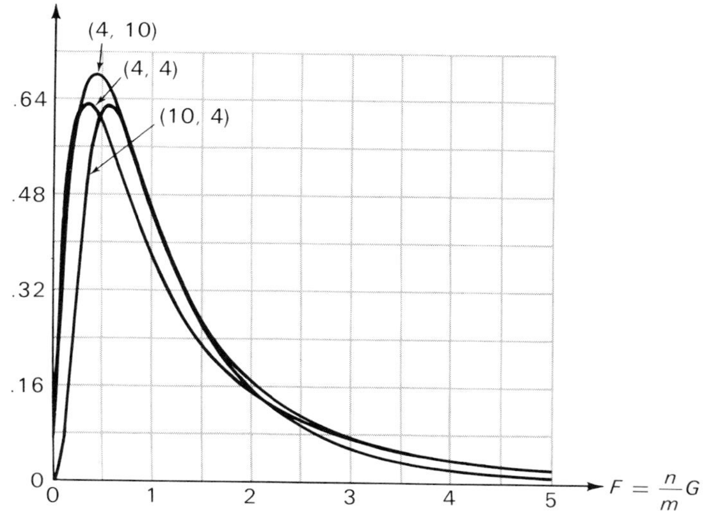

FIGURE 2.23
$F(m, n)$ distribution; $(m, n) = (4, 10), (4, 4), (10, 4)$.

Applications: With repetitions $z_1, z_2, \ldots$ from the standard normal, we will see that $G = \Sigma_1^m z_i^2 / \Sigma_{m+1}^{m+n} z_j^2$ is canonical $F(m, n)$, and $F = (\Sigma_1^m z_i^2/m)/(\Sigma_{m+1}^{m+n} z_j^2/n)$ is standard $F(m, n)$ described below.

Show that h is a probability density function; $m, n > 0$. Use the transformation $G = u/(1 - u)$ and results from Problem 17. Note that the approach involving events in Section C shows that $F = (n/m)G$ has the *standard $F(m, n)$ distribution* with density function

$$g(F) = \frac{\Gamma\left(\frac{m+n}{2}\right)}{\Gamma\left(\frac{m}{2}\right)\Gamma\left(\frac{n}{2}\right)} \frac{\left(\frac{m}{n}F\right)^{(m/2)-1}}{\left(1 + \frac{m}{n}F\right)^{(m+n)/2}} \frac{m}{n} \qquad \text{if } 0 < F < \infty$$

and zero otherwise. Percentage points are recorded in Appendix II.

19 The *canonical Student(n) distribution* has density function

$$h(T) = \frac{\Gamma\left(\frac{n+1}{2}\right)}{\Gamma\left(\frac{1}{2}\right)\Gamma\left(\frac{n}{2}\right)} (1 + T^2)^{-(n+1)/2} \qquad -\infty < T < \infty.$$

Applications: With repetitions $z, z_1, \ldots, z_n$ from the standard normal, we will see that $T = z/(\Sigma_1^n z_i^2)^{1/2}$ is canonical Student(n) and $t = z/(\Sigma_1^n z_i^2/n)^{1/2}$ is standard Student(n) as defined below. The Student(n) also provides a pattern for variation with more probability in the tails than the normal.

(a) Show that $\int_{-\infty}^{0} h(T)\, dT = \int_0^\infty h(T)\, dT$.

(b) Show that h is a probability density function; $n > 0$. Use the transformation $G = T^2$ and the results from Problem 18. Note that the approach involving events in Section C shows that $t = \sqrt{n}T$ has the standard *Student(n) distribution* with density function

$$g(t) = \frac{\Gamma\left(\frac{n+1}{2}\right)}{\Gamma\left(\frac{1}{2}\right)\Gamma\left(\frac{n}{2}\right)} \left(1 + \frac{t^2}{n}\right)^{-(n+1)/2} \frac{1}{\sqrt{n}};$$

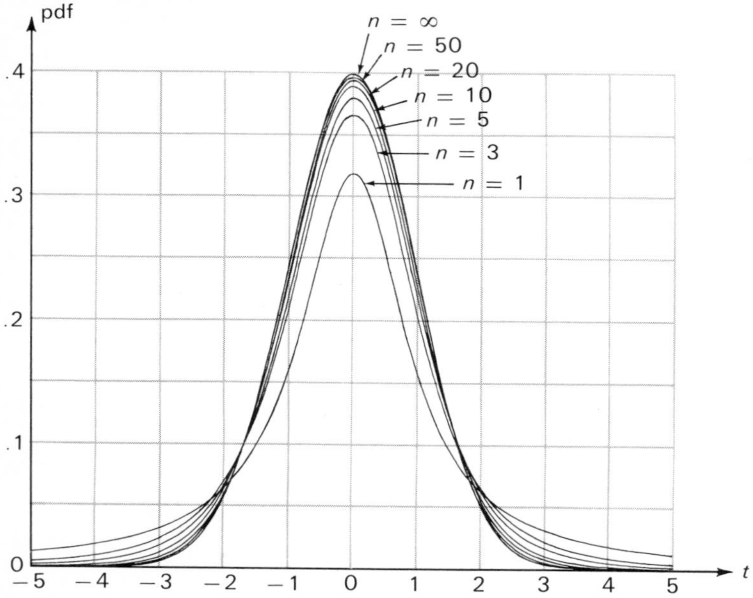

FIGURE 2.24
Student(n) distribution with $n = 1, 3, 5, 10, 20, 50, \infty$.

see Figure 2.24. Percentage points are recorded in Appendix II. The Student(n) distribution is symmetrical and bell-shaped like the normal, but it has more probability in the tails; it ranges from the Cauchy for $n = 1$ to the standard normal as $n \to \infty$; see Figure 2.25.

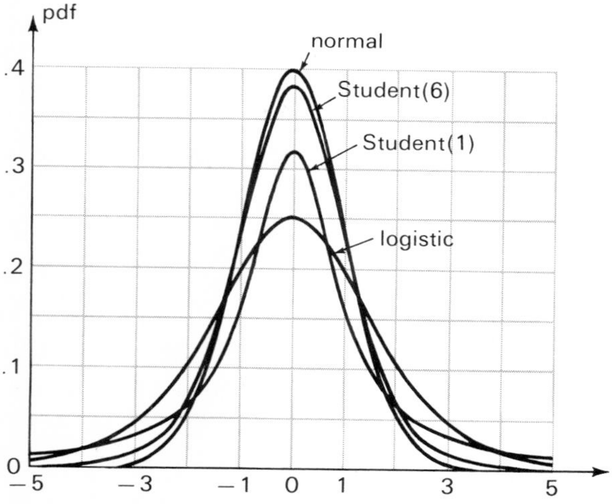

FIGURE 2.25
Normal(0, 1), Student(6), Student(1), and standard logistic distributions.

H PROBLEMS

20 The failure or mortality rate for a distribution is defined by $r(t) = f(t)/(1 - F(t))$.
(a) Give an explanation for the name.

(b) Show that the failure rate for the standard exponential is 1.
(c) Show that the failure rate for the standard Weibull(β) distribution is $\beta t^{\beta-1}$.

21 The gamma function Γ is defined by

$$\Gamma(p) = \int_0^\infty e^{-t} t^{p-1}\, dt$$

for $p > 0$. Show that
(a) $\Gamma(1) = 1$.
(b) $\Gamma(p + 1) = p\Gamma(p)$ by integrations by parts.
(c) $\Gamma(n + 1) = n!$ for $n = 1, 2, \ldots$; thus the Γ function provides a continuous version of the factorial function.
(d) Use the normed property of the normal to show that $\Gamma(1/2) = \sqrt{\pi}$.

22 The beta function B is defined by

$$\mathrm{B}(p, q) = \int_0^1 u^{p-1}(1 - u)^{q-1}\, du$$

for $p, q > 0$. Show that $\mathrm{B}(p, q) = \Gamma(p)\Gamma(q)/\Gamma(p + q)$ by making the change of variable $t = x + y$, $u = x/(x + y)$ in the double integral

$$\Gamma(p)\Gamma(q) = \int_0^\infty \int_0^\infty e^{-x-y} x^{p-1} y^{q-1}\, dx\, dy.$$

2.3

PROBABILITY ON THE PLANE

With many random systems we are interested in responses that have several real-valued components, say $y_1, \ldots, y_k$. For example, with a diet treatment we could measure y_1, the weight increase, and y_2, the blood cholesterol level. Or with a psychology investigation we could measure y_1, the time to reaction, and y_2, the intensity of the reaction.

The obvious sample space for a k-dimensional response $\mathbf{y} = (y_1, \ldots, y_k)'$ is

$$R^k = \{(y_1, \ldots, y_k) : y_i \in R\},$$

the k-fold Cartesian product of the reals.

On the space R^k we are clearly interested in rectangles such as

$$(a_1, b_1] \times (a_2, b_2] \times \cdots \times (a_k, b_k],$$

which would represent the intersection of the propositions $a_1 < y_1 \le b_1, \ldots, a_k < y_k \le b_k$. The smallest class of sets containing the rectangles and closed under the standard operations is the Borel class $\mathcal{B}^k$. Now consider a distribution on R^k as given by a measure P on the Borel sets. In this section we examine some alternative and simpler ways of describing the distribution on R^k.

A DISTRIBUTION FUNCTION ON R^k

We first define a distribution function that presents probability only for certain special subsets, the lower corners or lower rectangles of the form

$$(-\infty, y_1] \times \cdots \times (-\infty, y_k].$$

For this it is convenient to use vector notation $\mathbf{y}' = (y_1, \ldots, y_k)$ and write

$$(-\infty, \mathbf{y}] = (-\infty, y_1] \times \cdots \times (-\infty, y_k].$$

The distribution function is effectively the restriction of P to subsets of the preceding form:

DEFINITION 1

The **distribution function** F is a real-valued function on R^k such that

$$(1) \qquad F(\mathbf{y}) = P((-\infty, y_1] \times \cdots \times (-\infty, y_k]) = P((-\infty, \mathbf{y}]).$$

Consider two simple examples.

EXAMPLE 1

Heads with two symmetric coins: Consider tossing two symmetric coins and observing (x_1, x_2), where x_i is the indicator for heads on the ith coin. From symmetry we have probability 1/4 at each of the four possible sample points; the distribution function can then easily be calculated:

$$\begin{aligned} F(x_1, x_2) &= 0 && -\infty < x_1 < 0 \quad \text{or} \quad -\infty < x_2 < 0, \\ &= 1/4 && 0 \le x_1 < 1 \quad \text{and} \quad 0 \le x_2 < 1, \\ &= 1/2 && 0 \le x_1 < 1 \qquad\qquad 1 \le x_2 < \infty, \\ &= 1/2 && 0 \le x_2 < 1 \qquad\qquad 1 \le x_1 < \infty, \\ &= 1 && 1 \le x_1 < \infty \qquad\qquad 1 \le x_2 < \infty. \end{aligned}$$

This is plotted in Figure 2.26.

EXAMPLE 2

Orientation with two symmetric coins: Consider a first and second tossing of two symmetric coins and recording of (y_1, y_2), where y_i is the angle that the line joining them makes with the front edge of the table on the ith toss; see Example 2.1.2. Symmetry gives, for example,

$$\begin{aligned} P((0, 10] \times (0, 10]) &= P((0, 10] \times (10, 20]) = \cdots, \\ &= P((170, 180] \times (170, 180]) = \frac{10^2}{180^2}, \end{aligned}$$

and so on with finer partitions into squares. It follows, then, that

$$\begin{aligned} F(y_1, y_2) &= 0 && -\infty < y_1 < 0 \quad \text{or} \quad -\infty < y_2 < 0, \\ &= \frac{y_1 y_2}{180^2} && 0 \le y_1 < 180 \quad \text{and} \quad 0 \le y_2 < 180, \end{aligned}$$

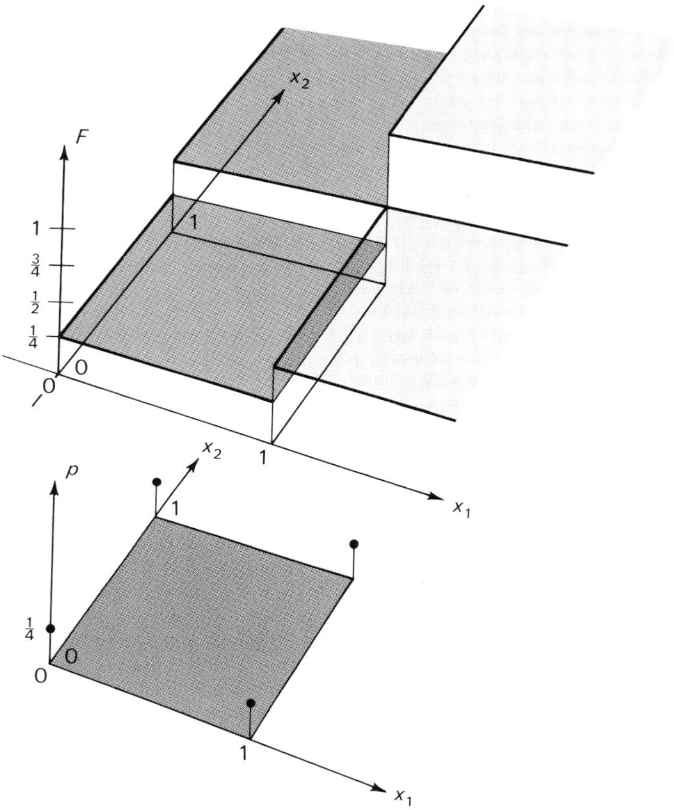

FIGURE 2.26
Distribution function F and probability function p for Example 1.

$$
\begin{aligned}
&= \frac{y_1}{180} && 0 \le y_1 < 180 && 180 \le y_2 < \infty, \\
&= \frac{y_2}{180} && 180 \le y_1 < \infty && 0 \le y_2 < 180, \\
&= 1 && 180 \le y_1 < \infty && 180 \le y_2 < \infty.
\end{aligned}
$$

This distribution has probability spread uniformly over the square $(0, 180] \times (0, 180]$ at the rate of $1/180^2$ per unit area; see Figure 2.27.

Now consider how the distribution function can be used to calculate some simple probabilities. For this we let $(a_1, b_1]_1$ be the difference operator in Section 2.1A as applied to the first argument of F:

$$
\begin{aligned}
(a_1, b_1]_1 F &= F(b_1, y_2, \ldots, y_k) - F(a_1, y_2, \ldots, y_k) \\
&= P((a_1, b_1] \times (-\infty, y_2] \times \cdots \times (-\infty, y_k]).
\end{aligned}
$$

Similarly, let $(a_2, b_2]_2$ be the difference operator $(a_2, b_2]$ as applied to the second argument of F:

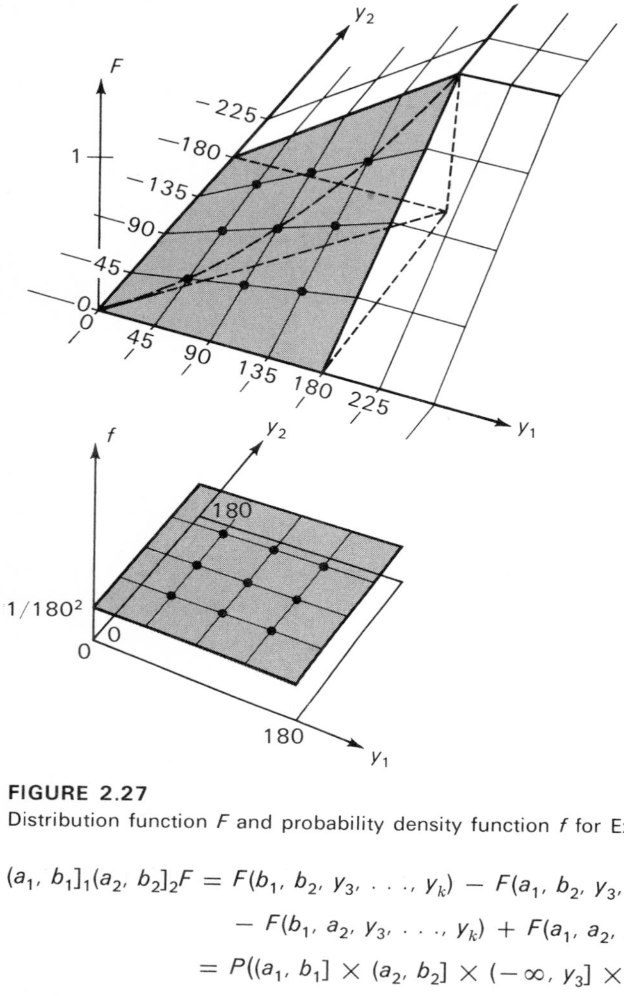

FIGURE 2.27
Distribution function F and probability density function f for Example 2.

$$
\begin{aligned}
(a_1, b_1]_1(a_2, b_2]_2F &= F(b_1, b_2, y_3, \ldots, y_k) - F(a_1, b_2, y_3, \ldots, y_k)\\
&\quad - F(b_1, a_2, y_3, \ldots, y_k) + F(a_1, a_2, y_3, \ldots, y_k);\\
&= P((a_1, b_1] \times (a_2, b_2] \times (-\infty, y_3] \times \cdots \times (-\infty, y_k]);
\end{aligned}
$$

the probability expression at this second stage follows directly from the probability expression at the first stage. Note that the order in which the operators are applied is irrelevant. Similarly, let $(a_i, b_i)_i$ be the corresponding operator for the ith argument. Then using the notation

$$(\mathbf{a}, \mathbf{b}] = (a_1, b_1]_1 \cdots (a_k, b_k]_k,$$

we can write

$$
\begin{aligned}
(2) \qquad (\mathbf{a}, \mathbf{b}]F &= \sum_{y=a,b} (-1)^{n(\mathbf{y})} F(y_1, \ldots, y_k)\\
&= P((\mathbf{a}, \mathbf{b}]),
\end{aligned}
$$

where the summation is over the 2^k terms obtained by replacing each y by an a or a b, and $n(\mathbf{y})$ is the number of a's in the argument. Thus by differencing F we can obtain the probability for any left-open right-closed rectangle.

B CHARACTERIZATION OF A DISTRIBUTION FUNCTION

The following are *necessary* properties of a distribution function F; they can be proved in a straightforward manner.

1. Generalized monotone: $(\mathbf{a}, \mathbf{b}]F \geq 0$ for all rectangles $(\mathbf{a}, \mathbf{b}]$.
2. Right continuous: $F(\mathbf{y} + \mathbf{0}) = F(\mathbf{y})$ for all $\mathbf{y}$.
3. Normed: $F(-\infty, y_2, \ldots, y_k) = \cdots = F(y_1, \ldots, y_{k-1}, -\infty) = 0$
   $F(+\infty, +\infty, \ldots, +\infty) = 1$.

Concerning the proofs, note the following: for property 1 the difference $(\mathbf{a}, \mathbf{b}]F$ denotes the probability in $(\mathbf{a}, \mathbf{b}]$. For property 2,

$$F(\mathbf{y} + \mathbf{0}) = \lim_{\delta_i \to 0} F(y_1 + \delta_1, \ldots, y_k + \delta_k),$$

where $\delta_1, \ldots, \delta_k > 0$; and the set $(-\infty, \mathbf{y} + \boldsymbol{\delta}] - (-\infty, \mathbf{y}]$ is a surface shell for the rectangle $(-\infty, \mathbf{y}]$ and is monotone decreasing to $\varnothing$ as $\boldsymbol{\delta}$ decreases to $\mathbf{0}$. And for property 3 the set $(-\infty, y_1] \times (-\infty, y_2] \times \cdots \times (-\infty, y_k]$ is monotone decreasing to $\varnothing$ as, say, $y_1 \to -\infty$, and is monotone increasing to R^k as all the $y_i \to \infty$.

The properties 1–3 are also sufficient for a function F to be the distribution function of a probability measure P. For certainly we can calculate the "probability" P for any left-open right-closed rectangle. The Extension Theorem recorded in Supplement Section 1.5D then gives a unique extension of P to all the sets in $\mathcal{B}^k$; compare with Section 2.1B.

C DISCRETE DISTRIBUTIONS

Consider a distribution over R^k with measure P and distribution function F:

DEFINITION 2

The probability function p is a real-valued function on R^k such that

(3) $p(\mathbf{y}) = P(\{\mathbf{y}\}) = (\mathbf{y} - \mathbf{0}, \mathbf{y}]F$.

Thus in Example 1 we have $p(\mathbf{y}) = 1/4$ at the four possible outcomes and zero elsewhere; and in Example 2 we have $p(\mathbf{y}) = 0$ everywhere. A point having $p(\mathbf{y}) > 0$ is called a *point with discrete probability*.

We now define a discrete distribution to be one that has all its probability at points with discrete probabilities:

DEFINITION 3

A distribution is a **discrete distribution** if its measure and distribution function have the form

$$(4)\qquad P(A) = \sum_{y\in A} p(\mathbf{y}), \qquad F(\mathbf{y}) = \sum_{t_1\leq y_1,\ldots,t_k\leq y_k} p(t_1, \ldots, t_k)$$

The *necessary* and *sufficient* conditions for a real-valued function p on R^k to be the probability function for a discrete distribution are

1. Nonnegative: $p(\mathbf{y}) \geq 0$ for all $\mathbf{y}$ in R^k.
2. Discrete: $p(\mathbf{y}) = 0$ except at a finite or countable number of points.
3. Normed: $\Sigma\, p(\mathbf{y}) = 1$.

EXAMPLE 3

Tossing two symmetric dice (Example 1.3.7): Two symmetric dice are tossed and (y_1, y_2) recorded, where y_i is the number of points on the ith die. Then

$$\begin{aligned} p(y_1, y_2) &= 1/36, & y_1, y_2 &= 1, 2, \ldots, 6, \\ &= 0, & &\text{elsewhere.} \end{aligned}$$

See Figure 2.26 and envisage $\{0, 1\}^2$ replaced by $\{1, 2, \ldots, 6\}^2$ and the value $1/4$ replaced by $1/36$.

D ABSOLUTELY CONTINUOUS DISTRIBUTIONS

Consider a distribution over R^k with measure P and distribution function F. We can check for a probability density limit at the point $\mathbf{y}$ by examining an open cube C_δ of side length δ containing the point $\mathbf{y}$, calculating the density per unit volume

$$\frac{P(C_\delta)}{\delta^k},$$

and then seeing if this approaches a fixed value for any sequence of cubes shrinking to $\varnothing$. We then define $f(\mathbf{y})$ to be this limit if it exists and to be equal to zero otherwise. A distribution is called "absolutely continuous" if all its probability is at points at which the probability density limit exists.

DEFINITION 4

A distribution is an **absolutely continuous distribution** if its measure and distribution function have the form

$$(5)\qquad P(A) = \int_A f(y_1, \ldots, y_k)\, dy_1 \cdots dy_k = \int_A f(\mathbf{y})\, d\mathbf{y},$$

$$F(\mathbf{y}) = \int_{(-\infty,\mathbf{y}]} f(\mathbf{t})\, d\mathbf{t} = \int_{-\infty}^{y_1} \cdots \int_{-\infty}^{y_k} f(t_1, \ldots, t_k)\, dt_1 \cdots dt_k,$$

where f is the probability density function defined above.

With an absolutely continuous distribution, probability for a set A is given by the volume over the set A beneath the function f and is obtained by integrating f over the set. Thus for the orientation of two symmetric coins as described in Example 2, we have

$$P((0, 10] \times (0, 10]) = \int_0^{10} \int_0^{10} \frac{1}{180^2} \, dy_1 \, dy_2 = \frac{10^2}{180^2}.$$

The following properties are necessary for the probability density of an absolutely continuous distribution:

1. Nonnegative: $f(\mathbf{y}) \geq 0$ for all $\mathbf{y}$ in R^k.
2. Measurable: $\{\mathbf{y} : f(\mathbf{y}) \leq c\} \in \mathcal{B}^k$ for all c.
3. Normed: $\int_{-\infty}^{\infty} \cdots \int_{-\infty}^{\infty} f(\mathbf{y}) \, d\mathbf{y} = 1$.

Properties 1 and 3 are basic; property 2 will be taken for granted.

The properties are also sufficient for a function f to define an absolutely continuous distribution. This follows easily by noting that

$$\int_{-\infty}^{y_1} \cdots \int_{-\infty}^{y_k} f(t_1, \ldots, t_k) \, dt_1 \cdots dt_k$$

satisfies the conditions for a distribution function as recorded in Section 2.3B. In peculiar cases a small technicality may arise—that such an f may differ from the probability density (as defined earlier) on a set having volume zero. For practical purposes we shall assume that f does give probability density.

If the distribution function F is a very nice regular function, then f can be calculated as

$$(6) \qquad f(\mathbf{y}) = \frac{\partial}{\partial y_k} \cdots \frac{\partial}{\partial y_1} F(y_1, \ldots, y_k)$$

$$= \frac{\partial}{\partial y_k} \cdots \frac{\partial}{\partial y_1} \int_{-\infty}^{y_1} \cdots \int_{-\infty}^{y_k} f(\mathbf{t}) \, dt_1 \cdots dt_k$$

by differentiating successively with respect to the coordinates. Note that if some function F has a derivative f as just defined, then the generalized monotone property 1 for a distribution function in Section B can be replaced by $f(\mathbf{y}) \geq 0$.

EXAMPLE 2 (*continuation*)

For points in the rectangle $(0, 180)^2$ we have

$$F(y_1, y_2) = \frac{y_1 y_2}{180^2},$$

$$f(y_1, y_2) = \frac{\partial}{\partial y_2} \frac{\partial}{\partial y_1} \frac{y_1 y_2}{180^2} = \frac{\partial}{\partial y_2} \frac{y_2}{180^2} = \frac{1}{180^2},$$

and for other points we have $f(y_1, y_2) = 0$. Now consider the probability for the event $y_1^2 + y_2^2 < c^2$ with $c < 180$:

$$P(y_1^2 + y_2^2 \leq c^2) = \int_{y_1^2+y_2^2 \leq c^2} f(\mathbf{t}) \, dt_1 \, dt_2 = \int_{A_c} \frac{1}{180^2} \, dy_1 \, dy_2,$$

where $A_c = \{(y_1, y_2) : 0 < y_1 < 180,\ 0 < y_2 < 180,\ y_1^2 + y_2^2 < c^2\}$ is the intersection of a disk and a square:

$$P(y_1^2 + y_2^2 \le c^2) = \frac{\pi c^2/4}{180^2},$$

which for this *uniform* distribution is *the area of the possible set for the event divided by the area of the possible set for the distribution;* compare with Exercise 1.3.10.

E MULTIVARIATE NORMAL DISTRIBUTION

Many response distributions on R^2 and R^3 are concentrated around a central value and drop off sharply in each direction in the bell-shaped pattern mentioned in Section 2.2C; consider the example for (y_1, y_2) mentioned at the beginning of this section. Sometimes different coordinates tend to move together in such a way that the probability is piled up in a ridge in some direction. The multivariate normal distribution can sometimes provide an approximation for such a pattern of variation. In the simplest rotationally symmetric case, we have the *standard normal distribution on* R^k, given by

$$g(z_1, \ldots, z_k) = (2\pi)^{-k/2} e^{-(z_1^2 + \cdots + z_k^2)/2};$$

see Figure 2.28.

We verify property 3 for a density function. The integral

$$\int_{R^k} g(\mathbf{z})\, d\mathbf{z} = \int_{R^k} (2\pi)^{-1/2} e^{-z_1^2/2} \cdots (2\pi)^{-1/2} e^{-z_k^2/2}\, dz_1 \cdots dz_k$$

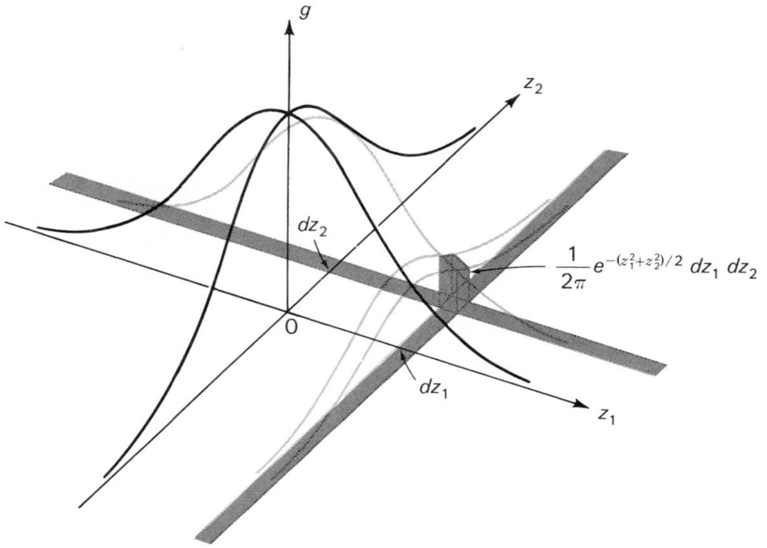

FIGURE 2.28
Density function for the standard bivariate normal.

has an integrand that factors so the variables separate and the range $R \times \cdots \times R$ factors as a product set. We can then integrate factor by factor:

$$\int_{R^k} g(\mathbf{z})\, d\mathbf{z} = \int_{-\infty}^{\infty} (2\pi)^{-1/2} e^{-z^2/2}\, dz \cdots \int_{-\infty}^{\infty} (2\pi)^{-1/2} e^{-z^2/2}\, dz;$$

and we are thus led back to the unproved property 3 for the standard normal on R. But consider the case with $k = 2$,

$$\int_{R^2} \frac{1}{2\pi} e^{-(z_1^2+z_2^2)/2}\, dz_1\, dz_2 = \int_0^{\infty} \int_0^{2\pi} \frac{1}{2\pi} e^{-r^2/2} r\, dr\, da,$$

where

$$z_1 = r\cos a, \qquad r^2 = z_1^2 + z_1^2,$$

$$z_2 = r\sin a, \qquad a = \tan^{-1}\frac{z_2}{z_1},$$

and where dr measures length radially at the point (z_1, z_2) and $r\,da$ measures length at right angles along the circle through (z_1, z_2); see Figure 2.29. Thus

$$\int_{R^2} \frac{1}{2\pi} e^{-(z_1^2-z_2^2)/2}\, dz_1\, dz_2 = \int_0^{\infty} e^{-r^2/2} r\, dr \cdot \int_0^{2\pi} \frac{1}{2\pi}\, da = 1.$$

And it follows that property 3 holds for $k = 2$, then $k = 1$, then all k.

The distribution function G for the standard normal on R^k is given by

$$\begin{aligned} G(z_1, \ldots, z_k) &= \int_{-\infty}^{z_1} \cdots \int_{-\infty}^{z_k} \frac{1}{\sqrt{2\pi}} e^{-t_k^2/2} \cdots \frac{1}{\sqrt{2\pi}} e^{-t_1^2/2}\, dt_1 \cdots dt_k \\ &= \int_{-\infty}^{z_1} \frac{1}{\sqrt{2\pi}} e^{-t_1^2/2}\, dt_1 \cdots \int_{-\infty}^{z_k} \frac{1}{\sqrt{2\pi}} e^{-t_k^2/2}\, dt_k \\ &= G(z_1) \cdots G(z_k) \end{aligned}$$

in terms of the distribution function for the standard normal on R.

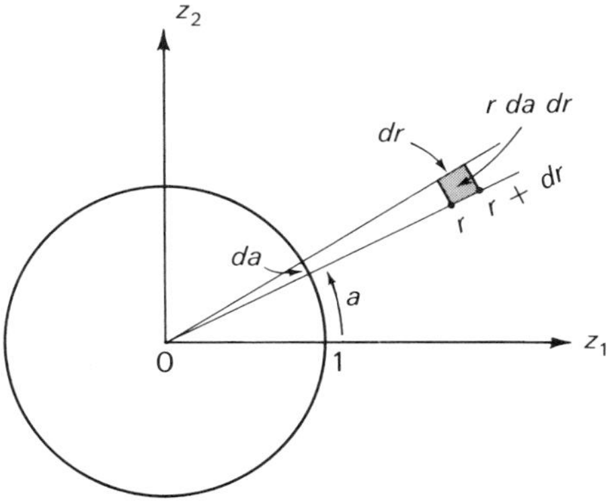

FIGURE 2.29
Area change under the transformation $z_1 = r\cos a$, $z_2 = r\sin a$.

F EXERCISES

1 Let $F(y_1, y_2) = F_1(y_1)F_2(y_2)$, where F_1 and F_2 are distribution functions on the real line.
(a) Calculate $(a_1, b_1]_1(a_2, b_2]_2F$ in terms of the component functions.
(b) Show that F is a distribution function.

2 A distribution on R^2 has probability function $p(y_1, y_2)$ with nonzero values as follows:

	y_2			
y_1	0	1	2	3
0	0	0	0.05	0.05
1	0.05	0.10	0.10	0.10
2	0.10	0.10	0.10	0.05
3	0.05	0.10	0.05	0

Calculate the values of the distribution function $F(y_1, y_2)$ at the points given in the table.

3 A function G on R^2 has the following form:

$$G(z_1, z_2) = 1 - (1 + z_1)^{-1} - (1 + z_2)^{-1} + (1 + z_1 + z_2)^{-1}, \qquad z_1, z_2 > 0$$

and zero otherwise.
(a) Calculate the function g of formula (6).
(b) Show that G is a distribution function; note the remark following formula (6).

4 Consider a uniform distribution for (y_1, y_2) over the unit disk centered at the origin.
(a) Calculate the probability $P(y_1^2 + y_2^2 \leq c^2)$.
(b) Plot the preceding probability as a function of c on the range $(0, \infty)$; and as a function of c^2 on the range $(0, \infty)$.

5 Consider the following function f on R^2: $f(y_1, y_2) = 4y_1y_2$ on $(0, 1)^2$ and zero elsewhere.
(a) Show that f is a probability density function.
(b) Calculate the probabilities: $P(y_1 \leq c_1)$; $P(y_2 \leq c_2)$; and $P(y_1 \leq c_1, y_2 \leq c_2)$.

6 A transistor and a condenser have lifetimes (y_1, y_2) with density function

$$f(y_1, y_2) = \frac{1}{6} \exp\left\{-\frac{y_1}{6} - y_2\right\}, \qquad y_1, y_2 > 0,$$

and zero otherwise. Verify that f is a probability density function.

7 (*continuation*) Calculate $P(y_1 > t, y_2 > t)$ for a positive number t. If the transistor and condenser are the only items subject to failure in a simple electronic system, calculate the probability that the system survives until time t.

8 Consider a uniform distribution for (y_1, y_2) on the square $(-1, +1)^2$. Calculate the probabilities (a) $P(y_1^2 \leq c^2)$, and (b) $P(y_1^2 + y_2^2 \leq c^2)$, both with $0 < c < 1$.

9 Consider a uniform distribution for (y_1, y_2, y_3) on the cube $(-1, +1)^3$. Calculate the following probabilities for $0 \leq c \leq 1$: (a) $P(y_1^2 \leq c^2)$; (b) $P(y_1^2 + y_2^2 \leq c^2)$; (c) $P(y_1^2 + y_2^2 + y_3^2 \leq c^2)$.

10 A coin of radius r is tossed on a smooth floor marked with east–west lines and with north–south lines at unit spacings. Let (y_1, y_2) with $(0 \leq y_i < 1)$ give the coordinates of its center in the square in which it falls.
(a) Record the probability density function based on symmetry.
(b) Calculate the probability that it will intersect an E–W line.
(c) Calculate the probability that it will intersect both an E–W line and a N–S line.
(d) Calculate the probability that it will not intersect any line.

11 A needle of length l is tossed onto the smooth floor described in Exercise 10. Let (y_1, y_2) give

the coordinates of its center (Exercise 10) and y_3 be its angle measured positively from E–W ($0 < y_3 < \pi$).

(a) Record the probability density for (y_1, y_2, y_3) based on symmetry.

(b) Calculate the probability that it does not intersect an E–W line; for simplicity assume that $l < 1$.

G PROBLEMS

12 The *Pareto*(α) *distribution* on R^2 has density function

$$g(z_1, z_2) = \alpha(\alpha + 1)(1 + z_1 + z_2)^{-(\alpha+2)}, \qquad z_1, z_2 > 0,$$
$$= 0, \qquad \text{otherwise.}$$

(a) Show that g is a probability density function; $\alpha > 0$.

(b) Calculate the distribution function G; compare with Exercise 3.

13 The *canonical Student*(n) *distribution* on R^2 has density function

$$g(z_1, z_2) = \frac{\Gamma\left(\frac{n+2}{2}\right)}{\pi\Gamma\left(\frac{n}{2}\right)}(1 + z_1^2 + z_2^2)^{-(n+2)/2}.$$

Show that g is a probability density function; $n > 0$. Use the integration results from Exercise 2.2.19.

14 The *Pareto*(α) *distribution* on R^k has density function

$$g(z_1, \ldots, z_k) = \alpha(\alpha + 1) \cdots (\alpha + k - 1)(1 + z_1 + \cdots + z_k)^{-(\alpha+k)}$$

for $z_1, \ldots, z_k > 0$, and zero otherwise. Show that g is a probability density function; $\alpha > 0$.

15 The *canonical Student*(n) *distribution* on R^k has density function

$$g(z_1, \ldots, z_k) = \frac{\Gamma\left(\frac{n+k}{2}\right)}{\pi^{k/2}\Gamma\left(\frac{n}{2}\right)}(1 + z_1^2 + \cdots + z_k^2)^{-(n+k)/2}.$$

Show that g is a probability density function; $n > 0$. Use the integration results of Exercise 2.2.19 and Problem 13.

16 The *Dirichlet*$(r_1, \ldots, r_{k+1})$ *distribution* on R^k has density

$$f(y_1, \ldots, y_k) = \frac{\Gamma(r_1 + \cdots + r_{k+1})}{\Gamma(r_1) \cdots \Gamma(r_{k+1})} y_1^{r_1-1} \cdots y_k^{r_k-1}\left(1 - \sum_1^k y_j\right)^{r_{k+1}-1}$$

on the set $\{\mathbf{y}: 0 < y_i, \Sigma_1^k y_i < 1\}$ and zero elsewhere. Show that f is a probability density function; $r_j > 0$. This is a generalized beta distribution; use Exercise 2.2.17.

17 The norming constant for the normal density can be verified without recourse to the bivariate and multivariate case. With

$$F(x) = \left(\int_0^x e^{-t^2/2}\, dt\right)^2, \qquad G(x) = \int_0^1 \frac{2e^{-x^2(1+t^2)/2}}{1 + t^2}\, dt$$

(a) Show that $F'(x) + G'(x) = 0$.

(b) Then show that $F(x) + G(x) = \pi/2$.

(c) Then consider $\lim_{x\to\infty} (F(x) + G(x))$ and thus obtain the norming constant.

2.4

CHANGE OF VARIABLES

There is often a considerable degree of arbitrariness in how we present the response variable of a random system. For example, in measuring the weight increase of an animal under diet therapy, we could use pounds or grams or some other units. In measuring the noise level at which hearing just occurs in a psychology test, we could use a convenient energy unit or we could take 10 times the logarithm of energy and use decibels. And in a catalyzed chemical reaction, we could measure the amount of converted material or we could use the ratio of converted material to unconverted material.

Often there is a preferred or natural way of presenting the response. This can relate to the simplicity with which response effects are determined by input variables. And it can relate to having a simple scaling for the variation, independent of the input variables.

In this section we examine the mathematics of changing the way we present a response, that is, of making a change of variable or of relabeling the response. The question of a natural labeling for the response will be mentioned briefly in Chapter 12.

A CHANGE OF VARIABLE FOR A REAL-VALUED RESPONSE

In Section 2.2C we introduced the standard normal density $g(z)$ for z and then obtained the general normal(μ, σ) density for $y = \mu + \sigma z$ by a simple argument in terms of events; see Figure 2.30. Now consider this change of variable more generally.

Let G be the distribution function for an initial response variable x. And suppose that $y = Y(x)$ is a one-to-one monotone-increasing function from R onto R. We investigate the distribution for this new or relabeled response $y = Y(x)$; let F_Y designate the distribution function.

For notation let $x = X(y)$ be the inverse function. We can then note that the event $(-\infty, y]$ for the new response is the same as the event $(-\infty, X(y)]$ for the original response; see Figure 2.31. Thus we obtain

(1) $$F_Y(y) = G(X(y)),$$

which expresses F_Y in terms of the original G. In this formula note that we merely substitute, expressing the original variable in terms of the new variable.

Formula (1) holds more generally. Suppose that $y = Y(x)$ is a one-to-one

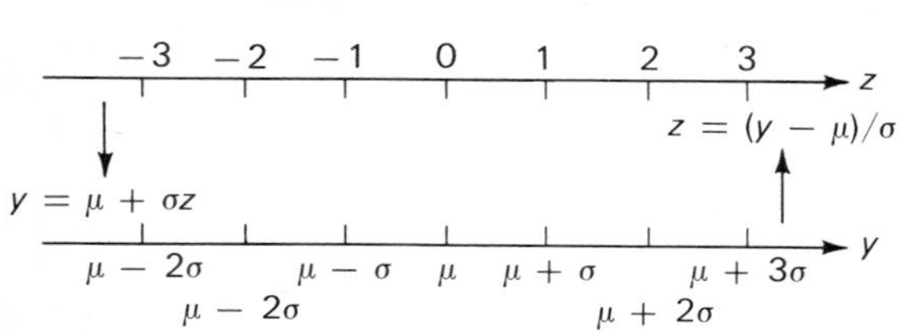

FIGURE 2.30
Transformation $y = \mu + \sigma z$ and inverse $z = (y - \mu)/\sigma$.

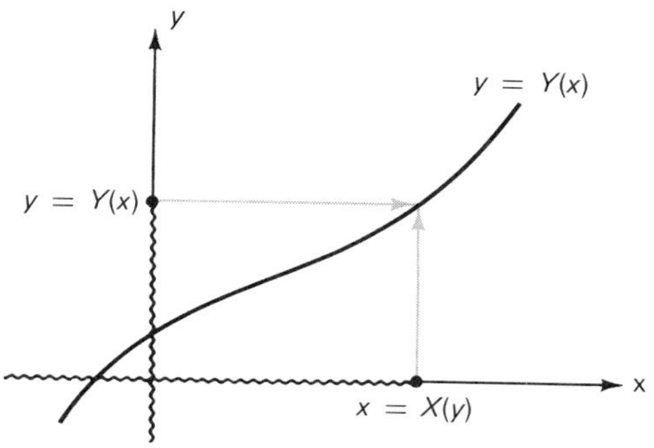

FIGURE 2.31
Transformation $y = Y(x)$ and inverse $x = X(y)$; the event $(-\infty, y]$ for the new response is the same as $(-\infty, X(y)]$ for the original response.

monotone-increasing function from an interval for the initial response to an interval for the new response. And suppose that all the probability for G is on the initial interval. Then formula (1) gives the new distribution function F_Y throughout the new interval.

EXAMPLE 1

Logistic and exponential growth: Consider the standard logistic distribution (Exercise 2.2.11) for an initial response z; recall Exercise 2.2.11 and Figure 2.17. The distribution function is easily derived and has the simple form

$$G(z) = \frac{1}{1 + e^{-z}} = \frac{e^z}{1 + e^z}, \qquad -\infty < z < \infty.$$

Now suppose that we take an exponential transformation $y = e^z$ of the initial response; this maps $(-\infty, \infty)$ into $(0, \infty)$. The inverse transformation is $z = \ln y$. Thus the distribution function for the new response is

$$F_Y(y) = \frac{e^{\ln y}}{1 + e^{\ln y}} = \frac{y}{1 + y}, \qquad 0 < y < \infty.$$

This is easily seen from Problem 2.2.18 to be canonical $F(2, 2)$. The logistic is often used to represent growth when there is an inhibiting upper limit; see Figure 2.32.

B IN TERMS OF THE PROBABILITY FUNCTION

For a discrete distribution the effects of relabeling are very simple if we use the probability function. Consider a discrete distribution for x with probability function p. And suppose that $y = Y(x)$ is a one-to-one transformation of R onto R (it suffices to have it one to one from possible values of x to possible values of y). Let p_Y be the probability function for the new $y = Y(x)$.

The probability function for the new response at the value y is, of course, the same as for the original response at $x = X(y)$; thus

(2) $\quad p_Y(y) = p(X(y)).$

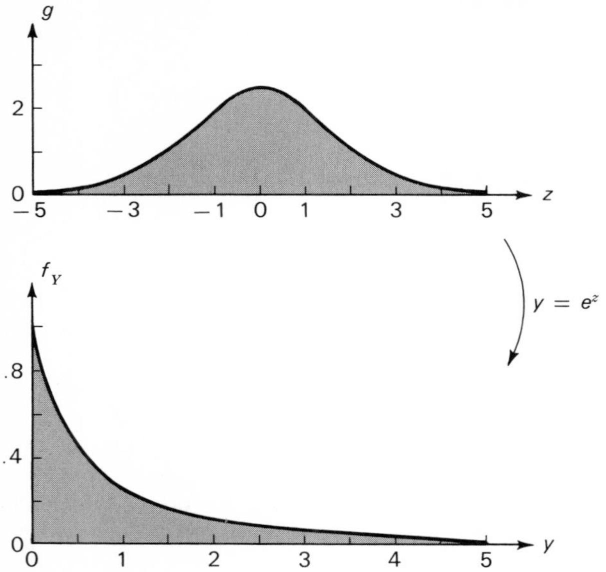

FIGURE 2.32
Logistic density function g for z. The density function f_Y for $y = e^z$ is canonical $F(2, 2)$.

We merely substitute expressing the original variable in terms of the new variable.

EXAMPLE 2

Heads with two symmetric coins: The distribution for the number, say x, of heads is available from Example 1.3.5:

$p(0) = 1/4, \quad p(1) = 1/2, \quad p(2) = 1/4.$

The probability function for the proportion $\hat{p} = Y(x) = x/2$ of heads is available immediately:

$p_Y(0) = 1/4, \quad p_Y(1/2) = 1/2, \quad p(1) = 1/4;$

see Figure 2.33.

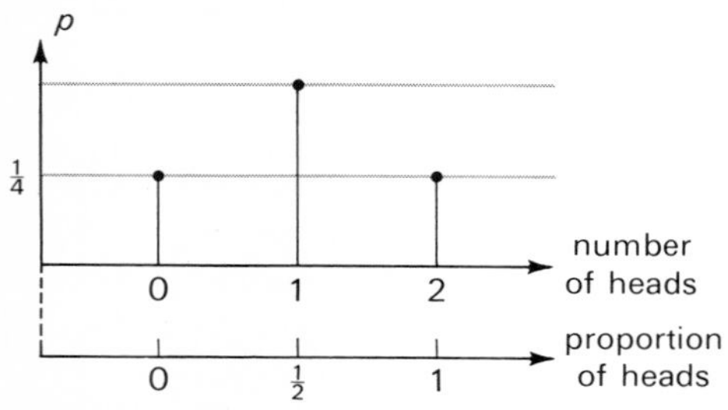

FIGURE 2.33
Probability function plotted against number of heads and against proportion of heads.

C IN TERMS OF THE PROBABILITY DENSITY FUNCTION

For an absolutely continuous distribution the effects of relabeling are quite simple if we use the probability density function. Consider an absolutely continuous distribution for x with probability density function g. And suppose that $y = Y(x)$ is a one-to-one transformation continuously differentiable each way from R onto R. Let f_Y be the density function for the new $y = Y(x)$ and F_Y be the corresponding distribution function.

Then in the monotone-increasing case we have

$$F_Y(y) = G(X(y)) = \int_{-\infty}^{X(y)} g(x)\, dx;$$

and differentiation with respect to y (using the chain rule) gives

$$(3) \qquad f_Y(y) = g(X(y)) \cdot \left|\frac{dx}{dy}\right|;$$

see Figure 2.34. It is easily seen that with the absolute-value sign the formula also holds in the monotone-decreasing case.

Formula (3), in fact, holds more generally. Suppose that $y = Y(x)$ has the continuity properties and maps an interval for the initial response into an interval for the new response. And suppose that all the probability for g is on the initial interval; then formula (3) gives the density function f_Y through the new interval.

Intuitively, it is more convenient to think of a chunk of probability

$$g(x)\, dx$$

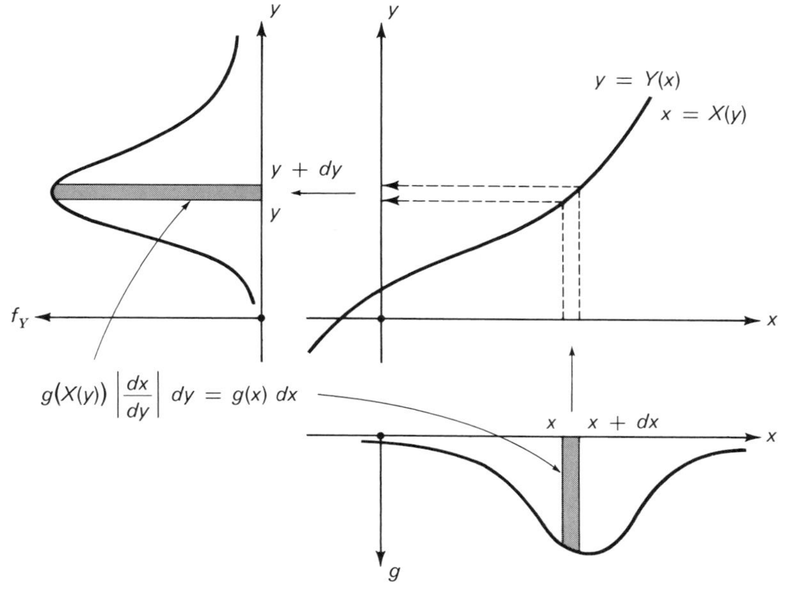

FIGURE 2.34
Density g for x transformed to the density f_Y for Y.

associated with the interval $(x, x + dx)$ for the initial variable. This same chunk of probability of course refers to the image $(y, y + dy)$ obtained from the mapping $y = Y(x)$. If the length of the increment dy is different from the length of dx, then the density must change inversely so that the area representing probability remains the same. This can be presented conveniently as follows:

$$g(x)\,dx = g(X(y))\,dX(y) = g(X(y)) \cdot \left|\frac{dX(y)}{dy}\right| dy \tag{4}$$
$$= f_Y(y)\,dy;$$

see Figure 2.34. More formally, this is an abbreviated way of presenting the essentials of the change-of-variable formula for an integral on the real line. Note the simplicity: *The density function is expressed in terms of the new variable and the differential is expressed in terms of the corresponding differential for the new variable.*

EXAMPLE 3

Consider a triangular distribution for the variable x:

$g(x)\,dx = 2x\,dx$ on $(0, 1)$

and zero elsewhere. And suppose that we are interested in $y = Y(x) = x^3$; this maps the region $(0, 1)$, where the probability is, into the region $(0, 1)$; see Figure 2.35.

The effect of the transformation on the differential is easily calculated:

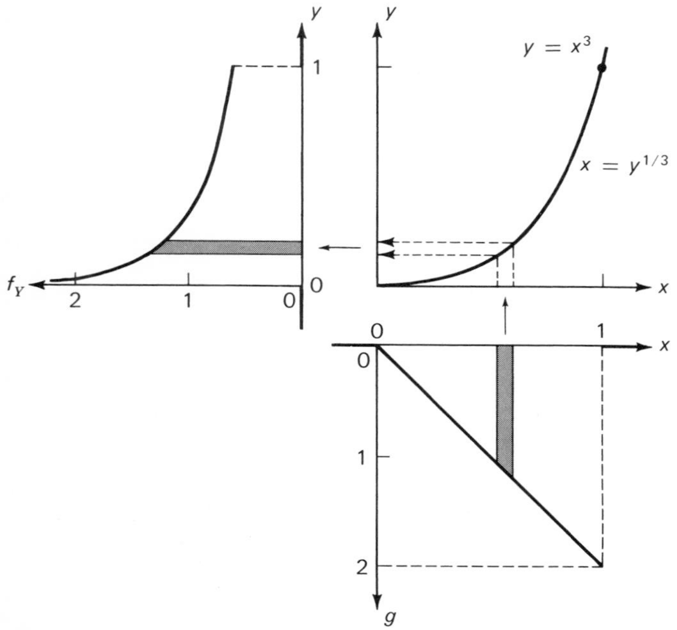

FIGURE 2.35
Triangular distribution for x, the transformation $y = x^3$, and the distribution f_Y for y.

$$x = X(y) = y^{1/3}, \qquad \frac{dx}{dy} = (1/3)y^{-2/3} \quad \text{on } (0,\ 1)$$

The density function for y is then obtained from (3):

$$g(X(y)) \cdot \left|\frac{dx}{dy}\right| = 2y^{1/3} \cdot (1/3)y^{-2/3} = (2/3)y^{-1/3} \qquad \text{on } (0,\ 1)$$

$$= \frac{\Gamma(5/3)}{\Gamma(2/3)\Gamma(1)} y^{(2/3)-1}(1 - y)^{1-1} \qquad \text{on } (0,\ 1),$$

which is the beta(2/3, 1) distribution in Problem 2.2.17.

EXAMPLE 4

Consider a gamma(1) distribution for the variable x:

$$g(x)\,dx = e^{-x}\,dx \qquad \text{on } (0,\ \infty)$$

and zero elsewhere. And suppose that we are interested in $u = U(x) = e^{-x}$; this maps the region $(0, \infty)$ into $(0, 1)$; see Figure 2.36.

The effect of the transformation on the differential is easily calculated:

$$x = X(u) = -\ln u, \qquad \frac{dx}{du} = \frac{-1}{u}.$$

The density function for u is then obtained from (3):

$$g(X(u)) \cdot \left|\frac{dx}{du}\right| = u \cdot \frac{1}{u} = 1$$

on the effective range $(0, 1)$; thus u has the uniform(0, 1) distribution. The details can be expressed more compactly in terms of differentials:

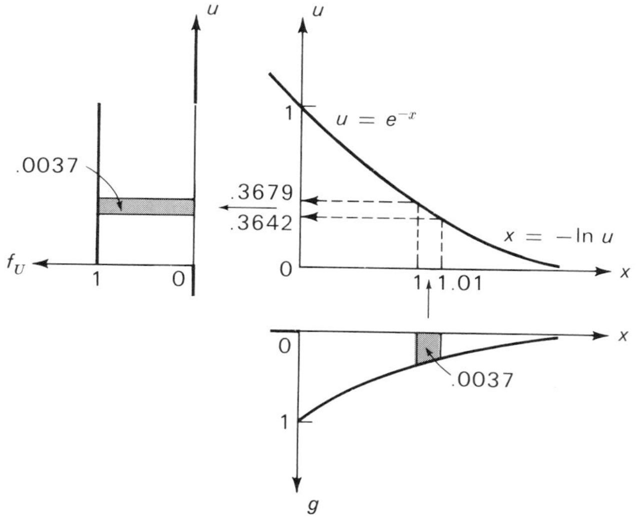

FIGURE 2.36
Gamma(1) distribution for x, the transformation $u = e^{-x}$, and the uniform(0, 1) distribution for u.

$$e^{-x}\,dx = u\left|\frac{d}{du}(-\ln u)\right|\,du = u\frac{1}{u}\,du = du$$

on the effective ranges $(0, \infty)$ and $(0, 1)$.

To illustrate the use of differentials to approximate probabilities, consider the interval (1, 1.01) for x and the corresponding interval (0.36422, 0.36788) for u. Then

$$P((1.00, 1.01)) = e^{-1}\cdot 0.01 = 0.00368,$$
$$P_U((0.36422, 0.36788)) = 0.00366.$$

D PROBABILITY INTEGRAL TRANSFORMATION

Consider some continuous distribution function G. And for simplicity in the discussion to follow suppose that G is one to one from an open interval for the initial variable, say x, to the interval (0, 1) for values, say u, of the function G; this avoids flat spots on the function G. We are going to consider a special transformation from x to u, the transformation $u = U(x) = G(x)$, and then the inverse transformation from u to x, say $x = X(u) = G^{-1}(u)$.

First consider an initial response x with distribution function G. And suppose that we define a new response $u = U(x) = G(x)$ by means of the given distribution function G; this is called the *probability integral transformation* (integral of density in the absolutely continuous case). In an experimental investigation, a certain theory may specify a distribution G. Then with an observed value x we could be interested in $u = G(x)$, the probability of a value x or smaller. The distribution describing possible u values under the theory would be of particular interest for assessing the theory; we now derive this distribution for u.

By formula (1) the distribution function for u is

$$(5)\qquad F_U(u) = G(X(u)) = G(G^{-1}(u)) = u, \qquad 0 < u < 1.$$

Thus $u = G(x)$ has the uniform(0, 1) distribution; see Figure 2.37. For convenience, let H be the distribution function for this uniform distribution: that is, $H(u) = u$ on (0, 1).

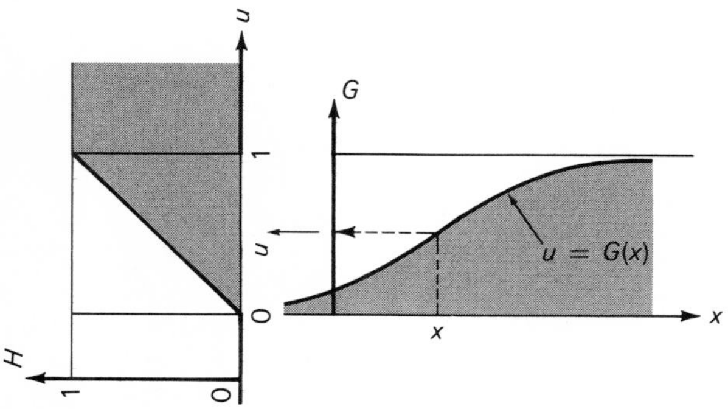

FIGURE 2.37
Probability integral transformation $u = G(x)$ gives the uniform(0, 1) distribution, and inverse probability integral transformation $X = G^{-1}(u)$ carries the uniform(0, 1) into the distribution with distribution function G.

Second, consider an initial response u that has a uniform(0, 1) distribution. And suppose that we define a new response $x = X(u) = G^{-1}(u)$, where G is a distribution function as described in the first paragraph. By formula (1) the distribution function for x is

$$F_X(x) = H(U(x)) = U(x) = G(x). \tag{6}$$

Thus $x = G^{-1}(u)$ has a distribution with distribution function G; see Figure 2.37. This is called the *inverse probability integral transformation*. It provides a means of carrying an initial variable with a uniform distribution into a new variable with a special distribution of interest. Values from a uniform distribution are easily produced on a computer; by an appropriate transformation, then, we can obtain values from any particular distribution of interest.

EXAMPLE 5

In Example 2.2.1 we noted how sample values can be obtained from the uniform(0, 1) distribution. Suppose that we are using three-figure accuracy and obtain the sequence

7 9 3 5 9 2 2 3 1

from a computer programmed to produce random digits or from a table of randomly produced digits. Then (the fourth-place 5 is added to center the values)

0.7935, 0.5925, 0.2315, . . .

are sample values (approximately) from the uniform(0, 1). For the logistic in Example 1, we can calculate the inverse function $G^{-1}(u) = \ln [u/(1 - u)]$ and from formula (6) obtain the sample values

$$\ln \frac{0.7935}{1 - 0.7935}, \quad \ln \frac{0.5925}{1 - 0.5925}, \quad \ln \frac{0.2315}{1 - 0.2315}, \quad \ldots$$

$$= \ln 3.843, \quad \ln 1.454, \quad \ln 0.3012, \quad \ldots$$

$$= 1.346, \quad 0.374, \quad -1.200, \quad \ldots;$$

these are sample values from the logistic distribution. For an initial sample value near 0 or near 1 we should add more digits to maintain accuracy under the transformation.

E CHANGE OF VARIABLE ON THE PLANE

Consider a distribution for an initial response (x_1, x_2). And suppose that $(y_1, y_2) = (Y_1(x_1, x_2), Y_2(x_1, x_2))$ is a one-to-one mapping of R^2 onto R^2. We investigate the distribution for the new response (y_1, y_2). For notation let $(x_1, x_2) = (X_1(y_1, y_2), X_2(y_1, y_2))$ be the inverse function. As an example, y_1 could be the average yield $(x_1 + x_2)/2$ on two plots of land and y_2 could be the difference, $x_2 - x_1$, in yields.

For the case of a discrete distribution, the values of the probability function are reassigned to the corresponding image points under the mapping; see Section B.

For the absolutely continuous case, let g be the density function for (x_1, x_2). And suppose that the mapping is continuously differentiable each way. Let $f_{Y_1Y_2}$ be the density function for the new response (y_1, y_2).

See Figure 2.38. The inverse mapping gives (x_1, x_2) from (y_1, y_2), and gives $(x_1 + dx_1, x_2 + dx_2)$ from $(y_1 + dy_1, y_2 + dy_2)$. By partial differentiation we can obtain the relationship between the increments dx_1, dx_2 and the increments dy_1, dy_2:

$$dx_1 = \frac{\partial X_1}{\partial y_1} dy_1 + \frac{\partial X_1}{\partial y_2} dy_2,$$

$$dx_2 = \frac{\partial X_2}{\partial y_1} dy_1 + \frac{\partial X_2}{\partial y_2} dy_2.$$

This can be rewritten in vector and matrix notation as

$$\begin{pmatrix} dx_1 \\ dx_2 \end{pmatrix} = \begin{pmatrix} \frac{\partial X_1}{\partial y_1} & \frac{\partial X_1}{\partial y_2} \\ \frac{\partial X_2}{\partial y_1} & \frac{\partial X_2}{\partial y_2} \end{pmatrix} \begin{pmatrix} dy_1 \\ dy_2 \end{pmatrix}.$$

The matrix is called the *Jacobian matrix*. The preceding equation shows how the inverse mapping produces the increment vector (dx_1, dx_2) from the increment vector (dy_1, dy_2); see Figure 2.38.

In particular, we record in Table 2.3 the increment vectors (dx_1, dx_2) that are obtained from the increment vectors $(dy_1, dy_2) = (1, 0)$ and $= (0, 1)$ (these correspond to the two axes on the y space). From this we see that a parallelogram in the initial space is obtained from a square in the new space; see Figure 2.38. The area of the parallelogram is obtained as the positive value of the determinant of the two vectors

$$\begin{vmatrix} \frac{\partial X_1}{\partial y_1} & \frac{\partial X_1}{\partial y_2} \\ \frac{\partial X_2}{\partial y_1} & \frac{\partial X_2}{\partial y_2} \end{vmatrix}_+ = \begin{vmatrix} \frac{\partial Y_1}{\partial x_1} & \frac{\partial Y_1}{\partial x_2} \\ \frac{\partial Y_2}{\partial x_1} & \frac{\partial Y_2}{\partial x_2} \end{vmatrix}_+^{-1},$$

which is, of course, the positive value of the Jacobian determinant; the second expression is available from advanced calculus. The area of the square is unity. Thus we obtain

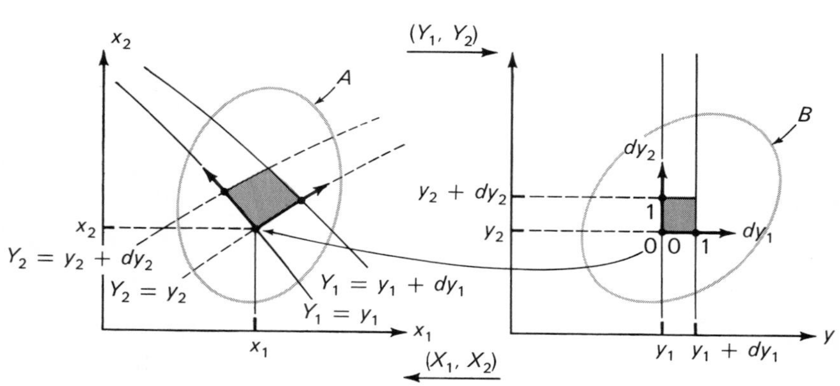

FIGURE 2.38
Inverse mapping (X_1, X_2) gives (x_1, x_2) from (y_1, y_2) and gives the parallelogram with vertex (x_1, x_2) from the square with vertex (y_1, y_2).

TABLE 2.3
Increment vectors

Initial space increment $\left[\begin{pmatrix} dx_1 \\ dx_2 \end{pmatrix}\right.$	$\xleftarrow[\text{mapping}]{\text{inverse}}$	New space increment$\left.\right]$ $\begin{pmatrix} dy_1 \\ dy_2 \end{pmatrix}$
$\begin{pmatrix} \dfrac{\partial X_1}{\partial y_1} \\ \dfrac{\partial X_2}{\partial y_1} \end{pmatrix}$		$\begin{pmatrix} 1 \\ 0 \end{pmatrix}$
$\begin{pmatrix} \dfrac{\partial X_1}{\partial y_2} \\ \dfrac{\partial X_2}{\partial y_2} \end{pmatrix}$		$\begin{pmatrix} 0 \\ 1 \end{pmatrix}$

the ratio of areas as expressed by

$$(7) \qquad d\mathbf{x} = \begin{vmatrix} \dfrac{\partial X_1}{\partial y_1} & \dfrac{\partial X_1}{\partial y_2} \\ \dfrac{\partial X_2}{\partial y_1} & \dfrac{\partial X_2}{\partial y_2} \end{vmatrix}_+ , \qquad d\mathbf{y} = \left|\frac{\partial X_1, X_2}{\partial y_1, y_2}\right|_+ d\mathbf{y}.$$

This intuitive description is presented to motivate the standard change-of-variable formula for a two-dimensional integral.

Let B be a set for the new response and A be the corresponding event for the original response. Then

$$P((y_1, y_2) \in B) = \int_A g(x_1, x_2)\, dx_1\, dx_2$$

$$= \int_B g(X_1(y_1, y_2), X_2(y_1, y_2)) \left|\frac{\partial X_1, X_2}{\partial y_1, y_2}\right|_+ dy_1\, dy_2.$$

Thus the density function for (y_1, y_2) is

$$(8) \qquad f_{Y_1Y_2}(y_1, y_2) = g(X_1(y_1, y_2), X_2(y_1, y_2)) \cdot \left|\frac{\partial X_1, X_2}{\partial y_1, y_2}\right|_+ .$$

Formula (8), in fact, holds more generally. Suppose that the mapping is one to one from a region for the initial response to a region for the new response. And suppose that all the probability for g is on the initial region. Then formula (8) gives the new density throughout the new region; it is zero otherwise.

EXAMPLE 6

Consider a uniform distribution for (x_1, x_2) over the unit square $(0, 1)^2$ and suppose that we are interested in a new variable (y_1, y_2) obtained by the affine transformation

$$\begin{aligned} y_1 &= 1 + x_1 - x_2, \\ y_2 &= 2 + 2x_1 + x_2 \end{aligned} \qquad \begin{pmatrix} y_1 \\ y_2 \end{pmatrix} = \begin{pmatrix} 1 \\ 2 \end{pmatrix} + \begin{pmatrix} 1 & -1 \\ 2 & 1 \end{pmatrix} \begin{pmatrix} x_1 \\ x_2 \end{pmatrix}.$$

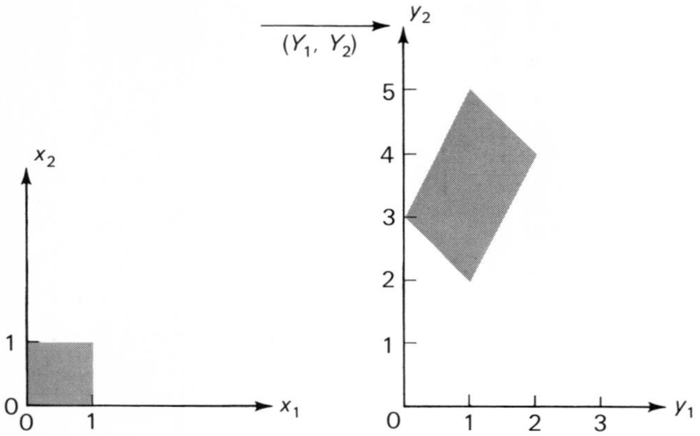

FIGURE 2.39
Uniform distribution on the unit square $(0, 1)^2$ is mapped into a uniform distribution for (y_1, y_2) on the parallelogram.

Note that this maps the square $(0, 1)^2$ on the initial space into the parallelogram with vertices (1, 2), (2, 4), (0, 3), (1, 5); see Figure 2.39. The probability density function g for the initial variable (x_1, x_2) is

$$g(x_1, x_2) = 1 \qquad \text{on } (0, 1)^2$$

and zero otherwise. For this example we do not need to calculate the inverse transformation. We can calculate the Jacobian for the given transformation

$$\begin{pmatrix} \dfrac{\partial Y_1}{\partial x_1} & \dfrac{\partial Y_1}{\partial x_2} \\ \dfrac{\partial Y_2}{\partial x_1} & \dfrac{\partial Y_2}{\partial x_2} \end{pmatrix} = \begin{pmatrix} 1 & -1 \\ 2 & 1 \end{pmatrix},$$

which has determinant 3. Thus from formula (8) we obtain

$$f_{Y_1Y_2}(y_1, y_2) = 1 \cdot 3^{-1} = 1/3$$

on the parallelogram formed by (1, 2), (2, 4), (0, 3), (1, 5) and zero otherwise. This is a uniform distribution on that parallelogram.

EXAMPLE 7

Consider a distribution for (x_1, x_2) with density function g. And suppose that we are interested in a new variable (r, θ) given by

$$r = (x_1^2 + x_2^2)^{1/2} \qquad x_1 = r\cos\theta,$$

$$\theta = \tan^{-1}\left(\frac{x_2}{x_1}\right) \qquad x_2 = r\sin\theta;$$

the inverse mapping is recorded on the right (see Figure 2.29). Note that the transformation carries R^2 less the origin into $(0, \infty) \times [0, 2\pi)$. The Jacobian determinant of the inverse mapping is easily calculated:

$$\begin{vmatrix} \dfrac{\partial x_1}{\partial r} & \dfrac{\partial x_1}{\partial \theta} \\ \dfrac{\partial x_2}{\partial r} & \dfrac{\partial x_2}{\partial \theta} \end{vmatrix}_+ = \begin{vmatrix} \cos\theta & -r\sin\theta \\ \sin\theta & r\cos\theta \end{vmatrix}$$

$$= r(\cos^2\theta + \sin^2\theta) = r.$$

Thus from formula (8) we obtain

$$f_{R\Theta}(r, \theta) = g(r\cos\theta, r\sin\theta)r$$

on $(0, \infty) \times [0, 2\pi)$ and zero otherwise. In the special case that g is rotationally symmetric, $g(x_1, x_2) = h(\sqrt{x_1^2 + x_2^2})$, then

$$f_{R\Theta}(r, \theta) = h(r)r$$

on $(0, \infty) \times [0, 2\pi)$ and zero otherwise. And in the particular case that g is the standard normal density (Section 2.3E),

$$f_{R\Theta}(r, \theta) = \frac{1}{2\pi} e^{-r^2/2} r$$

on $(0, \infty) \times [0, 2\pi)$ and zero otherwise. The simplicity of these last two cases will be examined under independence in Section 3.3.

F CHANGE OF VARIABLE FOR VECTOR RESPONSES

Now consider the general case of a distribution on R^k with density function g. And suppose that we have a function $\mathbf{Y}' = (Y_1, \ldots, Y_k)$ which maps R^k onto R^k and is one to one and continuously differentiable each way. We investigate the distribution for the new response $\mathbf{y} = \mathbf{Y}(\mathbf{x})$. For notation let $\mathbf{x} = \mathbf{X}(\mathbf{y})$ designate the inverse function.

Then in the pattern in the preceding section, we have

$$f_{\mathbf{Y}}(\mathbf{y}) = g(\mathbf{X}(\mathbf{y})) \left|\frac{\partial \mathbf{x}}{\partial \mathbf{y}}\right|_+ \tag{9}$$

where

$$\left|\frac{\partial \mathbf{x}}{\partial \mathbf{y}}\right| = \begin{vmatrix} \dfrac{\partial X_1}{\partial y_1} & \cdots & \dfrac{\partial X_1}{\partial y_k} \\ \vdots & & \vdots \\ \dfrac{\partial X_k}{\partial y_1} & \cdots & \dfrac{\partial X_k}{\partial y_k} \end{vmatrix}$$

is the Jacobian determinant. This can be presented more conveniently in terms of differentials:

$$g(\mathbf{x})\, d\mathbf{x} = g(\mathbf{X}(\mathbf{y})) \cdot \left|\frac{\partial \mathbf{x}}{\partial \mathbf{y}}\right|_+ d\mathbf{y} = f_{\mathbf{Y}}(\mathbf{y})\, d\mathbf{y}. \tag{10}$$

Formula (9) holds more generally. Suppose that the mapping is one to one from a region for the initial response to a region for the new response. And suppose that all the probability for g is in the initial region. Then formula (9) gives the new density function throughout the new region. (For details, see a text on advanced calculus.)

EXAMPLE 8

Consider a distribution for $\mathbf{x}' = (x_1, \ldots, x_k)$ with density function

$$g(\mathbf{x}) = exp\,\{-\Sigma\, x_i\} \qquad \text{if each } x_i > 0$$

and zero otherwise. And suppose that we are interested in $\mathbf{y}' = (y_1, \ldots, y_k)$, given by

$$\begin{aligned} y_1 &= \frac{x_1}{\Sigma\, x_i} & x_1 &= y_1 y_k \\ &\vdots & &\vdots \\ y_{k-1} &= \frac{x_{k-1}}{\Sigma\, x_i} & x_{k-1} &= y_{k-1} y_k \\ y_k &= \Sigma\, x_i & x_k &= (1 - y_1 - \cdots - y_{k-1}) y_k; \end{aligned}$$

the inverse transformation is recorded on the right. Note that the transformation carries the region $(0, \infty)^k$, where the probability for $\mathbf{x}$ is onto the set

$$S_k = \{(y_1, \ldots, y_{k-1}) : 0 < y_i, \Sigma_1^{k-1}\, y_i < 1\} \times \{y_k : 0 < y_k < \infty\}.$$

The Jacobian determinant is easily calculated:

$$\begin{vmatrix} \dfrac{\partial x_1}{\partial y_1} & \cdots & \dfrac{\partial x_1}{\partial y_k} \\ \vdots & & \vdots \\ \dfrac{\partial x_k}{\partial y_1} & \cdots & \dfrac{\partial x_k}{\partial y_k} \end{vmatrix} = \begin{vmatrix} y_k & & 0 & y_1 \\ & \ddots & & \vdots \\ 0 & & y_k & y_{k-1} \\ -y_k & \cdots & -y_k & (1 - \Sigma_1^{k-1}\, y_i) \end{vmatrix} = y_k^{k-1}.$$

Thus from formula (9) we obtain

$$f_Y(\mathbf{y}) = \exp\,\{-y_k\} y_k^{k-1}$$

on the set S_k in R^k and zero elsewhere. The simplicity of this last formula will become transparent in Section 3.3.

G MULTIVARIATE NORMAL

Consider the standard normal distribution for $\mathbf{z} = (z_1, \ldots, z_k)'$ on R^k:

$$g(z_1, \ldots, z_k) = (2\pi)^{-k/2} e^{-(z_1^2 + \cdots + z_k^2)/2},$$

as recorded in Section 2.3E. Suppose that we picture $\mathbf{z}$ as describing in standard form the variation for a k-dimensional response. And suppose the response itself is obtained by the relabeling,

$$\begin{aligned} y_1 &= \mu_1 + \gamma_{11} z_1 + \cdots + \gamma_{1k} z_k \\ &\vdots \\ y_k &= \mu_k + \gamma_{k1} z_1 + \cdots + \gamma_{kk} z_k, \end{aligned}$$

or more briefly in vector and matrix notation,

$$\mathbf{y} = \boldsymbol{\mu} + \Gamma \mathbf{z}$$

where $\boldsymbol{\mu} = (\mu_1, \ldots, \mu_k)'$ is the column vector giving the relocation and

$$\Gamma = \begin{pmatrix} \gamma_{11} & \cdots & \gamma_{1k} \\ \vdots & & \vdots \\ \gamma_{k1} & \cdots & \gamma_{kk} \end{pmatrix}$$

is a matrix giving the linear scaling and combining of components; see Figure 2.40.

For a one-to-one transformation we assume that $|\Gamma| \neq 0$. The Jacobian matrix of $\mathbf{y}$ with respect to $\mathbf{z}$ is just Γ. Thus

$$d\mathbf{y} = |\Gamma|_+ \, d\mathbf{z}; \qquad d\mathbf{z} = |\Gamma|_+^{-1} \, d\mathbf{y}.$$

The probability at an element $d\mathbf{z}$ for the initial variable is given by

$$(2\pi)^{-k/2} \exp\{-(1/2)(z_1^2 + \cdots + z_k^2)\} \, d\mathbf{z} = (2\pi)^{-k/2} \exp\{-(1/2)\mathbf{z}'\mathbf{z}\} \, d\mathbf{z},$$

where we use the matrix product of the row vector $\mathbf{z}'$ and the column vector $\mathbf{z}$ to give Σz_i^2. By expressing this in terms of the new labels, we obtain

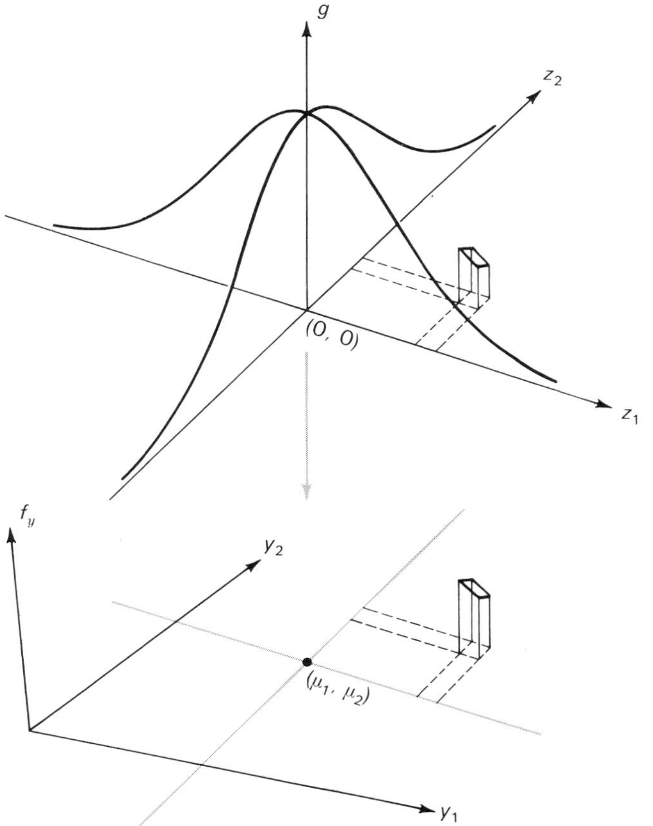

FIGURE 2.40
Standard normal g. The general normal(μ; Σ) located at $\boldsymbol{\mu}$ and scaled by Σ. The volume of the rectangular cylinders is the same; if the base area is smaller, the height is proportionately greater.

$$f(\mathbf{y})\,d\mathbf{y} = (2\pi)^{-k/2} \exp\{-(1/2)(\mathbf{y}-\boldsymbol{\mu})'\Gamma'^{-1}\Gamma^{-1}(\mathbf{y}-\boldsymbol{\mu})\}|\Gamma|_+^{-1}\,d\mathbf{y}$$
$$= (2\pi)^{-k/2} \exp\{-(1/2)(\mathbf{y}-\boldsymbol{\mu})'\Sigma^{-1}(\mathbf{y}-\boldsymbol{\mu})\}|\Sigma|^{-1/2}\,d\mathbf{y},$$

where the matrix

$$\Sigma = \Gamma\Gamma' = \begin{pmatrix} \sigma_{11} & \cdots & \sigma_{1k} \\ \vdots & & \vdots \\ \sigma_{k1} & \cdots & \sigma_{kk} \end{pmatrix}$$

has typical component $\sigma_{ij} = \gamma_{i1}\gamma_{j1} + \cdots + \gamma_{ik}\gamma_{jk}$, which is the inner product of the ith row vector and the jth row vector of Γ; for some matrix details, see Supplement Section 2.5. The probability differential for y can be expressed more fully as

$$f(\mathbf{y})\,d\mathbf{y} = (2\pi)^{-k/2}|\Sigma|^{-1/2} \exp\left\{-(1/2)\sum_{i,j=1}^{k} \sigma^{ij}(y_i-\mu_i)(y_j-\mu_j)\right\} d\mathbf{y},$$

where the inverse matrix

$$\Sigma^{-1} = \begin{pmatrix} \sigma^{11} & \cdots & \sigma^{1k} \\ \vdots & & \vdots \\ \sigma^{k1} & \cdots & \sigma^{kk} \end{pmatrix}$$

has components σ^{ij}. This is the *multivariate normal*$(\boldsymbol{\mu}; \Sigma)$ *distribution,* normal with mean $\boldsymbol{\mu}$ and variance Σ, as we shall see later. The multivariate normal is used in archeology, medicine, education—wherever there is a need to describe reasonably behaved multiple responses.

H EXERCISES

1 The distribution function G is equal to y^2 on $(0, 1)$ (Exercise 2.2.1). Find the distribution function of $w = y^2$.

2 A distribution function H is equal to $x^2/4$ on $(0, 2)$. Find the distribution function for $w = x^3$.

3 Show that the distribution of $y = e^z$ in Example 1 is canonical $F(2, 2)$; see Problem 2.2.18.

4 A symmetrical die is tossed, yielding y points. Record the probability function for $w = y^2$.

5 The distribution of cells on a slide has λ per unit area. It can be shown that the distance d from a cell to its nearest neighbor has density

$$g(d) = 2\pi\lambda d \exp\{-\lambda\pi d^2\}, \qquad d > 0.$$

The area of the largest free disk centered on a cell has area $A = \pi d^2$. Find the density function for A.

6 Let G be the distribution function for x, and let $y = Y(x)$ with inverse $x = X(y)$ be one to one monotone decreasing from R to R. Show that the distribution function of y is such that

$$F_Y(y-0) = 1 - G(X(y)).$$

7 (*continuation*) Consider the gamma(1) distribution for x. Find the distribution function for $y = -x$.

8 (*continuation*) Show that $u = e^{-x}$ is the probability integral transformation for the modified response $-x$; compare with Example 4.

9 Let z be Pareto(α) (Exercise 2.2.10). Show that $y = (1 + z)^p - 1$ with $p > 0$ is Pareto(α/p).

10 Let z be Weibull(β) (Exercise 2.2.13). Show that $y = z^\beta$ is exponential(1); and that $y = z^p$ is Weibull(β/p) with $p > 0$.

11 Let z be extreme value (Exercise 2.2.14). Show that $y = e^{-z}$ is exponential(1).

I EXERCISES: INVERSE PROBABILITY INTEGRAL TRANSFORMATION

12 Show that the inverse probability integral transformation (IPIT) for the exponential(θ) distribution (Exercise 2.2.7) is

$$y = -\theta \ln (1 - u).$$

13 (*continuation*) If u is uniform(0, 1), deduce that $-\ln u$ is exponential(1) or gamma(1); and that $-2 \ln u$ is chi square(2); see Exercises 2.2.15 and 2.2.16.

14 Show that the IPIT for the Cauchy distribution (Exercise 2.2.8) is

$$z = \tan\left(\pi u - \frac{\pi}{2}\right).$$

15 Show that the IPIT for the Pareto(α) distribution (Exercise 2.2.10) is

$$z = (1 - u)^{-1/\alpha} - 1.$$

16 Show that IPIT for the Weibull(β) distribution (Exercise 2.2.13) is

$$z = [-\ln (1 - u)]^{1/\beta}.$$

17 Show that IPIT for the extreme-value distribution (Exercise 2.2.14) is

$$z = -\ln (-\ln u).$$

J EXERCISES: COMPUTER USAGE

Find a computer program that gives reliable random digits (Exercise 2.1.8) or a table that contains randomly obtained digits. Use the inverse probability integral transformation IPIT to obtain a sample of 25 from each of the following distributions and see where the values fall in relation to the density function.

18 Exponential(3) distribution; Exercise 12.
19 Chi-square(2) distribution; Exercise 13.
20 Cauchy distribution; Exercise 14.
21 Pareto(2) distribution; Exercise 15.
22 Weibull(2) distribution; Exercise 16.
23 Extreme-value distribution; Exercise 17.

K PROBLEMS

Many problems involving changes of variable will be found in Sections 3.2 and 3.3, where some additional concepts will make the calculations very fruitful.

24 Consider the standard normal distribution for (z_1, z_2) and define (y_1, y_2) by

$$\begin{pmatrix} y_1 \\ y_2 \end{pmatrix} = \begin{pmatrix} \mu_1 \\ \mu_2 \end{pmatrix} + \begin{pmatrix} \sigma_1 & 0 \\ \rho\sigma_2 & \sqrt{1 - \rho^2}\,\sigma^2 \end{pmatrix} \begin{pmatrix} z_1 \\ z_2 \end{pmatrix},$$

where $-1 < \rho < 1$.

(a) Determine the elements of the matrix Σ.
(b) Determine the elements of the inverse matrix Σ^{-1}.
(c) Show that (y_1, y_2) has the bivariate normal density $(\mu_1, \mu_2, \sigma_1, \sigma_2, \rho)$ as given by

$$f(y_1, y_2) = \frac{1}{2\pi\sigma_1\sigma_2\sqrt{1 - \rho^2}} \exp\left\{\frac{-1}{2(1 - \rho^2)}\left[\frac{(y_2 - \mu_2)^2}{\sigma_2^2} - 2\rho\frac{(y_1 - \mu_1)(y_2 - \mu_2)}{\sigma_1\sigma_2} + \frac{(y_2 - \mu_2)^2}{\sigma_2^2}\right]\right\}.$$

25 *Rotationally symmetrical distribution:* Let $g(r^2)$ be the density function for $(z_1, \ldots, z_k)$, where for brevity $r^2 = (z_1^2 + \cdots z_k^2)$. Consider the transformation $\mathbf{y} = \boldsymbol{\mu} + \Gamma\mathbf{z}$, where $|\Gamma| \neq 0$ and $\Sigma = \Gamma\Gamma'$.

(a) Show that the density function f for $\mathbf{y}$ is given by

$$f(\mathbf{y}) = |\Sigma|^{-1/2} g((\mathbf{y} - \boldsymbol{\mu})'\Sigma^{-1}(\mathbf{y} - \boldsymbol{\mu})).$$

(b) Record the density function for the case that g describes the canonical Student(n) distribution.

SUPPLEMENTARY MATERIAL

Supplement Section 2.5 is on pages 520–525.

A brief survey of *vector algebra* is given and some discussion of *inner products* is included. The continuous and discrete components of a distribution have been discussed rather fully in preceding sections; the remaining component, the *singular component,* is examined for some interesting details.

NOTES AND REFERENCES

For general material on probability theory, see the Notes and References in Chapter 1. Simple notions from vector algebra are used intermittently through the chapters on probability and statistics; results needed for this are collected in Supplement Section 2.5.

In this chapter we have discussed ways of presenting probability distributions on the line, plane, and higher-dimensional real spaces. A familiar approach to this topic is in terms of *random variables,* a random variable being a real-valued function on a probability space (a more general definition is given in Chapter 3). With this other approach, the distributions on the line, plane, and higher-dimensional real spaces are viewed as coming from an antecedent probability space by means of functions (or random variables), although the functions as such are almost never presented. As our subject matter is the distributions themselves, we have chosen to discuss them directly. In Chapter 3 we shall examine how probability is mapped from one space to another by functions—by *functions on a probability space.*

The use of lowercase letters such as x and y for convenient reference to distributions should not be taken as suggesting that they be treated as identity or projection functions so as to conform to the language of random variables. Rather, they should be treated as the small ω that is used with the probabilists' probability space $(\Omega, \mathcal{A}, P)$, ω being thought of as an arbitrary point in the space Ω. The substantive material discussed will, of course, always be events and probabilities for events; the use of the lowercase letters provides a notationally convenient way of presenting the events as subsets on appropriate spaces.

In Chapter 1 we developed the basic language for handling probabilities. Then in Chapter 2 we examined various distributions of probability on the line, plane, and higher dimensions. This gives us the wherewithal to describe the common real and vector response variables in applications—in science, engineering, and everyday life.

Now consider a large batch of data in an application. It is common practice to calculate one, two, or several statistics that represent salient features of the data. For example, a psychologist with 30 interreaction time measurements might calculate the average time interval. Or a biologist with 125 lengths for 1-year-old fish might calculate the average length representing the location of the values and a standard deviation representing how spread out the values are. In the first sections of Chapter 3 we examine various probability aspects of such data ***reduction.***

We will then be in a position to discuss the ***statistical independence*** *of two different responses. This important concept of probability and statistical theory is examined in the third section of Chapter 3. We will also see that independence gives us the means for building larger models from small initial models; call this* ***compounding.***

With the concepts of reduction and compounding in hand we will have the needed theory to derive the common discrete distributions—the ***binomial,*** *the* ***multinomial,*** *and so on. These distributions are then derived in the concluding sections of Chapter 3.*

3

MARGINAL PROBABILITY

3.1

DATA REDUCTION

In the preface to this chapter we commented on the practice of simplifying a large batch of data by calculating one, two, or several statistics that represent salient features of the data. In this section we examine some general ideas connected with this *data reduction*.

The process of reduction is, in fact, the reverse of that of observing more, as discussed in Section 1.1C. In this section we examine reduction briefly for the probability model in terms of the pure Boolean events and then in more detail using a sample space.

A REDUCTION IN TERMS OF EVENTS: AN EXAMPLE

Consider the tossing of two symmetric coins and suppose that we are interested initially in the face showing for each coin. The standard operations in Section 1.1 give the class $\mathcal{B} = \mathcal{B}_2$ of 16 events and symmetry gives the corresponding 16 probabilities, shown in Table 3.1.

Now suppose that we restrict our interest to events connected with the number of heads. Let E_j be the event "exactly j heads." We then have the obvious events

$$\varnothing, \quad E_0 = T_1T_2, \quad E_1 = H_1T_2 \vee T_1H_2, \quad E_2 = H_1H_2, \quad \mathfrak{U},$$

and we obtain the events

$$E_1 \vee E_2 = (T_1T_2)^c, \quad E_2 \vee E_0 = H_1H_2 \vee T_1T_2, \quad E_0 \vee E_1 = (H_1H_2)^c$$

by the standard operations. These 8 events form a class $\mathcal{B}_*$ closed under the standard

TABLE 3.1
Probabilities for the initial events

$\mathcal{B}$	$\mathcal{B}_*$	P	$\mathcal{B}$	$\mathcal{B}_*$	P	$\mathcal{B}$	$\mathcal{B}_*$	P
$\varnothing$	$\varnothing$	0	H_1		1/2	$(H_1H_2)^c$	E_2^c	3/4
H_1H_2	E_2	1/4	T_1		1/2	$(H_1T_2)^c$		3/4
H_1T_2		1/4	H_2		1/2	$(T_1H_2)^c$		3/4
T_1H_2		1/4	T_2		1/2	$(T_1T_2)^c$	E_0^c	3/4
T_1T_2	E_0	1/4	$H_1H_2 \vee T_1T_2$	E_1^c	1/2	$\mathfrak{U}$	$\mathfrak{U}$	1
			$H_1T_2 \vee T_1H_2$	E_1	1/2			

operations. Thus our reduced interest has taken us from the original class $\mathcal{B}$ of 16 events to the subclass $\mathcal{B}_*$ of the 8 events still of interest, and correspondingly it has restricted our attention to the probabilities for those 8 events.

More formally, our reduced interest has taken us from the original model $(\mathcal{B}, P)$ to the *marginal model* $(\mathcal{B}_*, P)$, where the values are as given in Table 3.2. Note that $\mathcal{B}_*$ is a Boolean σ algebra using the standard operations, and P is the original measure P but now restricted to the subclass $\mathcal{B}_*$ of $\mathcal{B}$.

TABLE 3.2
Probabilities for the special events of interest

$\mathcal{B}_*$	P	$\mathcal{B}_*$	P
$\varnothing$	0	E_0^c	3/4
E_0	1/4	E_1^c	1/2
E_1	1/2	E_2^c	3/4
E_2	1/4	$\mathcal{U}$	1

Consider the reduction process more generally. Let $(\mathcal{B}, P)$ be an initial probability model with Boolean σ-algebra $\mathcal{B}$ and probability measure P. Suppose that we are interested only in certain events that form a subclass $\mathcal{B}_*$ closed under the standard operations. Then our reduced interest takes us from the initial model $(\mathcal{B}, P)$ to the marginal model $(\mathcal{B}_*, P)$, where P is the original measure but now restricted to the subclass $\mathcal{B}_*$.

B REDUCTION IN TERMS OF SAMPLE SPACES: AN EXAMPLE

Consider the coin-tossing example again and let $s = (x_1, x_2)$ designate the outcome, where x_i is the number of heads for the ith coin. Then the sample space is given by

(1) $$\mathcal{S} = \{(x_1, x_2) : x_i = 0, 1\};$$

and the probability measure for sample points is recorded on the left of Table 3.3.

Now suppose that we restrict our attention to events connected with the number of heads. Let y designate the number of heads and Y be the function that calculates the number of heads,

$$y = Y(x_1, x_2) = x_1 + x_2.$$

TABLE 3.3
Probabilities for sample points

s	$P(\{s\}) = p(s)$	$\xrightarrow{Y}$	y	$P_Y(\{y\}) = p_Y(y)$
(0, 0)	1/4	$\longmapsto$	0	1/4
(1, 0) (0, 1)	1/4 1/4	$\longmapsto$	1	1/2
(1, 1)	1/4	$\longmapsto$	2	1/4

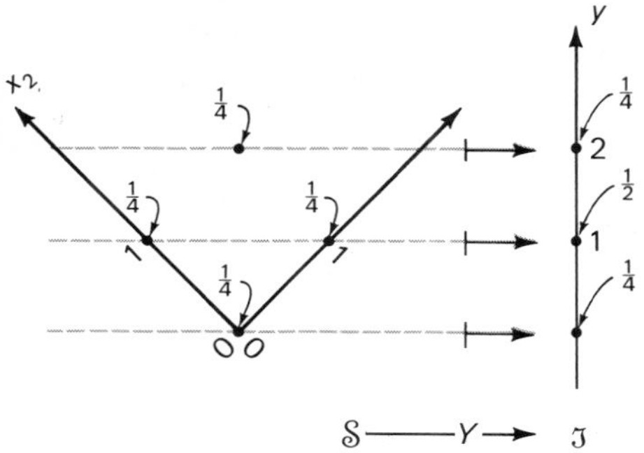

FIGURE 3.1
Mapping Y from sample space $\mathcal{S}$ to sample space $\mathfrak{I}$.

The sample space for y is

(2) $\mathfrak{I} = \{0, 1, 2\}$.

Note that y is the *sum function* mapping $\mathcal{S}$ into $\mathfrak{I}$ or even R^2 into R; see Figure 3.1 and Table 3.3.

TABLE 3.4
Some events for s and y

Event for s		Event for y
$\{(0, 0)\}$	$\leftrightarrow$	$\{0\}$
$\{(1, 0), (0, 1)\}$	$\leftrightarrow$	$\{1\}$
$\{(1, 1)\}$	$\leftrightarrow$	$\{2\}$

Table 3.4 records some events in terms of new sample points y and in terms of original sample points s. The probability for an event in terms of new sample points is, of course, the probability for the event as recorded in terms of original sample points. The probabilities for the new sample points are recorded on the right in Table 3.3. This is the marginal distribution for Y or for $y = Y(s)$; see Figure 3.1.

Reduction is more complicated when sample spaces are used. But, sample spaces are very useful. In the remainder of this section we examine reduction from one sample space to another.

C REDUCTION

Consider a random system that produces an outcome s, say the vector of interaction times in the psychologist's investigation. And suppose that we have a probability model $(\mathcal{S}, \mathcal{A}, P)$, where $\mathcal{S}$ is the sample space, $\mathcal{A}$ the class of events expressed as subsets, and P the probability measure.

Now for the random system, suppose that we are interested only in events based on a certain function Y. Let $\mathfrak{I}$ be the sample space for $y = Y(s)$; then Y is a function that maps $\mathcal{S}$ into $\mathfrak{I}$; see Figure 3.2.

The set of values for the function Y, the set $\{Y(s) : s \in \mathcal{S}\}$ may not be the whole of the space $\mathfrak{I}$: let C be the remaining portion of $\mathfrak{I}$; then $Y^{-1}(C) = \varnothing$. For the present it is convenient to ignore the set C and assume that Y maps $\mathcal{S}$ onto $\mathfrak{I}$.

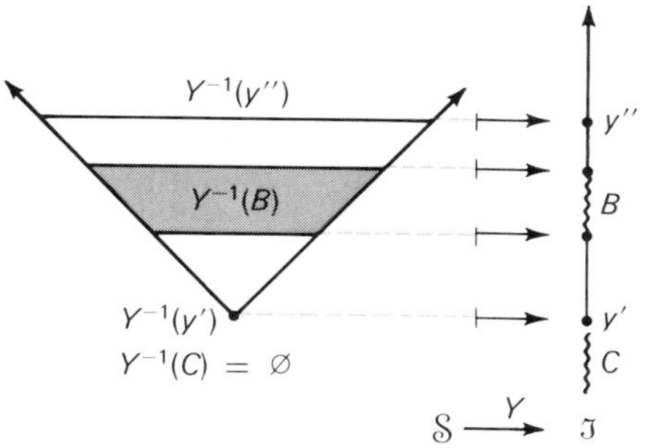

FIGURE 3.2
Mapping Y from space $\mathcal{S}$ into space $\mathfrak{J}$. The event "y in B" is the same as "s in $Y^{-1}(B)$."

Consider the assertion that the function Y has the value y; this is equivalent to the assertion that the original sample point is in the set

(3) $\quad Y^{-1}(y) = \{s : Y(s) = y\},$

called the *preimage* of y. And consider the assertion that the function Y has value in B; this is equivalent to the assertion that the original sample point is in the set

(4) $\quad Y^{-1}(B) = \{s : Y(s) \in B\},$

called the preimage of B. Thus the event expressed as B on the new space is equivalent to the event expressed by $Y^{-1}(B)$ on the initial space; call it an *event for the function Y*.

Now consider repetitions of the random system. Values of the function Y will fall in a set B exactly as original sample points fall in the preimage set $Y^{-1}(B)$. Thus the probability that attaches to the set B on the new space must be the probability as given for the preimage set $Y^{-1}(B)$ on the initial space. This leads us to the following definition:

DEFINITION 1

The **marginal probability** for the set B of values for the function Y is

(5) $\quad P_Y(B) = P(Y^{-1}(B)).$

We have neglected a small technical problem. We might be interested in a very reasonable set B on the new space $\mathfrak{J}$ and yet have a very peculiar function Y such that the preimage set $Y^{-1}(B)$ was not a set for which a probability was given. Clearly we want our function Y to be respectable so that such an embarrassment does not occur. Accordingly, we require that the function Y be measurable:

DEFINITION 2

A function Y from the space $\mathcal{S}$ with σ-algebra $\mathcal{A}$ to the space $\mathfrak{J}$ with σ-algebra $\mathcal{B}$ is **measurable** if for each B in $\mathcal{B}$, $Y^{-1}(B)$ is in $\mathcal{A}$.

Such a function on a probability space ($\mathcal{S}$, $\mathcal{A}$, P) is also called a *random variable*. All ordinary continuous functions and derived functions obtained by limits are measurable functions. For some details, see Supplement Section 3.6G.

We can now present the *marginal* probability model for the measurable function Y. The model is

(6) $(\mathcal{J}, \mathcal{B}, P_Y)$,

where

(7) $P_Y = PY^{-1}$

is given by Definition 1; this is the marginal model for Y or for $y = Y(s)$. Of course we must be sure that P_Y is a probability measure; but this is primarily a notational matter and is covered by some details in Section D.

EXAMPLE 1

Uniform distribution on $(0, 1)^2$: Consider the uniform distribution for (x_1, x_2) on the unit square $(0, 1)^2$; see Figure 3.3. Find the probability that $y = x_1 + x_2 \leq 0.5$. The set $\{y \leq 0.5\}$ on the image space is equivalent to the set $\{(x_1, x_2) : x_1 + x_2 \leq 0.5\}$ on the initial space. This is the set of points to the lower left from the line $x_1 + x_2 = 0.5$, and within the square it is just the triangle formed by (0, 0), (0.5, 0), (0, 0.5). With the uniform distribution the probability can be calculated as a ratio of areas: $\frac{1}{2}(0.5)(0.5)/1$, which is 0.125. Find the probability that $0.9 < y < 1.2$. The set $\{0.9 < y < 1.2\}$ on the image space is equivalent to the set $\{(x_1, x_2) : 0.9 < x_1 + x_2 < 1.2\}$ on the initial space. Calculated as a ratio of areas we obtain

$$(1 - \tfrac{1}{2}(0.9)(0.9) - \tfrac{1}{2}(0.8)(0.8))/1 = 0.275.$$

D SOME PROPERTIES OF A FUNCTION

Throughout our applications we will be interested in many different kinds of functions; as examples, consider the average $\bar{y}$ and sample variance $s_y^2 = \Sigma (y_i - \bar{y})^2/(n - 1)$. Consider a function Y from a space $\mathcal{S}$ into a space $\mathcal{J}$; see Figure 3.2. The set $\{Y(s) : s \in \mathcal{S}\}$ of values of the function Y may not be the whole of the

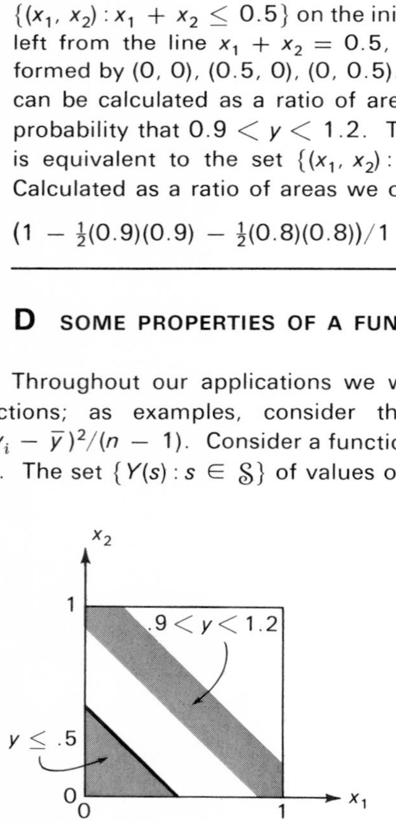

FIGURE 3.3
Sample space $S = (0, 1)^2$. The events $y \leq 0.5$ and $0.9 < y < 1.2$, where $y = x_1 + x_2$.

space $\mathfrak{Y}$. Let C be the remaining portion of $\mathfrak{Y}$; then $Y^{-1}(C) = \varnothing$. Again it is convenient to ignore the set C and assume that Y maps $\mathcal{S}$ onto $\mathfrak{Y}$.

Consider points in the space $\mathfrak{Y}$, and the corresponding preimage sets:

$$Y^{-1}(y') \leftrightarrow y',$$

$$Y^{-1}(y'') \leftrightarrow y''.$$

If $y' \neq y''$, then $Y^{-1}(y')$ and $Y^{-1}(y'')$ are clearly disjoint. Also, each point in $\mathcal{S}$ lies in some set, $Y^{-1}(y)$. Thus the different sets $Y^{-1}(y)$ form a partition of $\mathcal{S}$, called the *partition induced by* Y. And the sets in this partition are in one-to-one correspondence with the points of $\mathfrak{Y}$; see Figure 3.2.

As an example, consider the sum function given by $y = x_1 + x_2$ mapping R^2 into R. A line $x_1 + x_2 = c$ is mapped into the point c. Thus R^2 is partitioned into the class of dashed lines as indicated in Figure 3.1. Knowing the value of $x_1 + x_2$ is equivalent to knowing that (x_1, x_2) is somewhere on the corresponding line; thus calculating the value of the function is equivalent to *losing* the information as to where the point (x_1, x_2) is on the line and thus to making a *reduction* on the information contained in the original outcome.

Note that the average function $\bar{x} = (x_1 + x_2)/2$ induces the same partition as does the sum function $x_1 + x_2$. For clearly the values of the two functions can be put into one-to-one correspondence: $\bar{x} = (1/2)(x_1 + x_2)$ and $(x_1 + x_2) = 2\bar{x}$.

Now consider sets on the image space $\mathfrak{Y}$ and the corresponding preimage sets on the original space. We have the one-to-one correspondence

$$(8) \qquad Y^{-1}(B) \leftrightarrow B.$$

These sets are describing the same proposition; that is, $Y(s) \in B$ is equivalent to $s \in Y^{-1}(B)$. Thus it is not surprising to find that the following relations hold:

$$(9) \qquad Y^{-1}(\cap_1^\infty B_i) = \cap_1^\infty Y^{-1}(B_i),$$

$$Y^{-1}(\cup_1^\infty B_i) = \cup_1^\infty Y^{-1}(B_i),$$

$$Y^{-1}(B^c) = (Y^{-1}(B))^c;$$

thus Y^{-1} *commutes with the standard operations*. The proofs are straightforward and can be put in the abbreviated form used in Section 1.2D; for example,

$$s \in Y^{-1}(\cap_1^\infty B_i) \;\leftrightarrow\; Y(s) \in \cap_1^\infty B_i \;\leftrightarrow\; Y(s) \in B_i \text{ for each } i$$

$$\leftrightarrow\; s \in Y^{-1}(B_i) \text{ for each } i \;\leftrightarrow\; s \in \cap_1^\infty Y^{-1}(B_i).$$

Now consider the sets B that form a σ-algebra $\mathcal{B}$ on the image space $\mathfrak{Y}$. The relations (9) show that the preimage sets form a class that is closed under countable union, countable intersection, and complementation. Designate the class of preimage sets by

$$Y^{-1}(\mathcal{B}) = \{Y^{-1}(B) : B \in \mathcal{B}\};$$

then $Y^{-1}(\mathcal{B})$ is a σ-algebra on the original space. Indeed in terms of propositions, the two σ-algebras are equivalent; they are just expressed on different spaces by the one-to-one correspondence

$$Y^{-1}(B) \leftrightarrow B.$$

We can then rephrase Definition 2 as:

DEFINITION 3

A function Y from $\mathcal{S}$ with σ-algebra $\mathcal{A}$ to $\mathcal{T}$ with σ-algebra $\mathcal{B}$ is **measurable** if $Y^{-1}(\mathcal{B}) \subset \mathcal{A}$.

In other words, Y is measurable if it does not introduce any new "events."

For a measurable function Y the events B or, equivalently, $Y^{-1}(B)$ will be called *the events for the function Y*.

We are now in a position to see that P_Y in Section C is a probability measure on $\mathcal{B}$. We know that P is a probability measure on $\mathcal{A}$ and hence on the smaller σ-algebra $Y^{-1}(\mathcal{B})$. But P on this smaller σ-algebra is equivalent,

$$P(Y^{-1}(B)) = P_Y(B),$$

to P_Y on the σ-algebra $\mathcal{B}$. Thus P_Y is a probability measure.

E EXERCISES

1 Three symmetric coins are tossed and $s = (x_1, x_2, x_3)$ is observed, where x_i is the indicator for heads for the ith coin.
(a) Tabulate s and $P(\{s\}) = p(s)$.
(b) Let Y give the number of heads: $Y(s) = x_1 + x_2 + x_3 = y$. Tabulate y and $P_Y(\{y\}) = p_Y(y)$.

2 Two symmetric dice are tossed and $s = (x_1, x_2)$ is observed, where x_i is the number of points on the ith die.
(a) Partially tabulate s and $p(s)$ in a convenient order for part (b).
(b) Let Y give the total points: $Y(s) = x_1 + x_2 = y$. Tabulate y and $P_Y(\{y\}) = p_Y(y)$.

3 Consider the uniform distribution for (x_1, x_2) on the unit square $(0, 1)^2$. Let Y be the sum function: $Y(x_1, x_2) = x_1 + x_2 = y$.
(a) Calculate the distribution function $F_Y(y)$ for $0 \le y \le 1$; and for $1 \le y \le 2$.
(b) Calculate the density function f_Y.

4 A biologist makes two measurements on a growth rate and obtains (y_1, y_2).
(a) Determine the preimage of $y_1 + y_2 = c$; of $y_1^2 + y_2^2 = d$; and of $(\Sigma y_i, \Sigma y_i^2) = (c, d)$.
(b) Let $y_{(1)}$ be the smallest of (y_1, y_2), and $y_{(2)}$ the second smallest. Determine the preimage of $y_{(1)} = c'$; of $y_{(2)} = d'$; and of $(y_{(1)}, y_{(2)}) = (c', d')$.

5 (*continuation*) Argue that $(\Sigma y, \Sigma y^2)$ is a function equivalent to $(y_{(1)}, y_{(2)})$.

6 A physicist makes n measurements on a current flow and obtains $(y_1, \ldots, y_n)$. The physicist is interested in the sample mean $\bar{y}$ and sample variance s_y^2:

$$\bar{y} = \frac{\sum y_i}{n}, \qquad s_y^2 = \frac{1}{n-1} \sum (y_i - \bar{y})^2.$$

Compare with Exercise 1.2.17. Note that $\bar{y}\mathbf{1} = (\bar{y}, \ldots, \bar{y})$ is the projection of $\mathbf{y}$ on the line $\mathcal{L}(\mathbf{1}) = \{a\mathbf{1} : a \in R\}$ and that $(n-1)s_y^2$ is the squared distance of $\mathbf{y}$ from the line $\mathcal{L}(\mathbf{1})$.
(a) For $n = 3$ determine the preimage of $\bar{y} = c$ and show it on a sketch.
(b) For $n = 3$ determine the preimage of $s_y^2 = d$ and show it on a sketch.
(c) For $n = 3$ determine the preimage of $(\bar{y}, s_y^2) = (c, d)$ and show it on a sketch.

F PROBLEMS

7 Let Y be a function from $\mathcal{S}$ into $\mathcal{T}$. Use the method in Section D to prove that $Y^{-1}(B^c) = (Y^{-1}(B))^c$.

8 (*continuation*) Prove that $Y^{-1}(\cup_1^\infty B_i) = \cup_1^\infty Y^{-1}(B_i)$.

9 Consider the following three functions from R^n into R^n: $(y_{(1)}, \ldots, y_{(n)})$, where $y_{(i)}$ is the ith smallest of $y_1, \ldots, y_n$; $(\Sigma y_i, \Sigma y_i^2, \ldots, \Sigma y_i^n)$; and $(\Sigma y_i, \Sigma_{i<j} y_iy_j, \Sigma_{i<j<k} y_iy_jy_k, \ldots, y_1 \cdots y_n)$. The algebra of symmetric functions shows that the last two functions are equivalent: in fact, if we know the value for say $(\Sigma y_i, \Sigma y_i^2)$, then we can calculate the value of $(\Sigma y_i, \Sigma_{i<j} y_iy_j)$, and conversely. Use an algebraic result concerning the roots of a polynomial to show that the first and third are equivalent.

10 A chemist makes two determinations on the logarithm of a concentration and obtains (y_1, y_2). For simplicity assume that (y_1, y_2) has the standard bivariate normal. Let W be the sum function $W(y_1, y_2) = y_1 + y_2 = w$.
   (a) Determine the set $W^{-1}((-\infty, w])$ and sketch it on the plane.
   (b) Use the rotational symmetry of the normal to calculate $P(W^{-1}(-\infty, w])$ and express it in terms of the standard normal distribution function G.
   (c) Determine the distribution function F_W for $w = y_1 + y_2$. What distribution is this?

11 The relationships(9) are not restricted to the case with a countable index set. Show that $Y^{-1}(\cap B_\alpha) = \cap Y^{-1}(B_\alpha)$ and $Y^{-1}(\cup B_\alpha) = \cup Y^{-1}(B_\alpha)$, where α ranges over some index set I.

3.2

MARGINAL PROBABILITY ON THE LINE AND PLANE

Most scientific and industrial data come in the form of real numbers, a moderate-to-large collection of real numbers. Analysis of such data often involves a reduction to one, two, or several real numbers that represent significant features of the full data. Correspondingly, a probability description of possibilities for the full data leads to a probability description of possibilities for the reduced data, the marginal distribution for such data. In this section we examine some notation and methods for marginal distributions on real spaces. We organize this in terms of distribution functions, probability functions, and probability density functions. Recall the basic formulas (3.1.5) and (3.1.7) for marginal probabilities: $P_Y = PY^{-1}$.

A MARGINAL DISTRIBUTION FUNCTIONS

Consider a basic distribution on a space $\mathcal{S}$ with probabilities given by P, say the distribution for $(x_1, \ldots, x_{10})$ on R^{10}, where x_i is the yield of barley on an ith plot of land. Suppose first that we are interested in the marginal distribution for a real-valued function X, say the sum function $X(x_1, \ldots, x_{10}) = \Sigma_1^{10} x_i$. Let F_X be the distribution function:

$$(1) \qquad F_X(x) = P_X((-\infty, x]) = P(X^{-1}(-\infty, x]) = P(\{s : -\infty < X(s) \leq x\}).$$

Thus to evaluate the distribution function we must calculate the probability for the set

$$\{s : -\infty < X(s) \leq x\},$$

the preimage set of $(-\infty, x]$; this is usually accomplished by summing values of a probability function or integrating a probability density function.

Now suppose that we are interested in the marginal distribution for a pair of functions (X, Y). Let F_{XY} be the distribution function:

$$(2)\qquad \begin{aligned} F_{XY}(x, y) &= P_{XY}((-\infty, x] \times (-\infty, y]) \\ &= P(X^{-1}(-\infty, x] \cap Y^{-1}(-\infty, y]) \\ &= P(\{s : -\infty < X(s) \leq x,\ -\infty < Y(s) \leq y\}). \end{aligned}$$

This is usually evaluated by summation or by integration.

EXAMPLE 1

Uniform distribution on $(0, 1)^2$: Consider a uniform distribution for (x_1, x_2) over the unit square $(0, 1)^2$. And suppose that we want the distribution of $y = Y(x_1, x_2) = x_1 + x_2$. Let F_Y be the marginal distribution function. Then following the pattern of calculation in Example 3.1.1, we obtain

$$\begin{aligned} F_Y(y) &= 0 && \text{if} \quad y < 0, \\ &= \frac{y^2}{2} && 0 \leq y < 1, \\ &= 1 - \frac{(2-y)^2}{2} && 1 \leq y < 2, \\ &= 1 && 2 \leq y. \end{aligned}$$

See Figure 3.4. The density function can be obtained by differentiation:

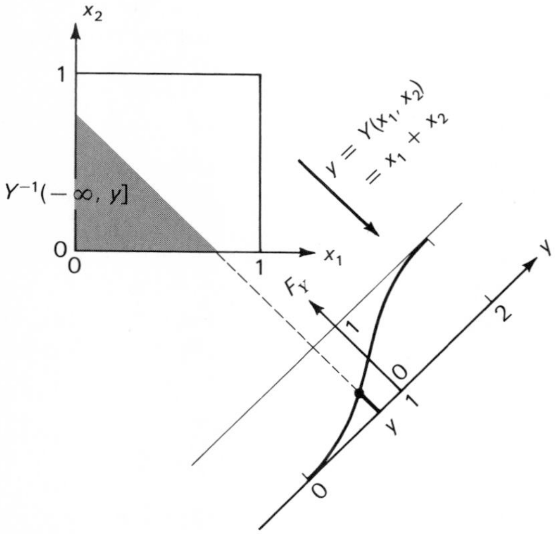

FIGURE 3.4
Probability in the square $(0, 1)^2$ is projected by $y = Y(x_1, x_2) = x_1 + x_2$ onto the real line. The distribution function at y, $F_Y(y)$, gives the probability for the pre-image set $Y^{-1}(-\infty, y]$ as marked.

$$f_Y(y) = y \qquad \text{if } 0 \le y < 1,$$
$$= 2 - y \qquad 1 \le y < 2,$$
$$= 0 \qquad \text{otherwise.}$$

This is a triangular distribution on (0, 2) with mode or peak probability at $y = 1$.

EXAMPLE 2

Square of a standard normal: Consider a standard normal distribution for z on R. And suppose that we want the distribution of $y = Y(z) = z^2$. Let F_Y be the marginal distribution function; then

$$F_Y(y) = P_Y((-\infty, y]) = P(Y^{-1}(-\infty, y])$$
$$= P([-\sqrt{y}, \sqrt{y}]) \qquad \text{if } y \ge 0.$$

See the function $y = z^2$ in Figure 3.5 and note that the function is two to one generally and that the preimage of $(-\infty, y]$ is $[-\sqrt{y}, \sqrt{y}]$ for $y \ge 0$. We can calculate directly from the standard normal distribution:

$$F_Y(y) = \int_{-\sqrt{y}}^{+\sqrt{y}} \frac{1}{\sqrt{2\pi}} e^{-t^2/2}\,dt = 2\int_0^{\sqrt{y}} \frac{1}{\sqrt{2\pi}} e^{-t^2/2}\,dt \qquad \text{if } y \ge 0.$$

The density function can be obtained by differentiation:

$$f_Y(y) = 2\frac{1}{2}y^{-1/2}\frac{1}{\sqrt{2\pi}}e^{-y/2} = \Gamma^{-1}\left(\frac{1}{2}\right)\left(\frac{y}{2}\right)^{(1/2)-1} e^{-y/2}\frac{1}{2} \qquad \text{if } 0 < y,$$
$$= 0 \qquad y \le 0.$$

Thus $y/2$ is gamma(1/2) and y is chi square(1); see Exercises 2.2.15 and 2.2.16.

Sometimes we are interested in very simple functions, coordinate projection functions. Consider a distribution on the plane with distribution function F; then

$$F(x, y) = P((-\infty, x] \times (-\infty, y]).$$

And suppose that we are interested in only the first coordinate, x. As a function of (x, y), x is the *projection* of $R \times R$ onto the first space in the product; by restricting

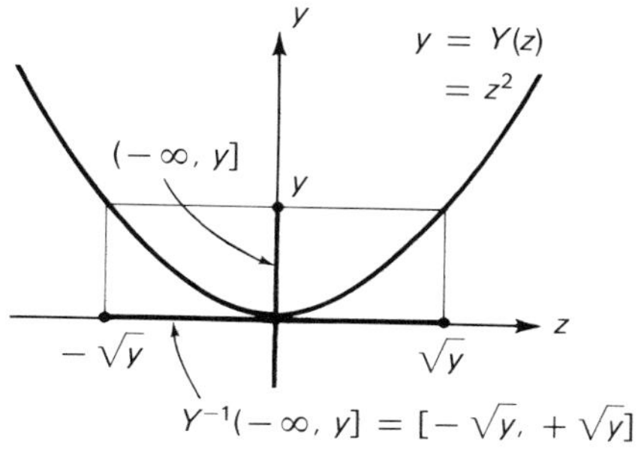

FIGURE 3.5
Function $y = Y(z) = z^2$. The preimage of $(-\infty, y]$ is $[-\sqrt{y}, +\sqrt{y}]$ for $y \ge 0$.

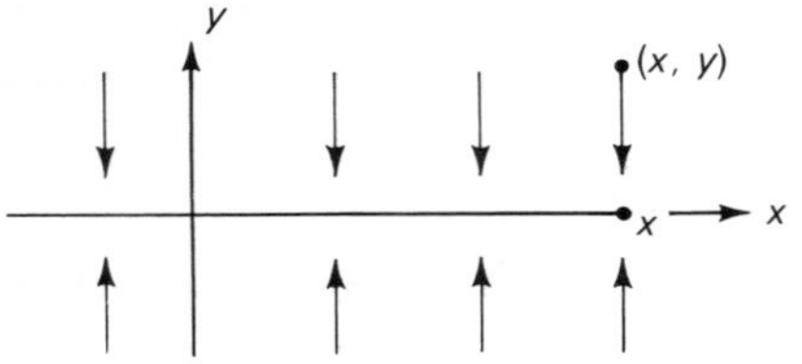

FIGURE 3.6
Marginal distribution for x; the probability on the plane is projected onto the first axis.

attention to the first coordinate, x, we are directly ignoring the second coordinate, y. Correspondingly, the probability distribution on the plane is effectively collapsed or projected onto the first axis; see Figure 3.6. Let F_1 be the marginal distribution function for the first coordinate x; then

(3) $$F_1(x) = P((-\infty, x] \times R) = \lim_{y\to\infty} F(x, y) = F(x, \infty).$$

Analogously, let F_2 be the marginal distribution function for the second coordinate y; then

(4) $$F_2(y) = F(\infty, y).$$

Now consider a distribution on R^k with distribution function F; then

$$F(y_1, \ldots, y_k) = P((-\infty, y_1] \times \cdots \times (-\infty, y_k]).$$

And suppose first that we are interested in just a single coordinate, say, the first, y_1. The first coordinate y_1 as a function of $(y_1, \ldots, y_k)$ is the projection of $R \times R \times \cdots \times R$ onto the first space in the product. Let F_1 be the marginal distribution function for the first coordinate y_1; then

(5) $$F_1(y_1) = P((-\infty, y_1] \times R \times \cdots \times R) = F(y_1, \infty, \ldots, \infty).$$

And suppose now that we are interested in just two coordinates, say the first two, y_1 and y_2. The first coordinate pair (y_1, y_2) as a function of $(y_1, \ldots, y_k)$ is the projection of $(R \times R) \times \cdots \times R$ onto the first space $R \times R$. Let F_{12} be the marginal distribution function for the first two coordinates (y_1, y_2); then

(6) $$\begin{aligned} F_{12}(y_1, y_2) &= P((-\infty, y_1] \times (-\infty, y_2] \times R \times \cdots \times R) \\ &= F(y_1, y_2, \infty, \ldots, \infty). \end{aligned}$$

And analogously, we can obtain, for example, the marginal distribution function F_{137} for the coordinate triple (y_1, y_3, y_7).

EXAMPLE 3

Orientation with two symmetric coins: Consider the angles (y_1, y_2) when a pair of symmetric coins are tossed twice (Example 2.3.2). The marginal distribution function $F_1(y_1)$ can be obtained from $F(y_1, y_2)$ by taking y_2 to ∞; this gives

$$\begin{aligned} F_1(y_1) &= 0 && \text{if} \quad y_1 < 0, \\ &= \frac{y_1}{180} && 0 \le y_1 < 180, \\ &= 1 && 180 \le y_1. \end{aligned}$$

This is the uniform(0, 180) distribution.

B MARGINAL PROBABILITY FUNCTIONS

The most convenient way of handling discrete distributions is almost always in terms of the probability function that records the discrete chunks of probability. Consider a basic discrete distribution on a space $\mathcal{S}$ with probability measure P and probability function p. Suppose first that we are interested in the marginal distribution for a real-valued function X. Let p_X be the probability function

$$(7) \qquad p_X(x) = P_X(\{x\}) = P(X^{-1}(x)) = \sum_{x=X(s)} p(s),$$

where the summation is, of course, over all points s with the given value for $X(s)$.

Now suppose that we are interested in the marginal distribution for a pair of functions (X, Y). Then

$$(8) \qquad p_{XY}(x, y) = P(X^{-1}(x) \cap Y^{-1}(y)) = \sum_{\substack{x=X(s)\\ y=Y(s)}} p(s).$$

EXAMPLE 4

Tossing two dice: Consider the tossing of two symmetric dice and the recording of (y_1, y_2), where y_i is the number of points on the ith die (Example 2.3.3). Then

$$p(y_1, y_2) = 1/36 \qquad \text{on } \{1, 2, 3, 4, 5, 6\}^2$$

and zero elsewhere. And suppose that we are interested only in the total points $t = T(y_1, y_2) = y_1 + y_2$ for the two dice; then

$$p_T(t) = p(T^{-1}(t)),$$

which is the summation of $1/36$ for each possible sample point having total t. As our initial distribution is symmetric, we can calculate by counting sample points:

$$p_T(t) = \frac{\text{number of points } (y_1, y_2) \text{ having } y_1 + y_2 = t}{\text{number of points } (y_1, y_2)},$$

where y_1 and y_2 take values 1, 2, . . ., 6. The probabilities are easily enumerated, giving the results shown in Table 3.5.

TABLE 3.5
Probabilities for t

t	$p_T(t)$	t	$p_T(t)$
2	1/36	8	5/36
3	2/36	9	4/36
4	3/36	10	3/36
5	4/36	11	2/36
6	5/36	12	1/36
7	6/36		1

The preceding is a special case of the convolution formula:

EXAMPLE 5

Convolution formula: Consider a discrete distribution on R^2 with probability function p. The probability function for the sum $t = T(y_1, y_2) = y_1 + y_2$ is given by

$$(9) \qquad p_T(t) = \sum_{t=y_1+y_2} p(y_1, y_2) = \sum_{y_2} p(t - y_2, y_2).$$

Sometimes we are interested in very simple functions—coordinate projection functions. Consider a discrete distribution on the plane with probability function p giving the probability $p(x, y)$ at the point (x, y). The marginal probability functions are then given by

$$(10) \qquad p_1(x) = \sum_{y} p(x, y),$$

$$p_2(y) = \sum_{x} p(x, y);$$

for example, $p_1(x)$ is obtained by summing the probabilities for all the discrete points (x, y) having first coordinate x; see Figure 3.6.

Now consider a discrete distribution on R^k with probability function p. The marginal probability function for the first coordinate is then given by

$$(11) \qquad p_1(y_1) = \sum_{y_2 \cdots y_k} p(y_1, \ldots, y_k)$$

and for the first two coordinates by

$$(12) \qquad p_{12}(y_1, y_2) = \sum_{y_3 \cdots y_k} p(y_1, \ldots, y_k).$$

Thus to obtain the marginal probability function we merely sum out the coordinates that are not of interest.

C MARGINAL PROBABILITY DENSITY FOR A PROJECTION

Consider an absolutely continuous distribution on the plane with probability density function f; then

$$P(A) = \int_A f(x, y)\, dx\, dy.$$

And suppose that we are interested in only the first coordinate, x. The first coordinate, x, is the projection of $R \times R$ onto the first factor, R; thus our interest is in the probability on the plane as projected or collapsed onto the first coordinate axis; see Figure 3.6 again. Let F_1 be the distribution function and f_1 the density function; then

$$F_1(x) = \int_{(-\infty, x]\times R} f(t, y)\, dt\, dy = \int_{-\infty}^{x} \left[\int_{-\infty}^{\infty} f(t, y)\, dy \right] dt,$$

and thus

$$(13) \qquad f_1(x) = \int_{-\infty}^{\infty} f(x, y)\, dy.$$

With a continuous density function f we can think of the probability in the strip between x and $x + dx$; see Figure 3.7. The *volume* above this strip can be integrated and represented as *area* above the interval $(x, x + dx)$; this gives

$$(14) \qquad f_1(x)\, dx = \int_{-\infty}^{\infty} f(x, y)\, dy \cdot dx.$$

More formally, we take this to be a probability differential, the marginal probability differential for the first coordinate x.

Analogously, let f_2 be the probability density function for the second coordinate y. The probability density is given by

$$(15) \qquad f_2(y) = \int_{-\infty}^{\infty} f(x, y)\, dx,$$

and the probability differential by

$$(16) \qquad f_2(y)\, dy = \int_{-\infty}^{\infty} f(x, y)\, dx \cdot dy.$$

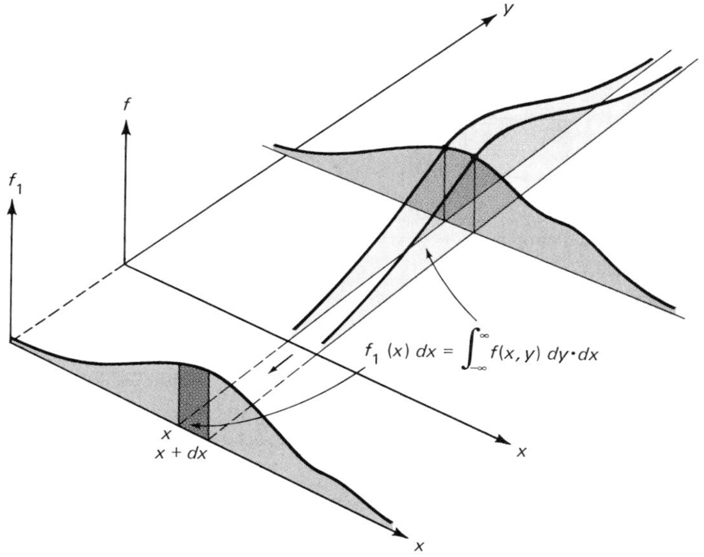

FIGURE 3.7
Joint probability density f; marginal density f_1 for x by integrating out y.

Now consider an absolutely continuous distribution on R^k with probability density function f. Then the marginal density for the first coordinate is given by

$$(17) \qquad f_1(y_1) = \int_{-\infty}^{\infty} \cdots \int_{-\infty}^{\infty} f(y_1, y_2, \ldots, y_k)\, dy_2 \cdots dy_k$$

and for the first two coordinates, say, by

$$(18) \qquad f_{12}(y_1, y_2) = \int_{-\infty}^{\infty} \cdots \int_{-\infty}^{\infty} f(y_1, y_2, y_3, \ldots, y_k)\, dy_3 \cdots dy_k.$$

EXAMPLE 6

Bivariate distribution: Consider an absolutely continuous distribution for (x, y) with density function

$$f(x, y) = 6x \qquad \text{if } 0 < x < 1, \quad 0 < y < 1 - x,$$

and zero otherwise. This is a distribution effectively over the triangle (0, 0), (1, 0), (0, 1); see Figure 3.8. The marginal densities are

$$(19) \qquad f_1(x) = \int_0^{1-x} 6x\, dy = 6xy \Big|_{y=0}^{1-x} = 6x(1 - x), \qquad 0 < x < 1,$$

and zero otherwise [this is the beta(2, 2) distribution], and

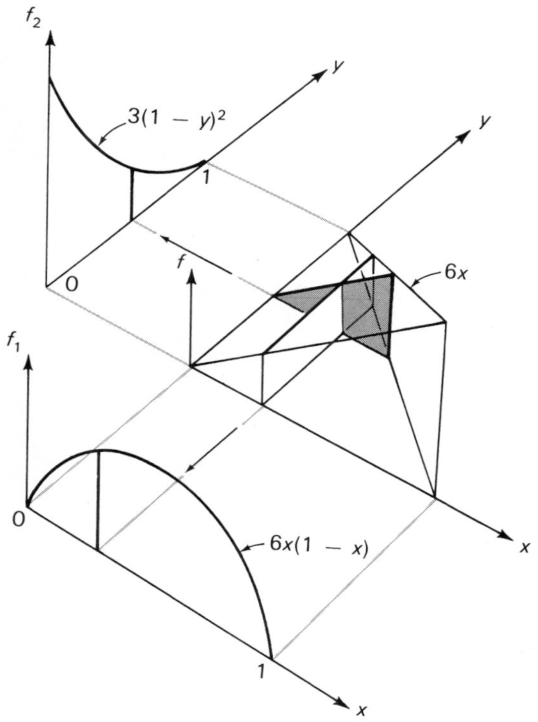

FIGURE 3.8
Joint and marginal densities in Example 6.

$$(20) \qquad f_2(y) = \int_0^{1-y} 6x\,dx = 3x^2\Big|_{x=0}^{1-y} = 3(1-y)^2, \qquad 0 < y < 1,$$

and zero otherwise [this is the beta(1, 3) distribution].

Now suppose that the distribution on R^2 has come from some initial space $\mathcal{S}$ with probability measure P. Let (X, Y) be the function that gives the distribution on R^2; then

$$P_{XY}(A) = P(\{s : (X(s), Y(s)) \in A\}) = \int_A f_{XY}(x, y)\,dx\,dy.$$

If the marginal density f_{XY} is available, then the marginal densities for X and Y alone are

$$f_X(x) = \int_{-\infty}^{\infty} f_{XY}(x, y)\,dy, \qquad f_Y(y) = \int_{-\infty}^{\infty} f_{XY}(x, y)\,dx.$$

D MARGINAL PROBABILITY DENSITY FOR A GENERAL FUNCTION

We now examine probability densities for functions more complicated than the simple coordinate projections just discussed. First consider an absolutely continuous distribution for x on the real line R with density function g. And suppose that we are interested in the distribution for $y = Y(x)$ produced by the function Y. If Y is a monotone, continuously differentiable one-to-one function, then we are in the change of variable (or relabeling) situation discussed in Section 2.4C and there is no reduction. Recall the density function:

$$(21) \qquad f_Y(y) = g(X(y))\left|\frac{dx}{dy}\right|_+,$$

where X is the inverse function Y^{-1}.

If the function Y is a several-to-one mapping, then several different neighborhoods for x can map into a single neighborhood for y; see Figure 3.9. The probability for the neighborhood of y is then calculated by adding the probabilities for the corresponding neighborhoods of x; with differentiability, formula (21) can then be applied component by component. Rather than present here a general formula that can be somewhat

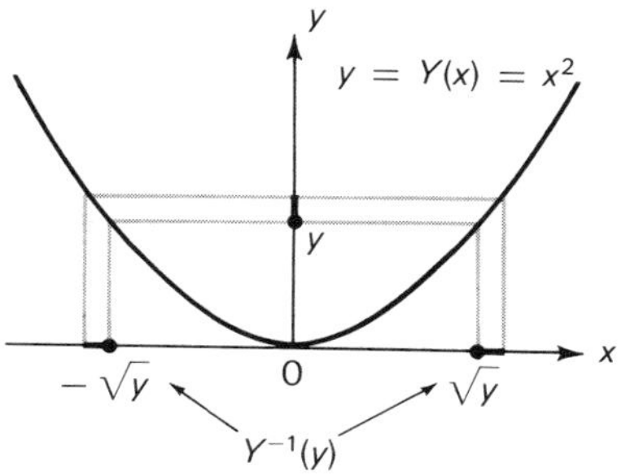

FIGURE 3.9
The probability for a neighborhood of y is the sum of the probabilities for the corresponding two neighborhoods for x.

cumbersome, we shall present a simple example that illustrates the use of the principles from Section 3.1 and the change of variable in formula (21).

EXAMPLE 7

Square function: Consider a distribution for x on the real line R with density function g. And suppose that we want the distribution of $y = Y(x) = x^2$; compare with Example 2 and see Figure 3.9. The preimage of y (>0) consists of two points, $-\sqrt{y}$ and $+\sqrt{y}$. The probability at y from the first of these points is

$$g(-\sqrt{y})\left|\frac{d-\sqrt{y}}{dy}\right| dy = g(-\sqrt{y})(1/2)y^{-1/2}\,dy,$$

and the probability at y from the second of the points is

$$g(\sqrt{y})\left|\frac{d\sqrt{y}}{dy}\right| dy = g(\sqrt{y})(1/2)y^{-1/2}\,dy.$$

Thus the total probability at y is

$$f_Y(y)\,dy = (g(\sqrt{y}) + g(-\sqrt{y}))(1/2)y^{-1/2}\,dy. \tag{22}$$

If g is standard normal, then substitution gives the result obtained in Example 2 by the distribution function route.

Now consider an absolutely continuous distribution for a multiple response $(x_1, \ldots, x_k)$ on R^k with density function g; for notational convenience we examine the case $k = 2$: for example, reaction time and reaction intensity in a psychology experiment.

Suppose that we are interested in the distribution for $(y_1, y_2) = (Y_1(x_1, x_2), Y_2(x_1, x_2))$ produced by the function (Y_1, Y_2). If (Y_1, Y_2) is a one-to-one continuously differentiable function, then we are in the change of variable (or relabeling) situation in Section 2.4E and there is no reduction. Recall the density function

$$f_{Y_1Y_2}(y_1, y_2) = g(X_1(y_1, y_2), X_2(y_1, y_2))\left|\frac{\partial x_1, x_2}{\partial y_1, y_2}\right|_+, \tag{23}$$

where (X_1, X_2) is the inverse function. If the function (Y_1, Y_2) is a several-to-one mapping, then several neighborhoods for (x_1, x_2) can map into a single neighborhood for (y_1, y_2). The probability for this neighborhood is then calculated by adding the probabilities for the corresponding neighborhoods for (x_1, x_2); the procedure follows as in Example 7.

Now suppose that we are interested in the distribution for $y = Y(x_1, x_2)$ produced by the function Y from R^2 into R. To have the value y for Y is to know that (x_1, x_2) is somewhere on the preimage contour $Y^{-1}(x_1, x_2)$ as indicated in Figure 3.10. In fact, the function Y partitions the plane into the class of such preimage contours. If the function Y has reasonable properties, then the contours will be continuous curves, as in Figure 3.10.

The probability for an interval $(y, y + dy)$ is given by the probability for the corresponding range of preimage contours, the curved band, as shown in Figure 3.10. To calculate this probability it is convenient to find a second function Y_2 whose preimage contours are cross sectional to those for Y. Thus we try to complete the initial

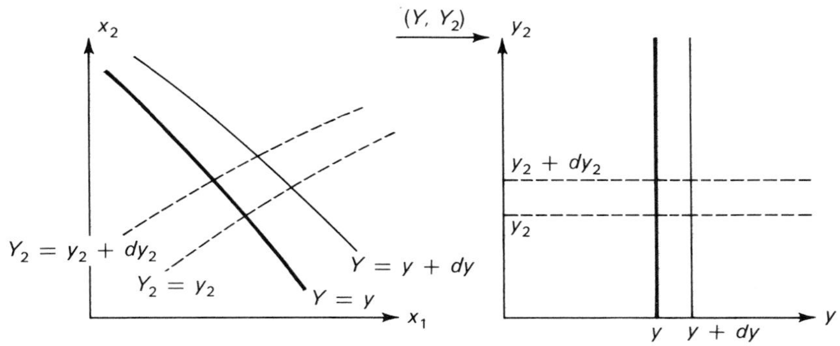

FIGURE 3.10
Contours for the function Y and contours for the complementing function Y_2.

function Y so as to obtain a function (Y, Y_2) with the one-to-one differentiable properties needed for formula (23). We then integrate out the unwanted coordinate as done in Section C:

$$(24) \qquad f_Y(y) = \int_{-\infty}^{\infty} g(X_1(y, y_2), X_2(y, y_2)) \left| \frac{\partial x_1, x_2}{\partial y, y_2} \right|_+ dy_2.$$

The method generalizes immediately to finding the density for several functions on R^k.

EXAMPLE 8

Convolution formula: Consider a distribution for (x_1, x_2) with density function g on R^2. And suppose that we want the distribution of $t = T(x_1, x_2) = x_1 + x_2$. A simple choice of second variable to complete this as a one-to-one transformation is $u = x_2$; thus

$$t = x_1 + x_2, \qquad x_1 = t - u,$$
$$u = x_2, \qquad x_2 = u.$$

The Jacobian matrix is

$$\begin{pmatrix} \dfrac{\partial x_1}{\partial t} & \dfrac{\partial x_1}{\partial u} \\ \dfrac{\partial x_2}{\partial t} & \dfrac{\partial x_2}{\partial u} \end{pmatrix} = \begin{pmatrix} 1 & -1 \\ 0 & 1 \end{pmatrix},$$

which is a constant (the transformation is linear); the determinant is 1. Thus

$$g(x_1, x_2)\, dx_1\, dx_2 = g(t - u, u)1\, dt\, du,$$

and the marginal density for $t = x_1 + x_2$ is

$$(25) \qquad f_T(t) = \int_{-\infty}^{\infty} g(t - u, u)\, du.$$

This is called the *convolution formula* and it gives the density for $t = x_1 + x_2$ in terms of the joint density for (x_1, x_2).

EXAMPLE 9

Consider the standard normal distribution for (z_1, z_2) on R^2. And suppose that we are interested in r and also interested in a, where

$$r = (z_1^2 + z_2^2)^{1/2}, \qquad z_1 = r \cos a,$$

$$a = \tan^{-1}\left(\frac{z_2}{z_1}\right), \qquad z_2 = r \sin a$$

for $0 < r < \infty$ and $0 < a < 2\pi$. The differential transformation is

$$\begin{pmatrix} dz_1 \\ dz_2 \end{pmatrix} = \begin{pmatrix} \cos a & -r \sin a \\ \sin a & r \cos a \end{pmatrix} \begin{pmatrix} dr \\ da \end{pmatrix},$$

the Jacobian determinant is $r(\cos^2 a + \sin^2 a) = r$, and thus

$$dz_1\, dz_2 = r\, dr\, da;$$

compare this with the geometrical derivation in Section 2.3E and recall Example 2.4.7. Thus

$$\frac{1}{2\pi} e^{-(z_1^2+z_2^2)/2}\, dz_1\, dz_2 = \frac{1}{2\pi} e^{-r^2/2} r\, dr\, da = e^{-r^2/2} r\, dr \cdot \frac{1}{2\pi}\, da$$

on the product set $(0, \infty) \times (0, 2\pi)$ and zero elsewhere. The marginal distributions are then available immediately:

$$(26) \qquad f_R(r)\, dr = e^{-r^2/2} r\, dr, \qquad f_A(a) = \frac{1}{2\pi}\, da$$

on the appropriate ranges. Note that $r^2/2$ is exponential(1) (Exercise 2.2.7), that r^2 is chi-square(2) (Exercise 2.2.16), and that a is uniform(0, 2π) (Example 2.2.1).

E EXERCISES: DISTRIBUTION FUNCTIONS

1 Let H_1 and H_2 be distribution functions on R. Then by Exercise 2.3.1, $F(y_1, y_2) = H_1(y_1)H_2(y_2)$ is a distribution function on R^2. Calculate the marginal distribution functions F_1 and F_2.

2 A distribution on R^2 has probability function p as given in Exercise 2.3.2. Calculate the marginal distribution function F_1 at $y_1 = 0, 1, 2, 3$. Calculate F_2 at $y_2 = 0, 1, 2, 3$.

3 A distribution on R^2 has distribution function F given by

$$F(y_1, y_2) = 1 - (1 + y_1)^{-1} - (1 + y_2)^{-1} + (1 + y_1 + y_2)^{-1}, \qquad y_1, y_2 \geq 0,$$

and zero otherwise.

(a) Calculate the marginal distribution functions F_1 and F_2.

(b) Calculate the marginal density functions.

4 The standard normal distribution on R^k has distribution function F given by $F(z_1, \ldots, z_k) = G(z_1) \cdots G(z_k)$, where G is the standard normal distribution function on R.

(a) Calculate the marginal distribution function F_1.

(b) Calculate the marginal distribution function F_{12}.

5 A distribution on R^k has distribution function F given by $F(y_1, \ldots, y_k) = H_1(y_1) \cdots H_k(y_k)$, where the H_i are distribution functions on R. Calculate the marginal distribution functions F_1, F_{12}, and F_{123}.

6 A distribution on R^2 has distribution function F given by

$$F(y_1, y_2) = 0, \qquad y_2 < 0,$$
$$= (1/2)G(y_1), \qquad 0 \le y_2 < 1,$$
$$= (1/2)G(y_1) + (1/2)H(y_1), \qquad 1 \le y_2 < \infty,$$

where G and H are distribution functions on R.
(a) Find the marginal distribution function F_1.
(b) Find the marginal distribution function F_2.

7 Let F be a distribution function on R^2. And let W be the function that carries (y_1, y_2) into $w = \max(y_1, y_2)$ and X be the function that carries (y_1, y_2) into $x = \min(y_1, y_2)$. Show that the marginal distribution functions F_W and F_X satisfy

$$F_W(w) = F(w, w), \qquad F_X(x) = F(x, \infty) + F(\infty, x) - F(x, x).$$

8 Let F be a distribution function on R^k. And let W be the function that carries $(y_1, \ldots, y_k)$ into $\max y_i$. Show that the marginal distribution function F_W satisfies $F_W(w) = F(w, \ldots, w)$.

9 A distribution on R^2 has $f(y_1, y_2) = \pi^{-1}$ for $y_1^2 + y_2^2 < 1$ and zero otherwise.
(a) Calculate the distribution function G for $w = y_1^2 + y_2^2$.
(b) Calculate the distribution function H for $r = (y_1^2 + y_2^2)^{1/2}$.

10 Consider a uniform distribution inside the sphere $y_1^2 + y_2^2 + y_3^2 = 1$.
(a) Calculate the distribution function G for $w = y_1^2 + y_2^2 + y_3^2$.
(b) Calculate the distribution function H for $r = (y_1^2 + y_2^2 + y_3^2)^{1/2}$.

11 The Pareto(α) distribution function on R^2 is given by ($\alpha > 0$)

$$F(y_1, y_2) = 1 - (1 + y_1)^{-\alpha} - (1 + y_2)^{-\alpha} + (1 + y_1 + y_2)^{-\alpha}, \qquad y_i \ge 0,$$

and zero otherwise. Determine the marginal distribution functions F_1 and F_2; verify that each is Pareto(α) on the real line R.

F EXERCISES: PROBABILITY FUNCTIONS

12 A symmetric die and a symmetric coin are tossed and (x_1, x_2) is recorded, where x_1 is the number of points on the die and x_2 is the number of heads showing. Record the symmetric probability function for (x_1, x_2). Calculate the probability function p_Y for the total $y = x_1 + x_2$.

13 A distribution on R^2 has probability function p as given in Exercise 2.3.2. Calculate the marginal probability functions p_1 and p_2.

14 Suppose that f is a symmetric density on R^2: $f(y_1, y_2) = f(y_2, y_1)$.
(a) Show that $P(y_1 < y_2) = P(y_2 < y_1) = 1/2$.
(b) Let $(r_1, r_2) = (1, 2)$ if $y_1 < y_2$ and $= (2, 1)$ if $y_1 > y_2$. Show that the probability function p for (r_1, r_2) is given by

$$p(r_1, r_2) = 1/2 \qquad \text{at } (1, 2) \text{ and } (2, 1)$$

and zero otherwise.

15 Suppose that f is a density on R^k and is symmetric $f(y_1, \ldots, y_k) = f(y_{i_1}, \ldots, y_{i_k})$, where $(i_1, \ldots, i_k)$ is any permutation of $(1, \ldots, k)$.
(a) Show that $P(y_{i_1} < \cdots < y_{i_k}) = 1/k!$.
(b) Let $\mathbf{r}(\mathbf{y}) = (r_1, \ldots, r_k)'$ be a function of $\mathbf{y}$ such that y_i is the r_ith smallest of $y_1, \ldots, y_k$. The vector $\mathbf{r}$ is called the *rank statistic;* determine the probability function for $\mathbf{r}$.

G EXERCISES: PROBABILITY DENSITY FUNCTIONS

16 Consider a distribution in R^2 with density f given by

$$f(y_1, y_2) = 18e^{-6y_1-3y_2}, \qquad y_1 > 0, y_2 > 0,$$

and zero otherwise. Determine the marginal densities f_1 and f_2.

17 Consider a distribution on R^2 with density f given by

$$f(y_1, y_2) = 2e^{-y_1-y_2}, \qquad 0 < y_1 < y_2 < \infty,$$

and zero otherwise. Determine the marginal densities f_1 and f_2.

18 Let f be the density function for the uniform distribution over the disk $y_1^2 + y_2^2 < 1$ in R^2. Calculate the marginal densities f_1 and f_2.

19 Let f be the density function for the uniform distribution over the unit ball $y_1^2 + y_2^2 + y_3^2 < 1$ in R^3.
(a) Determine the marginal density f_{12}.
(b) Determine the marginal density f_1.

20 Consider the density f on R^k given by $f(y_1, \ldots, y_k) = h_1(y_1) \cdots h_k(y_k)$, where each h_i is a density on R. Determine the marginal densities f_1, f_{12}, and f_{123}.

21 Consider the Pareto(α) distribution on R^2 with density f given by ($\alpha > 0$)

$$(27) \qquad f(y_1, y_2) = \alpha(\alpha + 1)(1 + y_1 + y_2)^{-\alpha-2}, \qquad y_1 > 0, y_2 > 0,$$

and zero otherwise. Show that each of the marginal densities f_1 and f_2 is Pareto(α) on R.

22 Consider the canonical Student(n) distribution for (y_1, y_2) on R^2 (Problem 2.3.13). Show that the marginal density f_1 is canonical Student(n) on R^1.

23 Consider a uniform distribution for (u_1, u_2) over the unit square $(0, 1)^2$. Use the convolution formula to obtain the distribution of $y = u_1 + u_2$.

H PROBLEMS

24 Consider the Pareto(α) distribution for $(y_1, \ldots, y_k)$ on R^k. Show that the marginal distribution of $(y_1, \ldots, y_{k-1})$ is Pareto(α) on R^{k-1}. Deduce that (y_1, y_2) is Pareto(α) on R^2; and that y_1 is Pareto(α) on R. See Problem 2.3.14.

25 Consider the canonical Student(n) distribution for $(y_1, \ldots, y_k)$ on R^k. Show that the marginal distribution of $(y_1, \ldots, y_{k-1})$ is canonical Student(n) on R^{k-1}. Deduce that (y_1, y_2) is canonical Student(n) on R^2; and that y_1 is canonical Student(n) on R. See Problem 2.3.15.

26 Consider the Dirichlet($r_1, \ldots, r_{k+1}$) distribution for $(y_1, \ldots, y_k)$ on R^k. Prove the convolution property: the marginal distribution of $(y_1 + y_2, y_3, \ldots, y_k)$ is Dirichlet($r_1 + r_2, r_3, \ldots, r_{k+1}$).

27 *Sum of chi-squares:* Consider the distribution for (χ_1^2, χ_2^2) with density f on R^2 given by

$$f(\chi_1^2, \chi_2^2) = \Gamma^{-1}\left(\frac{m}{2}\right) e^{-\chi_1^2/2} \left(\frac{\chi_1^2}{2}\right)^{(m/2)-1} \cdot \frac{1}{2} \cdot \Gamma^{-1}\left(\frac{n}{2}\right) e^{-\chi_2^2/2} \left(\frac{\chi_2^2}{2}\right)^{(n/2)-1} \cdot \frac{1}{2}$$

for $\chi_1^2, \chi_2^2 > 0$ and zero otherwise. Use the convolution formula and Problem 2.2.22 to show that the marginal distribution of $\chi^2 = \chi_1^2 + \chi_2^2$ is chi square($m + n$).

28 Consider the uniform distribution over the unit ball $y_1^2 + y_2^2 + y_3^2 < 1$ in R^3. Let $y_1 = r \cos a \sin b$, $y_2 = r \cos a \cos b$, $y_3 = r \sin a$ define distance r from the origin in $(0, \infty)$, latitude a in $(-\pi/2, \pi/2)$, and longitude b in $(0, 2\pi)$.
(a) Determine the density function for (r, a, b).
(b) Determine the marginal density of r, of a, and of b.

I PROBLEMS: OTHER CONVOLUTION FORMULAS

29 *Convolution: a difference:* Consider the distribution $f(x, y)$ for (x, y) on R^2, and derive the following expression for the marginal density of $x - y$:

$$f_{X-Y}(t) = \int_{-\infty}^{\infty} f(t + y, y)\, dy.$$

30 *Convolution: a ratio:* Consider a distribution with density f on R^2 with $f(x, y) = 0$ for $y \leq 0$. Show that the density function for $v = x/y$ is given by

$$(28) \qquad f_V(v) = \int_0^\infty f(vy, y)y\, dy.$$

31 *Convolution: a proportion:* Consider a distribution with density f on R^2 with $f(x, y) = 0$ unless $x > 0$, $y > 0$. Show that the density function for $u = x/(x + y)$ is given by

$$(29) \qquad f_U(u) = \int_0^\infty f(ut, (1 - u)t)t\, dt$$

on (0, 1) and zero otherwise.

32 *Convolution: linear combination:* Let $w = ax + by$ with $a \neq 0$. Derive the following expression for the marginal density of w:

$$f_W(w) = |a|^{-1} \int_{-\infty}^{\infty} f\left(\frac{w - by}{a}, y\right) dy.$$

33 *(continuation of Problem 27)* Use Problem 30 to show that $G = \chi_1^2/\chi_2^2$ is canonical $F(m, n)$; recall Problem 2.2.21. Thus $F = (\chi_1^2/m)/(\chi_2^2/n)$ is standard $F(m, n)$.

34 *(continuation of Problem 27)* Use Problem 31 to show that $B = \chi_1^2/(\chi_1^2 + \chi_2^2)$ is beta$(m/2, n/2)$; recall Problem 2.2.21.

35 Consider the distribution for (z, χ) with density f on R^2 given by

$$f(z, \chi) = (2\pi)^{-1/2}e^{-z^2/2} \cdot \Gamma^{-1}\left(\frac{n}{2}\right) e^{-\chi^2/2} \left(\frac{\chi^2}{2}\right)^{(n/2)-1} \chi$$

for $\chi > 0$ and zero otherwise. Use Problem 30 to show that $T = z/\chi$ is canonical Student(n); recall Problem 2.2.21.

3.3

STATISTICAL INDEPENDENCE

In science and industry it is common practice to take repeated observations on a random system operating under essentially constant conditions. At one extreme this can involve duplicating measurements or determinations on some physical quantity: for example, measurements on the lifetime of high-pressure sodium light bulbs made by a particular industrial process; or measurements of the weight increase in rabbits under a certain diet therapy. At another extreme this can involve repeating or replicating a basic design of response determinations coming from specified input conditions: for example, a basic design has three fertilizers applied respectively to three plots of land in a block and the corresponding yields observed; this design pattern is then replicated on several blocks of land.

Consider the process of taking repeated observations on a random system operated under constant conditions. The probability pattern for a performance will depend on the conditions. With constant conditions the pattern will be independent of results from antecedent performances—or otherwise the conditions would not, in fact, be constant.

This can be examined empirically by considering a pair of separate performances and then repeating the *pair*. If the performances in a pair are kept separate, then under repetitions the performance pattern for the pair is found to exhibit a symmetry or independence: the performance pattern for one element of the pair appears symmetrically with respect to possibilities for the other element.

In this section we examine this symmetry or independence for a pair of performances under constant conditions. And more generally we examine this symmetry or independence for a pair of different systems operated independently.

Independence is an important and fundamental property. It is useful for the probability analysis of given probability models; and it allows us to build large probability models from simple models that describe separate components of a system.

A INDEPENDENT EVENTS

Consider the tossing of a symmetric coin; let H be the event "heads for the coin": $P(H) = 1/2$. Also consider the tossing of a symmetric die; let A be the event "ace": $P(A) = 1/6$. Now consider the tossing of the coin and die in a thorough way to ensure symmetry among the pairings of a face for the coin with a face for the die. The symmetry among the 2×6 pairings then gives

$$P(H)P(A) = (1/2) \cdot (1/6) = 1/12,$$

which we note is equal to

$$P(HA) = 1/12.$$

A basic way to ensure symmetry is to have the coin tossing physically separate or independent of the tossing of the die. Accordingly, with

$$P(HA) = P(H)P(A),$$

we say that the event H is "statistically independent" of the event A.

Now consider a sample space $\mathcal{S}$ with probabilities given by P. Then we define a generalized symmetry relating an event A to an event B as follows:

DEFINITION 1

A is **statistically independent** of B if

(1) $P(AB) = P(A)P(B)$.

As empirical support for the term "statistically independent" we note that, with events A and B coming from physically independent systems, then for a large number of performances of the combined system, the estimated probabilities approximate the factorization presented in the definition; this is of course an expression of the generalized symmetry just mentioned.

The statistical independence of A and B implies the independence of events simply related to A and B.

PROPOSITION 2

If A is statistically independent of B, then each event in $\{\varnothing, A, A^c, \mathcal{S}\}$ is statistically independent of each event in $\{\varnothing, B, B^c, \mathcal{S}\}$.

Proof The proof is straightforward. Consider the pair A and $\varnothing$:

$$P(A\varnothing) = P(\varnothing) = 0,$$
$$P(A)P(\varnothing) = P(A) \cdot 0 = 0.$$

And consider, for example, the pair A and B^c:

$$P(AB^c) = P(A) - P(AB) = P(A) - P(A)P(B),$$
$$P(A)P(B^c) = P(A)(1 - P(B)) = P(A) - P(A)P(B).$$

Consider an example where the events clearly belong to a single random system:

EXAMPLE 1

One card from a deck: A deck of 52 playing cards is thoroughly shuffled and one card is dealt face up. Let S be the event ''spade'' and A be the event ''ace.'' Then

$$P(S) = 13/52 = 1/4, \qquad P(A) = 4/52 = 1/13,$$
$$P(SA) = 1/52 = P(S)P(A).$$

Thus the events S and A are statistically independent. It then follows from Proposition 2 that, say, S^c and A are statistically independent.

$$P(S^cA) = 3/52 = (3/4) \cdot (1/13) = P(S^c)P(A).$$

B INDEPENDENT ALGEBRAS

With independent events A and B, Proposition 2 shows that each event in $\{\varnothing, A, A^c, \mathcal{S}\}$ is independent of each event in $\{\varnothing, B, B^c, \mathcal{S}\}$. We now discuss independence between *classes* of events. Consider a sample space $\mathcal{S}$ with σ-algebra $\mathcal{A}$ of events and probability measure P. Let $\mathcal{A}_1$ be a class of events, say referring to certain characteristics of a system being investigated, and $\mathcal{A}_2$ be another class, say referring to other characteristics of the system. There are some advantages to thinking of the classes as being closed under the standard operations $\cup_1^\infty$, $\cap_1^\infty$, ^{c}; accordingly, we will assume that they are σ-algebras.

DEFINITION 3

A σ-algebra $\mathcal{A}_1$ is **statistically independent** of a σ-algebra $\mathcal{A}_2$ if

$$P(A_1A_2) = P(A_1)P(A_2)$$

for each A_1 in $\mathcal{A}_1$ and A_2 in $\mathcal{A}_2$.

EXAMPLE 2

One card from a deck: One card is dealt from a well-shuffled deck (Example 1). We have seen that S, ''spade,'' and A, ''ace,'' are statistically independent. By Proposition 2 it follows that $\{\varnothing, S, S^c, \mathcal{S}\}$ and $\{\varnothing, A, A^c, \mathcal{S}\}$ are statistically independent. Indeed, it should be clear that any σ-algebra describing suits is independent of any σ-algebra describing denominations.

C INDEPENDENT FUNCTIONS

Now consider independence between two functions X and Y defined on the sample space $\mathcal{S}$. Of course, we assume that the functions are respectable, that is, measurable, as discussed in Section 3.1.

DEFINITION 4

The functions X and Y are **statistically independent** if each event for X is statistically independent of each event for Y.

Note that the probability statement for the independence of an event B_1 for X and an event B_2 for Y can be expressed directly as

(2) $$P_{XY}(B_1 \times B_2) = P_X(B_1)P_Y(B_2),$$

or in terms of preimage events as

(3) $$P(X^{-1}(B_1) \cap Y^{-1}(B_2)) = P(X^{-1}(B_1))P(Y^{-1}(B_2)).$$

The independence of X and Y can be related to the independence of algebras in Definition 3: let $\mathcal{B}_1$ be the events for the function X and $\mathcal{B}_2$ be the events for the function Y; then independence says that $X^{-1}(\mathcal{B}_1)$ is statistically independent of $Y^{-1}(\mathcal{B}_2)$.

EXAMPLE 3

One card from a deck: One card is dealt from a well-shuffled deck (Examples 1 and 2). Let X be the indicator function for the event S and Y be the indicator for the event A. Then X and Y are statistically independent; this uses Proposition 2 with the results in Example 2. This simple example has been used repeatedly; some substantial examples are given in Sections E and F.

Note that when we talk of two functions X and Y being statistically independent, we are saying very little about X and Y as functions. Rather the property centers on the probability measure on the sample space and the functions X and Y merely delineate events on that sample space.

We now present a simple proposition that emphasizes an important aspect of independence, an aspect that is almost trivial as based on the definitions we have used:

PROPOSITION 5

If X and Y are statistically independent and if h_1 and h_2 are measurable functions on the ranges of X and Y, respectively, then $h_1(X)$ and $h_2(Y)$ are statistically independent.

Proof: From Definition 4 we have that each event for X is statistically independent of each event for Y. Now consider $h_1(X)$ and $h_2(Y)$. An event for $h_1(X)$ is necessarily an event for the antecedent function X; see Definition 3.1.2 for a measurable function. Similarly, an event for $h_2(Y)$ is necessarily an event for Y. Thus from the initial statement we have trivially that each event for $h_1(X)$ is independent of each event for $h_2(Y)$. Note that the functions h_1 and h_2 just cut down on the number of events being considered.

D INDEPENDENCE ON THE LINE AND PLANE

On real spaces we have seen the importance of distribution functions, probability functions, and density functions. Now consider independence in terms of these functions.

PROPOSITION 6

Real-valued functions X and Y are statistically independent if and only if

(4) $$F_{XY}(x, y) = F_X(x)F_Y(y)$$

for all x and y on the real line R.

Proof If X and Y are statistically independent, then the events $(-\infty, x] \times R$ and $R \times (-\infty, y]$ are statistically independent on the space R^2 for (X, Y). This gives

$$\begin{aligned} F_{XY}(x, y) &= P_{XY}((-\infty, x] \times (-\infty, y]) = P_X((-\infty, x])P_Y((-\infty, y]) \\ &= F_X(x)F_Y(y). \end{aligned}$$

The proof of the converse is recorded in Supplement Section 3.6; it is more complicated and uses the Extension Theorem, 1.5.5 in Supplement Section 1.5.

PROPOSITION 7

For a discrete distribution, X and Y are statistically independent if and only if

(5) $$p_{XY}(x, y) = p_X(x)p_Y(y)$$

for all x and y on the real line.

Proof The proof follows immediately from Proposition 6. For consider

$$F_{XY}(x, y) = \sum_{t \le x, u \le y} p_{XY}(t, u),$$

$$F_X(x)F_Y(y) = \sum_{t \le x} p_X(t) \sum_{u \le y} p_Y(u) = \sum_{t \le x, u \le y} p_X(t)p_Y(u).$$

If X and Y are independent, then by Proposition 6 the left sides are equal; but a distribution function determines the probability function; it follows that

$p_{XY} = p_X p_Y$. Conversely, if $p_{XY} = p_X p_Y$, then the right sides are equal and thus the left sides are equal; independence follows from Proposition 6.

PROPOSITION 8

For an absolutely continuous distribution, X and Y are statistically independent if and only if

(6) $f_{XY}(x, y) = f_X(x) f_Y(y)$

almost everywhere on the plane (except perhaps on a set that has area measure zero).

Proof The proof parallels that for the preceding proposition but uses

$$F_{XY}(x, y) = \int_{-\infty}^{x} \int_{-\infty}^{y} f_{XY}(t, u)\, dt\, du,$$

$$F_X(x)F_Y(y) = \int_{-\infty}^{x} f_X(t)\, dt \int_{-\infty}^{y} f_Y(u)\, du = \int_{-\infty}^{x} \int_{-\infty}^{y} f_X(t) f_Y(u)\, dt\, du.$$

Factorization so that the variables separate is the essential part of the condition in each of these propositions. More flexible versions of these propositions are recorded as Exercises 1, 2, and 3.

E COMBINING INDEPENDENT SYSTEMS

Now consider the use of independence to build probability models. Suppose that we have two random systems that are physically independent of each other. For the first system with response s_1, let $\mathcal{S}_1$ be the sample space, $\mathcal{A}_1$ be the class of events, and P_1 be the probability measure. And for the second system with response s_2, let $\mathcal{S}_2$ be the sample space, $\mathcal{A}_2$ be the class of events, and P_2 be the probability measure.

Now suppose that we are interested in a performance of the combined system. The combined response is (s_1, s_2); thus the sample space $\mathcal{S}$ is the product $\mathcal{S}_1 \times \mathcal{S}_2$. We would certainly be interested in events given by a product set $A_1 \times A_2$. Accordingly, we take $\mathcal{A}$ to be the smallest σ-algebra that includes the product sets $A_1 \times A_2$.

And if the systems are independent, we would have $P(A_1 \times A_2) = P_1(A_1)P_2(A_2)$ on the basis of events for the first system being statistically independent of events for the second system. The measure for the product sets then extends uniquely to a probability measure for the class $\mathcal{A}$ of events; this uses the Extension Theorem, 1.5.5, with details as in Section 2.3B.

Thus we obtain the *product space* $\mathcal{S}$, the *product algebra* $\mathcal{A}$ of events, and the *product probability measure P*. In this way we can build progressively larger models by introducing components one by one. We can thus produce quite complicated models from quite simple ingredients.

EXAMPLE 4

An engineer measures the lifetime y_1 of an electronic component; it is known to be exponential(θ). He also measures the lifetime y_2 for an independent electronic component; it is known to be exponential(2θ). Construct the model for the combined system with outcome (y_1, y_2). The density function for (y_1, y_2) is available from Proposition 8:

$$f(y_1, y_2) = \frac{1}{\theta} e^{-y_1/\theta} \frac{1}{2\theta} e^{-y_2/2\theta}$$

$$= \frac{1}{2\theta^2} e^{-(2y_1+y_2)/2\theta} \qquad \text{if } y_1, y_2 > 0$$

and zero otherwise. See Exercise 3.2.16.

F SAMPLES FROM A DISTRIBUTION

Consider a random system with response y. Sometimes we may be interested in a single performance of the system. Other times we may be interested in a sequence of independent performances yielding a combined response $(y_1, \ldots, y_n)$. If we have a model for a single performance, we can build a model for the combined performance by using the given model for each component and then combining by independence as in Section E.

DEFINITION 9

If $y_1, \ldots, y_n$ are statistically independent and if each y_i has the same distribution with measure P, or with probability function p, or with density function f, then $\mathbf{y} = (y_1, \ldots, y_n)'$ is a **sample** from the distribution given by P, or by p, or by f.

Alternatively, we can say that $y_1, \ldots, y_n$ are IID (independent and identically distributed) according to P, p, or f.

In the discrete case the joint probability function for the sample $\mathbf{y}$ on R^n is

(7) $$p(\mathbf{y}) = p(y_1) \cdots p(y_n);$$

and in the absolutely continuous case, the joint probability density for the sample $\mathbf{y}$ on R^n is

(8) $$f(\mathbf{y}) = f(y_1) \cdots f(y_n).$$

EXAMPLE 5

Sample from Bernoulli(p): A certain drug therapy has a probability p of success; let x be the indicator for success. The probability function is

$$p(x) = p^x q^{1-x} \qquad \text{if } x = 0, 1$$

and zero otherwise ($q = 1 - p$). This is the Bernoulli(p) distribution. Now consider n repetitions on different animals obtained randomly from a large population; let $(x_1, \ldots, x_n)$ be the vector of indicators. The probability function is obtained by Proposition 7:

$$p(x_1, \ldots, x_n) = p^{x_1} q^{1-x_1} \cdots p^{x_n} q^{1-x_n}$$

$$= p^{\Sigma x_i} q^{n-\Sigma x_i}$$

on the points $\{0, 1\}^n$ in R^n and zero elsewhere.

EXAMPLE 6

Sample from Poisson(λ): Consider a count y from a radioactive source having a Poisson distribution

$$p(y) = \frac{\lambda^y e^{-\lambda}}{y!} \qquad \text{if } y = 0, 1, 2, \ldots$$

and zero otherwise. And suppose that we wish to consider n independent performances of the system. Let $y_1, \ldots, y_n$ designate the responses for 1st, . . ., nth performances. The independence among the various performances then prescribes statistical independence, and by Proposition 7 we obtain

$$\begin{aligned} p(y_1, \ldots, y_n) &= p(y_1) \cdots p(y_n) \\ &= \frac{\lambda^{y_1} e^{-\lambda}}{y_1!} \cdots \frac{\lambda^{y_n} e^{-\lambda}}{y_n!} \qquad (9) \\ &= \frac{\lambda^{\Sigma y_i} e^{-n\lambda}}{\Pi y_i!} \qquad y_i = 0, 1, 2, \ldots \end{aligned}$$

and zero otherwise. We thus obtain a discrete distribution on R^n, and it describes the combined response $\mathbf{y} = (y_1, \ldots, y_n)'$.

In an application we may not know the true value of λ that gives the correct Poisson distribution for the response. We may then make n performances of the system and obtain an observed sample $(y_1, \ldots, y_n)$. In order to determine plausible values for λ, we will, in the statistics chapters, see how the observed sample relates to the distribution $p(\cdot, \ldots, \cdot)$ prescribed by each of the possible values for λ.

EXAMPLE 7

Sample from the normal(μ, σ): Consider a voltage measurement y having a normal distribution(μ, σ):

$$f(y) = \frac{1}{\sqrt{2\pi}\sigma} \exp\left\{-\frac{1}{2\sigma^2}(y-\mu)^2\right\}.$$

Now suppose that we wish to consider n independent performances of the system. Let $y_1, \ldots, y_n$ designate the responses for the 1st, . . ., nth performances. Then $(y_1, \ldots, y_n)$ is a sample from the normal(μ, σ) distribution; the probability density function is available from Proposition 8:

$$\begin{aligned} (10) \qquad f(y_1, \ldots, y_n) &= f(y_1) \cdots f(y_n) \\ &= (2\pi)^{-n/2}\sigma^{-n} \exp\left\{-\frac{1}{2\sigma^2}\sum_1^n (y_i - \mu)^2\right\}. \end{aligned}$$

Thus $\mathbf{y} = (y_1, \ldots, y_n)'$ has a rotationally symmetric normal distribution in R^n; it is located at some point $\mu\mathbf{1} = \mu(1, \ldots, 1)' = (\mu, \ldots, \mu)'$ on the line $\mathcal{L}(\mathbf{1})$.

In an application we may not know the values for μ and σ. We may then make n performances of the system and obtain an observed sample $(y_1, \ldots, y_n)$. In order to determine plausible values for (μ, σ), we will, in the statistics chapters, see how the observed sample relates to the distribution $f(\cdot, \ldots, \cdot)$ prescribed by each of the possible values for (μ, σ).

G EXERCISES: CRITERIA FOR INDEPENDENCE

1 Prove: X and Y are statistically independent if and only if $F_{XY}(x, y) = g(x)h(y)$ factors so that the variables separate. Use Proposition 6.

2 Prove for the discrete case: X and Y are statistically independent if and only if $p_{XY}(x, y) = g(x)h(y)$ factors so that the variables separate. Use Proposition 7.

3 Prove for the absolutely continuous case: X and Y are statistically independent if and only if $f_{XY}(x, y) = g(x)h(y)$ factors so that the variables separate, except perhaps on a set *that has* area measure zero. Use Proposition 8.

H EXERCISES

4 Consider the standard normal distribution for (z_1, z_2) on R^2; see Example 3.2.9. Show that r and a are statistically independent, where $z_1 = r \cos a$ and $z_2 = r \sin a$.

5 *Sum of chi squares:* Let $\chi_1^2, \ldots, \chi_k^2$ be statistically independent and χ_i^2 have a chi-square(m_i) distribution. Problem 3.2.27 shows that $\chi_1^2 + \chi_2^2$ is chi square$(m_1 + m_2)$. Deduce by induction that $\chi^2 = \chi_1^2 + \cdots + \chi_k^2$ is chi square $(m_1 + \cdots + m_k)$. This is called the *reproductive property* of the chi-square distribution.

6 Let $(z_1, \ldots, z_k)$ be a sample from the standard normal. Show that $\Sigma_1^k z_i^2$ is chi square(k); recall Example 3.2.2 and Exercise 5.

I PROBLEMS

7 Three symmetric dice are tossed and $s = (x_1, x_2, x_3)$ is observed, where x_i is the number of points on the ith die. Determine the probability function for W, where $W(x_1, x_2, x_3) = x_1 + x_2 + x_3$; recall Examples 3.2.4 and 3.2.5.

8 Let χ_1^2 and χ_2^2 be statistically independent and χ_i^2 have a chi-square(m_i) distribution. Show that $T = \chi_1^2 + \chi_2^2$ and $G = \chi_1^2/\chi_2^2$ are statistically independent, and deduce that T is chi square$(m_1 + m_2)$ and G is canonical $F(m_1, m_2)$.

9 Let χ_1^2 and χ_2^2 be statistically independent and χ_i^2 have a chi-square(m_i) distribution. Show that $T = \chi_1^2 + \chi_2^2$ and $P = \chi_1^2/(\chi_1^2 + \chi_2^2)$ are statistically independent, and deduce that T is chi square$(m_1 + m_2)$ and P is beta$(m_1/2, m_2/2)$.

10 Let (z_1, z_2) be a sample of 2 from the standard normal. Use Problem 3.2.32 to show that $W = z_1 + bz_2$ is normal $(0, (1 + b^2)^{1/2})$.

11 *(continuation)* Let y_1 and y_2 be statistically independent and y_i be normal (μ_i, σ_i). Represent y_1 and y_2 in terms of standard normal z_1 and z_2 and show that $y_1 + y_2$ is normal $(\mu_1 + \mu_2, (\sigma_1^2 + \sigma_2^2)^{1/2})$.

J PROBLEMS: COMPUTER USAGE

Some computer exercises are recorded in Sections 2.4H and 2.4I. Use the results there for the following theoretical and applied problems.

12 Let u_1 and u_2 be statistically independent and u_i be uniform on (0, 1), and define $(z_1, z_2) = ((-2 \ln u_1)^{1/2} \cos 2\pi u_2, (-2 \ln u_1)^{1/2} \sin 2\pi u_2)$. Show that (z_1, z_2) is a sample of 2 from the standard normal; recall Example 3.2.9.

13 *(continuation)* Use a computer program for random digits to obtain a sample of 50 from the standard normal; and see where the values fall in relation to the density function.

3.4

COUNTING SAMPLE POINTS

Many qualitative scientific investigations are concerned with the frequency of occurrence of an event in a sequence of repetitions of a random system. In Section 3.5 we shall determine the marginal distributions appropriate to such frequencies. Typically, the underlying distribution for a sequence of repetitions has a variety of symmetries, symmetries that lead to the equality of some or all of the underlying probabilities. Accordingly, the derivation of various marginal probabilities can simplify to the counting of sample points. For example, with a symmetric distribution on a set $\mathcal{S}_0$, we have

$$P(A) = \frac{c(A \cap \mathcal{S}_0)}{c(\mathcal{S}_0)},$$

where c is the cardinality measure. In this section we develop formulas for counting sample points in a set—for determining the cardinality of a set.

A COUNTING SEQUENCES: INDIVIDUAL POPULATIONS

Suppose that we have several finite sets, or *populations,* $\mathfrak{I}_1, \mathfrak{I}_2, \ldots, \mathfrak{I}_k$, and let $c(\mathfrak{I}_i) = N_i$ be the number of elements in the set $\mathfrak{I}_i$. Consider sequences of the form

$$(x_1, \ldots, x_k), \qquad x_i \in \mathfrak{I}_i.$$

And then consider the sample space $\mathcal{S}_1$ consisting of such sequences:

$$\mathcal{S}_1 = \{(x_1, \ldots, x_k) : x_i \in \mathfrak{I}_i\};$$

this is the Cartesian product of the spaces $\mathfrak{I}_1, \ldots, \mathfrak{I}_k$.

The cardinality of the space $\mathcal{S}_1$ is

$$(1) \qquad c(\mathcal{S}_1) = c(\mathfrak{I}_1) \cdots c(\mathfrak{I}_k) = N_1 \cdots N_k.$$

For the first x can be any of the N_1 elements in $\mathfrak{I}_1$; and given the first, the second can be any of the N_2 elements in $\mathfrak{I}_2$; and given the first two, the third can be any of the N_3 elements in $\mathfrak{I}_3$, and so on; thus the total number is $N_1 \cdots N_k$.

EXAMPLE 1

***n* tosses of a coin:** A symmetric coin is tossed n times. Let x_i be the indicator for heads on the ith toss; the space for x_i is

$$\mathfrak{I}_i = \{0, 1\}; \qquad c(\mathfrak{I}_i) = 2.$$

Then for the sequence of n tosses we have

$$(x_1, \ldots, x_n), \qquad x_i \in \mathfrak{I}_i,$$

giving the sample space

$$\mathcal{S}_1 = \{(x_1, \ldots, x_n) : x_i \in \mathfrak{I}_i\}.$$

The cardinality of $\mathcal{S}_1$ is $c(\mathcal{S}_1) = 2^n$. We then have the probabilities

$p(0, \ldots, 0) = 1/2^n$,

$p(1, \ldots, 1) = 1/2^n$,

$P(r$ 1s and $n - r$ 0s in a specified order$) = 1/2^n$.

B COUNTING SEQUENCES: A SINGLE POPULATION

Suppose that we have a finite set or population $\mathcal{S}$, and let $c(\mathcal{S}) = N$ be the number of elements in the set $\mathcal{S}$. Consider sequences of the form

$$(x_1, \ldots, x_n), \qquad x_i \in \mathcal{S},\ x_i \neq x_j \quad (i \neq j),$$

where if a particular element of the population is in a position x_i, then it is unavailable for the remaining positions in the sequence. And then consider the sample space $\mathcal{S}_2$ consisting of such sequences:

$$\mathcal{S}_2 = \{(x_1, \ldots, x_n);\ x_i \in \mathcal{S},\ x_i \neq x_j \quad (i \neq j)\}.$$

The cardinality of the space $\mathcal{S}_2$ is

(2) $$c(\mathcal{S}_2) = N^{(n)} = N(N-1)\cdots(N-n+1) = \frac{N!}{(N-n)!}.$$

For the first x can be any of the N elements in $\mathcal{S}$; and given the first, the second can be any of the $N - 1$ remaining elements, and so on for the n positions in the sequence; thus the total number is $N(N-1)\cdots(N-n+1)$. Note that $N^{(n)}$ is called *N descending factorial n;* it has n factors, commencing with N.

EXAMPLE 2

Thirteen cards from a deck: A deck of 52 playing cards is thoroughly shuffled and 13 cards are dealt in succession face up. Let x_i designate the ith card dealt. Then for the sequence of 13 cards we have

$$(x_1, \ldots, x_{13}), \qquad x_i \in \mathcal{S}, \quad x_i \neq x_j \quad (i \neq j),$$

giving the sample space

$$\mathcal{S}_2 = \{(x_1, \ldots, x_{13}) : x_i \in \mathcal{S},\ x_i \neq x_j \quad (i \neq j)\}.$$

The cardinality of $\mathcal{S}_2$ is $52^{(13)} = 52 \cdot 51 \cdot \cdots \cdot 40$. We then have

$$p(x_1, \ldots, x_{13}) = 1/52^{(13)}$$

for each possible sequence $(x_1, \ldots, x_{13})$ of 13 cards from 52. Now suppose that we are interested in the probability of 13 spades. The number of sequences of 13 spades is $13^{(13)} = 13!$ Thus

$$P(13 \text{ spades}) = 13!/52^{(13)}.$$

C COUNTING SETS

Suppose that we have a finite set or population $\mathcal{S}$ and let $c(\mathcal{S}) = N$ be the number of elements in the set $\mathcal{S}$. Consider subsets of $\mathcal{S}$ that contain n elements:

$$\{x_1, \ldots, x_n\}, \qquad x_i \in \mathcal{S},\ x_i \neq x_j \quad (i \neq j).$$

And then consider the sample space $\mathcal{S}_3$ consisting of such subsets:

$$\mathcal{S}_3 = \{\{x_1, \ldots, x_n\} : x_i \in \mathcal{S},\ x_i \neq x_j \quad (i \neq j)\}.$$

The cardinality of the space $\mathcal{S}_3$ *is*

$$(3) \qquad c(\mathcal{S}_3) = \binom{N}{n} = \frac{N^{(n)}}{n!} = \frac{N!}{n!(N-n)!}.$$

For we can form a mapping Y from the preceding space $\mathcal{S}_2$ onto the present space $\mathcal{S}_3$ such that

$$(x_1, \ldots, x_n) \mapsto \{x_1, \ldots, x_n\};$$

that is, the sequence $(x_1, \ldots, x_n)$ is carried into the set $\{x_1, \ldots, x_n\}$. This is an $n!$ to one mapping; thus

$$c(\mathcal{S}_3) = \frac{c(\mathcal{S}_2)}{n!} = \frac{N^{(n)}}{n!} = \binom{N}{n}.$$

Note that $\binom{N}{n}$ is called *N choose n*.

EXAMPLE 3

Thirteen cards from a deck; see Example 2: Thirteen cards are dealt from a well-shuffled deck $\mathcal{S}$ and are collected together as a *hand,* or set of 13 cards, without regard for the order in which they were dealt. The number of different hands is

$$c(\mathcal{S}_3) = \binom{52}{13} = \frac{52^{(13)}}{13!} = 635{,}013{,}559{,}600.$$

We then have (recall the mapping Y)

$$p_Y(\{x_1, \ldots, x_{13}\}) = 1 \big/\!\!\big/ \binom{52}{13}$$

for possible hands $\{x_1, \ldots, x_{13}\}$ of 13 cards. The probability for 13 spades is

$$P_Y(13 \text{ spades}) = 1 \big/\!\!\big/ \binom{52}{13} = \frac{13!}{52^{(13)}},$$

which agrees with the value calculated earlier on the basis of sequences of cards. The probability of a hand containing 5 spades, 3 hearts, 2 diamonds, and 3 clubs can now be calculated using, in addition, the results from Section A:

P(5 spades, 3 hearts, 2 diamonds, 3 clubs)

$$= \frac{\binom{13}{5}\cdot\binom{13}{3}\cdot\binom{13}{2}\cdot\binom{13}{3}}{\binom{52}{13}}.$$

Subsets from a space $\mathcal{S}$ can be described in another way by placing an indicator function on the population $\mathcal{S}$—to indicate whether an element is in the sample or not. Let w_j be equal to b if the jth population element is in the sample and equal to a if it is not. Then the sample space $\mathcal{S}_3$ can be expressed alternatively as

$$\mathcal{S}_3' = \left\{(w_1, \ldots, w_N) : \begin{matrix} n & \text{of the } w\text{'s are equal to } b \\ N - n & \text{of the } w\text{'s are equal to } a \end{matrix}\right\}.$$

The cardinality of $\mathcal{S}_3'$ is of course the number of ways in which we can choose n places for the b's.

D COUNTING SEQUENCES OF SETS

Consider a finite set or population $\mathcal{S}$ and let $c(\mathcal{S}) = N$ be the number of elements in the set $\mathcal{S}$. And suppose that we form a first subset $\{x_1, \ldots, x_{n_1}\}$ that contains n_1 elements, a second subset $\{x_{n_1+1}, \ldots, x_{n_1+n_2}\}$ that contains n_2 elements, . . ., and a last subset $\{x_{N-n_k+1}, \ldots, x_N\}$ that contains n_k elements ($\Sigma\, n_j = N$); and then consider

$$(\{x_1, \ldots, x_{n_1}\}, \ldots, \{x_{N-n_k+1}, \ldots, x_N\}), \qquad x_i \in \mathcal{S},\ x_i \neq x_j.$$

The sample space $\mathcal{S}_4$ is the set of such elements.

The cardinality of the space $\mathcal{S}_4$ *is*

$$(4) \qquad c(\mathcal{S}_4) = \binom{N}{n_1 \cdots n_k} = \frac{N!}{n_1! \cdots n_k!} = \binom{N}{n_1}\binom{N-n_1}{n_2}\cdots\binom{n_k}{n_k}.$$

For consider the space $\mathcal{S}_2$ with $n = N$; let W be the mapping from the space $\mathcal{S}_2$ to the present space $\mathcal{S}_4$ such that

$$(x_1, \ldots, x_N) \mapsto (\{x_1, \ldots, x_{n_1}\}, \ldots, \{x_{N-n_k+1}, \ldots, x_N\});$$

that is, the sequence $(x_1, \ldots, x_N)$ is carried into the sequence of sets of size $n_1, \ldots,$ size n_k. This is an $n_1! \cdots n_k!$ to one mapping; thus

$$c(\mathcal{S}_4) = \frac{c(\mathcal{S}_2)}{n_1! \cdots n_k!} = \frac{N!}{n_1! \cdots n_k!}.$$

Alternatively, we can form the first set in $\binom{N}{n_1}$ different ways, then the second set in $\binom{N-n_1}{n_2}$ different ways, and so on, giving the right-hand expression (4). Note that $\binom{N}{n_1 \cdots n_k}$ *is called N choose* $n_1, \ldots, n_k$.

EXAMPLE 4

A deck $\mathcal{S}$ of 52 playing cards is thoroughly shuffled and 13 cards are dealt to E, 13 to S, 13 to W, and 13 to N. We thus obtain a sequence of four hands each of

13 cards. The number of such *bridge distributions* is

$$c(\mathcal{S}_4) = \frac{52!}{13!\,13!\,13!\,13!}.$$

The probability for any particular bridge distribution is

$$p_W(\{x_1, \ldots, x_{13}\}, \ldots, \{x_{40}, \ldots, x_{52}\}) = \frac{(13!)^4}{52!}.$$

A sequence of subsets from a space $\mathcal{S}$ can be described in another way by placing an indicator function on the population $\mathcal{S}$—to indicate for any element which of the k samples it goes in. Let $w_j = b_1, \ldots, b_k$ according as the jth element of $\mathcal{S}$ is in the first, ..., kth subset, respectively. Then the sample space $\mathcal{S}_4$ can be expressed alternatively as

$$\mathcal{S}_4' = \left\{ (w_1, \ldots, w_N) : \begin{matrix} n_1 \text{ of the } y\text{'s are equal to } b_1 \\ \vdots \\ n_k \text{ of the } y\text{'s are equal to } b_k \end{matrix} \right\}$$

The cardinality of $\mathcal{S}_4'$ is, of course, the number of ways we can choose n_1 places for b_1's, then n_2 places for b_2's, and so on.

E EXTENDED FACTORIALS

The factorial expressions in this section can occur in various summation formulas, and it is convenient to extend their definition and thus avoid clumsy restrictions on the summations.

The *factorial function* $n! = n(n-1) \cdots 2 \cdot 1$ is defined initially for positive integers n. This is extended to 0 by defining $0! = 1$.

The *descending factorial* $N^{(n)} = N(n-1) \cdots (N-n+1)$ is defined for all real N and initially for positive integers n. This is extended to $n = 0$ by defining $N^{(0)} = 1$.

The *combinatorial function* is then defined by the formula

$$(5) \qquad \binom{N}{n} = \frac{N^{(n)}}{n!}$$

for all real N and for integers $n = 0, 1, 2, \ldots$; and is extended by

$$\binom{N}{0} = 1, \qquad \binom{N}{n} = 0, \qquad n = -1, -2, \ldots.$$

Then note that for integer N, n

$$\binom{N}{n} = \binom{N}{N-n}$$

The *binomial theorem* can then be expressed in the form

$$(6) \qquad (1+t)^N = 1 + \binom{N}{1} t + \binom{N}{2} t^2 + \cdots$$

for all real N provided that $|t| < 1$ and for positive integer N without restriction on t. Note that for positive integers N the series terminates; otherwise, it is a power series.

The *generalized combinatorial function*

(7) $$\binom{N}{n_1 \cdots n_k} = \frac{N!}{n_1! \cdots n_k!}$$

is defined initially for positive integers $n_1, \ldots, n_k$ such that $\sum n_i = N$. The definition of 0! gives an immediate extension to nonnegative integers $n_1, \ldots, n_k$ such that $\sum n_i = N$. We extend it to arbitrary integers $n_1, \ldots, n_k$ and positive integer N by defining

$$\binom{N}{n_1 \cdots n_k} = 0$$

if any n_i is negative or if $\sum n_i \neq N$.

The multinomial theorem can be expressed in the form

(8) $$(t_1 + \cdots + t_r)^N = \sum \binom{n}{n_1 \cdots n_r} t_1^{n_1} \cdots t_r^{n_r}$$

for positive integer N. The summation effectively is over all sequences $(n_1, \ldots, n_r)$ with integers $n_i \geq 0$ and $\sum n_i = N$.

The values of the factorial function can be approximated for large n by *Stirling's formula,*

$$\Gamma(N + 1) = n! \approx \sqrt{2\pi}\, n^{n+(1/2)} e^{-n} = \sqrt{2\pi n}\left(\frac{n}{e}\right)^n$$

F COMBINATORIAL EXERCISES

1 Prove algebraically the following recursion relation for the combinatorial symbol (real n, integer r):

$$\binom{n}{r-1} + \binom{n}{r} = \binom{n+1}{r}.$$

Repeated use of the relation generates the Pascal triangle recording values of $\binom{n}{r}$.

	r				
n	0	1	2	3	4
0	1				
1	1	1			
2	1	2	1		
3	1	3	3	1	
4	1	4	6	4	1
	...				

2 Prove the preceding recursion relation in terms of the number of different sets from a population (positive integer n, $0 \leq$ integer $r \leq n$).

3 Prove the following relations (positive integer r):

$$\binom{-1}{r} = (-1)^r, \qquad \binom{-2}{r} = (-1)^r(r + 1).$$

4 For a nonnegative integer n prove that

$$\binom{n}{0} + \binom{n}{1} + \cdots + \binom{n}{n} = 2^n.$$

$$\binom{n}{0} - \binom{n}{1} + \cdots + (-1)^n \binom{n}{n} = 0.$$

5 For a positive integer n prove that

$$\binom{n}{1} + 2\binom{n}{2} + \cdots + n\binom{n}{n} = n2^{n-1},$$

$$\binom{n}{1} - 2\binom{n}{2} + \cdots (-1)^{n-1} n \binom{n}{n} = 0.$$

G COMBINATORIAL PROBLEMS

6 For positive integers $n > r$, use Exercise 1 to prove that

$$\binom{n}{r} = \binom{n-1}{r} + \binom{n-2}{r-1} + \cdots + \binom{n-r-1}{0}$$

$$= \binom{n-1}{r-1} + \binom{n-2}{r-1} + \cdots + \binom{r-1}{r-1}.$$

7 Examine the expansion of $(1+t)^m(1+t)^n$ and prove that

$$\binom{m}{0}\binom{n}{r} + \binom{m}{1}\binom{n}{r-1} + \cdots + \binom{m}{r}\binom{n}{0} = \sum_t \binom{m}{t}\binom{n}{r-t} = \binom{m+n}{r}$$

for real m, n and integer r.

8 For real n and integer r, prove that

$$\binom{-n}{r} = (-1)^r \binom{n+r-1}{r}, \qquad \binom{2r}{r} 2^{-2r} = (-1)^r \binom{-1/2}{r}.$$

9 Use Problems 7 and 8 to prove that

$$\binom{n-1}{r} = \binom{n}{r} - \binom{n}{r-1} + \cdots \pm \binom{n}{1} \mp 1,$$

$$\binom{n-k}{n-r} = \sum_x (-1)^x \binom{k}{x}\binom{n-x}{r}$$

for integers n, k, and r.

10 Use Problems 7 and 8 to prove that

$$\binom{m+n+r-1}{r} = \sum_{x=0}^{r} \binom{m+r-x-1}{r-x}\binom{n+x-1}{x}$$

for real m, n and integer r.

11 Use Stirling's formula to show that

$$\lim_{n\to\infty} \frac{\Gamma\left(\frac{n+1}{2}\right)}{\sqrt{\frac{n}{2}}\,\Gamma\left(\frac{n}{2}\right)} = 1.$$

H PROBABILITY PROBLEMS

Some of the familiar probability problems that go with the combinatorial formulas are rephrased as distribution problems and presented in the appropriate context at the end of Section 5.

12 A symmetrical die is rolled 5 times. Find the probability of 5 aces; of 5 even numbers; and of 5 nonaces.

13 Random digits are defined in Problem 2.1.8, the uniform distribution on $\mathcal{S}_0 = \{0, 1, \ldots, 9\}$. An n-digit random number is a sample from such a uniform distribution. Find the probability that a 5-digit number contains 5 even digits; and exactly 2 even digits.

14 Under conventional assumptions find the probability that a family with 5 children has 5 girls; and exactly 2 girls.

15 A hand of poker contains 5 cards from a well-shuffled deck of 52 playing cards. Find the probability that the cards are all spades.

16 Find the probability that the cards in a 5-card poker hand are all of different denominations.

17 Find the probability that a 5-card poker hand is a
(a) Full house (3 cards of one denomination and 2 of another denomination).
(b) Four of a kind (4 cards of one denomination and 1 other card).

18 Find the probability that a 5-card poker hand is a
(a) Royal flush (10, J, Q, K, A in a single suit).
(b) Straight flush (5 cards in sequence in a single suit, other than a royal flush).
(c) Flush (5 cards in a single suit, other than a straight or royal flush).

I PROBLEMS FROM PHYSICS

19 Suppose that n identical balls are distributed among k different cells. Let y_i be the number of balls in the ith cell. Show that the number of different arrangements $(y_1, \ldots, y_k)$ is

$$A_{n:k} = \binom{n+k-1}{n} = \binom{n+k-1}{k-1}.$$

Note: A die with k faces is tossed n times and $(y_1, \ldots, y_k)$ is recorded where y_i is the frequency for the ith face; $A_{n:k}$ is the number of different vectors $(y_1, \ldots, y_k)$.

20 Consider n different balls and k different cells. Suppose that the balls are independently tossed in the cells and the probabilities are equal for the different cells: the symmetrical probability measure on k^n points. Calculate the probability function $p(y_1, \ldots, y_k)$, where y_i is the number of balls in the ith cell. *Note:* If each of n particles can be in any of k different states, the preceding Maxwell–Boltzmann probabilities have been proposed as a description of arrangements.

21 (*continuation*) Suppose that $n = k$ balls are distributed randomly among k different cells.
(a) Find the probability that all cells are occupied.
(b) Find the probability that exactly one cell is unoccupied.

22 (*continuation*) Physical particles do not seem to behave as balls symmetrically into cells. The Bose–Einstein model is a symmetrical model based on different arrangements $(y_1, \ldots, y_k)$ as sample points: thus probability $1/A_{n:k}$ for each arrangement. Consider 6 particles and 7 states. Find the probability that some state has 2 particles, some other 4 states have 1 particle each, and the remaining states have none—without specifying which states.

23 (*continuation*) The Fermi–Dirac model is a symmetric model based on arrangements—but only on those arrangements with at most one particle in any state. Show that the probability for a permissible arrangement is $1/\binom{k}{n}$. Some particles seem to follow Bose–Einstein and others seem to follow Fermi–Dirac.

24 (*continuation*) Suppose that $n = k$ particles are distributed among k states:
(a) Find the probability that all states are occupied (Bose–Einstein).
(b) Find the probability that all states are occupied (Fermi–Dirac).

J PROBLEM FROM ANALYSIS

25 Verify the following steps in a proof of Stirling's formula.

(a) $\Gamma(n+1) = n^{n+(1/2)}e^{-n} \int_{-\sqrt{n}}^{\infty} \exp\{n \log(1 + v/\sqrt{n}) - \sqrt{n}\,v\}\, dv.$

(b) $\lim_{n\to\infty} \dfrac{n^{n+(1/2)}e^{-n}}{\Gamma(n+1)} \int_{-\sqrt{n}}^{\sqrt{n}} \exp\{n \log(1 + v/\sqrt{n}) - \sqrt{n}\,v\}\, dv = 1.$

(c) $\lim_{n\to\infty} \exp\{n \log(1 + v/\sqrt{n}) - \sqrt{n}v\} = \exp\{-v^2/2\}$ uniformly on $[-\sqrt{n}, \sqrt{n}]$.

(d) $\lim_{n\to\infty} \dfrac{n^{n+(1/2)}e^{-n}}{\Gamma(n+1)} (2\pi)^{1/2} = 1.$

3.5

MARGINAL DISTRIBUTIONS FOR FREQUENCIES

We are now in a position to derive various marginal distributions describing frequencies. In one direction we shall be concerned with frequencies such as the number of reactions when a new drug is administered to a sequence of animals from a certain genetic strain; or the number of sterile water samples when a sequence of samples is taken from a water source; or the number of aces with a sequence of tosses of a die; or the number of heads with a sequence of tosses of a coin. In these examples the frequency is the number of occurrences of an event in a sequence of independent repetitions of a random system and the event can occur or not occur on each repetition. The marginal distribution for such frequencies is examined in Sections A and B: binomial distribution and multinomial distribution.

In another direction we shall be concerned with frequencies such as the number of tagged fish in a sample caught from the fish in a particular lake (care needed to ensure equal probability for each possible sample), the number of people favoring a certain proposal in a sample of adult voters in a certain city (care needed to ensure equal probability for each possible sample), the number of defective tires in a sample of 50 from an incoming shipment of 1000 (thorough mixing or random numbers needed to ensure equal probability for each possible sample), or the number of spades in a hand of 13 from a well-shuffled deck of 52 playing cards. In these examples there is a finite population of people, tires, or cards, and random sampling is used to ensure that each possible sample has the same probability of being drawn, and the frequency is the number of sample elements with a particular property. The marginal distribution for such frequencies in a random sample from a finite population is examined in Sections C and D: hypergeometric distribution and multivariate hypergeometric distribution. The marginal distribution for a general frequency is examined in Section E. The derivations use the marginal concepts of Sections 3.1 and 3.2, the independence of Section 3.3, and the counting formulas of Section 3.4.

A BINOMIAL(n, p) DISTRIBUTION

Consider the Bernoulli(p) distribution for an indicator x:

$$p(1) = p, \qquad p(0) = q,$$

or, equivalently,

$$p(x) = p^x q^{1-x}, \qquad x = 0, 1,$$

and zero otherwise. Note that the letter p is being used for the probability at the point $x = 1$ and quite separately is being used for the probability function; this slight notational overlap allows consistencies to be maintained elsewhere.

The indicator x could register the occurrence of a reaction when a drug is administered to an animal from a certain genetic strain, or could register sterility of a water sample from a certain source, or could register the occurrence of an ace with the toss of a die, or could register the occurrence of a head with the toss of a coin.

Now consider n independent performances of the system being investigated. More formally, let $(x_1, \ldots, x_n)$ be a sample from the Bernoulli(p) distribution. Then by the independence example (Example 3.5),

$$\begin{aligned} p(x_1, \ldots, x_n) &= p^{x_1} q^{1-x_1} \cdots p^{x_n} q^{1-x_n}, \qquad & x_j = 0, 1, \\ &= p^y q^{n-y}, & x_j = 0, 1, \end{aligned}$$

and zero otherwise, where we write $y = \Sigma\, x_j$ for the frequency of 1's in the sample $(x_1, \ldots, x_n)$.

Now consider the marginal distribution of $y = x_1 + \cdots + x_n$, the frequency of 1's in the sample $(x_1, \ldots, x_n)$. Of course, a value of y occurs with any vector $(x_1, \ldots, x_n)$ that contains y 1's and $n - y$ 0's. The number of such vectors is given by $\binom{n}{y}$ in Section 3.4C; thus the marginal probability function is

$$(1) \qquad \begin{aligned} p_Y(y) &= \binom{n}{y} p^y q^{n-y}, \qquad & y = 0, 1, \ldots, n. \\ &= 0, & \text{otherwise.} \end{aligned}$$

This is the *binomial(n, p) distribution* (see Example 1.3.5 and Exercise 2.1.10).

Consider some examples of the binomial distribution. The binomial(4, 1/2) and binomial(4, 3/4) are given in Table 3.6 and plotted in Figure 3.11. For comparison

TABLE 3.6
Binomial distributions

Binomial(4, 1/2)			Binomial(4, 3/4)		
y	$p_Y(y)$	$F_Y(y)$	y	$p_Y(y)$	$F_Y(y)$
0	1/16	1/16	0	0.0039	0.0039
1	4/16	5/16	1	0.0469	0.0508
2	6/16	11/16	2	0.2109	0.2617
3	4/16	15/16	3	0.4219	0.6836
4	1/16	1	4	0.3164	1.0000

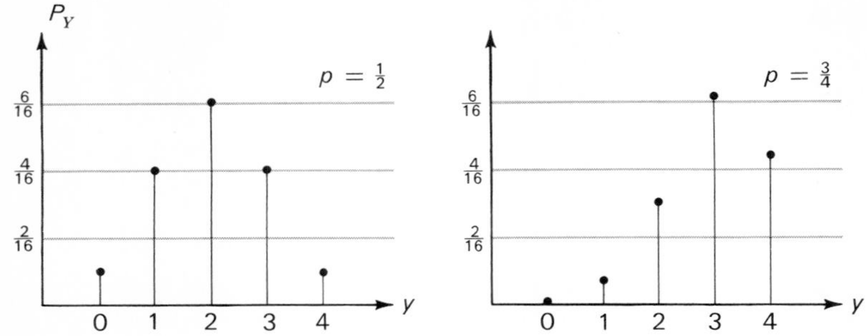

FIGURE 3.11
Binomial(4, 1/2) and binomial(4, 3/4) distributions.

purposes we record a few values for the corresponding distributions with $n = 4$ replaced by $n = 40$, Table 3.7.

TABLE 3.7
Binomial distributions

Binomial(40, 1/2)			Binomial(40, 3/4)		
y	$p_Y(y)$	$F_Y(y)$	y	$p_Y(y)$	$F_Y(y)$
0	0.0000	0.0000	0	0.0000	0.0000
⋮			⋮		
10	0.0008	0.0011	10	0.0000	0.0000
⋮			⋮		
20	0.1254	0.5627	20	0.0004	0.0006
⋮			⋮		
30	0.0008	0.9997	30	0.1444	0.5605
⋮			⋮		
40	0.0000	1.0000	40	0.0000	1.0000

In Section 1.1B we tossed a coin 100 times and obtained the observed frequency 53 for heads. If the probability for heads is p, then the distribution describing possible frequencies is binomial (100, p). This can be tabulated for various values of p. We could then examine these distributions to see which ones are harmonious with the observed frequency 53.

B MULTINOMIAL($n, p_1, \ldots, p_k$) DISTRIBUTION

We have been considering a dichotomy for the response of the random system being investigated: for example, {reaction, no reaction} for the drug to an animal, {sterile, nonsterile} for a water sample, {ace, nonace} for the toss of a die, $\{H, T\}$ for the toss of a coin. Now consider more generally some partition $\{B_1, \ldots, B_k\}$ for the response of the random system being investigated.

Let $p_1, \ldots, p_k$ be the probabilities corresponding to $B_1, \ldots, B_k$ and let $x_1, \ldots, x_k$ be the corresponding indicators, Table 3.8; note that exactly one $x(= 1)$ in each row is different from zero. This gives the *multivariate Bernoulli*($p_1, \ldots, p_k$) *distribution* for the vector $\mathbf{x} = (x_1, \ldots, x_k)'$:

$$p(\mathbf{x}) = p_1^{x_1} \cdots p_k^{x_k}, \qquad x_i = 0, 1; \sum x_i = 1,$$

TABLE 3.8
Indicators and probabilities

	Indicators				
Event	x_1	x_2	$\cdots$	x_k	Probability
B_1	1	0	$\cdots$	0	p_1
B_2	0	1	$\cdots$	0	p_2
$\vdots$					$\vdots$
B_k	0	0	$\cdots$	1	p_k

and zero otherwise, where $p_i \geq 0$ and $\Sigma\ p_i = 1$.

The indicators $x_1, \ldots, x_k$ could register occurrence for "type A reaction," "type B reaction," or "no reaction" for the drug to an animal ($k = 3$), or could register occurrence for "sterility," "contaminated (nonfecal)," or "contaminated (fecal)" for the water sample, or could register occurrence for "ace," "two or three," or "four or more" for the toss of a die, or could register "H with orientation in $(0, 90°)$," "H with orientation in $[90°, 180°]$," or "T" for the toss of a coin.

Now consider n independent performances of the system being investigated. More formally, let $\mathbf{x}_1, \ldots, \mathbf{x}_n$ be a sample from the preceding multivariate Bernoulli $(p_1, \ldots, p_k)$ distribution and write

$$X = (\mathbf{x}_1, \ldots, \mathbf{x}_n) = \begin{pmatrix} x_{11} & \cdots & x_{1n} \\ \vdots & & \vdots \\ x_{k1} & \cdots & x_{kn} \end{pmatrix},$$

where the jth column records the indicator values for the jth sample element. The joint probability function is then given by formula (3.3.7):

$$\begin{aligned} p(X) &= (p_1^{x_{11}} \cdots p_k^{x_{k1}}) \cdots (p_1^{x_{1n}} \cdots p_k^{x_{kn}}) \qquad \text{if } x_{ij} = 0, 1 \text{ and } \sum_i x_{ij} = 1 \\ &= p_1^{y_1} \cdots p_k^{y_k} \end{aligned}$$

and zero otherwise, where we write $y_1 = \Sigma_j\, x_{1j}, \ldots, y_k = \Sigma_j\, x_{kj}$ for the frequencies of the events $B_1, \ldots, B_k$, respectively.

Now consider the marginal distribution for the frequencies $(y_1, \ldots, y_k)$. A particular value for $(y_1, \ldots, y_k)$ will occur for any response sequence that involves y_1 B_1's, $\ldots$, y_k B_k's for the n positions in the sequence. The number of such sequences is available from the concluding remarks in Section 3.4D; the number is

$$\binom{n}{y_1 \cdots y_k},$$

which is the number of ways of selecting y_1 places for the B_1's, y_2 places for the B_2's, and so on. Thus the marginal probability function is

$$(2) \qquad p_Y(y_1, \ldots, y_k) = \binom{n}{y_1 \cdots y_k} p_1^{y_1} \cdots p_k^{y_k} \qquad \text{if } y_i = 0, 1, \ldots;\ \Sigma\, y_i = n$$

and zero otherwise. This is the *multinomial*$(n, p_1, \ldots, p_k)$ *distribution*. Note directly that the probabilities add to 1 (multinomial theorem in Section 3.4E).

EXAMPLE 1

Tossing a modified die: Consider the tossing of a modified symmetrical die that has one face marked 1, two faces marked 2, and three faces marked 3. The marginal distribution for the frequencies (y_1, y_2, y_3) is multinomial(5, 1/6, 2/6, 3/6); the probability function is

$$p_Y(y_1, y_2, y_3) = \binom{5}{y_1 y_2 y_3}(1/6)^{y_1}(2/6)^{y_2}(3/6)^{y_3} \qquad \text{if } y_i = 0, 1, \ldots; \Sigma y_i = 5$$

and zero otherwise; this is shown in Table 3.9 and plotted in Figure 3.12.

TABLE 3.9
Multinomial(5, 1/6, 2/6, 3/6)

y_1	y_2	y_3	$7776p_Y$	y_1	y_2	y_3	$7776p_Y$
0	0	5	243	3	1	1	120
1	0	4	405	2	2	1	360
0	1	4	810	1	3	1	480
2	0	3	270	0	4	1	240
1	1	3	1080	5	0	0	1
0	2	3	1080	4	1	0	10
3	0	2	90	3	2	0	40
2	1	2	540	2	3	0	80
1	2	2	1080	1	4	0	80
0	3	2	720	0	5	0	32
4	0	1	15				7776

C HYPERGEOMETRIC(N, n, p) DISTRIBUTION

Consider a collection or population of N elements, each being either a type B element or a type G element. Let Np be the number of type B elements and Nq be the number of type G elements; $p + q = 1$. Now consider a random sample of n from the population: a sequence of n elements such that each of the $N^{(n)}$ possible sequences has the same probability $1/N^{(n)}$. Let y be the number of type B elements in the sample.

For example, y could be the number of tagged fish in a sample of 200 from a lake containing 1000 tagged fish and 9000 unmarked fish, or y could be the number of people favoring a proposal in a sample of 100 from the 150,000 voters in a certain city, or y could be the number of defective tires in a sample of 50 from a shipment of 1000, or y could be the number of spades in a hand of 13 from a deck of 52.

Consider briefly how this differs from the binomial or "infinite" population situation. Suppose that we think of the sample being formed one element at a time. For the first element in the sample we must have by symmetry that $P(\text{B}) = p$ and $P(\text{G}) = q$. After we have the first element, we can consider the possibilities for the second element. The probabilities for B and G will be slightly changed and will depend, of course, on which element was obtained for the first position. Thus successive draws are not statistically independent and to calculate probabilities we would need the concept of conditional probability which is developed in Chapter 4.

The marginal probability distribution for the frequency y can be derived from the

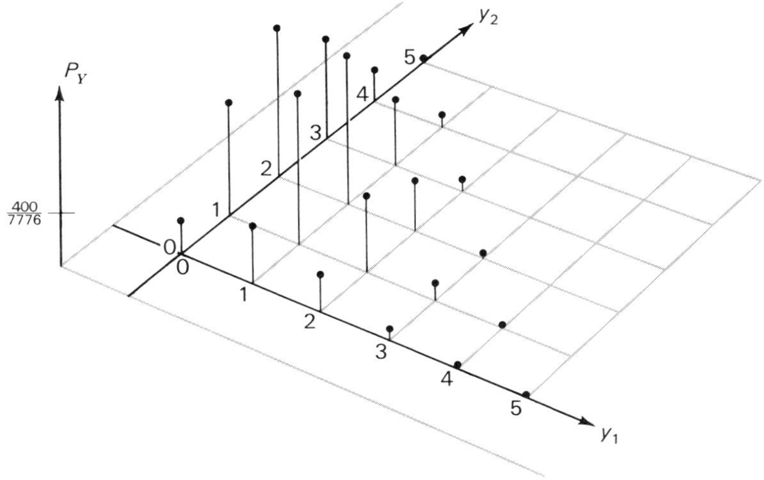

FIGURE 3.12
Multinomial(5, 1/6, 2/6, 3/6) distribution plotted against (y_1, y_2); recall that $y_3 = 5 - y_1 - y_2$ or at points on $\Sigma y_i = 5$ in R^3.

symmetric distribution assigning probability $1/N^{(n)}$ to each possible sequence or equivalently from the symmetric distribution assigning probability $1/\binom{N}{n}$ to each possible sample set. We use the second of these symmetric distributions and note that the number of different sets that contain y B elements and $n - y$ G elements is available from Sections 3.4A and 3.4C:

$$\binom{Np}{y}\binom{Nq}{n-y}.$$

Thus we obtain the marginal probability function:

$$p_Y(y) = \frac{\binom{Np}{y}\binom{Nq}{n-y}}{\binom{N}{n}} \qquad \text{for } y \text{ an integer} \tag{3}$$

and zero otherwise. This is the *hypergeometric(N, n, p) distribution*. See Exercise 2.1.12 and note the effective range of the distribution.

Consider the numerical examples listed in Table 3.10 and plotted in Figure 3.13; notice how the distribution is more concentrated than the corresponding binomial distribution in Figure 3.11. The hypergeometric probability function can be expressed in the alternative form:

$$\begin{aligned} p_Y(y) &= \binom{n}{y}\frac{(Np)^{(y)}(Nq)^{(n-y)}}{N^{(n)}}, && y = 0, 1, \ldots, n, \\ &= 0 && \text{otherwise,} \end{aligned}$$

which can be justified directly by considering sequences rather than sets.

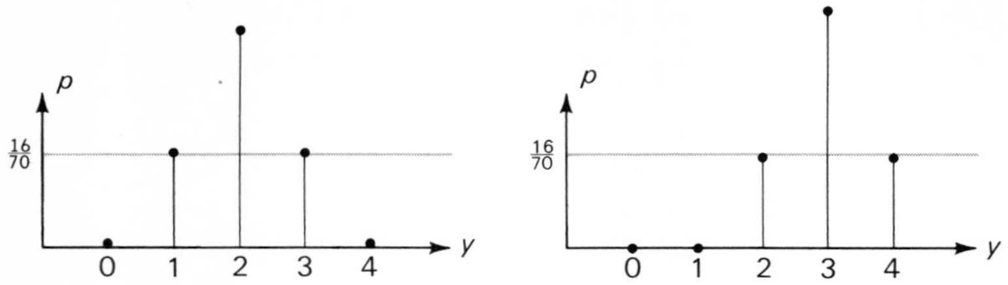

FIGURE 3.13
Hypergeometric(8, 4, 1/2) and hypergeometric(8, 4, 3/4) distributions. Compare with Figure 3.11.

TABLE 3.10
Example distribution

Hypergeometric(8, 4, 1/2)		Hypergeometric(8, 4, 3/4)	
y	$p_Y(y)$	y	$p_Y(y)$
0	1/70	0	0/70
1	16/70	1	0/70
2	36/70	2	15/70
3	16/70	3	40/70
4	1/70	4	15/70

EXAMPLE 2

Sample of desk calculators: A random sample of 4 units is taken from a shipment of 8 used desk calculators. Let y be the number of defective calculators in the sample. The marginal distribution of y is then hypergeometric $(8, 4, D/8)$, where D is the number of defectives in the lot. The distribution is plotted in Figure 3.13 for the cases $D = 4$ and $D = 8$.

D MULTIVARIATE HYPERGEOMETRIC $(N, n, p_1, \ldots, p_k)$ DISTRIBUTION

Consider the finite population further. In the preceding section we assumed that each element was either a type B element or a type G element, a *dichotomy*. Now suppose that an element can be a B_1 element, a B_2 element, . . ., or a B_k element, a *partition* of the possible descriptions for an element.

More specifically consider a population of N elements consisting of Np_1 elements marked B_1, . . ., Np_k elements marked B_k, where $\Sigma p_i = 1$. And suppose that we sample n elements in succession randomly so that each of the $N^{(n)}$ sequences of elements has the same probability $1/N^{(n)}$. Let y_1 be the number of B_1 elements in the sample, y_2 be the number of B_2 elements in the sample, . . ., and y_k the number of B_k elements. Note of course that $\Sigma y_i = n$.

The marginal probability distribution for the frequencies $(y_1, \ldots, y_k)$ can be derived from the symmetric distribution assigning probability $1/N^{(n)}$ to each sequence or from the symmetric distribution assigning probability $1/\binom{N}{n}$ to each possible sample set. We use the second of these and note that the number of different sets that contain y_1 B_1 elements, . . ., y_k B_k elements is available from Sections 3.4A and 3.4C:

$$\binom{Np_1}{y_1} \cdots \binom{Np_k}{y_k}.$$

Thus we obtain the marginal probability function:

$$(4) \qquad p_Y(y_1, \ldots, y_k) = \frac{\binom{Np_1}{y_1} \cdots \binom{Np_k}{y_k}}{\binom{N}{n}}, \qquad y_i = \text{integer}, \quad \Sigma\, y_i = n$$

and zero otherwise. This is the *multivariate hypergeometric*$(N, n, p_1, \ldots, p_k)$ *distribution*.

EXAMPLE 3

Sample of outboard motors: A random sample of 3 units is taken from a shipment of 6 used outboard motors. Let y_1 be the number of unrepairable units, y_2 the number of repairable units, and y_3 the number of satisfactory units. The marginal distribution of (y_1, y_2, y_3) is then multivariate hypergeometric$(6, 3, p_1, p_2, p_3)$. The distribution for the case $6p_1 = 1$, $6p_2 = 2$, $6p_3 = 3$ is given in Table 3.11 and plotted in Figure 3.14.

TABLE 3.11
Hypergeometric(6, 3, 1/6, 2/6, 3/6)

y_1	y_2	y_3	p_Y
0	0	3	1/20
0	1	2	6/20
0	2	1	3/20
1	0	2	3/20
1	1	1	6/20
1	2	0	1/20

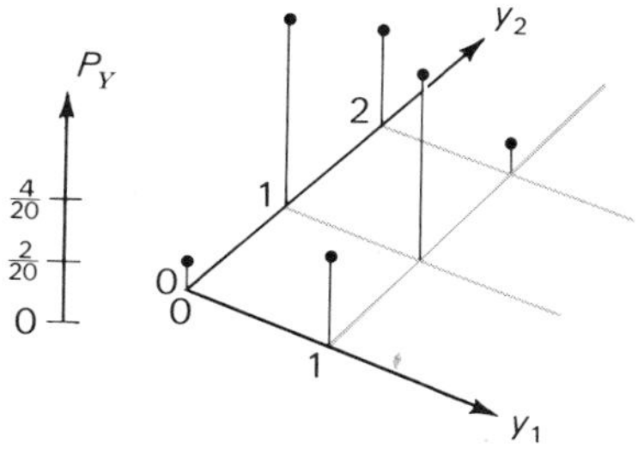

FIGURE 3.14
Multivariate hypergeometric(6, 3, 1/6, 2/6, 3/6) distribution plotted against (y_1, y_2); recall that $y_3 = 3 - y_1 - y_2$ or on points on $\Sigma\, y_i = 3$ in R^3.

E GENERAL FREQUENCY DISTRIBUTION

Consider a sample space $\mathcal{S}$ with class $\mathcal{A}$ of events and probability measure P. And suppose that an investigator is interested in certain events $A_1, A_2, \ldots, A_n$, specifically, in the number of these events that occur in any performance of the system. Let $y = Y(s)$ be the number of events $A_1, \ldots, A_n$ that occur with the sample point s,

$$Y(s) = \text{number of } A_j \text{ such that } s \in A_j.$$

The probability function for Y is available from Problem 1.5.3:

$$(5) \qquad p_Y(y) = S_y - \binom{y+1}{y} S_{y+1} + \cdots + (-1)^{n-y}\binom{n}{y} S_n,$$

where $S_0 = 1$ and

$$S_r = \sum_{i_1 < \cdots < i_r} P(A_{i_1} \cdots A_{i_r})$$

is the sum of the probabilities for the rth-order intersection sets. The probability for at least r of the events is recorded in Problem 1.5.5.

Consider the classical example for this general distribution:

EXAMPLE 4

A well-shuffled deck of $N = 52$ playing cards is dealt successively as the dealer calls AS, KS, QS, . . ., 3C, 2C. Let A_i be the event the ith call correctly identifies the ith card dealt. And let y be the number of cards called correctly.

$$P(A_1) = \frac{(N-1)!}{N!} = \frac{1}{N} = P(A_i)$$

$$P(A_1A_2) = \frac{(N-2)!}{N!} = \frac{1}{N(N-1)} = P(A_iA_j), \qquad i \neq j,$$

$$P(A_1A_2A_3) = \frac{(N-3)!}{N!} = \frac{1}{N(N-1)(N-2)} = P(A_iA_jA_k), \qquad i \neq j \neq k \neq i.$$

Thus

$$S_0 = 1,$$

$$S_1 = \sum P(A_i) = 1,$$

$$S_2 = \sum \frac{1}{N(N-1)} = \binom{N}{2}\frac{1}{N(N-1)} = \frac{1}{2!},$$

$$S_3 = \binom{N}{3}\frac{1}{N(N-1)(N-2)} = \frac{1}{3!}.$$

The distribution can now be given, Table 3.12. The Poisson(1) probabilities provide a convenient approximation:

$$p_Y(y) \approx \frac{1}{y!}e^{-1}.$$

TABLE 3.12
Distribution of cards

y	$P_Y(y)$
0	$1\left(1 - 1 + \frac{1}{2!} - \cdots + \frac{1}{52!}\right)$
1	$1\left(1 - 1 + \frac{1}{2!} - \cdots - \frac{1}{51!}\right)$
2	$\frac{1}{2!}\left(1 - 1 + \frac{1}{2!} - \cdots + \frac{1}{50!}\right)$
3	$\frac{1}{3!}\left(1 - 1 + \frac{1}{2!} - \cdots - \frac{1}{49!}\right)$
$\vdots$	
51	$\frac{1}{51!}(1 - 1)$
52	$\frac{1}{52!}(1)$

F EXERCISES: SAMPLE FROM A RANDOM SYSTEM

1 Tabulate the binomial(3, 1/2) probability function.

2 Tabulate the binomial(5, 1/2) probability function.

3 Let y be the number of even digits in a three-digit random number; tabulate the probability function for y.

4 Let y be the number of zeros in a three-digit random number; tabulate the probability function for y.

5 Tabulate the multinomial(3, 1/3, 1/3, 1/3) probability function.

6 Tabulate the multinomial(3, 1/10, 3/10, 6/10) probability function.

7 Consider a symmetric die and suppose that two faces are labeled 1, two are labeled 2, and two are labeled 3. Consider three tosses and tabulate the probability function for (y_1, y_2, y_3), where y_i is the number of faces that show i.

8 Consider a four-digit random number and suppose that (y_1, y_2, y_3) is reported, where y_1 is the number of 1s, y_2 is the number of 2s, and y_3 is the number of other digits. Record the probability function for (y_1, y_2, y_3).

9 An article from production can be acceptable, repairable, or scrap. Suppose that production is stable and p_1, p_2, and p_3 are the respective probabilities for the condition of an article. And suppose that articles are collected in lots of 100. Give the probability function for (y_1, y_2, y_3), where y_1, y_2, and y_3 are the number of items of the respective conditions in a lot.

10 A certain pair of black mice produce an offspring that is brown with probability 1/4 and black with probability 3/4. Let y be the number of black mice in a litter of 5. Record the probability function for y; ignore the possibility of twins.

11 A bioassay procedure exposes 5 animals to a first dosage and 5 animals to a second dosage of a drug. Let y_1 be the number with a reaction in the first group and y_2 the number in the second group. Let p_1 and p_2 be the probability of reaction at the two dosages. Record the probability function for (y_1, y_2). Assume independence between animals.

12 A coin may be biased. You decide to test the coin by making n tosses ($= 5$ to simplify arithmetic; too small realistically). As a procedure for making a tentative judgment, you decide to accept unbiasedness if the number of heads is not equal to the extremes 0 and 5 and reject unbiasedness if the extremes occur. Let $p = P(H)$ be the probability for the

coin. If $p = 1/2$, what is the probability of rejection? If $p = 3/4$, what is the probability of rejection? Let $\mathcal{P}(p)$ be the probability of rejection as a function of p. Calculate the power function $\mathcal{P}$ of this mechanized judgment procedure. Plot it. Is the procedure very reliable at this sample size?

G EXERCISES: SAMPLE FROM A FINITE POPULATION

13 Tabulate the hypergeometric(6, 3, 1/2) probability function.

14 Tabulate the hypergeometric(6, 5, 1/2) probability function.

15 Three coils of cable are randomly sampled from a store of 6 coils. Let y be the number of sampled coils with defects. Tabulate the probability function for y if half the coils in the store contain defects.

16 Record the probability function for y, the number of aces in a hand of 13 from a well-shuffled deck.

17 Record the probability function for z, the number of aces in the combined hands (13 each) for N and S in a bridge distribution of 13 cards to each of 4 players.

18 Consider an urn containing 5 balls: 1 ball numbered 1, 1 ball numbered 2, and 3 balls numbered 3. The urn is thoroughly stirred and 3 balls are drawn at random. And (y_1, y_2, y_3) is reported; y_i is the number of i balls in the sample. Record the probability function and tabulate.

19 A hand of 13 cards is dealt from a well-shuffled standard deck of 52 cards. Calculate the probability function

(a) $p_1(x)$, where x is the number of spades in the hand.

(b) $p_2(y)$, where y is the number of hearts in the hand.

(c) $p_3(x, y)$, where x and y are as given in parts (a) and (b).

20 A hand of 13 cards is dealt from a well-shuffled deck; let x, y, z, and w be the number of spades, hearts, diamonds, clubs, respectively. Record the probability function $p(x, y, z, w)$.

21 An urn contains R red balls, W white balls, B blue balls, where $N = R + W + B$. The urn is thoroughly mixed and a random sample of n balls is drawn; let r, w, and b be the number of red, white, and blue balls, respectively. Record the probability function for (r, w, b).

22 A production lot of 100 items contains D defectives. A sample of 5 is taken at random from the lot and the number of defectives y is reported.

(a) Record the probability function for y.

(b) Tabulate the probability function for the case $D = 0$; the case $D = 1$; and the case $D = 2$.

(c) A quality-control procedure is to reject the lot if any defectives are found; the operating characteristic function Q of the procedure records $Q(D)$, the probability of acceptance when the lot contains D defectives. Partially tabulate and sketch the operating characteristic Q.

H COMPUTER EXERCISES

23 Obtain a sample of 1000 random digits and tabulate the proportion of 0's, 1's, . . ., 9's. Does the empirical distribution seem reasonably symmetrical?

24 Let y be the number of even digits in a block of five random digits. For a sample of 1000 blocks, tabulate the proportion with $y = 0$, $y = 1$, $y = 2$, $y = 3$, $y = 4$, and $y = 5$. Compare with the theoretical distribution in Exercise 2.

25 Let y_1, y_2, and y_3 be, respectively, the number of 0's, the number that are multiples of 3, and the number of other digits in a block of three random digits. For a sample of 1000 blocks,

tabulate the proportion for each value for (y_1, y_2, y_3). Compare with the theoretical distribution in Exercise 6.

PROBLEMS

26 Binomial(n, p) probabilities can be calculated in sequence. Show that $r(y) = p(y)/p(y-1) = (n-y+1)p/yq$, where $p(y)$ is the binomial probability function.

27 *(continuation)* The mode of a discrete distribution is the sample point having maximum probability. Show that the mode of the binomial distribution is $[(n+1)p]$, where $[x]$ is the largest integer $\le x$. However, if $(n+1)p$ is an integer, there is a double mode, at $(n+1)p - 1$ and at $(n+1)p$.

28 Let F be the distribution function for the binomial(n, p) distribution. Show that at the integers $y = 0, 1, \ldots, n-1$,

$$F(y) = \sum_{x=0}^{y} \binom{n}{x} p^x q^{n-x} = \frac{\Gamma(n+1)}{\Gamma(y+1)\Gamma(n-y)} \int_p^1 t^y (1-t)^{n-y-1}\, dt;$$

the integral is an incomplete beta integral.

29 *Sum of binomials:* Let y_1 and y_2 be independent and y_i be binomial(n_i, p). Use the convolution formula (3.2.9) to show that $y_1 + y_2$ is binomial$(n_1 + n_2, p)$.

30 *(continuation)* Represent each y_i in terms of a sample from the Bernoulli(p) and thus show that $y_1 + y_2$ is binomial $(n_1 + n_2, p)$.

31 Poisson(λ) probabilities can be calculated in sequence: calculate $r(y) = p(y)/p(y-1)$.

32 *(continuation)* Show that the mode of the Poisson(λ) is $[\lambda]$; compare with Problem 27. However, if λ is an integer, there is a double mode, at $\lambda - 1$ and at λ.

33 By appropriate differentiation and integration, show that the Poisson distribution function can be reexpressed as an incomplete gamma integral:

$$F(y) = \sum_{0 \le x \le y} \frac{e^{-\lambda}\lambda^x}{x!} = \frac{1}{y!} \int_{\lambda}^{\infty} e^{-t} t^y\, dt$$

for $y = 0, 1, 2, \ldots$.

34 *Sum of Poisson:* Let y_1 and y_2 be independent and y_i be Poisson(λ_i). Use the convolution formula (3.2.9) to show that $y_1 + y_2$ is Poisson$(\lambda_1 + \lambda_2)$.

35 Let $p(y{:}p)$ be the binomial(n, p) probability function. Let $\lambda = np$ be fixed; then $p = \lambda/n$; show that $\lim p(y{:}p)$ as $n \to \infty$ is given by the Poisson(λ) probability function.

36 *Negative binomial(r, p) distribution:* Consider a sequence of independent indicators, each Bernoulli(p). Let y be the number of 0's until r 1's have occurred. Show that the marginal distribution of y is negative binomial(r, p); recall Exercise 2.1.11.

37 *Combining multinomial coordinates:* Let (y_1, y_2, y_3) be multinomial(n, p_1, p_2, p_3), and consider $(y_1, y_2 + y_3)$. Show that y_1 is binomial(n, p_1).

38 *(continuation)* Let $(y_1, \ldots, y_k)$ be multinomial$(n, p_1, \ldots, p_k)$. Show that the marginal distribution of $(y_1, \ldots, y_{k-2}, y_{k-1} + y_k)$ is multinomial$(n, p_1, \ldots, p_{k-2}, p_{k-1} + p_k)$.

39 *(continuation)* Let $(y_1, \ldots, y_k)$ be multinomial$(n, p_1, \ldots, p_k)$. And let w_1 be the sum of the first r_1 y's, w_2 the sum of the next r_2 y's, $\ldots$, w_s the sum of the last r_s y's. And let P_1 be the sum of the first r_1 p's, and so on for $P_2, \ldots, P_s$. Use Problem 38 to justify that the marginal of $(w_1, \ldots, w_s)$ is multinomial$(n, P_1, \ldots, P_s)$.

40 Let $p(y_1, \ldots, y_k)$ be the multinomial$(n, p_1, \ldots, p_k)$ probability function. Let $\lambda_1 = np_1, \ldots, \lambda_{k-1} = np_{k-1}$ be fixed; then $p_i = \lambda_i/n$ for $i = 1, \ldots, k-1$. Show that as $n \to \infty$, the probability function for $y_1, \ldots, y_{k-1}$ converges to the probability function for $k-1$ independent Poissons with parameters $\lambda_1, \ldots, \lambda_{k-1}$.

41 Let $(y_1, \ldots, y_k)$ be multivariate hypergeometric$(N, n, p_1, \ldots, p_k)$. Show that the marginal distribution of $(y_1, \ldots, y_{k-2}, y_{k-1} + y_k)$ is multivariate hypergeometric $(N, n, p_1, \ldots, p_{k-2}, p_{k-1} + p_k)$. Use Problem 3.4.7.

J PROBLEMS: GENERAL FREQUENCY DISTRIBUTION

42 Use the general frequency distribution(5) to derive the binomial(n, p) probability function for the number of 1s in a sample of n from the Bernoulli(p).

43 Derive the probability function for the number y of ace–king pairs in a bridge hand.

44 Derive the probability that a bridge hand contains at least one card from each suit.

45 Derive the probability that at least one of the four hands at bridge contains a complete suit.

46 *General bivariate frequency distribution:* Consider events $A_1, \ldots, A_n$ and events $B_1, \ldots, B_m$ for a random system. Let x be the number of events A that occur and y be the number of events B that occur. Show that

$$p_{XY}(x, y) = S_{x,y} - \binom{x+1}{x} S_{x+1,y} - \binom{y+1}{y} S_{x,y+1} \cdots$$

$$= \sum_{\alpha,\beta=0}^{\infty} (-1)^{\alpha+\beta} \binom{x+\alpha}{x}\binom{y+\beta}{y} S_{x+\alpha,\, y+\beta},$$

where

$$S_{x,y} = \sum_{\substack{i_1<\cdots<i_x \\ j_1<\cdots<j_y}} P(A_{i_1} \cdots A_{i_x} B_{j_1} \cdots B_{j_y}), \qquad x \le n,\; y \le m,$$

$$= 0, \qquad \text{otherwise.}$$

Use Problem 2 of Supplement Section 1.5.

K PROBLEMS: MULTINOMIALS OVER OTHER DISTRIBUTIONS

47 Two symmetric coins are tossed 4 times and (y_1, y_2, y_3) is observed, where y_1 is the number of double heads, y_2 is the number of double tails, and y_3 is the number of mixed tosses. Derive the probability function for (y_1, y_2, y_3).

48 (*continuation of Exercise 18*) The urn is stirred, a ball is drawn and replaced; the urn is stirred, a ball is drawn and replaced; the urn is stirred, a ball is drawn and replaced; (y_1, y_2, y_3) is reported, where y_i is the number of i balls in the sample. Derive the probability function.

49 (*continuation of Exercise 16*) For 10 successive hands of bridge with thorough shuffling at each step, find the probability function for $(y_0, y_1, y_2, y_3, y_4)$, where y_i is the number of hands with i aces.

SUPPLEMENTARY MATERIAL

Supplement Section 3.6 is on pages 526–537.

The proof is given for the following result: $F_{XY} = F_X F_Y$ *is a necessary and sufficient condition for the independence of X and Y.*

The normal theory basis for *chi square, Student t, and F* is an application of the theory in this chapter and is presented in detail. The *noncentral theory* is also presented, in text and problems.

Some *order statistic distributions* illustrate the theory in this chapter and are developed in detail.

Product spaces are the means by which bigger models are built from smaller initial models; the formation of product spaces, as touched on in Section 3.3, is examined in detail.

Various results for *measurable functions* are recorded; for example, the composition of measurable functions is measurable; and the limit of a sequence of measurable functions is measurable.

NOTES AND REFERENCES

For some references to other texts, see the Notes and References in Chapter 1. In this chapter we have organized the material in terms of two basic probability concepts, *marginal probability* and *independence,* and in terms of a major application, the derivation of the standard frequency distributions, the binomial and multinomial. These distributions need the two probability concepts, and we have chosen to examine the concepts first, in special detail. We feel that the two probability concepts are very important, and have separated the material accordingly. The common separation by type of distribution, discrete or continuous, has been deemphasized, and appears just as a minor difference in formulas, Σ or $\int$.

In Chapter 3 we examined how probability is mapped from one space to another by means of functions (or random variables). And we discussed various ways of describing the resulting ***marginal distributions.***

With marginal probability as the basis we then examined the important concept of ***statistical independence.*** *On the surface this is concerned with how a large model can sometimes be split into essentially independent components. In fact, however, the concept is more concerned with combining simple models to form larger more complex models—with model building.*

We then used independence and marginal probability to derive the important ***binomial, multinomial,*** *and* ***hypergeometric distributions.***

All our examples so far have involved a random system about to be performed or a random system that has been performed with the outcome concealed from the investigator. These two situations represent, in fact, two extremes, and we are often somewhere in between: a random system has been performed and the outcome is partially concealed from the investigator. Conditional probability is concerned with this intermediate situation involving partial information concerning an outcome.

In Chapter 4 we examine this very important and fundamental concept, ***conditional probability:*** *Section 4.1 is concerned with the general definition and interpretation; and Section 4.2 is concerned with formulas and methods on real spaces.*

4

CONDITIONAL PROBABILITY

4.1

DEFINITION AND INTERPRETATION

Conditional probability is concerned with describing a real-world system when there is partial information concerning the outcome of the system. In this section we examine the definition and interpretation of conditional probability. Then in Section 4.2 we examine various formulas and methods for calculating conditional probability on real spaces.

Conditional probability is the most important concept of probability theory—after probability itself. We have delayed its introduction to this point in order to have a broad contact with probability methods and be able to examine the concept in depth.

Consider an engineer investigating reliability. A control system has two primary components in series. The first component is a new unit and the second component has been in service and survived 900 hours. The distribution for the life of new components is known to be approximately normal(1000, 100). For the first component we can use this normal distribution directly. However, for the second component we are interested in residual lifetime given survival for 900 hours: we *know* that the realized lifetime for the second component will be at least 900 hours.

Consider an actuary investigating 1-year term insurance policies for 40-year-olds. Actuarial investigations lead to life tables that contain (implicitly) the distribution of lifetime for a person just born in a certain racial and social context. However, for 40-year-olds we know that the realized lifetime is at least 40; in fact, the outcomes of interest have been *selected* to be those with lifetime at least 40 and not yet 41.

Consider a geneticist examining a certain strain of brown and black mice. A gene for color can be dominant B for black or recessive b for brown. A mouse has two genes for color and can be of the following three genotypes: BB, Bb, bb. A mouse bb without a dominant B is brown; a mouse with a dominant B is black. Now consider the mating of a Bb mouse with a Bb mouse. Genetic theory prescribes the following probability for each kind of offspring:

BB	Bb	bb
1/4	1/2	1/4

But suppose that the geneticist is interested in a particular offspring that is black. Clearly, the case bb is excluded. He is interested in outcomes of the random system that have been *selected* on the basis of having a dominant black gene.

A SELECTION OF PERFORMANCES: EXAMPLE

Consider a random system involving the determination of the sex of an offspring. Let M denote male and F denote female; then in the standard context we have the distribution

M	F
1/2	1/2

Or consider a system in which repetitions are more simply visualized: the tossing of a coin and observing the upward face. Let H denote heads and T denote tails; then in the standard context we have the distribution

H	T
1/2	1/2

For this case suppose that we have an operator who tosses the coin twice and records (x_1, x_2), where x_i is the indication for heads on the ith toss; then we have

(0, 0)	(0, 1)	(1, 0)	(1, 1)
1/4	1/4	1/4	1/4

Now suppose that the operator selects certain performances and ignores the remaining performances. Specifically, suppose that the operator selects those performances on which C, ''at least one head,'' occurs and ignores those performances on which C^c occurs; see Figure 4.1.

We now discuss the probability model appropriate to such selected performances. Consider a large number, say 1000, of repetitions of the initial random system involving two tosses. The performances yielding (0, 1), (1, 0), (1, 1) will pass selec-

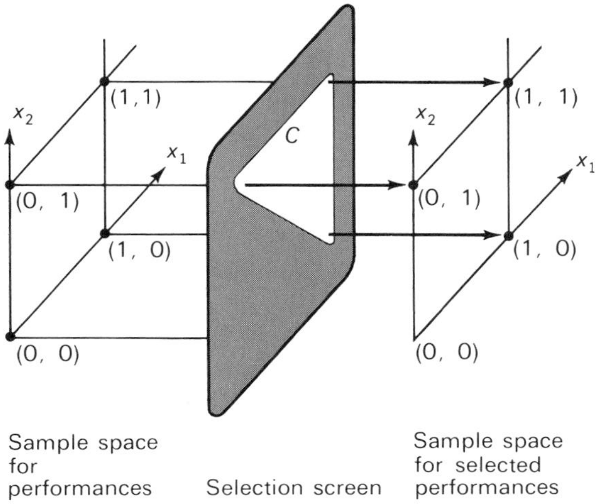

FIGURE 4.1
For two tosses, selection by the event C, ''at least one head.'' For 1000 initial performances, approximately 750 will pass selection. The 750 will separate approximately as 250 at each of the points (0, 1), (1, 0), (1, 1).

tion; there will be approximately 250 for each of these outcomes. The performances yielding (0, 0) will not pass selection; there will be approximately 250 of these rejected outcomes. Among selected performances we have the proportions

$$250/750 \qquad 250/750 \qquad 250/750$$

for the three possible outcomes

$$(0, 1) \qquad (1, 0) \qquad (1, 1).$$

Accordingly, we write

$$P((0, 1):C) = 1/3, \qquad P((1, 0):C) = 1/3, \qquad P((1, 1):C) = 1/3$$

for the estimated probabilities given the event C. These are called "conditional probabilities given the event C" and are interpreted in terms of performances of the system selected on the basis of the event C.

B DEFINITION OF CONDITIONAL PROBABILITY

Now consider in general a random system with sample space $\mathcal{S}$, with class $\mathcal{A}$ of events, and with probability measure P. And suppose that performances of the system are selected in accord with an event C; thus any performance yielding an outcome in C is reported to an investigator and the remaining outcomes in C^c are ignored. We investigate probabilities that describe selected performances; see Figure 4.2.

Consider a large number N of performances of the basic system. The number of outcomes that pass selection is approximately $NP(C)$. Suppose now that we are interested in some event A. The number of outcomes that pass selection and yield the event A is approximately $NP(AC)$. Thus for the investigator examining selected performances, the proportion or estimated probability for the event A is

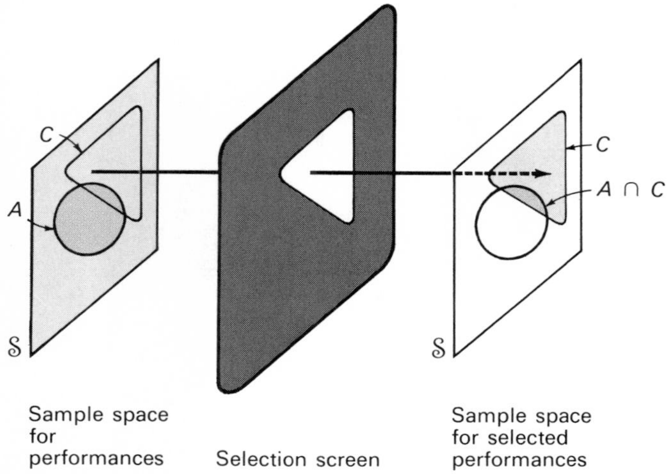

FIGURE 4.2
Selection in accord with event C. The proportion for A given selection is $P(A:C) = P(AC)/P(C)$.

$$\frac{NP(AC)}{NP(C)} = \frac{P(AC)}{P(C)}.$$

This leads us to the following formal definition:

DEFINITION 1

The **conditional probability** of A given C is

(1) $$P_C(A) = P(A:C) = \frac{P(AC)}{P(C)}$$

provided that $P(C) \neq 0$.

The interpretation of $P_C(A)$ is in terms of the long-run proportions for the event A among outcomes that have passed selection based on C.

Consider the construction of a model for a random system with selection. Let $(\mathcal{S}, \mathcal{A}, P)$ be the model for the unselected system, and let C be the event for selection. For the selected system it is convenient to use the same sample space $\mathcal{S}$ even though certain sample points cannot occur. And it is convenient to use the same class $\mathcal{A}$ of events. The function P_C given by Definition 1 gives the probability for any event A given the condition C; it is straightforward to show that P_C is a probability measure (Problem 7). Thus we obtain the model

$$(\mathcal{S}, \mathcal{A}, P_C)$$

describing the selected system. Informally, we shall refer to the *conditional distribution* P_C for the C selected performances of the system.

A conditional distribution P_C is used to describe a partially concealed realization from a random system, the available information being represented by C. Some examples have been mentioned earlier in this section; further examples are given in Section C. A conditional distribution is the appropriate distribution for a great many applications.

The definition of conditional probability gives us an alternative condition for independence:

PROPOSITION 2

If $P(C) \neq 0$, then the events A and C are statistically independent if and only if the conditional probability of A given C is equal to the marginal probability of A, that is, $P(A:C) = P(A)$.

The proof follows immediately from Definition 3.3.1.

C EXAMPLES

Consider some examples with coins, urns, and cards, simple examples that focus clearly on certain important aspects of conditional probability.

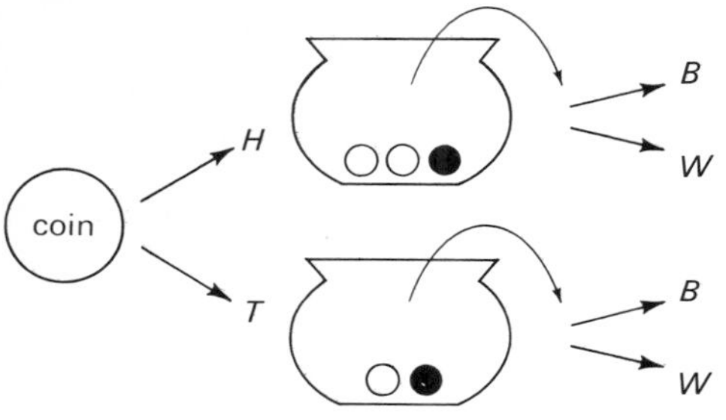

FIGURE 4.3
A coin is tossed. If H, then go to H urn, which has 2 W and 1 B balls; draw one ball. If T, then go to T urn, which has 1 W and 1 B balls; draw one ball.

EXAMPLE 1

A priori probability: Suppose that a technician tosses a symmetric coin. If the coin shows H, he mixes the balls in an H urn containing two white balls and one black ball, and then draws one ball and notes its color. If the coin shows T, he mixes the balls in a T urn containing one white ball and one black ball, and then draws one ball and notes its color. An outcome from these operations gives the face for the coin and the color for the ball. This is presented schematically in Figure 4.3 and as a probability tree in Figure 4.4.

As a first situation, suppose that the technician tosses the coin and reports the result to the investigator before drawing from the appropriate urn. The investigator receives the report H and is concerned with the probability for B, ''black ball.'' The probability refers to a future draw from the H urn and is

$$P(B : H) = \frac{1/6}{3/6} = \frac{1}{3}.$$

This is an a priori probability describing an outcome in the future. Note that the selected performances all involve drawing a single ball from the H urn, and thus refer to an ordinary performance of a component system; the probability can be calculated directly from the component system.

EXAMPLE 2

A posteriori probability: Consider a second situation for the preceding coin–urn example. Suppose that the technician tosses the coin, draws from the urn, and then reports the face showing on the coin but nothing concerning the color of the ball. The investigator receives the report H and is concerned with the probability for B, ''black ball.'' The probability refers to a realization of the full system. For the selected performances, however, we are concerned essentially with the same H urn as in Example 1; we have

$$P(B : H) = \frac{1/6}{3/6} = \frac{1}{3}.$$

This is an a posteriori probability describing a realized outcome that is partially concealed from the investigator. The interpretation is in terms of a large number of

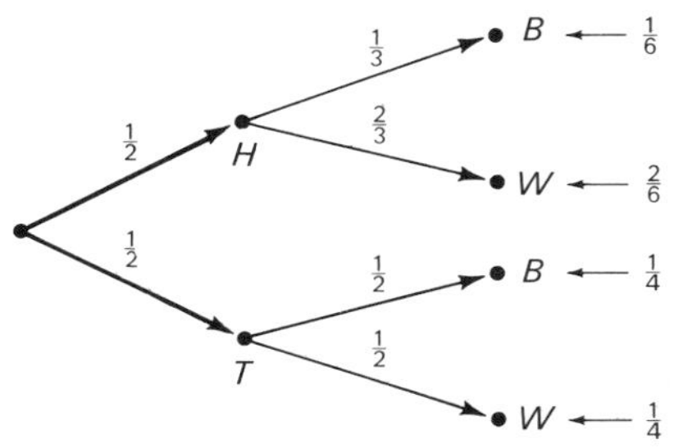

FIGURE 4.4
A coin is tossed and a ball is drawn accordingly from the H urn or the T urn. Branch probabilities and terminal probabilities are recorded.

observations on the selected system, selected to agree with the information available concerning the realization.

EXAMPLE 3

A posteriori probability for an antecedent event: Consider a third situation for the coin–urn example. Suppose that the technician tosses the coin, draws from the urn, and then reports the color of the ball but nothing concerning the face on the coin. The investigator receives the report B, "black ball," and is concerned with the probability for H. The probability refers to a realized toss of the coin, indeed to a coin toss that was antecedent in time to the drawing of the black ball:

$$P(H : B) = \frac{1/6}{1/6 + 1/4} = \frac{2}{5}.$$

This is also called an a posteriori probability. Indeed, it is a special a posteriori probability, a probability for an event H that would be antecedent in time to the event B representing the available information.

EXAMPLE 4

Card distributions at bridge: A deck of 52 playing cards is dealt so that 13 cards go to each of E, S, W, and N. Suppose that N can see the cards in his own hand and also those in the hand of S. And suppose that he observes 10 spades and is interested in the probability that the remaining 3 spades are split 2 to E and 1 to W. Let C be the event 10 spades in the combined hands of N and S, or equivalently 3 spades in the hands of E and W. And let A be the event 2 spades in the hand of E and 1 spade in the hand of W. Then

$$P(A : C) = \frac{P(AC)}{P(C)} = \frac{\binom{13}{2}\binom{39}{11} \cdot \binom{11}{1}\binom{28}{12} \Big/ \binom{52}{13}\binom{39}{13}}{\binom{13}{3}\binom{39}{23} \Big/ \binom{52}{26}};$$

this follows the pattern of calculation in Section 3.4: for the numerator, a set of 13 for E and then a set of 13 for W; for the denominator, a set of 26 for E and W.

For applications we need to be very careful when considering such a probability. Perhaps certain bidding has taken place, bidding that to some degree partially eliminates possibilities in C; for example, certain card distributions might require a different pattern of play and we would not be faced with a concern for the spade split. The probability as just calculated could then be inappropriate.

D MULTIPLICATION FORMULAS

So far we have been thinking of conditional probability as giving a description of a system that has been selected, or giving a description of a realization from such a system where there is partial information concerning the realization. Conditional probability, however, can also be used for computational purposes.

The definition of conditional probability can be rewritten as a *multiplication formula,*

$$(2) \qquad P(AC) = P(C)P(A:C) \qquad \text{if } P(C) \neq 0.$$

This can be used for calculating probabilities if we have available some direct method for calculating the conditional probability; symmetry can often be used for this.

The multiplication formula can be applied repeatedly. For example, with $P(BC) \neq 0$ we have

$$(3) \qquad P(ABC) = P(C) \cdot \frac{P(BC)}{P(C)} \cdot \frac{P(ABC)}{P(BC)} = P(C)P(B:C)P(A:BC).$$

Thus if appropriate conditional probabilities are available, say from symmetry, then the probability for a compound event can be calculated by multiplying together the appropriate conditional probabilities.

EXAMPLE 5

Randomization: Treatments T_1, T_2, T_3, and T_4 are to be randomly assigned to guinea pigs P_1, P_2, P_3, and P_4. Four cards with designations T_1, T_2, T_3, and T_4 are thoroughly shuffled and dealt one by one to positions P_1, P_2, P_3, and P_4; the treatments are given accordingly to the animals. Find the probability that the treatments are assigned T_1, T_2, and T_3, respectively, to P_1, P_2, and P_3. The probability that T_1 goes to P_1 is $1/4$, as T_1 is one of 4 symmetrical possibilities for P_1; the probability that T_2 goes to P_2 given T_1 to P_1 is $1/3$, as T_2 is one of the 3 remaining symmetrical possibilities for P_2. The probability that T_3 goes to P_3 given T_1, T_2, respectively, to P_1, P_2 is $1/2$, as T_3 is one of 2 remaining symmetrical possibilities for P_3. Thus the required probability is

$$(1/4) \cdot (1/3) \cdot (1/2) = 1/24.$$

This particular probability, however, is, of course, easily calculated directly.

EXAMPLE 6

Industrial acceptance sampling: Consider a lot containing $N = 10$ articles of which $D = 3$ are defective and $N - D = 7$ are good. Suppose that the articles are thoroughly mixed and then tested one by one. Find the probability that the last defective is found on the fifth test. Let C be the event "2 defectives in the first 4 tested," and A be "a defective on the fifth test." The required probability is then

$$P(AC) = P(C) \cdot P(A:C) = \frac{\binom{3}{2}\binom{7}{2}}{\binom{10}{4}} \cdot \frac{1}{6}.$$

The first factor is calculated from symmetry for a set of 4; the second factor from symmetry for a set of 1 but from the lot as modified by the first 4 removed.

E MARGINAL PROBABILITIES AND INDEPENDENCE

Sometimes conditional probabilities can be used in a somewhat different way to calculate a marginal probability, say $P(A)$. Suppose that we have available the conditional probabilities $P(A:C)$ and $P(A:C^c)$ and we find that

(4) $\qquad P(A:C) = P(A:C^c).$

Then

(5) $$\begin{aligned} P(A) &= P(AC) + P(AC^c) = P(C)P(A:C) + P(C^c)P(A:C^c) \\ &= P(A:C)(P(C) + P(C^c)) = P(A:C); \end{aligned}$$

thus we have calculated the probability $P(A)$.

Note: If equation (4) holds, then we can deduce (5) and obtain by Proposition 2 the independence of A and C. Conversely, if A and C are independent [with $P(C) \neq 0 \neq P(C^c)$], then Proposition 2 gives equation (4). This proves the essentials of the following proposition, presenting independence in terms of conditional probability.

PROPOSITION 3

A and C are statistically independent if and only if

$$P(A:C) = P(A:C^c)$$

wherever defined; that is, the conditional probability does not depend on the condition.

EXAMPLE 7

Coin and two urns: A symmetric coin is tossed. If the coin shows H, "heads," then a ball is drawn from an H urn containing 1 black ball and 2 white balls; if the coin shows T, "tails," then a ball is drawn from a T urn containing 2 black balls and 4 white balls. Clearly from symmetry we have

$$P(B:H) = 1/3, \qquad P(B:T) = 2/6 = 1/3.$$

Then by the preceding proposition we have that B and H are statistically independent and that

$$P(B) = P(B:H) = 1/3.$$

More generally suppose that we have a partition $\{C_1, \ldots, C_r\}$ of the sample space $\mathcal{S}$ and find that

(6) $P(A : C_i)$ is functionally independent of $i = 1, \ldots, r$.

Then

$$(7) \quad P(A) = \sum_1^r P(AC_i) = \sum_1^r P(C_i)P(A : C_i)$$

$$= P(A : C_1) \sum_1^r P(C_i) = P(A : C_1);$$

thus we have calculated the probability $P(A)$.

Note: If equation (6) holds, then we can deduce (7) and then obtain by Proposition 2 the independence of A and each C_i in the partition; it follows that A is independent of the σ-algebra generated by the C's. Conversely, if A is independent of the C_i's, then the conditional probabilities $P(A : C_i)$ do not depend on C_i. This proves the following:

PROPOSITION 4

A is statistically independent of the partition $\{C_1, \ldots, C_r\}$ *if and only if*

$P(A : C_i)$

wherever defined is functionally independent of i; that is, the conditional probability is functionally independent of the condition.

F PROBABILITIES FOR ANTECEDENT EVENTS

Example 3 records a simple example of an a posteriori probability for an antecedent event. Consider this more generally. Suppose that we have a system with two stages in sequence. For the first stage, let $H_1, \ldots, H_r$ be disjoint events with probabilities $P(H_1), \ldots, P(H_r)$. And for the second stage, let A be the observable event and suppose that its probability $P(A : H_i)$ depends on the outcome H_i from the first stage; see Figure 4.5.

Now suppose that we have observed the outcome from the second stage, say the occurrence of A; and suppose that we are interested in probabilities concerning antecedent events from the first stage. From Definition 1, we have

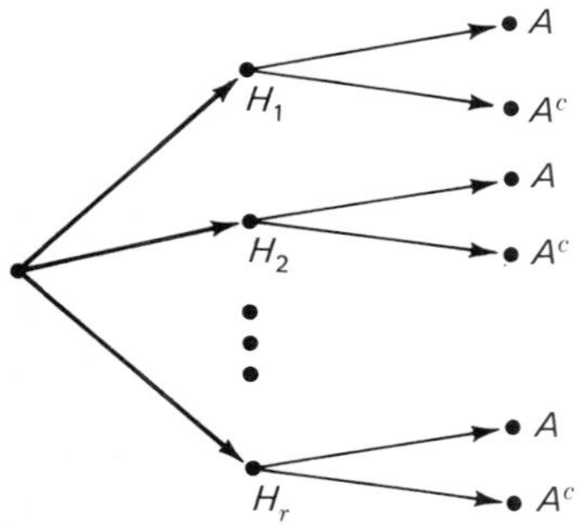

FIGURE 4.5
First stage, giving $H_1, \ldots, H_r$; and second stage, giving A or A^c.

$$(8) \qquad P(H_i : A) = \frac{P(H_i A)}{P(A)} = \frac{P(H_i)P(A : H_i)}{\sum_1^r P(H_j)P(A : H_j)}.$$

This is the ordinary expression for conditional probability, but note that we are expressing the probability for an antecedent event, stage 2 to stage 1, in terms of the forward probabilities, stage 1 to stage 2. Formula (8) is sometimes called Bayes formula. For a simple example, see the earlier Example 3.

Now consider a system that can be in any one of several conditions $H_1, \ldots, H_r$. And suppose that the system is performed giving an outcome characterized by the event A. The investigator, of course, is interested in which condition H_i applies to the system. Certainly he can examine the probability for what has occurred; the probability $P(A : H_i)$ as a function of the condition H_i; call it the *likelihood* of H_i. Some statistical analysis from this point of view is developed in Chapter 8.

Some statisticians feel that it is appropriate to attach numbers $P(H_1), \ldots, P(H_r)$ to the possible conditions $H_1, \ldots, H_r$; the numbers are chosen to represent in some way feelings concerning the plausibility of the various $H_1, \ldots, H_r$. They then treat these numbers as probabilities for a hypothetical initial system and call them *prior probabilities*. The use of formula (8) then gives the *posterior probabilities*

$$P(H_i : A) = \frac{P(H_i)P(A : H_i)}{\sum_1^r P(H_j)P(A : H_j)}, \qquad i = 1, \ldots, r.$$

These "probabilities" have meaning at most to the degree that it is reasonable or proper to describe feelings by means of numbers $P(H_1), \ldots, P(H_r)$. This *Bayesian analysis* will be discussed briefly in Supplement Section 9.6.

G CAUTIONS WITH CONDITIONAL PROBABILITY

In Example 4 we indicated briefly certain hazards that can arise in the application of conditional probabilities. We chose a card-dealing example for this as the basic calculations are easy and yet the potential for complications in applications is very large. The example involved a distribution of 52 cards into bridge hands, the observation of hands N and S, and the question of how cards are split between E and W.

The picture becomes more complicated and indeed realistic if we think of initial bidding and then play until a small number of cards remains in each hand. A probability calculation can be made as in Example 4; it would typically be based on symmetry and on the cards actually observed. There are a variety of other factors, however: the pattern of bidding can partially or completely eliminate some of the possibilities in the formal condition; the pattern of play can also partially or completely eliminate some of the possibilities; and indeed the players can use randomness in their choice of play. In short, there can be many complications that can invalidate the simplistic calculation based on symmetry among the various possibilities. However, if the conventions of bidding and play and any randomness used therein are clearly specified, then a conditional probability can be calculated from the corresponding, more complex model.

The interpretation of conditional probability is in terms of a selected system. This interpretation, in turn, provides the key to the proper application of the concept. Three conditions are sufficient for its application to a partially concealed realization of a random system.

The investigator receives information concerning the performance of the system; let C be the set of outcomes for which the information is true. The investigator must know that:

1. *One element of C has occurred.* In effect, this says that he knows that the information concerning the performance is true.

The investigator must also know something about the source of the information concerning the realization, that is, how it was generated and how it was selected for transmission to the investigator.

2. *Each element in C could have occurred.* In effect, this says that there must be no explicit selection: all the apparent possibilities in C must be real possibilities and not excluded by the way the information was generated or transmitted to the investigator.

And the generation and transmission of the information must be such that an identical situation would be obtained for each possibility in C.

3. *Each element in C would lead to the same information.* In effect, this says that there must be no implicit selection: the full information to the investigator must be identical to the information that the outcome is in C.

These conditions require that the system and the generation of the information be equivalent to a selected system as described in Section B.

H EXERCISES

A technician tosses a symmetrical copper coin and a symmetrical steel coin and then reports to an investigator, "the number of heads is at least 1." Determine the probability that both are heads if the technician's reporting pattern is as follows:

1 The technician observes both coins and reports "all tails" if the number of heads is zero; reports "the number of heads is at least 1" if the number of heads is 1 or 2.

2 The technician observes both coins and reports "all tails" if the number of heads is zero; reports "the number of heads is at least 1" if the number of heads is 1; reports "all heads" if the number of heads is 2.

3 The technician observes the copper coin and reports "the number of heads is at least 1" if the copper coin shows heads; reports "the number of tails is at least 1" if the copper coin shows tails.

I PROBLEMS

4 A bridge player finds exactly 9 spades in his own and partner's hands. Give a combinatorial expression for the probability that the remaining 4 spades split 4–0 in the hands left–right; split x–y in the hands left–right where $x + y = 4$.

5 A bridge player finds exactly 8 diamonds in his hand. Give an expression for the probability that his partner has 0 diamonds; at least 1 diamond; and at least 2 diamonds.

6 From family background it is known that with probability 1/2 a woman is a carrier of hemophilia. If she is a carrier, then with probability 1/2 a male offspring has the disease; transmission is independent for different offspring. If she is not a carrier, then a male offspring does not inherit the disease. Determine the probability that she is a carrier if her first son is normal; and if her first two sons are normal.

7 Show that P_C in Definition 1 is a probability measure.

8 A family has two children. It is reported that there is at least one boy. Under three reporting patterns as indicated by Exercises 1, 2, and 3, with "heads" replaced by "boys," determine the probability that both children are boys.

9 As a first approximation, assume that birthdays are uniformly distributed through a year of 365 days. Give an expression for the probability that n people sampled at random have different birthdays. Obtain a rough approximation for the probability. Find the smallest n for which the probability is less than 1/2.

10 The student organizer wants to choose a class representative for a class of 25. He thoroughly shuffles a deck of 52 cards and then proceeds through the class dealing one card to each student. The student receiving the first ace becomes the class representative. Give an expression for the probability that
  (a) The first student becomes class representative on the first run through the class.
  (b) The ith student in the class becomes class representative on the first run through the class.
  (c) No class representative is chosen on the first run through the class.
  (d) The first student becomes class representative.

11 In the game of craps, one of the players throws a pair of symmetric dice. On the first throw he wins if he gets 7 or 11 points and he loses if he gets 2, 3, or 12 points. Otherwise, he continues to toss until he obtains either the original number of points or 7 points. If he duplicates his original points, he wins; if he gets a 7, he loses. Find the probability that the player with the dice wins.

4.2

CONDITIONAL PROBABILITY ON THE PLANE

We have noted before the importance of real spaces for probability theory. In this section we examine conditional probability on the plane and more generally on real spaces.

A CONDITIONAL PROBABILITY FUNCTION

Consider a distribution for $(x_1, \ldots, x_k)$ on R^k with probability function p. And suppose that we want the conditional distribution given that a certain function Y takes the value y. Then directly from Definition 4.1.1 we obtain

$$(1) \qquad p(x_1, \ldots, x_k : Y = y) = \frac{p(x_1, \ldots, x_k)}{p_Y(y)} \qquad \text{if } Y(x_1, \ldots, x_k) = y$$

and zero otherwise. Note that the denominator is purely a norming constant that scales up in value the original probabilities along the contour $Y(x_1, \ldots, x_k) = y$; thus the relative conditional probabilities for points on the contour are just given by the original probability function. For example, in a safety investigation where x_i is the annual total number of accidents in plant i, we could be interested in the conditional distribution of $(x_1, \ldots, x_4)$ *given* the annual total accidents $y = x_1 + \cdots + x_4$ for a company.

EXAMPLE 1

Conditioning two Poissons: Consider a Poisson(λ_1) distribution for y_1 and an independent Poisson(λ_2) distribution for y_2. Then by Section 3.3E the joint probability function on the plane is given by

$$p(y_1, y_2) = \frac{\lambda_1^{y_1} e^{-\lambda_1}}{y_1!} \frac{\lambda_2^{y_2} e^{-\lambda_2}}{y_2!} \qquad \text{if } y_1, y_2 = 0, 1, 2, \ldots$$

and zero otherwise. And suppose that we want the conditional distribution given the value of the total $T(y_1, y_2) = y_1 + y_2 = t$. The marginal probability function for T is available from formula (3.2.9),

$$\begin{aligned} P_T(t) &= \sum_{y_2} \frac{\lambda_1^{t-y_2} e^{-\lambda_1}}{(t-y_2)!} \frac{\lambda_2^{y_2} e^{-\lambda_2}}{y_2!} \\ &= \frac{(\lambda_1 + \lambda_2)^t e^{-\lambda_1 - \lambda_2}}{t!} \sum_{y_2=0}^{t} \frac{t!}{y_2!(t-y_2)!} \left(\frac{\lambda_2}{\lambda_1 + \lambda_2}\right)^{y_2} \left(\frac{\lambda_1}{\lambda_1 + \lambda_2}\right)^{t-y_2} \\ &= \frac{(\lambda_1 + \lambda_2)^t e^{-\lambda_1 - \lambda_2}}{t!}, \end{aligned}$$

where a sum of binomial probabilities [formula (3.5.1)] is 1; note that T is Poisson$(\lambda_1 + \lambda_2)$. Then by division using formula (1) we obtain the conditional distribution given the total $y_1 + y_2 = t$:

$$(2) \qquad \begin{aligned} p(y_1, y_2 : t) &= \frac{\lambda_1^{y_1} e^{-\lambda_1} \lambda_2^{y_2} e^{-\lambda_2} / y_1! y_2!}{(\lambda_1 + \lambda_2)^t e^{-\lambda_1 - \lambda_2} / t!} & \text{if } & y_1, y_2 = 0, 1, 2, \ldots, \\ & & & y_1 + y_2 = t, \\ &= \binom{t}{y_2} \left(\frac{\lambda_2}{\lambda_1 + \lambda_2}\right)^{y_2} \left(\frac{\lambda_1}{\lambda_1 + \lambda_2}\right)^{t-y_2} & & y_2 = 0, 1, \ldots, t, \\ & & & y_1 + y_2 = t, \end{aligned}$$

and zero otherwise. Note that this distribution along a possible contour $y_1 + y_2 = t$ is binomial$(t, \lambda_2/(\lambda_1 + \lambda_2))$ in terms of y_2, and similarly is binomial $(t, \lambda_1/(\lambda_1 + \lambda_2))$ in terms of y_1; see Figure 4.6. Thus if we are counting radioactive particles (with a Poisson distribution) over two time intervals and obtain a particular total count, then the events segregate between the two time intervals in accord

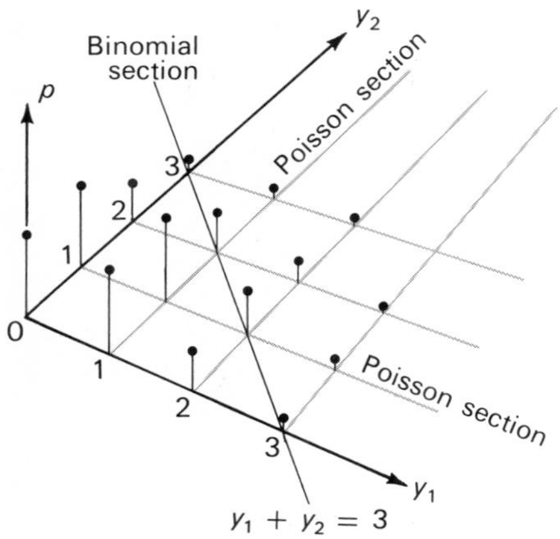

FIGURE 4.6
Poisson(λ_1) for y_1; and independent Poisson(λ_2) for y_2. The $y_1 + y_2 = 3$ section has the binomial$(t, \lambda_1/(\lambda_1 + \lambda_2))$ distribution for y_1. Illustration has $\lambda_1 = \lambda_2 = 1$.

with the binomial distribution; this conditional distribution can be useful for making statistical tests for the equality of λ_1 and λ_2 in an application.

Sometimes we are interested in very simple functions—coordinate projection functions.

Consider a discrete distribution on the plane R^2 with probability function p. And suppose that we want the conditional distribution for the second coordinate y given the value of the first coordinate x. A condition that specifies the value of the first coordinate x is a condition that specifies the value of the *projection function* onto the first axis; see Figure 4.7. Note that the second coordinate conveniently labels the points on the resulting x section.

Consider a first-coordinate value x having $p_1(x) \neq 0$. Then the *conditional probability function* for y given x is obtained directly from Definition 4.1.1:

$$(3) \qquad p_2(y:x) = \frac{p(x, y)}{p_1(x)}.$$

Note the form of $p_2(y:x)$: as a function of y it comes entirely from the numerator $p(x, y)$; the denominator is present only as a norming constant, so summation over y gives unity. For example, we might be interested in the distribution of the number of white blood cells on a slide given the number of red blood cells on the slide.

The definition (3) for $p_2(y:x)$ can be rearranged to give the multiplication formula

$$(4) \qquad p(x, y) = p_1(x)p_2(y:x),$$

where the second factor is undefined for x values having $p_1(x) = 0$.

The preceding multiplication formula, together with Exercise 3.3.2, proves the following proposition:

PROPOSITION 1

For the discrete case, x and y are statistically independent if and only if $p_2(y:x)$, where defined, is functionally independent of x.

Note the simple consequence: if $p_2(y:x)$ does not depend on x, then $p_2(y:x)$ is also the *marginal* probability function for y.

In a similar manner we can define the conditional probability function for x given y:

$$(5) \qquad p_1(x:y) = \frac{p(x, y)}{p_2(y)}$$

provided that $p_2(y) \neq 0$. And if the probability on the plane R^2 has come from some antecedent space, then the conditional probability function for Y given X can be presented as

$$p_{Y:X}(y:x) = \frac{p_{XY}(x, y)}{p_X(x)}$$

provided that $p_X(x) \neq 0$.

Analogous formulas exist for a distribution on R^k with probability function p. For example, the conditional probability function for $(y_4, \ldots, y_k)$ given (y_1, y_2, y_3) is

(6) $$p_{4\cdots k}(y_4, \ldots, y_k : y_1, y_2, y_3) = \frac{p(y_1, \ldots, y_k)}{p_{123}(y_1, y_2, y_3)}$$

provided that the denominator is nonzero. Note that the numerator gives the functional dependence and the denominator provides just the norming constant.

EXAMPLE 2

Conditioning the multinomial: First consider the multinomial Example 3.5.1: tossing a symmetric die with one face marked 1, two faces marked 2, and three faces marked 3; record (y_1, y_2, y_3), where y_i is the number of i faces in 5 tosses. The probability function can be expressed in terms of (y_1, y_2) since $y_3 = 5 - y_1 - y_2$; the values of $7776p_Y$ as retabulated from Example 3.5.1 are recorded in Table 4.1.

TABLE 4.1
$7776p_Y$ Values

	y_1					
y_2	0	1	2	3	4	5
0	243	405	270	90	15	1
1	810	1080	540	120	10	
2	1080	1080	360	40		
3	720	480	80			
4	240	80				
5	32					
			1250			

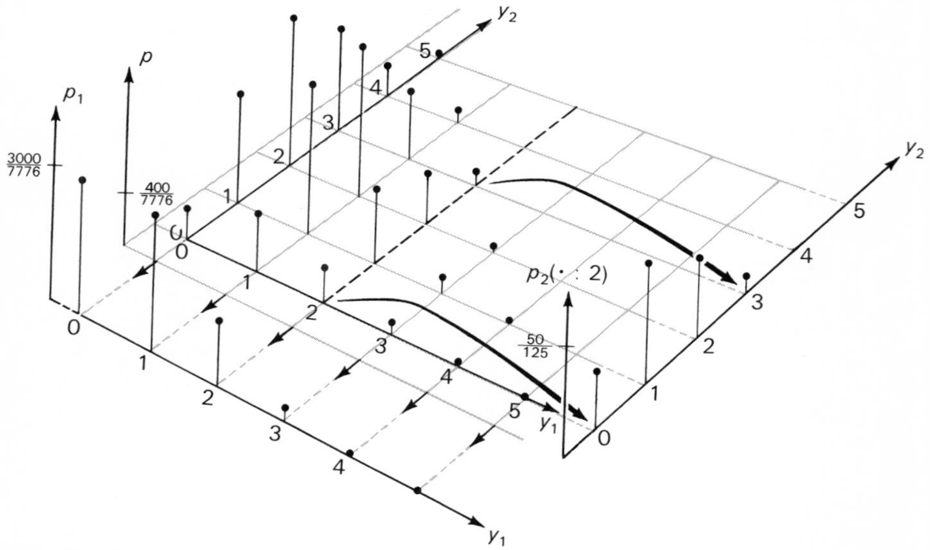

FIGURE 4.7
Multinomial(5, 1/6, 2/6, 3/6). The marginal for y_1 is binomial(5, 1/6). The conditional for y_2 given $y_1 = 2$ is binomial(3, 2/5).

Now suppose that we want the conditional distribution given that $y_1 = 2$. The marginal probability for $y_1 = 2$ is obtained by summing over y_2; we get 1250/7776. The conditional distribution for y_2, given $y_1 = 2$, is then obtained by division from values in the $y_1 = 2$ column; we obtain

y_2	0	1	2	3
$p_2(y_2 : 2)$	$\frac{27}{125}$	$\frac{54}{125}$	$\frac{36}{125}$	$\frac{8}{125}$

which is plotted in Figure 4.7; this is binomial(3, 2/5).

Now in general consider the multinomial(n, p_1, p_2, p_3) distribution for (y_1, y_2, y_3); this is effectively a distribution for (y_1, y_2) in the plane, since $y_3 = n - y_1 - y_2$. The probability function is

$$p(y_1, y_2) = \frac{n!}{y_1!\, y_2!\, (n - y_1 - y_2)!} p_1^{y_1} p_2^{y_2} p_3^{n-y_1-y_2}$$

on the points $y_i = 0, 1, \ldots$ with $y_1 + y_2 \leq n$. The marginal probability function for y_1 can be obtained by summing out y_2:

$$(7) \qquad \begin{aligned} p_1(y_1) &= \sum_{y_2} p(y_1, y_2) \\ &= \frac{n!}{y_1!\, (n - y_1)!} p_1^{y_1} (p_2 + p_3)^{n-y_1} \\ &\quad \cdot \sum \frac{(n - y_1)!}{y_2!\, (n - y_1 - y_2)!} \left(\frac{p_2}{p_2 + p_3}\right)^{y_2} \left(\frac{p_3}{p_2 + p_3}\right)^{n-y_1-y_2} \\ &= \binom{n}{y_1} p_1^{y_1} (p_2 + p_3)^{n-y_1} \end{aligned}$$

where a sum of binomial probabilities [formula (3.5.1)] is 1; note that the marginal distribution of y_1 is binomial(n, p_1). The conditional distribution for y_2 given y_1 can then be calculated from formula (3):

$$(8) \qquad \begin{aligned} p_2(y_2 : y_1) &= \frac{\binom{n}{y_1\ y_2\ y_3} p_1^{y_1} p_2^{y_2} p_3^{n-y_1-y_2}}{\binom{n}{y_1} p_1^{y_1} (p_2 + p_3)^{n-y_1}} \\ &= \binom{n - y_1}{y_2} \left(\frac{p_2}{p_2 + p_3}\right)^{y_2} \left(\frac{p_3}{p_2 + p_3}\right)^{n-y_1-y_2} \end{aligned}$$

for $y_2 = 0, 1, \ldots, n - y_1$ and zero elsewhere; note that this conditional distribution for y_2 is binomial($n - y_1, p_2/(p_2 + p_3)$). For n repetitions where y_1, y_2, and y_3 count the number of events B_1, B_2, and B_3, we see that with y_1 repetitions giving the first kind of event, there are $n - y_1$ repetitions that can be either the second or third kind of event; the formula above shows that the two kinds of events then segregate according to the binomial distribution with probabilities $p_2/(p_2 + p_3)$ and $p_3/(p_2 + p_3)$ for the second and third kinds of events.

B CONDITIONAL PROBABILITY DENSITY GIVEN A PROJECTION

Consider an absolutely continuous distribution on the plane R^2 with probability density function f. And suppose that we are interested in the conditional distribution of the second coordinate y given the value of the first coordinate x, for example, say, the distribution for the height of eldest son given the height of the father.

The probability for a particular value x of the first coordinate is of course 0. Thus we cannot directly use the Definition 4.1.1 for conditional probability and must reconsider the selection interpretation discussed in Section 4.1B. If we consider the practical meaning of the first coordinate being equal to x we are led to consider an interval $(x - h, x + h)$ and to contemplate the distribution for the second coordinate y given that the first coordinate lies in the interval $(x - h, x + h)$; see Figure 4.8.

Let G_h be the distribution function for the conditional distribution of y given that x lies in the interval $(x - h, x + h)$. Then from Definition 4.1.1 we obtain

$$G_h(y) = \frac{\int_{x-h}^{x+h} \int_{-\infty}^{y} f(t, u)\, dt\, du}{\int_{x-h}^{x+h} \int_{-\infty}^{\infty} f(t, u)\, dt\, du} = \frac{\int_{x-h}^{x+h} \int_{-\infty}^{y} f(t, u)\, dt\, du/2h}{\int_{x-h}^{x+h} \int_{-\infty}^{\infty} f(t, u)\, dt\, du/2h}.$$

Now consider the limit as $h \to 0$. Both the numerator and denominator produce a derivative with respect to a limit of integration; we obtain

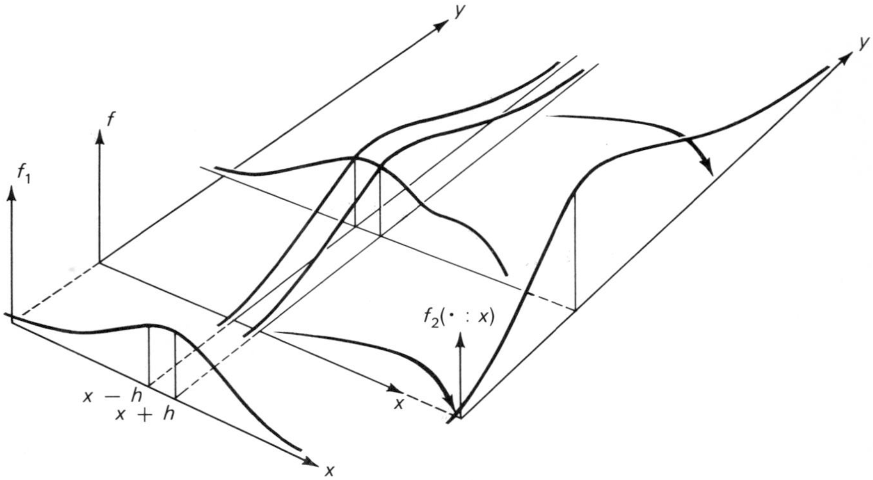

FIGURE 4.8
Joint density f; marginal density f_1 for x; and conditional density $f_2(\cdot : x)$ for y given x.

(9) $$G(y) = \lim_{h \to 0} G_h(y) = \frac{\int_{-\infty}^{y} f(x, u)\, du}{\int_{-\infty}^{\infty} f(x, u)\, du} = \frac{\int_{-\infty}^{y} f(x, u)\, du}{f_1(x)}$$

This is undefined if $f_1(x) = 0$; and it is undefined more generally if the marginal density is not given by a limit (but such points form a set with length measure zero). Now by differentiation we obtain the conditional probability density function for y given x:

(10) $$f_2(y : x) = \frac{f(x, y)}{f_1(x)}.$$

Note the form of $f_2(y : x)$: as a function of y it comes entirely from the numerator $f(x, y)$; the denominator is present only as a norming constant. Thus the form of the conditional distribution is given by the x section through the joint density function.

Formula (10) for $f_2(y : x)$ can be rearranged to give the multiplication formula

(11) $$f(x, y) = f_1(x) f_2(y : x),$$

where the second factor may be undefined for points x in a set having probability zero.

The preceding multiplication formula, together with Exercise 3.3.3, proves the following proposition:

PROPOSITION 2

x and y are statistically independent if and only if $f_2(y : x)$ is functionally independent of x (except possibly on a set having probability zero).

Note the simple consequence: if $f_2(y : x)$ does not depend on x, then $f_2(y : x)$ is also the marginal density function for y.

In a similar manner we can define the conditional probability density function for x given y:

(12) $$f_1(x : y) = \frac{f(x, y)}{f_2(y)}.$$

And if the probability on R^2 has come from some antecedent space, then the conditional probability density function for Y given X can be presented as

(13) $$f_{Y:X}(y : x) = \frac{f_{XY}(x, y)}{f_X(x)}$$

provided that $f_X(x) \neq 0$.

Analogous formulas exist for a distribution on R^k with probability density function f. For example, the conditional probability density function for $(y_4, \ldots, y_k)$ given (y_1, y_2, y_3) is

(14) $$f_{4\ldots k}(y_4, \ldots, y_k : y_1, y_2, y_3) = \frac{f(y_1, \ldots, y_k)}{f_{123}(y_1, y_2, y_3)}.$$

A conditional density function can be used for calculations. For example, to find the probability in a set A (Figure 4.9) we can write

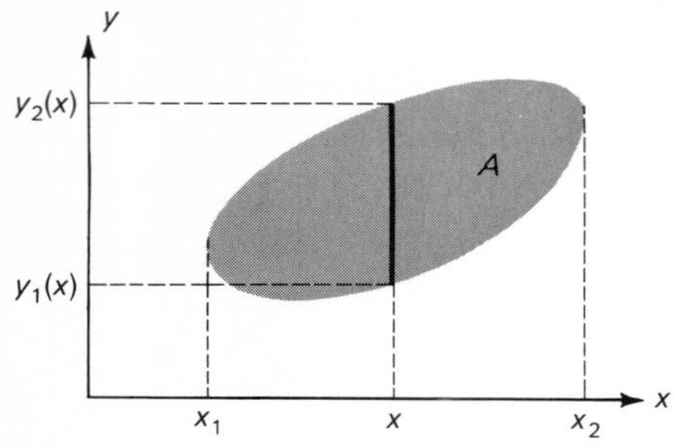

FIGURE 4.9
Set A and its x section $(y_1(x), y_2(x))$.

(15) $$P(A) = \int_{x_1}^{x_2} \left(\int_{y_1(x)}^{y_2(x)} f_2(y:x)\, dy \right) f_1(x)\, dx,$$

where the conditional probability

$$\int_{y_1(x)}^{y_2(x)} f_2(y:x)\, dy$$

is multiplied by the marginal probability $f_1(x)\,dx$ and integrated. Note that the expression (15) is just a rearrangement of the standard expression

$$P(A) = \int_{x_1}^{x_2} \int_{y_1(x)}^{y_2(x)} f(x, y)\, dx\, dy.$$

Perhaps the most convenient way to remember the conditional density is as part of a probability differential

(16) $$f(x, y)\, dx\, dy = f_1(x)\, dx \cdot f_2(y:x)\, dy,$$

in which the probability $f_1(x)\,dx$ for an increment for the first coordinate is multiplied by the conditional probability $f_2(y:x)\,dy$ for an increment for the second coordinate.

EXAMPLE 3

Bivariate distribution: Consider further the bivariate Example 3.2.6 with density function

$$f(x, y) = 6x \qquad \text{if } 0 < x < 1,\ 0 < y < 1 - x,$$

and zero otherwise. The marginal density for x was calculated in Example 3.2.6:

$$f_1(x) = 6x(1 - x) \qquad \text{if } 0 < x < 1$$

and zero otherwise. The conditional density for y can then be obtained from formula (10):

(17) $$f_2(y:x) = (1 - x)^{-1} \qquad \text{if } 0 < y < 1 - x$$

and zero otherwise. These are sketched in Figure 4.10.

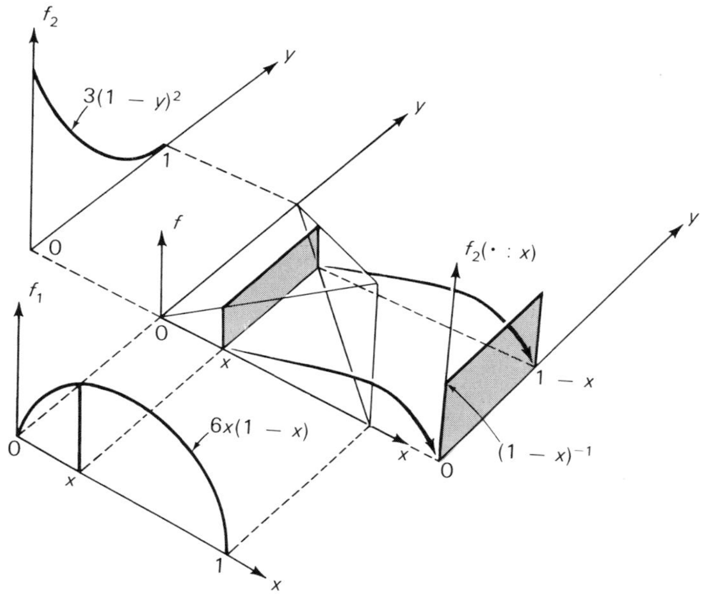

FIGURE 4.10
Conditional density $f_2(\cdot : x)$ for y given x.

Note that the initial density $6x$ *apparently* factors into a function $6x$ of the first variable and a function 1 of the second variable. It is *not,* however, a factorization of the function f which is defined on R^2; in particular, the region of positive density is not a product set. Thus x and y are not statistically independent; certainly this is evident from the conditional density $f_2(y : x)$, which is uniform $(0, 1-x)$ and *does* depend on x.

The marginal density for y was calculated in Example 3.2.6:

$$f_2(y) = 3(1-y)^2 \qquad \text{if } 0 < y < 1$$

and zero otherwise. The conditional density for x can then be obtained from formula (10):

$$(18) \qquad f_1(x : y) = \frac{2x}{(1-y)^2} \qquad \text{if } 0 < x < 1-y$$

and zero otherwise; this is a triangular distribution of $(0, 1-y)$.

C CONDITIONAL PROBABILITY DENSITY GIVEN A GENERAL FUNCTION

Consider a distribution for $(x_1, \ldots, x_k)$ on R^k with density function g. And suppose that we are interested in the conditional distribution given that a certain function $Y(x_1, \ldots, x_k)$ has the value y; for example, the conditional distribution of the yields $x_1, \ldots, x_7$ on plots $1, \ldots, 7$ given the total yield $x_1 + \cdots + x_7$ for the seven plots. The basic procedure is to make a change of variable on R^k so as to separate

$Y(x_1, \ldots, x_k)$ as one of the new variables and then to use the results appropriate to projections from Section B. We examine the details for the case $k = 2$.

Consider a distribution for (x_1, x_2) on R^2 with density function g. And suppose that we want the conditional distribution given a value for some function Y_1. For most reasonable functions Y_1, we can find a complementing function Y_2 such that the mapping

$$(y_1, y_2) = (Y_1(x_1, x_2), Y_2(x_1, x_2))$$

is one to one and continuously differentiable each way; let

$$(x_1, x_2) = (X_1(y_1, y_2), X_2(y_1, y_2))$$

designate the inverse mapping; see Figure 3.10.

In terms of differentials we can present the distribution for the original variables (x_1, x_2) and for the new variables (y_1, y_2):

$$g(x_1, x_2)\, dx_1\, dx_2 = g(X_1(y_1, y_2), X_2(y_1, y_2)) \left|\frac{\partial X_1, X_2}{\partial y_1, y_2}\right|_+ dy_1\, dy_2.$$

The conditional density for y_2 given y_1 is then available from formula (10) for density given a projection:

$$\frac{g(X_1(y_1, y_2), X_2(y_1, y_2)) \left|\dfrac{\partial X_1, X_2}{\partial y_1, y_2}\right|_+}{\displaystyle\int_{-\infty}^{\infty} g(X_1(y_1, u), X_2(y_1, u)) \left|\dfrac{\partial X_1, X_2}{\partial y_1, u}\right|_+ du}. \tag{19}$$

Note that y_2 labels the points along the y_1 contour on the original space; and note that

$$\left|\frac{\partial X_1, X_2}{\partial y_1, y_2}\right|_+$$

allows for variation in the width between the contours y_1 and $y_1 + dy_1$.

EXAMPLE 4

Symmetric normal distribution on R^2: Consider a sample (x_1, x_2) from the central normal$(0, \sigma)$ distribution, and suppose that we want the conditional distribution given the sample sum $t = T(x_1, x_2) = x_1 + x_2$. Some preimage contours for the sum function are indicated in Figure 4.11; note that they are lines orthogonal to the 1-vector $\mathbf{1} = (1, 1)'$. Thus we are interested in the conditional distribution along a strip such as that indicated in the figure.

For our basic procedure we need a second function whose preimage contours are cross sectional to those just described. A convenient set of contours is given by the lines parallel to the 1-vector and thus orthogonal to the first set of lines; a function that describes these contours is the difference function $x_2 - x_1$.

We now choose simple coordinates that measure Euclidean distance along the two sets of contours:

$$y_1 = \frac{x_1}{\sqrt{2}} + \frac{x_2}{\sqrt{2}},$$

$$y_2 = -\frac{x_1}{\sqrt{2}} + \frac{x_2}{\sqrt{2}},$$

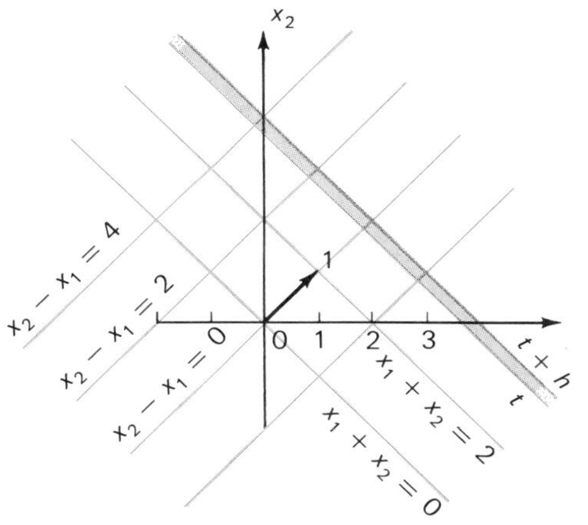

FIGURE 4.11
Preimage contours for the sum function; and preimage contours for the difference function.

or, in matrix form,

$$\begin{pmatrix} y_1 \\ y_2 \end{pmatrix} = \begin{pmatrix} \frac{1}{\sqrt{2}} & \frac{1}{\sqrt{2}} \\ -\frac{1}{\sqrt{2}} & \frac{1}{\sqrt{2}} \end{pmatrix} \begin{pmatrix} x_1 \\ x_2 \end{pmatrix} \quad \text{or} \quad \mathbf{y} = A\mathbf{x}.$$

This transformation is an *orthogonal transformation*. Note that the row vectors in A have unit length and are mutually orthogonal; that is, the inner product matrix $AA' = I$ is the identity matrix; for details see Supplement Section 2.5. (In general, a square matrix A is called orthogonal if it has the property $AA' = I$ or $A'A = I$, and correspondingly the transformation $\mathbf{y} = A\mathbf{x}$ is called "orthogonal"; note that such a transformation does not change the sum of squares

$$\sum y_i^2 = \mathbf{y}'\mathbf{y} = \mathbf{x}'A'A\mathbf{x} = \mathbf{x}'\mathbf{x} = \sum x_i^2;$$

thus Euclidean distance is preserved.) The Jacobian determinant is

$$\left|\frac{\partial \mathbf{y}}{\partial \mathbf{x}}\right| = \begin{vmatrix} \frac{1}{\sqrt{2}} & \frac{1}{\sqrt{2}} \\ -\frac{1}{\sqrt{2}} & \frac{1}{\sqrt{2}} \end{vmatrix}_+ = 1.$$

The initial distribution can now be expressed in terms of the new variables

$$\frac{1}{2\pi\sigma^2} \exp\left\{-\frac{x_1^2 + x_2^2}{2\sigma^2}\right\} dx_1\, dx_2 = \frac{1}{2\pi\sigma^2} \exp\left\{-\frac{y_1^2 + y_2^2}{2\sigma^2}\right\} \cdot 1 \cdot dy_1\, dy_2;$$

hence the distribution of (y_1, y_2) is that of a sample of 2 from the central normal$(0, \sigma)$ distribution. Thus the marginal distribution of y_1 is normal$(0, \sigma)$, and the conditional of y_2 given y_1 is normal$(0, \sigma)$. In conclusion, note that the conditional distribution of $(x_1, x_2) = (y_1/\sqrt{2} - y_2/\sqrt{2}, y_1/\sqrt{2} + y_2/\sqrt{2})$ is normal *on* the contour with the given value for $x_1 + x_2 = \sqrt{2}y_1$; it is located at $(y_1/\sqrt{2}, y_1/\sqrt{2})$ and has scaling σ.

D EXERCISES: PROBABILITY FUNCTIONS

1 Consider the multinomial(5, 1/6, 2/6, 3/6) for (y_1, y_2) as tabulated in Example 2.

(a) Tabulate the conditional distribution of y_2 given $y_1 = 1$; verify that it is binomial (4, 2/5).

(b) Tabulate the conditional distribution of y_2 given $y_1 = 0$; verify that it is binomial (5, 2/5).

2 Automatic transmissions received for repair are unrepairable with probability 1/6, repairable with probability 2/6, and actually in satisfactory condition with probability 3/6; see Example 2. Unrepairable units are quickly identified. A consignment of 5 units contains 1 unrepairable unit; determine the distribution of y_2, the number of units that are repairable.

3 Consider the multivariate hypergeometric(6, 3, 1/6, 2/6, 3/6) in Example 3.5.3; tabulate the probability function in terms of (y_1, y_2), where $y_3 = 3 - y_1 - y_2$.

(a) Tabulate the conditional distribution of y_2 given $y_1 = 1$; verify that it is hypergeometric (5, 2, 2/5).

(b) Tabulate the conditional distribution of y_2 given $y_1 = 0$; verify that it is hypergeometric (5, 3, 2/5).

4 *Sample of outboard motors* (Example 3.5.3). The dealer reports that the sample of 3 contains 1 unrepairable unit; determine the distribution of y_2, the number of repairable units.

5 *Industrial acceptance sampling* (Example 4.1.6). A lot contains $N = 10$ articles of which $D = 3$ are defective and $N - D = 7$ are good. The articles are tested successively. Verify the expression

$$\binom{4}{2}\frac{3^{(3)}7^{(2)}}{10^{(5)}}$$

for the probability that the last defective is found on the fifth test by considering possible sequences of 5 articles for testing.

6 (*continuation*) Verify the expression

$$\frac{\binom{3}{2\ 1\ 0}\binom{7}{2\ 0\ 5}}{\binom{10}{4\ 1\ 5}}$$

for the probability just described by considering a sequence of 3 sets of size 4, 1, 5.

E PROBLEMS: PROBABILITY FUNCTIONS

7 Consider independent $y_1, \ldots, y_k$, where y_i is Poisson(λ_i). Show that the conditional distribution of $(y_1, \ldots, y_k)$ given the total $T(y_1, \ldots, y_k) = y_1 + \cdots + y_k = y$ is multinomial$(y, \lambda_1/\lambda, \ldots, \lambda_k/\lambda)$, where $\lambda = \lambda_1 + \cdots + \lambda_k$. Use Problem 3.5.34.

8 Let $(y_1, \ldots, y_k)$ be multinomial $(n, p_1, \ldots, p_k)$. Show that the conditional distribution of $(y_{r+1}, \ldots, y_k)$ given $(y_1, \ldots, y_r)$ is multinomial$(n - y_1 - \cdots - y_r, p_{r+1}/p, \ldots, p_k/p)$, where $p = p_{r+1} + \cdots + p_k$. Use Problem 3.5.38.

9 Let $(y_1, \ldots, y_k)$ be multivariate hypergeometric$(N, n, p_1, \ldots, p_k)$. Show that the conditional distribution of $(y_{r+1}, \ldots, y_k)$ given $(y_1, \ldots, y_r)$ is multivariate hypergeometric$(Np, n - y_1 - \cdots - y_r, p_{r+1}/p, \ldots, p_k/p)$, where $p = p_{r+1} + \cdots + p_k$. Use Problem 3.5.41.

F EXERCISES: PROBABILITY DENSITY FUNCTIONS

10 Consider a distribution on R^2 with density

$$f(y_1, y_2) = 18e^{-6y_1 - 3y_2}, \qquad y_1 > 0, y_2 > 0$$

and zero otherwise. Determine the conditional density function for y_2 given y_1; see Exercise 3.2.16.

11 Consider a distribution on R^2 with density

$$f(y_1, y_2) = 2e^{-y_1-y_2}, \qquad 0 < y_1 < y_2 < \infty,$$

and zero otherwise. Determine the conditional density function for y_2 given y_1; see Exercise 3.2.17.

12 Let f be the density function for the uniform distribution over the disk $y_1^2 + y_2^2 < 1$. Calculate the conditional density $f_2(\cdot : y_1)$ for y_2 given y_1; see Exercise 3.2.18.

13 Consider the uniform distribution over the unit ball $y_1^2 + y_2^2 + y_3^2 < 1$. Determine the conditional distribution of y_3 given (y_1, y_2); and the conditional distribution of (y_2, y_3) given y_1; see Exercise 3.2.19.

14 Consider the density on R^k given by $f(y_1, \ldots, y_k) = h_1(y_1) \cdots h_k(y_k)$, where each h_i is a density on R. Determine the conditional density of $(y_4, \ldots, y_k)$ given (y_1, y_2, y_3); see Exercise 3.2.20.

15 Consider the Pareto(α) distribution on R^2 with density f given by ($\alpha > 0$)

$$f(y_1, y_2) = \alpha(\alpha + 1)(1 + y_1 + y_2)^{-\alpha-2}, \qquad 0 < y_1, 0 < y_2,$$

and zero otherwise. Determine the conditional distribution of y_2 given y_1; see Exercise 3.2.21.

G PROBLEMS: PROBABILITY DENSITY FUNCTIONS

16 Consider the canonical Student(n) distribution for (y_1, y_2) on R^2. Determine the conditional density for y_2 given y_1; see Exercise 3.2.22. Verify that the conditional distribution can be described as: $y_2/(1 + y_1^2)^{1/2}$ is canonical Student($n + 1$). Note that the conditional distribution of y_2 is more spread out with larger values of $|y_1|$.

17 Consider the uniform distribution for (x_1, x_2) over the unit square $(0, 1)^2$. Determine the distribution of (y_1, y_2) where $y_1 = x_1 + x_2$, $y_2 = x_2$ and sketch the region of positive density. Determine the conditional distribution of x_2 given $x_1 + x_2 = t$.

18 (*continuation of Example 3.2.8*) Consider a distribution for (x_1, x_2) with density g. Determine the conditional distribution of x_2 given $t = x_1 + x_2$.

19 Consider the Pareto(α) distribution for $(y_1, \ldots, y_k)$ on R^k; see Problem 2.3.14. Show that the conditional distribution given y_1 can be described as: $(y_2/(1 + y_1), \ldots, y_k/(1 + y_1))$ is Pareto($\alpha + 1$); see Problem 3.2.24.

20 Consider the canonical Student(n) distribution for $(y_1, \ldots, y_k)$. Show that the conditional distribution given y_1 can be described as: $(y_2/(1 + y_1^2)^{1/2}, \ldots, y_k/(1 + y_1^2)^{1/2})$ is canonical Student($n + 1$); see Problem 3.2.25.

21 Consider the Dirichlet($r_1, \ldots, r_{k+1}$) distribution for $(y_1, \ldots, y_k)$. Show that the conditional distribution given $(y_1, \ldots, y_p)$ can be described as: $(y_{p+1}/Y, \ldots, y_k/Y)$ is Dirichlet($r_{p+1}, \ldots, r_{k+1}$), where $Y = 1 - y_1 - \cdots - y_p$; see Problem 3.2.26.

22 Consider a distribution for (x, y) on R^2 with density f. Determine the conditional density for y given $t = x - y$; see Problem 3.2.29.

23 Consider a distribution for (x, y) on R^2 with density f, where $f(x, y) = 0$ for $x < 0$ and for $y \leq 0$. Determine the conditional density for $t = x + y$ given $v = x/y$; see Problem 3.2.30.

24 A distribution for lifetime t on the interval $(0, \infty)$ has a constant failure rate:

$$\frac{f(t)}{1 - F(t)} = c;$$

note that there is no "aging." Show that t is exponential $(1/c)$.

H PROBLEMS: ACCEPTANCE SAMPLING

A lot contains N articles, of which $D = Np$ are defective and $N - D = Nq$ are good. The articles are thoroughly mixed and are sampled and tested one by one until n have been tested. Let

r be the number of defectives in the sample and p_1 be the probability function. From Section 3.5C we have

$$p_1(r) = \frac{\binom{D}{r}\binom{N-D}{n-r}}{\binom{N}{n}}.$$

25 Verify the following expression directly; see Example 4.1.6.

$$p_1(r) = \binom{n}{r}\frac{D^{(r)}(N-D)^{(n-r)}}{N^{(n)}}.$$

26 (*continuation*) Verify the following expression directly:

$$p_1(r) = \binom{n}{r}\binom{N-n}{D-r}\frac{D!\,(N-D)!}{N!}.$$

27 (*continuation*) Verify the following expression directly:

$$p_1(r) = \binom{N-n}{D-r}\frac{D^{(D-r)}(N-D)^{(N-n-D+r)}}{N^{(N-n)}}.$$

28 (*continuation*) Verify the following expression directly:

$$p_1(r) = \frac{\binom{D}{r \quad D-r}\binom{n-D}{n-r \quad N-D-n+r}}{\binom{N}{n \quad N-n}}.$$

Consider a sequence of two sets of size n, $N-n$.

Alternatively, the articles are thoroughly mixed and are sampled and tested one by one until r defectives have been obtained. Let n be the number of items tested and p_2 be the probability function.

29 In the pattern of Example 4.1.6, verify the following expression:

$$p_2(n) = \frac{\binom{D}{r-1}\binom{N-D}{n-r}}{\binom{N}{n-1}}\frac{D-r+1}{N-n+1}.$$

30 (*continuation*) Verify the following expression directly:

$$p_2(n) = \binom{n-1}{r-1}\frac{D^{(r)}(N-D)^{(n-r)}}{N^{(n)}}.$$

Consider a sequence of N articles for testing.

31 Verify the following expressions directly:

$$p_2(n) = \binom{n-1}{r-1}\binom{N-n}{D-r}\frac{D!\,(N-D)!}{N!} = \binom{N-n}{D-r}\frac{D^{(D-r+1)}(N-D)^{(N-n-D+r)}}{N^{(N-n+1)}}.$$

32 Verify the following expressions directly:

$$p_2(n) = \frac{\binom{D}{r-1 \quad 1 \quad D-r}\binom{N-D}{n-r \quad 0 \quad N-D-n+r}}{\binom{N}{n-1 \quad 1 \quad N-n}}$$

$$= \frac{\binom{D}{r-1}\binom{D-r+1}{1}\binom{N-D}{n-r}}{\binom{N}{n-1}\binom{N-n+1}{1}} = \frac{\binom{D}{D-r}\binom{r}{1}\binom{N-D}{N-D-n+r}}{\binom{N}{N-n}\binom{n}{1}}.$$

I PROBLEMS: COMPUTER USAGE

The elements in a finite population have serial numbers 101, 102, . . ., 685. A random sample of 25 can be obtained from a computer that produces random digits: the digits are taken in blocks of three; if a three-digit block is a serial number and has not been obtained previously, it is accepted; continue until 25 serial numbers have been accepted. This process is justified on the basis of conditional probability and symmetry.

33 Engines produced by an automobile plant in a particular year have serial numbers 474400001, 474400002, . . ., 474438756. Use a random-digit generator to obtain a random sample of 30.

34 The students in a freshman class have student numbers A0001, A0002, . . ., A2387. Use a random-digit generator to obtain a random sample of 30.

Consider a density function f on R which is zero outside the interval (a, b) and has maximum value M. Let v_1 and v_2 be independent, v_1 uniform(a, b), and v_2 uniform$(0, M)$. Sample values for (v_1, v_2) can be obtained by relocating and rescaling values from the uniform(0, 1); these, in turn, can be approximated by taking batches of random digits. A sample value of (v_1, v_2) is tested to see if $v_2 \le f(v_1)$: if so, v_1 is accepted; if not, v_1 is rejected. Conditional probability shows that this produces sample values from the distribution with density f.

35 Use random digits to obtain a sample of 25 from the triangular distribution with $f(x) = 2x$ on $(0, 1)$ and zero otherwise.

36 Use random digits to obtain a sample of 25 from the beta(3, 3) distribution.

J PROBLEMS: NORMAL DISTRIBUTION

37 Consider the bivariate normal$(\mu_1, \mu_2, \sigma_1, \sigma_2, \rho)$ density f for (y_1, y_2) in Problem 2.4.24. Let h_1 be the normal(μ_1, σ_1) density and $h_2(\cdot : y_1)$ be the normal$(\mu_2 + \rho\sigma_2(y_1 - \mu_1)/\sigma_1, \sigma_2(1 - \rho^2)^{1/2})$ density. Show that

$$f(y_1, y_2) = h_1(y_1)h_2(y_2 : y_1)$$

and deduce that the marginal distribution of y_1 is normal(μ_1, σ_1) and the conditional of y_2 is normal$(\mu_2 + \rho\sigma_2(y_1 - \mu_1)/\sigma_1, \sigma_2(1 - \rho^2)^{1/2})$.

38 Consider the multivariate normal$(\boldsymbol{\mu}; \Sigma)$ distribution for $\mathbf{y}$ as given in Section 2.4G and with new notation in Problem 3.6.9. Verify the following factorization of the density function,

$$f(\mathbf{y}) = \frac{|\Sigma_{11}|^{-1/2}}{(2\pi)^{k/2}} \exp\left\{-\frac{1}{2}(\mathbf{y}_1 - \boldsymbol{\mu}_1)'\Sigma_{11}^{-1}(\mathbf{y}_1 - \boldsymbol{\mu}_1)\right\}$$

$$\cdot \frac{|\Sigma^{22}|^{1/2}}{(2\pi)^{l/2}} \exp\{-(1/2)(\mathbf{y}_2 - \boldsymbol{\mu}_2 - \Sigma_{21}\Sigma_{11}^{-1}(\mathbf{y}_1 - \boldsymbol{\mu}_1))'\Sigma^{22}(\mathbf{y}_2 - \boldsymbol{\mu}_2 - \Sigma_{21}\Sigma_{11}^{-1}(\mathbf{y}_1 - \boldsymbol{\mu}_1))\}$$

by using matrix identities derived from $\Sigma\Sigma^{-1} = I$, such as $\Sigma^{11} - \Sigma^{12}(\Sigma^{22})^{-1}\Sigma^{21} = \Sigma_{11}^{-1}$. Deduce that the marginal distribution of $\mathbf{y}_1$ is normal$(\boldsymbol{\mu}_1; \Sigma_{11})$, and that the conditional distribution of $\mathbf{y}_2$ given $\mathbf{y}_1$ is normal$(\boldsymbol{\mu}_2 + \Sigma_{21}\Sigma_{11}^{-1}(\mathbf{y}_1 - \boldsymbol{\mu}_1); (\Sigma^{22})^{-1})$.

SUPPLEMENTARY MATERIAL

Supplement Section 4.3 is on pages 537–544.

Some basic notation for *stochastic processes* is presented, and a brief introduction is given for *Poisson processes, Markov processes,* and *normal processes*. The *Polya urn,* which has applications in biology and medicine, is also described.

NOTES AND REFERENCES

In this chapter we have given detailed attention to the concept of conditional probability. Conditional probability is usually examined for an isolated probability $P(A:C)$. By contrast we have emphasized the *measure* $P(\cdot : C)$ or P_C; the measure gives a full probability description of a system performance on the basis of information represented by C. We have also provided a selection model for the interpretation of this conditional probability measure. Conditional probability is a very fundamental concept for applications.

For some further reading on the supplemental topic of stochastic processes, see the following texts:

Karlin, S. (1966). *A First Course in Stochastic Processes*. New York: Academic Press, Inc.
Parzen, E. (1962). *Stochastic Processes*. San Francisco: Holden-Day, Inc.

We have now examined the basic methods for handling probabilities themselves. In Chapter 1 we introduced the probability model and in Chapter 2 its detailed form on the line and plane. Then in Chapter 3 we examined marginal probability for functions of initial responses, and independence for combining responses. And in Chapter 4 we discussed conditional probability for the case of partial information concerning a response. A model of course has probabilities, but some other characteristics can also be very important.

To investigate the yield of a new variety of wheat, an agricultural statistician will collect data from a moderately large number of plots and will average the values as an estimate of the mean, or long-run average yield. A chemist investigating a new process for producing an isotope will examine a moderately large number of runs on the process and will average the values as an estimate of the mean, or long-run average yield.

In Chapter 5 we discuss the ***mean value*** *for a probability distribution on the real line. Then by examining the mean for functions on the line and plane we obtain some powerful tools for describing location, scaling, and other properties of distributions.*

5

MEAN VALUE FOR REAL AND VECTOR DISTRIBUTIONS

5.1

DEFINITION OF THE MEAN

In this chapter we discuss the **mean value**—the concept itself and its use to describe and predict important characteristics of probability distributions.

In this Section 5.1 we introduce the mean value and record some simple properties. In Section 5.2 we use the concept to describe the location of a distribution on the real line, and then, by simple vector extension, the location of a distribution on the plane. In Section 5.3 we use it to describe the scaling of a distribution on the line—from very compact through to very diffuse—and then by vector extension the scaling of a distribution on the plane. Matrix notation carries the initial cases on the real line into the vector extension in a simple routine way.

In Section 5.4 we examine location and scaling for a sample from a distribution and we derive important formulas that predict these characteristics from corresponding characteristics of the initial distribution. Again matrix notation provides the vector extension immediately from the initial case on the real line.

Conditional characteristics are examined in Section 5.5, moment sequences in Section 5.6, and moment-generating functions in Section 5.7. These provide a powerful collection of methods for predicting characteristics of various probability distributions.

A DEFINITION

Consider the rolling of a symmetric die and the recording of the number y of points showing. First suppose that we have found someone who will give us \$1 for each point showing. If this game were played a large number of times, say N, then $y = 1$ would occur approximately $N/6$ times, . . ., $y = 6$ would occur approximately $N/6$ times. The total amount received would be approximately

$$1 \cdot (1/6)N + 2 \cdot (1/6)N + \cdots + 6 \cdot (1/6)N$$

dollars. Thus the mean or long-run average per play would be

$$E(y) = 1 \cdot (1/6) + 2 \cdot (1/6) + \cdots + 6 \cdot (1/6) = 3\tfrac{1}{2}$$

dollars. Alternatively, suppose that we have found someone who will give us y^2 dollars on any toss; then, in a similar way, the mean or long-run average per play would be

$$E(y^2) = 1^2 \cdot (1/6) + 2^2 \cdot (1/6) + \cdots + 6^2 \cdot (1/6) = 15\tfrac{1}{6}.$$

Now, more generally, consider a random system with sample space $\mathcal{S}$, with events $\mathcal{A}$, and with probability measure P. And suppose that we are interested in a real-valued function Y giving the derived response $y = Y(s)$ from the basic response s. For example, we might be interested in the mean power Y in an electronic system in a case where $Y = x_1^2 + \cdots + x_{10}^2$ in terms of terms of underlying variables $x_1, \ldots, x_{10}$.

First suppose that Y is a simple function, a function taking only a finite number of values; let $y_1, \ldots, y_m$ be the values. Then, in direct analogy to the preceding example, we define the mean value of Y written variously as

$$E(Y) = E(Y(s)) = E(y) = \mu_Y$$

by the summation

$$(1) \qquad E(Y) = \sum_1^m y_j p_Y(y_j) = \sum_1^m y_j P(\{s : Y(s) = y_j\}).$$

This is the *summation of values y_j for the function weighted by corresponding probabilities*. Note that we have used several different notations for the mean, each being convenient in appropriate contexts; the Greek letter μ is often used for mean and a subscript indicates the distribution being examined.

Now suppose that Y is a bounded real-valued function: $-M_1 < Y(s) < M_2$ for all s in $\mathcal{S}$; see Figure 5.1. And suppose that we subdivide the range $(-M_1, M_2)$ by points $-M_1 = a_0 < a_1 < \cdots < a_n = M_2$ such that all the differences $a_i - a_{i-1}$ are less than some small number ϵ. Then we can approximate Y from above by the simple function

$$Y_n = \sum_1^n a_j I_{A_j},$$

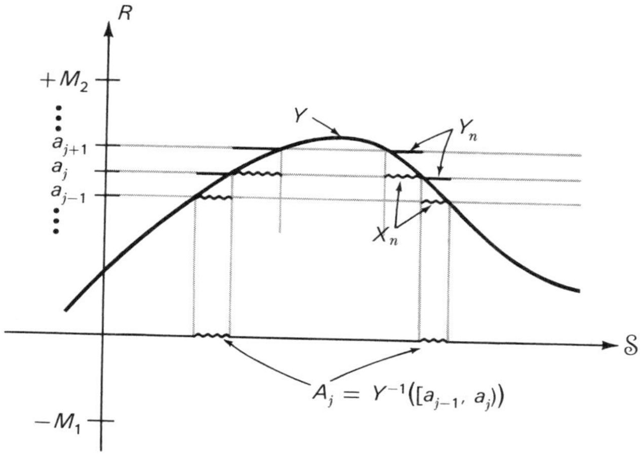

FIGURE 5.1
Set A_j, on which the function takes values in $[a_{j-1}, a_j)$; the function Y_n approximates Y from above; the function X_n approximates Y from below.

where $A_j = Y^{-1}([a_{j-1}, a_j))$ is the preimage of the interval $[a_{j-1}, a_j)$ and I_{A_j} is the corresponding indicator function; see Figure 5.1. This simple function has mean value

$$E(Y_n) = \sum_1^n a_j P(A_j).$$

Similarly, we can approximate Y from below by the simple function

$$X_n = \sum_1^n a_{j-1} I_{A_j},$$

which has mean value

$$E(X_n) = \sum_1^n a_{j-1} P(A_j).$$

If we take a sequence of values of ϵ going to zero, then the simple functions Y_n and X_n converge to Y from above and below. Consider how the simple functions differ in their mean values:

$$E(Y_n) - E(X_n) \leq \sum_1^n \epsilon P(A_j) = \epsilon;$$

thus they can be at most ϵ apart. Accordingly, we define $E(Y)$ to be the limiting value as the upper approximation Y_n and lower approximation X_n converge to Y:

DEFINITION 1

The **mean value** $E(Y)$ of a bounded function Y is defined by

(2) $$E(Y) = \lim_{\epsilon \to 0} \sum_1^n a_j P(A_j) = \lim_{\epsilon \to 0} \sum_1^n a_{j-1} P(A_j)$$

and denoted by

(3) $$E(Y) = \int Y(s)\, dP(s),$$

where the maximum subdivision interval ϵ goes to zero.

Now suppose that Y is an arbitrary real-valued function. We obtain a bounded function $Y_{M_1M_2}$ by flattening Y off at an upper limit M_2 and at a lower limit $-M_1$:

$$\begin{aligned} Y_{M_1M_2}(s) &= M_2 && \text{if} \quad M_2 \leq Y(s) < \infty, \\ &= Y(s) && \quad -M_1 < Y(s) < M_2, \\ &= -M_1 && \quad -\infty < Y(s) \leq -M_1. \end{aligned}$$

The mean value of $Y_{M_1M_2}$ is available from Definition 1; we obtain the mean value of Y by taking M_1 and M_2 to ∞:

DEFINITION 2

The **mean value** $E(Y)$ of a function Y is defined by

$$(4) \qquad E(Y) = \lim_{\substack{M_1 \to \infty \\ M_2 \to \infty}} E(Y_{M_1 M_2})$$

and written

$$(5) \qquad E(Y) = \int Y(s)\, dP(s),$$

provided the limit exists as M_1 and M_2 separately approach ∞; otherwise, we say that $E(Y)$ does not exist or $E(Y)$ diverges.

We, of course, assume that any function Y respects our events and is a measurable function in accord with Definition 3.1.2. Some theory concerning the preceding general definition of an integral will be discussed in Supplement Section 5.8.

B CALCULATIONS

First consider the case of a discrete function Y or even a more general function Y provided its distribution is discrete; let $y_1, y_2, \ldots$ be the points where there is probability. Then directly from Definition 2 we obtain the mean value of Y as

$$(6) \qquad E(Y) = E(Y(s)) = \sum y_j P(Y^{-1}(y_j))$$

provided that the summation converges absolutely; otherwise, we say that $E(Y)$ does not exist or $E(Y)$ diverges.

EXAMPLE 1

Consider 4 tosses of a symmetric coin giving $(x_1, \ldots, x_4)$, where x_i is the indicator for heads on the ith toss. And let $y = Y(x_1, \ldots, x_4)$ be the number of heads. By Section 3.5A the response y is binomial(4, 1/2) with distribution

y	0	1	2	3	4
$p_Y(y)$	1/16	4/16	6/16	4/16	1/16

The mean value of y is

$$\begin{aligned} E(y) &= \sum y p_Y(y) \\ &= 0 \cdot (1/16) + 1 \cdot (4/16) + 2 \cdot (6/16) + 3 \cdot (4/16) + 4 \cdot (1/16) \\ &= 2. \end{aligned}$$

Thus with four tosses of a symmetric coin the mean number of heads is 2.

Now consider the case of a function Y defined on R^1 or R^k and suppose that the distribution on R^1 or R^k has density function f. Then Definition 2 (with Proposition 7 of Supplement Section 5.8) gives the mean value of Y as

$$(7)\qquad E(Y) = E(Y(x_1, \ldots, x_k)) = \int_{R^k} Y(x_1, \ldots, x_k) f(x_1, \ldots, x_k)\, dx_1 \cdots dx_k$$

provided that the integral converges absolutely; otherwise, we say that $E(Y)$ does not exist or that $E(Y)$ diverges.

EXAMPLE 2

Minimum lifetime: Consider an electronic control and suppose that the engineer has found that the lifetime y of the control is exponential(800). We evaluate the mean lifetime:

$$\begin{aligned} E(y) &= \int_0^\infty y(1/800)e^{-y/800}\, dy \\ &= 800 \int_0^\infty ze^{-z}\, dz, \qquad \text{where } z = y/800 \\ &= 800 \left(-ze^{-z} \Big|_0^\infty + \int_0^\infty e^{-z}\, dz \right) \\ &= 800; \end{aligned}$$

the mean lifetime is 800.

Now briefly consider the general case of a real-valued function Y. Definition 2 gives the mean value of Y as

$$(8)\qquad E(Y) = E(Y(s)) = \int_{\mathcal{S}} Y(s)\, dP(s)$$

in terms of the underlying distribution on $\mathcal{S}$. Alternatively, formula (20) of Supplement Section 5.8 gives the mean value of Y as

$$(9)\qquad E(Y) = E(Y(s)) = \int_R y\, dP_Y(y) = \int_R y\, dF_Y(y)$$

in terms of the marginal distribution of Y. Formulas (8) and (9) apply provided that the integral exists; otherwise, $E(Y)$ does not exist or diverges.

All our examples, however, will involve just the summations (6) *or the ordinary integrals* (7); some theory for the general case is recorded in Supplement Section 5.8.

Now, more generally, consider a vector-valued function $(Y_1, \ldots, Y_k)$ taking values in R^k. Then we define the mean vector to be

$$E(Y_1, \ldots, Y_k) = (E(Y_1), \ldots, E(Y_k)),$$

provided that each coordinate exists.

C PROPERTIES

We now record some properties of the mean value for functions defined on a space $\mathcal{S}$ with probability P; these important properties will be used freely in succeeding sections.

1. Normed: The mean value of the constant function c is equal to c:

(10) $\quad E(c) = c.$

The proof follows trivially by noting that the preimage of the value c is, of course, the whole sample space $\mathcal{S}$.

The remaining properties follow directly from propositions in Supplement Section 5.8; for each property the needed proposition is cited.

2. Linear: If $E(Y_1)$ and $E(Y_2)$ exist, then

(11) $\quad E(aY_1 + bY_2) = aE(Y_1) + bE(Y_2)$

for real values a and b: mean linear combinations equals linear combination means (Proposition 5.8.3).

3. Continuity: If $Y_1, Y_2, \ldots$ is a sequence of functions such that $\lim_{i\to\infty} Y_i(s)$ exists and if $|Y_i(s)| < M(s)$ where $E(M)$ exists, then (Proposition 5.8.8)

(12) $\quad \lim_{i\to\infty} E(Y_i) = E\left(\lim_{i\to\infty} Y_i\right).$

4. Independence: If Y_1 and Y_2 are statistically independent, then

(13) $\quad E(Y_1Y_2) = E(Y_1)E(Y_2):$

mean independent product equals product mean (Proposition 5.8.10).

EXAMPLE 3

Suppose that we have found someone who will pay us in dollars the sum plus the product of the points on two good dice; what is a fair price to pay to play this game? Let Y_1 and Y_2 be the points on the first and second dies, respectively. Then the payoff is $y_1 + y_2 + y_1y_2$ and it has mean value

$$E(y_1 + y_2 + y_1y_2) = E(y_1) + E(y_2) + E(y_1)E(y_2)$$
$$= 3.5 + 3.5 + (3.5)^2 = 19.25.$$

If a person pays \$19.25, the net payoff is $y_1 + y_2 + y_1y_2 - 19.25$, and it has mean value

$$E(y_1 + y_2 + y_1y_2 - 19.25) = 19.25 - 19.25 = 0.$$

Thus if a person pays \$19.25 to play the game, the mean exchange on a play is 0 and the game is called *fair*.

D EXERCISES

1 A modified slot machine pays \$50 with probability 1/100, \$1 with probability 10/100, and zero otherwise. Calculate the mean payoff. If a person pays \$1 to play, determine the mean net payoff.

2 A person rolls a die and receives \$5 if the ace comes up; how much should he pay otherwise so that the game is fair?

3 A distribution for y has density function $3y^2$ on $(0, 1)$ and zero elsewhere. Determine $E(y)$ and $E(y^2)$.

4 A distribution for y has density $12y^2(1-y)$ on $(0, 1)$ and zero elsewhere. Determine $E(y)$ and $E(y^2)$.

5 A symmetrical coin is tossed until the first head occurs; let x be number of tosses. Consider the payoff $y = 2^x$. Show that $E(y)$ does not exist.

6 Consider the standard exponential distribution for x. Show that $E(y)$ does not exist for the payoff $y = e^x$.

5.2

THE LOCATION OF A DISTRIBUTION

A statistician has a distribution that describes the lifetime of high-pressure sodium lamps. Of course, he is interested in the location of the distribution, a location point about which realized values tend to be distributed. Also, he is interested in the scaling of the distribution, a distance from the location point that represents the variability in realized values. In this section we consider the mean value as a possible measure of the location of a distribution on the real line R^1 or on a real space R^k. We then examine scaling in the next section.

A LOCATION ON R^1

Consider a distribution for x on the real line R as given by the probability measure P, or as given by the probability function p in the discrete case or the probability density function f in the absolutely continuous case; for example, x could be the lifetime for a certain type of printed circuit. Consider the mean value for such a variable.

DEFINITION 1

The mean of the distribution is the point μ,

$$(1) \qquad \mu = \mu_x = E(x) = \int_R x\, dP(x)$$

$$= \sum xp(x) \qquad \text{if discrete,}$$

$$= \int_{-\infty}^{\infty} xf(x)\, dx \qquad \text{absolutely continuous}$$

provided that the integral exists (absolute convergence; Definition 5.1.2); otherwise, the mean diverges or does not exist; several notations are recorded.

EXAMPLE 1

The uniform(a, b) distribution: The density function for this uniform distribution is equal to $1/(b-a)$ on (a, b) and zero elsewhere. The mean of the distribution is

$$\mu = \int_{-\infty}^{\infty} xf(x)\,dx = \int_a^b x\frac{1}{b-a}\,dx = \frac{b^2/2 - a^2/2}{b-a} = \frac{a+b}{2}.$$

Note that the distribution is symmetric and that the mean is at the point of symmetry.

The mean value of a distribution is presented in Figure 5.2 as a point μ on the space of the distribution. If we consider the deviations $x - \mu$ of values x from μ, then some deviations will be positive and some negative, but the mean deviation from μ is zero:

(2) $$E(x-\mu) = E(x) - E(\mu) = \mu - \mu = 0;$$

use properties 1 and 2 in Section 5.1C. If we think of probability as unit mass distributed on R, then μ is the center of gravity, the point about which the mass distribution balances: the equation states that the moment of the force of gravity about μ is zero.

Consider a distribution that is symmetrical about some point m, and suppose that the mean μ of the distribution exists. Symmetry can be expressed by saying that $x - m$ and $-(x - m)$ have the same distribution and thus the same mean: $E(x - m) = E(-(x - m))$. Thus (properties 1 and 2)

$$E(x) - m = -E(x) + m,$$
$$E(x) = m.$$

Hence the mean μ is at the point m of symmetry.

Consider an affine transformation $y = a + cx = Y(x)$ carrying the initial response x into the new response y; for example, such a transformation might correspond to a change from Celsius to Fahrenheit. We have (properties 1 and 2)

(3) $$E(y) = E(a + cx) = a + cE(x),$$
$$\mu_Y = a + c\mu.$$

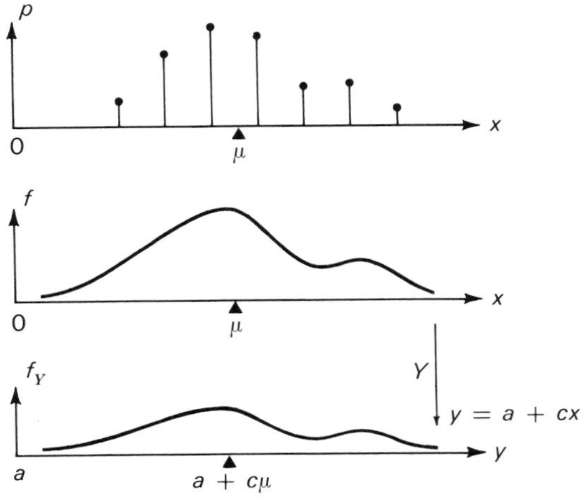

FIGURE 5.2
Mean of a discrete distribution; mean of an absolutely continuous distribution; mean after a transformation $y = a + cx$; illustration has $c > 0$.

Thus the transformation also carries the initial mean into the new mean; this is illustrated in Figure 5.2 in terms of a relabeling of the points on R.

Now consider a distribution for (x_1, x_2) on the plane and suppose that we form a new response $y = x_1 + x_2 = Y(x_1, x_2)$; then

$$(4) \qquad \mu_Y = E(y) = E(x_1 + x_2) = E(x_1) + E(x_2) = \mu_1 + \mu_2.$$

Thus: if we add the variables, then we add the means.

Now suppose that the probability distribution has come from some initial space $\mathcal{S}$ with class $\mathcal{A}$ and measure P. Let X be a real-valued measurable function; then

$$(5) \qquad \mu_X = E(X) = \int_{\mathcal{S}} X(s)\, dP(s) = \int_R x\, dP_X(x),$$

using formula (5.1.9). And if $Y = a + cX$, then $\mu_Y = a + c\mu_X$.

B SOME EXAMPLES ON R^1

EXAMPLE 2

The normal distribution: Let us examine a standard normal distribution for z. The distribution is symmetric about 0; and the mean clearly exists because of the exponential tails: thus $E(z) = 0$. Indeed, we have

$$(6) \qquad E(z) = \int_{-\infty}^{\infty} z \frac{1}{\sqrt{2\pi}} e^{-z^2/2}\, dz = 0,$$

as clearly the integrand is an odd function. Now suppose that we generate the general normal(μ, σ) distribution by the transformation $y = \mu + \sigma z$; then

$$E(y) = E(\mu + \sigma z) = \mu + \sigma E(z) = \mu.$$

Thus the μ we have used in defining the general normal is, in fact, the mean of the distribution.

EXAMPLE 3

The standard Cauchy distribution: This distribution is symmetrical, but the integral for the mean

$$\int_{-\infty}^{\infty} z \frac{1}{\pi} \frac{1}{1 + z^2}\, dz$$

diverges (recall Definition 5.1.2); thus the mean of the Cauchy does not exist. In Section A we saw some of the attractive properties of the mean as a measure of location, and we now see one of the unattractive properties—that it does not exist for some mathematically reasonable distributions. We need other measures of location for the Cauchy; see Problems 14 and 17.

EXAMPLE 4

The Poisson(λ) distribution: For a discrete example we examine the Poisson(λ) distribution:

$$E(y) = \sum_{y=0}^{\infty} y \frac{\lambda^y e^{-\lambda}}{y!} = \sum_{y=1}^{\infty} \frac{\lambda^y e^{-\lambda}}{(y-1)!} = \lambda \cdot \sum_{y=1}^{\infty} \frac{\lambda^{y-1} e^{-\lambda}}{(y-1)!} = \lambda,$$

where the sum of Poisson probabilities is 1. We have now verified that the mean of the Poisson(λ) is just λ; recall the large-sample data on α-particles in Section 2.1E.

Note: The first term with $y = 0$ can be omitted; the y cancels with the first factor in $y!$; the adjusted summation is a summation of Poisson probabilities and gives 1. This calculation method recurs with many discrete distributions: the expression being averaged cancels into the functional form of the distribution; with appropriate factors removed, the residual summation is a sum of probabilities ($= 1$) for some value of the parameter of the distribution. The binomial(n, p) distribution provides another illustration and we obtain $E(y) = np$.

C WHAT DOES μ SAY ABOUT P?

If we know that a distribution has mean equal to μ, then what constraints does this place on how the probability is distributed on the line R? We obtain a rather interesting relation if we ask this for a distribution with all probability on the positive axis:

$$\text{(7)} \qquad \mu = \int_0^{\infty} y\, dP(y) \geq \int_t^{\infty} y\, dP(y) \geq \int_t^{\infty} t\, dP(y) = tP([t, \infty)).$$

Note: The integrand is nonnegative and we reduce the range of integration; we replace y by the lower bound t; see Figure 5.3. Thus we obtain *Chebyshev's inequality:*

$$\text{(8)} \qquad P(y \geq t) \leq \frac{\mu}{t} \quad \text{or} \quad P(y \geq k\mu) \leq \frac{1}{k}$$

for a distribution on $[0, \infty)$. Note, for example, that there is at most 1/3 probability beyond 3μ.

D LOCATION ON R^k

Now we examine the parallel development on R^k. Consider a distribution for $\mathbf{x} = (x_1, \ldots, x_k)'$ on R^k as given by the probability measure P, or as given by the

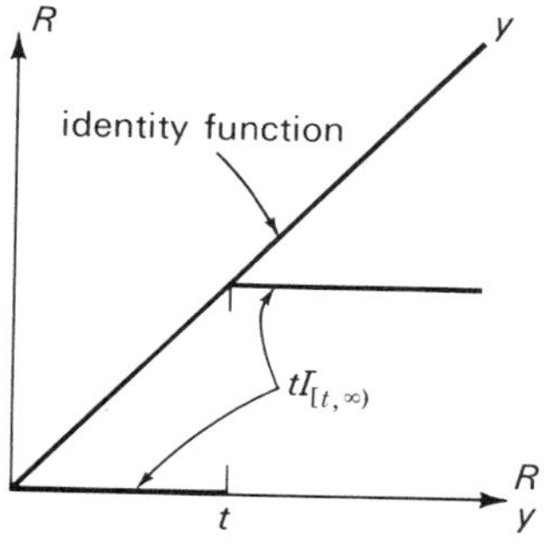

FIGURE 5.3
On $[0, \infty)$, the identity function is greater than or equal to the function $tI_{[t,\infty)}$.

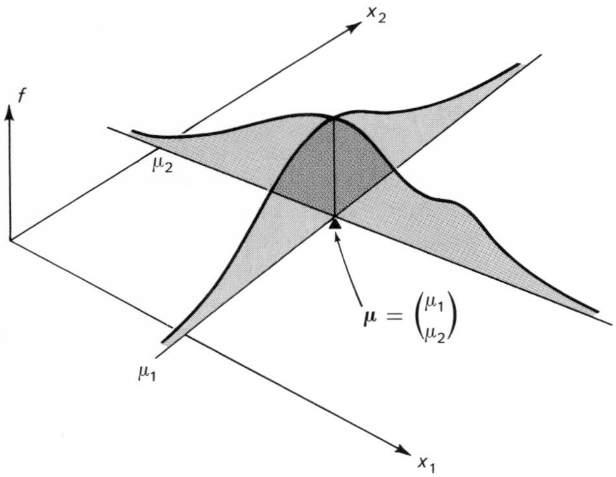

FIGURE 5.4
Mean (μ_1, μ_2) of a distribution on the plane is the center of gravity.

probability function p in the discrete case or the probability density function f in the absolutely continuous case. For example, x_1, x_2, and x_3 could be the time, intensity, and direction of the reaction to a stimulus in psychological testing.

DEFINITION 2

The mean of the distribution is the point $\boldsymbol{\mu} = (\mu_1, \ldots, \mu_k)'$,

(9) $$\boldsymbol{\mu} = E(\mathbf{x}) = (E(x_1), \ldots, E(x_k))',$$

in R^k provided that all the coordinate means exist; otherwise, we say that the mean diverges or does not exist.

See Figure 5.4. In general, if E is applied to a vector or to a matrix, then by definition it gives the vector or matrix formed by applying E to each element; the linearity properties of E make this a fruitful definition. Note that a coordinate mean can be calculated in a variety of ways:

(10) $$E(x_i) = \int_{R^k} x_i \, dP(\mathbf{x})$$

$$= \sum x_i p(\mathbf{x}) = \sum x_i p_i(x_i) \qquad \text{if discrete,}$$

$$= \int_{R^k} x_i f(\mathbf{x}) \, d\mathbf{x} = \int_R x_i f_i(x_i) \, dx_i \qquad \text{absolutely continuous.}$$

EXAMPLE 5

A distribution on R^2: Consider the absolutely continuous distribution for (x, y) with density f in Example 3.2.6:

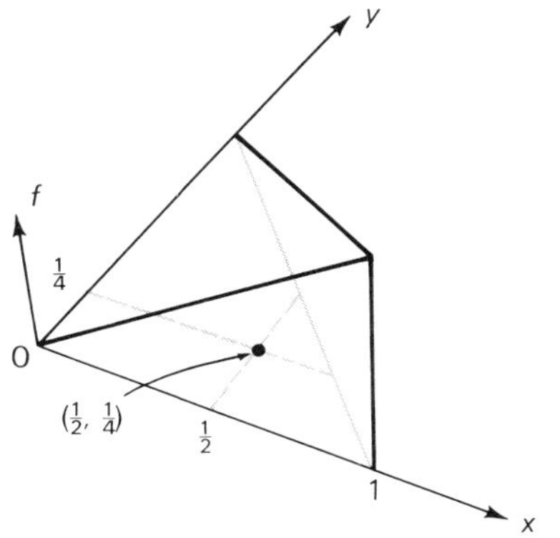

FIGURE 5.5
Mean $\boldsymbol{\mu} = (1/2, 1/4)'$ of the distribution in Example 5.

$$f(x, y) = 6x \qquad \text{if } 0 < x < 1,\ 0 < y < 1 - x$$

and zero elsewhere. The marginal densities are

$$f_1(x) = 6x(1 - x) \qquad \text{if } 0 < x < 1,$$
$$f_2(y) = 3(1 - y)^2 \qquad 0 < y < 1$$

and zero elsewhere. Thus we calculate

$$\mu_1 = \int_{R^2} xf(x, y)\,dx\,dy = \int_R xf_1(x)\,dx = \int_0^1 x \cdot 6x(1 - x)\,dx = 1/2,$$

$$\mu_2 = \int_{R^2} yf(x, y)\,dx\,dy = \int_R yf_2(y)\,dy = \int_0^1 y \cdot 3(1 - y)^2\,dy = 1/4.$$

Thus the mean $\boldsymbol{\mu}$ is $(1/2, 1/4)'$, as indicated in Figure 5.5. Note that the integrals are available from Problem 2.2.22 or indirectly from Problem 2.2.17.

The mean of a distribution is presented in Figure 5.4 as a point in R^k. If we consider the deviation vector $\mathbf{x} - \boldsymbol{\mu}$, then its values center at $\mathbf{0}$:

$$E(\mathbf{x} - \boldsymbol{\mu}) = E(\mathbf{x}) - E(\boldsymbol{\mu}) = \boldsymbol{\mu} - \boldsymbol{\mu} = \mathbf{0}.$$

If we think of probability as unit mass distributed on the plane R^2($k = 2$ case), then the mass balances about the line $x_1 = \mu_1$; it also balances about $x_2 = \mu_2$; and it balances about any line $\Sigma\, l_i x_i = \Sigma\, l_i \mu_i$ through the point $\boldsymbol{\mu}$; in fact, it balances on the *point* $\boldsymbol{\mu}$.

Consider an affine transformation $\mathbf{y} = \mathbf{a} + C\mathbf{x}$, where $\mathbf{x}$ in R^k is mapped to $\mathbf{y}$ in R^r and $\mathbf{a}$ is $r \times 1$ and C is $r \times k$; then the mean $\boldsymbol{\nu}$ of $\mathbf{y}$ is

$$\text{(11)} \qquad \boldsymbol{\nu} = E(\mathbf{y}) = E(\mathbf{a} + C\mathbf{x}) = \mathbf{a} + CE(\mathbf{x}) = \mathbf{a} + C\boldsymbol{\mu}.$$

Thus the transformation also carries the original mean into the new mean.

Now consider a distribution for $(\mathbf{x}_1, \mathbf{x}_2)$, where each $\mathbf{x}_i$ is in R^k, and suppose that we form a new response $\mathbf{y} = \mathbf{x}_1 + \mathbf{x}_2$. Then

$$\text{(12)} \qquad E(\mathbf{y}) = E(\mathbf{x}_1 + \mathbf{x}_2) = E(\mathbf{x}_1) + E(\mathbf{x}_2).$$

Thus, if we add vector variables, we add the vector means.

E SOME EXAMPLES ON R^k

EXAMPLE 6

A sample from the Bernoulli(p) distribution: Consider a sample $(x_1, \ldots, x_n)$ from the Bernoulli(p) distribution. Then

$$E(x_i) = 1 \cdot p + 0 \cdot q = p;$$

thus the mean of the distribution is

$$E(\mathbf{x}) = (p, \ldots, p)' = p\mathbf{1},$$

where $\mathbf{1} = (1, \ldots, 1)'$ is the 1-vector.

Now suppose that we form the binomial variable $y = \Sigma_1^n x_i$; then

$$E(y) = E(x_1 + \cdots + x_n) = p + \cdots + p = np;$$

thus the mean of the binomial(n, p) distribution is np. Note, for example, that with 100 tosses of a symmetric coin, the mean number of heads is 50.

EXAMPLE 7

The normal distribution on R^k: Consider a standard normal distribution for $\mathbf{z}$ on R^k; note that we can consider $\mathbf{z} = (z_1, \ldots, z_k)'$ as a sample from the standard normal on R^1. Using Example 2, we obtain the mean

$$E(\mathbf{z}) = (E(z_1), \ldots, E(z_k))' = (0, \ldots, 0)' = \mathbf{0}.$$

Now suppose that we generate the general normal by the transformation $\mathbf{y} = \boldsymbol{\mu} + \Gamma\mathbf{z}$, where $\boldsymbol{\mu}$ is $k \times 1$ and Γ is $k \times k$. Then the multivariate normal(μ; Σ) distribution has mean

$$E(\mathbf{y}) = E(\boldsymbol{\mu} + \Gamma\mathbf{z}) = \boldsymbol{\mu} + \Gamma E(\mathbf{z}) = \boldsymbol{\mu}.$$

Thus the $\boldsymbol{\mu}$ we have used to locate the general normal is, in fact, the mean of the distribution.

F EXERCISES

1 The binomial(4, 1/2) distribution is tabulated in Section 3.5A. Calculate the mean from the tabulation.

2 The binomial(4, 3/4) distribution is tabulated in Section 3.5A. Calculate the mean from the tabulation.

3 The hypergeometric(8, 4, 1/2) distribution is tabulated in Section 3.5C. Calculate the mean from the tabulation.

4 The hypergeometric(8, 4, 3/4) distribution is tabulated in Section 3.5C. Calculate the mean from the tabulation.

5 Consider a distribution for y with probability function $p(y) = 2^{-y}$ on $y = 1, 2, \ldots$ and zero elsewhere. Calculate $E(y)$.

6 A distribution on R has density function f given by $f(y) = 1 - y/2$ on $(0, 2)$ and zero elsewhere. Calculate the mean.

7 A distribution on R has density function f given by $f(y) = 12y(1 - y)^2$ on $(0, 1)$ and zero elsewhere. Calculate the mean.

8 The hypergeometric(6, 3, 1/6, 2/6, 3/6) distribution is tabulated in Section 3.5D. Calculate the mean (μ_1, μ_2, μ_3) from the tabulation.

9 The multinomial(5, 1/6, 2/6, 3/6) distribution for (y_1, y_2, y_3) is tabulated in Section 3.5B and in Section 4.2A. Calculate $E(y_1)$ and $E(y_2)$ from the tabulation; calculate $E(y_3)$ from $y_1 + y_2 + y_3 = 5$; record the mean of the distribution for (y_1, y_2, y_3).

10 Calculate the mean of the uniform distribution on the unit square $(0, 2)^2$.

G PROBLEMS

Additional problems concerning means will be found at the end of Section 5.3.

11 For a discrete distribution on the nonnegative integers, show that

$$E(y) = \sum_{j=0}^{\infty} P(y > j).$$

12 Consider a distribution on R with $E(|y|^r) < \infty (r > 0)$; show that for $t > 0$,

$$P(|y| \geq t) \leq \frac{E(|y|^r)}{t^r}.$$

13 Consider a distribution on R with $E(\phi(y)) < \infty$, where $\phi(y) \geq 0$ and $\phi(y) \geq k > 0$ for y in A; show that

$$P(A) \leq \frac{E(\phi(y))}{k}.$$

14 The median is often used as a location measure; the median $\tilde{\mu}$ is a solution of the relation $F(x - 0) \leq 1/2 \leq F(x)$ when F is the distribution function. Note that if F is strictly increasing, then $\tilde{\mu}$ is unique. If a new response is obtained by $y = h(x)$, where h is strictly increasing and continuous, then show that the median $\tilde{\nu}$ of y is given by the same transformation: $\tilde{\nu} = h(\tilde{\mu})$.

15 Let $a_1 < a_2 < \cdots < a_n$ be real numbers. Show that $\Sigma\, |a_i - t|$ is a minimum if t is a median $\tilde{a}$ of the n numbers; that is, the number of $a_i < \tilde{a}$ is equal to the number $a_i > \tilde{a}$.

16 (*continuation*) Consider a distribution for y on the line R. Show that $E(|y - t|)$ is a minimum if t is a median of the distribution.

17 The mode of a distribution with density f on the real line is a value μ^0 such that $f(\mu^0) \geq f(x)$ for all x. If a new response y is obtained by $y = a + cx$ $(c \neq 0)$, then show that the mode ν^0 of y is given by the same transformation $\nu^0 = a + c\mu^0$.

5.3

THE SCALING OF A DISTRIBUTION

We have mentioned the statistician who has a distribution describing the lifetime of high-pressure sodium lamps, and is interested in its location and scaling. In the preceding section we discussed the mean as a measure of the location of a distribution on the real line. In this section we examine the scaling of a distribution on the real line, scaling as measured by the **variance** or by its square root the **standard deviation**. We also examine the scaling of a distribution on the plane or on R^k; the formulas for this follow by a straightforward matrix extension of those presented for the simple case on the real line.

The scaling can be of interest in itself. For example, how variable is the lifetime for a certain type of printed circuit? Or, more frequently, it can be of interest as a measure of precision—the degree with which a response value will approximate the location of the distribution; but more on this in the later chapters on statistics.

A SCALING ON R^1

Consider a distribution for x on the real line R as given by the probability measure P, or as given by the probability function p in the discrete case or the probability density function f in the absolutely continuous case.

DEFINITION 1

The **variance** of the distribution is

$$(1) \qquad \sigma^2 = \sigma_x^2 = \text{var}(x) = E((x-\mu)^2) = \int_R (x-\mu)^2\, dP(x)$$

$$= \sum (x-\mu)^2 p(x) \qquad \text{if discrete,}$$

$$= \int_{-\infty}^{\infty} (x-\mu)^2 f(x)\, dx \qquad \text{absolutely continuous}$$

provided that the integral exists; otherwise, the variance diverges or does not exist; alternative notations are recorded.

The variance is the mean-squared deviation from the location or center of the distribution. The variance is measured in squared units. The square root of the variance is measured in the same units as the original variable for the distribution; it is called the standard deviation (or RMS, root mean square).

DEFINITION 2

The **standard deviation** of the distribution is

$$(2) \qquad \sigma = \sigma_x = \text{SD}(x) = \sqrt{\text{var}(x)} = [E((x-\mu)^2)]^{1/2}.$$

EXAMPLE 1

The uniform(0, 1) distribution: The uniform(0, 1) distribution has mean $\mu = 1/2$; see Example 5.2.1. Thus we can calculate the variance as follows:

$$\sigma^2 = \int_{-\infty}^{\infty} (x-1/2)^2 f(x)\, dx = \int_0^1 (x-1/2)^2\, dx = \frac{1}{12}.$$

The standard deviation is $\sigma = 1/\sqrt{12} = 0.2887$.

The standard deviation is presented in Figure 5.6 as a *scaling unit* or as a *statistical distance* from the mean μ. For the preceding uniform distribution we see

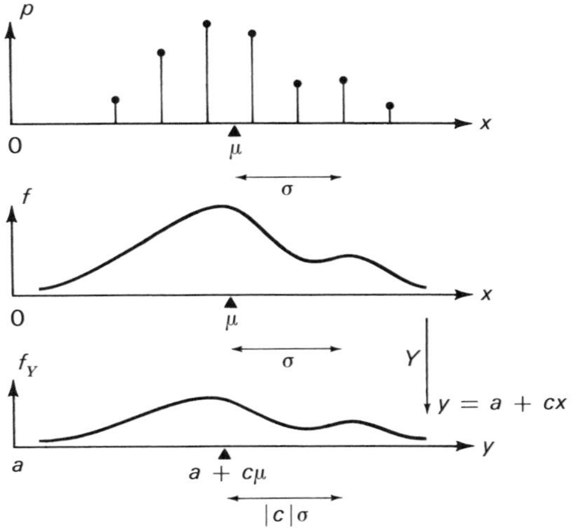

FIGURE 5.6
Standard deviation σ of a discrete distribution. The standard deviation σ of an absolutely continuous distribution. The standard deviation after a transformation $y = a + cx$; illustration has $c > 0$.

that $\mu \pm \sigma$ contains about 58 percent of the probability; for the normal the corresponding figure is 68.26 percent. In general, for a distribution with a central hump of probability and not-too-long tails, we find that $\mu \pm \sigma$ contains about 2/3 of the probability.

Some alternative formulas for the variance can often be more convenient for calculations. The formula

$$(3) \qquad \sigma^2 = E((x - \mu)^2) = E(x^2 - 2x\mu + \mu^2) = E(x^2) - 2\mu^2 + \mu^2$$
$$= E(x^2) - (E(x))^2 = \text{MS} - \text{SM},$$

which presents the variance as *mean square minus square mean,* can be useful with continuous distributions. And the formula

$$(4) \qquad \sigma^2 = E(x(x - 1)) - \mu(\mu - 1) = E(x^{(2)}) - \mu^{(2)}$$

can be useful for discrete distributions; recall the calculation method in Example 5.2.4.

Consider an affine transformation $y = a + cx$ carrying the initial response x with mean μ into the new response y with mean ν. Then

$$y = a + cx, \qquad \nu = a + c\mu,$$
$$y - \nu = c(x - \mu),$$
$$E((y - \nu)^2) = c^2 E((x - \mu)^2).$$

Thus

$$(5) \qquad \text{var}\,(a + cx) = c^2\,\text{var}\,(x), \qquad \text{SD}\,(a + cx) = |c|\,\text{SD}\,(x).$$

The transformation changes the scale by the factor $|c|$ and correspondingly changes the standard deviation by the same factor; see Figure 5.6. In particular, note that *relocation alone does not change the variance or standard deviation.*

Now consider a distribution for (x_1, x_2) on the plane with mean (μ_1, μ_2) and suppose that we form a new response $y = x_1 + x_2$ with mean ν:

$$y = x_1 + x_2, \qquad \nu = \mu_1 + \mu_2,$$

$$y - \nu = x_1 - \mu_1 + x_2 - \mu_2$$

$$\begin{aligned}\text{var}\,(y) &= E((x_1 - \mu_1 + x_2 - \mu_2)^2) \\ &= E((x_1 - \mu_1)^2 + 2(x_1 - \mu_1)(x_2 - \mu_2) + (x_2 - \mu_2)^2) \\ &= \text{var}\,(x_1) + 2E((x_1 - \mu_1)(x_2 - \mu_2)) + \text{var}\,(x_2).\end{aligned}$$

Now suppose that x_1 and x_2 are statistically independent. Then by formula (5.1.13), the middle term,

(6) $$E((x_1 - \mu_1)(x_2 - \mu_2)) = E(x_1 - \mu_1)E(x_2 - \mu_2) = 0 \cdot 0 = 0,$$

is a product of mean deviations and thus is equal to zero. Hence for independent variables we have

(7) $$\text{var}\,(x_1 + x_2) = \text{var}\,(x_1) + \text{var}\,(x_2).$$

Thus: if we add independent variables, we add the variances.

Now suppose that the probability distribution has come from some initial space $\mathcal{S}$ with class $\mathcal{A}$ and measure P. Let X be a real-valued measurable function; then

$$\sigma_X^2 = E((X - \mu_X)^2) = \int (X(s) - \mu_X)^2 \, dP(s).$$

And if $Y = a + cX$, then $\sigma_Y^2 = c^2\sigma_X^2$.

B SOME EXAMPLES ON R^1

EXAMPLE 2

The uniform(a, b) distribution: The uniform(0, 1) distribution has mean $\mu = 1/2$; we recalculate the variance:

$$E(x^2) = \int_{-\infty}^{\infty} x^2 f(x)\, dx = \int_0^1 x^2\, dx = 1/3;$$

$$\text{var}\,(x) = (1/3) - (1/2)^2 = 1/12.$$

Now suppose that we generate the uniform(a, b) distribution by the transformation $y = a + (b - a)x$. Then

$$\mu_y = a + (b - a) \cdot (1/2) = \frac{a + b}{2},$$

$$\sigma_y^2 = (b - a)^2 \cdot (1/12), \qquad \sigma_y = \frac{b - a}{\sqrt{12}} = 0.2887(b - a).$$

EXAMPLE 3

The normal distribution: The standard normal distribution for z has $E(z) = 0$. We calculate the variance:

$$E(z^2) = \int_{-\infty}^{\infty} z^2 \frac{1}{\sqrt{2\pi}} e^{-z^2/2}\,dz = \frac{2}{\sqrt{2\pi}} \int_0^{\infty} z \cdot e^{-z^2/2} z\,dz$$

$$= \frac{-2}{\sqrt{2\pi}} \left(z \cdot e^{-z^2/2} \Big|_0^{\infty} - \int_0^{\infty} e^{-z^2/2}\,dz \right) = 1;$$

thus

$$\text{var}\,(z) = E(z^2) - (E(z))^2 = 1 - 0 = 1.$$

Now, suppose that we generate the general normal(μ, σ) distribution by the transformation $y = \mu + \sigma z$; then

$$\text{var}\,(y) = \sigma^2 \cdot 1 = \sigma^2.$$

Thus the σ we have used in defining the general normal is, in fact, the standard deviation of the distribution. Recall from Section 2.2C that $\mu \pm \sigma$ contains 68.26 percent of the probability.

EXAMPLE 4

The Poisson(λ) distribution: In Example 5.2.4 we saw that the mean is λ. For the variance we first calculate

$$E(y^{(2)}) = \sum_{y=0}^{\infty} y(y-1) \frac{\lambda^y e^{-\lambda}}{y!} = \sum_{y=2}^{\infty} \frac{\lambda^y e^{-\lambda}}{(y-2)!} = \lambda^2,$$

following the pattern in the earlier example. Thus the variance is

$$\text{var}\,(y) = E(y^{(2)}) - \lambda^{(2)} = \lambda^2 - \lambda(\lambda - 1) = \lambda.$$

And the standard deviation is $\sqrt{\lambda}$; note that this is the square root of the mean.

EXAMPLE 5

A sample from the Bernoulli(p) distribution: In Example 5.2.6 we saw that the mean of the Bernoulli(p) distribution is p. We calculate the variance:

$$E(x^2) = 1^2 \cdot p + 0^2 \cdot q = p, \qquad\qquad E(x^{(2)}) = 0 \cdot p + 0 \cdot q = 0,$$

or

$$\text{var}\,(x) = p - p^2 = p(1 - p) = pq. \qquad \text{var}\,(x) = 0 - p^{(2)} = pq.$$

Now consider a sample $(x_1, \ldots, x_n)$ from the Bernoulli(p) distribution. We form the binomial variable y and from formula (7) obtain

$$\text{var}\,(y) = \text{var}\,(x_1) + \cdots + \text{var}\,(x_n) = npq.$$

Thus the variance of the binomial(n, p) distribution is npq. Note, for example, that with 100 tosses of a symmetric coin the variance of the number of heads is 25 and the standard deviation is 5.

C WHAT DOES σ SAY ABOUT P?

If we know that a distribution has variance σ^2, then what constraints does this place on how the probability is distributed on the line R?

The variance has a rather powerful null property: *the variance* var $(x) = 0$ *if and only if x is equal to a constant with probability* 1 (all probability at a point). For consider the variance

$$\text{var}(x) = \int_{-\infty}^{\infty} (x - \mu)^2 \, dP(x).$$

By formula (24) of Supplement Section 5.8 this variance can be zero if and only if $(x - \mu)^2 = 0$ with probability 1, that is $x = \mu$ with probability 1. This can be a very useful criterion for determining if a distribution is concentrated with all probability at a point.

Now consider a distribution for x with mean μ and variance σ^2 and apply the result in Section 5.2C with y replaced by $(x - \mu)^2$ and t replaced by t^2:

$$P((x - \mu)^2 \geq t^2) \leq \frac{\sigma^2}{t^2}. \tag{8}$$

The alternative form,

$$P(|x - \mu| \geq k\sigma) \leq \frac{1}{k^2}, \tag{9}$$

is another version of Chebyshev's inequality. Thus there is at most 1/9 probability outside the interval $(\mu - 3\sigma, \mu + 3\sigma)$.

In practice the variance is more specific. For distributions with a central pile of probability and not too much probability in the tails, the interval $(\mu - \sigma, \mu + \sigma)$ contains about 2/3 of the probability.

D SCALING ON R^k

Consider a distribution for $\mathbf{x} = (x_1, \ldots, x_k)'$ on R^k as given by the probability measure P—or as given by the probability function p in the discrete case, or the probability density function f in the absolutely continuous case. For example, x_1, x_2, and x_3 could be time, intensity, and direction for the reaction to a stimulus in the psychological testing mentioned earlier. Let $\boldsymbol{\mu} = (\mu_1, \ldots, \mu_k)'$ be the mean. We first define the covariance between two coordinates:

DEFINITION 3

The **covariance** between x_i and x_j is

$$\begin{aligned} \sigma_{ij} = \text{cov}(x_i, x_j) &= E((x_i - \mu_i)(x_j - \mu_j)) = \int_{R^k} (x_i - \mu_i)(x_j - \mu_j) \, dP(\mathbf{x}) \\ &= \sum_{\infty} (x_i - \mu_i)(x_j - \mu_j) p_{ij}(x_i, x_j) \qquad \text{if discrete,} \\ &= \iint_{-\infty} (x_i - \mu_i)(x_j - \mu_j) f_{ij}(x_i, x_j) \, dx_i \, dx_j \qquad \text{absolutely continuous} \end{aligned} \tag{10}$$

if the integral exists; otherwise, we say that the covariance diverges or does not exist.

Note: If x_i and x_j are statistically independent, then cov $(x_i, x_j) = 0$; see formula (6).

The covariances can be assembled to form a variance matrix:

DEFINITION 4

The **variance matrix** of a distribution is

$$\Sigma = \text{VAR}\,(\mathbf{x}) = \begin{pmatrix} \sigma_{11} & \cdots & \sigma_{1k} \\ \vdots & & \vdots \\ \sigma_{k1} & \cdots & \sigma_{kk} \end{pmatrix} \tag{11}$$

$$= E\begin{pmatrix} (x_1 - \mu_1)(x_1 - \mu_1) & \cdots & (x_1 - \mu_1)(x_k - \mu_k) \\ \vdots & & \vdots \\ (x_k - \mu_k)(x_1 - \mu_1) & \cdots & (x_k - \mu_k)(x_k - \mu_k) \end{pmatrix}.$$

The variance matrix can be expressed more compactly using matrix multiplication:

$$\Sigma = E(\begin{pmatrix} x_1 - \mu_1 \\ \vdots \\ x_k - \mu_k \end{pmatrix} (x_1 - \mu_1, \ldots, x_k - \mu_k)) = E((\mathbf{x} - \boldsymbol{\mu})(\mathbf{x} - \boldsymbol{\mu})'). \tag{12}$$

The matrix may seem unwieldy, but we will see later that it is very useful. Note that Σ is a symmetric matrix; also it can be shown that Σ is positive semidefinitive (Problem 32).

EXAMPLE 6

The uniform distribution on (0, 1)^2: Consider the uniform distribution for (x_1, x_2) over the unit square $(0, 1)^2$. The marginal distribution of x_1 is uniform(0, 1); thus $E(x_1) = 1/2$, $\text{var}\,(x_1) = 1/12$ by Example 1. Similarly, $E(x_2) = 1/2$, $\text{var}\,(x_2) = 1/12$. We calculate

$$\text{cov}\,(x_1, x_2) = \int_{R^2} (x_1 - 1/2)(x_2 - 1/2) f(x_1, x_2)\, dx_1\, dx_2 = \left(\int_0^1 (x - 1/2)\, dx\right)^2 = 0$$

by the factorization formula (5.1.13). Thus the mean and variance matrix are

$$\boldsymbol{\mu} = \begin{pmatrix} 1/2 \\ 1/2 \end{pmatrix}, \qquad \Sigma = \begin{pmatrix} 1/12 & 0 \\ 0 & 1/12 \end{pmatrix}.$$

Some alternative formulas can often be convenient for calculations. For the covariance we have

$$\begin{aligned} \sigma_{ij} &= E((x_i - \mu_i)(x_j - \mu_j)) = E(x_i x_j - \mu_i x_j - x_i \mu_j + \mu_i \mu_j) \\ &= E(x_i x_j) - \mu_i \mu_j = \text{MP} - \text{PM}, \end{aligned} \tag{13}$$

which presents the covariance as *mean product minus product mean*. The analogous result for the variance matrix itself is equally simple:

$$\begin{aligned} \Sigma = \text{VAR}\,(\mathbf{x}) &= E((\mathbf{x} - \boldsymbol{\mu})(\mathbf{x} - \boldsymbol{\mu})') = E(\mathbf{x}\mathbf{x}' - \boldsymbol{\mu}\mathbf{x}' - \mathbf{x}\boldsymbol{\mu}' + \boldsymbol{\mu}\boldsymbol{\mu}') \\ &= E(\mathbf{x}\mathbf{x}') - \boldsymbol{\mu}\boldsymbol{\mu}' = \text{MP} - \text{PM}, \end{aligned} \tag{14}$$

which is mean product minus product mean, where the product is given by matrix multiplication.

Now consider a distribution for $\mathbf{x}' = (x_1, \ldots, x_k, x_{k+1}, \ldots, x_{k+l}) = (\mathbf{x}_1', \mathbf{x}_2')$ on R^{k+l}; note that the first k coordinates are grouped as a vector $\mathbf{x}_1$ and the last l as $\mathbf{x}_2$. The variance matrix Σ for $\mathbf{x}$ can be partitioned into the variance matrices Σ_{11} for $\mathbf{x}_1$ and Σ_{22} for $\mathbf{x}_2$ and the covariance matrix $\Sigma_{12} = \mathrm{COV}\,(\mathbf{x}_1, \mathbf{x}_2)$ between $\mathbf{x}_1$ and $\mathbf{x}_2$; we have

$$(15) \qquad \Sigma = \begin{pmatrix} \sigma_{11} & \cdots & \sigma_{1k} & \sigma_{1k+1} & \cdots & \sigma_{1k+l} \\ \vdots & & \vdots & \vdots & & \vdots \\ \sigma_{k1} & \cdots & \sigma_{kk} & \sigma_{kk+1} & \cdots & \sigma_{kk+l} \\ \sigma_{k+11} & \cdots & \sigma_{k+1k} & \sigma_{k+1k+1} & \cdots & \sigma_{k+1k+l} \\ \vdots & & \vdots & \vdots & & \vdots \\ \sigma_{k+l1} & \cdots & \sigma_{k+lk} & \sigma_{k+lk+1} & \cdots & \sigma_{k+lk+l} \end{pmatrix} = \begin{pmatrix} \Sigma_{11} & \Sigma_{12} \\ \Sigma_{21} & \Sigma_{22} \end{pmatrix}.$$

Note that the covariance matrix

$$(16) \qquad \Sigma_{12} = \mathrm{COV}\,(\mathbf{x}_1, \mathbf{x}_2) = E((\mathbf{x}_1 - \boldsymbol{\mu}_1)(\mathbf{x}_2 - \boldsymbol{\mu}_2)') = E(\mathbf{x}_1, \mathbf{x}_2') - \boldsymbol{\mu}_1\boldsymbol{\mu}_2'$$

records the covariances between coordinates of $\mathbf{x}_1$ and coordinates of $\mathbf{x}_2$; and $\Sigma_{21} = \Sigma_{12}'$.

Consider an affine transformation $\mathbf{y} = \mathbf{a} + C\mathbf{x}$, where $\mathbf{x}$ in R^k is mapped to $\mathbf{y}$ in R^r and $\mathbf{a}$ is $r \times 1$ and C is $r \times k$. Let $\boldsymbol{\mu}$ and Σ be the mean and variance matrix of $\mathbf{x}$; and let $\boldsymbol{\nu}$ be the mean for $\mathbf{y}$. Then

$$\mathbf{y} = \mathbf{a} + C\mathbf{x}, \qquad \boldsymbol{\nu} = \mathbf{a} + C\boldsymbol{\mu}, \qquad \mathbf{y} - \boldsymbol{\nu} = C(\mathbf{x} - \boldsymbol{\mu}).$$

Thus the variance matrix for $\mathbf{y}$ is

$$(17) \qquad \begin{aligned} \mathrm{VAR}\,(\mathbf{y}) &= E((\mathbf{y} - \boldsymbol{\nu})(\mathbf{y} - \boldsymbol{\nu})') = E(C(\mathbf{x} - \boldsymbol{\mu})(\mathbf{x} - \boldsymbol{\mu})'C') \\ &= CE((\mathbf{x} - \boldsymbol{\mu})(\mathbf{x} - \boldsymbol{\mu})')C' = C\Sigma C' = C\,\mathrm{VAR}\,(\mathbf{x})C'. \end{aligned}$$

In particular, we obtain

$$(18) \qquad \mathrm{var}\left(a_1 + \sum_j c_{1j}x_j\right) = \sum_{j,j'=1}^{k} c_{1j}c_{1j'}\sigma_{jj'},$$

$$(19) \qquad \mathrm{cov}\left(a_1 + \sum_j c_{1j}x_j,\ a_2 + \sum_j c_{2j}x_j\right) = \sum_{j,j'=1}^{k} c_{1j}c_{2j'}\sigma_{jj'}$$

for the $(1, 1)$ and $(1, 2)$ elements in $\mathrm{VAR}\,(\mathbf{y})$.

Now consider a distribution for $(\mathbf{x}_1, \mathbf{x}_2)$ where each $\mathbf{x}_i$ is in R^k, and assume that $\mathbf{x}_1$ is statistically independent of $\mathbf{x}_2$. Let $\boldsymbol{\mu}_1$ and Σ_{11} be the mean and variance matrix for $\mathbf{x}_1$, and $\boldsymbol{\mu}_2$ and Σ_{22} for $\mathbf{x}_2$; note that the covariance matrix $\Sigma_{12} = 0$ by independence. Suppose that we form a new variable $\mathbf{y} = \mathbf{x}_1 + \mathbf{x}_2$ with mean $\boldsymbol{\nu}$:

$$\mathbf{y} = \mathbf{x}_1 + \mathbf{x}_2, \qquad \boldsymbol{\nu} = \boldsymbol{\mu}_1 + \boldsymbol{\mu}_2, \qquad \mathbf{y} - \boldsymbol{\nu} = \mathbf{x}_1 - \boldsymbol{\mu}_1 + \mathbf{x}_2 - \boldsymbol{\mu}_2.$$

Thus the variance matrix for $\mathbf{y}$ is

$$(20) \qquad \begin{aligned} \mathrm{VAR}\,(\mathbf{y}) &= E((\mathbf{x}_1 - \boldsymbol{\mu}_1 + \mathbf{x}_2 - \boldsymbol{\mu}_2)(\mathbf{x}_1 - \boldsymbol{\mu}_1 + \mathbf{x}_2 - \boldsymbol{\mu}_2)') \\ &= E((\mathbf{x}_1 - \boldsymbol{\mu}_1)(\mathbf{x}_1 - \boldsymbol{\mu}_1)' + (\mathbf{x}_1 - \boldsymbol{\mu}_1)(\mathbf{x}_2 - \boldsymbol{\mu}_2)' \\ &\qquad + (\mathbf{x}_2 - \boldsymbol{\mu}_2)(\mathbf{x}_1 - \boldsymbol{\mu}_1)' + (\mathbf{x}_2 - \boldsymbol{\mu}_2)(\mathbf{x}_2 - \boldsymbol{\mu}_2)') \\ &= \Sigma_{11} + \Sigma_{12} + \Sigma_{21} + \Sigma_{22} \\ &= \Sigma_{11} + \Sigma_{22}. \end{aligned}$$

Thus if we add independent vectors, we add the variance matrices.

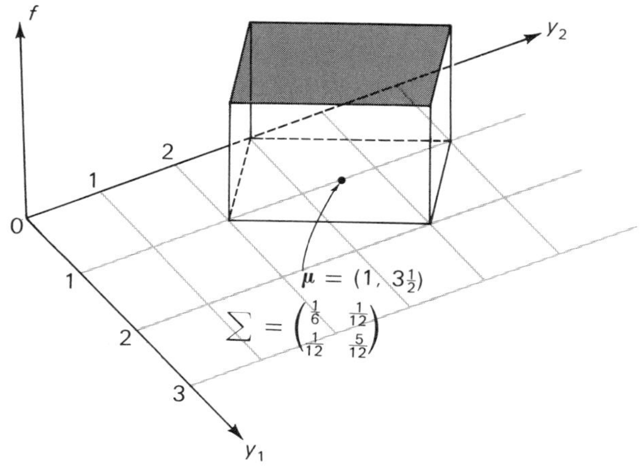

FIGURE 5.7
Mean μ and variance Σ for the uniform distribution on a parallelogram.

E SOME EXAMPLES

EXAMPLE 7

Uniform distribution on a parallelogram: First consider a uniform distribution for (x_1, x_2) on the unit square $(0, 1)^2$; the mean and variance matrix are recorded in Example 6. Now suppose that we form a new variable (y_1, y_2) by the transformation

$$\begin{aligned} y_1 &= 1 + x_1 - x_2 \\ y_2 &= 2 + 2x_1 + x_2 \end{aligned}, \qquad \begin{pmatrix} y_1 \\ y_2 \end{pmatrix} = \begin{pmatrix} 1 \\ 2 \end{pmatrix} + \begin{pmatrix} 1 & -1 \\ 2 & 1 \end{pmatrix}\begin{pmatrix} x_1 \\ x_2 \end{pmatrix};$$

it is straightforward to see that $\mathbf{y}$ is uniformly distributed over the parallelogram with vertices (1, 2), (2, 4), (0, 3), (1, 5); see Example 2.4.6. The mean and variance matrix of $\mathbf{y}$ are

$$\mu = \begin{pmatrix} 1 \\ 2 \end{pmatrix} + \begin{pmatrix} 1 & -1 \\ 2 & 1 \end{pmatrix}\begin{pmatrix} 1/2 \\ 1/2 \end{pmatrix} = \begin{pmatrix} 1 \\ 3\frac{1}{2} \end{pmatrix}$$

and

$$\Sigma = \begin{pmatrix} 1 & -1 \\ 2 & 1 \end{pmatrix}\begin{pmatrix} 1/12 & 0 \\ 0 & 1/12 \end{pmatrix}\begin{pmatrix} 1 & 2 \\ -1 & 1 \end{pmatrix} = \begin{pmatrix} 1/6 & 1/12 \\ 1/12 & 5/12 \end{pmatrix}.$$

See Figure 5.7.

EXAMPLE 8

A distribution on R^2: Consider the absolutely continuous distribution for (x, y) with density f as given in Example 3.2.6:

$$f(x, y) = 6x \qquad \text{if } 0 < x < 1,\ 0 < y < 1 - x$$

and zero elsewhere. The marginal densities are

$$f_1(x) = 6x(1 - x) \qquad \text{if } 0 < x < 1$$
$$f_2(y) = 3(1 - y)^2 \qquad 0 < y < 1$$

and zero elsewhere. The mean $\boldsymbol{\mu} = (1/2, 1/4)'$ was calculated in Example 5.2.5. Now consider calculations toward obtaining the variance matrix:

$$E(x^2) = \int_0^1 x^2 \cdot 6x(1 - x)\, dx = 6/20, \quad \sigma_{11} = (6/20) - (1/2)^2 = 1/20,$$

$$E(y^2) = \int_0^1 y^2 \cdot 3(1 - y)^2\, dy = 1/10, \quad \sigma_{22} = (1/10) - (1/4)^2 = 3/80,$$

$$E(xy) = \int_0^1 \left(\int_0^{1-x} xy \cdot 6x\, dy \right) dx = \int_0^1 6x^2 \frac{(1 - x)^2}{2}\, dx = 1/10,$$

$$\sigma_{12} = (1/10) - (1/2) \cdot (1/4) = -1/40.$$

Thus the mean and variance matrix are

$$\boldsymbol{\mu} = \begin{pmatrix} 1/2 \\ 1/4 \end{pmatrix}, \qquad \Sigma = \begin{pmatrix} 1/20 & -1/40 \\ -1/40 & 3/80 \end{pmatrix}.$$

EXAMPLE 9

A sample from the Bernoulli(p) distribution: Let $(x_1, \ldots, x_n)$ be a sample from the Bernoulli(p) distribution. From Example 5.2.6 we have $E(\mathbf{x}) = p\mathbf{1}$. From Example 5 we have var $(x_i) = pq$ and from the independence of coordinates we have cov $(x_i, x_j) = 0$ for $i \neq j$; this gives us the variance matrix

$$\Sigma = \begin{pmatrix} pq & & 0 \\ & \ddots & \\ 0 & & pq \end{pmatrix}.$$

Now suppose that we form the binomial variable $y = \Sigma\, x_i = \mathbf{1}'\mathbf{x}$. We have calculated its variance in Example 5; we now calculate it again using formula (17):

$$\text{var}\,(y) = (1, \ldots, 1) \begin{pmatrix} pq & & 0 \\ & \ddots & \\ 0 & & pq \end{pmatrix} \begin{pmatrix} 1 \\ \vdots \\ 1 \end{pmatrix} = (pq, \ldots, pq) \begin{pmatrix} 1 \\ \vdots \\ 1 \end{pmatrix} = npq.$$

EXAMPLE 10

The normal distribution: Consider a standard normal distribution for $\mathbf{z}$ on R^k. The mean $E(\mathbf{z}) = \mathbf{0}$ was calculated in Example 5.2.7. We note that the $z_1, \ldots, z_n$ are statistically independent and each is standard normal; thus

$$E(z_i^2) = 1, \quad E(z_i z_j) = E(z_i)E(z_j) = 0, \qquad i \neq j.$$

It follows that the variance matrix is the $k \times k$ identity matrix I.

Now suppose that we generate the general normal by the transformation $\mathbf{y} = \boldsymbol{\mu} + \Gamma\mathbf{z}$, where $\boldsymbol{\mu}$ is $k \times 1$ and Γ is $k \times k$. We have calculated the mean $E(\mathbf{y}) = \boldsymbol{\mu}$ in Example 5.2.7. We now calculate the variance matrix using formula (17):

$$\text{(21)} \qquad \text{VAR}\,(\mathbf{y}) = \Gamma\, \text{VAR}\,(\mathbf{z})\Gamma' = \Gamma I \Gamma' = \Gamma\Gamma' = \Sigma;$$

this is the matrix Σ that was defined in Section 2.4G for a special case, $|\Sigma| \neq 0$. Thus the normal $(\boldsymbol{\mu}; \Sigma)$ distribution has mean $\boldsymbol{\mu}$ and variance matrix Σ.

F WHAT DOES Σ SAY ABOUT P?

Consider the case $k = 2$; the variance matrix is

$$\Sigma = \begin{pmatrix} \sigma_{11} & \sigma_{12} \\ \sigma_{21} & \sigma_{22} \end{pmatrix}.$$

The diagonal element σ_{11} is the variance σ_1^2 of the first coordinate x_1; it describes the scaling of the marginal distribution of x_1, that is, of the probability on the plane as projected on the first axis. Similarly, σ_{22} is the variance σ_2^2 of the second coordinate x_2; it describes the scaling of the probability as projected onto the second axis. Now consider the covariance between the coordinates:

$$\sigma_{12} = \sigma_{21} = E((x_1 - \mu_1)(x_2 - \mu_2)).$$

Suppose that we standardize it with respect to the scaling σ_1 for x_1 and the scaling σ_2 for x_2 (assume that σ_1 and $\sigma_2 > 0$); we then obtain the *correlation coefficient*

$$(22) \qquad \rho_{12} = \frac{\sigma_{12}}{\sigma_1\sigma_2} = E\left(\frac{x_1 - \mu_1}{\sigma_1}\frac{x_2 - \mu_2}{\sigma_2}\right)$$

$$= \int \frac{x_1 - \mu_1}{\sigma_1}\frac{x_2 - \mu_2}{\sigma_2}\, dP(\mathbf{x}) = \operatorname{cov}\left(\frac{x_1 - \mu_1}{\sigma_1}, \frac{x_2 - \mu_2}{\sigma_2}\right);$$

this is the covariance between the standardized variables. The correlation coefficient is a measure of how two variables tend to move in the same direction.

From the Schwarz inequality (Problem 17 of Supplement Section 5.8), we obtain the *correlation inequality,*

$$(23) \qquad -1 \leq \rho_{12} \leq 1,$$

with equality on the right (*left*) *if and only if*

$$\frac{x_2 - \mu_2}{\sigma_2} = \frac{x_1 - \mu_1}{\sigma_1} \qquad \left(\frac{x_2 - \mu_2}{\sigma_2} = -\frac{x_1 - \mu_1}{\sigma_1}\right)$$

holds with probability 1. Thus $\rho_{12} = +1$ means that all probability is on a line with positive slope, and $\rho_{12} = -1$ means that all probability is on a line with negative slope. In practice with reasonable variables, a large positive correlation means a distribution with a compressed shape tilted up to the right and a large negative correlation means a distribution with a compressed shape tilted down to the right.

Note that independence implies that $\sigma_{12} = 0$ and $\rho_{12} = 0$. But $\sigma_{12} = 0$ or $\rho_{12} = 0$ does not imply independence (see Problem 33).

The Schwarz inequality (Problem 17 of Supplement Section 5.8) can be applied directly to $(x_1 - \mu_1)$ and $(x_2 - \mu_2)$, giving the *covariance inequality*

$$(24) \qquad (\sigma_{12})^2 \leq \sigma_{11}\sigma_{22}$$

with equality if and only if $x_1 - \mu_1$ *and* $x_2 - \mu_2$ *are linearly related with probability 1.*

G EXERCISES

1 The binomial(4, 1/2) distribution is tabulated in Section 3.5A. Calculate the variance for the tabulation.

2 The binomial(4, 3/4) distribution is tabulated in Section 3.5A. Calculate the variance for the tabulation.

3 The hypergeometric(8, 4, 1/2) distribution is tabulated in Section 3.5C. Calculate the variance from the tabulation.

4 The hypergeometric(8, 4, 3/4) distribution is tabulated in Section 3.5C. Calculate the variance from the tabulation.

5 A distribution on R has density f given by $f(y) = 1 - y/2$ on (0, 2) and zero elsewhere. Calculate the variance.

6 A distribution on R has density f given by $f(y) = 12y(1 - y)^2$ on (0, 1) and zero elsewhere. Calculate the variance.

7 The hypergeometric(6, 3, 1/6, 2/6, 3/6) distribution is tabulated in Section 3.5D. Calculate the variance matrix Σ; record the covariance matrix $\Sigma_{12} = \text{COV}\,(y_1, (y_2, y_3))$ between y_1 and (y_2, y_3).

8 Calculate the mean $\boldsymbol{\mu}$ and variance matrix Σ for the uniform distribution on the square with vertices $(-1, 0)$, $(0, -1)$, $(1, 0)$, $(0, 1)$.

H PROBLEMS FOR DISCRETE DISTRIBUTIONS

For each of the following distributions, verify the corresponding expressions for

$$E(y), \quad E(y^{(2)}), \quad \text{var}\,(y).$$

9 *The uniform distribution on the integers 1, 2, . . ., N:*

$$\frac{N+1}{2}, \quad \frac{N^2-1}{3}, \quad \frac{N^2-1}{12}.$$

10 *Poisson(λ) distribution:*

$$\lambda, \quad \lambda^2, \quad \lambda.$$

11 *Geometric(p) distribution:*

$$\frac{q}{p}, \quad 2\left(\frac{q}{p}\right)^2, \quad \frac{q}{p^2}.$$

12 *Binomial(n, p) distribution:*

$$np, \quad n^{(2)}p^2, \quad npq.$$

13 *Negative binomial(k, p) distribution:*

$$k\frac{q}{p}, \quad (k+1)k\left(\frac{q}{p}\right)^2, \quad k\frac{q}{p^2}.$$

14 *Hypergeometric distribution(N, n, p):*

$$np, \quad n^{(2)}p\,\frac{Np-1}{N-1}, \quad np(1-p)\,\frac{N-n}{N-1}.$$

For each of the following distributions, verify the corresponding expressions for

$$E(y_i), \quad E(y_i^{(2)}), \quad \text{var}\,(y_i),$$

$$E(y_i y_j), \quad \text{cov}\,(y_i, y_j), \quad i \neq j.$$

15 *Multinomial*$(n, p_1, \ldots, p_k)$ *distribution:*

$$np_i, \qquad n^{(2)}p_i^2, \qquad np_i(1 - p_i),$$

$$n^{(2)}p_ip_j, \qquad -np_ip_j.$$

16 *Multivariate hypergeometric*$(N, n, p_1, \ldots, p_k)$ *distribution:*

$$np_i, \qquad n^{(2)}p_i\frac{Np_i - 1}{N - 1}, \qquad np_i(1 - p_i)\frac{N - n}{N - 1},$$

$$n^{(2)}p_ip_j\frac{N}{N - 1}, \qquad -np_ip_j\frac{N - n}{N - 1}.$$

PROBLEMS: FOR ABSOLUTELY CONTINUOUS DISTRIBUTIONS

For each of the following distributions, verify the corresponding expressions for

$$E(y), \qquad E(y^2), \qquad \text{var}(y).$$

The more difficult problems are marked with an asterisk.

17 *Standard exponential distribution:*

$$1, \qquad 2, \qquad 1.$$

18 *Pareto*(α) *distribution:*

$$\frac{1}{\alpha - 1} \quad (\alpha > 1), \qquad \frac{2}{(\alpha - 2)(\alpha - 1)} \quad (\alpha > 2), \qquad \frac{\alpha}{(\alpha - 1)^2(\alpha - 2)} \quad (\alpha > 2).$$

19 *Standard Laplace distribution:*

$$0, \qquad 2, \qquad 2.$$

*20 *Standard logistic distribution:*

$$0, \qquad \pi^2/3, \qquad \pi^2/3.$$

21 *Standard log-normal*(τ) *distribution:*

$$e^{\tau^2/2}, \qquad e^{2\tau^2}, \qquad e^{\tau^2}(e^{\tau^2} - 1).$$

22 *Gamma*(p) *distribution:*

$$p, \qquad (p + 1)p, \qquad p.$$

23 *Chi-square*(m) *distribution:*

$$m, \qquad (m + 2)m, \qquad 2m.$$

24 *Beta*(p, q) *distribution:*

$$\frac{p}{p + q}, \qquad \frac{p(p + 1)}{(p + q)(p + q + 1)}, \qquad \frac{pq}{(p + q)^2(p + q + 1)}.$$

25 *Canonical F*(m, n) *distribution:*

$$\frac{m}{n - 2} \quad (n > 2), \qquad \frac{m(m + 2)}{(n - 2)(n - 4)} \quad (n > 4), \qquad \frac{2m(m + n - 2)}{(n - 2)^2(n - 4)} \quad (n > 4).$$

26 *Canonical Student*(n) *distribution:*

$$0 \quad (n > 1), \qquad \frac{1}{n - 2} \quad (n > 2), \qquad \frac{1}{n - 2} \quad (n > 2).$$

27 *Standard Weibull(β) distribution:*

$$\Gamma(\beta^{-1} + 1), \qquad \Gamma(2\beta^{-1} + 1), \qquad \Gamma(2\beta^{-1} + 1) - \Gamma^2(\beta^{-1} + 1).$$

For the following distribution, verify the corresponding expressions for

$$E(y_i), \qquad E(y_i^2), \qquad \text{var}\,(y_i),$$
$$E(y_i y_j), \qquad \text{cov}\,(y_i, y_j), \qquad i \neq j.$$

28 *Dirichlet($r_1, \ldots, r_{k+1}$) distribution:*

$$\frac{r_i}{\sum r_\alpha} \qquad \frac{(r_i + 1)r_i}{\left(\sum r_\alpha + 1\right)\sum r_\alpha}, \qquad \frac{r_i\left(\sum r_\alpha - r_i\right)}{\left(\sum r_\alpha\right)^2\left(\sum r_\alpha + 1\right)},$$

$$\frac{r_i r_j}{\left(\sum r_\alpha + 1\right)\sum r_\alpha} \qquad \frac{-r_i r_j}{\left(\sum r_\alpha\right)^2\left(\sum r_\alpha + 1\right)}.$$

J SOME GENERAL PROBLEMS

29 Show that the mean square deviation from t, $E((x - t)^2)$, is a minimum if $t = \mu$.

30 For a sample from the Bernoulli(p) distribution, show that the proportion $\hat{p} = \Sigma_1^n x_i/n = y/n$ has mean p and variance pq/n. Consider a possibly biased coin: how many times should the coin be tossed to be 95 percent sure that the proportion $\hat{p}$ is within 0.01 of the probability p? (Use Chebyshev and note that pq has maximum value 1/4.)

31 *Bivariate normal in Problem 2.4.24:* Show that the bivariate normal($\mu_1, \mu_2, \sigma_1, \sigma_2, \rho$) has mean and variance matrix,

$$\begin{pmatrix}\mu_1\\ \mu_2\end{pmatrix} \quad \text{and} \quad \begin{pmatrix}\sigma_1^2 & \sigma_1\sigma_2\rho\\ \sigma_1\sigma_2\rho & \sigma_2^2\end{pmatrix},$$

and correlation coefficient ρ. See Problem 4.2.37.

32 Consider a distribution for $\mathbf{x}$ on R^k with variance matrix Σ. Show that the variance of $\Sigma_i t_i x_i = \mathbf{t}'\mathbf{x}$ is given by $\mathbf{t}'\Sigma\mathbf{t}$. Hence deduce that Σ is positive semidefinite; use the results in Supplement Section 2.5C.

33 Consider a standard normal distribution for z and define a distribution for (z_1, z_2) on R^2 by $(z_1, z_2) = (z, |z|)$. Calculate the mean and variance matrix for (z_1, z_2). Calculate the correlation coefficient ρ_{12}; are z_1 and z_2 statistically independent?

34 For the uniform distribution over the disk $x_1^2 + x_2^2 < 4$, show that the mean and variance matrix are

$$\begin{pmatrix}0\\ 0\end{pmatrix}, \qquad \begin{pmatrix}1 & 0\\ 0 & 1\end{pmatrix}.$$

Note that these are the same as the standard normal on R^2.

35 (*continuation*) Show that the bivariate normal in Problem 31 has the same mean and covariance matrix as the uniform distribution over the elliptical disk

$$\frac{1}{1-\rho^2}\left[\frac{(y_1 - \mu_1)^2}{\sigma_1^2} - 2\rho\frac{(y_1 - \mu_1)(y_2 - \mu_2)}{\sigma_1\sigma_2} + \frac{(y_2 - \mu_2)^2}{\sigma_2^2}\right] < 4.$$

36 For the uniform distribution over the ball $\Sigma_1^k x_i^2 < k + 2$, show that the mean and variance matrix are $\mathbf{0}$ and I, where $\mathbf{0}$ is the 0 vector in R^k and I is the $k \times k$ identity matrix. *Method:* Use the intuitively reasonable fact that $r = (x_1^2 + \cdots + x_k^2)^{1/2}$ has a marginal distribution with density cr^{k-1} on $(0, \sqrt{k+2})$.

37 (*continuation*) Show that the multivariate normal($\boldsymbol{\mu}$; Σ) with $|\Sigma| \neq 0$ as given in Section 2.4G has the same mean and variance matrix as the uniform distribution on the elliptical ball $(\mathbf{y} - \boldsymbol{\mu})'\Sigma^{-1}(\mathbf{y} - \boldsymbol{\mu}) < k + 2$.

5.4

THE LOCATION AND SCALING OF A SAMPLE

Consider a chemist with a sample of n determinations of the yield for a new isotope process. He would certainly be interested in the mean yield for the process and perhaps in the variance of the process. With the sample of n values in hand he would naturally wonder to what degree the location and scaling for the sample represented the true mean and variance for the process.

In this section we develop the powerful formulas mentioned at the beginning of this chapter, formulas that relate the location and scaling of a sample to the location and scaling of the distribution being sampled. The applications are far-reaching, both in probability theory and in statistics.

We first examine some functions that describe the location and scaling of a sample of n values on the real line R, then functions that describe location and scaling on R^k. We then consider the distribution of a sample from an initial distribution and investigate some distribution properties of these location and scaling functions.

A SAMPLE LOCATION ON R

Consider a sequence $(x_1, \ldots, x_n)$ of values on the real line R. A simple measure of the location of this sample is given by the *sample average*

(1) $$\bar{x} = \frac{x_1 + \cdots + x_n}{n} = \frac{1}{n}\sum x_i.$$

If we consider unit masses at each of the points x_i, then $\bar{x}$ is the center of gravity, the point about which the array balances; see Figure 5.8. For we have

(2) $$\sum_1^n (x_i - \bar{x}) = \sum x_i - n\bar{x} = 0.$$

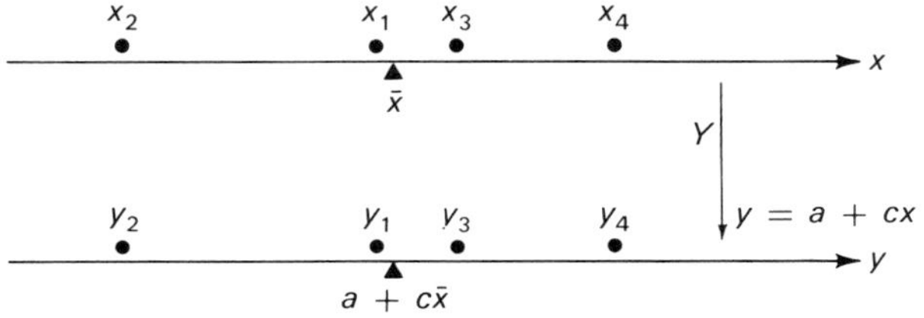

FIGURE 5.8
Sample (x_1, x_2, x_3, x_4) with average $\bar{x}$; corresponding sample (y_1, y_2, y_3, y_4) with average $\bar{y} = a + c\bar{x}$.

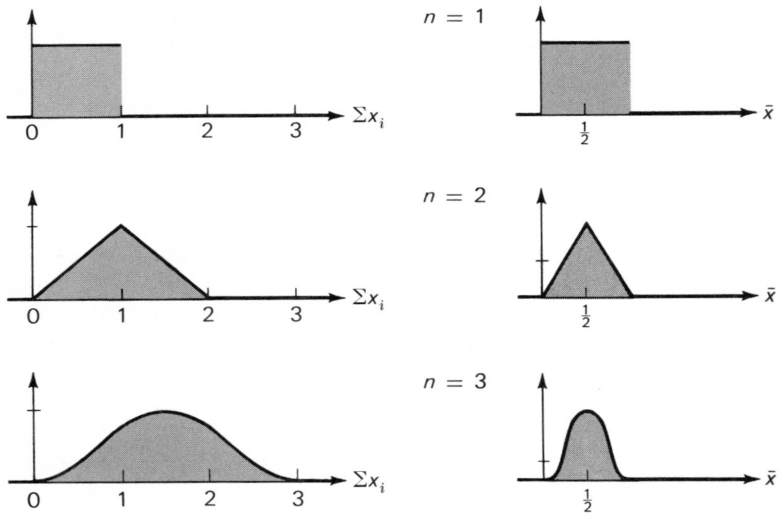

FIGURE 5.9
Distribution of $\Sigma_1^n x_i$ and $\bar{x}$ for samples of $n = 1, 2, 3$ from the uniform(0, 1) distribution.

If we obtain a new sequence $(y_1, \ldots, y_n)$ from an initial sequence $(x_1, \ldots, x_n)$ by an affine transformation $y = a + cx$, then

(3) $$\bar{y} = \frac{1}{n} \sum (a + cx_i) = a + c\bar{x}.$$

Now let $(x_1, \ldots, x_n)$ be a sample from a distribution on R with mean μ and variance σ^2. We then have a distribution for $\mathbf{x} = (x_1, \ldots, x_n)'$ on R^n in which the coordinates are statistically independent and each has the same given distribution with mean μ and variance σ^2. Consider some characteristics of the marginal distribution of the function $\bar{x}$; note that $\bar{x}$ is a function from R^n to R^1.

From formulas (5.2.4) and (5.3.7) we obtain

(4) $$E\left(\sum x_i\right) = n\mu, \qquad \operatorname{var}\left(\sum x_i\right) = n\sigma^2.$$

Then from formulas (5.2.3) and (5.3.5), we obtain

(5) $$E(\bar{x}) = \mu, \qquad \operatorname{var}(\bar{x}) = \frac{\sigma^2}{n}.$$

Note that the distribution of $\bar{x}$ has the same location μ as the distribution being sampled, but its variance is reduced by a factor $1/n$, and its standard deviation is reduced by a factor $1/\sqrt{n}$. For an example, see Figures 5.9 and 5.10.

B SAMPLE SCALING ON *R*

Consider again a sequence $(x_1, \ldots, x_n)$ of values on the real line R. The residual vector $(x_1 - \bar{x}, \ldots, x_n - \bar{x})$ shows how the values are spread out with respect to the sample average $\bar{x}$. First, some geometrical motivation.

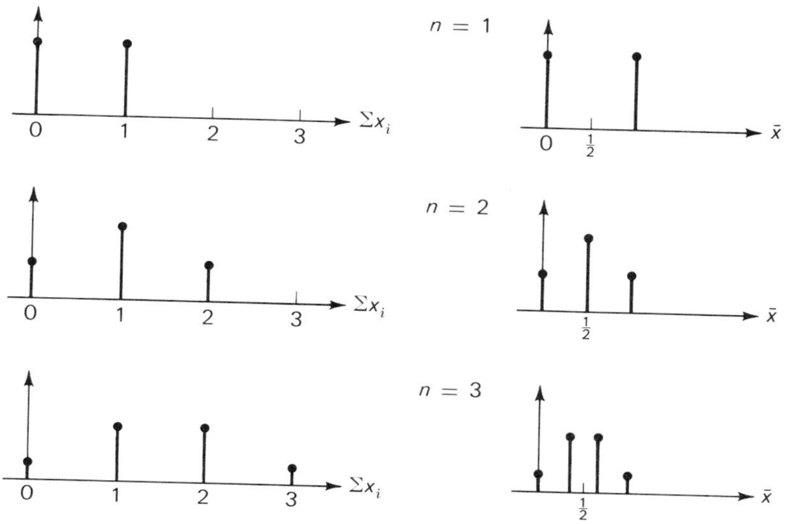

FIGURE 5.10
Distribution of $\Sigma_1^n x_i$ and $\bar{x}$ for samples of $n = 1, 2, 3$ from the Bernoulli(1/2) distribution.

We consider $\mathbf{x} = (x_1, \ldots, x_n)'$ as a point in R^n; see Figure 5.11. The point $\bar{x}\mathbf{1} = (\bar{x}, \ldots, \bar{x})'$ is the *fitted point* with all coordinates equal and yet with coordinates that add to the same total as for the original vector; that is, $\bar{x}\mathbf{1}$ lies on $\mathfrak{L}(\mathbf{1})$ and also lies on the plane $\Sigma x_i = c$ that passes through the original $\mathbf{x}$. The point $\mathbf{x} - \bar{x}\mathbf{1}$ is the

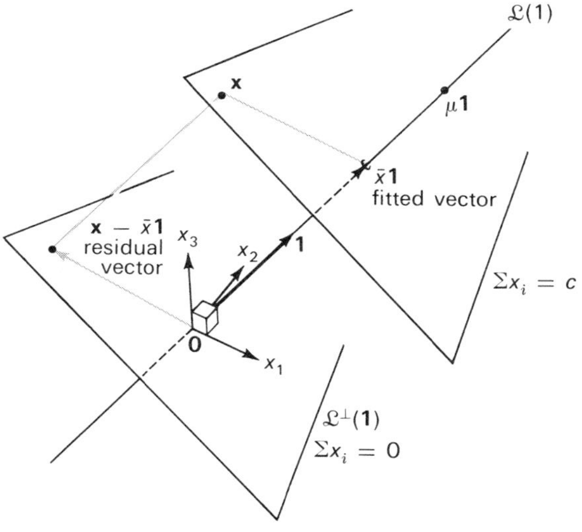

FIGURE 5.11
Vector $\mathbf{x}$: its projection on the extended 1-vector $\mathfrak{L}(\mathbf{1})$ gives the fitted vector $\bar{x}\mathbf{1}$; its projection on the orthogonal complement $\mathfrak{L}^{\perp}(\mathbf{1})$ of the vector $\mathbf{1}$ gives the residual vector $\mathbf{x} - \bar{x}\mathbf{1}$. Note that the sum of these two components reconstructs the original vector $\mathbf{x}$.

residual vector or the deviation vector from the fitted $\bar{x}\mathbf{1}$. Note that these vectors $\bar{x}\mathbf{1}$ and $\mathbf{x} - \bar{x}\mathbf{1}$ are orthogonal: the inner product is

$$(\bar{x}\mathbf{1}, \mathbf{x} - \bar{x}\mathbf{1}) = \sum_1^n \bar{x}(x_i - \bar{x}) = \bar{x}\sum_1^n (x_i - \bar{x}) = 0.$$

Also note that they add to give the original vector,

$$\bar{x}\mathbf{1} + (\mathbf{x} - \bar{x}\mathbf{1}) = \mathbf{x};$$

and in addition that their squared lengths add to give the squared length of the original vector,

$$(6) \qquad n\bar{x}^2 + \sum (x_i - \bar{x})^2 = \sum x_i^2,$$

an easily checked consequence of the orthogonality. Thus we see that $\bar{x}\mathbf{1}$ is the *projection* of $\mathbf{x}$ onto the line $\mathcal{L}(\mathbf{1})$; that is, it has an orthogonal residual. And we see that $\mathbf{x} - \bar{x}\mathbf{1}$ is the *projection* of $\mathbf{x}$ onto the plane $\mathcal{L}^{\perp}(\mathbf{1}) = \{\mathbf{y} : \Sigma y_i = 0\}$ consisting of all vectors orthogonal to $\mathbf{1}$; that is, it has an orthogonal residual which is just $\bar{x}\mathbf{1}$.

We will see that an ''average'' squared deviation, called the *sample variance,*

$$(7) \qquad s_x^2 = \frac{1}{n-1}\sum (x_i - \bar{x})^2 = \frac{1}{n-1}\left(\sum x_i^2 - n\bar{x}^2\right),$$

provides a very reasonable measure of scaling in squared units, and that root-average-squared deviation, the *sample standard deviation,* s_x, provides a reasonable measure of scaling in given units; see Figure 5.12.

If we obtain a new sequence $(y_1, \ldots, y_n)$ from a given sequence $(x_1, \ldots, x_n)$ by an affine transformation $y = a + cx$, it is easily checked that

$$(8) \qquad s_y^2 = c^2 s_x^2, \qquad s_y = |c| s_x.$$

Now let $(x_1, \ldots, x_n)$ be a sample from a distribution on R with mean μ and variance σ^2. We then have a distribution for $\mathbf{x} = (x_1, \ldots, x_n)'$ on R^n with independent coordinates each having the given distribution with mean μ and variance σ^2. Consider some characteristics of the marginal distribution of the function s_x^2: note that s_x^2 is a function from R^n to R^1.

If we replace x_i by $x_i - \mu$ in the identity (6) and rearrange, we obtain

$$\sum (x_i - \bar{x})^2 = \sum (x_i - \mu)^2 - n(\bar{x} - \mu)^2.$$

Then using $E(x_i - \mu)^2 = \text{var}(x_i) = \sigma^2$ and $E(\bar{x} - \mu)^2 = \text{var}(\bar{x}) = \sigma^2/n$, we obtain

$$(9) \qquad \begin{aligned} E\left(\sum (x_i - \bar{x})^2\right) &= n\sigma^2 - \frac{n\sigma^2}{n} = (n-1)\sigma^2, \\ E(s_x^2) &= \sigma^2. \end{aligned}$$

Thus the mean value of the sample variance is equal to the variance of the distribution being sampled.

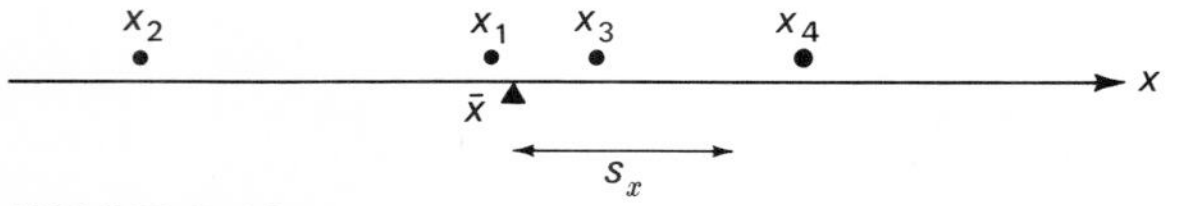

FIGURE 5.12
Sample average $\bar{x}$ and sample standard deviation s_x.

C SAMPLE LOCATION ON R^k

Consider a sequence $(\mathbf{x}_1, \ldots, \mathbf{x}_n)$ of n points in R^k. For example, an $\mathbf{x}$ might record (length, girth, age)′ for a fish in a sample of n fish. We can average the jth coordinates of these vectors,

$$\bar{x}_j = \frac{x_{j1} + \cdots + x_{jn}}{n} = \frac{1}{n} \sum_{i=1}^{n} x_{ji},$$

and then form the *sample average vector,*

$$(10) \qquad \bar{\mathbf{x}} = \begin{pmatrix} \bar{x}_1 \\ \vdots \\ \bar{x}_k \end{pmatrix} = \frac{1}{n} \sum_{i=1}^{n} \mathbf{x}_i,$$

as a point in R^k, the sample average vector $\bar{\mathbf{x}}$ gives the center of gravity of the n points $\mathbf{x}_1, \ldots, \mathbf{x}_n$; see Figure 5.13. For the example mentioned earlier $\bar{\mathbf{x}}$ records (average length, average girth, average age)′.

Now let $(\mathbf{x}_1, \ldots, \mathbf{x}_n)$ be a sample from a distribution on R^k with mean $\boldsymbol{\mu}$ and variance matrix Σ. Consider some characteristics of the marginal distribution of the function $\bar{\mathbf{x}}$; note that $\bar{\mathbf{x}}$ is a function from $(R^k)^n = R^{kn}$ into R^k.

From formulas (5.2.12) and (5.3.20) we obtain

$$(11) \qquad E\left(\sum_{i=1}^{n} \mathbf{x}_i\right) = n\boldsymbol{\mu}, \qquad \text{VAR}\left(\sum_{i=1}^{n} \mathbf{x}_i\right) = n\Sigma.$$

Then from formulas (5.2.11) and (5.3.17), we obtain

$$(12) \qquad E(\bar{\mathbf{x}}) = \boldsymbol{\mu}, \qquad \text{VAR}\,(\bar{\mathbf{x}}) = n^{-1}\Sigma.$$

Thus, in particular, $E(\bar{x}_1) = \mu_1$, var $(\bar{x}_1) = \sigma_{11}/n$, and cov $(\bar{x}_1, \bar{x}_2) = \sigma_{12}/n$. Thus the distribution of $\bar{\mathbf{x}}$ has the same location as the distribution being sampled, but its variance matrix is reduced by a factor $1/n$.

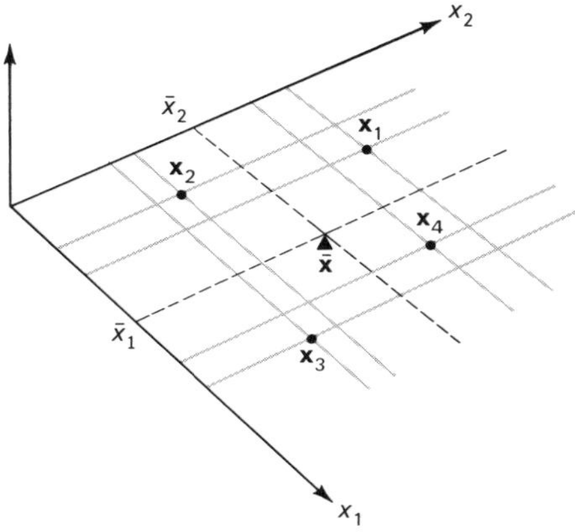

FIGURE 5.13
Sample $(\mathbf{x}_1, \mathbf{x}_2, \mathbf{x}_3, \mathbf{x}_4)$ on the plane. The sample average vector $\bar{x}$.

D SAMPLE SCALING ON R^k

Consider again a sequence $(\mathbf{x}_1, \ldots, \mathbf{x}_n)$ of n points in R^k, say the measurements for the n fish. We follow the pattern in Section B; for this we briefly reverse our geometry and consider k points in R^n. From the original first coordinates we form

$$(x_{11}, \ldots, x_{1n}),$$

and then the residual

$$(x_{11} - \bar{x}_1, \ldots, \bar{x}_{1n} - \bar{x}_1).$$

And from the original second coordinates we form

$$(x_{21}, \ldots, x_{2n})$$

and then the residual

$$(x_{21} - \bar{x}_2, \ldots, x_{2n} - \bar{x}_2);$$

see Figure 5.14. The identity

$$(13) \qquad \sum x_{1i}x_{2i} = \sum (x_{1i} - \bar{x}_1)(x_{2i} - \bar{x}_2) + n\bar{x}_1\bar{x}_2$$

is easily checked algebraically. A measure of how the first coordinate varies with respect to the second coordinate is given by an "average" product deviation, the *sample covariance,*

$$(14) \qquad s_{12} = \frac{1}{n-1} \sum (x_{1i} - \bar{x}_1)(x_{2i} - \bar{x}_2) = \frac{1}{n-1}\left(\sum x_{1i}x_{2i} - n\bar{x}_1\bar{x}_2\right).$$

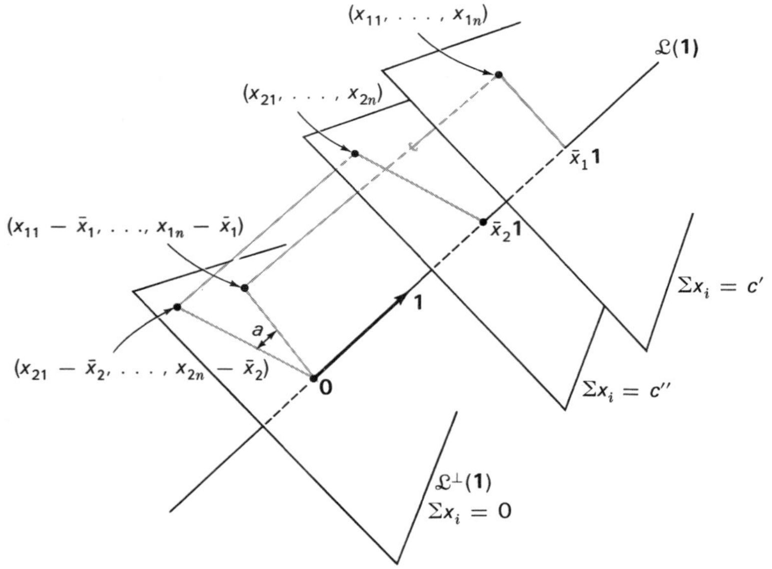

FIGURE 5.14
Vectors $(x_{11}, \ldots, x_{1n})$ and $(x_{21}, \ldots, x_{2n})$ in R^n. The corresponding residuals $(x_{11} - \bar{x}_1, \ldots, x_{1n} - \bar{x}_1)$ and $(x_{21} - \bar{x}_2, \ldots, x_{2n} - \bar{x}_2)$ in $\mathcal{L}^{\perp}(\mathbf{1})$. The angle a between the residuals; the correlation r_{12} is the cosine of the angle a.

Note that the sample covariance of the first coordinate with the first coordinate is just the sample variance, say s_{11}, for that coordinate. The sample covariances can be assembled as a *sample variance matrix*,

$$S = \begin{pmatrix} s_{11} & \cdots & s_{1k} \\ \vdots & & \vdots \\ s_{k1} & \cdots & s_{kk} \end{pmatrix} \tag{15}$$

$$= \frac{1}{n-1} \begin{pmatrix} \sum (x_{1i} - \bar{x}_1)^2 & \cdots & \sum (x_{1i} - \bar{x}_1)(x_{ki} - \bar{x}_k) \\ \vdots & & \vdots \\ \sum (x_{ki} - \bar{x}_k)(x_{1i} - \bar{x}_1) & \cdots & \sum (x_{ki} - \bar{x}_k)^2 \end{pmatrix}.$$

Note that $(n-1)S$ is the inner product matrix for the residual vectors.

Now let $(\mathbf{x}_1, \ldots, \mathbf{x}_n)$ be a sample from a distribution on R^k with mean $\boldsymbol{\mu}$ and variance matrix Σ. Consider some characteristics of the marginal distribution of the sample variance matrix.

If we replace each x_{1i} by $x_{1i} - \mu_1$ and each x_{2i} by $x_{2i} - \mu_2$ in the identity (13) and rearrange, we obtain

$$\sum (x_{1i} - \bar{x}_1)(x_{2i} - \bar{x}_2) = \sum (x_{1i} - \mu_1)(x_{2i} - \mu_2) - n(\bar{x}_1 - \mu_1)(\bar{x}_2 - \mu_2). \tag{16}$$

Then from $E((x_{1i} - \mu_1)(x_{2i} - \mu_2)) = \sigma_{12}$ and $E((\bar{x}_1 - \mu_1)(\bar{x}_2 - \mu_2)) = \operatorname{cov}(\bar{x}_1, \bar{x}_2) = \sigma_{12}/n$, we obtain

$$E(s_{12}) = \sigma_{12}. \tag{17}$$

Accordingly, we have

$$E(S) = \Sigma; \tag{18}$$

the mean value of the sample variance matrix is the variance matrix of the distribution being sampled.

E SAMPLE CORRELATION

Consider a sequence $(\mathbf{x}_1, \ldots, \mathbf{x}_n)$ of points in R^k, and for simplicity take $k = 2$. We follow the pattern in Sections B and D and briefly reverse our geometry and consider two points in R^n. From the first coordinates we obtain the deviation vector

$$(x_{11} - \bar{x}_1, \ldots, x_{1n} - \bar{x}_1),$$

and from the second coordinates we obtain the deviation vector

$$(x_{21} - \bar{x}_2, \ldots, x_{2n} - \bar{x}_2).$$

The cosine of the angle between these two deviation vectors (see Figure 5.14) is called the *sample correlation coefficient* between the first two coordinates:

$$r_{12} = \frac{\sum (x_{1i} - \bar{x}_1)(x_{2i} - \bar{x}_2)}{\left(\sum (x_{1i} - \bar{x}_1)^2\right)^{1/2}\left(\sum (x_{2i} - \bar{x}_2)^2\right)^{1/2}} = \frac{s_{12}}{(s_{11}s_{22})^{1/2}}.$$

If the two deviation vectors are pointing in almost the same direction, then r_{12} will be close to $+1$; if they are pointing in almost opposite directions, then r_{12} will be close to

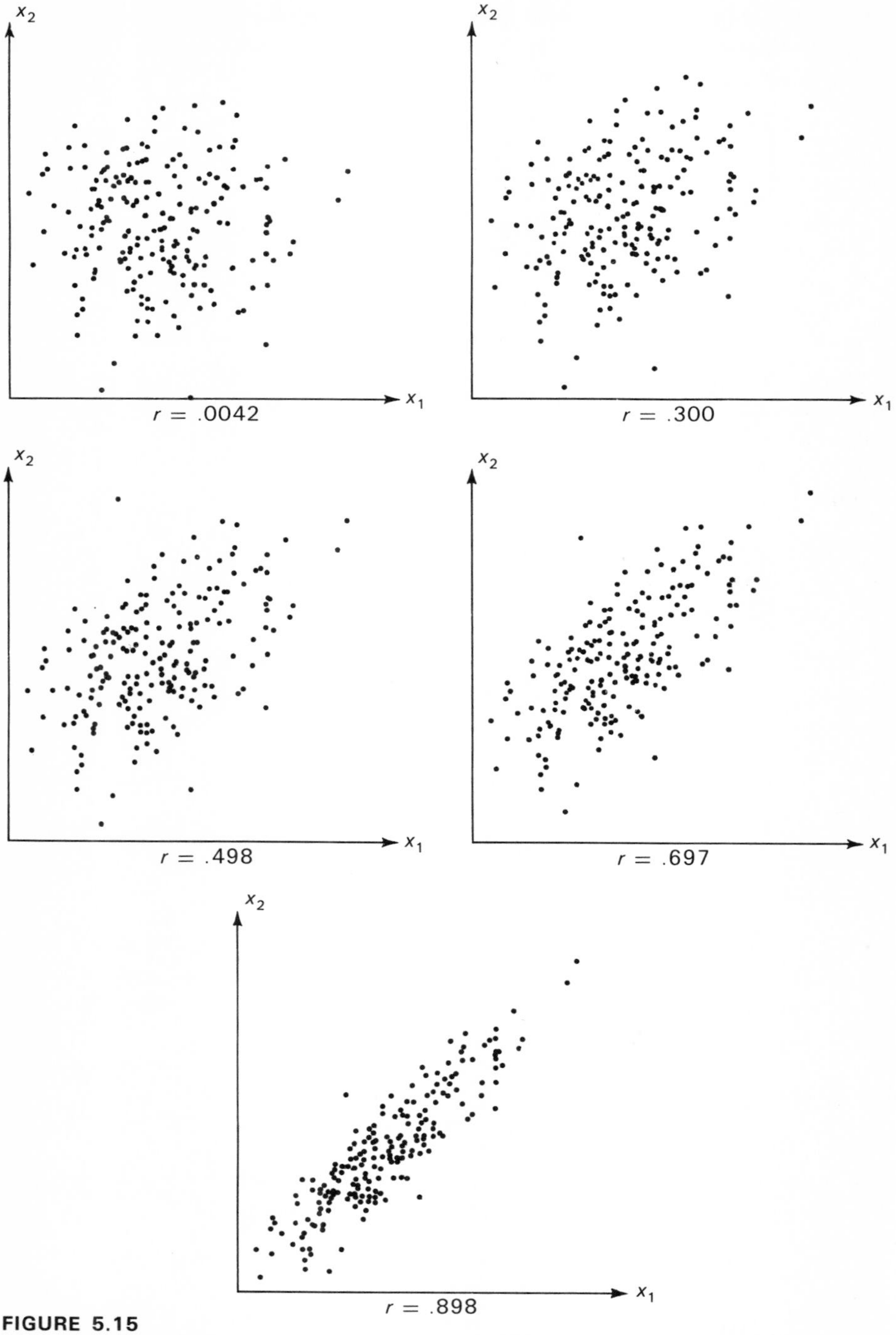

FIGURE 5.15
Samples of 100 plotted on the plane R^2 with correlations as marked. Samples were obtained by computer from bivariate normal distributions with $\sigma_1 = \sigma_2$ and $\rho = 0, 0.3, 0.5, 0.7, 0.9$.

-1; r_{12} measures how the values for the first coordinate tend to move with those for the second coordinate.

The Schwarz inequality (Problem 5.8.17) can be applied to the deviation vectors using a unit measure for each of the n coordinate positions; we obtain

$$\left(\sum (x_{1i} - \bar{x}_1)(x_{2i} - \bar{x}_2)\right)^2 \leq \sum (x_{1i} - \bar{x}_1)^2 \sum (x_{2i} - \bar{x}_2)^2$$

with equality if and only if one deviation vector is a multiple of the other; note equivalent forms of the inequality

$$s_{12}^2 \leq s_{11}s_{22}, \qquad -1 \leq r_{12} \leq 1.$$

Now we return to the original geometry of a sequence $(\mathbf{x}_1, \ldots, \mathbf{x}_n)$ of n points in R^2. In practice, with reasonable response vectors, a large positive correlation means a rather compressed array of points tilted up to the right, and a large negative correlation means a distribution with a compressed array of points tilted down to the right; see Figure 5.15.

F EXERCISES

1 For a new sample $(y_1, \ldots, y_n)$ obtained from a given sample $(x_1, \ldots, x_n)$ by the affine transformation $y = a + cx$, show that $s_y^2 = c^2 s_x^2$, $s_y = |c| s_x$.

2 Show that the sum of squares of deviations from t, $\Sigma (y_i - t)^2$, is a minimum if $t = \bar{y}$.

3 The following sample data were obtained in a chemical determination of sodium content: 17.1, 14.9, 15.2, 18.1, 16.4, 19.1, 18.5, 14.6, 17.3. Calculate the sample mean $\bar{y}$, the sample variance s_y^2, and the sample standard deviation s_y.

4 Ten determinations were made in a spectral line, giving the data

65.175	65.169	65.172	65.183	65.179
65.178	65.185	65.180	65.181	65.180

Calculate the sample mean $\bar{y}$, the sample variance s_y^2, and the sample standard deviation s_y: for easy arithmetic use formulas (3) and (8).

5 The following sample of 5 was obtained on the vector variable $\begin{pmatrix} x \\ y \end{pmatrix}$, where x is the amount of fertilizer and y is the yield

$$\left(\begin{pmatrix} 1 \\ 5 \end{pmatrix}, \begin{pmatrix} 2 \\ 7 \end{pmatrix}, \begin{pmatrix} 3 \\ 11 \end{pmatrix}, \begin{pmatrix} 4 \\ 10 \end{pmatrix}, \begin{pmatrix} 5 \\ 13 \end{pmatrix}\right).$$

Calculate the sample mean and sample variance matrix

$$\begin{pmatrix} \bar{x} \\ \bar{y} \end{pmatrix}, \qquad S = \begin{pmatrix} s_{xx} & s_{xy} \\ s_{yx} & s_{yy} \end{pmatrix}.$$

Calculate the sample correlation coefficient r_{xy}. In this example the values of x were chosen by design; thus there would not be a distribution for (x, y) on R^2.

6 The following sample of 7 was obtained on the vector $(y_1, y_2)'$, where y_1 is the tensile strength and y_2 the Brinell hardness of cold-drawn copper.

$$\left(\begin{pmatrix} 106.1 \\ 40.4 \end{pmatrix}, \begin{pmatrix} 106.3 \\ 40.8 \end{pmatrix}, \begin{pmatrix} 104.4 \\ 39.5 \end{pmatrix}, \begin{pmatrix} 103.8 \\ 37.0 \end{pmatrix}, \begin{pmatrix} 101.5 \\ 33.2 \end{pmatrix}, \begin{pmatrix} 100.6 \\ 29.9 \end{pmatrix}, \begin{pmatrix} 106.2 \\ 40.6 \end{pmatrix}\right).$$

Calculate the sample mean, the sample variance matrix, and the sample correlation coefficient. For easier arithmetic, relocate and rescale the y_1 values, relocate and rescale the

y_2 values, then use appropriate correction formulas—for example, simplified versions of those in Problem 7.

G PROBLEMS

7 For a new sample $(\mathbf{y}_1, \ldots, \mathbf{y}_n)$ obtained from a given sample $(\mathbf{x}_1, \ldots, \mathbf{x}_n)$ by the affine transformation $\mathbf{y} = \mathbf{a} + C\mathbf{x}$, show that

$$\bar{\mathbf{y}} = \mathbf{a} + C\bar{\mathbf{x}}, \qquad S_{\mathbf{yy}} = CS_{\mathbf{xx}}C',$$

where $S_{\mathbf{yy}}$ $(S_{\mathbf{xx}})$ is the sample variance matrix for the $\mathbf{y}_i$'s $(\mathbf{x}_i$'s).

8 The sequence $(\mathbf{x}_1, \ldots, \mathbf{x}_n)$ of n points in R^k can be presented as a matrix,

$$X = \begin{pmatrix} x_{11} & \cdots & x_{1n} \\ \vdots & & \vdots \\ x_{k1} & \cdots & x_{kn} \end{pmatrix}.$$

(a) Check that the residual vectors described in Section D are given by the row vectors of

$$X - \bar{\mathbf{x}}\mathbf{1}',$$

where $\mathbf{1}$ is the 1-vector in R^n.

(b) Show that the sample variance matrix is given by

$$S = (n-1)^{-1}(X - \bar{\mathbf{x}}\mathbf{1}')(X - \bar{\mathbf{x}}\mathbf{1}')' = (n-1)^{-1}(XX' - n\bar{\mathbf{x}}\bar{\mathbf{x}}').$$

9 Show that the sum of squared deviations $\Sigma\,(y_i - bx_i)^2$ of $\mathbf{y}$ from a point $b\mathbf{x}$ in $\mathcal{L}(\mathbf{x})$ is minimized by choosing $b = \Sigma\, x_i y_i / \Sigma\, x_i^2$. *Method:* The expression is quadratic in b; complete the square in b, obtaining an expression $a(b - b_0)^2 + d$.

10 Let $(u_1, \ldots, u_n)$ be a sample from the uniform(0, 1) distribution. And let u and v be the smallest and largest in the sample. The results in Problem 3.6.12 can be specialized to give the following probability density for (u, v):

$$f(u, v) = n(n-1)(v-u)^{n-2} \qquad \text{if } 0 < u < v < 1$$

and zero otherwise. Determine the mean and variance matrix of this distribution.

11 (*continuation*) Show that the interorder statistic distances $c_1, c_2, \ldots, c_{n+1}$ all have mean $1/(n+1)$; a very short proof is possible using Problem 3.6.10. Thus show that the order statistic $u_{(i)}$ has mean $i/(n+1)$.

12 (*continuation*) The joint density of $u_{(r)}, u_{(r+s)}$ can be obtained from Problem 3.6.12:

$$f(u_{(r)}, u_{(r+s)}) = \frac{n!}{(r-1)!(s-1)!(n-r-s)!}\, u_{(r)}^{r-1}(u_{(r+s)} - u_{(r)})^{s-1}(1 - u_{(r+s)})^{n-r-s}$$

on $0 < u_{(r)} < u_{(r+s)} < 1$ and zero elsewhere. Determine $E(u_{(r)}u_{(r+s)})$ and deduce $\operatorname{cov}(u_{(r)}, u_{(r+s)})$. Record the mean and variance matrix for $(u_{(1)}, \ldots, u_{(n)})$.

H PROBLEMS: COMPUTER USAGE

Program the computer to calculate as follows: (a) from an incoming sample of n, calculate $\bar{y}$ and s_y^2; (b) from a large number N of repetitions on $\bar{y}$ and s_y^2, calculate the long-run average $\hat{\mu}_{\bar{y}} = \Sigma\, \bar{y}/N$ for $\bar{y}$, the long-run variance $\hat{\sigma}_{\bar{y}}^2 = \Sigma\,(\bar{y} - \hat{\mu}_{\bar{y}})^2/(N-1)$ for $\bar{y}$, and the long-run average $\hat{\mu}_{s_y^2} = \Sigma\, s_y^2/N$ for s_y^2.

13 Obtain a sample of $n = 10$ from the uniform distribution on $\{0, 1, \ldots, 9\}$, and calculate $\bar{y}$ and s_y^2. For $N = 1000$ repetitions, calculate the long-run averages,

$$\hat{\mu}_{\bar{y}}, \qquad \hat{\sigma}_{\bar{y}}^2, \qquad \hat{\mu}_{s_y^2}.$$

Compare with the theoretical values $E(\bar{y})$, Var $(\bar{y})$, and $E(s_y^2)$.

14 Obtain a sample of 5 from the triangular distribution with $f(y) = 2y$ on (0, 1) and zero elsewhere (use Problem 4.2.35) and calculate $\bar{y}$ and s_y^2. For $N = 1000$ repetitions, calculate

$$\hat{\mu}_{\bar{y}}, \qquad \hat{\sigma}_{\bar{y}}^2, \qquad \hat{\mu}_{s_y^2}.$$

Compare with the theoretical values $E(\bar{y})$, var $(\bar{y})$, and $E(s_y^2)$.

5.5

CONDITIONAL LOCATION AND SCALING

We noted in Chapter 4 that partial information is sometimes available concerning a realization and that conditional probability then provides the appropriate description. Certainly the location and scaling of such a conditional distribution would be of prime importance. Consider some examples.

An early example involved the measurement of (x, y), where x is the height of a father and y the height of the eldest son. A large batch of data led to a bivariate normal as a reasonable approximation. For a father of height x with a young son, the conditional distribution of y given x would describe the future adult height of the son. The mean and variance of this distribution would be of particular interest.

Or consider a distribution describing (x, y), where x is the length and y the age for a fish of a certain species. Length is easy to determine and y is difficult to determine. For a fish of length x, the conditional distribution of y, given x, provides an estimating procedure for age. The mean and variance of this distribution would be of particular importance.

A CONDITIONAL MEANS AND VARIANCES ON R^2

Consider a distribution for (x, y) on the plane R^2 with probability function p in the discrete case or probability density function f in the absolutely continuous case. Let p_1 or f_1 describe the marginal distribution for x; and $p_2(\cdot : x)$ or $f_2(\cdot : x)$ describe the conditional distribution for y given x.

We examine the conditional distribution of y given x; see Figure 5.16. For this distribution, let $\mu_2(x)$ be the mean and $\sigma_2^2(x)$ be the variance; assuming they exist we have

(1) $$\mu_2(x) = E(y : x) = \sum_y y p_2(y : x) \qquad \text{if discrete,}$$
$$= \int_{-\infty}^{\infty} y f_2(y : x)\, dy \qquad \text{absolutely continuous,}$$

(2) $$\sigma_2^2(x) = \text{var}\,(y : x) = \sum_y (y - \mu_2(x))^2 p_2(y : x) \qquad \text{if discrete,}$$
$$= \int_{-\infty}^{\infty} (y - \mu_2(x))^2 f_2(y : x)\, dy \qquad \text{absolutely continuous.}$$

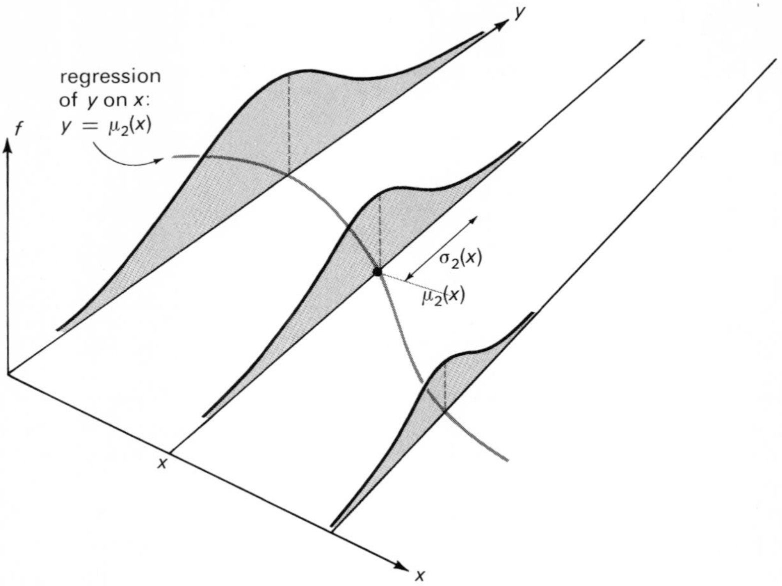

FIGURE 5.16
Mean $\mu_2(x)$ and standard deviation $\sigma_2(x)$ of the conditional distribution of y given x. The regression curve of y on x is given by $y = \mu_2(x)$.

EXAMPLE 1

A distribution on R^2: Consider the distribution for (x, y) with density f as examined in Example 4.2.3:

$$f(x, y) = 6x \qquad \text{if } 0 < x < 1, \quad 0 < y < 1 - x$$

and zero otherwise. The mean and variance matrix have been calculated in Example 5.3.8. The conditional density for y given x is

$$f_2(y : x) = (1 - x)^{-1} \qquad \text{if } 0 < y < 1 - x$$

and zero otherwise; this is the uniform(0, $1 - x$) distribution. The mean and variance for the uniform distribution are available from Example 5.3.2; thus

$$\mu_2(x) = \frac{1 - x}{2}, \quad \sigma_2^2(x) = \frac{(1 - x)^2}{12} \qquad \text{for } 0 < x < 1.$$

Suppose that we view $\mu_2(x)$ and $\sigma_2^2(x)$ as functions of x. Then we define:

DEFINITION 1

The **regression** of y on x is the function $\mu_2(x)$ recording the conditional mean for each value of x.

DEFINITION 2

The **variance about regression** on x is the function $\sigma_2^2(x)$ recording the conditional variance for each value of x.

See Figure 5.16.

Now suppose that we view $\mu_2(x)$ as a function of (x, y), a function on R^2 that happens not to depend on the second coordinate y. Then we have:

LEMMA 3

The marginal mean is equal to the mean value of the conditional mean:

(3) $\quad \mu_2 = E(\mu_2(x)) \quad \text{or} \quad E(y) = E(E(y:x)).$

Proof We record the proof for the absolutely continuous case:

$$E(\mu_2(x)) = \int_{-\infty}^{\infty}\left(\int_{-\infty}^{\infty} yf_2(y:x)\,dy\right) f_1(x)\,dx = \iint_{-\infty}^{\infty} yf_2(y:x)f_1(x)\,dx\,dy$$
$$= \iint_{-\infty}^{\infty} yf(x, y)\,dx\,dy = \mu_2.$$

This lemma says that we can calculate a mean value in two stages: first the conditional mean over a section of the distribution; then the mean of that conditional mean.

Now suppose that we view $\sigma_2^2(x)$ as a function of (x, y), a function on R^2 that happens not to depend on the second coordinate. Then we have:

LEMMA 4

The marginal variance is equal to the mean conditional variance plus the variance of the conditional mean:

$$\sigma_2^2 = E(\sigma_2^2(x)) + \text{var}\,(\mu_2(x))$$
(4)
$$\text{var}\,(y) = E(\text{var}\,(y:x)) + \text{var}\,(E(y:x)).$$

In brief, variance is mean variance about regression plus variance of regression.

Proof By formula (5.3.3) we have $E(w^2) = \text{var}\,(w) + (E(w))^2$; let w have the conditional distribution of $y - \mu_2$, given x:

$$E((y - \mu_2)^2 : x) = \text{var}\,(y:x) + (E(y:x) - \mu_2)^2.$$

Then, using Lemma 3, we obtain

$$\text{var}\,(y) = E((y - \mu_2)^2) = E(E((y - \mu_2)^2 : x)) = E(\text{var}\,(y:x)) + \text{var}\,(E(y:x)).$$

For an interpretation, see Figure 5.16 again. The term var (y) is the variance along the second axis when all the probability is projected onto that axis. Now suppose that the probability is collapsed along x sections and placed on the curve $\mu_2(x)$; in doing this, we relinquish $E(\text{var}\,(y:x))$, the mean variance about regression. And, indeed, we are left with just the variance along the second axis of this collapsed distribution.

B SOME EXAMPLES ON R^2

EXAMPLE 2

The multinomial(n, p_1, p_2, p_3) distribution: Consider the multinomial(n, p_1, p_2, p_3) distribution for (y_1, y_2, y_3). As noted before $y_3 = n - y_1 - y_2$; thus, effectively, we have a distribution for (y_1, y_2) on R^2. The marginal distribution for y_1 is binomial(n, p_1); see Problem 3.5.37. The conditional distribution for y_2 given y_1 is binomial($n - y_1, p_2/(p_2 + p_3)$); see Example 4.2.2. The mean and variance of the binomial distribution are given in Examples 5.2.6 and 5.3.5; thus

$$(5) \qquad \mu_2(y_1) = (n - y_1)\frac{p_2}{p_2 + p_3}, \qquad \sigma_2^2(y_1) = (n - y_1)\frac{p_2}{p_2 + p_3}\frac{p_3}{p_2 + p_3}.$$

The regression of y_2 on y_1 is linear, decreasing to zero at $y_1 = n$. The variance about regression decreases to zero as y_1 increases to the maximum value n. See Figure 5.17.

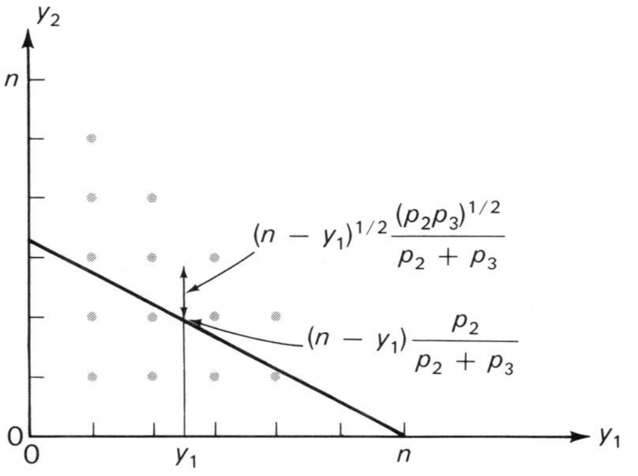

FIGURE 5.17
Regression of y_2 on y_1 for the multinomial(n, p_1, p_2, p_3).

EXAMPLE 3

The bivariate normal distribution: Consider the bivariate normal($\mu_1, \mu_2, \sigma_1, \sigma_2, \rho$) for (y_1, y_2); see Problem 4.2.37. The marginal for y_1 is normal(μ_1, σ_1); and the conditional of y_2 given y_1 is normal($\mu_2 + \rho\sigma_2(y_1 - \mu_1)/\sigma_1, \sigma_2(1 - \rho^2)^{1/2}$). Thus we obtain

$$(6) \qquad \mu_2(y_1) = \mu_2 + \rho\frac{\sigma_2}{\sigma_1}(y_1 - \mu_1), \qquad \sigma_2^2(y_1) = \sigma_2^2(1 - \rho^2);$$

see Figure 5.18. The regression of y_2 on y_1 is linear, tilted up to the right if ρ is positive and down to the right if ρ is negative. The variance about regression is constant; note that ρ^2 is the proportion of σ_2^2 attributable to regression and $1 - \rho^2$ is the proportion attributable to variation about regression. Thus if ρ^2 is large, the distribution is very tight about the regression line; compare with the remarks in Section 5.3F; see Figure 5.18.

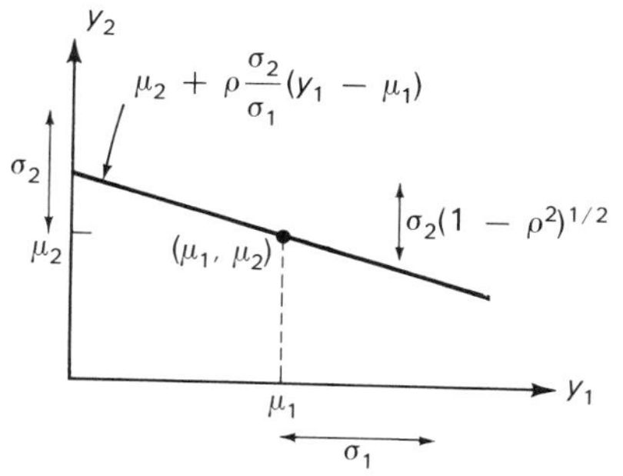

FIGURE 5.18
Regression of y_2 on y_1 for the bivariate normal$(\mu_1, \mu_2, \sigma_1, \sigma_2, \rho)$ for $\rho < 0$.

C CONDITIONAL MEANS AND VARIANCES ON R^{k+l}

Consider a distribution on R^{k+l}. For notation, let

$$\mathbf{y}' = (y_1, \ldots, y_k, y_{k+1}, \ldots, y_{k+l}) = (\mathbf{y}_1', \mathbf{y}_2'),$$

where we have partitioned the vector into two component vectors corresponding to the first k and the last l coordinates.

We examine the conditional distribution of the last l coordinates as given by $\mathbf{y}_2$ given the first k coordinates as given by $\mathbf{y}_1$. For this distribution let $\boldsymbol{\mu}_2(\mathbf{y}_1)$ be the mean and $\Sigma_{22}(\mathbf{y}_1)$ be the variance matrix, assuming they exist:

$$(7) \qquad \boldsymbol{\mu}_2(\mathbf{y}_1) = E(\mathbf{y}_2 : \mathbf{y}_1) = \sum_{\mathbf{y}_2} \mathbf{y}_2 p_{k+1\cdots k+l}(\mathbf{y}_2 : \mathbf{y}_1) \qquad \text{if discrete,}$$

$$= \int \overset{\infty}{\underset{-\infty}{\cdots}} \int \mathbf{y}_2 f_{k+1\cdots k+l}(\mathbf{y}_2 : \mathbf{y}_1)\, d\mathbf{y}_2 \qquad \text{absolutely continuous,}$$

$$(8) \qquad \Sigma_{22}(\mathbf{y}_1) = \text{VAR}\,(\mathbf{y}_2 : \mathbf{y}_1) = E((\mathbf{y}_2 - \boldsymbol{\mu}_2(\mathbf{y}_1))(\mathbf{y}_2 - \boldsymbol{\mu}_2(\mathbf{y}_1))' : \mathbf{y}_1).$$

Suppose that we view $\boldsymbol{\mu}_2(\mathbf{y}_1)$ and $\Sigma_{22}(\mathbf{y}_1)$ as functions of $\mathbf{y}_1$. Then we define

DEFINITION 5

The **regression** of $\mathbf{y}_2$ on $\mathbf{y}_1$ is the function $\boldsymbol{\mu}_2(\mathbf{y}_1)$ recording the conditional mean for each value of $\mathbf{y}_1$.

DEFINITION 6

The **variance matrix about regression** on $\mathbf{y}_1$ is the matrix-valued function $\Sigma_{22}(\mathbf{y}_1)$ recording the conditional variance matrix for each value of $\mathbf{y}_1$.

Now suppose that we view $\boldsymbol{\mu}_2(\mathbf{y}_1)$ as a function of $\mathbf{y}$, a function on R^{k+l} that happens not to depend on the last l coordinates. Then we have:

LEMMA 7

The marginal mean is equal to the mean value of the conditional mean:

(9) $\mu_2 = E(\mu_2(\mathbf{y}_1))$ or $E(\mathbf{y}_2) = E(E(\mathbf{y}_2 : \mathbf{y}_1))$.

Proof Essentially the same as for Lemma 3.

Now suppose that we view $\Sigma_{22}(\mathbf{y}_1)$ as a function of $\mathbf{y}$, a function on R^{k+l} that happens not to depend on the last l coordinates. Then we have

LEMMA 8

The marginal variance matrix is equal to the mean conditional variance matrix plus the variance matrix of the conditional mean:

$$(10) \qquad \begin{aligned} \Sigma_{22} &= E(\Sigma_{22}(\mathbf{y}_1)) + \text{VAR}\ (\mu_2(\mathbf{y}_1)) \\ \text{VAR}\ (\mathbf{y}_2) &= E(\text{VAR}\ (\mathbf{y}_2 : \mathbf{y}_1)) + \text{VAR}\ (E(\mathbf{y}_2 : \mathbf{y}_1)). \end{aligned}$$

In brief, VARiance is mean VARiance about regression plus VARiance of regression.

Proof By formula (5.3.14) we have

$E(\mathbf{w}\mathbf{w}') = \text{VAR}\ (\mathbf{w}) + E(\mathbf{w})E(\mathbf{w}')$;

let $\mathbf{w}$ have the conditional distribution of $\mathbf{y}_2 - \mu_2$ given $\mathbf{y}_1$:

$E((\mathbf{y}_2 - \mu_2)(\mathbf{y}_2 - \mu_2)' : \mathbf{y}_1) = \text{VAR}\ (\mathbf{y}_2 : \mathbf{y}_1) + (\mu_2(\mathbf{y}_1) - \mu_2)(\mu_2(\mathbf{y}_1) - \mu_2)'$.

Then, using Lemma 7, we obtain

$\text{VAR}\ (\mathbf{y}_2) = E((\mathbf{y}_2 - \mu_2)(\mathbf{y}_2 - \mu_2)') = E(\Sigma_{22}(\mathbf{y}_1)) + \text{VAR}\ (\mu_2(\mathbf{y}_1))$.

D AN EXAMPLE ON R^{k+l}

EXAMPLE 4

Multivariate normal(μ; Σ) distribution: Consider the multivariate normal distribution on R^{k+l} with mean and variance matrix,

$$\mu = \begin{pmatrix} \mu_1 \\ \mu_2 \end{pmatrix}, \qquad \Sigma = \begin{pmatrix} \Sigma_{11} & \Sigma_{12} \\ \Sigma_{21} & \Sigma_{22} \end{pmatrix},$$

partitioned corresponding to the first k and the last l coordinates. By Problem 4.2.38 the multivariate normal density can be factored into a normal(μ_1; Σ_{11}) density for $\mathbf{y}_1$ and a normal($\mu_2 + \Sigma_{21}\Sigma_{11}^{-1}(\mathbf{y}_1 - \mu_1)$; $(\Sigma^{22})^{-1}$) density for $\mathbf{y}_2$ given $\mathbf{y}_1$; note that Σ^{22} is a component of the partitioned inverse matrix

$$\Sigma^{-1} = \begin{pmatrix} \Sigma^{11} & \Sigma^{12} \\ \Sigma^{21} & \Sigma^{22} \end{pmatrix}.$$

It follows then that

$$(11) \qquad \mu_2(\mathbf{y}_1) = \mu_2 + \Sigma_{21}\Sigma_{11}^{-1}(\mathbf{y}_1 - \mu_1), \qquad \Sigma_{22}(\mathbf{y}_1) = (\Sigma^{22})^{-1}.$$

Note that the regression surface is an affine function of $\mathbf{y}_1$, and that the variance about regression is constant.

Lemma 7 gives the obvious

$$\boldsymbol{\mu}_2 = \boldsymbol{\mu}_2 + \Sigma_{21}\Sigma_{11}^{-1}(\boldsymbol{\mu}_1 - \boldsymbol{\mu}_1).$$

Lemma 8 gives the less-than-obvious

$$(12) \qquad \Sigma_{22} = (\Sigma^{22})^{-1} + \text{VAR}(\boldsymbol{\mu}_2(\mathbf{y}_1)) = (\Sigma^{22})^{-1} + \Sigma_{21}\Sigma_{11}^{-1}\Sigma_{11}\Sigma_{11}^{-1}\Sigma_{12}$$
$$= (\Sigma^{22})^{-1} + \Sigma_{21}\Sigma_{11}^{-1}\Sigma_{12}$$

or the equivalent,

$$(13) \qquad (\Sigma^{22})^{-1} = \Sigma_{22} - \Sigma_{21}\Sigma_{11}^{-1}\Sigma_{12}.$$

This is a property of partitioned positive definite matrices and can be checked directly in a mechanical way; see Problem 7.

E EXERCISES

1 Calculate $\mu_1(y)$ and $\sigma_1^2(y)$ for Example 1.

2 Verify that $\mu_2 = E(\mu_2(x))$ and $\sigma_2^2 = E(\sigma_2^2(x)) + \text{var}(\mu_2(x))$ hold for Example 1.

3 Consider the uniform distribution on the disk $x_1^2 + x_2^2 < 1$; see Problem 4.2.12. Determine $\mu_2(x_1)$ and $\sigma_2^2(x_1)$.

4 Let u and v be the smallest and largest in a sample $(u_1, \ldots, u_n)$ from the uniform (0, 1); see Problem 5.4.10. For the distribution of (u, v), calculate $\mu_2(u)$ and $\sigma_2^2(u)$.

F PROBLEMS

5 Consider the order statistic $(u_{(1)}, \ldots, u_{(n)})$ for a sample from the uniform(0, 1). The distribution of $(u_{(r)}, u_{(r+s)})$ is recorded in Problem 5.4.12. Determine the conditional mean and variance of $u_{(r+s)}$ given $u_{(r)}$.

6 Consider the Dirichlet$(r_1, \ldots, r_{k+1})$ distribution; see Problems 4.2.21 and 5.3.28. Determine the conditional mean and variance matrix given $(y_1, \ldots, y_p)$.

7 Consider Σ and Σ^{-1} as partitioned in Example 4. Multiply out $\Sigma\Sigma^{-1} = I$ by partitioned matrix multiplication. Use the resulting equations to show that

$$(\Sigma^{22})^{-1} = \Sigma_{22} - \Sigma_{21}\Sigma_{11}^{-1}\Sigma_{12}.$$

8 *Kolmogorov inequality:* An extension of the Chebyshev inequality in Section 5.3C. Let $(y_1, \ldots, y_n)$ be nonnegative and suppose that $E(y_k : y_1, \ldots, y_r) \geq y_r$ for $k > r$; then

$$P(\max\{y_1, \ldots, y_n\} \geq t) \leq \frac{E(y_n)}{t}$$

for any $t > 0$. *Method:* Let A_j be the event $\{y_j \geq t, y_1 < t, \ldots, y_{j-1} < t\}$; then $\{A_1, \ldots, A_n\}$ is a partition of $\{\max\{y_1, \ldots, y_n\} \geq t\}$; examine the integral of y_n over $\cup_1^n A_j$.

9 (*continuation*) An extension of the Chebyshev inequality in Section 5.3C. Let $x_1, \ldots, x_n$ be statistically independent with $E(x_i) = 0$ and $\text{var}(x_i) = \sigma_i^2$; and let $S_k = x_1 + \cdots + x_k$; $L_n = \max\{|S_1|, \ldots, |S_n|\}$. Then

$$P(L_n \geq t) \leq \frac{E(S_n^2)}{t^2} = \frac{\sum_1^n \sigma_i^2}{t^2}.$$

Method: From Problem 8, in the way that the second Chebyshev inequality was obtained from the first Chebyshev inequality in Section 5.3.

10 Consider a distribution for $\mathbf{y}' = (\mathbf{y}_1', \mathbf{y}_2') = (y_1, \ldots, y_k, y_{k+1}, \ldots, y_{k+l})$ on R^{k+l}. Let $\boldsymbol{\mu}' = (\boldsymbol{\mu}_1', \boldsymbol{\mu}_2')$ be the mean and

$$\Sigma = \begin{pmatrix} \Sigma_{11} & \Sigma_{12} \\ \Sigma_{21} & \Sigma_{22} \end{pmatrix}$$

be the variance matrix as partitioned in Example 4. Assume that Σ_{11} is positive definite. The *linear regression* of $\mathbf{y}_2$ on $\mathbf{y}_1$ is defined by the affine function,

$$\mathbf{m}_2 = \boldsymbol{\mu}_2 + B(\mathbf{y}_1 - \boldsymbol{\mu}_1),$$

of $\mathbf{y}_1$ where B is $l \times k$ and is chosen to "minimize" the variance matrix of the residual $\mathbf{y}_2 - \mathbf{m}_2$; the *residual* is given by

$$\mathbf{y}_2 - \mathbf{m}_2 = \mathbf{y}_2 - \boldsymbol{\mu}_2 - B(\mathbf{y}_1 - \boldsymbol{\mu}_1).$$

(a) Verify that the VARiance of the residual is

$$\begin{aligned} \text{VAR}(\mathbf{y}_2 - \mathbf{m}_2) &= \Sigma_{22} - B\Sigma_{12} - \Sigma_{21}B' + B\Sigma_{11}B' \\ &= (B - \Sigma_{21}\Sigma_{11}^{-1})\Sigma_{11}(B - \Sigma_{21}\Sigma_{11}^{-1})' + (\Sigma_{22} - \Sigma_{21}\Sigma_{11}^{-1}\Sigma_{12}). \end{aligned}$$

(b) The first matrix expression on the last line is positive semidefinite but can be made zero by choosing $B = \Sigma_{21}\Sigma_{11}^{-1}$; this gives the following VARiance for the residual from linear regression:

$$\text{VAR}(\mathbf{y}_2 - \mathbf{m}_2) = \Sigma_{22} - \Sigma_{21}\Sigma_{11}^{-1}\Sigma_{12}.$$

Deduce that this variance matrix is the zero matrix if and only if $\mathbf{y}_2$ is an affine function of $\mathbf{y}_1$ with probability 1.

5.6

MOMENTS OF A DISTRIBUTION

The mean and standard deviation give some indication as to how probability is distributed along the real line R. Perhaps by having further characteristics, we can pin down more precisely how the probability is distributed—perhaps even pin it down exactly and determine the distribution. In this section we consider some common sequences of characteristics. The results can be useful in many areas of applications—they provide a powerful method for determining needed marginal distributions.

A THE MOMENT SEQUENCE

The mean $\mu = E(x)$ of a distribution on the real line is sometimes called the *first moment,* by analogy with the moment of the force of gravity acting on unit mass. In a similar way, $E(x^2)$ is called the *second moment* and corresponds to the moment of inertia. We generalize this notion:

DEFINITION 1

The ***r*th moment** of a distribution for x with probability measure P is

$$(1)\qquad \mu_r = E(x^r) = \int_{-\infty}^{\infty} x^r\, dP(x) \qquad r = 0, 1, 2, \ldots,$$
$$= \sum x^r p(x) \qquad \text{if discrete,}$$
$$= \int_{-\infty}^{\infty} x^r f(x)\, dx \qquad \text{absolutely continuous,}$$

provided that the integral exists; otherwise, we say that the moment does not exist.

Note that $\mu_0 = 1$ and $\mu_1 = \mu$, the ordinary mean. Also note that if μ_r exists, then μ_s exists for $s \leq r$; this follows from $|x^s| \leq \max(1, |x^r|)$ and from Problem 5.8.16 in Supplement Section 5.8K.

It is convenient to assemble the moments and form a moment sequence or vector:

DEFINITION 2

The **moment sequence** of a distribution on the real line is given by

$$(2)\qquad \mathbf{m} = (1, \mu_1, \mu_2, \ldots)'$$

and the **truncated moment sequence** by

$$(3)\qquad \mathbf{m}_r = (1, \mu_1, \mu_2, \ldots, \mu_r)'.$$

EXAMPLE 1

The uniform(0, 1) distribution: For the uniform distribution, we have

$$\mu_r = \int_{-\infty}^{\infty} x^r f(x)\, dx = \int_0^1 x^r\, dx = \frac{1}{r+1};$$

thus

$$\mathbf{m} = (1, 1/2, 1/3, \ldots)', \qquad \mathbf{m}_r = (1, 1/2, 1/3, \ldots, 1/(r+1))'.$$

Suppose that a distribution for y is obtained from a distribution for x by the affine transformation $y = a + cx$. Then the moments ν_r for y are easily calculated from the moments μ_r for x; for example,

$$\nu_3 = E(y^3) = E((a+cx)^3) = E(a^3 + 3a^2cx + 3ac^2x^2 + c^3x^3)$$
$$= a^3 + 3a^2c\mu_1 + 3ac^2\mu_2 + c^3\mu_3.$$

These relations can be presented easily by means of matrix multiplication:

$$(4)\qquad \begin{pmatrix} 1 \\ \nu_1 \\ \nu_2 \\ \nu_3 \\ \vdots \end{pmatrix} = \begin{pmatrix} 1 & 0 & 0 & 0 & \cdots \\ a & c & 0 & 0 & \\ a^2 & \binom{2}{1} ac & c^2 & 0 & \\ a^3 & \binom{3}{1} a^2c & \binom{3}{2} ac^2 & c^3 & \\ \vdots & & & & \end{pmatrix} \begin{pmatrix} 1 \\ \mu_1 \\ \mu_2 \\ \mu_3 \\ \vdots \end{pmatrix};$$

note the terms of the binomial expansions of $(a + c)$, $(a + c)^2$, $(a + c)^3$, ...

Now suppose that x and y are independent and we want the moment sequence for the sum $x + y$. Let μ and ν refer to the moments of x and y, respectively; then

$$(5) \qquad E((x + y)^r) = E\left(x^r + \binom{r}{1} x^{r-1}y + \cdots + y^r\right)$$
$$= \mu_r\nu_0 + \binom{r}{1} \mu_{r-1}\nu_1 + \cdots + \mu_0\nu_r;$$

note that we have used "mean product is product mean" given independence; see formula (5.1.13).

B THE CENTRAL MOMENT SEQUENCE

The ordinary moments change drastically under a simple location change of a distribution. We can avoid this by examining moments of the deviation $x - \mu$ from the center of the distribution:

DEFINITION 3

The ***r*th central moment** of a distribution for x with probability measure P is

$$(6) \qquad \bar{\mu}_r = E((x - \mu)^r) = \int_{-\infty}^{\infty} (x - \mu)^r \, dP(x) \qquad r = 0, 1, 2, \ldots,$$
$$= \sum (x - \mu)^r p(x) \qquad \text{if discrete,}$$
$$= \int_{-\infty}^{\infty} (x - \mu)^r f(x) \, dx \qquad \text{absolutely continuous,}$$

where $\mu = E(x)$, provided that the integral exists.

Note that $\bar{\mu}_0 = 1$, $\bar{\mu}_1 = 0$, and $\bar{\mu}_2 = \sigma^2$.

DEFINITION 4

The **central moment sequence** of a distribution on the real line is given by

$$(7) \qquad \bar{\mathbf{m}} = (1, 0, \bar{\mu}_2, \bar{\mu}_3, \ldots)'$$

and the **truncated central moment sequence** by

$$(8) \qquad \bar{\mathbf{m}}_r = (1, 0, \bar{\mu}_2, \bar{\mu}_3, \ldots, \bar{\mu}_r)'.$$

EXAMPLE 2

Uniform(0, 1) distribution: The uniform(0, 1) distribution has mean 1/2; thus the deviation $y = x - 1/2$ is uniform$(-1/2, 1/2)$. The rth central moment of x is

$$E((x - 1/2)^r) = \int_{-1/2}^{1/2} y^r \, dy = 2\frac{(1/2)^{r+1}}{r + 1} = \frac{1}{(r + 1)2^r} \qquad r \text{ even,}$$
$$= 0 \qquad r \text{ odd.}$$

Thus,

$\bar{\mathbf{m}} = (1, 0, 1/3 \cdot 2^2, 0, 1/5 \cdot 2^4, \ldots)$.

We can easily see that a location change from x to $x + a$ does not affect the central moments, and that a scale change from x to cx changes $\bar{\mu}_r$ to $c^r\bar{\mu}_r$.

The ordinary moments can be calculated from the central moments plus the mean. We use the transformation $x = \mu + (x - \mu)$ and equation (4):

$$(9) \qquad \begin{pmatrix} 1 \\ \mu \\ \mu_2 \\ \mu_3 \\ \vdots \end{pmatrix} = \begin{pmatrix} 1 & 0 & 0 & 0 & \cdots \\ \mu & 1 & 0 & 0 & \\ \mu^2 & \binom{2}{1}\mu & 1 & 0 & \\ \mu^3 & \binom{3}{1}\mu^2 & \binom{3}{2}\mu & 1 & \\ \vdots & & & & \end{pmatrix} \begin{pmatrix} 1 \\ 0 \\ \bar{\mu}_2 \\ \bar{\mu}_3 \\ \vdots \end{pmatrix}.$$

The central moments can be calculated from the ordinary moments. We use the transformation $x - \mu = -\mu + x$ and equation (4):

$$(10) \qquad \begin{pmatrix} 1 \\ 0 \\ \bar{\mu}_2 \\ \bar{\mu}_3 \\ \vdots \end{pmatrix} = \begin{pmatrix} 1 & 0 & 0 & 0 & \cdots \\ -\mu & 1 & 0 & 0 & \\ \mu^2 & -\binom{2}{1}\mu & 1 & 0 & \\ -\mu^3 & \binom{3}{1}\mu^2 & -\binom{3}{2}\mu & 1 & \\ \vdots & & & & \end{pmatrix} \begin{pmatrix} 1 \\ \mu \\ \mu_2 \\ \mu_3 \\ \vdots \end{pmatrix}$$

Note that the truncated moment sequence $\mathbf{m}_r$ can be obtained from the mean μ and the truncated central moment sequence $\bar{\mathbf{m}}_r$, and conversely. Thus we see that $\mathbf{m}_r$ and $(\mu, \bar{\mathbf{m}}_r)$ provide the same information concerning the distribution.

Also note the simple coefficients in the two matrices, which indeed are inverses of each other.

EXAMPLE 3

Standard normal distribution: Consider the standard normal distribution for z:

$$\mu_r = E(z^r) = \int_{-\infty}^{\infty} z^r \frac{1}{\sqrt{2\pi}} e^{-z^2/2}\, dz.$$

If r is odd, then $\mu_r = 0$. If r is even, then the integral can be adjusted into a gamma integral; alternatively, integration by parts gives

$$\int_0^{\infty} z^{2m} e^{-z^2/2}\, dz = -\int_0^{\infty} z^{2m-1}\, de^{-z^2/2} = (2m - 1) \int_0^{\infty} z^{2m-2} e^{-z^2/2}\, dz;$$

thus $\mu_{2m} = (2m - 1)(2m - 3) \cdots 3.1$. Hence

$\mathbf{m} = (1, 0, 1, 0, 3, 0, 5 \cdot 3, 0, 7 \cdot 5 \cdot 3, \ldots)$.

C THE FACTORIAL MOMENT SEQUENCE

A modification of the ordinary moments has substantial computational advantage for the common discrete distributions that have factorials in the denominator of the probability function.

DEFINITION 5

The rth factorial moment of a distribution for x with probability measure P is

$$(11)\qquad \mu_{(r)} = E(x^{(r)}) = \int_{-\infty}^{\infty} x(x-1)\cdots(x-r+1)\,dP(x), \qquad r = 0, 1, \ldots,$$

$$= \sum x^{(r)} p(x) \qquad \text{if discrete,}$$

$$= \int_{-\infty}^{\infty} x^{(r)} f(x)\,dx \qquad \text{absolutely continuous,}$$

provided that the integral exists.

DEFINITION 6

The **factorial moment sequence** of a distribution on the real line is given by

$$(12)\qquad \mathbf{m}^* = (1, \mu_{(1)}, \mu_{(2)}, \mu_{(3)}, \ldots).$$

The factorial moments can be calculated from the ordinary moments; we use the relations $x^{(1)} = x$, $x^{(2)} = x(x-1) = x^2 - x$, $x^{(3)} = x^3 - 3x^2 + 2x, \ldots$:

$$(13)\qquad \begin{pmatrix} 1 \\ \mu \\ \mu_{(2)} \\ \mu_{(3)} \\ \vdots \end{pmatrix} = \begin{pmatrix} 1 & 0 & 0 & 0 & \cdots \\ 0 & 1 & 0 & 0 & \\ 0 & -1 & 1 & 0 & \\ 0 & 2 & -3 & 1 & \\ \vdots & & & & \end{pmatrix} \begin{pmatrix} 1 \\ \mu \\ \mu_2 \\ \mu_3 \\ \vdots \end{pmatrix}.$$

The ordinary moments can be calculated from the factorial moments:

$$(14)\qquad \begin{pmatrix} 1 \\ \mu \\ \mu_2 \\ \mu_3 \\ \vdots \end{pmatrix} = \begin{pmatrix} 1 & 0 & 0 & 0 & \cdots \\ 0 & 1 & 0 & 0 & \\ 0 & 1 & 1 & 0 & \\ 0 & 1 & 3 & 1 & \\ \vdots & & & & \end{pmatrix} \begin{pmatrix} 1 \\ \mu \\ \mu_{(2)} \\ \mu_{(3)} \\ \vdots \end{pmatrix};$$

the coefficients can be obtained by simple rules (Problem 9).

Note that the truncated moment sequence $\mathbf{m}_r$ can be obtained from the truncated factorial moment sequence $\mathbf{m}_r^*$, and conversely.

Thus, in summary, we have that $\mathbf{m}_r$, $(\mu, \bar{\mathbf{m}}_r)$, and $\mathbf{m}_r^*$ provide the same information concerning the distribution.

EXAMPLE 4

The Poisson(λ) distribution: Consider the factorial moments of the Poisson(λ) distribution:

$$\mu_{(r)} = E(y^{(r)}) = \sum_{y=0}^{\infty} y^{(r)} \frac{\lambda^y e^{-\lambda}}{y!} = \sum_{y=r}^{\infty} \frac{\lambda^y e^{-\lambda}}{(y-r)!} = \lambda^r,$$

$$\mathbf{m}^* = (1, \lambda, \lambda^2, \lambda^3, \ldots),$$

$$\sigma^2 = \mu_{(2)} - \lambda^{(2)} = \lambda^2 - \lambda(\lambda - 1) = \lambda.$$

D MOMENTS ON R^k

Consider a distribution for $\mathbf{x}$ on R^k. The $(r_1, \ldots, r_k)$th ordinary moment is

$$\mu_{r_1 \cdots r_k} = E(x_1^{r_1} \cdots x_k^{r_k}), \qquad r_i = 0, 1, 2, \ldots;$$

in particular, $\mu_{10\cdots 0} = \mu_1, \ldots, \mu_{0\cdots 01} = \mu_k$. The $(r_1, \ldots, r_k)$th central moment is

$$\bar{\mu}_{r_1 \cdots r_k} = E((x_1 - \mu_1)^{r_1} \cdots (x_k - \mu_k)^{r_k}), \qquad r_i = 0, 1, 2, \ldots;$$

in particular, $\bar{\mu}_{10\cdots 0} = \cdots = \bar{\mu}_{0\cdots 01} = 0$, and $\bar{\mu}_{20\cdots 0} = \sigma_1^2, \ldots, \bar{\mu}_{0\cdots 02} = \sigma_k^2$, and $\mu_{110\cdots 0} = \sigma_{12}$. And the $(r_1, \ldots, r_k)$th factorial moment is

$$\mu^*_{r_1 \cdots r_k} = \mu_{(r_1)\cdots(r_k)} = E(x_1^{(r_1)} \cdots x_k^{(r_k)}), \qquad r_i = 0, 1, 2, \ldots.$$

The ordinary moments provide the same information concerning the distribution as the central moments plus the mean, and the same as the factorial moments.

E EXERCISES

1 Calculate the rth moment of the triangular distribution with density $f(x) = 2x$ on (0, 1), and zero otherwise.

2 Calculate the rth moment of the uniform(a, b) distribution.

3 Calculate the rth moment and the rth central moment for the Bernoulli(p) distribution.

4 Calculate the central moment sequence for the normal(0, σ).

F PROBLEMS

Additional problems on the calculation of moments will be found at the end of Section 5.7.

5 For the standard normal distribution, calculate $E(z^{2m})$ by adjusting the integral into the form of a gamma function.

6 Determine the rth factorial moment for the uniform distribution on the integers 1, . . ., N. *Method:* Use the easily verified relation

$$\sum_{x=a}^{b} x^{(r)} = \frac{1}{r+1}((b+1)^{(r+1)} - a^{(r+1)}).$$

7 Let $(x_1, \ldots, x_n)$ be a sample from a distribution with moments $1, \mu_1, \mu_2, \ldots$. The sample mean $\bar{x}$ is an analog of μ_1; and the sample rth moment $m_r = \Sigma_1^n x_i^r/n$ is an analog of μ_r. We know that $E(\bar{x}) = \mu_1$, var $(\bar{x}) = \sigma^2/n$. Show that $E(m_r) = \mu_r$ and var $(m_r) = (\mu_{2r} - \mu_r^2)/n$.

8 (*continuation*) The sample rth central moment is $\bar{m}_r = \Sigma(x_i - \bar{x})^r/n$. Express the ordinary moments $1, m_1, m_2, \ldots$ in terms of the central moments $1, \bar{m}_1, \bar{m}_2, \ldots$, and $\bar{x}$; and conversely. Show that $E\bar{m}_2 \neq \bar{\mu}_2$, except in a degenerate case.

9 The connection between ordinary and factorial moments. Let Δ be the finite-difference operator and $E = 1 + \Delta$ be the advancing operator:

$$\Delta f(x) = f(x + 1) - f(x) = (E - 1)f(x),$$

$$\Delta^2 f(x) = f(x + 2) - 2f(x + 1) + f(x) = (E - 1)^2 f(x),$$

$$\Delta^r f(x) = \Delta \cdot \Delta \cdots \Delta f(x) = (E - 1)^r f(x) = f(x + r) - \binom{r}{1} f(x + r - 1) + \cdots + (-1)^r f(x),$$

$$f(x) = (1 + \Delta)^x f(0) = f(0) + x^{(1)} \Delta f(0) + \frac{x^{(2)}}{2!} \Delta^2 f(0) + \frac{x^{(3)}}{3!} \Delta^3 f(0) + \cdots.$$

If f is a polynomial of the rth degree, then the series will terminate at the rth difference; the series is a polynomial that agrees with the original polynomial at the integers, and hence for all x. Thus

$$x^r = x^{(1)} \Delta 0^r + \frac{x^{(2)}}{2!} \Delta^2 0^r + \cdots + \frac{x^{(r)}}{r!} \Delta^r 0^r,$$

where, for example, $\Delta^2 0^r = 2^r - 2.1^r + 0^r$ is the second difference of x^r at $x = 0$; and we obtain

$$\mu_r = \Delta 0^r \mu_{(1)} + \frac{\Delta^2 0^r}{2!} \mu_{(2)} + \cdots + \frac{\Delta^r 0^r}{r!} \mu_{(r)}.$$

Show that

$$\Delta^n 0^r = n(\Delta^{n-1} 0^{r-1} + \Delta^n 0^{r-1}).$$

and extend the following table to $r = 4$:

r	$\Delta 0^r$	$\Delta^2 0^r$	$\Delta^3 0^r$	$\cdots$
1	1	0		
2	1	2	0	

5.7

MOMENT-GENERATING AND CHARACTERISTIC FUNCTIONS

As we have noted, the mean and standard deviation give some indication as to how probability is distributed along the real line R. Moment sequences were defined with the hope that additional characteristics could pin down more precisely how the probability is distributed. We now develop an analytically more useful way of presenting the information content of a moment sequence. And we discuss how this information content pins down exactly how the probability is distributed on the line R. Again, the results can be useful in many areas of applications—they provide a powerful method for determining needed marginal distributions.

A MOMENT-GENERATING FUNCTIONS

Consider a distribution for x on the real line R with probability measure P. To calculate the moments, we take the mean, E, of

$$1 \quad x \quad x^2 \quad x^3 \quad \ldots;$$

To do this collectively, we could take the mean, E, of

$$e^{xt} = 1 + xt + x^2\frac{t^2}{2!} + x^3\frac{t^3}{3!} + \cdots.$$

DEFINITION 1

The **moment-generating function** m of a distribution for x with probability measure P is given by

$$(1) \qquad m(t) = E(e^{xt}) = \int_{-\infty}^{\infty} e^{xt}\, dP(x)$$

$$= \sum e^{xt}p(x) \qquad \text{if discrete,}$$

$$= \int_{-\infty}^{\infty} e^{xt} f(x)\, dx \qquad \text{absolutely continuous}$$

for those values of t for which the integral exists. The function m is said to exist if the integral (1) exists for t in an open interval containing the origin 0.

The integral can be shown to exist for *an* interval containing 0; if the distribution has a lot of probability in the tails, then the integral may exist only for the trivial $t = 0$. In analysis, m is called the Laplace transform of the distribution P.

A closely related complex-valued function always exists:

DEFINITION 2

The **characteristic function** c of a distribution for x with probability measure P is given by

$$(2) \qquad c(t) = E(e^{ixt}) = E(\cos xt + i \sin xt)$$

$$= \int_{-\infty}^{\infty} \cos xt\, dP(x) + i\int_{-\infty}^{\infty} \sin xt\, dP(x)$$

for real values of t.

The integrands $\cos xt$, $\sin xt$ are bounded by the function 1, a function that can be integrated with respect to the probability measure P. Then from Supplement Problem 5.8.16 it follows that the characteristic function exists for all real values of t. In analysis, the function c is called the Fourier transform of the distribution P.

EXAMPLE 1

The binomial(n, p) distribution: The moment-generating function for the binomial distribution is

$$(3) \qquad m(t) = \sum_0^n e^{yt}\binom{n}{y}p^y q^{n-y} = \sum_0^n \binom{n}{y}(pe^t)^y q^{n-y} = (pe^t + q)^n.$$

Similarly, the characteristic function is

$c(t) = (pe^{it} + q)^n.$

EXAMPLE 2

The standard normal: The moment-generating function is

$$(4) \qquad m(t) = \int_{-\infty}^{\infty} e^{zt}\frac{1}{\sqrt{2\pi}}e^{-z^2/2}\,dz = e^{t^2/2}\int_{-\infty}^{\infty}\frac{1}{\sqrt{2\pi}}e^{-(z-t)^2/2}\,dz = e^{t^2/2}.$$

The characteristic function is more difficult to derive; it can be obtained by contour integration in the complex plane:

$c(t) = e^{-t^2/2}.$

Now, suppose that the distribution on R has come from some initial space $\mathcal{S}$ with class $\mathcal{A}$ and measure P. Let X be a real-valued function; then

$$m_X(t) = E(e^{X(s)t}) = \int_{\mathcal{S}} e^{X(s)t}\,dP(s),$$

$$c_X(t) = E(e^{iX(s)t}) = \int_{\mathcal{S}} e^{iX(s)t}\,dP(s).$$

It is of some formal interest that the functions m and c defined by (1) and (2) are two sections of the function $E(e^{xw})$ for w in the complex plane.

B PROPERTIES OF MOMENT-GENERATING FUNCTIONS

Two different distributions cannot have the same moment-generating function or the same characteristic function. Thus the moment-generating function or the characteristic function contains enough information to fully determine the corresponding distribution. This uniqueness result is established in Supplement Section 5.8.

The uniqueness result can be presented in terms of one-to-one correspondences as illustrated in Figure 5.19. Certain operations on distribution functions or on density functions may be difficult to complete, and yet the corresponding operations on moment-generating functions may be very simple. Thus if we can move back and forth easily between distributions and generating functions, then we have available a simple round-about route for completing the operations. We now consider some operations that are simple in terms of the generating functions.

Consider a distribution for x with moment-generating function m and characteristic

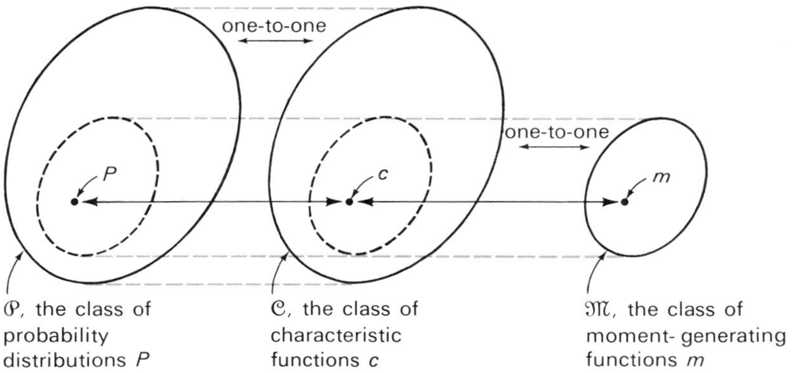

FIGURE 5.19
One-to-one correspondence between 𝒫 and 𝒞; one-to-one correspondence between a subclass of 𝒫 and ℳ.

function c. And suppose that we obtain a distribution for y by the affine transformation $y = a + kx = Y(x)$. Then the corresponding functions for y are

$$(5) \qquad \begin{aligned} m_Y(t) &= E(e^{(a+kx)t}) = e^{at}E(e^{kxt}) = e^{at}m(kt), \\ c_Y(t) &= E(e^{i(a+kx)t}) = e^{iat}E(e^{ikxt}) = e^{iat}c(kt). \end{aligned}$$

Consider a distribution for (x_1, x_2) and suppose that x_1 is *independent* of x_2 and that x_i has moment-generating function m_i and characteristic function c_i. Then the corresponding functions for the sum $y = x_1 + x_2 = Y(x_1, x_2)$ are

$$(6) \qquad \begin{aligned} m_Y(t) &= E(e^{yt}) = E(e^{x_1t}e^{x_2t}) = m_1(t)m_2(t), \\ c_Y(t) &= E(e^{iyt}) = E(e^{ix_1t}e^{ix_2t}) = c_1(t)c_2(t), \end{aligned}$$

where we have used formula (5.1.13). Thus *to add independent variables, we multiply the moment-generating functions and we multiply the characteristic functions*. Note that we can write $y = x_1 + x_2$, but to calculate its distribution directly we must perform an integration; recall the convolution formulas (3.2.9) and (3.2.25)!

Now consider a moment-generating function that exists for the closed interval $[-\delta, +\delta]$. Then we can write

$$(7) \qquad m(t) = E(e^{xt}) = E\left(1 + xt + x^2\frac{t^2}{2!} + \cdots\right) = 1 + \mu_1 t + \mu_2\frac{t^2}{2!} + \cdots,$$

using formula (5.1.12) and the fact that a partial sum $1 + xt + \cdots + x^i t^i/i!$ is bounded by the integrable function $e^{x\delta} + e^{-x\delta}$. The preceding formula gives grounds for the name "moment-generating function."

And, in a corresponding way, if we follow the pattern for a Taylor expansion and take the rth derivative at $t = 0$, we obtain the rth moment:

$$(8) \qquad \left.\frac{d^r m(t)}{dt^r}\right|_{t=0} = \mu_r.$$

Now suppose that the rth moment μ_r exists. Then by Propositions 8 and 9 of Supplement Section 5.8, we obtain

$$(9) \qquad c(t) = 1 + i\mu t - \mu_2 \frac{t^2}{2!} - i\mu_3 \frac{t^3}{3!} + \cdots + (i)^r \mu_r \frac{t^r}{r!} + o(t^r)$$

$$(10) \qquad \left. \frac{d^r c(t)}{dt^r} \right|_{t=0} = E(i^r x^r e^{ixt}) \Big|_{t=0} = i^r \mu_r.$$

C SOME EXAMPLES INVOLVING MOMENT-GENERATING FUNCTIONS

EXAMPLE 3

The normal(μ, σ) distribution: Consider the standard normal distribution and suppose that we expand the moment-generating function in a power series in accord with formula (7):

$$m(t) = e^{t^2/2} = 1 + \left(\frac{t^2}{2}\right) + \left(\frac{t^2}{2}\right)^2 \frac{1}{2!} + \left(\frac{t^2}{2}\right)^3 \frac{1}{3!} + \cdots$$

$$= 1 + \frac{t^2}{2!} + 3 \cdot 1 \frac{t^4}{4!} + 5 \cdot 3 \cdot 1 \frac{t^6}{6!} + \cdots.$$

Then we obtain

$$(11) \qquad \mu_{2m+1} = 0, \qquad \mu_{2m} = (2m-1)(2m-3) \cdots 3 \cdot 1.$$

Now suppose that we form the general normal(μ, σ) distribution by the affine transformation $y = \mu + \sigma z = Y(z)$. Then by formula (5) we obtain

$$(12) \qquad m_Y(t) = e^{\mu t} e^{(\sigma t)^2/2} = e^{\mu t + \sigma^2 t^2/2},$$

$$c_Y(t) = e^{i\mu t} e^{-(\sigma t)^2/2} = e^{i\mu t - \sigma^2 t^2/2}.$$

EXAMPLE 4

Convolution of the normal: Consider a distribution for $(x_1, \ldots, x_n)$ on R^n and suppose that the $x_1, \ldots, x_n$ are statistically independent and x_i is normally distributed (μ_i, σ_i). By Example 3 the moment-generating function of x_i is

$$m_i(t) = e^{\mu_i t + \sigma_i^2 t^2/2}.$$

Now suppose that we form the sum variable $y = x_1 + \cdots + x_n = Y(x_1, \ldots, x_n)$. Then by formula (6) we obtain

$$m_Y(t) = m_1(t) \cdots m_n(t) = \prod_1^n e^{\mu_i t + \sigma_i^2 t^2/2}$$

$$= e^{\Sigma \mu_i t + \Sigma \sigma_i^2 t^2/2}.$$

Note that this is the moment-generating function for the normal $(\Sigma \mu_i, (\Sigma \sigma_i^2)^{1/2})$ distribution. By the uniqueness property discussed in Section B, it follows that y is normal $(\Sigma \mu_i, (\Sigma \sigma_i^2)^{1/2})$. Thus if we add independent normal variables, we obtain a normal variable.

D MOMENT-GENERATING FUNCTIONS ON R^k

Consider a distribution for $\mathbf{x}$ on R^k.

DEFINITION 3

The moment-generating function m of a distribution for $\mathbf{x}$ on R^k with probability measure P is given by

$$(13) \qquad m(\mathbf{t}) = m(t_1, \ldots, t_k) = E(e^{\Sigma x_i t_i}) = E(e^{\mathbf{x}'\mathbf{t}}) = \int e^{\mathbf{x}'\mathbf{t}}\, dP(\mathbf{x})$$

provided that the integral exists. The function m is said to exist if the integral (13) exists for $\mathbf{t}$ in an open rectangle containing the origin $\mathbf{0}$. The characteristic function is defined by

$$(14) \qquad c(\mathbf{t}) = c(t_1, \ldots, t_k) = E(e^{i\Sigma x_i t_i}) = \int e^{i\mathbf{x}'\mathbf{t}}\, dP(\mathbf{x}).$$

Again, on R^k, two different distributions cannot have the same moment-generating function or the same characteristic function. The moment-generating function or the characteristic function thus *contains enough information* to fully determine the corresponding distribution. See the earlier remarks in Section B; the formal uniqueness results will be recorded in Supplement Section 5.8E.

Consider a distribution for $\mathbf{x}$ with moment-generating function m. Then the moment-generating function for $\mathbf{y} = \mathbf{a} + C\mathbf{x} = \mathbf{Y}(\mathbf{x})$ is

$$(15) \qquad m_Y(\mathbf{t}) = E(e^{\mathbf{y}'\mathbf{t}}) = E(e^{\mathbf{a}'\mathbf{t}+\mathbf{x}'C'\mathbf{t}}) = e^{\mathbf{a}'\mathbf{u}}m(C'\mathbf{t}).$$

Consider a distribution for $(\mathbf{x}_1, \ldots, \mathbf{x}_n)$ and suppose that the $\mathbf{x}_i$ are independent and $\mathbf{x}_i$ has moment-generating function m_i. Then the sum $\mathbf{y} = \Sigma\, \mathbf{x}_i = \mathbf{Y}(\mathbf{x}_1, \ldots, \mathbf{x}_n)$ has moment-generating function

$$(16) \qquad m_{\mathbf{Y}}(\mathbf{t}) = E(e^{\Sigma \mathbf{x}_i'\mathbf{t}}) = m_1(\mathbf{t}) \cdots m_n(\mathbf{t}).$$

Now suppose that a moment-generating function m is expanded in a series about $\mathbf{0}$:

$$(17) \qquad m(\mathbf{t}) = \sum \mu_{r_1 \cdots r_k} \frac{t_1^{r_1}}{r_1!} \cdots \frac{t_k^{r_k}}{r_k!}.$$

The moments can be obtained by differentiation:

$$(18) \qquad \left.\frac{\partial^{r_1}}{\partial t_1^{r_1}} \cdots \frac{\partial^{r_k}}{\partial t_k^{r_k}} m\right|_{\mathbf{t}=0} = \mu_{r_1 \cdots r_k}.$$

EXAMPLE 5

The multivariate normal(μ; Σ): Consider $\mathbf{z} = (z_1, \ldots, z_k)'$ with a standard normal distribution on R^k. From the independence among the coordinates we obtain

$$m(t_1 \cdots t_k) = E(e^{z_1 t_1 + \cdots + z_k t_k}) = e^{t_1^2/2} \cdots e^{t_k^2/2} = e^{\mathbf{t}'\mathbf{t}/2}.$$

Suppose that we form the general normal($\boldsymbol{\mu}$; Σ) distribution by the affine transformation $\mathbf{y} = \boldsymbol{\mu} + \Gamma\mathbf{z} = \mathbf{Y}(\mathbf{z})$, where $\mathbf{Y}$ is in R^l. Then by (15) we obtain the moment-generating function for the general normal($\boldsymbol{\mu}$; Σ),

$$m_{\mathbf{Y}}(\mathbf{t}) = e^{\boldsymbol{\mu}'\mathbf{t}+\mathbf{t}'\Gamma\Gamma'\mathbf{t}/2} = e^{\boldsymbol{\mu}'\mathbf{t}+\mathbf{t}'\Sigma\mathbf{t}/2},$$

where $\Sigma = \Gamma\Gamma'$. Formulas (15) and (17) can then show that $\boldsymbol{\mu}$ is the mean and Σ is the variance matrix; see Problem 42.

E MARGINAL DISTRIBUTIONS

In Chapter 2 we developed a variety of methods for determining marginal distributions. With the uniqueness theorems in this section we now have a new range of methods for determining marginal distributions. Often we can find fairly easily the characteristic function, or the moment-generating function, or the moments of some marginal variable of interest. Then if we recognize the characteristic function, or use the inversion formula, we can obtain the corresponding distribution. These new methods are largely restricted to linear operations on the initial variables. See Example 4 in Section C; also the problems in Section K below.

F PROBABILITY-GENERATING FUNCTIONS

For discrete distributions on the nonnegative integers we can define a very useful function closely related to the moment-generating function. It provides yet another way of determining a marginal distribution—for responses involving counts.

DEFINITION 4

The **probability-generating function** r of a distribution for x on the nonnegative integers is given by

$$(19) \qquad r(t) = E(t^x) = t^0p(0) + t^1p(1) + t^2p(2) + \cdots$$

for those values of t for which the summation converges; the summation converges for an interval that includes $t = 0$ and $t = 1$.

For certain purposes connected with moments, we find it convenient to say that *r exists* if the summation (19) exists for an *open* interval containing $t = 1$.

Note the obvious connection with the moment-generating function:

$$(20) \qquad m(t) = r(e^t); \qquad r(t) = m(\ln t), \qquad t > 0.$$

Also note the fundamental property that the coefficients of the powers of t are the corresponding probabilities of the distribution.

For an affine transformation $y = a + kx = Y(x)$, we have

$$(21) \qquad r_Y(t) = t^a r(t^k).$$

For the sum $y = x_1 + x_2 = Y(x_1 + x_2)$ where x_1 and x_2 are independent with probability-generating functions r_1 and r_2, we have

(22) $r_Y(t) = r_1(t)r_2(t)$.

Now suppose that a probability-generating function is differentiated s times and evaluated at $t = 1$:

$$\frac{d}{dt}r(t) = \sum xt^{x-1}p(x), \qquad \frac{d^s r(t)}{dt^s} = \sum x^{(s)}t^{x-s}\,p(x);$$

(23) $$\left.\frac{d}{dt}\,r(t)\right|_{t=1} = \mu, \qquad \left.\frac{d^s r(t)}{dt^s}\right|_{t=1} = E(x^{(s)}) = \mu_{(s)};$$

thus we obtain the factorial moments.

EXAMPLE 6

Sample from the Bernoulli(p) distribution: Consider a sample from the Bernoulli(p) distribution. For an individual x_i we have

$$r_i(t) = q + pt.$$

Then for the sum $y = x_1 + \cdots + x_n$, we have

$$r_Y(t) = (q + pt)^n = q^n + \binom{n}{1}pq^{n-1}t + \cdots + \binom{n}{r}p^r q^{n-r}t^r + \cdots + p^n t^n.$$

The distribution of y is of course binomial(n, p); note that we have obtained the binomial probabilities as coefficients in the expansion of r_Y.

The factorial moments for the binomial distribution can be calculated by evaluating the derivatives at $t = 1$. We can accomplish this equivalently by expanding in a series about $t = 1$. For this let $u = t - 1$; then

$$\begin{aligned} r(t) &= r(1 + u) = (q + p(1 + u))^n = (1 + pu)^n \\ &= 1 + npu + n^{(2)}p^2\frac{u^2}{2!} + n^{(3)}p^3\frac{u^3}{3!} + \cdots + p^n u^n; \end{aligned}$$

thus

$$\mu_{(s)} = E(y^{(s)}) = n^{(s)}p^s.$$

The analogous definition for the probability-generating function of a distribution for $\mathbf{x}$ on R^k is given by

$$r(t_1, \ldots, t_k) = E(t_1^{x_1} \cdots t_k^{x_k}) = m(\ln t_1, \ldots, \ln t_k), \qquad t_i > 0.$$

Differentiation at the point **1** gives

$$\left.\frac{\partial^{s_1}}{\partial t_1^{s_1}} \cdots \frac{\partial^{s_k}}{\partial t_k^{s_k}}\, r\right|_{\mathbf{t}=\mathbf{1}} = E(x_1^{(s_1)} \cdots x_k^{(s_k)}) = \mu_{(s_1)\cdots(s_k)},$$

the $(s_1, \ldots, s_k)$th factorial moment.

G EXERCISES

1 Calculate the moment-generating function for the uniform distribution on the integers 1, 2, 3. Use the moment-generating function to deduce the mean and second moment.

2 The binomial(4, 1/2) distribution is tabulated in Section 3.5A. Calculate the moment-generating function.

3 Show that the moment-generating function for the uniform(0, 1) distribution has $m(t) = (e^t - 1)/t$. Use the moment-generating function to deduce the mean.

4 Calculate the moment-generating function for the triangular distribution with density $f(x) = 2x$ on (0, 1) and zero elsewhere.

5 Show that the moment-generating function for the uniform distribution on the square $(0, 1)^2$ has $m(t_1, t_2) = (e^{t_1} - 1)(e^{t_2} - 1)/t_1t_2$.

H PROBLEMS: STATISTICAL INDEPENDENCE

6 Consider a distribution for (x_1, x_2) on the plane R^2 with moment-generating function $m(t_1, t_2)$ and characteristic function $c(t_1, t_2)$. Show that the moment-generating function for x_1 is $m_1(t_1) = m(t_1, 0)$ and for x_2 is $m_2(t_2) = m(0, t_2)$. Correspondingly, show that $c_1(t_1) = c(t_1, 0)$, $c_2(t_2) = c(0, t_2)$.

7 (*continuation*) Show that x_1 and x_2 are statistically independent if and only if $c(t_1, t_2) = g(t_1)h(t_2)$ factors so that the variables separate. Use the uniqueness discussed in Section B.

8 (*continuation*) Suppose that the moment-generating function m exists. Then show that x_1 and x_2 are statistically independent if and only if $m(t_1, t_2) = g(t_1)h(t_2)$ factors so that the variables separate. Use the uniqueness described in Section B.

I PROBLEMS FOR DISCRETE DISTRIBUTIONS

For each of the following distributions, verify the corresponding expressions for

$$\mu_{(r)}, \qquad r(t), \qquad m(t).$$

9 *The uniform distribution on the integers* 1, 2, . . ., *N:*

$$\frac{(N+1)^{(r+1)}}{N(r+1)} \quad (r > 0), \qquad \frac{t^{N+1} - t}{N(t-1)}, \qquad \frac{e^{(N+1)t} - e^t}{N(e^t - 1)};$$

see Problem 5.6.6.

10 *Poisson*(λ) *distribution:*

$$\lambda^r, \qquad \exp\{\lambda t - \lambda\}, \qquad \exp\{\lambda e^t - \lambda\}.$$

11 *Geometric*(*p*) *distribution:*

$$\frac{r!\,q^r}{p^r}, \qquad \left(\frac{1 - qt}{p}\right)^{-1}, \qquad \left(\frac{1 - qe^t}{p}\right)^{-1}.$$

12 *Binomial*(*n*, *p*) *distribution:*

$$n^{(r)}p^r, \qquad (q + pt)^n \qquad (q + pe^t)^n.$$

13 *Negative binomial*(*k*, *p*) *distribution:*

$$\frac{(k + r - 1)^{(r)}q^r}{p^r}, \qquad \left(\frac{1 - qt}{p}\right)^{-k}, \qquad \left(\frac{1 - qe^t}{p}\right)^{-k}.$$

14 *Hypergeometric*(*N*, *n*, *p*) *distribution:*

$$\frac{n^{(r)}(Np)^{(r)}}{N^{(r)}}.$$

For each of the following distributions, verify the corresponding expressions for

$$\mu_{(s_1)\cdots(s_k)}, \qquad r(t_1, \ldots, t_k), \qquad m(t_1, \ldots, t_k).$$

15 *Multinomial*$(n, p_1, \ldots, p_k)$ *distribution:*

$$n^{(\Sigma s_i)} p_1^{s_1} \cdots p_k^{s_k}, \qquad (p_1 t_1 + \cdots + p_k t_k)^n, \qquad (p_1 e^{t_1} + \cdots + p_k e^{t_k})^n.$$

16 *Multivariate hypergeometric*$(N, n, p_1, \ldots, p_k)$ *distribution:*

$$\frac{n^{(\Sigma s_i)} (Np_1)^{(s_1)} \cdots (Np_k)^{(s_k)}}{N^{(\Sigma s_i)}}.$$

J PROBLEMS FOR ABSOLUTELY CONTINUOUS DISTRIBUTIONS

For each of the following distributions, verify the corresponding expressions for

$$\mu_r, \qquad m(t), \qquad c(t).$$

More difficult problems are marked with an asterisk.

17 *Standard exponential distribution:*

$$r!, \qquad (1 - t)^{-1}, \qquad (1 - it)^{-1}.$$

18 *Pareto*(α) *distribution for z; the distribution for* $1 + z$ *has moments*

$$\frac{\alpha}{\alpha - r} \quad (r < \alpha).$$

19 *Standard Laplace distribution:*

$$\begin{matrix} 0 & (r \text{ odd}) \\ r! & (r \text{ even}) \end{matrix}, \qquad (1 - t^2)^{-1}, \qquad (1 + t^2)^{-1}.$$

**20 *Standard logistic distribution:*

$$\begin{matrix} 0 & (r \text{ odd}) \\ 2\Gamma(r + 1)(1 - 2^{1-r})\zeta(r) & (r \text{ even}) \end{matrix} \qquad B(1 - t, 1 + t), \qquad \pi t \operatorname{cosech} \pi t,$$

where B is the beta function (Problem 2.2.22) and $\zeta(x) = \Sigma_{i=1}^{\infty} i^{-x}$ is the Riemann zeta function.

21 *Standard lognormal*(τ):

$$\exp\{(1/2) r^2 \tau^2\}.$$

22 *Gamma*(p) *distribution:*

$$\frac{\Gamma(p + r)}{\Gamma(p)}, \qquad (1 - t)^{-p}, \qquad (1 - it)^{-p}.$$

23 *Chi-square*(m) *distribution:*

$$m(m + 2) \cdots (m + 2r - 2), \qquad (1 - 2t)^{-m/2}, \qquad (1 - 2it)^{-m/2}.$$

24 *Beta*(p, q) *distribution:*

$$\frac{p^{[r]}}{(p + q)^{[r]}}, \qquad \sum_{r=0}^{\infty} \frac{p^{[r]}}{(p + q)^{[r]}} \frac{t^r}{r!}, \qquad \sum_{r=0}^{\infty} \frac{p^{[r]}}{(p + q)^{[r]}} \frac{(it)^r}{r!},$$

where $p^{[r]} = p(p + 1) \cdots (p + r - 1)$ is p ascending r factorial.

25 *Canonical F*(m, n) *distribution:*

$$\frac{(m/2)^{[r]}}{(n/2 - 1)^{(r)}}, \qquad \sum_{r=0}^{\infty} \frac{(m/2)^{[r]}}{(n/2 - 1)^{(r)}} \frac{t^r}{r!}.$$

26 *Canonical Student(n) distribution:*

$$\begin{cases} 0 & (r \text{ odd}) \\ \dfrac{(r-1)(r-3)\cdots 3\cdot 1}{(n-2)(n-4)\cdots(n-r)} & (r \text{ even}) \end{cases}$$

27 *Standard Weibull(β) distribution:*

$\Gamma(r\beta^{-1} + 1)$.

*28 *Standard Cauchy distribution has characteristic function*

$c(t) = \exp\{-|t|\}$.

Method: Use Problem 19 and the inversion formula in Supplement Section 5.8.

K PROBLEMS INVOLVING UNIQUENESS AND OTHER PROPERTIES OF GENERATING FUNCTIONS

29 Let $x_1, \ldots, x_k$ be statistically independent and x_i be binomial(n_i, p). Show that $y = \Sigma x_i$ is binomial$(\Sigma n_i, p)$.

30 Let $x_1, \ldots, x_k$ be statistically independent and x_i be Poisson(λ_i). Show that $y = \Sigma x_i$ is Poisson$(\Sigma \lambda_i)$.

31 Let z be standard normal. Determine the moment-generating function of $y = z^2$ directly from the normal density and then show that $y = z^2$ is chi-square(1).

32 Let $x_1, \ldots, x_k$ be statistically independent and x_i be gamma(p_i). Show that $y = \Sigma x_i$ is gamma(Σp_i).

33 Let $\chi_1^2, \ldots, \chi_k^2$ be statistically independent and χ_i^2 be chi-square (f_i). Show that $\chi^2 = \Sigma \chi_i^2$ is chi-square(Σf_i).

34 Let $(x_1, \ldots, x_k)$ be a sample from the standard Cauchy distribution. Show that $y = \Sigma x_i/k$ has the same standard Cauchy distribution!

35 Let $(x_1, \ldots, x_n)$ be a sample from the normal(μ, σ) distribution. Use moment-generating functions to show that $\bar{x} = \Sigma x_i/n$ is normal$(\mu, \sigma/\sqrt{n})$.

36 Let $(\mathbf{x}_1, \ldots, \mathbf{x}_n)$ be a sample from the multivariate normal$(\boldsymbol{\mu}; \Sigma)$. Show that $\bar{\mathbf{x}} = \Sigma \mathbf{x}_j/n$ is multivariate normal$(\boldsymbol{\mu}; n^{-1}\Sigma)$.

37 Let $\mathbf{x}$ be multivariate normal$(\boldsymbol{\mu}; \Sigma)$. Show that $\mathbf{y} = \mathbf{a} + C\mathbf{x}$ is multivariate normal$(\mathbf{a} + C\boldsymbol{\mu}; C\,\Sigma\,C')$.

38 Consider a distribution for $\mathbf{x}$ on R^k with probability measure P. If the marginal distribution of $\Sigma a_i x_i = \mathbf{a}'\mathbf{x}$ is given for all vectors $\mathbf{a}$, then show that the probability measure P is uniquely determined.

39 Consider a distribution for x with all probability on the nonnegative integers. Let r be the probability generating function and R be the distribution-generating function

$$R(t) = \sum_x F(x)t^x. \qquad R(t) = \frac{r(t)}{1-t};$$

prove the connecting relation between R and r.

40 Let $x_1, x_2, \ldots$ be statistically independent with a common discrete distribution with probability generating function $r(t)$. Let n be statistically independent with a discrete distribution on $\{1, 2, \ldots\}$ with probability generating function $g(t)$. Let $s = x_1 + \cdots + x_n$; show that its probability generating function is $g(r(t))$.

L FURTHER PROBLEMS

41 The moment-generating function for the multivariate normal $(\boldsymbol{\mu}; \Sigma)$ is recorded in Example 5. Specialize and record the moment-generating function for the bivariate normal$(\mu_1, \mu_2, \sigma_1, \sigma_2, \rho)$.

42 (*continuation*) For the general case expand in a series and show that $E(\mathbf{y}) = \boldsymbol{\mu}$. Determine the generating function for $\mathbf{y} - \boldsymbol{\mu}$; expand in a series and show that VAR $(\mathbf{y}) = \Sigma$.

43 The logarithm of the characteristic function is called the *cumulant-generating function k*,

$$k(t) = \ln c(t), \qquad c(t) = e^{k(t)}.$$

If $c(t)$ can be expanded to r terms (if the rth moment exists), then so can the cumulant-generating function; the coefficients (analogous to moments) are called cumulants and are designated $\kappa_1, \kappa_2, \kappa_3, \ldots$:

$$k(t) = \kappa_1(it) + \kappa_2\frac{(it)^2}{2!} + \kappa_3\frac{(it)^3}{3!} + \cdots + \kappa_r\frac{(it)^r}{r!} + o(t^r).$$

Use the exponential or logarithmic expansion to show that

$$\begin{aligned} &\kappa_1 = \mu && \mu_1 = \kappa_1 \\ &\kappa_2 = \sigma^2 && \mu_2 = \kappa_2 + \kappa_1^2 \\ &\kappa_3 = \bar{\mu}_3 && \mu_3 = \kappa_3 + 3\kappa_1\kappa_2 + \kappa_1^3 \\ &\vdots && \vdots \end{aligned}$$

44 If x_1 and x_2 are statistically independent with cumulant-generating functions k_1 and k_2, show that the cumulant-generating function for the convolution distribution of $x_1 + x_2$ is $k_1 + k_2$. This additivity often makes cumulants more convenient than moments.

45 Let $(x_1, \ldots, x_n)$ be a sample from a distribution with cumulants $\kappa_1, \kappa_2, \ldots$. Functions of $(x_1, \ldots, x_n)$, called k statistics, have been defined for use with cumulants:

$$k_1 = \frac{S_1}{n}, \qquad k_2 = \frac{nS_2 - S_1^2}{n(n-1)}, \qquad k_3 = \frac{n^2S_3 - 3nS_2S_1 + 2S_1^3}{n(n-1)(n-2)},$$

where $S_r = x_1^r + \cdots + x_n^r$. Show that $k_1 = \bar{x}$, $k_2 = s_x^2$. Show that $E(k_i) = \kappa_i$ for $i = 1, 2, 3$.

SUPPLEMENTARY MATERIAL

Supplement Section 5.8 is on pages 544–554. The *uniqueness theorem* for characteristic functions is proved; also the theorem for moment-generating functions. The uniqueness theorems for moment sequences are recorded; also the various theorems for the multivariate case.

A brief development of the *general integral* is given, and various properties and theorems are developed or presented as problems; for example, *linearity* and *limiting* properties. The Fubini theorem is recorded in a convenient form for later reference.

NOTES AND REFERENCES

This chapter assembles various concepts connected with the mean-value concept: mean, variance, covariance, moments, moment-generating functions, characteristic functions. These concepts all depend on the integral.

The first section presents the mean value in terms of the integral, examining the special cases of summation and the Riemann integral. Supplement Section 5.8 records various properties of the general integral.

The basic development is for distributions on the real line but always in a form suitable for matrix extension to distributions in several dimensions. This makes available some rather rich results for the vector case.

For some references to other texts, see the Notes and References for Chapter 1 and the following:

Pfeiffer, P. E. (1965). *Concepts of Probability Theory*. New York: McGraw-Hill Book Company.

We have examined probabilities themselves in Chapters 1–4 and mean-value characteristics in Chapter 5. These results lead to the following: the average $\bar{y}$ for a sample $(y_1, \ldots, y_n)$ from the normal(μ, σ) distribution has a normal $(\mu, \sigma/\sqrt{n})$ distribution. A proof can be based on routine probability methods (Problem 3.3.11) or on mean-value characteristics (Problem 5.7.35).

Empirically, however, we find a much broader result: The average $\bar{y}$ for a sample $(y_1, \ldots, y_n)$ from an arbitrary distribution with mean μ and standard deviation σ is approximately normal$(\mu, \sigma/\sqrt{n})$ as n becomes large. A theoretical verification of this can be based on the notion of limiting distribution. In Chapter 6 we examine ***limiting distributions:*** *the approach to a limit in Section 6.1, and the approach to a normal limit (Central Limit Theorems) in Section 6.2. The concept of* ***limiting function*** *is discussed in Supplement Section 6.3, together with some concluding comments (Section 6.3N) on the development of the probability model.*

6

LIMITING DISTRIBUTIONS AND LIMITING FUNCTIONS

6.1

LIMITING DISTRIBUTIONS

Many response variables in science and industry are approximately normally distributed. A theoretical argument supporting this runs as follows. There are many small sources of variation internal to the system being investigated; these components add to produce the variation in the response. The sum of bounded independent components has a limiting normal distribution. In this section we examine the notion of limiting distribution. Then in Section 6.2 we examine some Central Limit Theorems for limiting normal distributions. Supplement Section 6.3 examines limiting functions and records some concluding comments (Section 6.3N) on the probability model. The limiting distributions provide some very useful approximations.

A EXAMPLE OF A LIMITING DISTRIBUTION

Consider a classical example involving a limiting distribution. The Poisson distribution is often used in applications where we can envisage a large number of independent sources for an event and yet a very small probability that any particular source produces the event. For example, consider radioactive material with a large number of atoms, each having a small probability of producing an α-particle; recall Section 2.1E. The binomial distribution, however, is the obvious distribution for the total frequency with independent components having common probability for producing the event.

For this example we examine the binomial(n, p) distribution with n large and p small; specifically we take a fixed mean $\lambda = np$, write $p = \lambda/n$, and investigate the limit as $n \to \infty$. Let y_n be used to present events and probabilities for this binomial distribution.

Consider the binomial probability function at a point x:

$$P(y_n = x) = p_n(x) = \binom{n}{x}\left(\frac{\lambda}{n}\right)^x\left(1 - \frac{\lambda}{n}\right)^{n-x} \qquad x = 0, 1, 2, \ldots, n$$

$$= \frac{\lambda^x}{x!}\left(1 - \frac{\lambda}{n}\right)^n \cdot \frac{n^{(x)}}{n^x}\left(1 - \frac{\lambda}{n}\right)^{-x}$$

$$= \frac{\lambda^x}{x!}\left(1 - \frac{\lambda}{n}\right)^n \cdot 1\left(1 - \frac{1}{n}\right)\cdots\left(1 - \frac{x-1}{n}\right)\cdot\left(1 - \frac{\lambda}{n}\right)^{-x}.$$

Consider the limit as $n \to \infty$:

$$\lim_{n\to\infty} P(y_n = x) = \lim_{n\to\infty} p_n(x) = \frac{\lambda^x}{x!} e^{-\lambda} \cdot 1 \cdot 1 = \frac{\lambda^x}{x!} e^{-\lambda} = p(x),$$

where p is the probability function for the Poisson(λ) distribution. Thus for large n the binomial(n, λ/n) probability function approaches the Poisson(λ) probability function.

We now examine the binomial distribution function at a point x:

$$P(y_n \le x) = F_n(x) = p_n(0) + p_n(1) + \cdots + p_n([x]),$$

where $[x]$ is the largest integer $\le x$. Taking the limit of the finite sum we obtain

$$\lim_{n\to\infty} F_n(x) = p(0) + p(1) + \cdots + p([x]) = F(x),$$

where F is the Poisson(λ) distribution function. Thus the binomial(n, λ/n) distribution function approaches the Poisson(λ) distribution function at all points x on the real line.

For a numerical illustration suppose that we have 50 independent repetitions of a system that has a certain event that occurs with probability 1/50. Let y_{50} designate the number of occurrences of the event. The distribution of y_{50} is binomial(50, 1/50); some values for this binomial probability function and for the corresponding Poisson(1) probability function are given in Table 6.1.

TABLE 6.1
Probability function values

x	0	1	2	3	4	$\cdots$
$p_{50}(x)$	0.3642	0.3716	0.1858	0.0607	0.0145	
$p(x)$	0.3679	0.3679	0.1839	0.0613	0.0153	

B DEFINITION OF A LIMITING DISTRIBUTION

In Section A we showed that the binomial distribution function with $p = \lambda/n$ converges to the Poisson(λ) distribution function. This convergence of a sequence of distribution functions will be the basis for defining a limiting distribution. We first examine a small technical complication illustrated by the following example.

EXAMPLE 1

Consider a distribution for y_n with probability 1/2 at each of $1/n$ and $-1/n$. As $n \to \infty$ the two points, each with probability 1/2, converge together toward zero. Let F_n be the distribution function for y_n:

$$\begin{aligned} F_n(x) &= 0 && \text{if } -\infty < x < -1/n, \\ &= 1/2 && -1/n \le x < 1/n, \\ &= 1 && 1/n \le x < \infty, \end{aligned}$$

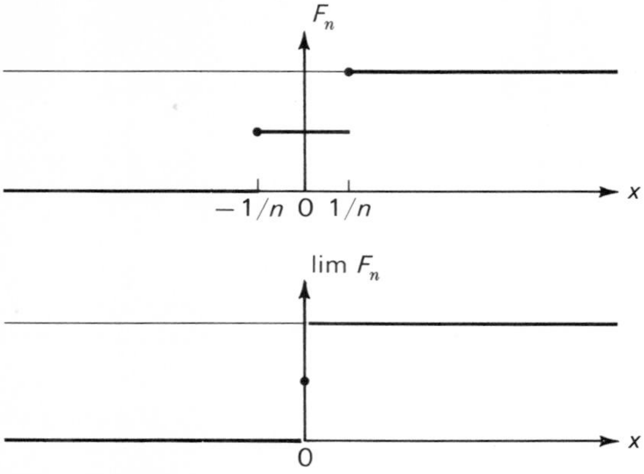

FIGURE 6.1
F_n is the distribution function for probability $1/2$ at $-1/n$ and at $1/n$; lim F_n is not a distribution function.

$$\lim_{n\to\infty} F_n(x) = 0 \qquad \text{if } -\infty < x < 0,$$
$$= 1/2 \qquad x = 0,$$
$$= 1 \qquad 0 < x < \infty.$$

See Figure 6.1 and note that lim F_n is *not* a distribution function.

Now consider intuitively the distribution for y_n. Clearly the two probability weights of $1/2$ are converging to the origin; and in a sense the total probability is piling up at the origin. Consider a distribution for y with all probability at the origin. Then we would want to say that $y_n \to y$ in a distribution sense.

Let H be the distribution function for y which has all probability at the point 0:

$$H(x) = 0,\ 1 \qquad \text{according as } x < 0,\ 0 \leq x.$$

Then we note that lim $F_n = H$ except at the point 0, where H has a discontinuity. Accordingly, we find it convenient to ignore nonconvergence at the points of discontinuity of the limit function.

We now formalize the concept of a sequence of distributions approaching a limiting distribution.

DEFINITION 1

A sequence of distributions with distribution functions F_n has limiting distribution with distribution function F if

(1) $$\lim_{n\to\infty} F_n(x) = F(x)$$

at all points x at which F is continuous; we write $F_n \to F$ or $y_n \xrightarrow{\mathfrak{D}} y$ or $P_n \xrightarrow{\mathfrak{D}} P$ and say that y_n converges to y in distribution.

For a substantial example, recall the binomial(n, λ/n) distribution converging to the Poisson(λ) distribution, as examined in Section A.

Now, in general, consider a sequence of distributions F_n with limiting distribution F. An interval $B = (a, b]$ on R is said to be an *interval of continuity* if the end points a and b are points of continuity of the distribution function F. Then, by differencing the distribution function, we obtain

$$\lim_{n\to\infty} P_n(B) = \lim_{n\to\infty} (F_n(b) - F_n(a)) = F(b) - F(a) = P(B).$$

Of course, the limit need not hold for other intervals or for most Borel sets. Accordingly, if we know a distribution approximately or if we are concerned with the limiting distribution of a sequence, we are essentially restricted to intervals—in fact, to rather special intervals. Thus we are withdrawing rather substantially from the Borel sets as the logical consequence of examining intervals.

C CRITERIA FOR LIMITING DISTRIBUTIONS

Sometimes limiting distributions can be determined directly from probability density functions.

PROPOSITION 2

If the distribution function F_n has density f_n and the distribution function F has density f, then

$$\lim_{n\to\infty} f_n(x) = f(x) \tag{2}$$

at all points x implies that $F_n \to F$.

For a method of proof, see Problem 17.

EXAMPLE 2

Student distribution: Consider the standard Student(n) distribution as examined in Problem 2.2.19. Does this distribution have a limiting distribution as $n \to \infty$? We calculate:

$$\begin{aligned}
\lim_{n\to\infty} f_n(t) &= \lim_{n\to\infty} \frac{\Gamma\left(\frac{n+1}{2}\right)}{\Gamma(1/2)\Gamma(n/2)}\left(1 + \frac{t^2}{n}\right)^{-(n+1)/2} \frac{1}{\sqrt{n}} \\
&= \lim_{n\to\infty} \frac{1}{\sqrt{2\pi}} \frac{\Gamma\left(\frac{n+1}{2}\right)}{\Gamma(n/2)\sqrt{n/2}}\left(1 + \frac{t^2}{n}\right)^{-n/2}\left(1 + \frac{t^2}{n}\right)^{-1/2} \\
&= \frac{1}{\sqrt{2\pi}} \cdot 1 \cdot e^{-t^2/2} \cdot 1 = \frac{1}{\sqrt{2\pi}} e^{-t^2/2};
\end{aligned} \tag{3}$$

this uses Problem 3.4.11. Thus the standard Student(n) distribution converges to the standard normal distribution as $n \to \infty$. See Figure 6.2.

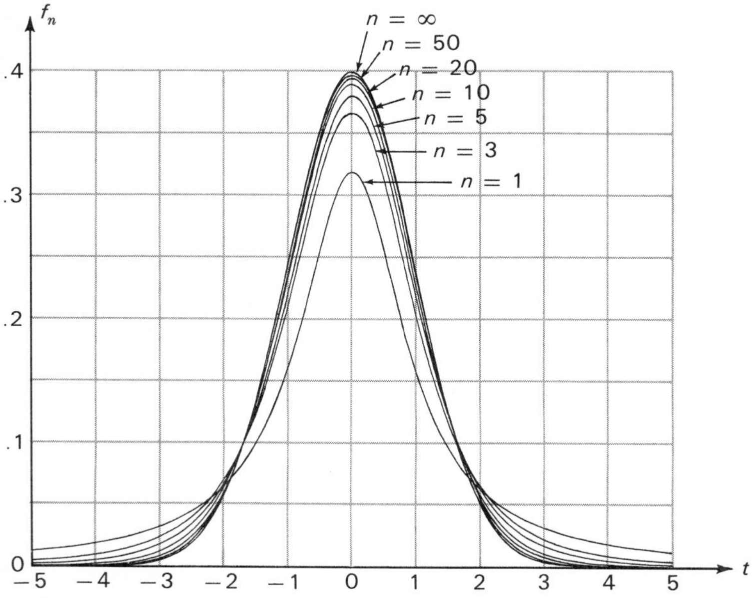

FIGURE 6.2
Standard Student(n) distribution for $n = 1, 3, 5, 10, 20, 50, \infty$. The limiting distribution is standard normal.

A very powerful way of determining limiting distributions is by means of moment-generating and characteristic functions. For this, we need the following continuity theorem:

THEOREM 3

If F_n has characteristic function c_n and F has characteristic function c, then $F_n \to F$ if and only if

(4) $$\lim_{n\to\infty} c_n(t) = c(t)$$

at all points t. If F_n has moment-generating function m_n and F has moment-generating function m, then $F_n \to F$ if and only if

(5) $$\lim_{n\to\infty} m_n(t) = m(t)$$

for an open interval containing the origin.

Proof The proof is recorded in Supplement Section 6.3H.

An analogous result holds for moment sequences—with a single qualification that the limiting moment sequence should correspond to a unique distribution.

Figure 6.3 presents the one-to-one mapping between distributions $\mathcal{P}$ and characteristic functions $\mathcal{C}$ and the one-to-one mapping between a subset of $\mathcal{P}$ and the moment-generating functions $\mathfrak{M}$. The theorem says that these one-to-one mappings are continuous; we refer to the theorem as the *continuity theorem*.

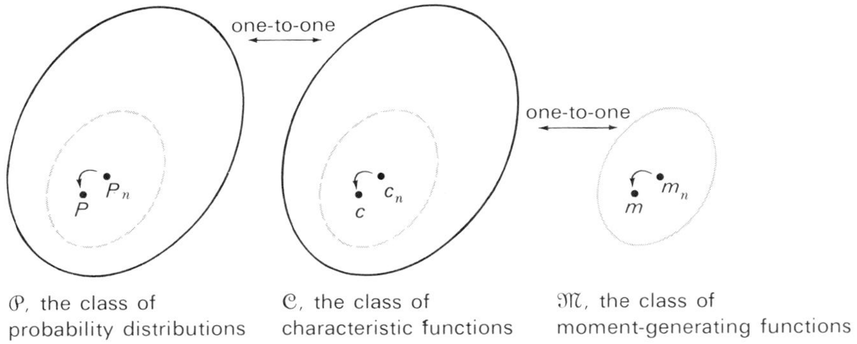

FIGURE 6.3
P_n converges to P if and only if c_n converges to c if and only if m_n converges to m; this assumes that each function belongs to the indicated class.

EXAMPLE 3

Poisson(λ) distribution: Consider the Poisson(λ) distribution for y. The mean of the distribution is λ and the standard deviation is $\sqrt{\lambda}$. As $\lambda \to \infty$, the distribution dilates and moves toward $+\infty$. Suppose that we examine the limiting distribution of the variable standardized with respect to location and scale. The moment-generating function of the Poisson is recorded in Problem 5.7.10. We now use properties (5.7.5) of the moment-generating function and calculate

Variable	y	$y - \lambda$	$(y - \lambda)/\sqrt{\lambda}$
Moment-generating function(t)	$\exp\{\lambda e^t - \lambda\}$	$\exp\{\lambda e^t - \lambda - \lambda t\}$	$\exp\{\lambda e^{t/\sqrt{\lambda}} - \lambda - \sqrt{\lambda}t\}$

Let $m_\lambda(t)$ be the moment-generating function of $(y - \lambda)/\sqrt{\lambda}$; then examining the logarithm we obtain

(6)
$$\begin{aligned}\lim_{\lambda\to\infty} \ln m_\lambda(t) &= \lim_{\lambda\to\infty} (\lambda e^{t/\sqrt{\lambda}} - \lambda - \sqrt{\lambda}t\} \\ &= \lim_{\lambda\to\infty}\left(\lambda\left(1 + \frac{t}{\sqrt{\lambda}} + \frac{t^2}{2!\,\lambda} + \frac{t^3}{3!\,\lambda^{3/2}} + \cdots\right) - \lambda - \sqrt{\lambda}t\right) \\ &= \lim_{\lambda\to\infty}\left(\frac{t^2}{2} + \frac{t^3}{3!\,\sqrt{\lambda}} + \frac{t^4}{4!\,\lambda} + \cdots\right) = \frac{t^2}{2};\end{aligned}$$

that is, $\lim m_\lambda(t) = e^{t^2/2}$, the moment-generating function for the standard normal. Thus *the standardized Poisson distribution converges to the standard normal distribution*.

D CLOSENESS OF THE POISSON APPROXIMATION

First consider the use of the preceding approximation to calculate a Poisson probability.

EXAMPLE 4

The number of fatal automobile accidents per year in a large city is known to approximate the Poisson(80) distribution. On the assumption that conditions remain essentially stable, find the probability that the number of accidents y for the coming calendar year is ≤ 110.

We use the normal approximation in Example 3 and refine it slightly. The frequency y has a discrete distribution on the integers; the normal is continuous. Thus for a frequency $y \leq 110$ it is perhaps reasonable to use the point $110 + 1/2$, half way between 110 and 111, as the dividing point. Let P_{80} designate probability calculated according to the Poisson(80) and P designate probability calculated according to the standard normal; we calculate

$$P_{80}(y \leq 110) = P_{80}\left(\frac{y - 80}{\sqrt{80}} \leq \frac{110 - 80}{\sqrt{80}}\right)$$

$$\approx P\left(z \leq \frac{110.5 - 80}{8.944}\right) = P(z \leq 3.410) = 0.99968.$$

The exact value for the Poisson(80) is 0.99929. Ordinarily, we would not depend on a limiting distribution for a calculation of probability far out on the tail of a distribution.

Now consider an alternative way to approximate Poisson(λ) probabilities on the real line R. From Example 3 we know that the limiting distribution of

$$z_1 = \frac{y - \lambda}{\sqrt{\lambda}}$$

is standard normal; note that y is rescaled by a different factor $1/\sqrt{\lambda}$ for each different λ value. A somewhat different standardization

$$z_2 = 2(\sqrt{y} - \sqrt{\lambda}) \tag{7}$$

can be more convenient: the scaling is the *same* for each λ. We now show that z_2 has a limiting standard normal distribution. For this let G_λ, H_λ be the distribution functions for z_1, z_2; then

$$H_\lambda(x) = P(z_2 \leq x) = P(2(\sqrt{y} - \sqrt{\lambda}) \leq x) = P(\sqrt{y} \leq \sqrt{\lambda} + x/2)$$

$$= P(y \leq \lambda + x\sqrt{\lambda} + x^2/4) = P\left(\frac{y - \lambda}{\sqrt{\lambda}} \leq x + \frac{x^2}{4\sqrt{\lambda}}\right)$$

$$= G_\lambda(x + x^2/4\sqrt{\lambda});$$

thus

$$\lim_{\lambda\to\infty} H_\lambda(x) = \lim_{\lambda\to\infty} G_\lambda(x + x^2/4\sqrt{\lambda}) = G(x),$$

where G is the standard normal distribution function. *Note:* $G_\lambda(x) < H_\lambda(x)$.

EXAMPLE 5

Consider again the Poisson(80) distribution for the accident frequency in Example 4. We calculate the Poisson probability $P_{80}(y \leq 110)$ using the square-root approximation z_2 in (7). Again it is reasonable to go to the appropriate half-integer point:

$$\begin{aligned} P_{80}(y \le 110) &= P_{80}(2(\sqrt{y} - \sqrt{80}) \le 2(\sqrt{110} - \sqrt{80})) \\ &\approx P(z_2 \le 2(\sqrt{110.5} - \sqrt{80})) \\ &= P(z_2 \le 3.135) = 0.99914. \end{aligned}$$

The exact value for the Poisson(80) is 0.99929. For this particular probability the square-root approximation is slightly better.

Now consider a comparison of the two approximations for the Poisson. Let F_λ be the distribution function for the Poisson(λ) and let G be the standard normal distribution function. A good approximation should yield a z value that accurately calculates the probability. The correct z value for the Poisson probability $F_\lambda(y)$ is

$$z = G^{-1}(F_\lambda(y)).$$

The two approximations give

$$z_1 = \frac{y + 1/2 - \lambda}{\sqrt{\lambda}}, \qquad z_2 = 2(\sqrt{y + 1/2} - \sqrt{\lambda}).$$

Consider the values for the Poisson(80) given in Table 6.2. The values in the last column are those found in Examples 4 and 5. See also Figure 6.4.

TABLE 6.2
Some Poisson(80) values

y	50	60	70	80	90	100	110
z_1	−3.30	−2.18	−1.06	0.06	1.17	2.29	3.41
z_2	−3.68	−2.33	−1.10	0.06	1.14	2.16	3.14
z	−3.52	−2.26	−1.07	0.07	1.17	2.22	3.24

In general, $z_2 < z_1$; this follows from $G_\lambda < H_\lambda$. An investigation of graphs such as Figure 6.4 indicates that $z_2 < z < z_1$, with z_2 approximating better than z_1; exceptions occur near the center, where, however, both approximations are very good. The second approximation z_2 given in (7) has, in addition, substantial advantages for statistics: *a single reexpression* $2\sqrt{y}$ *gives the scaled variable regardless of the value for* λ.

E CONVERGENCE TO A CONSTANT

If all the probability piles up at a point c, then it is reasonable to say that the distribution converges to the constant c; see Example 1. Consider a distribution function F_n for y_n on the real line R. Then as a special case of Definition 1, we say that y_n *converges in distribution to the constant* c, $y_n \xrightarrow{\mathfrak{D}} c$, if

$$(8) \qquad \begin{aligned} \lim_{n\to\infty} F_n(x) &= 0 \qquad x < c, \\ &= 1 \qquad x > c. \end{aligned}$$

An alternative criterion for this convergence to the constant c is that

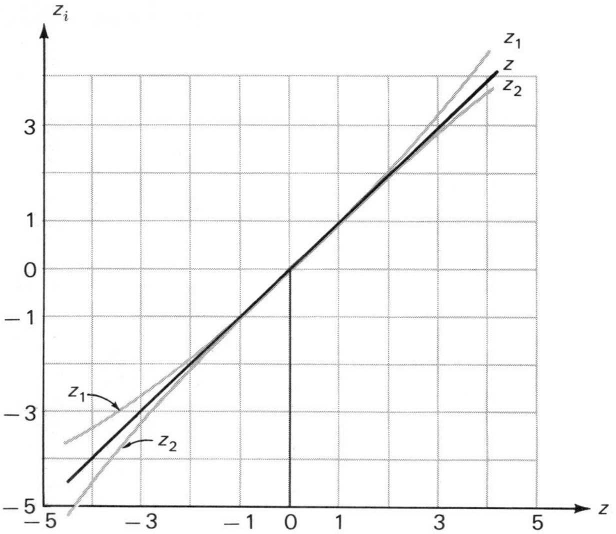

FIGURE 6.4
Approximations z_1 and z_2 plotted against the true z for the Poisson(20) distribution for y. As λ increases, the approximations become closer in the pattern indicated.

(9) $$P(|y_n - c| > \epsilon) \to 0 \qquad \text{for each } \epsilon > 0$$

as $n \to \infty$, or that

(10) $$P(|y_n - c| \leq \epsilon) \to 1 \qquad \text{for each } \epsilon > 0$$

as $n \to \infty$. These are just other ways of saying that the probability for y_n piles up at the point c.

Part of the importance of this special definition is indicated by the following theorem.

THEOREM 4

If $y_n \xrightarrow{\mathfrak{D}} y$ *and if* $x_n \xrightarrow{\mathfrak{D}} 0$, *then* $y_n + x_n \xrightarrow{\mathfrak{D}} y$.

Proof The proof is recorded in Supplement Section 6.31. For some generalizations see Problems 6.1.14–6.1.16.

We can interpret the theorem by thinking of x_n as a perturbation whose effects go to zero as $n \to \infty$. Thus we see that adding the perturbation x_n to y_n does not change the limiting distribution for y_n.

F EXERCISES

1 The number y of fatal accidents per year in a large city is known to be approximately Poisson(100). Calculate the probabilities

$P(90 \leq y \leq 110)$, $\quad P(80 \leq y \leq 120)$

using a normal approximation.

2 Use moment-generating or characteristic functions to show that the distribution of

$$\frac{\chi^2 - m}{\sqrt{2m}}$$

converges to the standard normal distribution where χ^2 has the chi-square(m) distribution.

3 Use Exercise 2 to show that the distribution of

$$\sqrt{2\chi^2} - \sqrt{2m - 1}$$

converges to the standard normal distribution where χ^2 has the chi-square(m) distribution.

4 Prove that formula (9) is equivalent to $y_n \xrightarrow{\mathfrak{D}} c$.

5 Show that if y_n converges in distribution to the constant c, then $g(y_n)$ converges in distribution to the constant $g(c)$, where g is a continuous function.

G PROBLEMS

6 Use moment-generating functions or characteristic functions to show that the distribution of

$$\frac{y - np}{\sqrt{npq}}$$

converges to the standard normal distribution, where y has the binomial(n, p) distribution.

7 Let $(u_{(1)}, \ldots, u_{(n)})$ be the order statistic for a sample of n from the uniform(0, 1) distribution.

(a) Show that $nu_{(1)}$ has a limiting standard exponential distribution. *Method:* Use the distribution or density function.

(b) Then deduce that $n(1 - u_{(n)})$ has a limiting standard exponential distribution.

8 Let $(y_{(1)}, \ldots, y_{(n)})$ be the order statistic of a sample of n from a distribution with continuous distribution function F. Use Problem 7 to show that $nF(y_{(1)})$ has a limiting standard exponential distribution; that $n(1 - F(y_{(n)}))$ has a limiting standard exponential distribution.

9 *Normal approximation to the Poisson; by density functions*. Consider a Poisson(λ) distribution for y and an independent uniform($-1/2$, $1/2$) distribution for u. Let $w = y + u$.

(a) Show that the density function for w is

$$f_\lambda(w) = \frac{\lambda^{(w)}e^{-\lambda}}{(w)!} \qquad \text{on } (-1/2, \infty),$$

where (w) is the nearest integer to w.

(b) Use Theorem 4 to show that as $\lambda \to \infty$

$$\frac{y - \lambda}{\lambda^{1/2}}, \qquad \frac{w - \lambda}{\lambda^{1/2}}$$

have the same limiting distribution, if any.

10 (*continuation*) Show that density for the standardized continuous Poisson $(w - \lambda)/\lambda^{1/2}$ converges to the standard normal density function as $\lambda \to \infty$; use Stirling's formula.

11 *Normal approximation to the binomial; by density functions*. Consider a binomial(n, p) distribution for y and an independent uniform($-1/2$, $1/2$) distribution for u. Let $w = y + u$. Show that the density function for w is

$$f_n(w) = \binom{n}{(w)} p^{(w)}q^{n-(w)} \qquad \text{on } (-1/2, n + 1/2),$$

where (w) is the nearest integer to w.

12 (*continuation*) Show that the density function for the standardized continuous binomial $(w - np)/(npq)^{1/2}$ converges to the standard normal density function as $n \to \infty$; use Stirling's formula and Theorem 4.

13 Let $(x_{(1)}, \ldots, x_{(n)})$ be the order statistic for a sample of n from a distribution with a continuous

density function f. For a given $p = 1 - q$ in (0, 1), let $x_{(np)}$ be the order statistic with subscript taken to be the largest integer in np, and ζ_p be such that $F(\zeta_p) = p$, where F is the distribution function. Show that if $f(\zeta_p) > 0$, then

$$\frac{x_{(np)} - \zeta_p}{(pq/nf^2(\zeta_p))^{1/2}}$$

has a limiting standard normal distribution as $n \to \infty$. *Method:* For given z, show that

$$x_{(np)} < \zeta_p + z\left(\frac{pq}{n}\right)^{1/2} \frac{1}{f(\zeta_p)}$$

is equivalent to $y \geq [np]$, where y is the number of observations less than $\zeta_p + z(pq/n)^{1/2}/f(\zeta_p)$, and also show that y has a binomial$(n, F(\zeta_p + z[pq/nf^2(\zeta_p)]^{1/2}))$ distribution. *Assume:* The standardized binomial distribution function approaches the standard normal distribution function uniformly in p with $0 < p_1 < p < p_2 < 1$.

14 If $y_n \xrightarrow{\mathfrak{D}} y$ and $x_n \xrightarrow{\mathfrak{D}} a$, show that $y_n + x_n \xrightarrow{\mathfrak{D}} y + a$. Thus if y_n has limiting distribution function $F(\cdot)$, then $y_n + x_n$ has limiting distribution function $G(\cdot)$, where $G(w) = F(w - a)$.

15 (*continuation*) If $a > 0$, then $w_n = x_n y_n$ has limiting distribution G, where $G(w) = F(w/a)$. *Hint:* As in Supplement Section 6.3I.

16 (*continuation*) If $a > 0$, then $w_n = y_n/x_n$ has limiting distribution H, where $H(w) = F(aw)$. *Hint:* As in Supplement Section 6.3I.

*17 Prove Proposition 2. *Method:* Apply the Fatou lemma in Supplement Problem 5.8.5 to a Borel set B and to the Borel set B^c.

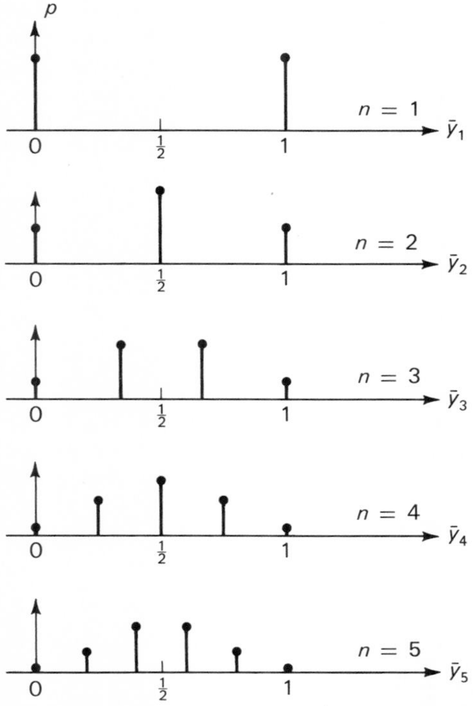

FIGURE 6.5
Distribution of the sample average $\bar{y}_n$ for a sample of n from the Bernoulli(1/2).

6.2

THE WEAK LAW OF LARGE NUMBERS AND THE CENTRAL LIMIT THEOREM

We have noted that many response variables in science and industry are approximately normally distributed—the explanation being that many small sources of variation add to produce normal variation in the response. We find also that many *derived* variables are very closely normally distributed. Notable among such variables of course is the average of a sample of independent repetitions on a random system; such samples are essential to experimental investigations. In Section A we examine two simple cases to see how the distribution of the sample average varies with the sample size. In Section B we examine the Weak Law of Large Numbers, which shows fairly generally that the sample average converges to the mean as the sample size increases. Then in Section C we examine some Central Limit Theorems that provide an explanation for the approximate normality for sums and averages and for many responses.

A EXAMPLES OF THE DISTRIBUTION OF THE SAMPLE AVERAGE

We now consider two distributions, one discrete and one continuous, and examine the distribution of the sample average $\bar{y}_n = \Sigma_1^n y_i/n$ as the sample size is increased.

For a discrete distribution we choose the simple Bernoulli(1/2). The distribution

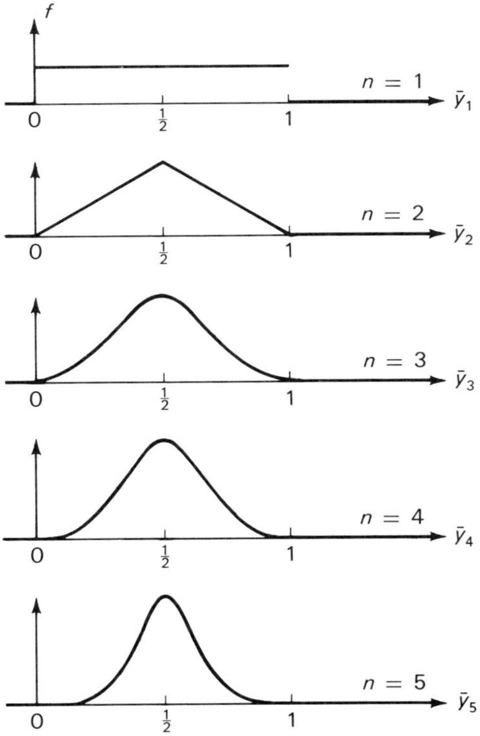

FIGURE 6.6
Distribution of the sample average $\bar{y}_n$ for a sample of n from the uniform(0, 1).

of $\Sigma_1^n y_i$ is binomial(n, 1/2). Correspondingly, the distribution of $\bar{y}$ is the binomial(n, 1/2) distribution but with the interval $[0, n]$ scaled into $[0, 1]$; see Figure 6.5.

For an absolutely continuous distribution we choose the uniform(0, 1). The distribution of $\Sigma_1^n y_i$ can be obtained by induction using the convolution formula (3.2.25); the density for $\bar{y}$ on (0, 1) is

$$f(t) = \frac{n}{(n-1)!}\left((nt)^{n-1} - \binom{n}{1}(nt-1)^{n-1} + \binom{n}{2}(nt-2)^{n-1} - \cdots (-1)^r\binom{n}{r}(nt-r)^{n-1}\right) \quad \text{if } \frac{r}{n} < t \le \frac{r+1}{n},$$

for $r = 0, 1, \ldots, n-1$ and zero otherwise. See Figure 6.6.

As n increases we notice in each case that the distribution of the sample average $\bar{y}_n$ is gradually concentrating around the mean μ of the distribution being sampled. This phenomenon in fact occurs in general provided only that the mean μ exists; it is covered by the Weak Law of Large Numbers discussed in Section B.

Now consider the shape of the distribution for the sample average $\bar{y}_n$. For this we need to remove the concentrating effect we have just described: the deviation from the mean is $\bar{y}_n - \mu$; the standard deviation of this deviation is $\sigma/\sqrt{n}$, where σ^2 is the variance of the distribution being sampled; we examine the distribution of $\sqrt{n}(\bar{y}_n - \mu)$. This distribution is recorded in Figure 6.7 for the Bernoulli and in Figure 6.8 for the uniform.

As n increases, we notice in each case that the distribution of the adjusted variable $\sqrt{n}(\bar{y}_n - \mu)$ is gradually coming toward the bell-shaped form of the normal distribution. This phenomenon occurs generally if the mean μ and variance σ^2 exist for the

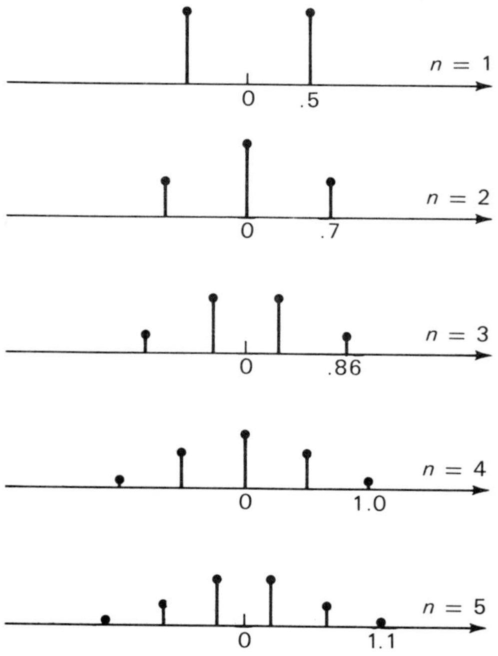

FIGURE 6.7
Distribution of $\sqrt{n}(\bar{y}_n - \mu)$ for a sample of n from the Bernoulli(1/2).

distribution being sampled; it is covered by the Central Limit Theorem discussed in Section C.

B THE WEAK LAW OF LARGE NUMBERS

Consider a distribution for y on the real line R with probability measure P and with mean μ, say the Bernoulli distribution for recovery under a surgical treatment. Let $\mathbf{y} = (y_1, \ldots, y_n)'$ be a sample: that is, with independent coordinates and with each coordinate having the same distribution P. The sample average $\bar{y}_n = (\Sigma_1^n y_i)/n$ is a function from R^n to R; it has a marginal distribution on R. In this section we see that the distribution of $\bar{y}_n$ concentrates or piles up at μ as the sample size $n \to \infty$.

THEOREM 1

Weak Law of Large Numbers: *If $(y_1, \ldots, y_n)$ is a sample from a distribution with mean μ, then*

(1) $$\lim_{n\to\infty} P(|\bar{y}_n - \mu| \leq \epsilon) = 1$$

for each $\epsilon > 0$; that is, $\bar{y}_n \xrightarrow{\mathfrak{D}} \mu$.

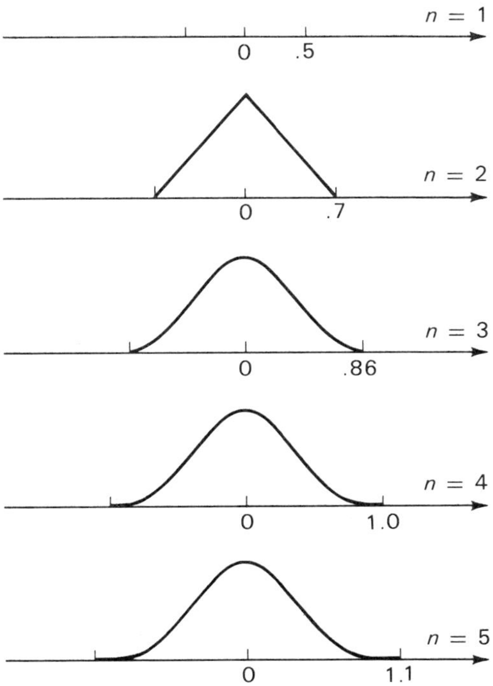

FIGURE 6.8
Distribution of $\sqrt{n}(\bar{y}_n - \mu)$ for a sample of n from the uniform(0, 1).

Proof We give the proof under the simplifying assumption that the variance σ^2 exists for the distribution being sampled; the more general proof is given in Section F. From formula (5.4.5) we have

$$E(\bar{y}_n) = \mu, \qquad \text{var}\,(\bar{y}_n) = \frac{\sigma^2}{n}.$$

The Chebyshev inequality (5.3.9) then gives

$$P(|\bar{y}_n - \mu| > \epsilon) \leq \frac{\sigma^2/n}{\epsilon^2},$$

$$P(|\bar{y}_n - \mu| \leq \epsilon) \geq 1 - \frac{\sigma^2/n}{\epsilon^2}.$$

Now consider the limit as $n \to \infty$:

$$\lim_{n\to\infty} P(|\bar{y}_n - \mu| \leq \epsilon) = 1;$$

and we have that $\bar{y}_n$ converges to μ in distribution, $\bar{y}_n \longrightarrow \mu$.

EXAMPLE 1

Sample from the Bernoulli(p): Consider a sample $(x_1, \ldots, x_n)$ from the Bernoulli(p) distribution. The function $y = x_1 + \cdots + x_n$ gives the frequency for n Bernoulli repetitions, and the function $\hat{p}_n = y/n = \bar{x}_n$ gives the corresponding proportion or estimated probability. For an application, consider some random system and a particular event of interest having probability p. For n independent repetitions the function $\hat{p}_n = \bar{x}_n$ is the proportion of occurrences for the event. We now examine a property of the distribution that describes this proportion. From the Weak Law of Large Numbers we obtain

$$(2) \qquad P(|\hat{p}_n - p| \leq \epsilon) \xrightarrow[n\to\infty]{} 1;$$

that is, the probability for $\hat{p}_n$ piles up at the true probability p. Our theory has produced, in fact, a model for the frequency interpretation of probability and the model is consistent with that frequency interpretation. This is an indication of the internal consistency of our theory.

EXAMPLE 2

Sample from the Cauchy distribution: Consider a sample $(y_1, \ldots, y_n)$ from the Cauchy distribution. From Example 5.2.3 we know that the mean does not exist. Thus it is not surprising to note from Problem 5.7.34 that the distribution for the sample average from the Cauchy distribution is just the same Cauchy distribution; thus

$$P(|\bar{y}_n| \leq \epsilon) = \int_{-\epsilon}^{\epsilon} \frac{1}{\pi(1 + y^2)}\, dy,$$

which is in sharp contrast to the Weak Law.

C THE CENTRAL LIMIT THEOREM

Consider a distribution for y on the real line R with probability measure P, with mean μ and with variance σ^2. Let $\mathbf{y} = (y_1, \ldots, y_n)'$ be a sample from the distribution. We have seen that the marginal distribution of the sample average $\bar{y}_n = (\Sigma_1^n y_i)/n$ concentrates at the mean μ. We now show that the shape of the marginal distribution approaches the normal; see for example, Figure 6.9.

From formulas (5.4.4) and (5.4.5) we have the location and scale characteristics of the sample sum $\Sigma_1^n y_i$ and the sample average $\bar{y}_n$:

$$E\left(\sum_1^n y_i\right) = n\mu, \qquad \operatorname{var}\left(\sum_1^n y_i\right) = n\sigma^2,$$

(3)

$$E(\bar{y}_n) = \mu, \qquad \operatorname{var}(\bar{y}_n) = \frac{\sigma^2}{n}.$$

We now form a new variable by standardizing and measuring departure from the mean in units of the standard deviation:

$$z_n = \frac{\sum_1^n y_i - n\mu}{\sqrt{n}\sigma} = \frac{\bar{y}_n - \mu}{\sigma/\sqrt{n}}.$$

The standardized variable has mean zero and variance 1.

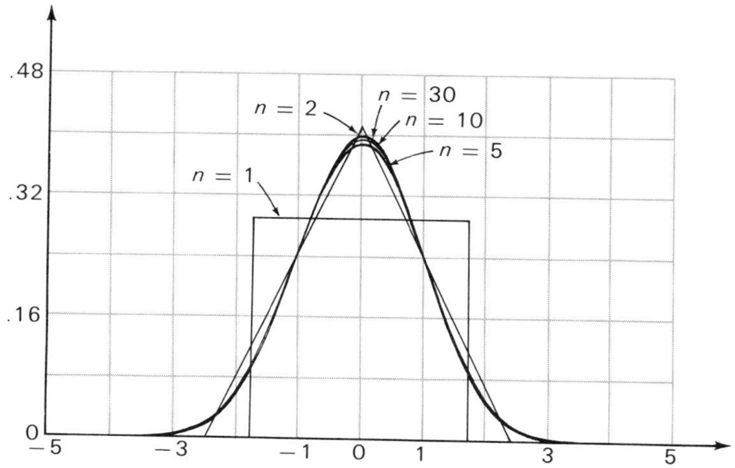

FIGURE 6.9
Distribution of $\sqrt{n}(\bar{y} - \mu)/\sigma$ for a sample of n from the uniform(0, 1); $n = 1, 2, 5, 10, 30$.

THEOREM 2

Central Limit Theorem: *If $(y_1, \ldots, y_n)$ is a sample from a distribution with mean μ and variance σ^2, then the distribution of*

$$z_n = \frac{\sum_1^n y_i - n\mu}{\sqrt{n}\sigma} = \frac{\bar{y}_n - \mu}{\sigma/\sqrt{n}} \tag{4}$$

converges to the standard normal distribution.

The theorem asserts that

$$\lim_{n\to\infty} P(z_n \le x) = \int_{-\infty}^{x} \frac{1}{\sqrt{2\pi}} e^{-z^2/2}\, dz;$$

and thus that

$$\lim_{n\to\infty} P(a < z_n \le b) = \int_{a}^{b} \frac{1}{\sqrt{2\pi}} e^{-z^2/2}\, dz. \tag{5}$$

The proof of the theorem is recorded in Section D.

EXAMPLE 3

Sample from the Bernoulli(p): Consider a sample $(x_1, \ldots, x_n)$ from the Bernoulli(p) distribution; the location and scaling of the Bernoulli are

$$E(x) = p, \qquad SD(x) = \sqrt{pq}.$$

The marginal distribution of the sample sum $y = \Sigma_1^n x_i$ is binomial(n, p) and the distribution of $\bar{x}_n = \hat{p}$ is a rescaled binomial.

In Example 1 we saw that the distribution of $\bar{x}_n = \hat{p}$ piled up at the value p. The present Theorem 2 says that the distribution of

$$z_n = \frac{y - np}{(npq)^{1/2}} = \frac{\hat{p} - p}{(pq/n)^{1/2}} \tag{6}$$

converges to the standard normal distribution.

Now suppose that we use this limiting distribution to calculate some approximate probabilities for the binomial(100, 1/2) distribution; we might, for example, be interested in the frequency y of heads for 100 tosses of an unbiased coin or in the number of 1's in a sample of 100 random bits; we have

$$E(y) = 50, \qquad \sigma_y = 5.$$

Before using the normal approximation we should note that the binomial differs from the normal in an obvious way: the binomial has probability at the integers and the normal has probability continuously on the line. Toward compensating for this we can proceed as in Example 6.1.4 and view the binomial as having come from a continuous distribution by lumping probability to the nearest integer. Thus to approximate the probability $P(45 \le y \le 55)$ we would use $P(44\tfrac{1}{2} \le y \le 55\tfrac{1}{2})$ with the limiting distribution. Thus we have

$$P(45 \le y \le 55) = P\left(\frac{45 - 50}{5} \le \frac{y - 50}{5} \le \frac{55 - 50}{5}\right)$$

$$\approx P\left(\frac{44\frac{1}{2} - 50}{5} \le z \le \frac{55\frac{1}{2} - 50}{5}\right) = P(-1.1 \le z < 1.1)$$

$$= 0.8643 - 0.1357 = 72.9\%.$$

The normal approximation gives reasonable accuracy for p near 0.5 and $n \ge 10$, for p in the range 0.4 to 0.6 and $n \ge 25$, and for p in the range 0.3 to 0.7 and $n \ge 100$. For p near the extreme values 0, 1, the Poisson approximation in Section 6.1A can be used.

D PROOF OF THE CENTRAL LIMIT THEOREM 2

The ordinary exponential limit $\lim_{n\to\infty} (1 + \delta/n)^n = e^\delta$ can be used to show that

(7) $$\lim_{n\to\infty}\left(1 + \frac{\delta + \delta_n}{n}\right)^n = e^\delta,$$

where $\{\delta_n\}$ is a null sequence (i.e., $\lim \delta_n = 0$).

The existence of the mean $\mu = E(y)$ and the inequality

$$|iye^{iyt}| \le |y|$$

used with Supplement Proposition 5.8.9 give the convergence of the integral

(8) $$c'(t) = \frac{dc(t)}{dt} = \int_{-\infty}^{\infty} iye^{iyt}\, dP(y).$$

The existence of the second moment $E(y^2)$ and the inequality

$$|i^2y^2e^{iyt}| \le y^2$$

used with Supplement Proposition 5.8.9 give the convergence of the integral

(9) $$c''(t) = \frac{d^2c(t)}{dt^2} = \int_{-\infty}^{\infty} i^2y^2e^{iyt}\, dP(y).$$

The continuity of the preceding integrand (as a function of t) used with Proposition 5.8.8 gives the continuity of $c''(t)$. Thus the Taylor expansion of $c(t)$ about the origin gives

(10) $$c(t) = 1 + c'(0)t + c''(\theta t)\frac{t^2}{2!} \qquad |\theta| < 1,$$

$$= 1 + i\mu t - \mu_2\frac{t^2}{2!} + o(t^2),$$

where $o(t^2)$ is a remainder that goes to zero as a proportion of t^2.

Now for the main part of the proof. We are concerned with the distribution of

$$z_n = \frac{\sum y_i - n\mu}{\sqrt{n}\sigma} = \frac{y_1 - \mu}{\sqrt{n}\sigma} + \cdots + \frac{y_n - \mu}{\sqrt{n}\sigma},$$

which is a sum of independent and identically distributed components. Note that only the deviations $y_i - \mu$ with $E(y_i - \mu) = 0$ are involved; it thus suffices to prove the theorem for the case $\mu = 0$. We use the expression (10) and the properties of the characteristic function (Table 6.3). We take the limit of the characteristic function of z_n:

TABLE 6.3
Distributions and corresponding characteristic functions

Variable	Characteristic function(t)
y_i	$1 - \sigma^2 \frac{t^2}{2!} + o(t^2)$
$\frac{y_i}{\sqrt{n}\sigma}$	$1 - \frac{t^2}{2n} + o\left(\frac{t^2}{n\sigma^2}\right)$
z_n	$\left[1 - \frac{t^2}{2n} + o\left(\frac{t^2}{n\sigma^2}\right)\right]^n$

$$\lim_{n\to\infty}\left(1 - \frac{t^2}{2n} + o\left(\frac{t^2}{n\sigma^2}\right)\right)^n = e^{-t^2/2},$$

which is the characteristic function of the standard normal. It follows then by the Continuity Theorem 6.1.3 that the limiting distribution of z_n is standard normal.

E MORE GENERAL FORMS OF THE CENTRAL LIMIT THEOREM

The Central Limit Theorem can be extended to cover independent components with different distributions. The following version is due to Lindeberg.

THEOREM 3

If $y_1, \ldots, y_n$ are statistically independent, if y_i has a distribution with probability measure P_i, mean μ_i, and variance σ_i^2, and if

$$\frac{\sum_1^n \int_{|t|\geq k\tau_n} (t - \mu_i)^2 \, dP_i(t)}{\tau_n^2} \to 0$$

for each $k > 0$, where $\tau_n^2 = \sigma_1^2 + \cdots + \sigma_n^2$, then the distribution of

$$z_n = \frac{\sum_1^n (y_i - \mu_i)}{\tau_n}$$

converges to the standard normal distribution.

A stronger condition due to Liapounov gives the following:

THEOREM 4

If $y_1, \ldots, y_n$ are statistically independent, if y_i has a distribution with mean μ_i, variance σ_i^2, and third absolute central moment $\rho_i^3 = E(|y_i - \mu_i|^3)$, and if

$$\frac{\rho_1^3 + \cdots + \rho_n^3}{\tau_n^3} \to 0,$$

where $\tau_n^2 = \sigma_1^2 + \cdots + \sigma_n^2$, then the distribution of

$$z_n = \frac{\sum_1^n (y_i - \mu_i)}{\tau_n}$$

has a limiting standard normal distribution.

Central Limit Theorems are also available for distributions on R^k. The following is the analog of Theorem 2.

THEOREM 5

If $(\mathbf{y}_1, \ldots, \mathbf{y}_n)$ is a sample from a distribution on R^k with mean $\boldsymbol{\mu}$ and variance matrix Σ, then

$$\mathbf{w}_n = n^{-1/2}\left(\sum_1^n \mathbf{y}_i - n\boldsymbol{\mu}\right) = n^{1/2}(\bar{\mathbf{y}} - \boldsymbol{\mu})$$

has a limiting multivariate normal$(\mathbf{0}; \Sigma)$ distribution [convergence $F_n \xrightarrow{\mathfrak{D}} F$: $F_n(\mathbf{x}) \to F(\mathbf{x})$ at points of continuity of F].

F GENERAL PROOF OF THE WEAK LAW OF LARGE NUMBERS

We now prove the Weak Law of Large Numbers assuming only that $\mu = E(y)$ exists. From the preliminaries in Section D we have

$$c(t) = 1 + c'(\theta t)t = 1 + i\mu t + o(t), \qquad |\theta| < 1,$$

where $o(t)$ is a remainder that goes to zero as a proportion of t.

We are concerned with the distribution of

$$\bar{y}_n = \frac{y_1}{n} + \cdots + \frac{y_n}{n},$$

which is a sum of independent identically distributed components. The properties of the characteristic function then give

Variable	y_i	y_i/n	$\bar{y}_n$
Characteristic function(t)	$1 + i\mu t + o(t)$	$1 + i\mu \frac{t}{n} + o\left(\frac{t}{n}\right)$	$\left[1 + i\mu \frac{t}{n} + o\left(\frac{t}{n}\right)\right]^n$

For any given t the characteristic function of $\bar{y}_n$ has by (7) the limit $e^{i\mu t}$. But this is the characteristic function of the distribution with all probability at the point μ. Thus by the Continuity Theorem 6.1.3, the distribution of $\bar{y}_n$ converges to μ; that is, $\bar{y} \xrightarrow{\mathfrak{D}} \mu$.

G EXERCISES

1 How many times should a possibly biased coin be tossed to be 95 percent sure that the proportion for heads is within 0.01 of the probability for heads? Use the Central Limit Theorem and compare with Problem 5.3.30.

2 Suppose that "1-pound" packages from a packaging machine have a weight distribution with mean 1.01 and standard deviation 0.01. Use reasonable assumptions to find the probability that a "100-pound" lot weighs less than 100 pounds.

3 Let $(y_1, \ldots, y_n)$ be a sample from a distribution with moments $\mu_1, \ldots, \mu_r$. Show that $m_s \xrightarrow{\mathfrak{D}} \mu_s$ $(s \leq r)$, where m_s is the sample sth moment defined in Problem 5.6.7.

4 (*continuation*) Let $(y_1, \ldots, y_n)$ be a sample from a distribution with moments $\mu_1, \ldots, \mu_{2r}$. Show that the standardized distribution of m_r converges to the standard normal.

5 Show that if $\mathbf{y}_n$ converges in distribution to the constant $\mathbf{c}$, then $g(\mathbf{y}_n)$ converges in distribution to the constant $g(\mathbf{c})$, where g is a continuous function; recall the definition in Theorem 6.2.5.

6 (*continuation of Exercise 3*) Let g be a continuous function of r arguments. Show that $g(m_1, \ldots, m_r) \xrightarrow{\mathfrak{D}} g(\mu_1, \ldots, \mu_r)$.

H PROBLEMS

7 Let $(y_1, \ldots, y_n)$ be a sample from a distribution with mean μ and variance σ^2. Show that

$$s_y^2 \xrightarrow{\mathfrak{D}} \sigma^2,$$

where s_y^2 is the sample variance; use Problem 6.

8 Let $((y_{11}, y_{21}), \ldots, (y_{1n}, y_{2n}))$ be a sample from a distribution with means μ_1, μ_2, variances σ_1^2, σ_2^2, and covariance σ_{12}. Show that

$$s_{12} \xrightarrow{\mathfrak{D}} \sigma_{12},$$

where s_{12} is the sample covariance; use Problem 6.

9 Show that $\bar{m}_s \xrightarrow{\mathfrak{D}} \bar{\mu}_s$ $(s \leq r)$, where $\bar{m}_s$ is the sample sth central moment defined in Problem 5.6.8; use Problem 6.

10 If μ_4 exists, show that the limiting distribution of $\sqrt{n}(s_y^2 - \sigma^2)$ is a central normal; see Problem 7 and recall Theorem 6.1.4.

11 If μ_{2r} exists, show that the limiting distribution of $\sqrt{n}(\bar{m}_r - \mu_r)$ is a central normal; see Problem 9 and recall Theorem 6.1.4.

12 Under the conditions of the Central Limit Theorem 2, show that $\sqrt{n}(g(\bar{y}) - g(\mu))$ has a limiting normal distribution $(0, g'(\mu)\sigma)$, where g is continuously differentiable in the neighborhood of μ.

13 Under the conditions of the Central Limit Theorem 5, show that the distribution of

$$\sqrt{n}(g(\bar{y}_1, \ldots, \bar{y}_k) - g(\mu_1, \ldots, \mu_k))$$

converges to the normal $(0, \mathbf{a}'\Sigma\mathbf{a})$ distribution, where g is continuously differentiable near $\boldsymbol{\mu}$ and where $\mathbf{a}' = (a_1, \ldots, a_k) = (\partial g/\partial\mu_1, \ldots, \partial g/\partial\mu_k)$ gives the derivatives taken at the mean $(\mu_1, \ldots, \mu_k)$; use an obvious generalization of Problem 6.1.15.

14 Let $(y_1, \ldots, y_k)$ have the multinomial $(n, p_1, \ldots, p_k)$ distribution. Use the Central Limit

Theorem 5 to show that the distribution of

$$(w_1, \ldots, w_k) = n^{-1/2}(y_1 - np_1, \ldots, y_k - np_k)$$

converges to the multivariate normal distribution $(0, \Sigma)$, where $\sigma_{ii} = p_i(1 - p_i)$ and $\sigma_{ij} = -p_i p_j (i \neq j)$.

15 *Alternative proof of the Weak Law of Large Numbers:* Show that it suffices to prove the law for the case $\mu = 0$.

16 (*continuation*) Assume that $\mu = 0$ and for $j = 1, \ldots, n$, define $w_j = y_j$ if $|y_j| \leq \delta n$ and zero otherwise; note that $E(w_j) = \mu^*$ depends on n. Show that $\mu^* \to 0$ as $n \to \infty$.

17 (*continuation*) Show that

$$P(|\bar{y}_n - \mu^*| > t) \leq P(|\bar{w}_n - \mu^*| > t) + P(\bar{y}_n \neq \bar{w}_n)$$
$$\leq \frac{\sigma^2_{w_n}}{nt^2} + nP(|y_n| > \delta n).$$

18 (*continuation*) Show that

$$\sigma^2_{w_n} \leq \delta n \int |x| \, dP(x)$$
$$P(|y_n| > \delta n) \leq \frac{1}{\delta n} \int_{|y_n| > \delta n} |x| \, dP(x).$$

19 (*continuation*) Prove that

$$P(|\bar{y}_n - \mu^*| > t) < \epsilon$$

for n sufficiently large. Then prove the Weak Law.

SUPPLEMENTARY MATERIAL

Supplement Section 6.3 is on pages 554–563.

Sequences of functions on a probability space are examined and several forms of convergence are defined: convergence in probability, convergence in mean(r), and convergence almost surely. Criteria are given for convergence almost surely. The Continuity Theorem from Section 6.1 is proved. The Strong Law of Large Numbers is discussed and proved.

A concluding section shows how we have come full circle. We have been able to build up the model so that it can describe large-sample proportions and large-sample averages, which are in fact the pragmatic basis for the initial definition of probabilities and means. Reassuringly, we find that the model describes the proportions and averages in a consistent and coherent way.

NOTES AND REFERENCES

The Weak Law of Large Numbers and the Central Limit Theorem in Section 6.2 are concerned primarily with sequences of distributions. The Strong Law and convergence in probability in Supplement Section 6.3 are concerned with sequences of functions. The material in this chapter has been separated in accordance with this difference.

For further reading, see the following:

Feller, W. (1968). *An Introduction to Probability Theory and Its Applications,* Vol. 1, 3rd ed. New York: John Wiley & Sons, Inc.

Feller, W. (1971). *An Introduction to Probability Theory and Its Applications,* Vol. 2, 2nd ed. New York: John Wiley & Sons, Inc.

Loève, M. (1963). *Probability Theory,* 3rd ed. New York: Van Nostrand Reinhold.

In Chapters 1–6 we developed introductory probability theory in a form suitable for further study of probability theory or for detailed study of statistics and its applications.

In Chapters 7–12 we shall develop statistical theory with applications. For this we assume a general background of introductory probability theory, such as that presented in the main sections of Chapters 1–6.

In Chapter 7 we shall introduce the basic ideas of statistical inference: ***estimation, tests of significance,*** *and* ***confidence intervals.*** *We introduce these concepts in a general context, without restrictive assumptions such as the familiar one of normal distribution form. And we use results immediately available to us from the preceding chapters: specifically, symmetric probability distributions, and limiting or large-sample distributions.*

By introducing the basic inference concepts in a general context we shall be able to see how inference can be applied to data without elaborate or specialized models or theory; we shall also have the freedom in later chapters to see statistical inference in a framework larger than the usual one involving the normal and related distributions. Certainly the normal distribution is important, but at best it is an approximation to a true distribution in an application. Our development acknowledges the need for methods suitable for other distribution forms and the need for methods suitable for a range of distribution forms.

7

STATISTICAL INFERENCE

7.1

THE STATISTICAL MODEL AND THE DATA

In this section we examine some simple examples of statistical models and then briefly consider some data from a biochemical investigation. Then in Section 7.2 we use large-sample theory to develop some basic inference methods; these methods are applied to the biochemical data. In Section 7.3 we examine *finite population sampling,* the elementary theory for sample surveys. In Section 7.4 we use symmetry to develop some standard *nonparametric* inference procedures. Supplement Section 7.5 looks briefly at *tests of randomness,* tests for the validity of a sample.

A PROBABILITY AND STATISTICS

In the preceding chapters we have been concerned with random systems as found throughout science and industry. We have seen how probability distributions can describe the variation that occurs with random systems. And we have developed in detail the theory for such probability distributions.

Probability theory is descriptive: it describes phenomena in the real world; specifically, the stable variation that occurs with random systems throughout science and industry.

A major question for any application of probability theory is—How do we determine the appropriate distribution? In a sense this is what statistics is about and is the subject matter for these next six chapters.

Statistical theory is prescriptive: it prescribes methods of determining the appropriate distribution for a random system. Specifically, it is concerned with the design of observational and experimental systems so that the things of interest are, in fact, those being investigated; it is concerned with performances of the system and the collection of the relevant data; and it is concerned with the analysis of the data and the drawing of conclusions concerning the things of interest.

How do we determine the appropriate distribution for a particular random system? In preceding chapters we have mentioned two methods. First, take a very large number of performances of the system, and use the observed proportions as estimates or approximations for the true probabilities. Second, examine the random system for symmetries, and use the available symmetries to calculate some or all of the true probabilities. In later chapters we shall develop a third basic method: check the background information for general properties that limit the range of possibilities for the distribution, and use this limited range of possibilities together with data from a small or moderate number of performances of the system.

In this chapter we examine the first two methods, involving large samples and symmetry. We use the theory that is available to us from the preceding chapters. And we develop the process of statistical inference from observed data to conclusions concerning particular unknowns of interest.

In practice, however, large samples and symmetry are not always available. For example, a large number of repetitions can be unrealistic or uneconomical. And enough symmetry occurs rather infrequently. In Chapters 8, 9, 10, and 11 we will see how certain general properties of the system being investigated can make statistical inference feasible for a small to moderate number of repetitions of the system. As an example, a general property of the system could be that the variation in the response is approximately normally distributed, or that the variation in the response is approximately Student(λ) with the degrees of freedom in a moderate range, say from 2 to 15.

Then in the concluding chapter, we examine the most important part of statistics, the design of experimental investigations. In the design of experiments, we consider the choice of values for input variables of a system and the randomization of such choice against any nonrandom sources of variation. The purpose of the design of experiments is to determine cause–effect relationships.

Thus from Chapters 7 through 12 we progress from the simple use of probability concepts for statistical inference through to the central concern of scientific investigation, the determination of cause–effect relationships.

B STATISTICAL INFERENCE FROM LARGE SAMPLES AND SYMMETRY

In this chapter we examine statistical inference using the theory available to us from preceding chapters, specifically the large-sample results from Chapter 6 and the symmetry results from Chapter 1.

This general approach does not prescribe any tight or strict model for an application. We are not, for example, saying that certain response variables are (exactly) normally distributed. In some ways this makes our initial approach to inference very open indeed. At times it may seem, in fact, to lack direction.

But this is as it should be. We do not wish to restrict our initial approach to inference to any narrow direction predicated by a particular model or method. Rather we wish the approach to inference to be broad and general. We will thus obtain an initial broad contact with the basic concepts of statistics; and we will see that a wide range of methods is available to us—just on the basis of large samples and symmetry.

In subsequent chapters, then, we will see how additional information and the corresponding specification for the model can lead to more directed and more specific inferences.

Our purpose now is to open as many directions for inference as possible; to have it as flexible and data- and application-based as possible; and to avoid conceptual commitment to any very specific form of modeling.

C THE UNKNOWNS OF A RANDOM SYSTEM

Consider a random system. If the distribution is known, we have a full description of the system as discussed in Chapters 1 to 6, and there is no statistical problem.

If the distribution is not known, a range of possibilities will exist for the system. Certain of the unknowns of the system may be of particular interest; consider some examples.

An investigator may be interested directly in certain probabilities of the system: for example, the probability of recovery from a certain disease using a particular drug therapy; or the probability of a lifetime greater than 5000 hours for a certain lighting unit.

Alternatively, the investigator may be interested in a characteristic of the system that is closely related to a probability: for example, the amount of the drug that gives a 50 percent probability of recovery from the particular disease; or the amount of gaseous additive that maximizes the probability of a lifetime greater than 5000 hours. These alternatives involve questions of experimental design and are examined in Chapter 12.

An investigator may be interested in characteristics that are derived from the probabilities for the system. He may, for example, be interested in the general level of a certain real-valued response variable y. The general level can often be measured by the mean $\mu = E(y)$. In other cases, if long tails are permitted for the response distribution, the general level may be measured more realistically by the median $\tilde{\mu}$, the value that divides the distribution 50–50. Or he may be interested in the magnitude of the variation of a certain real-valued response variable. Often the variation can be measured by the variance $\sigma^2 = \text{var}(y)$. In other cases, if long tails are permitted for the response distribution, the variation can be measured by the range that contains the central 50 percent of the distribution—the interquartile range; or it can be measured by the distance plus or minus from the median that contains 68.26 percent of the distribution (this gives 1 standard deviation in the case of the normal).

Alternatively, the investigator may be interested in a characteristic that is closely related to the preceding characteristics: for example, in the change in the general response level caused by a change in an input variable to the system. This alternative involves questions of cause–effect relationships and is examined in Chapter 12.

The statistical model for a stable system with unknowns is the *set of possible descriptions for the system*. In some cases the statistical model can be the set of possible distributions for the response of the system. If there is very little information concerning the system, the set may be very large and embrace, say, all the distributions on the particular space. Or if there is a lot of information, the set may be quite small and embrace, say, the normal distributions on the real line or the normal distributions with specified variance. With such a model, some one of the distributions is the true distribution and provides a reasonable approximation for the system. The problem of statistical inference, then, is to use *data* from a small or moderate number of repetitions to draw conclusions concerning *which distribution in the model is the true distribution, or to draw conclusions about a particular unknown of interest such as the mean or variance of the true distribution*.

In other cases the investigator can describe the variation in the system by means of a probability model. The statistical model then involves the set of possibilities for the response as a function of the underlying variation; with such a model some one of the functions is the true function that provides a reasonable approximation for the system. The problem of statistical inference, then, is to use data from a small or moderate number of repetitions to draw conclusions about *which function in the model is the true response function or to draw conclusions about some particular unknown of interest, say the mean or variance of the true response function*.

D EXAMPLES OF A STATISTICAL MODEL

Consider several examples of statistical models. The examples present a gradation in the amount of information available concerning the system being investigated.

EXAMPLE 1

Consider a biochemical system and let y designate a certain real-valued response of the system as operated under specified conditions. Suppose that y can be measured to a moderate number of figures so that the continuous range R of possible values provides a reasonable approximation for the sample space. We would certainly be interested in intervals, so the Borel sets provide a reasonable class of events. If there are no constraints on the possible distributions, then the model would be the set $\mathfrak{D} = \{P\}$ of all probability distributions P on the real line. If there is information concerning continuity of the response, then the model could be the set $\mathfrak{D} = \{f\}$ of all probability densities on the real line, or perhaps the set of all continuous probability densities on the real line. Note that we have shifted to a more convenient way of presenting the distribution, by density rather than measure. Also note that if we use, say, three figures to record the response, then the distribution is technically discrete, but a continuous distribution with a density function may be a very reasonable approximation. In some contexts there may be additional constraints on the distribution, say that the mean of the distribution exists, or that the mean and variance exist.

To obtain information on the system we would want observations on a sequence of independent repetitions of the system. In order to assess such observations we need the distribution for such a sequence of repetitions. Let $(y_1, \ldots, y_n)$ designate the vector response for a sequence of repetitions. A distribution for this sequence is the distribution of a sample from a distribution in $\mathfrak{D}$; see Section 3.3F. Our model for the sequence is

$$\{f(y_1) \cdots f(y_n) : f \in \mathfrak{D}\},$$

and we would have appropriate constraints on the possible densities f as discussed in the preceding paragraph. In this chapter we see how an observed sequence can be used to make inferences concerning the true f or concerning some relevant characteristic of the true f. Note in the display that we are concerned with the *functions* but that arguments have been inserted for clarity of presentation.

EXAMPLE 2

A new treatment is applied to animals of a certain breed—is applied to *experimental units*. Let y designate the response. The variation affecting the response has been investigated under various treatments and conditions and is known to be approximately normal. Under present conditions let μ designate the general level of the response and σ the scaling of the response. The statistical model for a treatment applied to a unit is

$$\left\{\frac{1}{\sqrt{2\pi}\sigma} \exp\left\{-\frac{1}{2\sigma^2}(y-\mu)^2\right\} : \begin{matrix}\mu \in R \\ \sigma \in R^+\end{matrix}\right\},$$

using densities to present distributions on the line R.

Now consider a sequence of n independent repetitions of the system, that is, the treatment applied to n experimental units. Let $(y_1, \ldots, y_n)$ designate the response sequence; the model for such a sequence is

$$\left\{\frac{1}{(2\pi\sigma^2)^{n/2}} \exp\left\{-\frac{1}{2\sigma^2}\sum (y_i-\mu)^2\right\} : \begin{matrix}\mu \in R \\ \sigma \in R^+\end{matrix}\right\},$$

using densities to present distributions on R^n. In Chapters 8, 9, and 10 we shall examine how an observed sequence can be used to make inferences concerning the true density f, that is, concerning the true values for the parameters μ and σ.

EXAMPLE 3

Consider the preceding example in somewhat greater detail. The distribution of the variation affecting the response has been identified. Let z designate this variation. In standard units the distribution for z is given by

$$\frac{1}{\sqrt{2\pi}} \exp\{-(1/2)z^2\}.$$

The model for the response is

$$\left\{y = \mu + \sigma z : \begin{matrix}\mu \in R \\ \sigma \in R^+\end{matrix}\right\},$$

where μ and σ designate the general level and the scaling of the response y.

Now consider a sequence of independent repetitions of the system. Let $(z_1, \ldots, z_n)$ designate the sequence for the variation and $(y_1, \ldots, y_n)$ designate the sequence for the response. The distribution for the variation is given by

$$\frac{1}{(2\pi)^{n/2}} \exp\left\{-(1/2)\sum z_i^2\right\}.$$

The model for the response is

$$\left\{\mathbf{y} = \mu\mathbf{1} + \sigma\mathbf{z} : \begin{matrix}\mu \in R \\ \sigma \in R^+\end{matrix}\right\},$$

where $\mathbf{1} = (1, \ldots, 1)'$ designates the 1-vector. In Chapter 11 we will examine how an observed response sequence $\mathbf{y}$ can be used to make inferences concerning the realized sequence $\mathbf{z}$ and concerning the true values for the parameters μ and σ. These inference methods work with arbitrary distributions for the variation.

E EXAMPLE OF DATA

In the biochemical investigation of a protein, the following $n = 9$ observations were obtained on a response y:

17.1 14.9 15.2 18.1 16.4

19.1 18.5 14.6 17.3.

We shall use these data for illustrations later in this chapter. Accordingly, it is convenient to record some simple calculations. The sample average is a simple analog

TABLE 7.1
Averages and vectors

$\mathbf{y}$	$\bar{y}\mathbf{1}$	$\mathbf{y} - \bar{y}\mathbf{1}$
17.1	16.8	0.3
14.9	16.8	−1.9
15.2	16.8	−1.6
18.1	16.8	1.3
16.4	16.8	−0.4
19.1	16.8	2.3
18.5	16.8	1.7
14.6	16.8	−2.2
17.3	16.8	0.5
151.2	151.2	0

$$\begin{aligned} \bar{y} &= 151.2/9 \\ &= 16.8, \\ \sum y_i^2 &= 2561.54 = \text{SS}, \\ \left(\sum y_i\right)^2/n &= (151.2)^2/9 \\ &= 2540.16 = \text{SS}_\text{A}, \\ \sum (y_i - \bar{y})^2 &= 21.38 = \text{SS}_\text{V}, \\ s_y^2 &= \sum (y_i - \bar{y})^2/(n-1) \\ &= 2.672, \\ s_y &= 1.635. \end{aligned}$$

of the mean of a distribution; the sample average is calculated on the right in Table 7.1. The average vector is recorded in the second column, and the residual vector recording deviations from the average is recorded in the third column; see Figure 7.1 and recall Figure 5.11. Note that the vector $\mathbf{y}$ is equal to the average vector $\bar{y}\mathbf{1}$ plus the residual vector $\mathbf{y} - \bar{y}\mathbf{1}$. Also note that the squared length SS of the vector $\mathbf{y}$ is equal to the squared length SS_A of the average vector plus the squared length SS_V of the residual vector:

$$(1) \qquad \sum y_i^2 = n\bar{y}^2 + \sum (y_i - \bar{y})^2 = \frac{\left(\sum y_i\right)^2}{n} + \sum (y_i - \bar{y})^2;$$

this is a consequence of the orthogonality of the average vector $\bar{y}\mathbf{1}$ and the residual

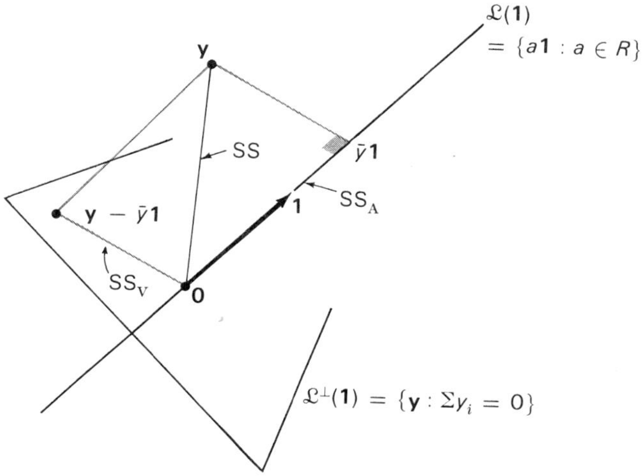

FIGURE 7.1
Orthogonal components $\bar{y}\mathbf{1}$ and $\mathbf{y} - \bar{y}\mathbf{1}$ add to give the vector $\mathbf{y}$. The corresponding squared lengths SS_A and SS_V add to give the squared length SS.

vector $\mathbf{y} - \bar{y}\mathbf{1}$. The sample variance and the sample standard deviation are calculated also.

Statistical inference is concerned with how to draw conclusions concerning unknowns: from (1) data such as just presented, and (2) a statistical model such as described in Section C, the model presenting the background information concerning the properties of the system being investigated.

F EXPERIMENTAL DESIGN

Usually an investigator must do much more than just collect data and analyze it with a statistical model. Factors that can affect the response must be controlled: some factors would have values chosen as part of seeing whether they affect the response; other factors would be kept constant. Other influences that can affect the response should be controlled as much as possible and randomization applied so that corresponding variation becomes random. Consider this further in the context of Example 2.

In Example 2 a treatment was applied to animals—to experimental units. The animals that receive the treatment should not be those "that did not get away." Rather, they should be representative of the collection or population of animals under study. Without information on all aspects of the animals, only probability theory seems able to give the appropriate balance and representativeness over the population. In its simplest form a symmetric distribution is used over the set of possible samples of a given size. This random sampling is discussed in Section 7.3.

A simple extension of Example 2 would have several different treatments. We would then have the problem of how to assign the various treatments to the experimental units that have been sampled as just described. An apparent difference in the effects of the treatments could be due to differences in the animals. Without information on all aspects of the animals, only probability theory seems able to give an appropriate balance among the treatments. In its simplest form a symmetric probability distribution is used over the set of possible assignments of treatments to experimental units, a specified number of experimental units for each treatment. This randomization is a fundamental ingredient of experimental design, of the process of determining cause–effect relationships. Other ingredients involve the control of the factors that could affect the response and the replication of a basic investigation so that treatment effects can be assessed over and above the inevitable variation from experimental unit to unit. Various aspects of experimental design are discussed in Chapter 12.

G EXERCISES

1 The resistance was determined for five pieces of wire sampled from production:

0.138 0.140 0.136 0.142 0.140.

Calculate the sample average, the sample average vector, the sample residual vector, and the sample variance.

2 The breaking load was determined for a sample of six from the production run of a certain plastic:

18.3 19.3 21.2 22.4 17.1 27.7.

Calculate the sample average, the sample average vector, the sample residual vector, and the sample variance.

3 The yield of a certain variety of corn was determined on seven plots of land:

87.7 87.2 92.3 94.2 93.6 85.6 93.5.

Calculate the sample average, the sample average vector, the sample residual vector, and the sample variance.

7.2

INFERENCE FROM LARGE SAMPLES

Consider a random system with a real-valued response y, for example, the biochemical system in the preceding section. Suppose that the investigator has had earlier experience with the system under various conditions and has grounds to assume that the response distribution is of reasonable form with not too much probability in the tails and with finite mean and variance. Also suppose that continuity properties of the system give grounds to assume that a continuous density function can provide a reasonable approximation to the true response distribution.

Consider an appropriate model. The sample space is clearly the real line, and the obvious class of events is the Borel class derived from the intervals. The statistical model can then be given in terms of density functions as $\mathfrak{M} = \{f\}$, where we restrict $\mathfrak{M}$ in a general way to continuous densities that do not have too much probability in the tails and have a *finite mean* and a *finite variance*.

Now suppose, as is typical, that the investigator is primarily interested in the general level of the response, say, as given by the mean $\mu = E(y)$. The variance of the response $\sigma^2 = \text{var}(y)$ can also be of interest—for what it says of course about variation in the response, but also for what it says about the accuracy of an estimate of the general level μ. Note that μ and σ^2 can be viewed as functions defined on $\mathfrak{M}$: Each f has a value for μ and a value for σ^2; such functions can be called *parameters* of the model. One of the distributions in $\mathfrak{M}$ is the true distribution; the corresponding values of μ and σ^2 are then the true values for these parameters.

Suppose that the investigator plans a sequence of n independent repetitions of the system in order to obtain information concerning the general level μ. Let $(y_1, \ldots, y_n)$ designate the compound response for the sequence of repetitions. The sample space then is R^n and the obvious class of events is the Borel class on R^n; the statistical model for the sequence of repetitions can be given in terms of density functions as

$$\{f(y_1) \cdots f(y_n) : f \in \mathfrak{M}\},$$

where $\mathfrak{M}$ is restricted as before.

EXAMPLE 1

Biochemical example: We shall illustrate various inference procedures with the biochemical data in Section 7.1E:

17.1 14.9 15.2 18.1 16.4

19.1 18.5 14.6 17.3

with sample average $\bar{y} = 16.8$, sample variance $s_y^2 = 2.672$, and sample standard deviation $s_y = 1.635$.

A ESTIMATE FOR THE MEAN

We have been considering the sample average $\bar{y}$ as a natural analog of the mean μ of the distribution being sampled. The sample average $\bar{y}$ is a *function defined on the sample space R^n;* such a function can be called a *statistic*. The sample average $\bar{y}$ has an attractive property [formula (5.4.5)] in relation to the mean μ:

(1) Unbiasedness: $E(\bar{y}) = \mu$.

Thus the distribution of $\bar{y}$ is located at the mean μ of the distribution being sampled: sometimes the values of $\bar{y}$ are larger than the true μ and sometimes smaller than the true μ, but the mean or long-run average value of $\bar{y}$ is equal to the mean μ, whatever value it may have.

For the biochemical illustration recorded in the introduction, the investigator can present the observed value $\bar{y} = 16.8$ as an estimate of the true mean response μ. He knows that the value of $\bar{y}$ may be smaller or larger than the true mean μ, but he gains some satisfaction from the property that the mean of the values he might have obtained is equal to the true mean response that he is interested in.

We record the general definition for unbiasedness:

DEFINITION 1

A function on the sample space is an **unbiased estimator** of a parameter of the model if the mean value of the function is equal to the value of the parameter, for each distribution in the model.

In Chapter 9 we shall discuss unbiased estimation in some detail, including how to choose a function on the sample space to obtain a good estimate.

A single value as an estimate of a parameter is not too useful unless we have some indication of its reliability. For this we think naturally of the scale of the distribution of $\bar{y}$, as given by, say, the variance or the standard deviation. For the sample average $\bar{y}$ we have [formula (5.4.5)]

$$(2) \qquad \text{var}(\bar{y}) = \frac{\sigma^2}{n}, \qquad \text{SD}(\bar{y}) = \frac{\sigma}{\sqrt{n}}.$$

These formulas are useful if we should happen to know the standard deviation σ; we will see later that they can be useful otherwise.

Sometimes earlier experience with the system under various conditions can give a reasonably accurate value for the standard deviation σ; designate this known value by σ_0. The various conditions may, however, allow the general level μ to be different and thus not give any reasonable indication as to the current value for μ. Suppose we know that the variance σ^2 of the distribution has the value σ_0^2. The investigator can then record the observed value of $\bar{y}$ together with the known standard deviation $\sigma_0/\sqrt{n}$ for the distribution of $\bar{y}$.

EXAMPLE 1 (*continuation*)

For the biochemical example, suppose that the standard deviation σ is known to be approximately equal to $\sigma_0 = 2.1$. The investigator can then record the observed $\bar{y} = 16.8$ together with the standard deviation

$$\text{SD}(\bar{y}) = \frac{2.1}{\sqrt{9}} = 0.7.$$

This standard deviation indicates how far the estimate can be away from the true mean μ. It does this in a weak way by the Chebyshev inequality in Section 5.3C. And it does so in a somewhat stronger way if a normal approximation is appropriate; we shall examine this in Section C.

B TEST FOR THE MEAN: VARIANCE KNOWN

Consider the random system as discussed in the introduction and suppose that earlier experience with the system has given the value σ_0^2 for the variance of the distribution. Sometimes in applications a value μ_0 for the general level μ may be suggested by some theoretical calculations or by some related empirical investigations with the same or a similar system. The investigator may then be interested in whether the value μ_0 applies to the system under current conditions. More specifically he may be interested in whether observed data are consistent with such a suggested value.

Note that we have used a subscript on σ_0 to denote a known value and the same subscript on μ_0 to denote a suggested or contemplated value. For such usages the context will provide the distinction.

The Central Limit Theorem 6.2.2 says that the sample average $\bar{y}$ appropriately standardized has a limiting standard normal distribution. And empirical investigations show that the approach to normality is quite fast, provided there is not too much probability in the tails of the distribution. Thus for the present model we assume that $\bar{y}$ is approximately normal$(\mu, \sigma_0/\sqrt{n})$ or that

$$z = \frac{\bar{y} - \mu}{\sigma_0/\sqrt{n}}$$

is approximately standard normal.

For our illustration a sample of size $n = 9$ is somewhat small and a size of $n = 20, 30, \ldots$ would be preferable. For a simple illustration, however, we assume that the normal distribution is applicable; that is,

$$z = \frac{\bar{y} - \mu}{0.7}$$

is approximately standard normal.

We now determine the kind of sample average $\bar{y}$ to expect if the mean $\mu = \mu_0$. If $\mu = \mu_0$, then the distribution of $\bar{y}$ is located at μ_0, is scaled by $\sigma_0/\sqrt{n}$, and has approximately normal form. Or, equivalently, if $\mu = \mu_0$, then the distribution of

$$(3) \qquad z = \frac{\bar{y} - \mu_0}{\sigma_0/\sqrt{n}}$$

is approximately standard normal. We can be very conservative and use the Chebyshev inequality, obtaining

$$(4) \qquad \begin{aligned} P(|z| \geq 2) &\leq 1/4, \\ P(|z| \geq 3) &\leq 1/9. \end{aligned}$$

Or, more realistically, we can accept the normal approximation, obtaining

$$(5) \qquad \begin{aligned} P(|z| \geq 1.96) &\approx 5\%, \\ P(|z| \geq 2.58) &\approx 1\%; \end{aligned}$$

the values are obtained from Appendix II.

We can now check whether observed data are consistent with the suggested or hypothesized value μ_0 for the mean. We calculate the observed value of

$$z = \frac{\bar{y} - \mu_0}{\sigma_0/\sqrt{n}}.$$

We compare it with the standard normal distribution, and we assess the hypothesis $\mu = \mu_0$ accordingly. If the observed value of z is reasonable for the distribution, then our data are in general conformity with the hypothesis $\mu = \mu_0$. If the observed value is an extreme, an unusual, or a virtually impossible value for the distribution, then we would have evidence against the hypothesis $\mu = \mu_0$ with corresponding strength.

EXAMPLE 1 (*continuation*)

For the biochemical example, some theoretical calculations have indicated a value 14.8 for the mean μ. We test this indicated value, and thus by implication test the theory that was used for the calculations. The observed value of $\bar{y}$ is 16.8, and the corresponding value for z is

$$z = \frac{16.8 - 14.8}{0.7} = 2.86;$$

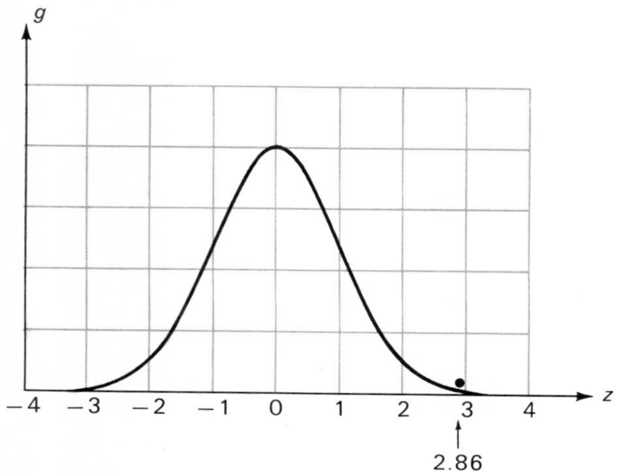

FIGURE 7.2
Standard normal distribution and observed value $z = 2.86$ under the hypothesis $\mu = 14.8$.

this value is compared with the standard normal distribution in Figure 7.2. The observed value z is not in the center of the distribution—in fact, it is well out on the right tail of the distribution. Thus the observed value of z is not the sort of value we could reasonably expect under the hypothesis, and it provides moderately strong evidence against the hypothesis. The operating principle in this argument is: if A implies B, then B^c implies A^c. One way of assessing the observed value is to calculate the probability, under the hypothesis, of a value as far or farther from the center of the distribution:

(6) $$P(|z| \geq 2.86) \approx 0.42\%.$$

This is called the *observed level of significance* and it measures the significance of the observed value in relation to the hypothesis $\mu = 14.8$.

In general, if the observed value of z in formula (3) is in the central part of the distribution, then, as assessed, the data are in accord with the hypothesis $\mu = \mu_0$. Of course this does not say that the hypothesis $\mu = \mu_0$ is true; in fact, we would not expect μ to be exactly equal to μ_0. Rather it is saying that the assessment does not give evidence against the value μ_0 as a reasonable approximation to the true μ.

On the other hand, if the observed value of z in formula (3) is very far out on a tail of the distribution, the data are substantially contradicting the hypothesis $\mu = \mu_0$, and we have substantial evidence against the hypothesis $\mu = \mu_0$. In summary, we would say that the observed $\bar{y}$ and the corresponding z are statistically significant in relation to the hypothesis $\mu = \mu_0$. However, in certain of these cases the value indicated for μ by the estimate $\bar{y}$ may, in fact, differ from μ_0 by an amount that makes no practical difference; in that case the statistical significance is of no practical significance.

One way of obtaining a numerical assessment of an observed value in relation to a distribution is to calculate the probability of a value as far or farther from the center of the distribution as that observed; this is called the *observed level of significance*. Such a calculation is reasonable typically with distributions that are symmetrical about a central point and drop off toward the tails.

C INTERVAL ESTIMATE FOR THE MEAN: VARIANCE KNOWN

Consider the random system discussed earlier and suppose that earlier experience with the system has given the value σ_0^2 for the variance σ^2 of the distribution.

In Section A we examined $\bar{y}$ as an estimate of μ and we noted that the standard deviation $\sigma_0/\sqrt{n}$ gives some indication of its reliability. We are now in a position to be more precise. We shall test various values for μ and see which ones are reasonable by the assessment procedures in Section B. Suppose that for each value for μ we determine a 95 percent interval from the distribution of the criterion

(7) $$\frac{\bar{y} - \mu}{\sigma_0/\sqrt{n}};$$

and then view this as an interval of acceptable values for our criterion. There is some moderate advantage in having the same interval for each hypothesized value; and the interval $(-1.96, +1.96)$ is a rather natural choice based on the approximate distribution.

EXAMPLE 1 (*continuation*)

Consider an interval estimate for μ in the biochemical example. A value μ is acceptable, as just discussed, if

$$-1.96 \leq \frac{16.8 - \mu}{2.1/\sqrt{9}} \leq 1.96$$

or, equivalently, if

$$16.8 - 1.96\frac{2.1}{\sqrt{9}} \leq \mu \leq 16.8 + 1.96\frac{2.1}{\sqrt{9}}.$$

Thus we see that the values of μ that are acceptable on the basis of the data form the interval

$$\left(16.8 \pm 1.96\frac{2.1}{\sqrt{9}}\right) = (15.4, 18.2). \tag{8}$$

Note that the interval is ($\bar{y} \pm 1.96$ standard deviations of $\bar{y}$). We thus have an *interval estimate* of the mean μ as based on a 95 percent acceptance level.

Now consider the general random system. A value μ is acceptable as discussed if

$$-1.96 \leq \frac{\bar{y} - \mu}{\sigma_0/\sqrt{n}} \leq 1.96$$

or, equivalently, if

$$\bar{y} - 1.96\frac{\sigma_0}{\sqrt{n}} \leq \mu \leq \bar{y} + 1.96\frac{\sigma_0}{\sqrt{n}}.$$

Thus we obtain the interval estimate

$$\left(\bar{y} \pm 1.96\frac{\sigma_0}{\sqrt{n}}\right) \tag{9}$$

for the mean as based on a 95 percent acceptance level.

We can obtain the same interval estimate by a different route that provides an interesting interpretation. We follow the same preliminaries as before but consider the *function* $\bar{y}$ rather than the observed value from the data. We can make the probability statement

$$P\left(-1.96 \leq \frac{\bar{y} - \mu}{\sigma_0/\sqrt{n}} \leq 1.96\right) \approx 95\%; \tag{10}$$

this holds for each value for the mean μ of the underlying distribution. We then rearrange the inequalities as before, obtaining

$$P\left(\bar{y} - 1.96\frac{\sigma_0}{\sqrt{n}} \leq \mu \leq \bar{y} + 1.96\frac{\sigma_0}{\sqrt{n}}\right) \approx 95\%. \tag{11}$$

Thus the probability is approximately 95 percent that the interval

$$\left(\bar{y} - 1.96\frac{\sigma_0}{\sqrt{n}}, \bar{y} + 1.96\frac{\sigma_0}{\sqrt{n}}\right) \tag{12}$$

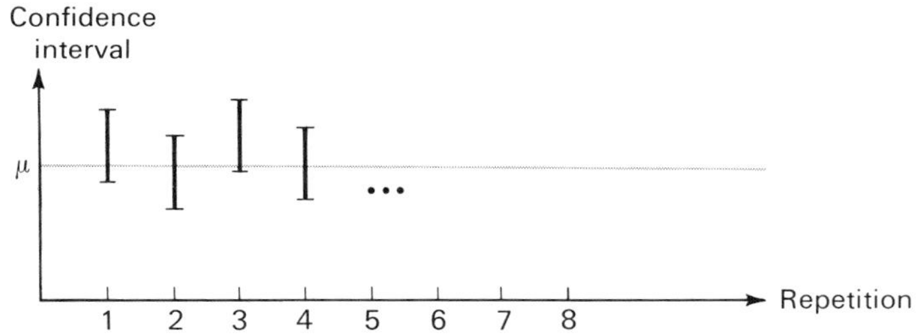

FIGURE 7.3
Repeated confidence intervals for the parameter μ; in the long run, 95 percent of them will intersect the line for the true mean μ.

brackets the true mean μ. Note that the interval is a function of the sample $(y_1, \ldots, y_n)$ and thus has a distribution; the probability statement is made on the basis of this distribution. For observed data we can then calculate the observed value of the interval (12); it is called a 95 percent *confidence interval* for the parameter μ.

EXAMPLE 1 (*continuation*)

For the biochemical data we obtain the interval

$$\left(16.8 \pm 1.96\frac{2.1}{\sqrt{9}}\right) = (15.4, 18.2)$$

as before; it is a 95 percent confidence interval for the true mean μ. Of course, either the interval brackets the true mean μ or it does not; the investigator does not know. But he gains confidence from the property that approximately 95 percent of the intervals he might have obtained would bracket the true mean μ.

A long-run interpretation can be given for this confidence-interval procedure. Suppose that we contemplate repeated samples of 9 from the biochemical system under investigation, and in each case calculate the observed value of the confidence interval (12); then we would obtain results as presented in Figure 7.3. The probability statement (11) asserts that in the long run, approximately 95 percent of these intervals will intersect the line representing the true value for the parameter μ. More on confidence intervals in Sections 8.3, 10.5, and 11.1.

D ESTIMATE FOR THE VARIANCE

Consider the random system as discussed in the introduction and suppose now that the variance σ^2 is unknown. The investigator may be interested in the variance σ^2 for what it says about the variation in the response or for what it says about the accuracy of an estimate such as $\bar{y}$ for the general level μ. We have considered the sample variance s_y^2 as a natural analog of the variance σ^2 of the distribution being sampled. The sample variance s_y^2 is a function defined on the sample space R^n; it has the attractive property (5.4.9) of unbiasedness in relation to the variance σ^2:

(13) Unbiasedness: $E(s_y^2) = \sigma^2$.

Thus the values for s_y^2 are sometimes larger and sometimes smaller than the true σ^2, but the mean value is equal to the variance σ^2 whatever value it may have.

EXAMPLE 1 (*continuation*)

For the biochemical example the investigator can present the observed value $s_y^2 = 2.672$ as an estimate of the true variance σ^2. He knows that the value may be smaller or larger than the true variance σ^2, but he gains some satisfaction from the unbiasedness property concerning what he might have obtained.

For the reliability of the estimator s_y^2 we might naturally turn to the scale of its distribution as given by, say, the variance: we have

$$\text{var}\,(s_y^2) = \frac{1}{n}\left(\bar{\mu}_4 - \frac{n-3}{n-1}\sigma^4\right) = \frac{1}{n}\left(\kappa_4 + \frac{2n}{n-1}\sigma^4\right) \tag{14}$$

(Problem 11). But this then requires information concerning the fourth central moment or the fourth cumulant.

For the standard deviation σ the investigator will probably use the sample standard deviation s_y. Unfortunately, however, it does not have the unbiasedness property; in fact,

$$\sigma > E(s_y), \tag{15}$$

as is easily seen from the following:

$$\text{var}\,(s_y) = E(s_y^2) - (E(s_y))^2 = \sigma^2 - (E(s_y))^2,$$
$$\sigma^2 - (E(s_y))^2 > 0.$$

What we are seeing here, in fact, is an unattractive property of unbiasedness—that unbiasedness is tied to a particular way of scaling or expressing a parameter of a distribution. In Chapters 9 and 10 we shall consider further some questions of estimation and reliability for the variance.

E TEST FOR THE MEAN: VARIANCE UNKNOWN

Consider the random system as discussed in the introduction to this section and suppose that the variance σ^2 is unknown. We now consider a test for the hypothesis $\mu = \mu_0$.

The Central Limit Theorem 6.2.2 says that

$$z = \frac{\bar{y} - \mu}{\sigma/\sqrt{n}}$$

has a limiting standard normal distribution. Problems 6.2.7 and 6.3.5 show that s_y converges to σ in probability and almost surely as $n \to \infty$. It follows, then, by Problem 6.1.16 that

$$t = \frac{\bar{y} - \mu}{s_y/\sqrt{n}} \tag{16}$$

has a limiting standard normal distribution as $n \to \infty$. Empirical investigations, however, show that the approach to normality can be quite slow: the approach is strongly influenced by the slowness of convergence of s_y to σ, particularly if the distribution is substantially different from normal form (see, for example, Section 8.3B). In the present context we assume that the sample size is sufficiently large that the normal distribution provides a reasonable approximation.

We now determine approximately the kind of sample average to expect under the hypothesis $\mu = \mu_0$. If $\mu = \mu_0$, then

$$(17) \qquad t = \frac{\bar{y} - \mu_0}{s_y/\sqrt{n}}$$

is approximately standard normal, and we have

$$P(|z| \geq 2) \approx 5\%,$$
$$P(|z| \geq 3) \approx 1\%.$$

Note that we acknowledge the rough approximation by using the simple values 2 and 3 rather than the values 1.96 and 2.58 for the normal.

We can now check whether observed data are consistent with the hypothesized value for μ_0. We calculate the observed value of t in formula (17), compare it with the standard normal, and assess the hypothesis accordingly.

EXAMPLE 1 (*continuation*)

For the biochemical example we have

$$t = \frac{16.8 - 14.8}{1.64/\sqrt{9}} = 3.66.$$

The observed value is well out on the right tail of the distribution, thus providing moderate evidence against the hypothesis: $\mu = 14.8$. The strength of such evidence, however, should be substantially qualified by the rough approximation involved with such a small sample size $n = 9$.

F INTERVAL ESTIMATE FOR THE MEAN: VARIANCE UNKNOWN

Consider further the random system as discussed in the introduction and suppose that the variance σ^2 is unknown. The distribution properties discussed in Section E give the following probability statement:

$$P\left(-2 \leq \frac{\bar{y} - \mu}{s_y/\sqrt{n}} \leq +2\right) \approx 95\%$$

or, equivalently,

$$(18) \qquad P\left(\bar{y} - 2\frac{s_y}{\sqrt{n}} \leq \mu \leq \bar{y} + 2\frac{s_y}{\sqrt{n}}\right) \approx 95\%.$$

Thus

(19) $(\bar{y} \pm 2s_y/\sqrt{n})$

is an approximate 95 percent confidence interval for the parameter μ.

EXAMPLE 1 (*continuation*)

For the biochemical example we have the observed interval

(20) $$\left(16.8 \pm 2\frac{1.64}{\sqrt{9}}\right) = (15.7, 17.9).$$

This is an approximate 95 percent confidence interval for the parameter μ; but we should be mindful of the rough approximation involved with such a small sample size. This interval can also be obtained by testing each possible parameter value μ against the data as in Section C.

For these approximate confidence intervals it is reasonable to use the simple values 2 and 3, corresponding to 95 percent and 99 percent confidence. More on confidence intervals in Sections 8.3, 10.5, and 11.1.

G INFERENCES OR DECISIONS

Consider briefly the testing of a hypothesis $\mu = \mu_0$. Some developments of statistical theory suggest that on the basis of the data a decision should be made—to accept the hypothesis or to reject the hypothesis. The theoretical problem, then, is a matter of choosing a decision rule that associates one or other decision with each possible set of data. For example, in Section B a rule might be to accept the hypothesis if the value of z falls in $(-1.96, +1.96)$ and to reject the hypothesis otherwise. With this rule the decision ''reject'' would be obtained for the biochemical data with $\mu_0 = 14.8$.

There is certainly some appeal to a categorical statement. However, as we have seen, the only instances where there would be substantial evidence to support a categorical statement would be those where the deviation was ''impossibly'' large, as assessed under the hypothesis. The use of a decision rule to ''decide'' on truths of hypothesis is perhaps due to the attractiveness of a mathematically very elegant theory; we shall discuss some aspects of this theory in Chapter 10. In practice, however, the use of such rules is essentially unscientific (making categorical assertions without supporting evidence), and the hazards of having scientific results asserted on such a basis are substantial. For some observations on how decision rules can interact with editorial policies of scientific journals, see Sterling (1959).

H EXERCISES

It is important to make accurate numerical calculations.

1 Sixteen measurements are made on a chemical response, yielding $\bar{y} = 285.3$. The response distribution is known to be of reasonable form with σ approximately 3.1. Test the hypothesis $\mu = 280$. Calculate a 95 percent confidence interval for μ. Calculate a 99 percent confidence interval for μ.

2 The resistances of five pieces of wire sampled from production are recorded in Exercise 7.1.1. The standard deviation is known to be approximately 0.002 and the distribution form

to be such that the limiting normal distribution is reasonable for this sample size. Test the hypothesis that $\mu = 0.135$.

3 The breaking loads for a sample of 6 are recorded in Exercise 7.1.2. The standard deviation is known to be approximately 2.5 and the distribution form to be such that the normal distribution is reasonable for this sample size. Calculate a 95 percent interval for the mean breaking load μ.

4 The yield for a certain variety of corn on 7 plots is recorded in Exercise 7.1.3. The distribution is known to be such that the normal approximation for t in formula (16) is roughly adequate. Test the hypothesis that $\mu = 90$. Calculate an approximate 95 percent confidence for the mean yield of the particular corn under the particular conditions.

5 A sample of 11 rats was obtained randomly from a population of rats. A blood viscosity test of each yielded 3.29, 3.91, 4.64, 3.55, 3.67, 4.18, 3.74, 4.67, 3.03, 4.61, and 3.84. On the assumption that the distribution has reasonable form, test the hypothesis: $\mu = 4.6$. Calculate an approximate 95 percent confidence interval for μ. Calculate an approximate 99 percent confidence interval for μ.

TWO-SAMPLE PROBLEMS

Consider a sample $(y_{11}, \ldots, y_{1m})$ from a first response and an independent sample $(y_{21}, \ldots, y_{2n})$ from a second response. Let μ_1 and σ_1^2 be the mean and variance for the first response and μ_2 and σ_2^2 be the mean and variance for the second response. And let $\bar{y}_1$ and s_1^2 be the sample average and variance for the first sample and $\bar{y}_2$ and s_2^2 be the sample average and variance for the second sample.

The difference $\bar{y}_2 - \bar{y}_1$ is the natural sample analog of the difference $\mu_2 - \mu_1$ between the mean responses. From Chapter 5 we know that $\bar{y}_2 - \bar{y}_1$ has mean $\mu_2 - \mu_1$ and variance $\sigma_1^2/m + \sigma_2^2/n$.

The results in Chapter 6 can be used to show that

$$\frac{\bar{y}_2 - \bar{y}_1 - (\mu_2 - \mu_1)}{(\sigma_1^2/m + \sigma_2^2/n)^{1/2}} \tag{21}$$

has a limiting standard normal distribution as $m, n \to \infty$ in a given ratio.

With some additional argument, the results in Chapter 6 show that

$$\frac{\bar{y}_2 - \bar{y}_1 - (\mu_2 - \mu_1)}{(s_1^2/m + s_2^2/n)^{1/2}} \tag{22}$$

has a limiting standard normal distribution.

If it is known that $\sigma_1 = \sigma_2$, then the sample variances can be combined in a natural way (discussed in Chapter 9) to give the combined estimate of variance,

$$s^2 = \frac{\sum_1^m (y_{1i} - \bar{y}_1)^2 + \sum_1^n (y_{2i} - \bar{y}_2)^2}{m + n - 2} = \frac{(m-1)s_1^2 + (n-1)s_2^2}{m + n - 2}. \tag{23}$$

The results in Chapter 6 can then be used to show that

$$\frac{\bar{y}_2 - \bar{y}_1 - (\mu_2 - \mu_1)}{s(1/m + 1/n)^{1/2}} \tag{24}$$

has a limiting standard normal distribution.

6 *(continuation of Exercise 1)* In addition, 16 measurements with a treatment input were made on the response yielding $\bar{y}_t = 288.5$. The treatment is known not to alter the response variance. Test the hypothesis: $\mu_t - \mu = 0$. Calculate an approximate 95 percent confidence interval for $\delta = \mu_t - \mu$.

7 *(continuation of Exercise 5)* In fact, 22 rats were randomly obtained from the population and

11 were assigned randomly as controls, yielding the data in Exercise 5, and the remaining were given an ephedrin treatment with data as follows: 3.45, 4.26, 4.71, 3.14, 3.45, 5.01, 4.43, 4.91, 4.22, 4.83, and 3.55. On the assumption that the treatment distribution has reasonable form, test the hypothesis: $\mu_t - \mu = 0$. Calculate an approximate 99 percent confidence interval for $\delta = \mu_t - \mu$.

8 (*continuation*) Assume that the variances, with and without treatment, are equal. Test the hypothesis that $\mu_t - \mu = 0$. Calculate an approximate 95 percent confidence interval for $\delta = \mu_t - \mu$.

9 Record the general formula for the confidence interval in Problem 7 for the case of samples of size m and n; with possibly different variances.

10 Record the general formula for the confidence interval in Problem 8 for the case of samples of size m and n; with equal variances.

J PROBLEMS

11 Prove formula (14) for var (s_y^2); use $\bar{\mu}_4 = \kappa_4 + 3\sigma^4$.

12 *Jensen inequality:* A real-valued function c on an open interval is called *convex* if $ac(s_1) + (1 - a)c(s_2) \geq c(as_1 + (1 - a)s_2)$ for $0 < a < 1$; the function is called *strictly convex* if the inequality is strict. Some analysis shows that any point $(s_0, c(s_0))$ on the graph of a convex function has a *line of support,* a line l through it such that $c(s) \geq l(s)$; and if c is strictly convex, then $c(s) > l(s)$ for $s \neq s_0$. Given a distribution for s on the open interval, show that $E(c(s)) \geq c(E(s))$: *mean convex function* is greater than or equal to *convex function of mean*. If c is strictly convex, show that the inequality is strict unless all probability is at a point.

13 Show that $\sigma^2 > (E(s_y))^2$ unless the distribution of s_y^2 has all probability at a point; use Problem 12.

14 Consider a distribution for y with all probability on the positive axis. Use Problem 12 to show that

$$E\left(\frac{1}{y}\right) \geq \frac{1}{E(y)}$$

with equality if and only if y has all probability at a point.

15 Consider a distribution for y on R with mean μ. Show that for $t \neq 0$,

$$E(e^{yt}) \geq e^{\mu t}$$

with equality if and only if y has all probability at a point.

7.3

INFERENCE FOR FINITE POPULATIONS

We now examine finite population sampling, an important area of inference that uses both large-sample and symmetry results from probability theory. Finite population sampling enters into almost all parts of statistics—scientific, industrial, commercial, and governmental. This ranges: From the scientist who samples experimental units for the application of treatments of interest; for example, people for a test of ascorbic acid effects, or mice for a test of cancer treatments, or fruit flies for a test of radiation effects. Through the industrial statistician who samples a day's production, or an incoming lot of a manufactured material. To the statistician in market research, opinion polls, or government information bureaus who wants information about people or production or resources.

A population may consist of the people of a given geographical district, or the houses in a certain city, or the rabbits of a particular genetic strain in a certain laboratory, or the transistors received on a particular contract, or the psychology students at a certain university.

Consider a population under investigation. The population must be well defined so that the elements are in some way clearly specified. As part of this, the size of the population is typically known to reasonable accuracy; let N designate the size. An element in the population will usually have several characteristics or variables of possible interest; typically one of the variables will be of central interest—we consider here the case of a single real variable. Let $c_1, \ldots, c_N$ designate the values of the variable for the N elements in the population. We then abstract and take the finite population to be the set

$$\Pi = \{c_1, \ldots, c_N\}.$$

In general an investigator will have some interest in knowing all the values in the population. In practice, however, he will be concerned with one or several derived characteristics. Almost certainly he will be interested in the location of the values in the population; for this we define the mean of Π to be the average value in the population:

$$\mu = \frac{\sum_1^N c_\alpha}{N}.$$

And typically he will also be interested in the scaling or variability in the population —for its own sake or for the information it gives concerning the reliability of an estimate of μ; for this we define the variance of Π:

$$\sigma^2 = \frac{1}{N-1}\sum_1^N (c_\alpha - \mu)^2 = \frac{1}{N-1}\left(\sum_1^N c_\alpha^2 - \left(\sum_1^N c_\alpha\right)^2 \Big/ N\right).$$

The division by $N - 1$ rather than N has certain computational advantages; and with any large population it makes no significant numerical difference. Note that as yet we have not mentioned any probability distribution.

An obvious way of obtaining information concerning μ and σ^2 is by means of a *census,* contacting and obtaining the value of each element in the population. A census of a large population can be very expensive and time consuming. And it can often be substantially inaccurate, owing to operational difficulties—such as that of obtaining skilled enumerators.

In some populations the elements are available in a serial order, such as in a directory or by geographical layout. Sometimes, then, systematic sampling is proposed in which every tenth or every fiftieth element is examined. With systematic sampling there are potential hazards, as, for example, if unexpected patterns exist between the variable of interest and the criterion underlying the serial ordering of the population.

In this section we consider *simple random sampling* from the population: we impose a symmetric probability distribution on the set of all possible samples of a given size.

A SIMPLE RANDOM SAMPLING

A random sample of size n from the population Π is a sequence $(y_1, \ldots, y_n)$ with probability function

$$p(y_1, \ldots, y_n) = 1/N^{(n)}$$

for each of the $N^{(n)}$ possible sequences of n elements from the population Π; note for this that we are treating the elements in the population as distinguishable even though some c_α values may be equal. This random sampling involves a symmetric distribution on the set $\mathcal{S}_2$ as defined in Section 3.4B.

The symmetry in the distribution for $(y_1, \ldots, y_n)$ leads to a simple method for performing the sampling. The first element y_1 can be any element in the population, each with the same probability $1/N$. And for given y_1, the second element y_2 can be any of the remaining elements in the population, each with the same probability $1/(N-1)$, and so on. Now suppose that serial numbers have been applied to the elements in the population. Random numbers can then be used to randomly sample the population: Use a block size of random numbers big enough to cover all the serial numbers; take blocks of random numbers until an actual serial number is obtained and then examine the corresponding population elements; continue in this way until a sample of size n is obtained. For some examples, see Section 4.2I.

The randomness in random sampling can be expressed explicitly by means of a matrix of indicators. Let $I_{i\alpha}$ be equal to 1 if the ith sample element is the αth element c_α in Π and zero otherwise; then

$$y_i = \sum_{\alpha=1}^{N} I_{i\alpha} c_\alpha$$

or in matrix form,

$$\begin{pmatrix} y_1 \\ \vdots \\ y_n \end{pmatrix} = \begin{pmatrix} I_{11} & \cdots & I_{1N} \\ \vdots & & \vdots \\ I_{n1} & \cdots & I_{nN} \end{pmatrix} \begin{pmatrix} c_1 \\ \vdots \\ c_N \end{pmatrix}.$$

Note that choosing the first sample element is equivalent to choosing the position for the single 1 in the first row; that choosing the second sample element is equivalent to choosing the position for the single 1 in the second row—any position except that corresponding to the 1 in the first row; and so on.

The indicators satisfy the obvious relations:

$$I_{i\alpha} I_{i\beta} = 0 = I_{i\alpha} I_{j\alpha}, \qquad i \neq j,\ \alpha \neq \beta,$$

$$I_{i\alpha} I_{i\alpha} = I_{i\alpha}.$$

The probability function for the indicator matrix has some simple marginal probabilities:

$$P(I_{i\alpha} = 1) = \frac{1}{N},$$

$$P(I_{i\alpha} I_{j\beta} = 1) = \frac{1}{N(N-1)}, \qquad i \neq j,\ \alpha \neq \beta,$$

and so on. For some complex calculations the indicators can make certain calculations mechanical. For the proofs here, however, we find it easier to argue directly.

B THE BASIC RESULTS FOR RANDOM SAMPLING

The basic results for random sampling can be assembled in the following theorem:

THEOREM 1

If $(y_1, \ldots, y_n)$ is a random sample from a population of size N with mean μ and variance σ^2, then

$$E(\bar{y}) = \mu, \qquad \text{var}(\bar{y}) = \frac{\sigma^2}{n}\left(1 - \frac{n}{N}\right),$$

$$E(s_y^2) = \sigma^2.$$

And if $n \to \infty$ and $N/n \to \infty$ together with mild regularity conditions for Π, then

$$\frac{\bar{y} - \mu}{\dfrac{\sigma}{\sqrt{n}}\left(1 - \dfrac{n}{N}\right)^{1/2}}, \qquad \frac{\bar{y} - \mu}{\dfrac{s_y}{\sqrt{n}}\left(1 - \dfrac{n}{N}\right)^{1/2}}$$

have a limiting standard normal distribution.

The proof of the first part of the theorem is given in Section C. The proof of the second part is more difficult and is omitted; it can be derived from the Wald–Wolfowitz–Hoeffding theorem for permutation distributions; see, for example, Fraser (1957, p. 239). The approximate normality is often adequate for medium to large samples.

EXAMPLE 1

A random sample of 25 from a population of 500 medical students gives the following reduced data:

$$\bar{y} = 126.0, \qquad s_y = 7.5.$$

Some related investigations in other professional groups suggest the value $\mu = 120$. We can test the hypothesis $\mu = 120$ by calculating the standardized deviation

$$z = \frac{126.0 - 120}{\dfrac{7.5}{\sqrt{25}}\left(1 - \dfrac{25}{500}\right)^{1/2}} = 4.10$$

and comparing it with the standard normal distribution. This value is an extreme value for the standard normal and gives moderately strong evidence against the hypothesis.

Now consider a confidence interval for μ. At the 95 percent level we have the interval $(-1.96, 1.96)$ for the standard normal. Thus

$$P\left(-1.96 \le \frac{\bar{y} - \mu}{\dfrac{s_y}{\sqrt{25}}\left(1 - \dfrac{25}{500}\right)^{1/2}} \le 1.96\right) \approx 95\%,$$

which can be rearranged to give

$$P\left(\bar{y} - 1.96\frac{s_y}{\sqrt{25}}\left(1 - \frac{25}{500}\right)^{1/2} \le \mu \le \bar{y} + 1.96\frac{s_y}{\sqrt{25}}\left(1 - \frac{25}{500}\right)^{1/2}\right) \approx 95\%.$$

Thus a 95 percent confidence interval for μ is

$$\left(126.0 \pm 1.96\frac{7.5}{\sqrt{25}}\left(1 - \frac{25}{500}\right)^{1/2}\right) = (123.1,\ 128.9).$$

C PROOFS

We now prove the first part of Theorem 1. For the sample average $\bar{y}$ we have

$$E(\bar{y}) = \frac{1}{n}\sum_{1}^{n} E(y_i) = \frac{1}{n}\cdot n \cdot \frac{\sum_{1}^{N} c_\alpha}{N} = \mu;$$

this uses the linearity of E and the fact that a y_i can take each value c_α with equal probability $1/N$.

For the next two formulas it is convenient to make a preliminary simplification. Suppose that we subtract μ from each element in Π and obtain a modified population $\Pi_0 = \{d_\alpha\} = \{c_\alpha - \mu\}$. Then any sample $(y_1, \ldots, y_n)$ from Π has a corresponding sample $(w_1, \ldots, w_n) = (y_1 - \mu, \ldots, y_n - \mu)$ from Π_0, and clearly

$$\bar{w} = \bar{y} - \mu, \qquad s_w^2 = s_y^2.$$

Thus var $(\bar{w})$ = var $(\bar{y})$ and $E(s_w^2) = E(s_y^2)$. It follows then that it suffices to prove the formulas for the case $\mu = 0$.

We assume that $\mu = 0$. As an immediate consequence, we obtain

$$\sum c_\alpha = 0, \qquad \sum c_\alpha^2 + \sum_{\alpha \neq \beta} c_\alpha c_\beta = 0.$$

Then for the variance of $\bar{y}$ we have

$$\begin{aligned}
\text{var}\,(\bar{y}) = E(\bar{y}^2) &= \frac{1}{n^2}E\Big(\sum y_i^2 + \sum_{i\neq j} y_i y_j\Big) \\
&= \frac{1}{n^2}\left(n\frac{\sum_\alpha c_\alpha^2}{N} + n(n-1)\frac{\sum_{\alpha\neq\beta} c_\alpha c_\beta}{N(N-1)}\right) \\
&= \frac{1}{n^2}\left(n\frac{(N-1)\sigma^2}{N} + n(n-1)\frac{-(N-1)\sigma^2}{N(N-1)}\right) = \frac{\sigma^2}{n}\left(1 - \frac{n}{N}\right).
\end{aligned}$$

The factor $(1 - n/N)$ is called the *finite population correction factor;* for example, it forces var $(\bar{y})$ to zero if the sample becomes a census $(n = N)$.

And for the mean of s_y^2, we have

$$\begin{aligned}
E(s_y^2) &= \frac{1}{n-1}E\Big(\sum y_i^2 - n\bar{y}^2\Big) \\
&= \frac{1}{n-1}\left(n\frac{(N-1)\sigma^2}{N} - n\frac{\sigma^2}{n}\left(1 - \frac{n}{N}\right)\right) = \sigma^2.
\end{aligned}$$

D EXERCISES

1 The registration records of a university were used to obtain a random sample of 100 undergraduates from the 12,352 full-time undergraduate students. Each was interviewed

and the value of a numerical quantity determined. The sample mean and standard deviation are $\bar{y} = 1452$, $s_y = 485$. Determine a 95 percent confidence interval for the population mean μ.

2 Consider a population of size N that consists of 0's and 1's only; the population mean μ then is P, where P is the proportion of 1's. Let $(y_1, \ldots, y_n)$ be a random sample; the sample mean $\bar{y}$ is the proportion of 1's in the sample. Specialize the general formulas and show that

$$E(\bar{y}) = P, \qquad \text{var}(\bar{y}) = \frac{PQ}{n}\frac{N-n}{N-1}.$$

Check these formulas by specializing the results for the hypergeometric distribution in Problem 5.3.14.

3 (*continuation*) A random sample of 150 was chosen from a population of 4085 invoices. Detailed examination of each invoice in the sample showed that 32 of them contained errors. Estimate the number of invoices in the population that contain errors; also record a 90 percent confidence interval.

4 A random sample of 30 households in a city having 14,848 households yielded the following numbers of persons per household:

4, 2, 4, 3, 4, 2, 1, 3, 3, 4, 3, 3, 4, 4, 5,

3, 4, 7, 2, 3, 4, 4, 3, 3, 3, 2, 3, 3, 6, 5.

Obtain a 90 percent confidence interval for the total number of people in the city.

5 A random sample of 20 pages from a 530-page text yielded the following number of typographical errors on an intensive examination:

3, 1, 0, 5, 3, 2, 4, 1, 0, 5,

3, 3, 2, 3, 0, 3, 1, 5, 3, 3.

Obtain a 90 percent confidence interval for the total number of typographical errors in the text.

6 Consider the population $\{1, 2, \ldots, N\}$. For a random sample of n, determine expressions for $E(\bar{y})$ and var $(\bar{y})$ as functions of N.

E PROBLEMS ON STRATIFIED SAMPLING

7 Sometimes it is possible to stratify a population into moderately homogeneous strata. Suppose that a population of N elements is stratified into k strata containing $N_1, \ldots, N_k$ elements, respectively. Let μ_j and σ_j^2 be the mean and variance in the jth stratum and μ and σ^2 be the mean and variance in the population. Show that

$$(N-1)\sigma^2 = \sum (N_j - 1)\sigma_j^2 + \sum N_j(\mu_j - \mu)^2.$$

8 (*continuation*) Now suppose that random samples are obtained in each stratum. Let $\bar{y}_j$ be the mean of a sample of n_j from the jth stratum. If $\Sigma_1^k a_j\bar{y}_j$ is necessarily an unbiased estimate of μ, then show that $a_j = N_j/N$. Let

$$\bar{y}_* = \sum_1^k \frac{N_j\bar{y}_j}{N}$$

be the unbiased estimate of μ; then show that the variance of $\bar{y}_*$ is

$$\text{var}(\bar{y}_*) = \sum_1^k \left(\frac{N_j}{N}\right)^2 \frac{\sigma_j^2}{n_j}\left(1 - \frac{n_j}{N_j}\right).$$

9 (*continuation*) If the sampling is proportional to stratum size, that is, $n_j/N_j = n/N$, where $n = \Sigma\, n_j$, and if all the strata have the same variance σ_*^2, then show that

$$\text{var}\,(\bar{y}_*) = \frac{\sigma_*^2}{n}\left(1 - \frac{n}{N}\right).$$

Thus if the variation within strata is substantially smaller than the overall variation, stratified sampling can give improvement in the precision of the estimate.

10 (*continuation*) Consider the minimization of var $(\bar{y}_*)$ subject to the constraint $\Sigma\, n_j = n$, where n is held fixed. Use partial differentiation with respect to n_j (taken to be continuous) and a Lagrange multiplier to accommodate the constraint, and show that minimum variance is attained with n_j proportional to $N_j\sigma_j$; that is, $n_j = nN_j\sigma_j/\Sigma\, N_i\sigma_i$. To optimize thus requires information concerning the relative sizes of the strata variances σ_j^2.

11 (*continuation*) Let V_1 be the variance of the estimate from a random sample of n; let V_2 be the variance of the estimate from a proportional sample of n; let V_3 be the variance from an optimum allocation (Problem 10) of a sample of n.

(a) If the stratum population sizes are large, so that the finite population correction factors can be ignored, show that

$$V_1 = \frac{\sigma^2}{n}, \qquad V_2 = \sum \frac{N_j}{N}\sigma_j^2/n, \qquad V_3 = \left(\sum \frac{N_j}{N}\sigma_j\right)^2/n.$$

(b) Under the same assumption, show that

$$V_1 = V_2 + \sum \frac{N_j}{N}(\mu_j - \mu)^2/n$$

$$V_2 = V_3 + \sum \frac{N_j}{N}(\sigma_j - \bar{\sigma})^2/n$$

where $\bar{\sigma} = \Sigma\, N_j\sigma_j/N$.

7.4

INFERENCE FROM SYMMETRY

We now examine some distribution-free or nonparametric methods of statistical inference, an important theoretical area of statistics that directly uses symmetry results from probability theory. Consider a random system with a real-valued response y, for example the biochemical system examined earlier in the chapter. And suppose there is continuity in the system and sufficient figures are used for the recorded response that a distribution without discrete probabilities is appropriate.

Consider an appropriate statistical model. The lack of discrete probabilities can conveniently be expressed in terms of continuity of the distribution function. Accordingly, we take the model to be $\mathfrak{M} = \{F\}$, where $\mathfrak{M}$ is restricted to continuous distribution functions. The theory we discuss in this section is sometimes called distribution-free—it does not depend on having a particular functional form for the distribution in the model, functional form such as the normal. The theory is also called nonparametric—the model is not given completely in terms of one, two, or several real parameters.

Suppose, as is typical, that the investigator is primarily interested in the general level of the response. For the model $\mathfrak{M}$ the mean μ does not exist for some of the distributions F. Accordingly, it is convenient to use the median ζ (or $\tilde{\mu}$) as a measure of the general level of the response. The median of a distribution given by the distribution function F is defined by

(1) $F(\zeta) = 1/2 = 1 - F(\zeta)$

provided the distribution does not have a central interval containing no probability; if there is such an interval, then there is an *interval of medians,* any one of which can be used for the analysis here.

Suppose that the investigator plans a sequence of n independent repetitions. Let $(y_1, \ldots, y_n)$ designate the compound response for the sequence. The sample space then is R^n and the statistical model can be given as

(2) $\{F(y_1) \cdots F(y_n) : F \in \mathfrak{M}\}.$

EXAMPLE 1

Biochemical example: We shall illustrate several inference procedures using the biochemical data from Section 7.1E. The data, with some simple calculations to be used later, are given in Table 7.2.

TABLE 7.2
Data for Example 1

y_i	$y_i - 14.8$	sgn $(y_i - 14.8)$
17.1	2.3	+1
14.9	0.1	+1
15.2	0.4	+1
18.1	3.3	+1
16.4	1.6	+1
19.1	4.3	+1
18.5	3.7	+1
14.6	−0.2	−1
17.3	2.5	+1
		$S = 7$

A SIGN SYMMETRY

The sign function sgn is defined by

sgn $(y) = +1, 0, -1$ according as $y > 0, = 0, < 0$.

Now consider y with a continuous distribution F and median ζ. The marginal distribution of sgn $(y - \zeta)$ has probability 1/2 at +1 and probability 1/2 at −1. This is a relocated, rescaled Bernoulli distribution. In fact, $(1/2)(\text{sgn}(y - \zeta) + 1)$ is Bernoulli(1/2); note that $(1/2)(\text{sgn}(y - \zeta) + 1)$ is the indicator function for a positive value for the deviation $y - \zeta$.

Now consider a sample $(y_1, \ldots, y_n)$ from the distribution F with median ζ. Then

(3) $S = \sum_1^n \text{sgn}(y_i - \zeta)$

has a distribution with probability at the points $-n, -n+2, \ldots, n-2, n$. This is a relocated, rescaled binomial(n, 1/2) distribution. In fact,

$$(4) \qquad K = (1/2)(S+n) = \sum_1^n (1/2)(\operatorname{sgn}(y_i - \zeta) + 1)$$

is binomial(n, 1/2); note that K is just the number of positive deviations $y_1 - \zeta, \ldots, y_n - \zeta$.

For a sample size $n = 9$, the distribution of S and K is as follows:

S	-9	-7	-5	$\cdots$	5	7	9
K	0	1	2	$\cdots$	7	8	9
p	1/512	9/512	36/512	$\cdots$	36/512	9/512	1/512

This is binomial(9, 1/2) for K.

For large samples the distribution of S and K can be approximated by treating

$$(5) \qquad \frac{K - n/2}{\sqrt{n}/2} = \frac{S}{\sqrt{n}}$$

as a standard normal variable; see Example 6.2.3.

B SIGN TEST

Consider the random system discussed in the introduction to this section and suppose that a value ζ_0 has been suggested for the median of the distribution. We consider a simple test for the hypothesis $\zeta = \zeta_0$.

The function

$$(6) \qquad S = \sum_1^n \operatorname{sgn}(y_i - \zeta_0)$$

gives the sum of the signs of the deviations from ζ_0, and the adjusted function

$$(7) \qquad K = (1/2)(S+n) = \sum_1^n (1/2)(\operatorname{sgn}(y_i - \zeta_0) + 1)$$

gives the number of positive deviations from ζ_0. This function indicates whether sample values $y_1, \ldots, y_n$ are distributed in a reasonable way about the supposedly central value ζ_0.

If $\zeta = \zeta_0$, then K is binomial(n, 1/2). On the other hand, if the true median ζ is different from ζ_0, then K is binomial(n, $1 - F(\zeta_0)$) and we would expect extreme values for K: a higher probability for large values of K if $F(\zeta_0)$ is small, and higher probability for small values of K if $F(\zeta_0)$ is large.

Consider the following test for the hypothesis $\zeta = \zeta_0$. We calculate the value of K in formula (7) and compare it with the binomial(n, 1/2) distribution. A central value for K is in conformity with the hypothesis. An extreme value on one or other tail of the distribution is an indication against the hypothesis.

EXAMPLE 1 (*continuation*)

Consider the biochemical data and suppose that we want to test the value $\zeta = 14.8$ as suggested by some theoretical calculations. The observed value of S is 7 and the corresponding value of K is 8. Note that this is the next-to-last point on the right tail of the distribution as tabulated at the end in Section A. We can, in part, assess the significance of this observed value by calculating the observed level of significance, the probability under the hypothesis of a value as far or farther from the center of the distribution as that observed:

$$\frac{1}{512} + \frac{9}{512} + \frac{9}{512} + \frac{1}{512} = \frac{20}{512} \approx 4\%.$$

This is a moderately small probability, and it gives a moderate indication against the hypothesis.

C SIGN CONFIDENCE INTERVAL

Now suppose that we want a confidence interval for the general level ζ. The distribution of

$$K = \sum_1^n (1/2)(\text{sgn}\,(y_i - \zeta) + 1)$$

is binomial(n, 1/2). From this binomial distribution we can determine a central 95 percent interval $[k_1, k_2]$ for K, central in the sense of approximately equal probability on each tail. We can then write

(8) $$P(k_1 \le K \le k_2) \approx 95\%$$
$$= P(k_1 \le (\text{number of } y_i > \zeta) \le k_2)$$
$$= P(y_{(n-k_2)} \le \zeta < y_{(n-k_1+1)}),$$

where $(y_{(1)}, \ldots, y_{(n)})$ is the order statistic that records the values $y_1, \ldots, y_n$ arranged in order from the smallest to the largest. Thus $[y_{(n-k_2)}, y_{(n-k_1+1)})$ is a 95 percent confidence interval for the median ζ. If n is large, we can use the normal approximation giving $[k_1, k_2] \approx (n/2 \pm z_{2\frac{1}{2}\%}\sqrt{n}/2)$, where z_γ is the value exceeded with probability γ according to the standard normal.

EXAMPLE 1 (*continuation*)

Consider the biochemical data. The distribution for K in Section A gives

(9) $$P(2 \le K \le 7) = 0.961$$
$$= P(y_{(2)} \le \zeta < y_{(8)}).$$

Thus $[y_{(2)}, y_{(8)}) = [14.9, 18.5)$ is a 96.1 percent confidence interval for the median ζ.

D MODEL WITH ADDITIONAL SYMMETRIES

For the random system discussed in the introduction, suppose there is the additional information that the response distribution is symmetrical about its median. As an appropriate statistical model based on continuity and the symmetry just mentioned, consider $\mathfrak{M} = \{F\}$, where the elements F are restricted to continuous distribution functions with the symmetry property

$$F(\zeta - d) = 1 - F(\zeta + d), \qquad d \geq 0,$$

where ζ is the median of F. The model then for a sequence of repetitions, $(y_1, \ldots, y_n)$, is

$$\{F(y_1) \cdots F(y_n) : F \in \mathfrak{M}\}.$$

EXAMPLE 1 (*continuation*)

We shall illustrate some additional inference procedures using the biochemical data. The data with some calculations of interest later are given in Table 7.3.

TABLE 7.3
Data for Example 1

y_i	$\lvert y_i - 14.8\rvert$	sgn $(y_i - 14.8)$	$\lvert y_i - 14.8\rvert$ sgn $(y_i - 14.8)$
17.1	2.3	+1	2.3
14.9	0.1	+1	0.1
15.2	0.4	+1	0.4
18.1	3.3	+1	3.3
16.4	1.6	+1	1.6
19.1	4.3	+1	4.3
18.5	3.7	+1	3.7
14.6	0.2	−1	−0.2
17.3	2.5	+1	2.5
	18.4		18.0

E CONDITIONAL SIGN SYMMETRY

Now consider y with a continuous distribution F that is symmetric about the median. The symmetry suggests that the conditional distribution of $y - \zeta$ given $|y - \zeta|$ has probability 1/2 at $+|y - \zeta|$ and probability 1/2 at $-|y - \zeta|$; see Figure 7.4 expressed in terms of a density function. For a formal statement of the distribution theory it is convenient to relocate at the median and in effect consider the case $\zeta = 0$.

LEMMA 1

If y has a continuous distribution that is symmetric about the origin, then $|y|$ and sgn (y) *are statistically independent and* sgn y *has probability* 1/2 *at* −1 *and at* +1.

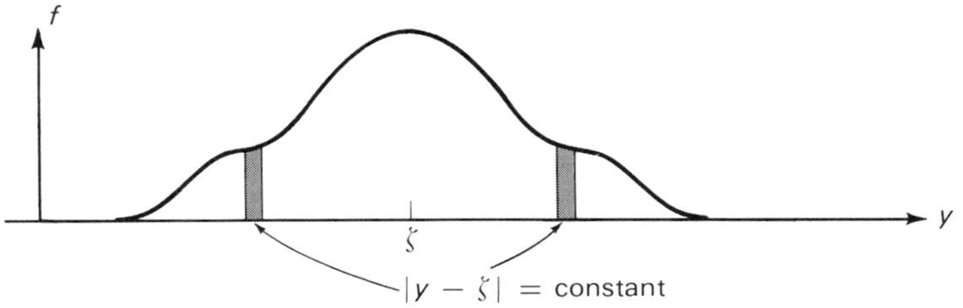

FIGURE 7.4
Conditional distribution of $y - \zeta$ given $|y - \zeta|$ has probability 1/2 at each of $+|y - \zeta|$, $-|y - \zeta|$.

Proof: We show independence by showing that the distribution function for $(|y|, \text{sgn}(y))$ factors, so the variables separate; see Proposition 3.3.6. The function sgn (y) takes essentially two values, -1 and $+1$; accordingly, it suffices to consider the distribution function for $(\text{sgn}(y), |y|)$ with sgn (y) at an intermediate point, say 0:

$$P(\text{sgn}(y) \leq 0, |y| \leq u) = P(-u \leq y \leq 0) = 1/2 - F(-u),$$
$$P(\text{sgn}(y) \leq 0) = P(y \leq 0) = 1/2,$$
$$P(|y| \leq u) = P(-u \leq y \leq u) = F(u) - F(-u).$$

And then, by symmetry, we have

$$(1/2) \cdot (F(u) - F(-u)) = (1/2) \cdot 2((1/2) - F(-u)) = (1/2) - F(-u).$$

Thus the product of the marginal distribution functions is equal to the joint distribution function, and it follows that $|y|$ and sgn (y) are statistically independent.

Now, in general, consider y with symmetric distribution function F and median ζ. Then by the lemma we have that $|y - \zeta|$ and sgn $(y - \zeta)$ are statistically independent and sgn $(y - \zeta)$ has probability 1/2 at -1 and at $+1$. Or, equivalently, we have that the conditional distribution of sgn $(y - \zeta)$ given $|y - \zeta|$ has probability 1/2 at -1 and at $+1$.

Now consider a sample $(y_1, \ldots, y_n)$ from the distribution F with median ζ. We investigate the conditional distribution of

$$(10) \qquad T = \sum_1^n |y_i - \zeta| \,\text{sgn}(y_i - \zeta) = \sum_1^n (y_i - \zeta)$$

given the absolute deviations $|y_1 - \zeta|, \ldots, |y_n - \zeta|$. This distribution can be expressed by

$$(11) \qquad \sum_1^n |y_i - \zeta| s_i,$$

where $(s_1, \ldots, s_n)$ is a sample from a distribution with probability 1/2 at -1 and at $+1$. For small and moderate samples this distribution can be calculated by hand or machine. And for large samples there is approximate normality provided the set of absolute deviations forms a reasonable array.

As an illustration consider a sample of size $n = 9$ and suppose the absolute deviations from the true median are

$$0.1,\quad 0.2,\quad 0.4,\quad 1.6,\quad 2.3,\quad 2.5,\quad 3.3,\quad 3.7,\quad 4.3.$$

The conditional distribution of T can be obtained by enumeration, easily done for the tails, but tedious for the center of the distribution.

T	-18.4	-18.2	-18.0	$\cdots$	18.0	18.2	18.4
P	1/512	1/512	1/512		1/512	1/512	1/512

The distribution is symmetric about 0. Some of the probabilities differ from 1/512, for example

$$P(T = 13.8) = 2/512,$$

as can be seen from

$$18.4 - 2(2.3) = 18.4 - 2(0.1 + 0.2 + 0.4 + 1.6) = 13.8;$$

thus the distribution is not flat or symmetric in the sense of Section 1.3E.

F SIGNED-DEVIATION OR FISHER TEST

Consider the random system discussed in Section D and suppose that ζ_0 has been suggested for the median of the distribution. We now consider the Fisher test for the hypothesis $\zeta = \zeta_0$.

The function

$$T = \sum_1^n |y_i - \zeta_0| \operatorname{sgn}(y_i - \zeta_0) = \sum_1^n (y_i - \zeta_0) \tag{12}$$

gives the sum of deviations from ζ_0. Given the absolute deviations $|y_1 - \zeta_0|, \ldots, |y_n - \zeta_0|$, the value of the function T provides an indication of how sample values $y_1, \ldots, y_n$ are distributed in relation to the supposedly theoretically central value ζ_0: the signs of the deviations from ζ_0 are weighted by the distances from ζ_0.

If $\zeta = \zeta_0$, then the distribution of T given $|y_1 - \zeta_0|, \ldots, |y_n - \zeta_0|$ is distributed as

$$\sum_1^n |y_i - \zeta_0| s_i, \tag{13}$$

where $(s_1, \ldots, s_n)$ is a sample from a distribution with probability 1/2 at -1 and at $+1$.

We can test the hypothesis $\zeta = \zeta_0$ by calculating the value of T in formula (12) and comparing it with the distribution expressed by (13). An extreme value on one or other tail of the distribution would be an indication against the hypothesis.

EXAMPLE 1 (*continuation*)

Consider the biochemical example as given in Section D and suppose that we want to test the hypothesis $\zeta = 14.8$. The observed value of T is 18.0. The distribution for T under the hypothesis $\zeta = 14.8$ is recorded in Section E. Note that the observed value 18.0 is the third point in on the right tail of the distribution. The observed level of significance, the probability of a value as far or farther from the center of the distribution as that observed, is

$$(1/512) + (1/512) + (1/512) + (1/512) + (1/512) + (1/512) = 6/512,$$

which is just slightly larger than 1 percent. This is not the kind of value that we would reasonably expect under the hypothesis, and in turn it provides moderate evidence against the hypothesis.

In conclusion, it is worth noting some equivalent functions that can be used for this conditional test. We can write

$$(14) \qquad T = \sum_{y_i > \zeta_0} (y_i - \zeta_0) + \sum_{y_i \le \zeta_0} (y_i - \zeta_0)$$

$$= 2 \sum_{y_i > \zeta_0} (y_i - \zeta_0) - \sum |y_i - \zeta_0|$$

$$= 2T^+ - k,$$

where $k = \Sigma\, |y_i - \zeta_0|$ is a constant; thus T and T^+ are monotone functions of each other (just as S and K were in Section B). In another direction we can write

$$(15) \qquad \frac{T^2/n}{\sum |y_i - \zeta_0|^2} = \frac{n(\bar{y} - \zeta_0)^2}{\sum (y_i - \zeta_0)^2}$$

$$= \frac{n(\bar{y} - \zeta_0)^2}{n(\bar{y} - \zeta_0)^2 + \sum (y_i - \bar{y})^2} = \frac{1}{1 + (n-1)/t^2},$$

where

$$(16) \qquad t = \frac{\bar{y} - \zeta_0}{s_y/\sqrt{n}}$$

is the t-function given by (7.2.17) in Section 7.2E. Thus $|T|$ can be put in one-to-one correspondence with $|t|$, and, in addition, the signs of T and t correspond; thus we could equivalently use t for the test discussed in this section.

G SIGNED-DEVIATION CONFIDENCE INTERVAL

Now suppose that we want a confidence interval for the general level ζ. The function

$$T = \sum_1^n |y_i - \zeta| \operatorname{sgn}(y_i - \zeta) = \sum_1^n (y_i - \zeta)$$

can be used together with the conditional distribution described by

$$\sum_1^n |y_i - \zeta| s_i,$$

where $(s_1, \ldots, s_n)$ has probability $1/2^n$ at each of the 2^n points formed by $s_i = \pm 1$. Consider the following equivalent statements:

$$\sum_1^n (y_i - \zeta) \leq m\text{th smallest of the } \sum_1^n (y_i - \zeta)s_i,$$

$$\sum_1^n (\zeta - y_i) \geq m\text{th largest of the } \sum_1^n (\zeta - y_i)s_i,$$

$$0 \geq m\text{th largest of the } \sum_1^n (s_i - 1)(\zeta - y_i),$$

$$0 \geq m\text{th largest of the } 2 \sum_{s_i=-1} (y_i - \zeta),$$

$$0 \geq m\text{th largest of the } \Big(\sum_{s_i=-1} y_i\Big)\Big/\Big(\sum_{s_i=-1} 1\Big) - \zeta,$$

$$\zeta \geq m\text{th largest of the } \Big(\sum_{s_i=-1} y_i\Big)\Big/\Big(\sum_{s_i=-1} 1\Big),$$

where, for example, $\Sigma_{s_i=-1} 1$ gives the number of coordinates of $(s_1, \ldots, s_n)$ that are equal to -1. Thus to form a confidence interval, we consider the $2^n - 1$ possible subsets of $\{y_1, \ldots, y_n\}$; we form the $2^n - 1$ subset averages

$$\Big(\sum_{s_i=-1} y_i\Big)\Big/\Big(\sum_{s_i=-1} 1\Big) = \sum_{s_i=-1} y_i/(\text{number of } s_i = -1);$$

we take the m_1st smallest and m_2nd largest of these subset averages; and we obtain a confidence interval for ζ having confidence level $1 - (m_1 + m_2)/2^n$.

EXAMPLE 1 (*continuation*)

Consider further the biochemical example as presented in Section D. The arithmetic is reasonable if we choose a 98 percent confidence interval: we exclude 1 percent on each tail; 1 percent of 512 is approximately 5; thus we form an interval from the 5th smallest to the 5th largest of the 511 subset averages. The data rearranged in accord with the order statistic are given in Table 7.4. For the subset averages on the left tail it is convenient to use 14.6 as a temporary origin:

Subset	{0}	{0, 0.3}	{0.3}	{0, 0.6}	{0, 0.3, 0.6}	...
Average	0	0.15	0.3	0.3	<u>0.3</u>	

TABLE 7.4
Data for example 1

$y_{(i)}$	$y_{(i)} - 14.6$	$19.1 - y_{(i)}$
14.6	0	4.5
14.9	0.3	4.2
15.2	0.6	3.9
16.4	1.8	2.7
17.1	2.5	2.0
17.3	2.7	1.8
18.1	3.5	1.0
18.5	3.9	0.6
19.1	4.5	0

For the subset averages on the right tail it is convenient to measure back from 19.1:

Subset	{0}	{0, 0.6}	{0, 1.0}	{0, 0.6, 1.0}	{0.6}	···
Average	0	0.3	0.5	0.533	<u>0.6</u>	

Thus an approximate 98 percent confidence interval for ζ is

$$(14.6 + 0.3,\ 19.1 - 0.6) = (14.9,\ 18.5).$$

Note that this happens to be the same interval as obtained in Section C, yet the confidence levels are different. The confidence level in each case, however, is determined in relation to intervals that "might have been obtained."

H EXERCISES

1 The resistances of five pieces of wire sampled from production are recorded in Exercise 7.1.1; assume continuity of the distribution function. Test the hypothesis that the median $\zeta = 0.135$. Form a 93.75 percent confidence interval for ζ.

2 The breaking loads for six samples of plastic are recorded in Exercise 7.1.2; assume continuity of the distribution function. Test the hypothesis that the median breaking strength is 27.0. Form a 96.875 percent confidence interval for the median ζ.

3 For Exercise 1, assume in addition that the distribution is symmetric; answer Exercise 1 using the signed-deviation function.

4 For Exercise 2, assume in addition that the distribution is symmetric; answer Exercise 2 using the signed-deviation function.

5 A sample of 11 rats was obtained randomly from a population of rats. A blood viscosity test of each yielded 3.29, 3.91, 4.64, 3.55, 3.67, 4.18, 3.74, 4.67, 3.03, 4.61, 3.84.
(a) Use the sign test to test the hypothesis: median $\zeta = 4.6$.
(b) Form a 98.8 percent confidence interval for the median ζ.

6 (*continuation*) Suppose there are grounds for the assumption that the distribution is symmetrical about the median ζ.
(a) Use the signed-deviation test to test the hypothesis: median $\zeta = 4.6$.
(b) Form a 99 percent confidence interval for the median ζ.

PROBLEMS

7 *On the sign test:* The sign test is based on $K = \Sigma_1^n c(y_i - \zeta_0)$, where c is the indicator function for the positive axis $(0, \infty)$. Some statistical theory suggests that *decisions* be made in a statistical investigation. From this point of view the conventional 5 percent (or α) sign test is to "reject" if $|K - n/2| \geq k$ and to "accept" otherwise; the constant k is chosen so that $P(|K - n/2| \geq k) \leq 5$ percent (or $\leq \alpha$) if $\zeta = \zeta_0$. The investigator then knows that the test will reject a true hypothesis with probability at most 5 percent. But what does the test do if $\zeta \neq \zeta_0$? Let p designate the probability $P(y > \zeta_0)$ for such a case. The investigator would then be interested in the *power* of the test: the probability $\beta(p) = P(|K - n/2| \geq k{:}p)$ of rejecting the hypothesis when the distribution has probability p to the right of ζ_0. For $p \neq 1/2$, the investigator would hope for high power, that is, high probability of rejecting the hypothesis. For a sample of $n = 7$, determine k for a 2 percent test. Calculate the power function β and plot against p.

8 Let $r_1, \ldots, r_n$ be the ranks of $y_1, \ldots, y_n$; recall Exercise 3.2.15. Show that

$$r_i = \sum_{j=1}^{n} c(y_i - y_j) + 1,$$

where c is the indicator function for the positive axis $(0, \infty)$. Assume that the y's are all different.

9 *Wilcoxon signed-rank test:* Consider the absolute deviations $|y_1 - \zeta|, \ldots, |y_n - \zeta|$ and let $r(|y_1 - \zeta|), \ldots, r(|y_n - \zeta|)$ be the corresponding ranks: thus $r(|y_i - \zeta|)$ is equal to r if $|y_i - \zeta|$ is the rth smallest of the absolute deviations. Under the assumptions in Section D we consider the conditional distribution of the function

$$T^+ = \sum_{y_i > \zeta} r(|y_i - \zeta|) = \sum r(|y_i - \zeta|)c(y_i - \zeta)$$

given the absolute deviations. The function c is defined in Problem 8.

(a) Tabulate the conditional distribution of T^+ for the case $n = 3$.

(b) Show that

$$E(T^+) = \frac{n(n+1)}{4}, \qquad \operatorname{var}(T^+) = \frac{n(n+1)(2n+1)}{24}.$$

10 (*continuation*) For the example in Exercises 5 and 6, use the preceding signed-rank test to test the hypothesis $\zeta = 4.6$ by:

(a) Calculating the necessary part of the distribution of T^+ directly; note that the distribution of T^+ is symmetrical.

(b) Using the limiting normal distribution: Theorem 6.2.4 shows that T^+, standardized with respect to its mean and standard deviation, has a limiting standard normal distribution.

11 (*continuation*) The continuity assumptions in Section D assure that there is zero probability for a tie in a sample $(y_1, \ldots, y_n)$ of size n. Thus assume that the ranks in Problem 8 are well defined.

(a) Use Problem 8 to show that the signed-rank test function can be reexpressed as

$$T^+ = \sum_{i=1}^{n} c(y_i - \zeta) + \sum_{i,j=1}^{n} c(y_i - \zeta)c(|y_i - \zeta| - |y_j - \zeta|).$$

(b) Let $S = \{(y_i + y_j)/2 : i \leq j = 1, \ldots, n\}$ be the set of $n(n+1)/2$ subset averages for subsets of sizes 1 and 2. Show that T^+ is the number of points in S that are larger than ζ.

12 (*continuation*) *The signed-rank confidence interval:* A value of T^+ can be associated with having ζ in one of the $n(n+1)/2 + 1$ intervals formed by the $n(n+1)/2$ points of S. Let $P(t_1 \leq T^+ \leq t_2) = 1 - \alpha$. Then show that a $1 - \alpha$ confidence interval is obtained by combining the intervals for which T^+ lies in the range $t_1 \leq T^+ \leq t_2$.

13 (*continuation*) For the example in Exercises 5 and 6, use the signed-rank test to form a central 99 percent confidence interval for ζ.

J TWO-SAMPLE PROBLEMS

14 *The median-test or Mood–Westenberg test:* Let $(y_{11}, \ldots, y_{1m})$ be a sample from a first distribution and $(y_{21}, \ldots, y_{2n})$ be a sample from a second distribution. Consider the case where the distributions have the same form; and suppose that we are interested in the difference in the locations of the two distributions. Specifically, let the first distribution be given by $F(y_1)$ and the second distribution be given by $F(y_2 - \theta)$, where F can be any continuous distribution function; and suppose that we are interested in testing the hypothesis $\theta = 0$. The median test is a partial analog of the sign test. It is performed in the following manner: The two samples are combined to form a larger "sample" of $m + n$ values; the $m + n$ values are arranged in order of magnitude from the smallest to the largest; the sequence is divided into two equal or nearly equal parts so that $[(m + n)/2]$ values are larger than the dividing point and $m + n - [(m + n)/2]$ values are smaller; each value in the sequence is replaced by an A if it originated from the first sample and by a B if it originated from the second sample; let S be the number of A's to the right of the dividing point. In other words, S is the number of first-sample values that are greater than the median of the combined samples. Under the hypothesis, show that the probability function of S is

$$p_N(S) = \binom{m}{S}\binom{n}{h - S}\Big/\binom{N}{h},$$

where $N = m + n$ and $h = [N/2]$.

15 (*continuation*) Under the hypothesis $\theta = 0$, obtain an expression for $E(S^{(r)})$ and show that

$$E(S) = \frac{mh}{N}, \qquad \text{var}(S) = \frac{mnh(N - h)}{N^2(N - 1)}.$$

If the two distributions have the same continuous distribution function, it can be shown (Problem 24) that $z = (S - E(S))/\sigma_S$ has a limiting standard normal distribution.

16 (*continuation*) For the data in Exercise 7.2.5 and Problem 7.2.7, test the hypothesis $\theta = 0$.

17 (*continuation*) *The median-test confidence interval:* Let s_1 and s_2 be chosen so that $P(s_1 \le S \le s_2) = 95$ percent (or $1 - \alpha$) using the hypothesis distribution of S. A 95 percent (or $1 - \alpha$) confidence "interval" consists of all those θ values for which the median test applied to $(y_{11}, \ldots, y_{1m}), (y_{21} - \theta, \ldots, y_{2n} - \theta)$ yields a value of S in the interval $[s_1, s_2]$. Let $(y_{1(1)}, \ldots, y_{1(m)})$ be the order statistic for the first sample and $(y_{2(1)}, \ldots, y_{2(n)})$ be the order statistic for the second sample.

(a) Show that the value S implies $y_{1(m-S+1)} > y_{2(m-h+S)} - \theta$ and $y_{2(m-h+S+1)} - \theta > y_{1(m-S)}$.

(b) Obtain the corresponding interval for θ.

(c) Show that the 95 percent (or $1 - \alpha$) confidence interval is $(y_{2(m-h+s_1)} - y_{1(m-s_1+1)}, y_{2(m-h+s_2+1)} - y_{1(m-s_2)})$.

18 *The Mann–Whitney test:* Assume the conditions given in Problem 14. The Mann–Whitney test is also a partial analog of the sign test. Let U be the number of pairs (y_{1i}, y_{2j}) having $y_{2j} > y_{1i}$:

$$U = \sum_{i=1}^{n}\sum_{j=1}^{m} c(y_{2i} - y_{1j}).$$

Under the hypothesis $\theta = 0$, show that

$$E(U) = \frac{mn}{2}, \qquad \text{var}(U) = \frac{mn(N + 1)}{12},$$

where $N = m + n$. A generalized central limit theorem for what are called U statistics can be used to show that U standardized with respect to its mean and standard deviation has a limiting standard normal distribution; see, for example, Fraser (1957, p. 225).

19 (*continuation*) For the data in Exercise 7.2.5 and Problem 7.2.7, test the hypothesis $\theta = 0$.

20 (*continuation*) *The Mann–Whitney confidence interval:* Suppose that $P(u_1 \le U \le u_2) = 95$ percent (or $1 - \alpha$) under the hypothesis distribution. The confidence interval is obtained by calculating U for $(y_{11}, \ldots, y_{1m})$ and $(y_{21} - \theta, \ldots, y_{2n} - \theta)$. Show that the 95 percent (or $1 - \alpha$) confidence interval is (d_{u_2+1}, d_{u_1}), where $d_{mn}, d_{mn-1}, \ldots, d_2, d_1$ are the mn differences $y_{2j} - y_{1i}$ arranged in order of magnitude from smallest to largest.

21 (*continuation*) *The Wilcoxon test:* The Wilcoxon test is performed in the following manner: the two samples are combined to form a larger "sample" of $m + n$ values; the $m + n$ values are arranged in order of magnitude from the smallest to the largest; the values in the sequence are replaced by their ranks from 1 to $m + n$; the sum W of the ranks corresponding to those values that come from the second sample is calculated. Use Problem 8 to show that

$$W = U + \frac{n(n+1)}{2}.$$

Thus the Wilcoxon test is equivalent to the Mann–Whitney test. The Mann–Whitney test was proposed several years after the Wilcoxon test and has some advantages for computation.

22 *The Pitman test:* The Pitman test is a two-sample analog of the Fisher test; assume the conditions given in Problem 14: a sample of m from a distribution $F(y_1)$ and a sample of n from a distribution $F(y_2 - \theta)$, where F can be any continuous distribution function. Having the same distribution form is the property that replaces the symmetry used in the Fisher test. The test is a conditional test given the $m + n$ values in the combined sample; the conditional distribution under the hypothesis gives equal probability $1\Big/\binom{m+n}{n}$ to each assignment of m values to the first sample set and the remaining n to the second sample set. The conditional test of the hypothesis: $\theta = 0$ given the set $\{y_{11}, \ldots, y_{1m}, y_{21}, \ldots, y_{2n}\}$ can be based on $\bar{y}_2$, or on $d = \bar{y}_2 - \bar{y}_1$, or on

$$t = (\bar{y}_2 - \bar{y}_1)\Big/\left(\frac{1}{m} + \frac{1}{n}\right)^{1/2} \left(\frac{\sum (y_{2j} - \bar{y}_2)^2 + \sum (y_{1j} - \bar{y}_1)^2}{m + n - 2}\right)^{1/2}$$

as defined in Section 7.2l. Show that the three functions are equivalent in the order they place on the $\binom{m+n}{n}$ segregations into possible first and second samples. The test is performed by calculating the observed value and comparing it with the conditional distribution based on the $\binom{m+n}{n}$ different segregations of the combined sample. For small samples the test can be performed easily using the function $\Sigma\, y_{2j}$. It involves calculating the possible values of $\Sigma\, y_{2j}$ under the various allocations of n values as possible second samples; often these can be calculated for just the extreme cases necessary for determining the level of significance.

23 (*continuation*) The following samples were obtained with two treatments applied to experimental units: (27.1, 25.3, 28.4); (28.3, 29.2, 28.5). Test the hypothesis $\theta = 0$.

24 (*continuation of Problems 14 and 15*) *Proof of the limiting normality:* Let $m = \alpha N$, $n = \beta N$ and let

$$f_N(z) = p_N(\mu_S + z\sigma_S)\sigma_S.$$

By substituting the expressions for the mean and standard deviation and using Stirling's formula for the factorials, show that $\lim_{N\to\infty} f_N(z) = \exp\{-z^2/2\}/\sqrt{2\pi}$. Then use Proposition 6.1.2 to prove the limiting normality; note the method in Problems 6.1.11 and 6.1.12.

SUPPLEMENTARY MATERIAL

Supplement Section 7.5 is on pages 563–572.

The inference methods in this chapter assume that there is randomness. The supplemental material is concerned with *tests of randomness*. A very powerful industrial tool, *quality-control charts,* is examined in some detail. And a simple *run test* for randomness is presented, partly for its quick convenient scientific uses and partly for the way it illustrates the application of some combinatorial formulas.

NOTES AND REFERENCES

The following are textbooks that develop introductory statistics:

Fraser, D. A. S. (1958). *Statistics, An Introduction.* New York: John Wiley & Sons, Inc.

Freund, J. E. (1971). *Mathematical Statistics.* Englewood Cliffs, N.J.: Prentice-Hall, Inc.

Hoel, P. G., S. C. Port, and C. J. Stone (1971). *Introduction to Statistical Theory.* Boston: Houghton Mifflin Company.

Kalbfleisch, J. G. (1971). *Probability and Statistical Inference.* Waterloo: J. G. Kalbfleisch, University of Waterloo.

Mendenhall, W., and R. L. Scheaffer (1972). *Mathematical Statistics with Applications.* North Scituate, Mass.: Duxbury Press.

Mood, A. M., F. A. Graybill, and D. C. Boes (1974). *Introduction to the Theory of Statistics.* New York: McGraw-Hill Book Company.

Roussas, G. G. (1973). *A First Course in Mathematical Statistics.* Reading, Mass.: Addison-Wesley Publishing Company, Inc.

This text introduces statistics in an open-ended frame presenting tests of significance, estimation, and confidence methods as integrated components of statistical inference. These methods are introduced directly from large-sample and symmetry results of probability theory. Thus the methods are not conceptualized in the context of the common normal distribution model. Rather, inference is seen as a process from data to conclusions, a process that depends on the particular model accepted for an application.

In later chapters this text progressively examines more specialized statistical models.

For some additional tests of randomness (Supplement Section 7.5), see Knuth (1969).

Fraser, D. A. S. (1957). *Nonparametric Methods in Statistics.* Ann Arbor, Mich.: Xerox Microfilms, OP64588.

Gibbons, J. D. (1971). *Nonparametric Statistical Inference.* New York: McGraw-Hill Book Company.

Knuth, D. E. (1969). *The Art of Computer Programming,* Vol. 2. Reading, Mass.: Addison-Wesley Publishing Company, Inc.

Sterling, T. D. (1959). Publication decisions and their possible effects on inferences drawn from tests of significance—or vice versa. *Journal of the American Statistical Association,* **54**, 30–34.

In this chapter we have introduced the basic concepts of statistical inference. Many of the fruitful ideas of contemporary inference are due to R. A. Fisher; the following three fundamental books of his are recorded for supplemental reading.

Fisher, R. A. (1973). *Statistical Methods for Research Workers,* 14th rev. ed. New York: Macmillan Publishing Co.

Fisher, R. A. (1971). *The Design of Experiments,* 9th ed. New York: Macmillan Publishing Co.

Fisher, R. A. (1973). *Statistical Methods and Scientific Inference,* 3rd ed. New York: Macmillan Publishing Co.

In Chapter 7 we introduced estimation, tests of significance, and confidence intervals, and showed how these closely related components of inference can be applied easily to data. The three methods of inference were examined in a context of applications—unlinked to particular models that would specify precise details of distribution form. The background theory used was the symmetry and large-sample distribution theory from the probability chapters.

In Chapter 8 we shall examine statistical models that do specify precise details of distribution form. In Section 8.1 we show that such a model together with data leads to an observed ***likelihood function,*** *a function that makes inferences immediately available concerning an unknown parameter value. For illustrations we then examine models that use the normal distribution: the distribution theory in Section 8.2, and the inference methods in Section 8.3. The process of going from a sample space to a likelihood function is investigated in Section 8.4. Certain easy results are then found to be restricted to the* ***exponential models*** *in Section 8.5 or to the* ***large-sample models*** *in Section 8.6.*

8

THE LIKELIHOOD FUNCTION IN STATISTICAL INFERENCE

8.1

THE OBSERVED LIKELIHOOD FUNCTION

We now investigate statistical inference based on a parametric statistical model. Such a model is a more structured model in which the form of the distribution is known except for one, two, or several real parameters. As possible applications, consider the following. The lifetime of aluminum crucibles is approximately normally distributed with unknown location and scale parameters μ and σ. The determination of potassium content in a solution is approximately Student(λ) with unknown location and scale parameters μ and σ and unknown parameter λ determining the *shape* of the distribution. The lifetime of an electronic component is approximately exponential(θ) with unknown scale parameter θ.

In this chapter we investigate what a parametric statistical model says about observed data. This leads us to the *likelihood function* in Section 8.1 and then to a detailed examination of inference for *normal models* in Sections 8.2 and 8.3. A theoretically important statistic based on likelihood analysis is examined in Section 8.4; this leads us to the exponential models in Section 8.5 and large-sample likelihood in Section 8.6.

A THE PARAMETRIC STATISTICAL MODEL

The term *parametric statistical model* is used for models in which the form of the distribution is known except for one, two, or several real parameters. For example, a determination of potassium content is known to be approximately of Student(λ) form with location μ and scale σ; for a single determination, the density function is

$$(1) \qquad f(y \mid \mu, \sigma, \lambda) = \frac{1}{\sigma} \frac{\Gamma\left(\frac{\lambda+1}{2}\right)}{\Gamma(1/2)\Gamma(\lambda/2)} \left(1 + \frac{(y-\mu)^2}{\sigma^2 \lambda}\right)^{-(\lambda+1)/2} \frac{1}{\sqrt{\lambda}}.$$

The full parameter is $\theta = (\mu, \sigma, \lambda)$ with values in the space $\Omega = R \times R^+ \times R^+$. One of the values of the parameter θ is the true value, and to a reasonable approximation it gives the distribution for a potassium determination. Note that θ for this example has dimension 3. Also note that the *shape* parameter λ determines the thickness of the tails of the distribution.

Now in general let $f(y|\theta)$ be the density function for the response y on a sample space $\mathcal{S}$. The parameter θ with values in a parameter space Ω indexes the various possible density functions for the response. Accordingly, the statistical model is

$$\text{(2)} \qquad \mathfrak{M} = \{f(\cdot\,|\theta) : \theta \in \Omega\},$$

as expressed in terms of density functions.

The sample space $\mathcal{S}$ will usually be an open set on R^1, R^2, or R^k, and the function f will be density with respect to length, area, or volume measure. Alternatively, we see from Problem 5.8.15 that the binomial and the Poisson can be presented in terms of density functions with respect to the cardinality measure (Example 1.3.2) on the integers. Accordingly, we use the notation $f(y|\theta)$ both for a density function in the absolutely continuous case and for the probability function in the discrete case.

The various possible distributions for the response are obtained as the parameter θ ranges over the parameter space Ω. If the model is a correct model for an application, then some value of θ in Ω is the true value that gives the appropriate distribution for the response.

In most applications we will need the model for a *sequence* of independent repetitions of the system. Let $\mathbf{y} = (y_1, \ldots, y_n)'$ designate the vector response. Then the density function for the compound $\mathbf{y}$ is

$$\text{(3)} \qquad f(\mathbf{y}|\theta) = f(y_1|\theta) \cdots f(y_n|\theta).$$

Note that it is convenient to use the letter f to designate both the single and the compound density function; the size of the response argument provides the distinction.

B THE LIKELIHOOD FUNCTION FROM OBSERVED DATA

Consider a response y that is known to be approximately normal(θ, 1). For a sample of n independent repetitions, the model is

$$\text{(4)} \qquad \left\{\frac{1}{(2\pi)^{n/2}} \exp\left\{-(1/2)\sum_1^n (y_i - \theta)^2\right\} : \theta \in R\right\}.$$

As an example, consider the lifetime in months of crucibles used in the electrolytic production of aluminum. For this application we should perhaps formally exclude negative values; for typical data, however, there is no difficulty in having the extended ranges.

For $n = 3$ repetitions, the following data were obtained:

$$\text{(5)} \qquad (y_1, y_2, y_3) = (10.5, 10.7, 9.4).$$

This is plotted in Figure 8.1.

Now consider what the statistical model (4) says about the observed data (5). The probability at the mathematical point (10.5, 10.7, 9.4) is of course 0. Hence to examine probabilities we must consider a small neighborhood N of the point (10.5, 10.7, 9.4); let v be the volume of N. The probability at the observed point (10.5, 10.7, 9.4) is then given approximately by

$$\text{(6)} \qquad P(N|\theta) = (2\pi)^{-3/2} \exp\left\{-(1/2)[(10.5 - \theta)^2 + (10.7 - \theta)^2 + (9.4 - \theta)^2]\right\}v,$$

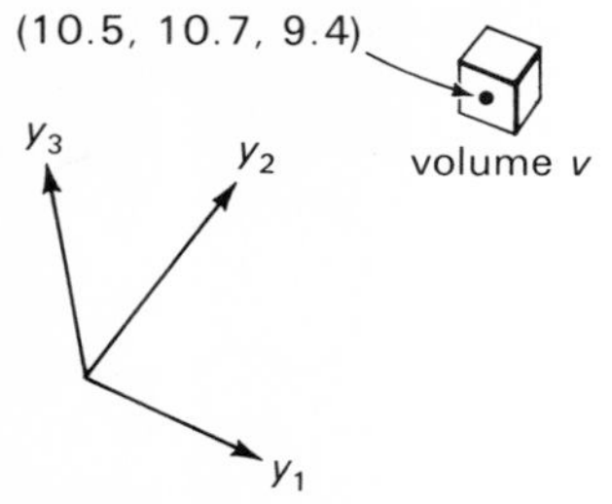

FIGURE 8.1
Point (10.5, 10.7, 9.4) and a neighborhood with volume v.

where θ takes values in $\Omega = R$. Of course, v could be arbitrarily small and certainly should be small for the approximation using the density function at the center of the neighborhood. Note that we write $P(N|\theta)$ for the probability in the set N as calculated from the distribution determined by the parameter value θ.

For any neighborhood of the observed data point with volume v, the probability (6) is a function of θ; it records the probability for what has been observed as a function of the possible values θ for the physical quantity being measured. The function is called the likelihood function (of θ) from the observed response $\mathbf{y} = (10.5, 10.7, 9.4)$ and designated $L(\mathbf{y}|\cdot)$:

(7) $$L(\mathbf{y}|\theta) = (2\pi)^{-3/2} \exp\left\{-(1/2)[(10.5 - \theta)^2 + (10.7 - \theta)^2 + (9.4 - \theta)^2]\right\} v$$

$$= c \exp\left\{-(1/2)\left[\sum_1^3 (y_i - 10.2)^2 + 3(10.2 - \theta)^2\right]\right\}$$

$$= c \exp\left\{-\frac{1}{2(1/3)}(10.2 - \theta)^2\right\};$$

for simplification we have used the standard identity (7.1.1) for sums of squares and we have incorporated some multiplicative constants into the arbitrary constant c derived from the arbitrary volume v.

The likelihood function (7) is plotted in Figure 8.2. The likelihood function is, in fact, not a single function but an equivalence class of similarly shaped curves; the class is generated by taking arbitrary values for v in (6) or for the constant c in (7); three typical curves in the class are plotted in Figure 8.2. Note that the likelihood function in this example has the shape of the normal density function as located at $\theta = 10.2$ and scaled by the factor $1/\sqrt{3} = 0.577$.

The likelihood function provides the basic probability assessment of an observed response; it is what the model says about the observed response.

C ASSESSMENT OF THE OBSERVED LIKELIHOOD FUNCTION

The likelihood function in Figure 8.2 records the probability for the observed response under various possible values for the parameter; it records what the model says about the observed response. For some values of θ close to 10.2, the probability for the observed response is relatively high; and for other values of θ, three or four

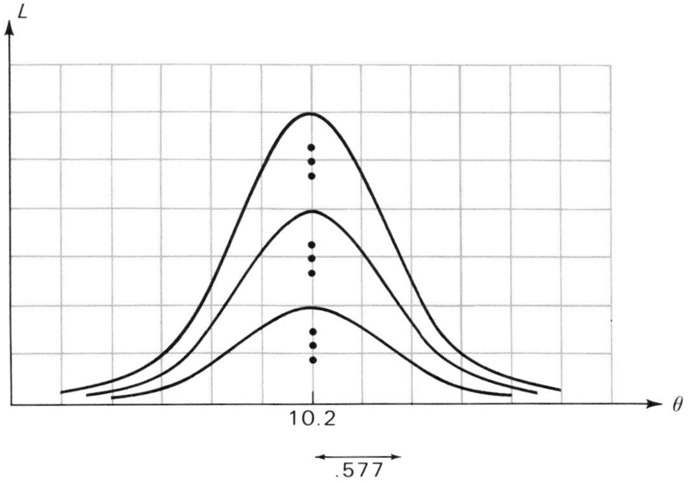

FIGURE 8.2
Likelihood function from the data in Section 8.1B.

scaling units from 10.2, the probability for the observed response is relatively very low.

Note that the likelihood function provides only a relative or comparative assessment of the various possible values for the parameter θ. This follows from the presence of the arbitrary multiplicative constant c or arbitrary volume v. Of course, each of the similarly shaped curves gives the same ratio of probabilities as determined for one θ value relative to another θ value; such a ratio is called the *likelihood ratio* for those θ values.

Our development makes clear the need for the arbitrary constant c: no particular value for the volume at the observed response has any special significance. Some texts neglect the arbitrary constant; this is a mistake both from the point of view of interpretation and of the theoretical results that depend on it (for example, the results in Sections 8.4 and 8.5).

Consider the observed likelihood function as recorded in Figure 8.2. The parameter value $\theta = \hat{\theta} = 10.2$ is the value for which the observed sample has maximum probability; it is called the maximum likelihood estimate (MLE) of the true parameter value. The likelihood function drops off from the maximum value and reaches the proportion

$$e^{-2^2/2} = e^{-2} = 0.135$$

of maximum value at two scaling units from the maximizing point. Accordingly, we form the interval (9.05, 11.35) for which the likelihood is greater than 0.135 of the maximum value; with nonnormal likelihood shapes, the fraction 0.135 gives a comparison in terms of the familiar normal shape.

Some developments of statistical theory have suggested a likelihood principle. The *likelihood principle* says that any inferences in a statistical investigation should use only the observed likelihood function; and that any other information about the data or about the statistical model should be suppressed. More specifically, it says that if two different statistical applications produce the same observed likelihood function, then inferences concerning the true parameter value should be the same. This is a severe principle in the amount that it excludes—particularly in the exclusion of all information concerning the statistical model beyond that implicitly contained in the observed likelihood function. The development of theory in this text does not endorse the likelihood principle. In fact, much of the remaining developments run counter to the likelihood principle.

For the present context we note that the likelihood function presents all that the model has to say about an observed response value. We are, of course, free to consider other possible response values and to calculate the corresponding likelihood functions. Indeed, we can even consider *the distribution of possible likelihood functions that would be obtained under any particular value for the parameter.*

D GENERAL DEFINITION OF THE LIKELIHOOD FUNCTION

Consider a statistical model

(8) $\{f(\cdot|\theta):\theta \in \Omega\},$

and let $\mathbf{y}$ be a possible value for the response. The probability for the response value $\mathbf{y}$ as calculated for a small neighborhood N with volume v is

(9) $L(\mathbf{y}|\theta) = f(\mathbf{y}|\theta)v.$

This gives us the *likelihood function* $L(\mathbf{y}|\cdot)$ from the observed response value $\mathbf{Y}$, where

(10) $L(\mathbf{y}|\theta) = cf(\mathbf{y}|\theta),$

and c is arbitrary in $(0, \infty)$; see Figure 8.3. The constant c generates an equivalence class of similarly shaped functions; several of the similarly shaped curves are plotted in the figure. Sometimes constants from $f(\mathbf{y}|\theta)$ can be combined with v; accordingly, we have used c for the arbitrary positive multiplicative constant.

Often if L does not take zero values, there is convenience in using the *log-likelihood function* $l(\mathbf{y}|\cdot)$ from the observed response $\mathbf{y}$, where

(11) $l(\mathbf{y}|\theta) = \ln f(\mathbf{y}|\theta) + a,$

and a is arbitrary in $(-\infty, \infty)$; see Figure 8.3.

And occasionally if L is differentiable with respect to a real parameter θ, there is convenience in using the *score function* $S(\mathbf{y}|\cdot)$ from the observed response $\mathbf{y}$, where

(12) $S(\mathbf{y}|\theta) = \frac{\partial}{\partial\theta} \ln f(\mathbf{y}|\theta) = \frac{\partial}{\partial\theta} l(\mathbf{y}|\theta);$

this records the *slope* of the log-likelihood function. See Figure 8.3 and note that $S(\mathbf{y}|\cdot)$ does not have an arbitrary constant.

If we have any one of the three functions $L(\mathbf{y}|\cdot)$, $l(\mathbf{y}|\cdot)$, $S(\mathbf{y}|\cdot)$, then we can calculate the others; for example, from $S(\mathbf{y}|\theta)$ we can integrate to obtain $l(\mathbf{y}|\cdot)$ and the additive constant a emerges as the arbitrary constant of integration. Thus the three functions convey the same information.

The arbitrary constant that generates the equivalence class of similarly shaped functions can occasionally be a nuisance. One way of avoiding the nuisance is to

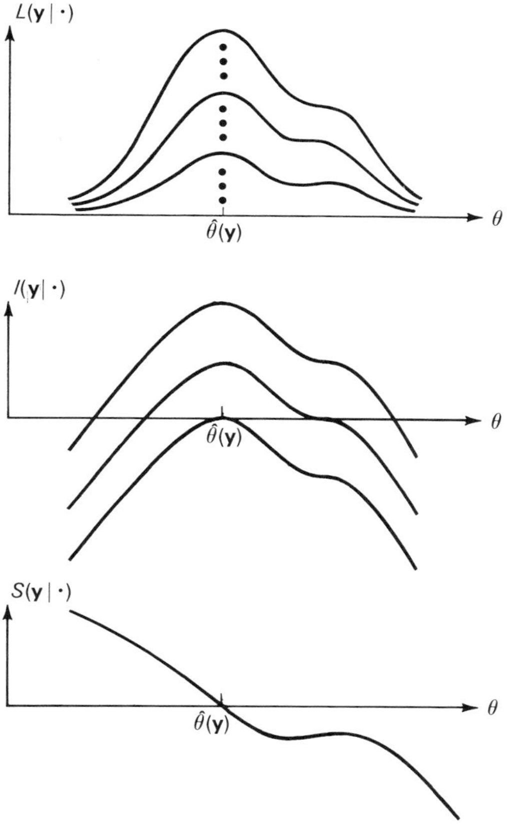

FIGURE 8.3
Likelihood, log-likelihood, and score functions from an observed response $\mathbf{y}$.

choose a representative from the equivalence class. For example, if there is a value θ_Δ for which $f(\mathbf{y}|\theta_\Delta) > 0$ for all possible sample points, then we could use the representative

$$(13) \qquad L_1(\mathbf{y}|\theta) = \frac{f(\mathbf{y}|\theta)}{f(\mathbf{y}|\theta_\Delta)}, \qquad l_1(\mathbf{y}|\theta) = \ln f(\mathbf{y}|\theta) - \ln f(\mathbf{y}|\theta_\Delta);$$

this amounts to choosing the L curve that has height 1 at θ_Δ; see Figure 8.4. Or if each likelihood curve has a finite maximum, we could use

$$(14) \qquad L_2(\mathbf{y}|\theta) = \frac{f(\mathbf{y}|\theta)}{f(\mathbf{y}|\widehat{\theta})}, \qquad l_2(\mathbf{y}|\theta) = \ln f(\mathbf{y}|\theta) - \ln f(\mathbf{y}|\widehat{\theta}),$$

where $\widehat{\theta} = \widehat{\theta}(\mathbf{y})$ is the parameter value at which the likelihood is a maximum; see Figure 8.4. Or if there are only two parameter values, that is, if $\Omega = \{\theta_1, \theta_2\}$, then we could use

$$(15) \qquad L_3(\mathbf{y}) = \frac{f(\mathbf{y}|\theta_2)}{f(\mathbf{y}|\theta_1)}, \qquad l_3(\mathbf{y}) = \ln f(\mathbf{y}|\theta_2) - \ln f(\mathbf{y}|\theta_1),$$

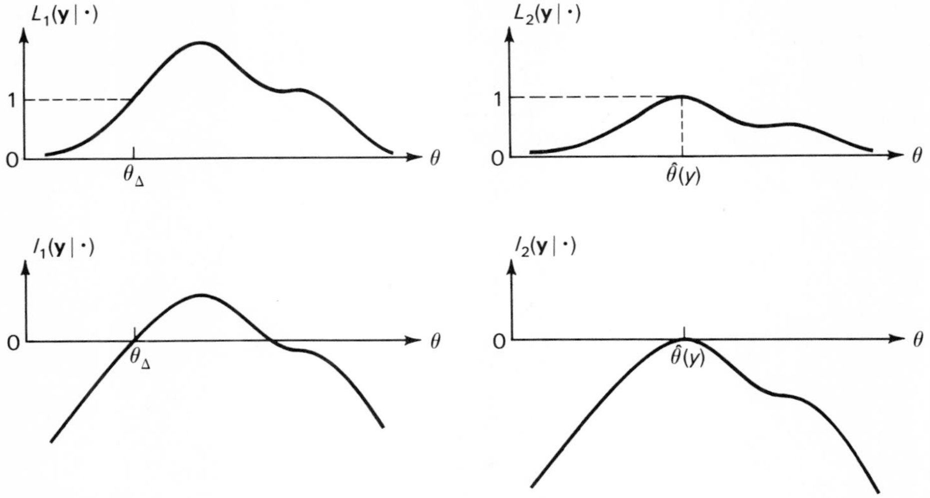

FIGURE 8.4
Likelihood and log-likelihood functions standardized (a) with respect to a value θ_Δ, (b) with respect to the maximum likelihood value $\widehat{\theta}$.

which is called the observed *likelihood ratio* (for θ_2 versus θ_1). Note that $L_3(\mathbf{y})$ is just a real number and hardly needs to be presented graphically.

E EXAMPLES

Consider three examples, two involving the normal distribution and one involving the binomial distribution.

EXAMPLE 1

The location normal: For example, the electrolytic crucibles in Section B. Let $\mathbf{y} = (y_1, \ldots, y_n)'$ be a sample from a normal(θ, σ_0) distribution with θ in $\Omega = R$; note that σ_0 designates a known value for the scaling σ. The likelihood function $L(\mathbf{y}\,|\,\cdot)$ from an observed response $\mathbf{y}$ is given by

$$L(\mathbf{y}\,|\,\theta) = c(2\pi\sigma_0)^{-n/2} \exp\left\{-\frac{1}{2\sigma_0^2}\sum_1^n (y_i - \theta)^2\right\}$$

$$= c \exp\left\{-\frac{1}{2\sigma_0^2}\sum_1^n (y_i - \bar{y})^2 - \frac{n}{2\sigma_0^2}(\bar{y} - \theta)^2\right\}$$

$$= c \exp\left\{-\frac{n}{2\sigma_0^2}(\bar{y} - \theta)^2\right\};$$

and, correspondingly, the log-likelihood $l(\mathbf{y}\,|\,\cdot)$ and score $S(\mathbf{y}\,|\,\cdot)$ functions are given by

$$l(\mathbf{y}\,|\,\theta) = -\frac{n}{2\sigma_0^2}(\bar{y} - \theta)^2 + a,$$

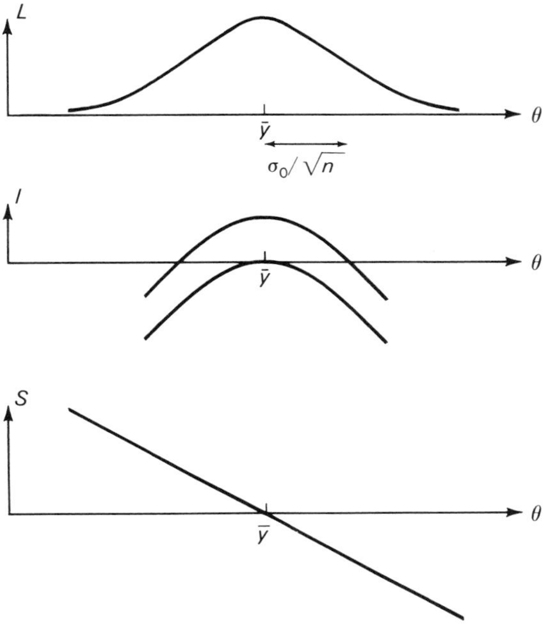

FIGURE 8.5
Functions L, l, and S for a sample from the location normal.

$$S(\mathbf{y}|\theta) = \frac{n}{\sigma_0^2}(\bar{y} - \theta).$$

See Figure 8.5 and note that the likelihood function has normal density shape located at $\bar{y}$ and scaled by $\sigma_0/\sqrt{n}$. Also note that the log-likelihood is quadratic with maximum at $\bar{y}$. And note that the score is linear with slope $-n/\sigma_0^2$. The value of θ that maximizes the likelihood can be obtained by solving

$$S(\mathbf{y}|\theta) = 0;$$

the resulting maximum likelihood estimate MLE for θ is

$$\hat{\theta} = \bar{y}.$$

As a point of particular interest, note that the same likelihood function is obtained from all the sample points $(y_1, \ldots, y_n)$ that have a particular value for the function $\bar{y}$. Thus *the model says the same thing about each of the sample points that have a particular value for the function* $\bar{y}$. Accordingly, for any analysis of observed data with this normal model, it suffices to have just the value of $\bar{y}$ together with the model. In more general situations, both location and scaling of the likelihood function may depend on the response.

EXAMPLE 2

The Bernoulli: Let $\mathbf{x} = (x_1, \ldots, x_n)'$ be a sample of n from the Bernoulli(p) distribution with p in [0, 1]. The likelihood function $L(\mathbf{x}|\cdot)$ from an observed response $\mathbf{x}$ is given by

$$L(\mathbf{x} \mid p) = cp^{\Sigma x_i} q^{n-\Sigma x_i}.$$

Now suppose that $n = 100$ and that from observed data we obtain $\Sigma x_i = 55$. Then $L(\mathbf{x} \mid \cdot)$, $l(\mathbf{x} \mid \cdot)$, and $S(\mathbf{x} \mid \cdot)$ are given by

$$L(\mathbf{x} \mid p) = cp^{55}(1 - p)^{45},$$

$$l(\mathbf{x} \mid p) = 55 \ln p + 45 \ln q + a,$$

$$S(\mathbf{x} \mid p) = \frac{55}{p} - \frac{45}{q}.$$

Note that the likelihood function has beta(56, 46) shape. The maximum likelihood estimate for p can be obtained by solving

$$S(\mathbf{x} \mid p) = \frac{55}{p} - \frac{45}{q} = 0,$$

which gives $\hat{p} = 55/100$; this is the proportion or estimated probability which we have discussed earlier.

As a point of particular interest, note that the same likelihood function is obtained from all the sample points $(x_1, \ldots, x_n)$ that have a particular value of the function Σx_i; thus the model says the same thing about each of the sample points with a particular value of the function Σx_i. Accordingly, for any analysis of observed data with this Bernoulli model, it suffices to have just the value of Σx_i, together with the model.

The arguments used for the definition of the likelihood function do not immediately apply to the case of a discrete distribution such as the Bernoulli. We can, however, always contemplate observing and modeling some additional response characteristic with a continuous distribution—for example, the orientation of a coin in addition to, say, heads or tails. The arguments would then apply and we would obtain the likelihood function as all that the model says about an observed response.

EXAMPLE 3

The location-scale normal: Let $(y_1, \ldots, y_n)$ be a sample from a normal(μ, σ) distribution with $\theta = (\mu, \sigma^2)$ in $R \times R^+$, for example the measurement of current flow in a conductor. The likelihood $L(\mathbf{y} \mid \cdot)$, log-likelihood $l(\mathbf{y} \mid \cdot)$, and score $\mathbf{S}(\mathbf{y} \mid \cdot)$ functions are given by

$$L(\mathbf{y} \mid \mu, \sigma^2) = c(2\pi\sigma^2)^{-n/2}\left\{\exp -\frac{1}{2\sigma^2} \sum (y_i - \mu)^2\right\}$$

$$= c(\sigma^2)^{-n/2} \exp\left\{-\frac{1}{2\sigma^2} \sum (y_i - \bar{y})^2\right\} \exp\left\{-\frac{n}{2\sigma^2}(\bar{y} - \mu)^2\right\};$$

$$l(\mathbf{y} \mid \mu, \sigma^2) = -\frac{n}{2} \ln \sigma^2 - \frac{\sum (y_i - \bar{y})^2}{2\sigma^2} - \frac{(\bar{y} - \mu)^2}{2\sigma^2/n} + a,$$

$$S_1(\mathbf{y} \mid \mu, \sigma^2) = \frac{\partial}{\partial\mu} l(\mathbf{y} \mid \mu, \sigma^2)$$

$$= \frac{n}{\sigma^2}(\bar{y} - \mu),$$

$$S_2(\mathbf{y}\,|\,\mu, \sigma^2) = \frac{\partial}{\partial\sigma^2} l(\mathbf{y}\,|\,\mu, \sigma^2)$$

$$= -\frac{n}{2\sigma^2} + \frac{\sum (y_i - \bar{y})^2 + n(\bar{y} - \mu)^2}{2\sigma^4}.$$

See Figure 8.6 and note that the likelihood function is defined over the upper half-plane $\Omega = R \times R^+$. It is a function that drops to zero at the boundary and rises to a hump at the maximum-likelihood estimate. Also note that the slope can be calculated in the direction μ, giving S_1, and in the direction σ^2, giving S_2. The maximum-likelihood estimate can be obtained by solving

$$S_1(\mathbf{y}\,|\,\mu, \sigma^2) = \frac{n}{\sigma^2}(\bar{y} - \mu) = 0,$$

$$S_2(\mathbf{y}\,|\,\mu, \sigma^2) = -\frac{n}{2\sigma^2} + \frac{\sum (y_i - \bar{y})^2 + n(\bar{y} - \mu)^2}{2\sigma^4} = 0;$$

we obtain

$$\hat{\mu} = \bar{y}, \qquad \hat{\sigma}^2 = \frac{\sum (y_i - \bar{y})^2}{n} = \frac{n-1}{n} s_y^2.$$

As a point of particular interest, note that the same likelihood function is obtained from all the sample points $(y_1, \ldots, y_n)$ that have a particular value for the function $(\bar{y}, s_y^2)$; thus the model says the same things about each of these

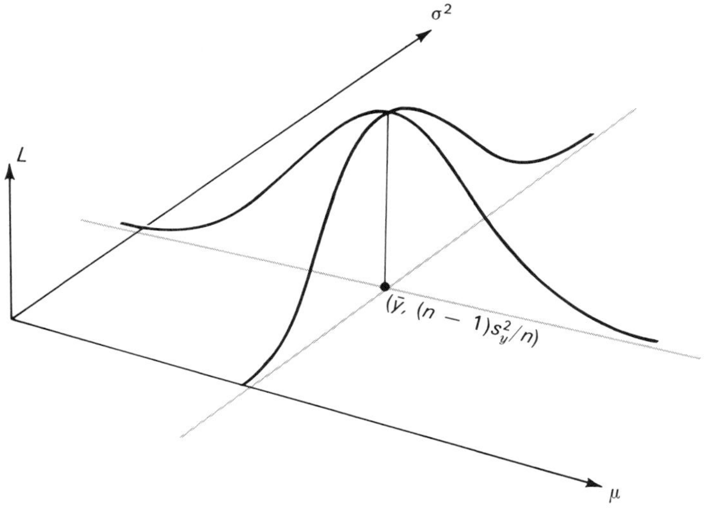

FIGURE 8.6
Likelihood function $L(\mathbf{y}\,|\,\cdot)$ from an observed $\mathbf{y}$; it is a function of (μ, σ^2) on $R \times R^+$.

sample points. Accordingly, for any analysis of observed data with this normal model it suffices to have just the value of $(\bar{y}, s_y^2)$ together with the model.

F EXERCISES

1 A new drug is administered to 100 infected animals. Assume that the responses of the animals are independent; let p be the probability of recovery. For the data in Example 2, sketch the likelihood, log-likelihood, and score functions.

2 The lifetime of a certain transducer is known to be exponential(θ) with θ in $(0, \infty)$. A sample of three yielded $\mathbf{y} = (2.1, 1.3, 2.9)'$.
(a) Determine and sketch the functions L, l, and S.
(b) Calculate the MLE estimate of θ.

3 The number of defects in a 1000-ft spool of cable is Poisson(θ) with θ in $(0, \infty)$. A sample of three yielded $\mathbf{y} = (3, 7, 5)'$.
(a) Determine and sketch the functions L, l, and S.
(b) Calculate the MLE estimate of θ.

4 Let $\mathbf{y} = (2.1, 3.9, 3.4)'$ be a sample of three from the location Laplace distribution with density $f(y|\theta) = \exp\{-|y - \theta|\}/2$.
(a) Determine and sketch the functions L and l.
(b) Calculate the MLE estimate of θ.
(c) Record a different sample with a *different* likelihood function and yet the same MLE estimate of θ.

5 Let $\mathbf{y} = (1.2, 4.7, 0.8, 3.5)'$ be a sample from the uniform$(0, \theta)$ distribution with θ in $(0, \infty)$.
(a) Determine and sketch the functions L and l; allow the value $-\infty$ for l.
(b) Calculate the MLE estimate of θ.

6 Let $\mathbf{y} = (y_1, \ldots, y_n)'$ be a sample from the normal(μ_0, σ) distribution with σ in $(0, \infty)$.
(a) Determine and sketch the functions L, l, and S for the parameter σ^2.
(b) Calculate the MLE estimate of σ^2.
(c) Show that the same likelihood function is obtained from all sample points that have a particular value for $\Sigma (y_i - \mu_0)^2$.

G PROBLEMS

7 Let $\mathbf{y} = (1.2, 4.7, 0.8, 3.5)'$ be a sample from the uniform(θ_1, θ_2) distribution with $-\infty < \theta_1 < \theta_2 < \infty$.
(a) Determine the likelihood function and sketch it.
(b) Find the MLE estimate of (θ_1, θ_2).

8 A certain electrical component is known to have an exponential(θ) lifetime. A sample of 10 was tested for 50 weeks: 3 survived the period; 7 failed during the test period at lifetimes 4, 27, 14, 24, 49, 4, and 33.
(a) Determine the likelihood function and sketch it.
(b) Find the MLE of θ.

9 An offspring in a breeding investigation can be A, B, or C with respective probabilities $(2 + p)/4$, $(1 - p)/2$, or $p/4$. For a sample of n, let a, b, and c be the number of progeny of types A, B, and C. The following data were obtained: $a = 58$, $b = 33$, and $c = 9$.
(a) Determine the likelihood function.
(b) In terms of a, b, and c, determine a quadratic equation for the MLE estimate of p.

8.2

DISTRIBUTION THEORY FOR THE NORMAL MODEL

The normal distribution has a prominent place in statistics. This is due partially to its success in providing a first approximation to the variation found in many applications, and due partially to some exceptional mathematical properties that make many theoretical calculations almost trivial for the normal—calculations that are virtually impossible for other forms for the variation. Indeed, the theory in Chapters 9 and 10 is essentially limited to the normal and certain related distribution forms. We will examine statistical inference for other forms for the variation in Chapter 11.

The exceptional property of the normal distribution is the rotational symmetry of the distribution for a sample $(y_1, \ldots, y_n)$ from a central normal$(0, \sigma)$ distribution. The density function is

$$f(\mathbf{y}) = (2\pi\sigma^2)^{-n/2} \exp\left\{-\frac{1}{2\sigma^2}\sum y_i^2\right\};$$

note that it depends only on the squared distance $r^2 = \Sigma\, y_i^2$ from the origin; see Figure 8.7. Thus if we should use orthogonally rotated axes, the normal distribution would look exactly the same.

In Section 8.1 we considered an observed sample in relation to a normal model with unknown location or unknown location and scale. And we saw that we needed from the data only the value of $\bar{y}$ in the first case and the value of $(\bar{y}, s_y^2)$ in the second case.

In this section we establish two theorems that formally express the rotational symmetry of the normal, and we use these theorems to obtain the marginal distributions for the functions $\bar{y}$ and s_y^2 based on a normal model.

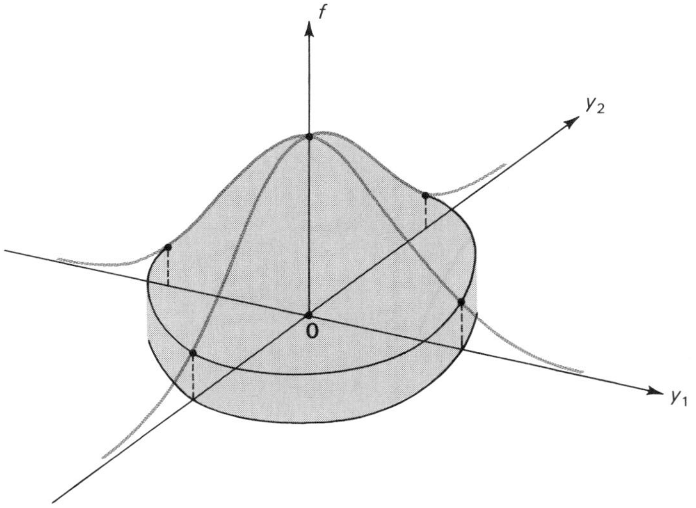

FIGURE 8.7
Distribution of a sample $(y_1, \ldots, y_n)$ from the normal$(0, \sigma)$ is rotationally symmetrical about the origin.

A SYMMETRY: SAMPLE FROM THE NORMAL(0, σ)

A sample $(y_1, \ldots, y_n)$ from the normal(0, σ) distribution has a rotationally symmetric distribution on R^n. Some aspects of this are given by the following theorem.

THEOREM 1

If $\mathbf{y} = (y_1, \ldots, y_n)'$ is a sample from the normal(0, σ) distribution and if C is an $n \times n$ orthogonal matrix, then $\mathbf{u} = C'\mathbf{y}$ also has the distribution of a sample from the normal(0, σ) distribution.

A special case of this theorem was proved as Example 4.2.4. The proof of the theorem is omitted, as it is a special case of Theorem 2 in Section B. We complete this section with a brief discussion of orthogonal transformations.

Consider an $n \times n$ matrix formed by n column vectors in R^n:

(1) $$C = (\mathbf{c}_1, \ldots, \mathbf{c}_n) = \begin{pmatrix} c_{11} & \cdots & c_{1n} \\ \vdots & & \vdots \\ c_{n1} & \cdots & c_{nn} \end{pmatrix}.$$

And suppose that the vectors $\mathbf{c}_1, \ldots, \mathbf{c}_n$ are each of length 1 and mutually orthogonal:

(2) $$(\mathbf{c}_i, \mathbf{c}_i) = \mathbf{c}_i'\mathbf{c}_i = \sum_{\alpha=1}^{n} c_{\alpha i}^2 = 1,$$
$$(\mathbf{c}_i, \mathbf{c}_j) = \mathbf{c}_i'\mathbf{c}_j = \sum_{\alpha=1}^{n} c_{\alpha i}c_{\alpha j} = 0, \qquad i \neq j.$$

We call such a set of n vectors an *orthonormal set*. The relations (2) can be expressed compactly by the following matrix equation:

(3) $$C'C = I,$$

where I is the $n \times n$ identity matrix. Note that $C'C$ is the *inner-product matrix* for the n vectors $\mathbf{c}_1, \ldots, \mathbf{c}_n$; see Supplement Section 2.5B. A matrix C satisfying property (2) or (3) is called an *orthogonal matrix*. Note from (3) that an orthogonal matrix has a very simple inverse—just the transpose

(4) $$C^{-1} = C'.$$

Consider an orthogonal matrix C and suppose that we use C' as a transformation matrix carrying $\mathbf{y}$ in R^n into $\mathbf{u}$ in R^n:

(5) $$\mathbf{u} = C'\mathbf{y},$$

or, equivalently,

(6) $$u_i = \mathbf{c}_i'\mathbf{y} = c_{1i}y_1 + \cdots + c_{ni}y_n.$$

Premultiplication by the matrix C gives

$$C\mathbf{u} = CC'\mathbf{y} = \mathbf{y};$$

thus

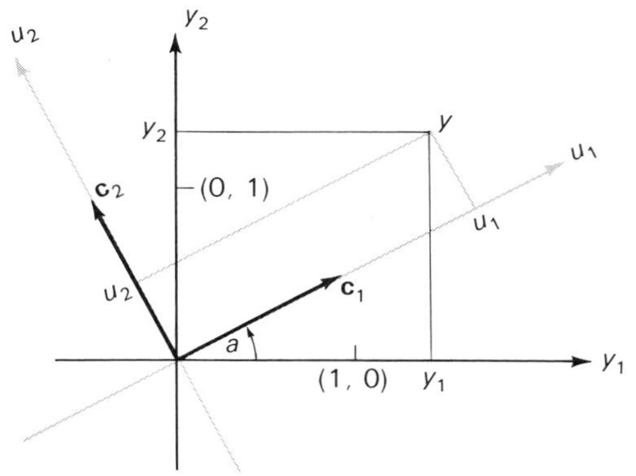

FIGURE 8.8
Coordinates u_1 and u_2 with respect to the new axes $\mathbf{c}_1 = (\cos a, \sin a)'$, $\mathbf{c}_2 = (-\sin a, \cos a)'$.

$$(7) \qquad \mathbf{y} = C\mathbf{u} = u_1\mathbf{c}_1 + \cdots + u_n\mathbf{c}_n.$$

Note that equation (7) expresses $\mathbf{y}$ as a sum of components in the various orthogonal directions $\mathbf{c}_1, \ldots, \mathbf{c}_n$; and that the coordinate u_i in the direction $\mathbf{c}_i$ is obtained as the inner product (6) of $\mathbf{y}$ with the unit vector $\mathbf{c}_i$; see Figure 8.8.

Consider the orthogonal transformation (5). The squared length of $\mathbf{y}$ can be expressed in terms of the new coordinates $u_1, \ldots, u_n$ by using relation (3):

$$(\mathbf{y}, \mathbf{y}) = \sum_1^n y_i^2 = \mathbf{y}'\mathbf{y} = \mathbf{u}'C'C\mathbf{u} = \mathbf{u}'\mathbf{u} = \sum u_i^2 = (\mathbf{u}, \mathbf{u}).$$

We thus have a decomposition of the vector $\mathbf{y}$ and of its squared length $\Sigma\, y_i^2$ according to the n orthogonal directions given by $\mathbf{c}_1, \ldots, \mathbf{c}_n$ (Table 8.1).

TABLE 8.1
Components and lengths of vector $\mathbf{y}$

Subspace	Dimension	Component	Squared Length
$\mathcal{L}(\mathbf{c}_1)$	1	$u_1\mathbf{c}_1$	u_1^2
$\mathcal{L}(\mathbf{c}_2)$	1	$u_2\mathbf{c}_2$	u_2^2
$\vdots$	$\vdots$	$\vdots$	$\vdots$
$\mathcal{L}(\mathbf{c}_n)$	1	$u_n\mathbf{c}_n$	u_n^2
	n	$\mathbf{y}$	$\sum y_i^2$

B SYMMETRY: VARIATION FROM THE NORMAL(0, σ)

Consider independent coordinates $y_1, \ldots, y_n$ and suppose that each coordinate has normal(0, σ) variation. An orthogonal transformation leaves these distribution properties unchanged; for an example, see Figure 8.9.

THEOREM 2

If $y_1, \ldots, y_n$ are statistically independent and y_i is normal(μ_i, σ) and if $\mathbf{u} = C'\mathbf{y}$, where C is an $n \times n$ orthogonal matrix, then $u_1, \ldots, u_n$ are statistically in-

dependent and u_i is normal(ν_i, σ), where the ν_i are given by the same orthogonal transformation, $\boldsymbol{\nu} = C'\boldsymbol{\mu}$.

Proof A change of variable for a probability density is best expressed in terms of a probability differential, the formal component of an integral expression for a probability. The probability differential for **y** is

$$(2\pi\sigma^2)^{-n/2} \exp\left\{-\frac{\sum (y_i - \mu_i)^2}{2\sigma^2}\right\} \prod dy_i = (2\pi\sigma^2)^{-n/2} \exp\left\{-\frac{(\mathbf{y} - \boldsymbol{\mu})'(\mathbf{y} - \boldsymbol{\mu})}{2\sigma^2}\right\} d\mathbf{y}.$$

The effect of the transformation on the volume differential is

$$(8) \qquad d\mathbf{y} = |C|_+ \, d\mathbf{u} = d\mathbf{u};$$

for clearly the Jacobian determinant of an orthogonal transformation is unity: $|C|^2 = |C'C| = |I| = 1$. The effect of the transformation on the squared distance in the exponent is obtained by using $\mathbf{y} = C\mathbf{u}$, $\boldsymbol{\mu} = C\boldsymbol{\nu}$; and thus $\mathbf{y} - \boldsymbol{\mu} = C(\mathbf{u} - \boldsymbol{\nu})$:

$$(9) \qquad (\mathbf{y} - \boldsymbol{\mu})'(\mathbf{y} - \boldsymbol{\mu}) = (\mathbf{u} - \boldsymbol{\nu})'C'C(\mathbf{u} - \boldsymbol{\nu}) = (\mathbf{u} - \boldsymbol{\nu})'(\mathbf{u} - \boldsymbol{\nu}).$$

The probability differential for **u** is then obtained by substituting (8) and (9) in the differential for **y**:

$$(2\pi\sigma^2)^{-n/2} \exp\left\{-\frac{(\mathbf{u} - \boldsymbol{\nu})'(\mathbf{u} - \boldsymbol{\nu})}{2\sigma^2}\right\} d\mathbf{u} = (2\pi\sigma^2)^{-n/2} \exp\left\{-\frac{\sum (u_i - \nu_i)^2}{2\sigma^2}\right\} d\mathbf{u};$$

this is the probability differential for the distribution cited in the theorem.

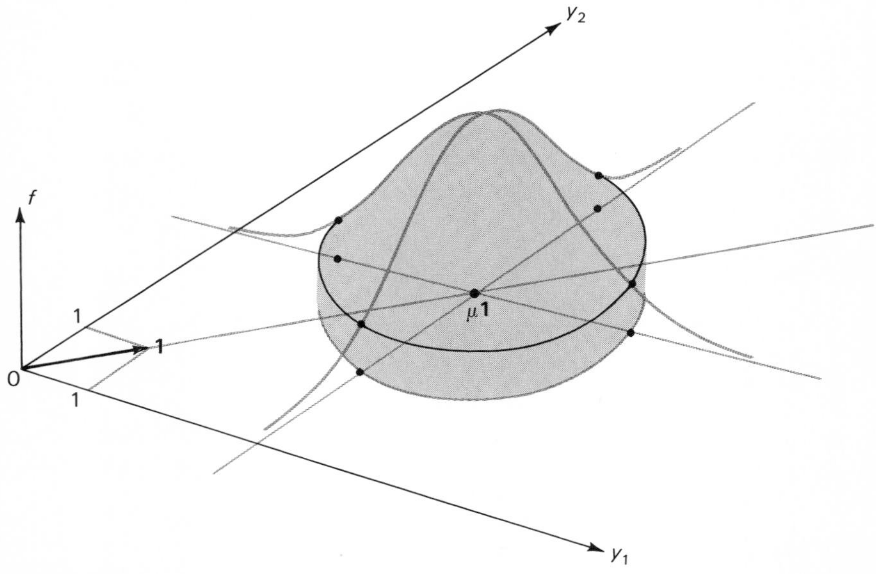

FIGURE 8.9
Distribution of a sample from the normal(μ, σ) distribution is rotationally symmetric about the central point $\mu\mathbf{1}$ on the line $\mathcal{L}(\mathbf{1})$.

Consider the interpretation of this theorem in terms of the orthogonal components discussed in Section A. The component of $\mathbf{y}$ in the direction $\mathbf{c}_i$ is $u_i\mathbf{c}_i$, and the component of $\boldsymbol{\mu}$ in the same direction is $\nu_i\mathbf{c}_i$:

$$u_i = \mathbf{c}_i'\mathbf{y}, \qquad \nu_i = \mathbf{c}_i'\boldsymbol{\mu}.$$

The theorem says that u_i is normally distributed about ν_i with standard deviation σ.

C THE DISTRIBUTION OF $\bar{y}$ AND s_y^2 FOR A NORMAL MODEL

Consider the case of a sample $(y_1, \ldots, y_n)$ from the normal (μ, σ) distribution. From Section 8.1 we know that with unknown location we need only the value of $\bar{y}$ and with unknown location and scale we need only $(\bar{y}, s_y^2)$. Thus $\bar{y}$ and s_y^2 are key functions for sampling from the normal.

In Section 7.1 we saw that the vector $\mathbf{y}$ can be expressed in terms of two components: the projection $\bar{y}\mathbf{1}$ onto the line $\mathcal{L}(\mathbf{1})$, and the projection $\mathbf{y} - \bar{y}\mathbf{1}$ onto the plane $\mathcal{L}^\perp(\mathbf{1})$ [the orthogonal complement of $\mathcal{L}(\mathbf{1})$]; and also that the squared length of $\mathbf{y}$ is the sum of the squared lengths of the two components:

$$\sum y_i^2 = n\bar{y}^2 + \sum (y_i - \bar{y})^2 = n\bar{y}^2 + (n - 1)s_y^2.$$

Recall Figure 7.1. This gives Table 8.2. We now obtain the distribution theory for $\bar{y}$ and s_y^2 by using new axes that align with the subspaces $\mathcal{L}(\mathbf{1})$ and $\mathcal{L}^\perp(\mathbf{1})$.

TABLE 8.2
Components of vector $\mathbf{y}$; corresponding squared lengths

Subspace	Dimension	Component	Squared Length
$\mathcal{L}(\mathbf{1})$	1	$\bar{y}\mathbf{1}$	$n\bar{y}^2$
$\mathcal{L}^\perp(\mathbf{1})$	$n - 1$	$\mathbf{y} - \bar{y}\mathbf{1}$	$\sum (y_i - \bar{y})^2$
	n	$\mathbf{y}$	$\sum y_i^2$

The sample average $\bar{y}$ measures location relative to the 1-vector; accordingly, as a first vector $\mathbf{c}_1$ for an orthogonal matrix C, we choose the unit vector in the direction of the 1-vector:

$$\mathbf{c}_1' = \frac{1}{\sqrt{n}}(1, \ldots, 1) = \left(\frac{1}{\sqrt{n}}, \ldots, \frac{1}{\sqrt{n}}\right).$$

An orthogonal transformation $\mathbf{u} = C'\mathbf{y}$ then has the form

$$(10) \qquad \begin{pmatrix} u_1 \\ u_2 \\ u_3 \\ \vdots \\ u_n \end{pmatrix} = \begin{pmatrix} \frac{1}{\sqrt{n}} & \frac{1}{\sqrt{n}} & \frac{1}{\sqrt{n}} & \cdots & \frac{1}{\sqrt{n}} \\ \frac{1}{\sqrt{2}} & \frac{-1}{\sqrt{2}} & 0 & \cdots & 0 \\ \frac{1}{\sqrt{6}} & \frac{1}{\sqrt{6}} & \frac{-2}{\sqrt{6}} & \ddots & 0 \\ \vdots & & & & \\ \frac{1}{\sqrt{n(n-1)}} & \cdots & & \frac{1}{\sqrt{n(n-1)}} & \frac{-(n-1)}{\sqrt{n(n-1)}} \end{pmatrix} \begin{pmatrix} y_1 \\ y_2 \\ y_3 \\ \vdots \\ y_n \end{pmatrix};$$

the remaining rows of C' are a convenient set of orthonormal vectors that span $\mathcal{L}^{\perp}(\mathbf{1})$.

Now consider the effect of the transformation C' on the components of $\mathbf{y} = \bar{y}\mathbf{1} + (\mathbf{y} - \bar{y}\mathbf{1})$:

$$(11) \qquad C'\begin{pmatrix} \bar{y} \\ \bar{y} \\ \vdots \\ \bar{y} \end{pmatrix} = \begin{pmatrix} \sqrt{n}\bar{y} \\ 0 \\ \vdots \\ 0 \end{pmatrix} = \begin{pmatrix} u_1 \\ 0 \\ \vdots \\ 0 \end{pmatrix}, \qquad C'\begin{pmatrix} y_1 - \bar{y} \\ y_2 - \bar{y} \\ \vdots \\ y_n - \bar{y} \end{pmatrix} = \begin{pmatrix} 0 \\ u_2 \\ \vdots \\ u_n \end{pmatrix};$$

this uses the orthogonality between the first row (representing the 1-vector) of C' and the remaining rows. Correspondingly, for the squared lengths we have

$$(12) \qquad n\bar{y}^2 = u_1^2, \qquad \sum (y_i - \bar{y})^2 = u_2^2 + \cdots + u_n^2.$$

We can now give a simple proof of the following theorem.

THEOREM 3

If $\mathbf{y} = (y_1, \ldots, y_n)$ is a sample from the normal(μ, σ) distribution, then $\bar{y}$ and $\Sigma (y_i - \bar{y})^2$ are statistically independent, $\bar{y}$ is normal$(\mu, \sigma/\sqrt{n})$, and $\Sigma (y_i - \bar{y})^2$ has the distribution of $\sigma^2\chi^2$, where χ^2 has the chi-square$(n - 1)$ distribution.

Proof Consider the orthogonal transformation $\mathbf{u} = C'\mathbf{y}$ as given by formula (10). From Theorem 2 we have that $u_1, \ldots, u_n$ are statistically independent and u_i is normal(ν_i, σ), where

$$(13) \qquad \boldsymbol{\nu} = C'\begin{pmatrix} \mu \\ \mu \\ \vdots \\ \mu \end{pmatrix} = \begin{pmatrix} n\mu/\sqrt{n} \\ 0 \\ \vdots \\ 0 \end{pmatrix} = \begin{pmatrix} \sqrt{n}\mu \\ 0 \\ \vdots \\ 0 \end{pmatrix};$$

the original mean is at $\mu\mathbf{1}$ on the 1-vector; the first new axis is in the direction of the 1-vector; accordingly, in terms of new coordinates, the mean is on the first new axis.

First, we have that $\sqrt{n}\bar{y} = u_1$, which is normal$(\sqrt{n}\mu, \sigma)$; thus $\bar{y}$ is normal$(\mu, \sigma/\sqrt{n})$.

Second, we have that $\Sigma(y_i - \bar{y})^2 = u_2^2 + \cdots + u_n^2$, which is the sum of squares of $n - 1$ normal$(0, \sigma)$ variables. Hence by Problem 3.3.6, or by Problems 5.7.31 and 5.7.33, or by the detailed development in Supplement Section 3.6B, it follows that $\Sigma(y_i - \bar{y})^2/\sigma^2$ has the chi-square$(n - 1)$ distribution or, equivalently, that s_y^2 is distributed as $\sigma^2\chi^2/(n - 1)$, where χ^2 is chi-square$(n - 1)$.

Third, the independence of u_1 and $(u_2, \ldots, u_n)$ implies the independence of $\bar{y}$ and s_y^2 (use Proposition 3.3.5).

We shall use these results for statistical inference in the next section.

D EXERCISES

1 Show that the inner product of two vectors is invariant under an orthogonal transformation.

Let $\mathbf{u}_1$ and $\mathbf{u}_2$ be the vectors obtained by applying the orthogonal transformation $\mathbf{u} = C'\mathbf{y}$ to $\mathbf{y}_1$ and $\mathbf{y}_2$; show that

$$(\mathbf{y}_1, \mathbf{y}_2) = (\mathbf{u}_1, \mathbf{u}_2).$$

2 For a sample of size $n = 2$, express s_y^2 in terms of $(y_1 - y_2)$; use the transformation (10). Let $(y_1, y_2) = (10.5, 10.9)$; calculate (u_1, u_2) as given by (10); use $1/\sqrt{2} = 0.707$.

3 For a sample of size $n = 3$, express s_y^2 in terms of $(y_1 - y_2)$ and $(y_1 + y_2 - 2y_3)$; use the transformation (10). Let $(y_1, y_2, y_3) = (5.50, 5.78, 5.31)$; calculate (u_1, u_2, u_3) as given by (10); use $1/\sqrt{2} = 0.707$, $1/\sqrt{3} = 0.577$, $1/\sqrt{6} = 0.408$.

4 Verify that the matrix C is orthogonal:

$$C' = \begin{pmatrix} 1/\sqrt{5} & 1/\sqrt{5} & 1/\sqrt{5} & 1/\sqrt{5} & 1/\sqrt{5} \\ 2/\sqrt{30} & 2/\sqrt{30} & 2/\sqrt{30} & -3/\sqrt{30} & -3/\sqrt{30} \\ 1/\sqrt{2} & -1/\sqrt{2} & 0 & 0 & 0 \\ 1/\sqrt{6} & 1/\sqrt{6} & -2/\sqrt{6} & 0 & 0 \\ 0 & 0 & 0 & 1/\sqrt{2} & -1/\sqrt{2} \end{pmatrix}.$$

5 Let $\mathbf{y} = (y_{11}, y_{12}, y_{13}, y_{21}, y_{22})'$, $\mathbf{y}_1 = (y_{11}, y_{12}, y_{13})'$, $\mathbf{y}_2 = (y_{21}, y_{22})'$, and $\mathbf{u} = C'\mathbf{y}$. Use the matrix C in Exercise 4 to show that

$$u_1 = \bar{y}/(1/5)^{1/2}, \qquad u_2 = (\bar{y}_1 - \bar{y}_2)/[(1/3) + (1/2)]^{1/2},$$

$$u_3^2 + u_4^2 + u_5^2 = \sum_1^3 (y_{1j} - \bar{y}_1)^2 + \sum_1^2 (y_{2j} - \bar{y}_2)^2.$$

E TWO-SAMPLE PROBLEMS

6 Let $(y_{11}, \ldots, y_{1m})$ be a sample from the normal(μ_1, σ) and $(y_{21}, \ldots, y_{2n})$ be a sample from the normal(μ_2, σ) with (μ_1, μ_2, σ) in $R^2 \times R^+$. Use Theorem 3 to show that $\bar{y}_1$, $\bar{y}_2$, and $S^2 = \Sigma (y_{1j} - \bar{y}_1)^2 + \Sigma (y_{2j} - \bar{y}_2)^2$ are statistically independent, and that $\bar{y}_1$ is normal$(\mu_1, \sigma/\sqrt{m})$, $\bar{y}_2$ is normal$(\mu_2, \sigma/\sqrt{n})$, and S^2 is $\sigma^2\chi^2$, where χ^2 has the chi-square$(m + n - 2)$ distribution; of course, assume independent samples.

7 Verify that the matrix C is orthogonal:

$$C' = \begin{pmatrix} \dfrac{1}{\sqrt{m+n}} & \cdots & \dfrac{1}{\sqrt{m+n}} & \dfrac{1}{\sqrt{m+n}} & \cdots & \dfrac{1}{\sqrt{m+n}} \\ \dfrac{\sqrt{n/m}}{\sqrt{m+n}} & \cdots & \dfrac{\sqrt{n/m}}{\sqrt{m+n}} & \dfrac{-\sqrt{m/n}}{\sqrt{m+n}} & \cdots & \dfrac{-\sqrt{m/n}}{\sqrt{m+n}} \\ & A_m & & & 0 & \\ & 0 & & & A_n & \end{pmatrix},$$

where A_n is the $(n - 1) \times n$ matrix obtained by deleting the first row of the matrix C' in formula (10).

8 Use Theorem 2 and the matrix in Problem 7 to prove the distribution results quoted in Problem 6.

9 Let $(y_{11}, \ldots, y_{1m})$ be a sample from the normal(μ_1, σ_1) and $(y_{21}, \ldots, y_{2n})$ be a sample from the normal(μ_2, σ_2) with $(\mu_1, \mu_2, \sigma_1, \sigma_2)$ in $R^2 \times (0, \infty)^2$. Let s_1^2 and s_2^2 be the respective sample variances. Use Problem 3.2.33 or the distribution results in Supplement Section 3.6B to show that if $\sigma_1^2 = \sigma_2^2$, then

$$F = \frac{s_1^2}{s_2^2}$$

has an $F(m - 1, n - 1)$ distribution and that in general $\sigma_2^2 F/\sigma_1^2$ has an $F(m - 1, n - 1)$ distribution.

8.3

STATISTICAL INFERENCE FOR THE NORMAL MODEL

We have noted that the normal distribution provides an approximation to the variation found in many applications and that the normal has some exceptional mathematical properties. In the preceding section we used these exceptional properties to derive the marginal distributions for the sample average $\bar{y}$ and the sample variance s_y^2.

In this section we use these marginal distribution results to examine inference methods for applications where the normal pattern for variation is appropriate. We then see that simple adjustments to the large-sample methods of Section 7.2 give methods that are exact for small samples with the normal model.

In practice, however, there is often more probability in the tails than is allowed by the normal distribution. The Student family, as indicated in Section 8.1A, can frequently provide a reasonable model for such cases. Inference methods for alternative distributions for variation will be examined in Chapter 11.

A INFERENCE FOR THE MEAN; VARIANCE KNOWN

Consider a random system with a response y known to be normal with known variance σ_0^2; for example, the treatment applied to animals in Example 7.1.2 but now with variance known from a related investigation. As a model for n repetitions, we have a sample $(y_1, \ldots, y_n)$ from the normal(θ, σ_0) distribution with θ in R. We now consider how an observed sample, together with this model, can produce inferences concerning the mean θ.

From Example 8.1.1 we know that it suffices to have just the value of $\bar{y}$ together with the model. And from Theorem 8.2.3 we know that $\bar{y}$ is normal$(\theta, \sigma_0/\sqrt{n})$ or that

$$z = \frac{\bar{y} - \theta}{\sigma_0/\sqrt{n}}$$

is standard normal.

First, we consider the problem of testing the hypothesis $\theta = \theta_0$. For an observed sample $(y_1, \ldots, y_n)$ we can calculate the value $\bar{y}$ and can compare the value with the hypothesis distribution which is normal$(\theta_0, \sigma_0/\sqrt{n})$. Or, equivalently, we can calculate the standardized deviation from θ_0,

$$(1) \qquad z = \frac{\bar{y} - \theta_0}{\sigma_0/\sqrt{n}},$$

and can compare the value with the standard normal distribution; see Appendix II. We then assess the hypothesis accordingly—as we have discussed in Section 7.2B. Thus we see that the large-sample test discussed in Section 7.2B is, in fact, an exact test for any sample size, provided that we have the restricted normal model.

Second, we consider the problem of forming a confidence interval for the location parameter θ. The distribution of the function

$$z = \frac{\bar{y} - \theta}{\sigma_0/\sqrt{n}}$$

is standard normal for each value of θ. Let $(-z_{\alpha/2}, z_{\alpha/2})$ be the central interval containing $1 - \alpha$ probability for the standard normal; then

$$P\left(-z_{\alpha/2} < \frac{\bar{y} - \theta}{\sigma_0/\sqrt{n}} < z_{\alpha/2}\right) = 1 - \alpha.$$

This can be rearranged as in Section 7.2C, giving

(2) $$P(\bar{y} - z_{\alpha/2}\sigma_0/\sqrt{n} < \theta < \bar{y} + z_{\alpha/2}\sigma_0/\sqrt{n}) = 1 - \alpha.$$

Thus $(\bar{y} \pm z_{\alpha/2}\sigma_0/\sqrt{n})$ is an exact $1 - \alpha$ confidence interval for the location θ. Thus we see that the large-sample confidence interval discussed in Section 7.2C is, in fact, an exact confidence interval for any sample size, provided that we have the restricted normal model. Recall the repeated sampling-frequency interpretation discussed in Section 7.2C.

EXAMPLE 1

An instrument for measuring current is unbiased and the error is approximately normal with standard deviation $\sigma = 0.025$. The following measurements were obtained for a constant current θ:

0.672, 0.619, 0.656, 0.651.

Some theory suggests the value $\theta = 0.700$; this can be tested by calculating

$$z = \frac{\bar{y} - 0.700}{0.025/\sqrt{4}} = \frac{0.6495 - 0.700}{0.0125} = -4.04$$

and comparing it with the standard normal; this is an extreme value for the standard normal, and it gives strong evidence against the hypothesis. A 95 percent confidence interval for θ can be obtained from formula (2):

$$\left(\bar{y} \pm 1.96\frac{0.025}{\sqrt{4}}\right) = (0.6495 \pm 0.0245) = (0.6250, 0.6740).$$

B INFERENCE FOR THE VARIANCE

Now consider a random system with a response y known only to have normal variation, for example, the treatment applied to animals in Example 7.1.2. As a model for n repetitions we have a sample $(y_1, \ldots, y_n)$ from the normal(μ, σ) with parameter $\theta = (\mu, \sigma)$ in $R \times R^+$. In this section we examine how an observed sample can be used with this model for inference concerning the variance σ^2.

From Example 8.1.3 we know that it suffices to have just the value of $(\bar{y}, s_y^2)$. And from Theorem 8.2.3 we know that $\bar{y}$ is normal$(\mu, \sigma/\sqrt{n})$, that $(n - 1)s_y^2$ is $\sigma^2\chi^2$, where χ^2 is chi-square$(n - 1)$, and that $\bar{y}$ and s_y^2 are statistically independent. The distribution of $\bar{y}$ has unknown location μ; accordingly, an observed value can give no information concerning the variance σ^2. We thus restrict our attention to s_y^2 for inferences concerning σ^2.

First, we consider the problem of testing the hypothesis $\sigma^2 = \sigma_0^2$. For an observed sample $(y_1, \ldots, y_n)$, we can calculate the value of $(n - 1)s_y^2 = \Sigma (y_i - \bar{y})^2$ and can compare it with the hypothesis distribution, which can be described by $\sigma_0^2\chi^2$. Or, equivalently, we can calculate

$$\text{(3)} \qquad \chi^2 = \frac{\sum (y_i - \bar{y})^2}{\sigma_0^2} = \frac{(n - 1)s_y^2}{\sigma_0^2}$$

and can compare it with the chi-square$(n - 1)$ distribution; see Appendix II. We then assess the hypothesis accordingly—in the pattern we have used before.

Second, we consider the problem of forming a confidence interval for the variance σ^2 or for the standard deviation σ. The distribution of the function

$$\chi^2 = \frac{\sum (y_i - \bar{y})^2}{\sigma^2} = \frac{(n - 1)s_y^2}{\sigma^2}$$

is chi-square$(n - 1)$. Let $(\chi_{1-\alpha/2}^2, \chi_{\alpha/2}^2)$ be a central interval containing $1 - \alpha$ probability for the chi-square$(n - 1)$ distribution; see Appendix II. The chi-square distribution is asymmetric; the interval just suggested is central in a probability sense, $\alpha/2$ probability excluded on each tail of the distribution. We then have

$$P\left(\chi_{1-\alpha/2}^2 < \frac{(n - 1)s_y^2}{\sigma^2} < \chi_{\alpha/2}^2\right) = 1 - \alpha$$

for each value for σ^2. The inequalities can be rearranged [take reciprocals and multiply through by $(n - 1)s_y^2$], giving

$$\text{(4)} \qquad P\left(\frac{(n - 1)s_y^2}{\chi_{\alpha/2}^2} < \sigma^2 < \frac{(n - 1)s_y^2}{\chi_{1-\alpha/2}^2}\right) = 1 - \alpha,$$

and it follows that

$$\text{(5)} \qquad \left(\frac{n - 1}{\chi_{\alpha/2}^2} s_y^2, \; \frac{n - 1}{\chi_{1-\alpha/2}^2} s_y^2\right)$$

is a $1 - \alpha$ confidence interval for the variance, and, of course,

$$\text{(6)} \qquad \left(\frac{(n - 1)^{1/2}}{\chi_{\alpha/2}} s_y, \; \frac{(n - 1)^{1/2}}{\chi_{1-\alpha/2}} s_y\right)$$

is a $1 - \alpha$ confidence interval for the standard deviation.

TABLE 8.3
Factors for s_y^2 and s_y

Factors for s_y^2			Factors for s_y	
Lower	Upper	n	Lower	Upper
0.59	2.09	21	0.77	1.45
0.67	1.64	41	0.82	1.28
0.77	1.35	101	0.88	1.16

There is some interest in seeing the 95 percent factors that must be applied to s_y^2 and s_y for various sample sizes (Table 8.3). Thus, even with a sample of $n = 101$, the factors allow a range of 12 percent low to 16 percent high for the standard deviation σ relative to s_y.

EXAMPLE 2

An instrument for measuring pH is unbiased and the error is approximately normal. A sample of five yielded $\bar{y} = 7.921$, $s_y = 0.017$. A 95 percent confidence interval for σ can be obtained from formula (6) with $n = 5$:

$$\left(\frac{\sqrt{4}}{\sqrt{11.143}} \cdot 0.017, \quad \frac{\sqrt{4}}{\sqrt{0.4844}} \cdot 0.017\right) = (0.010, 0.049);$$

the percentage points are available from Appendix II.

C INFERENCE FOR THE MEAN: VARIANCE UNKNOWN

Consider further a random system with a response y known only to have normal variation. As a model for n repetitions we have a sample $(y_1, \ldots, y_n)$ from the normal(μ, σ) with parameter $\theta = (\mu, \sigma)$ in $R \times R^+$. In this section we examine how an observed sample can be used with this model for inference concerning the mean μ.

From Example 8.1.3 we know that it suffices to have just the value of $(\bar{y}, s_y^2)$. And from the distribution theory recorded in Section B, we know that the deviation $\bar{y} - \mu$ of the average $\bar{y}$ from the true mean μ has a normal$(0, \sigma/\sqrt{n})$ distribution. The adjusted standard deviation $s_y/\sqrt{n}$ provides an estimate of $\sigma/\sqrt{n}$. Accordingly, we consider the deviation $\bar{y} - \mu$ in units of its estimated standard deviation:

$$t = \frac{\bar{y} - \mu}{s_y/\sqrt{n}} = \frac{\sqrt{n}(\bar{y} - \mu)/\sigma}{s_y/\sigma}.$$

Consider the distribution of t: the numerator of the right-hand expression is standard normal; the denominator is $\chi/\sqrt{n-1}$, where χ^2 is chi-square$(n-1)$; the numerator and denominator are independent. By Problem 3.2.35 for the canonical Student or by the details in Supplement Section 3.6B it follows that t has the Student$(n-1)$ distribution. We thus restrict our attention to the function t for inference concerning μ.

First, we consider the problem of testing the hypothesis $\mu = \mu_0$. We can measure the deviation of $\bar{y}$ from μ_0 by calculating

$$(7) \qquad t = \frac{\bar{y} - \mu_0}{s_y/\sqrt{n}};$$

note that if $\mu = \mu_0$, this t is Student$(n-1)$. For an observed sample $(y_1, \ldots, y_n)$, we can calculate the value of t in formula (7) and can compare it with the Student$(n-1)$ distribution; see Appendix II. We then assess the hypothesis accordingly—as we have discussed in Section 7.2E. Thus we see that the large-sample test discussed in Section 7.2E is easily adjusted to give an exact test for any sample size, provided that we have the normal model.

Note that if μ is different from μ_0, then the distribution of t in formula (7) is noncentral Student($n-1$) with noncentrality $\delta = \sqrt{n}(\mu - \mu_0)/\sigma$; see Problem 2 in Supplement Section 3.6.

Second, we consider the problem of forming a confidence interval for the location parameter μ. The distribution of the function

$$t = \frac{\bar{y} - \mu}{s_y/\sqrt{n}} \tag{8}$$

is the Student($n-1$) distribution for each value of $\theta = (\mu, \sigma)$ in $R \times R^+$; it is symmetric about 0. Let $(-t_{\alpha/2}, t_{\alpha/2})$ be the central interval containing $1-\alpha$ probability for the Student($n-1$) distribution; see Appendix II. We then have

$$P\left(-t_{\alpha/2} < \frac{\bar{y} - \mu}{s_y/\sqrt{n}} < t_{\alpha/2}\right) = 1 - \alpha$$

for each value of (μ, σ). The inequalities can be rearranged as in Section 7.2F, giving

$$P\left(\bar{y} - t_{\alpha/2}\frac{s_y}{\sqrt{n}} < \mu < \bar{y} + t_{\alpha/2}\frac{s_y}{\sqrt{n}}\right) = 1 - \alpha,$$

and it follows that $(\bar{y} \pm t_{\alpha/2}s_y/\sqrt{n})$ is a $1-\alpha$ confidence interval for θ.

TABLE 8.4
Comparison of 95 percent confidence intervals

Variance Known	n	Variance Estimated
$\bar{y} \pm 1.96\sigma_0/\sqrt{10}$	10	$\bar{y} \pm 2.26s_y/\sqrt{10}$
$\bar{y} \pm 1.96\sigma_0/\sqrt{20}$	20	$\bar{y} \pm 2.09s_y/\sqrt{20}$

There is some interest in comparing the form of the confidence intervals for the two cases: variance known, and variance unknown. For $n = 10$ and 20, the 95 percent confidence intervals for μ are as shown in Table 8.4. In Section 7.2F we used the factor ± 2 with an estimated standard deviation to form an approximate large-sample confidence interval; and we recommended caution. We now see that even with normal variation and with a moderately large sample, there is need for a factor bigger than 2. From Example 6.1.2 we know that the Student(n) distribution converges to the standard normal as $n \to \infty$. Thus for large samples the Student limits for the confidence interval approximate the normal limits proposed in Section 7.2F.

EXAMPLE 3

Consider further the data and model in Example 2. A 95 percent confidence interval for the mean pH is

$$\left(7.921 \pm 2.776\frac{0.017}{\sqrt{5}}\right) = (7.900, 7.942).$$

D EXERCISES

1 A strength measurement on steel castings has a normal(θ, 2.5) distribution with θ in R. A random sample of 10 castings yields

59.5 61.5 63.5 63.0 64.5

61.5 60.0 65.0 59.5 57.0.

(a) Test the hypothesis $\theta = 58.0$.
(b) Form 95 and 99 percent confidence intervals for θ.

2 A method of measuring temperature remotely has a normal(μ, σ) distribution with (μ, σ) in $R \times R^+$. The following seven determinations were made:

683, 688, 683, 687, 692, 687, 682.

(a) Test the hypothesis $\sigma = 2.0$.
(b) Form a 95 percent confidence interval for σ.

3 (*continuation*) (a) Test the hypothesis $\mu = 680$. (b) Form 95 and 99 percent confidence intervals for μ.

4 The breaking strength for samples of a certain yarn is approximately normal. The following data were obtained:

19.1, 22.8, 21.2, 26.5, 25.6, 20.7, 22.5.

(a) Calculate a 95 percent confidence interval for σ.
(b) Calculate a 95 percent confidence interval for μ.

5 Determinations of viscosity for gasoline are approximately normal. For a particular type of gasoline the following results were obtained for 25 determinations: $\bar{y} = 36.48$, $s_y = 11.37$. Calculate a 99 percent confidence interval for the mean viscosity μ.

6 The lifetime for certain light bulbs is approximately normal. A sample of 10 for a particular type yielded

$\bar{y} = 1495, \qquad s_y = 217.$

(a) Calculate a 95 percent confidence interval for σ.
(b) Calculate a 95 percent confidence interval for the mean lifetime μ.

E TWO-SAMPLE PROBLEMS

7 For samples from the normal(μ_1, σ) and normal(μ_2, σ), the distribution of

$$\frac{S^2}{\sigma^2} = \frac{\sum (y_{1j} - \bar{y}_1)^2 + \sum (y_{2j} - \bar{y}_2)^2}{\sigma^2} = \frac{(m + n - 2)s^2}{\sigma^2}$$

is chi-square($m + n - 2$), as obtained in Problem 8.2.6.
(a) Describe the procedure for testing the hypothesis $\sigma^2 = \sigma_0^2$.
(b) Determine a $1 - \alpha$ confidence region for σ^2.

8 (*continuation*) The lifetime for certain light bulbs is approximately normal with variance independent of the type of such bulbs. Ten bulbs of type *A* and 10 of type *B* yielded the following data:

A	1028	1270	1094	1466	1643	1340	1497	1614	1380	1293
B	1143	1017	1138	1021	1711	1092	1383	1065	1627	1061

Calculate a 95 percent confidence interval for the common variance σ^2.

9 For samples from the normal(μ_1, σ) and normal(μ_2, σ), the distribution of

$$t = \frac{\bar{y}_2 - \bar{y}_1 - (\mu_2 - \mu_1)}{s(1/m + 1/n)^{1/2}}$$

is the Student$(m + n - 2)$ distribution; see Problem 7.
(a) Use Problem 8.2.6 to justify the preceding distribution for t.
(b) Describe the procedure for testing the hypothesis: $\mu_1 = \mu_2$.
(c) Determine a $1 - \alpha$ confidence interval for $\mu_2 - \mu_1$.

10 (*continuation*) Use the data and model from Problem 8 to determine a 95 percent confidence interval for the difference $\mu_B - \mu_A$ in mean lifetimes.

11 Eleven pieces of material were sampled from a lot; five chosen at random were subjected to a first treatment, and the remaining six to a second treatment. The resulting response measurement yields

$$\bar{y}_1 = 0.275, \qquad \bar{y}_2 = 0.293,$$

$$s_1^2 = 0.00045, \qquad s_2^2 = 0.00039.$$

On the basis of the model in Problem 9,
(a) Test the hypothesis $\mu_1 = \mu_2$.
(b) Form a 95 percent confidence interval for $\delta = \mu_2 - \mu_1$.

12 For the model as given in Problem 8.2.9, we have that

$$\frac{\sigma_2^2}{\sigma_1^2}F = \frac{s_1^2/\sigma_1^2}{s_2^2/\sigma_2^2}$$

is the F distribution$(m - 1, n - 1)$.
(a) Describe the procedure for testing the hypothesis $\sigma_1^2 = \sigma_2^2$.
(b) Determine a $1 - \alpha$ confidence interval for $\gamma = \sigma_1^2/\sigma_2^2$.

13 (*continuation*) For the data in Problem 11, test the hypothesis $\sigma_1^2 = \sigma_2^2$.

8.4

AN IMPORTANT STATISTIC

For the normal(θ, σ_0) model we have seen (Example 8.1.1) that each sample point $(y_1, \ldots, y_n)$ that has a particular value for the function $\bar{y}$ gives the same likelihood function; accordingly, as we have noted, the model says the same thing about each of these sample points. As a result we restricted our attention for statistical inference to $\bar{y}$ and its marginal model.

For this example we can see that a particular value for $\bar{y}$ determines the particular likelihood function, and, indeed, a particular likelihood function in return determines the particular value for $\bar{y}$. Thus $\bar{y}$ indexes the possible likelihood functions. Such an essentially unique statistic that indexes the likelihood functions is called the *likelihood statistic*. The likelihood statistic is a fundamental part of the analysis of most statistical models; indeed, it guides the analysis in almost all areas of application.

Of course, for this example the function $\bar{y}$ is the maximum likelihood estimate (MLE). For some more general problems, however, we may need to know *more than* the MLE estimate in order to determine the particular likelihood function. Thus we should not expect generally that a maximum likelihood estimate will serve as the likelihood statistic; recall Exercise 8.1.4.

A THE LIKELIHOOD STATISTIC

First, consider the normal model with known variance σ_0^2. The likelihood function $L(\mathbf{y}|\cdot)$ from a possible sample $\mathbf{y} = (y_1, \ldots, y_n)'$ is given by

$$(1) \qquad L(\mathbf{y}\,|\,\theta) = c \exp\left\{-\frac{n}{2\sigma_0^2}(\bar{y}-\theta)^2\right\};$$

the arbitrary constant c generates the equivalence class of similarly shaped functions of θ. We can eliminate the constant by choosing, say, the representative that is standardized with respect to maximum likelihood:

$$(2) \qquad L_2(\mathbf{y}\,|\,\theta) = \exp\left\{-\frac{n}{2\sigma_0^2}(\bar{y}-\theta)^2\right\}.$$

Note that the likelihood function $L_2(\mathbf{y}\,|\,\cdot)$ has normal shape, is scaled by $\sigma_0/\sqrt{n}$, and has maximum value at the point $\theta = \bar{y}$.

Each sample point $(y_1, \ldots, y_n)$ that has $\bar{y}$ equal to k produces a particular likelihood function located at k, and each point that has $\bar{y}$ equal to a different k' produces a particular likelihood function located at a different point k'; see Figure 8.10. Thus the values for $\bar{y}$ can be put in one-to-one correspondence with the possible likelihood functions; that is, $\bar{y}$ *indexes* the possible likelihood functions. Any other function that indexes the likelihood functions is of course one-to-one equivalent to $\bar{y}$: for example $\Sigma_1^n y_i$, or $\bar{y}^3$, or $\exp\{\bar{y}\}$; recall Section 3.1D. Accordingly, we call $\bar{y}$ the *likelihood statistic* for the normal model with known variance; and indeed we can call $\bar{y}^3$ the likelihood statistic.

Now consider a statistical model on R^n with density function $f(\cdot\,|\,\theta)$ with θ in Ω; we assume some regularity, as discussed in Section 8.1A. The likelihood function from a sample point $\mathbf{y}$ is $L(\mathbf{y}\,|\,\cdot)$, where

$$(3) \qquad L(\mathbf{y}\,|\,\theta) = cf(\mathbf{y}\,|\,\theta);$$

the arbitrary constant c generates the equivalence class of similarly shaped functions of θ.

We now consider a particular likelihood function and we identify the preimage set of points that produce that likelihood function; in a similar way, we consider some other likelihood function and we identify the preimage set of points that produce that likelihood function; see Figure 8.11. Suppose that we can find a function s on R^n that indexes the likelihood functions; then the set of points that produce a particular likelihood function is the set of points for which $s(\mathbf{y})$ takes some value k.

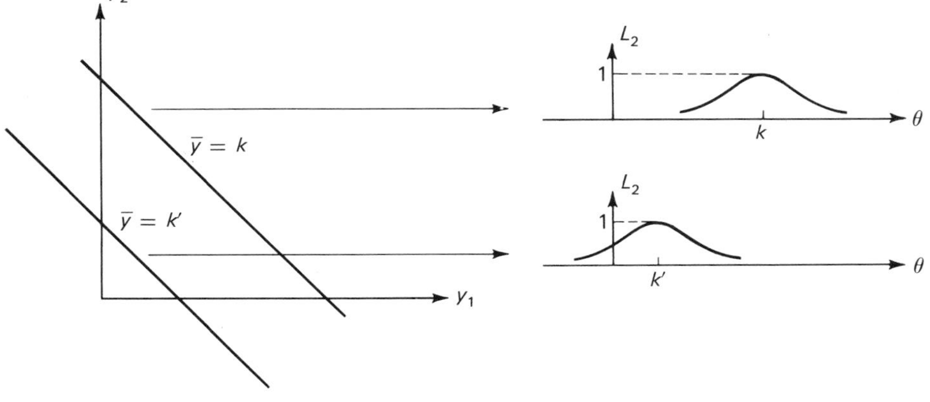

FIGURE 8.10
Likelihood function from points having $\bar{y} = k$; likelihood function from points having $\bar{y} = k'$.

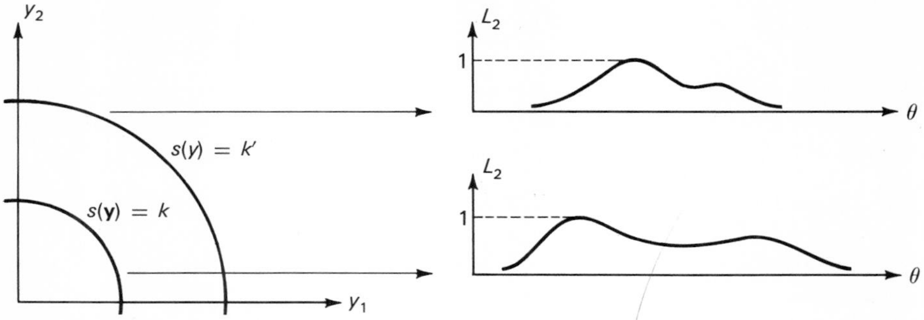

FIGURE 8.11
Likelihood function from points having $s(\mathbf{y}) = k$; likelihood function from points having $s(\mathbf{y}) = k'$.

DEFINITION 1

The **likelihood statistic** is a function on the sample space that indexes the possible likelihood functions; the likelihood statistic is essentially unique up to a one-to-one equivalence.

In conclusion we remark that this concept can be examined from a very general point of view. Let $\mathcal{L}$ be the set of possible likelihood functions. We can view $L(\mathbf{y} \mid \cdot)$ as a function that carries a point $\mathbf{y}$ in R^n to a particular likelihood function in $\mathcal{L}$; call this function the *likelihood map*. The likelihood map itself is an example of the likelihood statistic; in the preceding paragraph, however, we had in mind a simple function such as $s(\mathbf{y}) = \bar{y}$.

B EXAMPLES OF LIKELIHOOD STATISTICS

We now determine the likelihood statistic for several examples.

EXAMPLE 1

The location-scale normal: Let $(y_1, \ldots, y_n)$ be a sample from the normal(μ, σ) distribution with (μ, σ) in $R \times R^+$. From Example 8.1.3 we know that $(\bar{y}, s_y^2)$ indexes the likelihood functions. Thus $s(\mathbf{y}) = (\bar{y}, s_y^2)$ is the likelihood statistic. Note some equivalent forms: $(\bar{y}, \Sigma\,(y_i - \bar{y})^2)$, $(\Sigma\, y_i, \Sigma\, y_i^2)$.

EXAMPLE 2

The Bernoulli: Let $(x_1, \ldots, x_n)$ be a sample from the Bernoulli(p) with p in $[0, 1]$. From Example 8.1.2 we know that $\Sigma\, x_i$ indexes the likelihood functions. Thus $s(\mathbf{x}) = \Sigma\, x_i$ is the likelihood statistic. An equivalent statistic is the sample proportion $\hat{p} = \Sigma\, x_i/n$.

EXAMPLE 3

The uniform: Let $(y_1, \ldots, y_n)$ be a sample from a uniform$(0, \theta)$ distribution with θ in $(0, \infty)$. For notation, let ϕ be the indicator function for the interval $(0, 1)$.

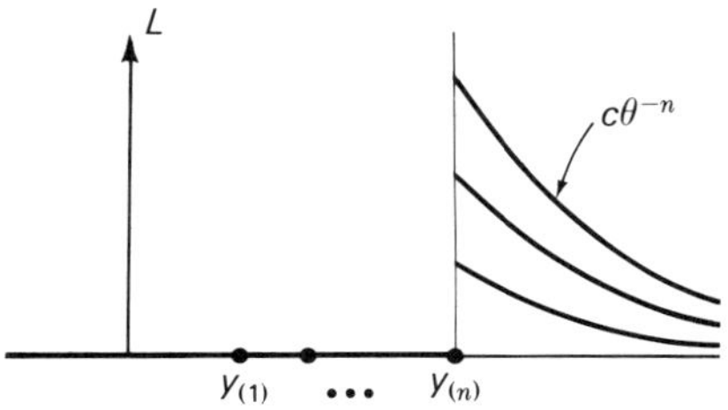

FIGURE 8.12
Likelihood function $L(\mathbf{y} \mid \theta)$ from a sample point $\mathbf{y}$.

Then for a single response we have

$$f(y \mid \theta) = \theta^{-1}\phi(y/\theta),$$

and for a sample of n we have

$$f(\mathbf{y} \mid \theta) = \theta^{-n} \prod_{1}^{n} \phi(y_i/\theta).$$

Thus the likelihood function from a sample point $\mathbf{y}$ is

$$(4) \qquad L(\mathbf{y} \mid \theta) = c\theta^{-n} \prod_{1}^{n} \phi(y_i/\theta).$$

This function has the form $c\theta^{-n}$ whenever the product of the ϕ's is equal to 1, that is, whenever θ is greater than all the y's or whenever θ is greater than $\max y_i = y_{(n)}$; and it is zero for $\theta \le y_{(n)}$. See Figure 8.12. Note that $y_{(n)}$ determines the likelihood function and the likelihood function determines $y_{(n)}$. Thus the function $s(\mathbf{y}) = y_{(n)}$ indexes the likelihood functions; it is the likelihood statistic. It follows that $y_{(n)}$ is all that is needed from an observed response.

EXAMPLE 4

Let $(y_1, \ldots, y_n)$ be a sample from a distribution given by $f(y \mid \theta)$ with θ in $\{\theta_1, \theta_2\}$. The likelihood function can be expressed conveniently by using the representative

$$(5) \qquad L_3(\mathbf{y}) = \frac{\prod_{1}^{n} f(y_i \mid \theta_2)}{\prod_{1}^{n} f(y_i \mid \theta_1)},$$

mentioned in Section 8.1D. For given $\mathbf{y}$ note that $L_3(\mathbf{y})$ is just a real number or $+\infty$ (nonzero numerator, zero denominator); it is called the *likelihood ratio* and it gives the ratio of the probability at $\mathbf{y}$ from θ_2 to that from θ_1. Note that $L_3(\mathbf{y})$ as a function on the sample space R^n maps into R or into $R \cup \{+\infty\}$. Thus $s(\mathbf{y}) = L_3(\mathbf{y})$ indexes the likelihood functions; it is a real-valued likelihood statistic.

C THE LIKELIHOOD STATISTIC WITH INDEPENDENCE

Now consider the likelihood statistic when independent models are combined. We see first that likelihood functions from independent systems combine in a very simple way.

Consider a statistical model for $\mathbf{y}_1$ given by $f'(\mathbf{y}_1|\theta)$ with θ in Ω and a statistical model for an independent $\mathbf{y}_2$ given by $f''(\mathbf{y}_2|\theta)$ with the same θ in Ω. The statistical model for the combined $(\mathbf{y}_1, \mathbf{y}_2)$ is then given by $f'(\mathbf{y}_1|\theta)f''(\mathbf{y}_2|\theta)$ with θ in Ω.

The likelihood functions from the components $\mathbf{y}_1$ and $\mathbf{y}_2$ are

$$L'(\mathbf{y}_1|\theta) = cf'(\mathbf{y}_1|\theta), \qquad L''(\mathbf{y}_2|\theta) = cf''(\mathbf{y}_2|\theta).$$

And the likelihood function from the combined $(\mathbf{y}_1, \mathbf{y}_2)$ is

(6) $$L(\mathbf{y}_1, \mathbf{y}_2|\theta) = cf'(\mathbf{y}_1|\theta)f''(\mathbf{y}_2|\theta) = L'(\mathbf{y}_1|\theta)L''(\mathbf{y}_2|\theta).$$

Thus *when independent models are combined, the likelihood functions are multiplied*. Correspondingly, *the log-likelihoods or the scores are added:*

(7) $$l(\mathbf{y}_1, \mathbf{y}_2|\theta) = l'(\mathbf{y}_1|\theta) + l''(\mathbf{y}_2|\theta),$$

(8) $$S(\mathbf{y}_1, \mathbf{y}_2|\theta) = S'(\mathbf{y}_1|\theta) + S''(\mathbf{y}_2|\theta).$$

Thus when two statistical models are combined, we have the simple formulas (6), (7), and (8) for combining the likelihood, log-likelihood, or score functions. Accordingly, we will almost always determine the likelihood statistic for a combined model directly from properties of the simply obtained likelihood function.

D THE LIKELIHOOD STATISTIC IS SUFFICIENT

We have seen that the likelihood function presents all that the model has to say about an observed response. We have also seen that a range of sample points can have the same likelihood function—that the model is saying the same thing about each of the points in the range. Accordingly, we used $\bar{y}$ with the normal(μ, σ_0) and we used $(\bar{y}, s_y^2)$ with the normal(μ, σ)—on the grounds that the model does not distinguish among the points having a particular value for the likelihood statistic. And, of course, any implications concerning the true parameter value would be the same from each of the points in the range.

We now consider a distribution property of the likelihood statistic, a property that is somewhat weaker than the preceding result. Suppose that we have an investigator who has collected data from a system being investigated and has reported to us the value of the likelihood statistic. With this value in hand we are, of course, free to contemplate the possibilities for the antecedent data, that is, to contemplate the conditional distribution given the value of the likelihood statistic. This conditional distribution cannot involve the parameter θ, for if it did, the model would have to say different things about various possibilities for the antecedent data. We examine this more specifically.

Consider a statistical model on R^n with density function $f(\cdot|\theta)$ with θ in Ω; we assume some regularity, as discussed in Section 8.1A. In the preceding paragraph we mentioned a conditional distribution being independent of a parameter; for this we introduce the following definition.

DEFINITION 2

A function t on the sample space R^n is a **sufficient statistic** if the conditional distribution given t does not depend on the parameter θ.

As support for the term "sufficient," suppose that we have been given the value $t(\mathbf{y}) = t'$ for the sufficient statistic in an application. Then it is reasonable to inquire about the antecedent data. The definition says that the distribution describing possible antecedent data does not involve θ. This suggests that the antecedent data would be of no additional use—that the statistic t is sufficient. Of course, t can be vector-valued.

Consider briefly an example.

EXAMPLE 5

Let (y_1, y_2) be a sample of $n = 2$ from the normal(θ, σ_0) with θ in R. From Section 8.2C we know that

$$u_1 = \frac{y_1 + y_2}{\sqrt{2}}, \qquad u_2 = \frac{y_1 - y_2}{\sqrt{2}}$$

are statistically independent, that u_1 is normal$(\sqrt{2}\theta, \sigma_0)$, and that u_2 is normal$(0, \sigma_0)$. We examine the conditional distribution given u_1: the conditional distribution is described by u_2 and is normal$(0, \sigma_0)$; it does not involve θ. Thus u_1 is sufficient. Of course, we note that $u_1 = \sqrt{2}\bar{y}$ is the likelihood statistic.

Our introductory remarks can now be formalized as the following theorem.

THEOREM 3

The likelihood statistic is sufficient.

Proof Our consideration of conditional distributions in Section 4.2 was limited to cases in which we could find a complementing function u such that there is a one-to-one correspondence, $\mathbf{y} \leftrightarrow (s(\mathbf{y}), u(\mathbf{y}))$; and in the absolutely continuous case, we require a continuous Jacobian each way. We record the proof for this absolutely continuous case; the discrete case is a direct analog without the Jacobians.

The likelihood function $L(\mathbf{y} \mid \cdot)$ is determined by the likelihood statistic $s(\mathbf{y})$. Let $L^*(s \mid \cdot)$ be a particular representative in the class of similarly shaped likelihood functions determined by $s(\mathbf{y}) = s$. Note that $f(\mathbf{y} \mid \theta)$ and $L^*(s(\mathbf{y}) \mid \theta)$ are similarly shaped functions of θ; let $a(\mathbf{y})$ give the proportionality constant,

(9) $$f(\mathbf{y} \mid \theta) = a(\mathbf{y})L^*(s(\mathbf{y}) \mid \theta).$$

Thus the probability differential for $\mathbf{y}$ can be written

(10) $$f(\mathbf{y} \mid \theta)\, d\mathbf{y} = a(\mathbf{y})L^*(s(\mathbf{y}) \mid \theta)\, d\mathbf{y}.$$

The joint distribution for (s, u) can be obtained from Section 2.4E. Let $J(s, u) = |\partial \mathbf{y}/\partial(s, u)|_+$ be the Jacobian determinant. Then substitution in (10) gives the probability differential for (s, u),

$$a(\mathbf{y})L^*(s \mid \theta)J(s, u)\, ds\, du. \tag{11}$$

The marginal distribution for s can be obtained by integrating out the variable u. The variable u occurs only in the factors $a(\mathbf{y})$ and $J(s, u)$; hence we obtain the marginal differential

$$L^*(s \mid \theta)k(s)\, ds, \tag{12}$$

where k is the obvious integral involving $a(\mathbf{y})$ and $J(s, u)$.

The conditional distribution for u given s can be obtained by dividing (11) by (12); we obtain

$$\frac{a(\mathbf{y})J(s, u)}{k(s)}\, du. \tag{13}$$

Note that this conditional distribution does not depend on θ; thus s is a sufficient statistic.

The preceding proof has an immediate corollary:

COROLLARY 4

The likelihood function obtained from an observed $\mathbf{y}$ *is the same as the likelihood function obtained from the likelihood statistic* $s(\mathbf{y})$ *and the marginal model for* s.

Proof The likelihood function from $s(\mathbf{y})$ and its marginal model is available from (12):

$$cL^*(s(\mathbf{y}) \mid \theta)k(s(\mathbf{y})) = cL^*(s(\mathbf{y}) \mid \theta). \tag{14}$$

But this is just the likelihood function from $\mathbf{y}$ as produced by (9).

E THE LIKELIHOOD STATISTIC IS MINIMAL SUFFICIENT

A sufficient statistic has an attractive property: it provides a reduction or simplification on the original response, and yet it records all that seems necessary concerning the original response. Clearly we would be interested in a sufficient statistic that makes the largest reduction—records as little as possible concerning the original response and yet records all that is necessary. For this we present the following definition:

DEFINITION 5

A sufficient statistic s is **minimal** if for any other sufficient statistic t we can find a function h so that $s(\mathbf{y}) = h(t(\mathbf{y}))$.

The definition says that a minimal sufficient statistic is a reduction on any other sufficient statistic. We now have the following property for the likelihood statistic.

THEOREM 6

The likelihood statistic is the minimal sufficient statistic.

Proof Consider a sufficient statistic t. We confine our proof as with Theorem 3 to the case where we have a complementing function u such that there is a one-to-one correspondence $\mathbf{y} \leftrightarrow (t(\mathbf{y}), u(\mathbf{y}))$; and in the absolutely continuous case we require a continuous Jacobian each way. We then obtain

(15) $$f(\mathbf{y}|\theta)\,d\mathbf{y} = f(\mathbf{y}|\theta)J(t, u)\,dt\,du = f_1(u:t)f_2(t|\theta)\,dt\,du,$$

where the middle expression comes by a change of variable and the last expression is an ordinary conditional-times-marginal expression in which the conditional density $f_1(u:t)$, by sufficiency, does not depend on θ. The likelihood function $L(\mathbf{y}|\cdot)$ can now be calculated and we have

(16) $$L(\mathbf{y}|\theta) = cf(\mathbf{y}|\theta) = cf_2(t(\mathbf{y})|\theta),$$

where we have combined $f_1(u:t)$ with the multiplicative constant c. Note that the likelihood function can be calculated from t alone, and thus the likelihood statistic can be expressed as a function of t.

The theorem states in effect that the likelihood statistic is the best sufficient statistic. The concept of a sufficient statistic has been widely used in statistical theory. The concept of the likelihood statistic, however, has recently emerged as a more appropriate concept: The likelihood statistic has the strong justification based on the likelihood function in Section 8.1, a justification that is substantially stronger than the distribution property of a sufficient statistic; the likelihood function itself admits direct interpretation and immediate usefulness in statistical inference; and, besides, the likelihood statistic is the fastest route to the minimal sufficient statistic.

F EXERCISES

1 *Poisson distribution:* Let $(y_1, \ldots, y_n)$ be a sample from the Poisson(λ) distribution with λ in $(0, \infty)$. Show that the likelihood statistic can be given by $\Sigma\, y_i$ or by $\bar{y}$.

2 *Binomial distribution:* Let $(x_1, \ldots, x_n)$ be a sample from the Bernoulli(p) distribution with p in $[0, 1]$. Show that the likelihood statistic can be given by $\Sigma\, x_i$ or by $\bar{x}$.

3 *Scale normal:* Let $(y_1, \ldots, y_n)$ be a sample from the normal(μ_0, σ^2) distribution with σ^2 in $(0, \infty)$. Show that the likelihood statistic is $\Sigma(y_i - \mu_0)^2$.

4 *Uniform* $(\theta, \theta + 1)$: Let $(y_1, \ldots, y_n)$ be a sample from the uniform($\theta, \theta + 1$) distribution with θ in R. Show that the likelihood statistic is $(y_{(1)}, y_{(n)})$.

5 *Scale-exponential:* Let $(y_1, \ldots, y_n)$ be a sample from the exponential(θ) with θ in $(0, \infty)$. Show that the likelihood statistic is $\Sigma\, y_i$.

6 *Scale gamma or chi-square:* Let $(y_1, \ldots, y_n)$ be a sample from the scale gamma having density $f(y|\theta) = \Gamma^{-1}(p)\theta^{-p}y^{p-1}\exp\{-y/\theta\}$ on $(0, \infty)$ and 0 otherwise, with θ in $(0, \infty)$. Show that the likelihood statistic is $\Sigma\, y_i$.

7 *Pareto:* Let $(y_1, \ldots, y_n)$ be a sample from the Pareto distribution $f(y|\theta) = \theta(1 + y)^{-\theta-1}$ on $(0, \infty)$ and 0 otherwise, with θ in $(0, \infty)$. Show that the likelihood statistic is $\Pi\,(1 + y_i)$.

8 *Order statistic:* Let $(y_1, \ldots, y_n)$ be a sample of n from an arbitrary continuous density on R. Determine the conditional distribution of $(y_1, \ldots, y_n)$ given $(y_{(1)}, \ldots, y_{(n)})$; and deduce that the order statistic $(y_{(1)}, \ldots, y_{(n)})$ is sufficient; see Section 3.6E.

G PROBLEMS

9 *Negative binomial distribution:* Let $(x_1, x_2, \ldots)$ be a sample sequence from the Bernoulli(p) distribution with p in (0, 1), and suppose the sequence is terminated when $\Sigma_1^n x_i$ first equals a given k. Show that the likelihood statistic is given by n.

10 *Hypergeometric:* Let $(x_1, \ldots, x_n)$ be a random sample from a finite population of D 0's and $N - D$ 1's with D in $\{0, 1, \ldots, N\}$. Show that the likelihood statistic is Σx_i (the hypergeometric variable).

11 *Multinomial:* Let $(\mathbf{y}_1, \ldots, \mathbf{y}_r)$ be a sample from the multinomial $(n, p_1, \ldots, p_k)$ with $\Sigma p_i = 1, p_i \geq 0$. Show that the likelihood statistic is $\Sigma_1^r \mathbf{y}_j$.

12 *Multivariate hypergeometric:* Let $(x_1, \ldots, x_n)$ be a random sample from a finite population containing Np_1 elements $b_1, \ldots, Np_k$ elements b_k, where the possible $p_1, \ldots, p_k$ satisfy $\Sigma p_i = 1, p_i \geq 0$. Show that the likelihood statistic is $(t_1, \ldots, t_k)$, where t_j is the number of b_j in $\{x_1, \ldots, x_n\}$.

13 *Variable carrier:* Let $(y_1, \ldots, y_n)$ be a sample from $f(y|\theta) = k(\theta)c(y - \theta)h(y)$, where c is the positive indicator function defined in Problem 7.4.8. Show that the likelihood statistic is $y_{(1)}$.

14 *Uniform*(θ_1, θ_2): Let $(y_1, \ldots, y_n)$ be a sample from the uniform distribution (θ_1, θ_2) with $\theta_1 < \theta_2$ and θ_i in R. Show that the likelihood statistic is $(y_{(1)}, y_{(n)})$.

15 *Location-scale exponential:* Let $(y_1, \ldots, y_n)$ be a sample from $f(y|\theta, \tau) = \tau^{-1} \exp\{-(y - \theta)/\tau\}$ on (θ, ∞) and 0 otherwise, with (θ, τ) in $R \times (0, \infty)$. Show that the likelihood statistic is $(y_{(1)}, \Sigma y_i)$.

16 A drug is administered to n animals at each of r doses $x_1, \ldots, x_r$. The "logistic" model is sometimes used for the probability p_i of reaction at dose x_i,

$$p_i = \frac{\alpha\beta^{x_i}}{1 + \alpha\beta^{x_i}},$$

with parameters α and β. Let $y_1, \ldots, y_r$ be the number of reactions, respectively, at doses $x_1, \ldots, x_r$. Show that the likelihood statistic is $(\Sigma y_i, \Sigma x_i y_i)$.

17 Let p_i be the probability of exactly i female children in a family having 3 children, assuming independence of successive births, and let θ be the probability of a female child. For a sample of n families having 3 children, let y_i be the number of families having i female children. Show that $y_1 + 2y_2 + 3y_3$ is the likelihood statistic.

18 (a) Let $(x_1, \ldots, x_n)$ be a sample from the Bernoulli(p) distribution and write $y = \Sigma x_i$. Determine the likelihood function.

(b) Let $x_1, x_2, \ldots$ be statistically independent with the Bernoulli(p) distribution. Let n be the coordinate number at which $\Sigma_1^n x_i = y$, where y is specified. Determine the likelihood function.

(c) If the values of (n, y) in (a) and (b) are the same, show that the likelihood functions are the same.

19 Consider a population containing D 1's and $N - D$ 0's with parameter D in $\Omega = \{0, 1, \ldots, N\}$.

(a) Let y be the number of 1's in a random sample of n. Determine the likelihood function from an observed y.

(b) Suppose that elements are sampled randomly until a specified number y of 1's has been obtained. Let n be the number of elements sampled. Determine the likelihood function from an observed n.

(c) If the values of (n, y) in (a) and (b) are the same, show that the likelihood functions are the same.

20 Suppose that the parameter space Ω is a connected open set in R^r. Then show that the score function $(S_1(y|\cdot), \ldots, S_r(y|\cdot))$ on Ω can be calculated from the log-likelihood, and conversely; the score components are

$$S_i(y|\boldsymbol{\theta}) = \frac{\partial}{\partial\theta_i} \ln f(y|\boldsymbol{\theta}).$$

21 From formula (16) we know that the likelihood function from an observed value of a sufficient statistic as calculated from the marginal model is the same as the likelihood function from the antecedent response as calculated from the full model. Let $(y_1, \ldots, y_n)$ be a sample from a distribution with density $f(y|\theta)$ with θ in $\{\theta_1, \theta_2\}$ and let $L_3(\mathbf{y})$ be the likelihood ratio given in formula (5). Then show that the likelihood function (third form) from the observed $L_3(\mathbf{y})$ is obtained from L_3 by the identity function.

22 Let $(T_1, \ldots, T_n)$ be a sample from the canonical Student(λ) distribution with parameter λ in $(0, \infty)$. Determine the likelihood statistic.

23 Let $(G_1, \ldots, G_n)$ be a sample from the canonical $F(\lambda_1, \lambda_2)$ distribution with parameter(λ_1, λ_2) in $(0, \infty) \times (0, \infty)$. Determine the likelihood statistic.

8.5

MODELS THAT HAVE EXPONENTIAL FORM

We have seen that, for a sample $(y_1, \ldots, y_{100})$ of 100 from the normal(μ, σ) model, it suffices to record the value of $(\bar{y}, s_y)$ for purposes of inference. This is a huge reduction—from a point in R^{100} to a point in R^2. In this section we investigate what kind of functional form for the model makes such reductions possible—in short, what makes the normal work.

Consider this normal example in detail. The density for a single y from the normal(μ, σ) can be expressed in the form

(1) $$f(y|\mu, \sigma) = (2\pi\sigma^2)^{-1/2} \exp\left\{-\frac{\mu^2}{2\sigma^2}\right\} \cdot \exp\left\{y\frac{\mu}{\sigma^2} - y^2\frac{1}{2\sigma^2}\right\}.$$

Note the simple form in the right-hand exponent—a linear combination of statistics and parameters. We obtain the density for a sample of n by adding the exponents:

(2) $$f(\mathbf{y}|\mu, \sigma) = (2\pi\sigma^2)^{-n/2} \exp\left\{-\frac{n\mu^2}{2\sigma^2}\right\} \cdot \exp\left\{\sum y_i \frac{\mu}{\sigma^2} - \sum y_i^2 \frac{1}{2\sigma^2}\right\}.$$

Recall that $(\Sigma y_i, \Sigma y_i^2)$ is the likelihood statistic.

Consider the Bernoulli(p) model. The density for a single x can be expressed in the form

(3) $$f(x|p) = p^x q^{1-x} = q \cdot \exp\{x(\ln p - \ln q)\}.$$

Note the simple form in the exponent—a linear combination of a statistic and a parameter. We obtain the density for a sample of n by adding the exponents:

(4) $$f(\mathbf{x}|p) = p^{\Sigma x_i} q^{n-\Sigma x_i} = q^n \cdot \exp\left\{\sum x_i(\ln p - \ln q)\right\}.$$

Recall that Σx_i is the likelihood statistic.

In this section we investigate statistical models that have this linear form in the exponent, and we see how this form relates to having a simple likelihood statistic regardless of the sample size.

A EXPONENTIAL FORM

The linear properties in the exponent lead to the following definition of exponential form.

DEFINITION 1

A statistical model $\{f(y|\theta) : \theta \in \Omega\}$ has exponential form if the density function can be expressed as

$$f(y|\theta) = \gamma(\theta) \exp \{a_1(y)\psi_1(\theta) + \cdots + a_r(y)\psi_r(\theta)\}h(y). \tag{5}$$

Note in particular that the range of the distribution does *not* depend on θ.

It is quite possible for such a model to be written with more ψ-functions than are really needed. For this the following definition is convenient:

DEFINITION 2

A model having exponential form is in a **reduced form** if the functions $1, \psi_1(\theta), \ldots, \psi_r(\theta)$ (on Ω) are linearly independent.

For the normal(μ, σ) example, it is easy to see that 1, μ/σ^2, and $1/2\sigma^2$ are linearly independent functions on $R \times R^+$. And for the Bernoulli(p) example, it is immediate that 1 and $\ln p - \ln q$ are linearly independent.

A model having exponential form can always be put in a reduced form. For if the ψ-functions are linearly dependent, we can solve for one in terms of the others, we can substitute in the density expression, and we can simplify, thus obtaining exponential form with one less ψ-function. This is repeated until a reduced form is obtained.

B THE LIKELIHOOD STATISTIC FOR A MODEL HAVING EXPONENTIAL FORM

The likelihood statistic is available immediately for a model having reduced exponential form.

THEOREM 3

If an exponential model (5) *is in a reduced form, then* $(a_1(y), \ldots, a_r(y))$ *is the likelihood statistic.*

Proof Consider an exponential model (5) in a reduced form. The log-likelihood $l(y|\cdot)$ from a sample point y has the form

$$l(y|\theta) = \ln \gamma(\theta) + a \cdot 1 + a_1(y)\psi_1(\theta) + \cdots + a_r(y)\psi_r(\theta). \tag{6}$$

Clearly the value of $(a_1(y), \ldots, a_r(y))$ determines the likelihood function; see Figure 8.13.

Now suppose that the values $(a_1', \ldots, a_r')$ and $(a_1'', \ldots, a_r'')$ for $(a_1(y), \ldots, a_r(y))$ both produce the same likelihood function. Then

$$\ln \gamma(\theta) + a' \cdot 1 + a_1'\psi_1(\theta) + \cdots + a_r'\psi_r(\theta)$$
$$= \ln \gamma(\theta) + a'' \cdot 1 + a_1''\psi_1(\theta) + \cdots + a_r''\psi_r(\theta),$$

and hence

$$a \cdot 1 + a_1\psi_1(\theta) + \cdots + a_r\psi_r(\theta) = 0,$$

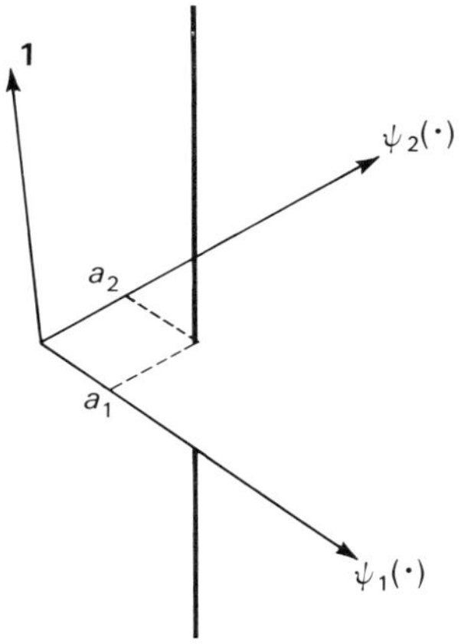

FIGURE 8.13
Adjusted log-likelihood function $l(y|\cdot) - \ln \gamma(\cdot) = a\mathbf{1} + a_1\psi_1(\cdot) + a_2\psi_2(\cdot)$ has coordinates a_1 and a_2 with respect to the functions ψ_1 and ψ_2; the arbitrary value for a produces the line as shown.

where $a = a' - a''$ and $a_i = a_i' - a_i''$. The linear independence then says that the $a_i = 0$ and hence that

$$(a_1', \ldots, a_r') = (a_1'', \ldots, a_r'').$$

Thus the likelihood function determines the value of $(a_1(y), \ldots, a_r(y))$. The two results show that $(a_1(y), \ldots, a_r(y))$ indexes the likelihood functions and is thus the likelihood statistic.

For the normal(μ, σ) example with density function given by (2), this theorem states that $(\Sigma y_i, \Sigma y_i^2)$ is the likelihood statistic. Note that for a sample of $n = 1$, this says that (y, y^2) is the likelihood statistic; of course, this is equivalent to y itself, since y^2 is redundant.

For the Bernoulli(p) example with density function given by (4), this theorem states that Σx_i is the likelihood statistic; this is, of course, equivalent to the proportion or estimated probability $\hat{p} = \Sigma x_i/n$.

C SAMPLING FROM A MODEL THAT HAS EXPONENTIAL FORM

A model that has exponential form has some very convenient properties under sampling. Consider a model for y that has exponential form with density

$$(7) \qquad f(y|\theta) = \gamma(\theta) \exp \{a_1(y)\psi_1(\theta) + \cdots + a_r(y)\psi_r(\theta)\}h(y).$$

Then the model for a sample $(y_1, \ldots, y_n)$ has the density function

$$(8) \qquad f(\mathbf{y}|\theta) = \gamma^n(\theta) \exp \left\{\sum a_1(y_i)\psi_1(\theta) + \cdots + \sum a_r(y_i)\psi_r(\theta)\right\} \prod h(y_i).$$

But this model also has exponential form and uses the same functions $\psi_1(\theta), \ldots, \psi_r(\theta)$.

Now suppose that the model (7) or (8) is in reduced exponential form. Then by

Theorem 3 it follows that $(\Sigma\, a_1(y_i), \ldots, \Sigma\, a_r(y_i))$ is the likelihood statistic for the sample $(y_1, \ldots, y_n)$. This establishes the following theorem:

THEOREM 4

If $(y_1, \ldots, y_n)$ is a sample from a model having reduced exponential form, then $(\Sigma_1^n\, a_1(y_i), \ldots, \Sigma_1^n\, a_r(y_i))$ is the likelihood statistic.

Consider a simple example.

EXAMPLE 1

Let $(y_1, \ldots, y_n)$ be a sample from a normal(μ_0, σ) distribution with σ in R^+. The density function for a single response y can be written in the form

$$f(y|\sigma) = (2\pi\sigma^2)^{-1/2} \exp\left\{-(y - \mu_0)^2 \frac{1}{2\sigma^2}\right\}.$$

This has reduced exponential form with one ψ-function $1/2\sigma^2$. Thus for a sample $(y_1, \ldots, y_n)$ it follows from Theorem 4 that $\Sigma_1^n (y_i - \mu_0)^2$ is the likelihood statistic; compare with Problem 8.4.3.

D WHEN DOES THE LIKELIHOOD STATISTIC HAVE FIXED DIMENSION?

For a sample from the normal(μ, σ) model which has two ψ-functions, we have seen that the likelihood statistic is $(\Sigma\, y_i, \Sigma\, y_i^2)$, which has just two dimensions regardless of how large the sample size is.

We have seen this property with another kind of model. Let $(y_1, \ldots, y_n)$ be a sample from the uniform$(0, \theta)$ distribution. Then by Example 8.4.3 the function $y_{(n)}$ is the likelihood statistic. This function has a single coordinate regardless of how large the sample size is. But note that this model exemplifies a rather different kind of model—a model in which the region of positive density varies with the parameter, a *variable carrier model*. However, such models are appropriate only in very special circumstances.

We now restrict our attention to statistical models that have a fixed region of positive density; that is, the limits for the distribution do not vary with θ. Let $f(y|\theta)$ be the density function relative to the length measure on R, and suppose that $\ln f(y|\theta)$ is continuously differentiable with respect to y on R; note that this implies that the logarithm exists and that f is nonzero on R. We now see that fixed dimension for the likelihood statistic occurs only for a model having exponential form:

THEOREM 5

If $\{f(y|\theta) : \theta \in \Omega\}$ is a statistical model with the regularity conditions in the preceding paragraph, then the likelihood statistic has maximum dimension r under sampling if and only if the model is exponential with r ψ-functions in a minimal reduced form.

The proof is recorded in Section E. Minimal means smallest value for r.

In Chapters 9 and 10 we will see that a large part of statistical theory works only if there is a likelihood statistic with one, two, or several coordinates—but not more than the number of parameter coordinates. By Theorem 5 this property occurs only for models that have exponential form. We of course then want to know what common statistical models have exponential form.

For continuous models we have seen that the normal(μ_0, σ) model and the normal(μ, σ) model have the exponential form just described. For discrete models we have seen that the binomial(p) model and the Poisson(θ) model have the exponential form. All the frequency-count models, however, are basically simple and hardly need the somewhat elaborate concepts that we have been discussing in this chapter.

This brings us back to asking what common *continuous* models have exponential form—other than the normal models we have just mentioned. A common statistical model is concerned with the general level of the response in the case where the distribution for variation is known; the response model is

$$(9)\qquad \{f(y-\theta):\theta \in R\},$$

with f known. It can be shown that such a model has exponential form with one ψ-function if and only if (a) y has a normal distribution ($= a + cz$), or (b) y has the distribution form of the logarithm of a χ^2 variable ($= a + c \ln \chi^2$). In effect, these are just the two normal examples that have a single parameter: the normal(μ, σ_0) with location μ; and a logarithmic transform of the normal(μ_0, σ) with location $\ln \sigma$ (this latter case does, of course, include the scale exponential and scale gamma).

An even more general statistical model is concerned with the general level and the scaling of the response in cases where the form of the variation is known; the response model is

$$(10)\qquad \left\{\frac{1}{\sigma} f\left(\frac{y-\mu}{\sigma}\right):(\mu, \sigma) \in R \times R^+\right\},$$

with f known. It can be shown that such a model has exponential form with two ψ-functions if and only if y has a normal distribution. This is the third of the three normal examples: the normal(μ, σ) with location μ and scale σ.

For other common statistical models we need some linear structure that is described in Chapter 9. For these models, also, it can be shown that exponential form occurs if and only if the distribution is normal.

It thus follows that a large part of the theory in the next two chapters works only for the three familiar normal examples, for the common frequency-count examples, and certain variable carrier examples.

E SUPPLEMENT: PROOF OF THEOREM 5

Consider a density function $f(y|\theta)$ satisfying the regularity conditions preceding Theorem 5. The log-likelihood function for a single response of y can be presented as

$$l(y|\theta) = \ln f(y|\theta) - \ln f(y|\theta_\Delta),$$

where we are using the form l_1 given by formula (8.1.13). The log-likelihood then for a sample $(y_1, \ldots, y_n)$ is

$$(11)\qquad l_n(\mathbf{y}|\theta) = l(y_1|\theta) + \cdots + l(y_n|\theta).$$

We examine the log-likelihood function for points in the neighborhood of a point $\mathbf{y}$:

$$(12) \qquad l_n(\mathbf{y} + \mathbf{t}|\theta) = l_n(\mathbf{y}|\theta) + \sum_1^n t_i \frac{\partial}{\partial y_i} l(y_i|\theta) + o(\mathbf{t}),$$

where $o(\mathbf{t})$ is a function that goes to zero as a proportion of $|\mathbf{t}|$; note that the derivative $\partial/\partial y_i$ applied to (11) catches only the term $l(y_i|\theta)$ on the right side. See Figure 8.14.

Now suppose that the likelihood statistic has maximum dimension r and that $\mathbf{y}$ is a point at which this dimension r is attained. The possible changes in the log-likelihood function as determined near $\mathbf{y}$ are given by

$$(13) \qquad \sum_1^n t_i \frac{\partial}{\partial y_i} l(y_i|\cdot),$$

a linear combination of n functions. But the likelihood statistic has maximum dimension r; hence the functions produced by formula (13) must have a basis consisting of just r functions, say $\psi_1(\cdot), \ldots, \psi_r(\cdot)$; accordingly, all the functions (13) can be generated as linear combinations of these r ψ's.

Now consider a sample $(y_1, \ldots, y_n, y)$ of $n + 1$; let $(y_1, \ldots, y_n)$ be the point described above and let y be arbitrary. Then, as above, the changes in the log-likelihood function as determined near this new point are given by

$$\sum_1^n t_i \frac{\partial}{\partial y_i} l(y_i|\cdot) + t \frac{\partial}{\partial y} l(y|\cdot).$$

But the likelihood statistic cannot have dimension greater than r. It has dimension r on the basis of just the first terms; hence the function

$$\frac{\partial}{\partial y} l(y|\cdot)$$

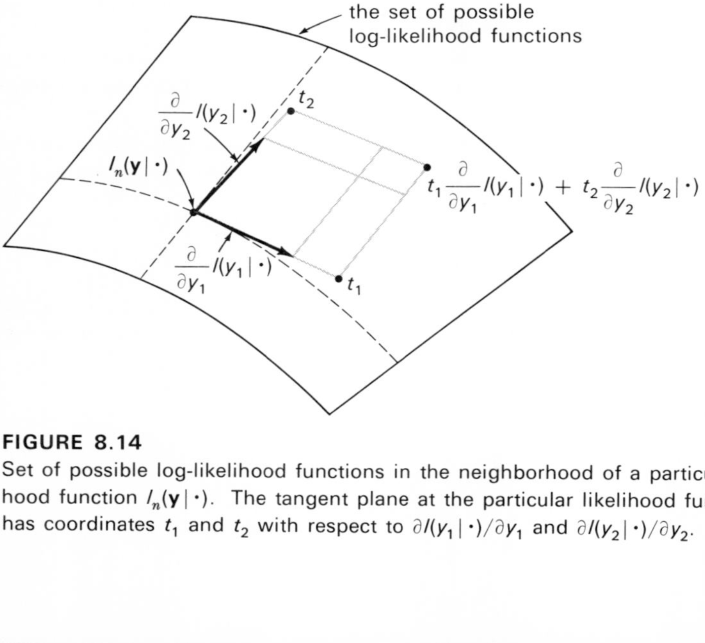

FIGURE 8.14

Set of possible log-likelihood functions in the neighborhood of a particular likelihood function $l_n(\mathbf{y}|\cdot)$. The tangent plane at the particular likelihood function has coordinates t_1 and t_2 with respect to $\partial l(y_1|\cdot)/\partial y_1$ and $\partial l(y_2|\cdot)/\partial y_2$.

must be one of those generated by the r ψ-functions.

The preceding paragraph shows that

$$\frac{\partial}{\partial y} l(y \mid \theta) = A_1(y)\psi_1(\theta) + \cdots + A_r(y)\psi_r(\theta).$$

Integration then gives

$$\ln \frac{f(y \mid \theta)}{f(y \mid \theta_\Delta)} = l(y \mid \theta) = a_1(y)\psi_1(\theta) + \cdots + a_r(y)\psi(\theta) + \phi(\theta).$$

Hence $f(y \mid \theta)$ has exponential form based on r ψ-functions.

The converse follows from Theorem 3 applied to a reduced exponential model with minimum value for r.

F EXERCISES

1 *Binomial distribution:* Let $(x_1, \ldots, x_n)$ be a sample from the Bernoulli(p) distribution. Use Theorem 4 to determine the likelihood statistic.

2 *Poisson distribution:* Let $(y_1, \ldots, y_n)$ be a sample from the Poisson(θ) distribution. Use Theorem 4 to determine the likelihood statistic.

3 *Scale gamma:* Let $(y_1, \ldots, y_n)$ be a sample from the scale gamma in Exercise 8.4.6. Use Theorem 4 to determine the likelihood statistic.

4 *Pareto:* Show that the Pareto distribution in Exercise 8.4.7 has exponential form. Use Theorem 4 to determine the likelihood statistic.

G PROBLEMS

5 *Multinomial:* Show that the multinomial($n, p_1, \ldots, p_k$) distribution with $\Sigma p_i = 1$ and $p_i \geq 0$ has exponential form. Determine the ψ-functions and for a sample verify the likehood statistic in Problem 8.4.11.

6 *Beta:* Let $(u_1, \ldots, u_n)$ be a sample from the beta(p, q) distribution with (p, q) in $R^+ \times R^+$. Show that the model has exponential form. Determine the ψ-functions and record a reasonable likelihood statistic.

7 *Normal linear model:* Consider a response vector $\mathbf{y}$ in R^n having a distribution with mean located in an r-dimensional subspace $\mathcal{L}(\mathbf{x}_1, \ldots, \mathbf{x}_r)$ and with variation $\sigma\mathbf{z}$, where $\mathbf{z}$ is a sample from the standard normal; that is, $\mathbf{y} = X\boldsymbol{\beta} + \sigma\mathbf{z}$, where $X = (\mathbf{x}_1, \ldots, \mathbf{x}_r)$ is an $n \times r$ matrix of rank r and the parameter $(\boldsymbol{\beta}, \sigma)$ is in $R^r \times R^+$. Show that the response density function is

$$f(\mathbf{y} \mid \boldsymbol{\beta}, \sigma) = (2\pi\sigma^2)^{-n/2} \exp\left\{-\frac{1}{2\sigma^2}(\mathbf{y} - X\boldsymbol{\beta})'(\mathbf{y} - X\boldsymbol{\beta})\right\}$$

and that it has exponential form. Determine a set of ψ-functions and show that $(\mathbf{y}'\mathbf{y}, \mathbf{y}'\mathbf{x}, \ldots, \mathbf{y}'\mathbf{x}_r) = (\mathbf{y}'\mathbf{y}, \mathbf{y}'X)$ is a likelihood statistic.

8 *Multivariate normal:* Consider the multivariate normal $(\boldsymbol{\mu}, \Sigma)$ with $\boldsymbol{\mu}$ in R^k and with essential coordinates of Σ in a region of $R^{k(k+1)/2}$ corresponding to positive definite matrices.

(a) Show that the model has exponential form. Determine the ψ-functions.

(b) Let $\mathbf{y}_1, \ldots, \mathbf{y}_n$ be a sample of n from the multivariate normal$(\boldsymbol{\mu}, \Sigma)$ as described. Show that $(\Sigma_1^n \mathbf{y}_i, \Sigma_1^n \mathbf{y}_i\mathbf{y}_i')$ is the likelihood statistic; it has essentially $k + k(k+1)/2$ coordinates.

8.6

INFERENCE FROM LARGE SAMPLES USING LIKELIHOOD

We have seen with a parametric model that the likelihood function from an observed response presents all that the model has to say about the observed response. In Section 8.1C we discussed some inference methods based on a direct assessment of the observed likelihood function.

We have also seen how to group together sample points that produce the same likelihood function. This led us in Section 8.4 to the definition of the likelihood statistic, a function that gives an important reduction on the sample space.

In this section we examine these two concepts together: for any contemplated parameter value we consider the distribution of possible likelihood functions obtained from samples. This *statistical model for possible likelihood functions* is central to our present examination of inference for parametric models. Recall the observed likelihood function for the electrolytic crucibles in Section 8.1B; it is natural to contemplate the likelihood functions that we might have obtained given any particular distribution for crucible life.

For a sample from the normal(θ, σ_0) model, the distribution for possible likelihood functions is easily derived. We examine this in detail in Section A.

For a sample from some other model, the distribution for possible likelihood functions may be complicated to derive or awkward to express. Fortunately, however, for large samples the distribution is easily obtained and is almost a copy of that for the normal(θ, σ_0) model. We examine this in Sections B and C.

We then examine some large-sample inference methods in Sections D and E. Our large-sample inference methods in Section 7.2 were developed for the mean of an essentially arbitrary distribution. The present methods provide the essential inference procedures for the general parameter of a parametric model. Indeed, they provide some basic procedures that work when other methods fail.

A THE NORMAL WITH KNOWN VARIANCE

Let $(y_1, \ldots, y_n)$ be a sample from a normal(θ, σ_0) distribution with θ in R. The likelihood, log-likelihood, and score functions are given by

$$L_n(\mathbf{y}\,|\,\theta) = c \exp\left\{-\frac{1}{2\sigma_0^2/n}(\bar{y}-\theta)^2\right\},$$

$$l_n(\mathbf{y}\,|\,\theta) = -\frac{1}{2\sigma_0^2/n}(\bar{y}-\theta)^2 + a,$$

$$S_n(\mathbf{y}\,|\,\theta) = \frac{\bar{y}-\theta}{\sigma_0^2/n};$$

the subscript n is used to indicate dependence on the sample size.

We are now going to consider the distribution of possible likelihood functions. For this we face a small notational problem. On the one hand, we have used θ for a parameter value that gives the distribution for the sample $(y_1, \ldots, y_n)$. And on the other hand, we have used θ as a free variable when examining the likelihood function from an observed response. Now that we are going to examine the distribution of possible likelihood functions, we must face these two roles for θ simultaneously. Accordingly, we will adopt distinguishing notations *where needed*. For the present section we will use θ for the free variable when examining the likelihood function. And then when we need to distinguish the parameter value that gives the distribution for the sample, we will use θ_* and refer to it as the *true value* of the parameter (of course, we can still use θ_* as a free variable when studying response distributions under various conditions).

We also face a small problem connected with the arbitrary constant in the expression for a likelihood function. The direct way of eliminating the constant is to choose a representative or standardize the likelihood function in some convenient way. We will do this, but we will do it in a very special and convenient way: we will standardize with respect to the true value θ_*. In an application we would not be able to do this, but for theoretical discussions here it is very convenient—and besides it is only a device for examining likelihood ratios or log-likelihood differences. Accordingly, in the present context we take the log-likelihood function to be

$$(1) \qquad l_n(\mathbf{y}|\theta) = -\frac{n/\sigma_0^2}{2}(\bar{y}-\theta)^2 + a = -\frac{n/\sigma_0^2}{2}(\bar{y}-\theta)^2 + \frac{n/\sigma_0^2}{2}(\bar{y}-\theta_*)^2$$
$$= -\frac{1/\sigma_0^2}{2}(\hat{\delta}-\delta)^2 + \frac{1/\sigma_0^2}{2}\hat{\delta}^2.$$

This log-likelihood function and the corresponding score function are plotted in Figure 8.15; note the adjusted parameter $\delta = \sqrt{n}(\theta - \theta_*)$ and the adjusted variable $\hat{\delta} = \sqrt{n}(\bar{y} - \theta_*)$.

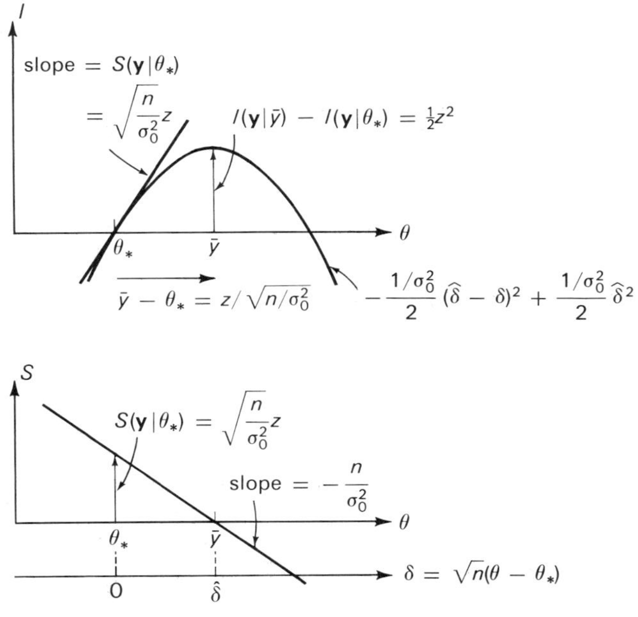

FIGURE 8.15
Normal(θ, σ_0). Distribution of the log-likelihood and score in terms of a standard normal variable z. Note that $\delta = \sqrt{n}(\theta - \theta_*)$ and $\hat{\delta} = \sqrt{n}(\bar{y} - \theta_*)$.

The distribution theory in Section 8.2 gives immediately the distribution theory for the log-likelihood function and for the score since both are functions of the likelihood statistic $\bar{y}$. The distribution of the likelihood statistic $\bar{y}$ is normal$(\theta_*, \sigma_0/\sqrt{n})$; thus we write $\bar{y} = \theta_* + (\sigma_0/\sqrt{n})z$ or $\hat{\delta} = \sigma_0 z$ in terms of a standard normal z. Now first we note that the distribution of the log-likelihood function (1) is given by its dependence on $\bar{y}$ or on $\hat{\delta}$. Then we note the distribution of several salient characteristics: the deviation from θ_* to the point of maximum log-likelihood is

(2) $$\bar{y} - \theta_* = \frac{\sigma_0}{\sqrt{n}} z, \qquad \text{or, as adjusted,} \qquad \hat{\delta} = \sigma_0 z;$$

the slope of the log-likelihood at the true θ_* is

(3) $$S_n(\mathbf{y}\,|\,\theta_*) = \frac{\bar{y} - \theta_*}{\sigma_0^2/n} = \frac{\sqrt{n}}{\sigma_0} z;$$

and the height to the maximum log-likelihood is

(4) $$l_n(\mathbf{y}\,|\,\bar{y}) - l_n(\mathbf{y}\,|\,\theta_*) = \frac{1}{2\sigma_0^2/n}(\bar{y} - \theta_*)^2 = (1/2)z^2;$$

see Figure 8.15. For large samples we will see that these same distribution properties are obtained with the general parametric model.

B THE DISTRIBUTION OF THE SCORE: MEANS AND VARIANCES

Consider a statistical model on a sample space $\mathcal{S}$ with density function $f(y\,|\,\theta)$ with θ in $\Omega = R$. We assume that $\ln f(y\,|\,\theta)$ can be differentiated several times with respect to θ and that this differentiation can be carried through a sample-space integration sign; in particular, this excludes variable-carrier models. The precise assumptions are recorded at the end of this section. Note the following expression for the derivative of the density function:

(5) $$\frac{\partial}{\partial\theta} f(y\,|\,\theta) = S(y\,|\,\theta) f(y\,|\,\theta).$$

First, we differentiate the integral

$$\int f(y\,|\,\theta)\, dy = 1$$

with respect to θ:

(6) $$\int \frac{\partial}{\partial\theta} f(y\,|\,\theta)\, dy = \int S(y\,|\,\theta) \cdot f(y\,|\,\theta)\, dy = 0.$$

Thus we obtain

(7) $$E(S(y\,|\,\theta)\,|\,\theta) = 0,$$

which says that *the mean value of the score function at the true value is zero,* or the mean slope of the log-likelihood at the true value is zero. This holds for all values for the parameter θ.

Consider the notation $E(h(y)\,|\,\theta)$, which designates the mean value of $h(y)$ calculated using the distribution with parameter value θ. The expression $E(S(y\,|\,\theta)\,|\,\theta)$ is the mean value of $h(y) = S(y\,|\,\theta)$, the score calculated at the true parameter value θ. Note here that the parameter argument for the score function is the *same* as the parameter of the distribution; thus we have no need for the specialized notation.

Second, we differentiate a second time working from the middle expression in formula (6) and again use the adjustment (5) in terms of the logarithmic derivative:

$$\int \frac{\partial}{\partial\theta} S(y|\theta) \cdot f(y|\theta)\, dy + \int S(y|\theta) \frac{\partial}{\partial\theta} f(y|\theta)\, dy = 0,$$

$$(8) \qquad \int S'(y|\theta) f(y|\theta)\, dy + \int S^2(y|\theta) f(y|\theta)\, dy = 0,$$

where the prime denotes a derivative with respect to θ. Thus using (7) we obtain

$$(9) \qquad \text{var}\,(S(y|\theta)|\theta) = E(S^2(y|\theta)|\theta) = -E(S'(y|\theta)|\theta),$$

which says that *the variance of the score at the true value is equal to the negative mean of the derivative of the score at the true value*—or to the negative mean of the second derivative of the log-likelihood at the true value. We write this

$$(10) \qquad \text{var}\,(S(y|\theta)|\theta) = -E(S'(y|\theta)|\theta) = j(\theta),$$

where $j(\theta)$ is called the *Fisher information* as obtained from a single response. The significance of the term "information" will appear with results later in this section.

From a single y from the normal(θ, σ_0), we have

$$S(y|\theta) = \frac{y - \theta}{\sigma_0^2}, \qquad j(\theta) = \text{var}\,[S(y|\theta)|\theta] = \frac{\sigma_0^2}{\sigma_0^4} = \frac{1}{\sigma_0^2}.$$

Third, we consider a sample $(y_1, \ldots, y_n)$ from the distribution given by $f(y|\theta)$. Then

$$\ln f(\mathbf{y}|\theta) = \ln f(y_1|\theta) + \cdots + \ln f(y_n|\theta),$$

$$(11) \qquad S_n(\mathbf{y}|\theta) = S(y_1|\theta) + \cdots + S(y_n|\theta),$$

$$S'_n(\mathbf{y}|\theta) = S'(y_1|\theta) + \cdots + S'(y_n|\theta).$$

We now examine mean values for the score and the score derivative at the true value θ. The terms on the right side of (11) are statistically independent; thus we obtain

$$(12) \qquad E(S_n(\mathbf{y}|\theta)|\theta) = 0,$$

$$(13) \qquad \text{var}\,(S_n(\mathbf{y}|\theta)|\theta) = J(\theta) = nj(\theta) = -E(S'_n(\mathbf{y}|\theta)|\theta),$$

where $J(\theta) = nj(\theta)$ is the Fisher information as obtained from a sample of n.

For a sample $(y_1, \ldots, y_n)$ from the normal(θ, σ_0), we have

$$S_n(\mathbf{y}|\theta) = \frac{\bar{y} - \theta}{\sigma_0^2/n}, \qquad J(\theta) = \frac{\sigma_0^2/n}{\sigma_0^4/n^2} = \frac{n}{\sigma_0^2}.$$

The following assumption supports the preceding analysis.

ASSUMPTION 1

The derivatives of the density function satisfy $|\partial f(y|\theta)/\partial\theta| < M_1(y)$, *and* $|\partial^2 f(y|\theta)/\partial\theta^2| < M_2(y)$, *where* $M_1(y)$, $M_2(y)$ *and*

$$\frac{\partial^2}{\partial\theta^2} \ln f(y|\theta) \cdot f(y|\theta)$$

are integrable on the sample space $\mathcal{S}$.

C THE DISTRIBUTION OF THE LARGE-SAMPLE LIKELIHOOD FUNCTION

Consider a statistical model with density $f(y|\theta)$ as discussed in the preceding section. Let $(y_1, \ldots, y_n)$ be a sample of n. We now examine the distribution of the large-sample likelihood function. We do this in terms of log-likelihood and as before we use θ_* to refer to the true value of the parameter. The distribution is presented in the following theorem based on Assumption 1 and on an Assumption 3 recorded with the proof in Section F.

THEOREM 2

As $n \to \infty$, the log-likelihood has a limiting quadratic shape

$$l_n(\mathbf{y}|\theta) = -\frac{nj(\theta_*)}{2}(\hat{\theta}(\mathbf{y}) - \theta)^2 + a = -\frac{j(\theta_*)}{2}(\hat{\delta} - \delta)^2 + \frac{j(\theta_*)}{2}\hat{\delta}^2;$$

this is the quadratic form (1) found in the normal$(0, \sigma_0)$ case but with the factor n/σ_0^2 replaced by $nj(\theta_*)$. Note the adjusted parameter $\delta = \sqrt{n}(\theta - \theta_*)$ and adjusted variable $\hat{\delta} = \sqrt{n}(\hat{\theta}(\mathbf{y}) - \theta_*)$.

The scaled deviation $\hat{\delta} = \sqrt{n}(\hat{\theta}(\mathbf{y}) - \theta_*)$ of the maximum likelihood estimate from the true value has a limiting central normal distribution with variance $1/j(\theta_*)$. The scaled slope $S_n(\mathbf{y}|\theta)/\sqrt{n}$ of the log-likelihood at the true value has a limiting central normal distribution with variance $j(\theta_*)$. The height $l_n(\mathbf{y}|\hat{\theta}) - l_n(\mathbf{y}|\theta_*)$ to the maximum log-likelihood has a limiting(1/2) chi-square(1) distribution.

The proof is given in Section F.

The limiting distribution can be presented in a more convenient and informal way. See Figure 8.16 and note the close similarities to Figure 8.15. Let z designate a standard normal variable; the limiting distribution can be presented as follows:

(14) $$\hat{\theta}(\mathbf{y}) - \theta_* \sim \frac{1}{\sqrt{nj(\theta_*)}}z, \qquad \text{or, as adjusted,} \qquad \hat{\delta} = \frac{1}{\sqrt{j(\theta_*)}}z,$$

(15) $$S_n(\mathbf{y}|\theta_*) \sim \sqrt{nj(\theta_*)}\, z,$$

(16) $$l_n(\mathbf{y}|\hat{\theta}(\mathbf{y})) - l_n(\mathbf{y}|\theta_*) \sim (1/2)z^2.$$

Note that the large-sample likelihood is the same as if we had taken a sample of n from a normal distribution with mean θ_* and reciprocal variance or *precision* $j(\theta_*)$; the maximum-likelihood estimate takes the place of the sample average. Thus we see that $j(\theta_*)$ is information in the sense of precision: each sample component contributes $j(\theta_*)$ to the precision $nj(\theta_*)$ of the estimate $\hat{\theta}(\mathbf{y})$, which locates the large-sample likelihood.

Also note in a parallel way that the limiting form of the log-likelihood is the same as would be obtained from the limiting marginal distribution of the maximum-likelihood estimate $\hat{\theta}$.

D TESTS AND CONFIDENCE INTERVALS FROM A LARGE SAMPLE

Theorem 2 shows that the large-sample likelihood depends on a single sample characteristic $\hat{\theta}(\mathbf{y})$. We can thus view $\hat{\theta} = \hat{\theta}(\mathbf{y})$ as the large-sample likelihood statistic and accordingly use it as a basis for tests and confidence intervals.

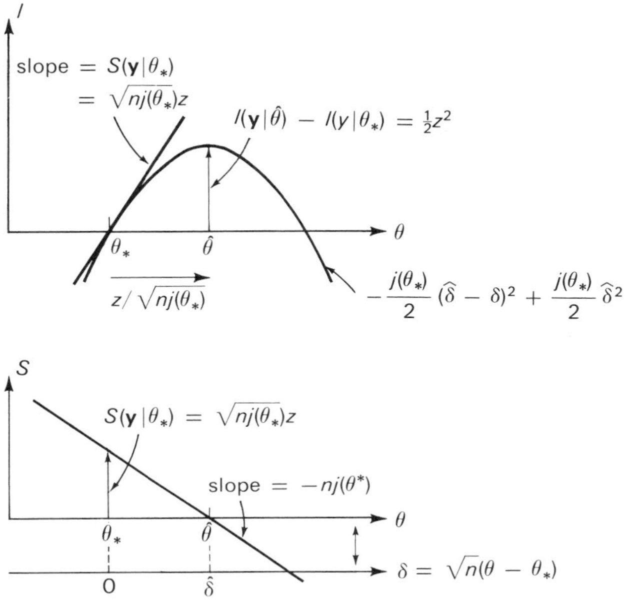

FIGURE 8.16

General case: limiting distribution of the log-likelihood and score in terms of a standard normal variable z. Note $\delta = \sqrt{n}(\theta - \theta_*)$ and $\widehat{\delta} = \sqrt{n}(\widehat{\theta}(\mathbf{y}) - \theta_*)$.

Let $(y_1, \ldots, y_n)$ be an observed large sample from a distribution satisfying Assumption 1 in Section B and Assumption 3 in Section F.

First, consider the hypothesis $\theta = \theta_0$. The hypothesis can be tested by calculating

$$(17) \qquad (nj(\theta_0))^{1/2}(\widehat{\theta}(\mathbf{y}) - \theta_0)$$

and comparing the value with the standard normal distribution. Alternatively, the hypothesis can be tested by calculating

$$(18) \qquad (nj(\theta_0))^{-1/2}S_n(\mathbf{y}\,|\,\theta_0)$$

and comparing the value with the standard normal distribution. Or it can be tested by calculating

$$(19) \qquad 2 \ln \frac{\prod f(y_i\,|\,\widehat{\theta}(\mathbf{y}))}{\prod f(y_i\,|\,\theta_0)}$$

and comparing the value with the chi-square(1) distribution; large values denote low likelihood at θ_0 and thus provide evidence against the hypothesis. In the limit these tests are equivalent; for moderate samples they may give slightly different results, and a choice will often be based on the ease with which computations can be made.

Second, consider the formation of a confidence interval for θ. A $1 - \alpha$ confidence interval can be based on

$$(20) \qquad P(-z_{\alpha/2} < (nj(\theta))^{1/2}(\widehat{\theta}(\mathbf{y}) - \theta) < z_{\alpha/2}) \approx 1 - \alpha,$$

or

$$(21) \qquad P\left(-z_{\alpha/2} < \frac{S_n(\mathbf{y}\,|\,\theta)}{(nj(\theta))^{1/2}} < z_{\alpha/2}\right) \approx 1 - \alpha.$$

or

$$(22) \qquad P\left(2 \ln \frac{\prod f(y_i \mid \widehat{\theta}(\mathbf{y}))}{\prod f(y_i \mid \theta)} \le \chi^2_\alpha\right) \approx 1 - \alpha.$$

The observed values would be substituted in the relations; the relations would then be rearranged (as we have discussed before) to give the interval of acceptable values, an approximate $1 - \alpha$ confidence interval for θ. An approximate $1 - \alpha$ confidence interval for θ can be obtained directly from the first expression (20):

$$(23) \qquad \widehat{\theta}(\mathbf{y}) \pm z_{\alpha/2}(nj(\widehat{\theta}(\mathbf{y})))^{-1/2};$$

compare this with the results in Section 7.2C and 7.2F.

E BINOMIAL TESTS AND CONFIDENCE INTERVALS

Let $(x_1, \ldots, x_n)$ be a sample from the Bernoulli(p) distribution with p in (0, 1). The density function for a single x is

$$f(x \mid p) = p^x q^{1-x}, \qquad x = 0, 1.$$

Thus

$$l(x \mid p) = x \ln p + (1 - x) \ln q,$$

$$S(x \mid p) = \frac{x}{p} - \frac{1 - x}{q},$$

$$S'(x \mid p) = -\frac{x}{p^2} - \frac{1 - x}{q^2},$$

from which we obtain

$$E(S(x \mid p) \mid p) = \frac{p}{p} - \frac{q}{q} = 0,$$

$$E(S'(x \mid p) \mid p) = -\frac{p}{p^2} - \frac{q}{q^2} = \frac{-1}{pq},$$

$$\text{var}\,(S(x \mid p) \mid p) = \frac{pq}{(pq)^2} = \frac{1}{pq} = j(p).$$

The maximum-likelihood estimate from the sample is obtained by solving $S_n(\mathbf{x} \mid p) = 0$; thus we obtain $\widehat{p} = \Sigma x_i/n$ as in Example 8.1.2.

The hypothesis $p = p_0$ can be tested by calculating, say (17),

$$(24) \qquad \frac{\widehat{p} - p_0}{(p_0 q_0/n)^{1/2}},$$

and comparing the value with the standard normal.

A confidence interval for p can be calculated on the basis of say (20). At the 95 percent level we have

$$-1.96 < \frac{\widehat{p} - p}{(pq/n)^{1/2}} < 1.96.$$

This can be rearranged to give

$$p_1(\hat{p}) < p < p_2(\hat{p}),$$

where $p_1(\hat{p})$ and $p_2(\hat{p})$ are the solutions of the equation

$$(\hat{p} - p)^2 = (1.96)^2 \frac{pq}{n}.$$

Thus

$$P(p_1(\hat{p}) < p < p_2(\hat{p})) = 95\% \tag{25}$$

and $(p_1(\hat{p}), p_2(\hat{p}))$ is a 95 percent confidence interval for p. At the $1 - \alpha$ level we have

$$-z_{\alpha/2} < \frac{\hat{p} - p}{(pq/n)^{1/2}} < z_{\alpha/2}.$$

This can be rearranged to give the $1 - \alpha$ confidence interval $(p_1(\hat{p}), p_2(\hat{p}))$, where $p_1(\hat{p})$ and $p_2(\hat{p})$ are the solutions of the equation

$$(\hat{p} - p)^2 = z^2_{\alpha/2} \frac{p(1 - p)}{n};$$

these solutions can be given explicitly as

$$p_i(\hat{p}) = \hat{p} + \frac{\dfrac{z^2_{\alpha/2}}{2n}(\hat{q} - \hat{p}) \pm \dfrac{z_{\alpha/2}}{\sqrt{n}}\left(\hat{p}\hat{q} + \dfrac{z^2_{\alpha/2}}{4n}\right)^{1/2}}{1 + z^2_{\alpha/2}/n}. \tag{26}$$

If n is reasonably large, the confidence interval can be approximated as

$$\hat{p} \pm z_{\alpha/2}\left(\frac{\hat{p}\hat{q}}{n}\right)^{1/2}.$$

F SUPPLEMENT: PROOF OF THEOREM 2

We now record the proof of Theorem 2. For this it is convenient to follow the notational device in Section A and standardize the likelihood function with respect to the *true value* θ_*. Thus the log-likelihood can be written without an arbitrary constant in the form

$$l_n(\mathbf{y} \mid \theta) = \ln f(\mathbf{y} \mid \theta) - \ln f(\mathbf{y} \mid \theta_*).$$

First, we expand the log-likelihood in a Taylor series about the true value θ_*:

$$l_n(\mathbf{y} \mid \theta) = (\theta - \theta_*) S_n(\mathbf{y} \mid \theta_*) + \frac{(\theta - \theta_*)^2}{2!} S'_n(y \mid \theta_*) + \frac{(\theta - \theta_*)^3}{3!} R \sum_1^n N(y_i),$$

where $|R| \le 1$. The coefficient $S_n(\mathbf{y} \mid \theta_*)$ has standard deviation $(nj(\theta_*))^{1/2}$; accordingly, we examine $\theta = \theta_* + \delta n^{-1/2}$, where δ is deviation in units of length $n^{-1/2}$:

$$l_n(\mathbf{y} \mid \theta_* + \delta n^{-1/2}) = \delta \frac{S_n(\mathbf{y} \mid \theta_*)}{n^{1/2}} + \frac{\delta^2}{2}\left(\frac{S'_n(\mathbf{y} \mid \theta_*)}{n} + \frac{\delta}{3\sqrt{n}} R \frac{\sum N(y_i)}{n}\right).$$

Note the correspondence between θ values and δ values:

θ	θ_*	$\widehat{\theta}$
δ	0	$\widehat{\delta}$

By the Central Limit Theorem 6.2.2 we have that

$$w = \frac{S_n(\mathbf{y} \mid \theta_*)}{n^{1/2}}$$

has a limiting normal(0, $\sqrt{j(\theta_*)}$) distribution. By the Strong Law of Large Numbers (Theorem 6.3.12), we have that

$$\frac{S_n'(y \mid \theta_*)}{n} + \frac{\delta R}{3\sqrt{n}} \frac{\sum N(y_i)}{n}$$

has limit $-j(\theta_*)$ with probability 1. Thus the limiting form of the log-likelihood is given by

$$(27) \qquad \delta w - \frac{\delta^2}{2} j(\theta_*) = -\frac{j(\theta_*)}{2} (\widehat{\delta} - \delta)^2 + \frac{j(\theta_*)}{2} \widehat{\delta}^2,$$

where the maximum-likelihood estimate $\widehat{\delta} = w/j(\theta_*) = \sqrt{n}(\widehat{\theta}(\mathbf{y}) - \theta_*) = z/\sqrt{j(\theta_*)}$ has a limiting normal(0, $1/\sqrt{j(\theta_*)}$) distribution. This is pictured in Figure 8.16. Note that $\delta = \sqrt{n}(\theta - \theta_*)$ uses a new origin at θ_* with scaling unit $n^{-1/2}$. This proof uses the following assumption together with Assumption 1:

ASSUMPTION 3

The information $j(\theta) > 0$ and $|\partial^3 \ln f(y \mid \theta)/\partial\theta^3| < N(y)$, where $N(y)f(y \mid \theta)$ is integrable on the sample space $\mathcal{S}$.

G SUPPLEMENT: LARGE-SAMPLE LIKELIHOOD WITH A VECTOR PARAMETER

Now consider the case of a vector parameter $\boldsymbol{\theta} = (\theta_1, \ldots, \theta_r)'$ in R^r or in an open set of R^r. The score function has a coordinate for each coordinate of $\boldsymbol{\theta}$:

$$S_1(y \mid \boldsymbol{\theta}) = \frac{\partial}{\partial\theta_1} \ln f(y \mid \boldsymbol{\theta}),$$
$$\vdots$$
$$S_r(y \mid \boldsymbol{\theta}) = \frac{\partial}{\partial\theta_r} \ln f(y \mid \boldsymbol{\theta}).$$

The score function is thus a vector-valued function $\mathbf{S}(y \mid \boldsymbol{\theta})$ and it records the gradient of the log-likelihood function. By the methods in Section B we obtain

$$(28) \qquad E(\mathbf{S}(y \mid \theta) \mid \boldsymbol{\theta}) = \mathbf{0}, \qquad \text{VAR}\,(\mathbf{S}(y \mid \boldsymbol{\theta}) \mid \boldsymbol{\theta}) = \mathbf{j}(\boldsymbol{\theta}),$$

where the matrix $\mathbf{j}(\boldsymbol{\theta})$ can also be calculated from

(29) $$\mathbf{j}(\boldsymbol{\theta}) = -E\left(\begin{pmatrix} \frac{\partial^2}{\partial\theta_1^2} \ln f(y|\boldsymbol{\theta}) & \cdots & \frac{\partial^2}{\partial\theta_1\,\partial\theta_r} \ln f(y|\boldsymbol{\theta}) \\ \vdots & & \vdots \\ \frac{\partial^2}{\partial\theta_r\,\partial\theta_1} \ln f(y|\boldsymbol{\theta}) & \cdots & \frac{\partial^2}{\partial\theta_r^2} \ln f(y|\boldsymbol{\theta}) \end{pmatrix} \middle| \boldsymbol{\theta}\right).$$

Now consider a large sample $\mathbf{y} = (y_1, \ldots, y_n)'$. By the Central Limit Theorem 6.2.5 we have that the scaled slope

(30) $$\mathbf{w} = \frac{\mathbf{S}(\mathbf{y}\,|\,\boldsymbol{\theta}_*)}{n^{1/2}}$$

of the log-likelihood at the true value has a limiting multivariate normal(0; $\mathbf{j}(\boldsymbol{\theta}_*)$) distribution.

Then by the methods in Section F we can show that the log-likelihood standarized with respect to the true value θ_* has the limiting form

(31) $$\boldsymbol{\delta}'\mathbf{w} - \frac{1}{2}\boldsymbol{\delta}'\mathbf{j}(\boldsymbol{\theta}_*)\boldsymbol{\delta} = \frac{1}{2}(\widehat{\boldsymbol{\delta}} - \boldsymbol{\delta})'\mathbf{j}(\boldsymbol{\theta}_*)(\widehat{\boldsymbol{\delta}} - \boldsymbol{\delta}) + \frac{1}{2}\widehat{\boldsymbol{\delta}}'\mathbf{j}(\boldsymbol{\theta}_*)\widehat{\boldsymbol{\delta}},$$

where $\boldsymbol{\delta} = \sqrt{n}(\boldsymbol{\theta} - \boldsymbol{\theta}_*)$ and where $\widehat{\boldsymbol{\delta}}$ is given by

(32) $$\widehat{\boldsymbol{\delta}} = \mathbf{j}^{-1}(\boldsymbol{\theta}_*)\mathbf{w} = \sqrt{n}(\widehat{\boldsymbol{\theta}}(\mathbf{y}) - \boldsymbol{\theta}_*).$$

From this it follows that the offset $\widehat{\boldsymbol{\delta}} = \mathbf{j}^{-1}(\boldsymbol{\theta}_*)\mathbf{w}$ to the point of maximum likelihood has a multivariate normal($\mathbf{0}$; $\mathbf{j}^{-1}(\theta_*)$) distribution; thus we can describe $\widehat{\boldsymbol{\theta}}(\mathbf{y})$ as multivariate normal located at the true $\boldsymbol{\theta}_*$ with variance matrix $J^{-1}(\boldsymbol{\theta}_*) = n^{-1}\mathbf{j}^{-1}(\boldsymbol{\theta}_*)$.

It also follows that the height to the maximum is distributed as $(1/2)\chi^2$ where χ^2 has a chi-square(r) distribution; thus the quadratic part $\widehat{\boldsymbol{\delta}}'J(\boldsymbol{\theta}_*)\widehat{\boldsymbol{\delta}}'$ of the multivariate normal density becomes $\Sigma_1^r z_1^2$ when related to standard normal variables.

The distribution properties for the likelihood can be summarized informally as

$\widehat{\theta}(\mathbf{y})$	multivariate normal($\boldsymbol{\theta}_*$; $n^{-1}\mathbf{j}^{-1}(\boldsymbol{\theta}_*)$),
$\mathbf{S}(\mathbf{y}\,|\,\boldsymbol{\theta}_*)$	multivariate normal($\mathbf{0}$; $n\mathbf{j}(\boldsymbol{\theta}_*)$),
$2 \ln \dfrac{\prod f(y_i\,|\,\widehat{\boldsymbol{\theta}}(\mathbf{y}))}{\prod f(y_i\,|\,\boldsymbol{\theta}_*)}$	chi-square(r).

H EXERCISES

1 A possibly biased coin was tossed 6400 times, yielding 3277 heads. Use the large-sample results in Sections C and E:
   (a) Test the hypothesis: $p = 1/2$.
   (b) Calculate a 95 percent confidence interval for p.

2 Calculate $S(y|\theta)$, $S'(y|\theta)$, and $j(\theta)$ for the Poisson(θ) distribution.

3 Assuming that traffic fatalities in a particular city are approximately Poisson, derive a 95 percent confidence interval for the mean annual rate θ given the observed count $y = 113$ for one year. Use Exercise 2 and the asymptotic results even though $n = 1$ (the annual count can be viewed as a sum of say 52 weekly counts).

4 A sample of nine from a Poisson(θ) distribution yielded $\bar{y} = 28\frac{1}{3}$. Use Exercise 2 and the large-sample results in Section D:
   (a) Test the hypothesis $\theta = 25$.
   (b) Form a 95 percent confidence interval for θ.

5 Let $(y_1, \ldots, y_n)$ be a sample from the normal(0, σ) with $\theta = \sigma^2$ in R^+. Calculate $S(\mathbf{y}|\sigma^2)$, $S'(\mathbf{y}|\sigma^2)$, and $nj(\sigma^2)$. Calculate $2 \ln (\Pi f(y_i|\hat{\sigma}^2)/\Pi f(y_i|\sigma_*^2))$ and quote its limiting distribution under the hypothesis.

I PROBLEMS

6 Calculate $S(y|\theta)$, $S'(y|\theta)$, and $j(\theta)$ for the location Cauchy $f(y|\theta) = \pi^{-1}(1 + (y - \theta)^2)^{-1}$ with θ in R; use $\int_{-\infty}^{\infty} (1 - t^2)(1 + t^2)^{-3}\, dt = \pi/4$. Determine σ_0 so that the normal(θ, σ_0) distribution has the same information as the location Cauchy.

7 Calculate $S(y|\theta)$, $S'(y|\theta)$, and $j(\theta)$ for the scale exponential $f(y|\theta) = \theta^{-1} \exp\{-y/\theta\}$ on R^+ with θ in R^+.

8 (*continuation*) For a large sample $(y_1, \ldots, y_n)$ determine a 95 percent confidence interval for θ based on the limiting likelihood.

9 Calculate $S(y|\theta)$, $S'(y|\theta)$, and $j(\theta)$ for the scale gamma $f(y|\theta) = \Gamma^{-1}(p)(y/\theta)^{p-1} \exp\{-y/\theta\}/\theta$ on R^+ with θ in R^+.

10 (*continuation*) For a sample of n, calculate $2 \ln (\Pi f(y_i|\hat{\theta})/\Pi f(y_i|\theta_*))$. Quote its limiting distribution from the results in this section.

11 Calculate $S(y|\theta)$, $S'(y|\theta)$, and $j(\theta)$ for the location(θ) logistic distribution. Determine the normal distribution (θ, σ_0) that has the same information as the location logistic.

12 Calculate $\mathbf{S} = (S_1(y|\mu, \sigma^2), S_2(y|\mu, \sigma^2))'$ for the normal(μ, σ) with parameter (μ, σ^2) in $R \times R^+$. Calculate the derivatives of S_1 and S_2 and thus determine the information matrix $\mathbf{j}(\mu, \sigma^2)$.

13 (*continuation*) For a sample of n, calculate $2 \ln (\Pi f(y_i|\hat{\mu}, \hat{\sigma}^2)/\Pi f(y_i|\mu_*, \sigma_*^2))$. Quote its limiting distribution from the results in this section.

14 Let $(y_1, \ldots, y_n)$ be a sample from a distribution with density $f(y|\theta)$ with θ in $\Omega = \{\theta_1, \theta_2\}$. Let μ_1 and σ_1^2 be the mean and variance of $l_3(y) = \ln f(y|\theta_2) - \ln f(y|\theta_1)$ under the distribution $f(y|\theta_1)$, and let μ_2 and σ_2^2 be the mean and variance under the distribution $f(y|\theta_2)$; suppose that $\sigma_1 > 0$ and $\sigma_2 > 0$. Show that $l_3(\mathbf{y})$ has a distribution with limiting normal form $(n\mu_i, \sqrt{n}\sigma_i)$ when the distribution being sampled is $f(y|\theta_i)$.

NOTES AND REFERENCES

Recently the observed likelihood function has been recognized as a basic means for assessing data. The acceptance of the likelihood function ranges from the Bayesians, who use only the likelihood function, to the traditional statisticians, who find that the likelihood concept provides feasible new ways of determining sufficiency and minimal sufficiency. In between there is a wide range of statisticians who use the likelihood function for some data assessment; for this in an extreme form, see Edwards (1972).

The likelihood function leads us to $(\bar{y}, s_y^2)$ for the normal(μ, σ) model. Sections 8.2 and 8.3 examine distribution theory and inference methods for the normal model.

In a similar way, the likelihood function leads to the likelihood statistic for a general statistical model; see Section 8.4. This provides a simple direct route to the usual sufficiency and minimal sufficiency results. Indeed, the likelihood statistic seems to replace the minimal sufficient statistic as the basic reduction method of statistics.

Sufficiency for exponential models follows almost trivially by the use of the likelihood statistic; compare, for example, with Barankin and Maitra (1963).

The chapter concludes with some large-sample results for the likelihood function. We see that the general statistical model using a large sample can be analyzed in a way quite similar to that for the normal.

The proper definition of the likelihood function is essential for these results. The reader is to be alerted to the importance of the "arbitrary multiplicative constant" in the definition of the likelihood function—its absence in many current statistical texts prevents the proper and full use of the concept.

On the likelihood function:

Edwards, A. W. F. (1972). *Likelihood.* New York: Cambridge University Press.

Fisher, R. A. (1922). On the mathematical foundations of theoretical statistics. *Phil. Trans. Roy. Soc. London,* **A222,** 309–368.

On large-sample likelihood:

Fraser, D. A. S. (1968). *The Structure of Inference,* Chap. 8. Huntington, N.Y.: Krieger Publishing Co.

On sufficiency:

Fisher, R. A. (1922). On the mathematical foundations of theoretical statistics. *Phil. Trans. Roy. Soc. London,* **A222,** 309–368.

Neyman, J. (1935). Sur un teorema concernente le cosidette statistiche sufficienti. *Giornale dell' Istituto Italiano degli Attuari,* **6,** 320–334.

On sufficiency and exponential models:

Barankin, E. W., and A. P. Maitra (1963). Generalization of the Fisher–Darmois–Koopman–Pitman theorem on sufficient statistics. *Sankhyā,* **A25,** 217–244.

Fraser, D. A. S. (1966). Sufficiency for regular models. *Sankhyā,* **A28,** 137–144.

In Chapter 7 we introduced estimation, tests of significance, and confidence intervals as the three basic concepts of statistical inference. We examined the concepts informally in that chapter, using the large-sample and symmetry results from the probability chapters. Then in Chapter 8, as part of examining likelihood with statistical models, we examined the three concepts in relation to the normal statistical model.

In Chapter 9 we shall examine ***estimation*** *in detail. First we investigate briefly three traditional methods of obtaining an estimate,* ***least squares, moments,*** *and* ***maximum likelihood*** *(Section 9.1). We then examine least squares in greater detail for models having a certain* ***linear structure*** *for location (Section 9.2). We investigate ways of combining direct estimates to obtain an overall estimate,* ***weighting by reciprocal variance,*** *and of combining indirect estimates, the* ***Gauss–Markov method*** *(Section 9.3). We investigate ways of obtaining best estimates with a one- or a several-parameter model, by the use of the* ***information inequality*** *(Section 9.4) and by use of* ***sufficiency and completeness*** *(Section 9.5). The* ***Bayesian method*** *of inference is examined in Supplement Section 9.6.*

9

ESTIMATION

9.1

SOME TRADITIONAL METHODS

A surveyor investigating the distance between two points may make several measurements or determinations of the distance and then combine the measurements in some suitable way to obtain an estimate of the distance. This is an example of the *direct measurement* of a physical quantity.

Or, more generally, a surveyor investigating the boundary of a piece of property may make a succession of measurements of angle and distance progressing from a starting point around the boundary and back to the starting point. Each measurement would directly refer to a physical quantity connected with the boundary. However, a relation exists among the physical quantities in the sense that the finishing point is, in fact, the starting point; the surveyed line has been closed. Accordingly, each measurement has some bearing on a range of physical quantities and there is thus some indirect measurement of physical quantities. Various tie lines or check lines could also be included in the survey and there would then be further indirect measurement on the physical quantities.

In this section we examine three traditional methods for obtaining an estimate or plausible value for a parameter. Direct measurement is somewhat straightforward; these methods are primarily concerned with indirect measurement.

A THE METHOD OF LEAST SQUARES

The method of least squares was proposed by Legendre and Gauss at the beginning of the nineteenth century. The method is concerned with measurements on physical quantities that satisfy a relationship or are expressed in terms of more elementary quantities. The method produces adjusted measurements that do satisfy the relationships and it produces estimates of the elementary quantities.

Consider an application with an observed response vector $\mathbf{y} = (y_1, \ldots, y_n)'$. Some related theory prescribes a location for the response vector: $\boldsymbol{\nu}(\boldsymbol{\theta}) = (\nu_1(\theta_1, \ldots, \theta_r), \ldots, \nu_n(\theta_1, \ldots, \theta_r))'$, where $\boldsymbol{\theta}$ is a parameter having r coordinates; these coordinates may be more elementary quantities of the system being investigated. For example, we might have measurements (y_1, y_2, y_3) for two sides and the hypotenuse of a right-angled triangle and we could have $\boldsymbol{\nu}(\boldsymbol{\theta}) = (\theta_1, \theta_2, \sqrt{\theta_1^2 + \theta_2^2})$. The method produces a fitted response vector $\hat{\mathbf{y}} = (\hat{y}_1, \ldots, \hat{y}_n)$ that satisfies the relations indicated by the location vector, and it produces an estimate $\hat{\boldsymbol{\theta}} = (\hat{\theta}_1, \ldots, \hat{\theta}_r)$ for the parameter.

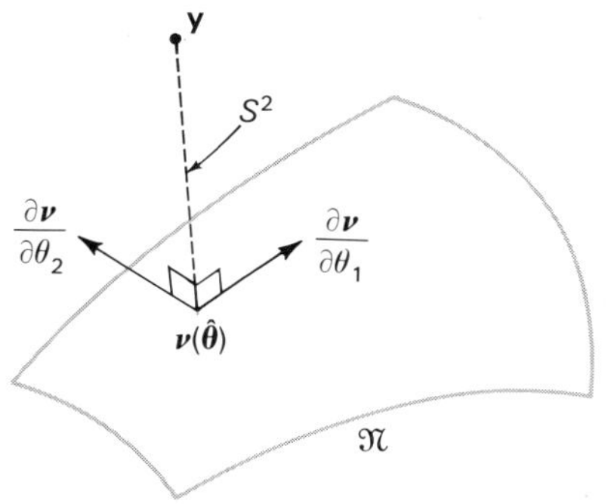

FIGURE 9.1
Observation $\mathbf{y}$. The set $\mathfrak{M}$ of possible locations $\boldsymbol{\nu}(\boldsymbol{\theta})$. The minimizing point $\boldsymbol{\nu}(\hat{\boldsymbol{\theta}})$ has deviation vector $\mathbf{y} - \boldsymbol{\nu}(\hat{\boldsymbol{\theta}})$ orthogonal to the tangent vectors in the direction of θ_1 change, of θ_2 change, ...

The least-squares method has been widely used in astronomy and geodesy, where a large number of measurements can be taken on a variety of physical dimensions. The geometry of the system gives relationships among the dimensions or presents the dimensions in terms of some independent components or parameters. The method of least squares can be viewed as adjusting or fitting the measurements in an orderly way so that they satisfy the geometry of the system.

The method of least squares also has important uses in statistical theory. In one direction it provides various computational procedures that are very useful. And in another direction it gives an estimation procedure that has certain optimum properties, depending on the form of the distribution.

We now describe the method. Let $\mathbf{y} = (y_1, \ldots, y_n)'$ be an observed response. And let $\boldsymbol{\nu}(\boldsymbol{\theta}) = (\nu_1(\boldsymbol{\theta}), \ldots, \nu_n(\boldsymbol{\theta}))'$ be a possible location for the response; the model for location is then the set

$$\mathfrak{M} = \{\boldsymbol{\nu}(\boldsymbol{\theta}) : \boldsymbol{\theta} \in \Omega\}$$

of possible response locations obtained as the parameter $\boldsymbol{\theta}$ ranges over the parameter space Ω. We can picture $\mathbf{y}$ and $\mathfrak{M}$ in Euclidean space as indicated in Figure 9.1:

$\mathbf{y}$	y_1	y_2	$\cdots$	y_n
$\boldsymbol{\nu}(\boldsymbol{\theta})$	$\nu_1(\boldsymbol{\theta})$	$\nu_2(\boldsymbol{\theta})$	$\cdots$	$\nu_n(\boldsymbol{\theta})$

DEFINITION 1

The **least-squares estimate** of $\boldsymbol{\theta}$ is the value $\hat{\boldsymbol{\theta}} = \hat{\boldsymbol{\theta}}(\mathbf{y})$ that minimizes the sum of squares

(1) $$S^2 = (y_1 - \nu_1(\boldsymbol{\theta}))^2 + \cdots + (y_n - \nu_n(\boldsymbol{\theta}))^2 = (\mathbf{y} - \boldsymbol{\nu}(\boldsymbol{\theta}))'(\mathbf{y} - \boldsymbol{\nu}(\boldsymbol{\theta})).$$

See Figure 9.1: S^2 is the squared distance from the point $\boldsymbol{\nu}(\boldsymbol{\theta})$ to the point $\mathbf{y}$; the method is to choose the value for $\boldsymbol{\theta}$ that makes $\boldsymbol{\nu}(\boldsymbol{\theta})$ closest to the observation $\mathbf{y}$. If the set $\mathfrak{M}$

is closed in R^n, then the minimizing parameter value will always exist: the closest point $\hat{\mathbf{y}} = \boldsymbol{\nu}(\hat{\boldsymbol{\theta}})$ is called the *fitted response vector*.

Note that we have not mentioned any response distribution. Of course with a specification for the response distribution, the method may or may not be appropriate; some aspects of this will be examined later.

Now suppose that $\boldsymbol{\nu}(\boldsymbol{\theta})$ is continuously differentiable with respect to $\boldsymbol{\theta}$. *If the minimizing value* $\hat{\boldsymbol{\theta}}$ *is an interior point* on the surface $\mathfrak{M}$, then it can be calculated by differentiating with respect to $\boldsymbol{\theta}$ and setting the derivative equal to zero:

$$(2) \qquad \begin{array}{ccc} \dfrac{\partial S^2}{\partial \theta_1} = 0, & & \sum \dfrac{\partial \nu_i(\boldsymbol{\theta})}{\partial \theta_1}(y_i - \nu_i(\boldsymbol{\theta})) = 0, \\ \vdots & \text{or} & \\ \dfrac{\partial S^2}{\partial \theta_r} = 0, & & \sum \dfrac{\partial \nu_i(\boldsymbol{\theta})}{\partial \theta_r}(y_i - \nu_i(\boldsymbol{\theta})) = 0; \end{array}$$

the differentiation produces factors 2 and -1; these have been factored from the right-hand expressions.

The first equation states that the deviation vector,

$$(y_1 - \nu_1(\boldsymbol{\theta}), \ldots, y_n - \nu_n(\boldsymbol{\theta})),$$

is orthogonal to the tangent vector,

$$\left(\frac{\partial \nu_1(\boldsymbol{\theta})}{\partial \theta_1}, \ldots, \frac{\partial \nu_n(\boldsymbol{\theta})}{\partial \theta_1}\right),$$

in the direction of θ_1 change; . . .; the last equation states that the deviation vector is orthogonal to the tangent vector,

$$\left(\frac{\partial \nu_1(\boldsymbol{\theta})}{\partial \theta_r}, \ldots, \frac{\partial \nu_n(\boldsymbol{\theta})}{\partial \theta_r}\right),$$

in the direction of θ_r change; see Figure 9.1.

In certain cases discussed in Section 9.2 these equations (2) are linear and can be solved directly. In other cases various iterative procedures can be used; for example Section 9.4D.

Consider a very simple example to illustrate the method; more substantial examples will be examined later.

EXAMPLE 1

Let $\mathbf{y} = (y_1, \ldots, y_n)'$ be an observation with theoretical location $\beta\mathbf{x}' = \beta(x_1, \ldots, x_n)' = (\beta x_1, \ldots, \beta x_n)'$, where the x_i are known in value. The least-squares method is to choose the value $\beta = \hat{\beta}$ so as to minimize

$$S^2 = (y_1 - \beta x_1)^2 + \cdots + (y_n - \beta x_n)^2,$$

$$(3) \qquad -\frac{1}{2}\frac{\partial S^2}{\partial \beta} = x_1(y_1 - \beta x_1) + \cdots + x_n(y_n - \beta x_n) = 0,$$

$$\sum x_i y_i - \beta \sum x_i x_i = \mathbf{x}'\mathbf{y} - \beta \mathbf{x}'\mathbf{x} = 0.$$

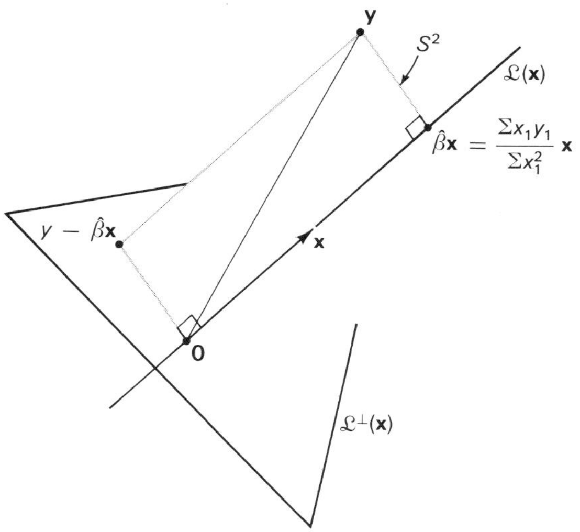

FIGURE 9.2
The point on the line $\mathcal{L}(\mathbf{x})$ closest to $\mathbf{y}$ is the projection of $\mathbf{y}$ onto that line.

Thus the least-squares estimate of β is

$$\hat{\beta} = \frac{\mathbf{x}'\mathbf{y}}{\mathbf{x}'\mathbf{x}} = (\mathbf{x}'\mathbf{x})^{-1}\mathbf{x}'\mathbf{y}, \tag{4}$$

and the fitted response vector is

$$\hat{\beta}\mathbf{x} = \mathbf{x}\hat{\beta} = \mathbf{x}(\mathbf{x}'\mathbf{x})^{-1}\mathbf{x}'\mathbf{y}. \tag{5}$$

Note that $\mathbf{x}(\mathbf{x}'\mathbf{x})^{-1}\mathbf{x}'$ is an $n \times n$ matrix that transforms the given response vector into the fitted response vector.

A clear picture of the geometry of least-squares procedures is particularly useful and informative. For the present example, see Figure 9.2 and recall the geometry in Sections 7.1E and 8.2C. The set $\mathfrak{N} = \{b\mathbf{x} : b \in R\}$ is the line generated by the vector $\mathbf{x}$. The point on the line $\mathcal{L}(\mathbf{x})$ that is closest to $\mathbf{y}$ is seen by equation (3) to be the projection $\hat{\beta}\mathbf{x}$ of $\mathbf{y}$ onto the line; the projection is, of course, the point whose residual vector $\mathbf{y} - \hat{\beta}\mathbf{x}$ is orthogonal to the line. Thus we have the orthogonal decomposition shown in Table 9.1.

TABLE 9.1
Components of vector $\mathbf{y}$; corresponding squared lengths

Subspace	Dimension	Component	Squared Length
$\mathcal{L}(\mathbf{x})$	1	$\hat{\beta}\mathbf{x}$	$\hat{\beta}^2 \sum x_i^2$
$\mathcal{L}^{\perp}(\mathbf{x})$	$n - 1$	$\mathbf{y} - \hat{\beta}\mathbf{x}$	$\sum (y_i - \hat{\beta}x_i)^2$
	n	$\mathbf{y}$	$\mathbf{y}'\mathbf{y}$

B THE METHOD OF MOMENTS

The method of moments is a very old method of estimation. A large part of early statistics was involved with fitting a distribution form to data involving a moderately large number of observations. Various distribution forms with a small number of parameters have been proposed; prominent among these is a Pearson family of curves that includes the normal, gamma, beta, and Student distributions. The method of moments provides a rather straightforward method of fitting such a distribution form to a large number of observations.

Let $(y_1, \ldots, y_n)$ be an observed sample from a distribution that has a parameter $\boldsymbol{\theta} = (\theta_1, \ldots, \theta_r)'$ with r coordinates. The moments of the distribution will, of course, depend on the parameter $\boldsymbol{\theta}$ of the distribution. The method of moments involves choosing as an estimate that parameter value $\boldsymbol{\theta}^*$ for which the first few moments of the distribution are equal to the corresponding moments of the sample; see Table 9.2.

TABLE 9.2
Sample and distribution moments

Sample Moment	Distribution Moment
$m_1 = \sum y_i/n$	$\mu_1(\boldsymbol{\theta}) = E(y\,\vert\,\boldsymbol{\theta})$
$m_2 = \sum y_i^2/n$	$\mu_2(\boldsymbol{\theta}) = E(y^2\,\vert\,\boldsymbol{\theta})$
$\vdots$	
$m_r = \sum y_i^r/n$	$\mu_r(\boldsymbol{\theta}) = E(y^r\,\vert\,\boldsymbol{\theta})$

DEFINITION 2

The **method-of-moments estimate** of $\boldsymbol{\theta}$ is the value $\theta^* = \theta^*(\mathbf{y})$ obtained by solving the equations

$$(6)\qquad \begin{aligned} m_1 &= \mu_1(\boldsymbol{\theta}), \\ &\vdots \\ m_r &= \mu_r(\boldsymbol{\theta}), \end{aligned}$$

using the first few moments.

Typically as many moments will be needed as there are coordinates for $\boldsymbol{\theta}$.

EXAMPLE 2

The location scale normal: Let $y_1, \ldots, y_n$ be a sample from a normal(μ, σ) distribution. The first two moments are shown in Table 9.3. Hence we obtain

$$\bar{y} = \mu, \quad \overline{y^2} = \mu^2 + \sigma^2; \qquad \mu^* = \bar{y}, \quad \sigma^{*2} = \frac{\sum y_i^2}{n} - \bar{y}^2 = \frac{n-1}{n} s_y^2.$$

Note that the method gives the sample second central moment as the estimate of variance.

TABLE 9.3
Sample and distribution moments

Sample Moment	Distribution Moment
$\bar{y} = \sum y_i/n$	μ
$\overline{y^2} = \sum y_i^2/n$	$\mu^2 + \sigma^2$

The method of moments can be very hazardous if more than just one or two parameters are estimated. Some indications for this come from the need to calculate higher powers of the observations; the higher powers are very sensitive to slight changes in the observations and the method can be unstable.

C THE MAXIMUM-LIKELIHOOD METHOD

The method of maximum likelihood can handle both vector location as with least squares and distribution form as with the method of moments. It was proposed by R. A. Fisher in the 1920s as an alternative to the method of moments; some aspects of the method had been used by Gauss in the early nineteenth century. The method is very widely used in almost all areas of statistical application.

Consider a statistical model

$$\{f(\cdot\,|\theta):\theta \in \Omega\}$$

and let $\mathbf{y}$ be an observed value. The likelihood function $L(\mathbf{y}|\cdot)$ and log-likelihood function $l(\mathbf{y}|\cdot)$ have been defined in Section 8.1D:

$$L(\mathbf{y}|\theta) = cf(\mathbf{y}|\theta), \qquad l(\mathbf{y}|\theta) = \ln f(\mathbf{y}|\theta) + a,$$

where c is in R^+ and a in R.

DEFINITION 3

The **maximum-likelihood estimate** (MLE) of θ is the value $\hat{\theta} = \hat{\theta}(\mathbf{y})$ that maximizes the likelihood function L:

$$(7) \qquad L(\mathbf{y}|\hat{\theta}) = \sup_{\theta\in\Omega} L(\mathbf{y}|\theta).$$

Note that the MLE estimate is the parameter value that maximizes the probability for the observed outcome.

If the likelihood function is continuously differentiable and if the maximum occurs on an open set of parameter values, then the MLE can be obtained as a solution of the score equation: for a real-valued parameter we have

$$(8) \qquad S(\mathbf{y}|\theta) = \frac{\partial \ln f(\mathbf{y}|\theta)}{\partial\theta} = 0;$$

and for a vector parameter $\boldsymbol{\theta} = (\theta_1, \ldots, \theta_r)'$ we have

$$
(9) \qquad
\begin{aligned}
S_1(\mathbf{y}\,|\,\boldsymbol{\theta}) &= \frac{\partial \ln f(\mathbf{y}\,|\,\boldsymbol{\theta})}{\partial \theta_1} = 0, \\
&\;\vdots \\
S_r(\mathbf{y}\,|\,\boldsymbol{\theta}) &= \frac{\partial \ln f(\mathbf{y}\,|\,\boldsymbol{\theta})}{\partial \theta_r} = 0.
\end{aligned}
$$

Some examples involving MLE estimates were given in Section 8.1E. The large-sample importance of the MLE method is closely tied to the large-sample methods discussed in Section 8.6.

We now show a rather close connection between the least-squares method and the maximum likelihood method—for the case of normally distributed variation.

THEOREM 4

If $y_1, \ldots, y_n$ are statistically independent and $y_i - \nu_i(\boldsymbol{\theta})$ is normal$(0, \sigma)$ with σ in R^+, then the maximum-likelihood estimate of $(\boldsymbol{\theta}, \sigma^2)$ is given by
(a) *the least-squares estimate $\widehat{\boldsymbol{\theta}}$ for $\boldsymbol{\theta}$ obtained by minimizing*

$$S^2 = \sum_1^n (y_i - \nu_i(\boldsymbol{\theta}))^2$$

and (b) *the average squared deviation*

$$s^2 = \frac{1}{n} \sum_1^n (y_i - \nu_i(\widehat{\boldsymbol{\theta}}))^2$$

for σ^2.

Proof The likelihood function from an observed $\mathbf{y}$ is given by

$$L(\mathbf{y}\,|\,\boldsymbol{\theta}, \sigma^2) = c(2\pi\sigma^2)^{-n/2} \exp\left\{-\frac{1}{2\sigma^2} S^2\right\}.$$

For given σ^2 this function is maximized by choosing $\widehat{\boldsymbol{\theta}}$ to minimize S^2; thus $\widehat{\boldsymbol{\theta}}$ is the least-squares estimate of θ.

Let S_0^2 be the minimized value of S^2. For maximization with respect to σ^2, we then have

$$L(\mathbf{y}\,|\,\widehat{\boldsymbol{\theta}}, \sigma^2) = c(\sigma^2)^{-n/2} \exp\left\{-\frac{1}{2\sigma^2} S_0^2\right\},$$

$$l(\mathbf{y}\,|\,\widehat{\boldsymbol{\theta}}, \sigma^2) = -\frac{n}{2} \ln \sigma^2 - \frac{1}{2\sigma^2} S_0^2 + a,$$

$$\frac{\partial l(\mathbf{y}\,|\,\widehat{\boldsymbol{\theta}}, \sigma^2)}{\partial \sigma^2} = -\frac{n}{2\sigma^2} + \frac{1}{2\sigma^4} S_0^2 = 0,$$

and hence $\widehat{\sigma}^2 = S_0^2/n$.

EXAMPLE 3

The location-scale normal: Let $(y_1, \ldots, y_n)$ be a sample from the normal(μ, σ) with (μ, σ^2) in $R \times R^+$. The least-squares estimate of μ is obtained by minimizing

$$S^2 = \sum_1^n (y_i - \mu)^2$$

and is of course $\hat{\mu} = \bar{y}$. Thus the maximum-likelihood estimate is $(\hat{\mu}, \hat{\sigma}^2) = \left(\bar{y}, \sum (y_i - \bar{y})^2/n\right)$.

D EXERCISES

1 The observed response vector $(2, 4, -1, -1)$ has theoretical location $\beta(1, 1, -1, -1) = \beta\mathbf{x}'$.
(a) Calculate the least-squares estimate of β.
(b) Calculate the fitted vector $\hat{\beta}\mathbf{x}$ and its squared length.
(c) Calculate the residual vector $\mathbf{y} - \hat{\beta}\mathbf{x}$ and its squared length.

2 (*continuation*) (a) The fitted vector $\hat{\beta}\mathbf{x}$ can be obtained by applying the projection matrix $\mathbf{x}(\mathbf{x}'\mathbf{x})^{-1}\mathbf{x}'$ to the response vector $\mathbf{y}$; calculate this 4×4 projection matrix. (b) The residual vector $\mathbf{y} - \hat{\beta}\mathbf{x}$ can be obtained by applying the projection matrix $I - \mathbf{x}(\mathbf{x}'\mathbf{x})^{-1}\mathbf{x}'$ to the response vector $\mathbf{y}$; calculate this 4×4 projection matrix.

3 The observed response vector $(2, 2, 4, 5)$ has theoretical location $\beta(1, 2, 3, 4)' = \beta\mathbf{x}$.
(a) Calculate the least-squares estimate of β.
(b) Calculate the fitted vector $\hat{\beta}\mathbf{x}$ and its squared length.
(c) Calculate the residual vector $\mathbf{y} - \hat{\beta}\mathbf{x}$ and its squared length.

4 (*continuation*)
(a) The fitted vector $\hat{\beta}\mathbf{x}$ can be obtained by applying the projection matrix $\mathbf{x}(\mathbf{x}'\mathbf{x})^{-1}\mathbf{x}'$ to the response vector $\mathbf{y}$; calculate the 4×4 projection matrix.
(b) The residual vector $\mathbf{y} - \hat{\beta}\mathbf{x}$ can be obtained by applying the projection matrix $I - \mathbf{x}(\mathbf{x}'\mathbf{x})^{-1}\mathbf{x}'$ to the response vector $\mathbf{y}$; calculate this 4×4 projection matrix.

5 Let $(y_1, \ldots, y_n)$ be a sample from the Poisson(λ) distribution with λ in R^+.
(a) Calculate the least-squares estimate of the location λ.
(b) Calculate the method-of-moments estimate of λ.
(c) Calculate the maximum-likelihood estimate of λ.

6 Let $(y_1, \ldots, y_k)$ be a sample from the binomial(n, p) distribution with p in $[0, 1]$.
(a) Calculate the least-squares estimate of the location np.
(b) Calculate the method-of-moments estimate of np.
(c) Calculate the maximum-likelihood estimate of np.

7 Let $(y_1, \ldots, y_n)$ be a sample from the normal(μ_0, σ) distribution with σ in R^+.
(a) Calculate the method-of-moments estimate of σ^2.
(b) Calculate the maximum-likelihood estimate of σ^2.

8 Let $(y_1, \ldots, y_n)$ be a sample from the uniform$(0, 2\theta)$.
(a) Calculate the least-squares estimate of θ.
(b) Calculate the method-of-moments estimate of θ.
(c) Calculate the maximum-likelihood estimate of θ.

E PROBLEMS

9 The vector $\beta\mathbf{x}$ was fitted by least squares in Example 1. Show that the squared length of the fitted vector is

$$\mathbf{y}'\mathbf{x}\hat{\beta} = \sum y_i x_i \hat{\beta}.$$

10 (*continuation*) Show that the squared length of the residual vector is

$$\sum (y_i - \hat{\beta}x_i)^2 = \sum y_i^2 - \sum y_i x_i \hat{\beta}.$$

11 Let $(y_{11}, \ldots, y_{1m}, y_{21}, \ldots, y_{2n})$ have location model $(\theta_1, \ldots, \theta_1, \theta_2, \ldots, \theta_2)$ with (θ_1, θ_2) in R^2. Find the least-squares estimate of (θ_1, θ_2).

12 Let m_1 and m_2 be the first and second moments for a sample from the scale gamma distribution

$$f(y \mid p, \beta) = \beta^{-p}\Gamma^{-1}(p)e^{-y/\beta}y^{p-1}, \qquad y > 0,$$

with (p, β) in $(0, \infty)^2$. Find the method-of-moments estimate of (p, β).

13 Let $(y_1, \ldots, y_n)$ have location given by $(m_1(\theta), \ldots, m_n(\theta))$. The least-absolute-deviations estimate is the value $\hat{\theta} = \hat{\theta}(\mathbf{y})$ that minimizes

$$\sum_1^n |y_i - m_i(\theta)|.$$

Prove the following analog of Theorem 4: *if $y_1, \ldots, y_n$ are statistically independent and $y_i - m_i(\theta)$ has the Laplace distribution*

$$f(v \mid \tau) = \frac{1}{2\tau} e^{-|v|/\tau}$$

with τ in R^+, then the maximum-likelihood estimate of (θ, τ) is (a) *the least-absolute-deviation estimate $\hat{\theta}$ for θ,* (b) *the average absolute deviation*

$$\hat{\tau} = \frac{1}{n}\sum_1^n |y_i - m_i(\hat{\theta})|$$

for τ. For example, see Problem 5.3.15.

9.2

LINEAR LEAST SQUARES

An application of least squares is concerned with a sequence of response values $y_1, \ldots, y_n$ and a corresponding sequence of theoretical locations $\nu_1(\boldsymbol{\theta}), \ldots, \nu_n(\boldsymbol{\theta})$; the locations can be adjusted toward fitting the response values by varying a parameter $\boldsymbol{\theta}$. Least squares is particularly simple if the locations $\nu_1(\boldsymbol{\theta}), \ldots, \nu_n(\boldsymbol{\theta})$ are linear in the parameter $\boldsymbol{\theta}$. In this section we examine this linear least squares.

We introduce this least-squares theory here in order to have some substantial statistical examples to illustrate the inference methods in the next few chapters. The approach to inference in Chapters 9 and 10 is concerned with response values that deviate from a true location by error from a normal distribution; the approach in Chapter 11 covers the case with error from other distribution forms.

A EXAMPLES WITH LINEAR LOCATION

Consider a response $\mathbf{y} = (y_1, \ldots, y_n)'$ with location $\boldsymbol{\nu}(\boldsymbol{\theta}) = (\nu_1(\boldsymbol{\theta}), \ldots, \nu_n(\boldsymbol{\theta}))$ that is linear in the parameter $\boldsymbol{\theta}$. We consider briefly some examples that illustrate the great wealth of generality available with this linear model for location.

The general level of yield for trefoil on certain plots is approximately affine in the amount of nitrogen added; see Figure 9.3. Or, in general, consider a random system

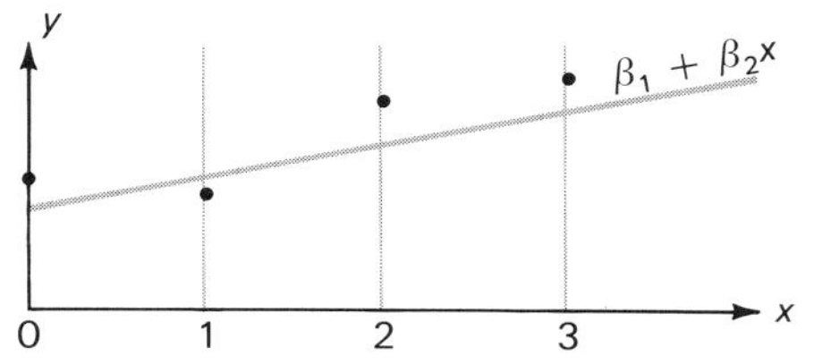

FIGURE 9.3
Amount of nitrogen x. True general level $\beta_1 + \beta_2 x$ for the yield y as a function of the amount of input x; in a typical application β_1 and β_2 would be unknown. Observed yields for $x = 0, 1, 2, 3$.

where the location of the response is known to be an affine function of an input variable x over a reasonable range. Then the location for the response would have the form

$$\beta_1 + \beta_2 x$$

with parameter (β_1, β_2). Correspondingly, the location for the response vector $\mathbf{y}$ with various input settings $x_1, \ldots, x_n$ would then be

$$\begin{aligned} \nu_1 &= \beta_1 + \beta_2 x_1, \\ &\vdots \\ \nu_n &= \beta_1 + \beta_2 x_n, \end{aligned} \qquad \boldsymbol{\nu}(\beta_1, \beta_2) = \beta_1 \mathbf{1} + \beta_2 \mathbf{x}.$$

More generally, it may be necessary to allow for possible quadratic dependence on the input variable x. Then the location for the response would have the form

$$\beta_1 + \beta_2 x + \beta_3 x^2$$

with parameter $(\beta_1, \beta_2, \beta_3)$. Correspondingly, the location for the response vector $\mathbf{y}$ with various input settings $x_1, \ldots, x_n$ would then be

$$\begin{aligned} \nu_1 &= \beta_1 + \beta_2 x_1 + \beta_3 x_1^2, \\ &\vdots \\ \nu_n &= \beta_1 + \beta_2 x_n + \beta_3 x_n^2, \end{aligned} \qquad \boldsymbol{\nu}(\beta_1, \beta_2, \beta_3) = \beta_1 \mathbf{x}_1 + \beta_2 \mathbf{x}_2 + \beta_3 \mathbf{x}_3,$$

where we write

$$\mathbf{x}_1 = \begin{pmatrix} 1 \\ \vdots \\ 1 \end{pmatrix}, \qquad \mathbf{x}_2 = \begin{pmatrix} x_1 \\ \vdots \\ x_n \end{pmatrix}, \qquad \mathbf{x}_3 = \begin{pmatrix} x_1^2 \\ \vdots \\ x_n^2 \end{pmatrix}.$$

The general level of sales of a certain commodity has an annual cyclic effect. Or, in general, consider a random system where the location of the response is a simple cyclic function of a position or time variable t. If t is conveniently scaled, the location for the response would have the form

$$\beta_1 + \beta_2 \sin t + \beta_3 \cos t = \beta_1 + \beta_4 \sin (t + \alpha)$$

with parameter $(\beta_1, \beta_2, \beta_3)$. Correspondingly, the location for the response vector $\mathbf{y}$ at various positions or times $t_1, \ldots, t_n$ would then be

$$\begin{aligned} \nu_1 &= \beta_1 + \beta_2 \sin t_1 + \beta_3 \cos t_1, \\ &\vdots \\ \nu_n &= \beta_1 + \beta_2 \sin t_n + \beta_3 \cos t_n, \end{aligned} \qquad \boldsymbol{\nu}(\beta_1, \beta_2, \beta_3) = \beta_1 \mathbf{x}_1 + \beta_2 \mathbf{x}_2 + \beta_3 \mathbf{x}_3,$$

where we write

$$\mathbf{x}_1 = \begin{pmatrix} 1 \\ \vdots \\ 1 \end{pmatrix}, \qquad \mathbf{x}_2 = \begin{pmatrix} \sin t_1 \\ \vdots \\ \sin t_n \end{pmatrix}, \qquad \mathbf{x}_3 = \begin{pmatrix} \cos t_1 \\ \vdots \\ \cos t_n \end{pmatrix}.$$

The general level of output from a chemical process is approximately quadratic in terms of operating pressure and amount of a particular catalyst. Or, in general, consider a system with two input variables x_1 and x_2 and suppose that it is necessary to allow for possible quadratic dependence of the response level on the two variables. The location for the response would then have the form

$$\beta_1 + \beta_2 x_1 + \beta_3 x_2 + \beta_4 x_1^2 + \beta_5 x_2^2 + \beta_6 x_1 x_2$$

with parameter $(\beta_1, \ldots, \beta_6)$. Correspondingly, the location for the response vector $\mathbf{y}$ with various input settings $(x_{11}, x_{21}), \ldots, (x_{1n}, x_{2n})$ would be

$$\boldsymbol{\nu}(\beta_1, \ldots, \beta_6) = \beta_1 \mathbf{x}_1 + \cdots + \beta_6 \mathbf{x}_6,$$

where $\mathbf{x}_1, \ldots, \mathbf{x}_6$ record successive values of the variables $1, x_1, x_2, x_1^2, x_2^2$, and $x_1 x_2$.

The general level of effectiveness of a language instruction program depends on which of three instructional methods is used. Or, in general, consider a system with a response y whose location depends on which of three treatments has been applied to the system. Let I_1, I_2, and I_3 be indicator variables for treatments 1, 2, and 3, respectively, and let β_1, β_2, and β_3 be the corresponding response locations. Then the location for the response can be presented as

$$\beta_1 I_1 + \beta_2 I_2 + \beta_3 I_3$$

with parameter $(\beta_1, \beta_2, \beta_3)$. Correspondingly, the location for the response vector $\mathbf{y}$ with, say, two observations using treatment 1, two observations using treatment 2, and two observations using treatment 3 would be

$$\begin{pmatrix} \nu_1 \\ \nu_2 \\ \nu_3 \\ \nu_4 \\ \nu_5 \\ \nu_6 \end{pmatrix} = \beta_1 \begin{pmatrix} 1 \\ 1 \\ 0 \\ 0 \\ 0 \\ 0 \end{pmatrix} + \beta_2 \begin{pmatrix} 0 \\ 0 \\ 1 \\ 1 \\ 0 \\ 0 \end{pmatrix} + \beta_3 \begin{pmatrix} 0 \\ 0 \\ 0 \\ 0 \\ 1 \\ 1 \end{pmatrix}, \qquad \boldsymbol{\nu}(\beta_1, \beta_2, \beta_3) = \beta_1 \mathbf{1}_1 + \beta_2 \mathbf{1}_2 + \beta_3 \mathbf{1}_3,$$

where $\mathbf{1}_1$, $\mathbf{1}_2$, and $\mathbf{1}_3$ are 1-vectors for the particular coordinates receiving the corresponding treatments.

DEFINITION 1

A model $\boldsymbol{\nu}(\boldsymbol{\beta})$ for the location of a response $\mathbf{y}$ is said to be **linear** if

$$\boldsymbol{\nu}(\boldsymbol{\beta}) = X\boldsymbol{\beta} = \beta_1 \mathbf{x}_1 + \cdots + \beta_r \mathbf{x}_r$$

and $\boldsymbol{\beta} \in R^r$; the vectors $\mathbf{x}_1, \ldots, \mathbf{x}_r$ are assembled side by side to form the $n \times r$ matrix X.

For some matrix details, see Supplement Section 2.5.

B THE LEAST-SQUARES SOLUTION

Now consider an observed response $\mathbf{y} = (y_1, \ldots, y_n)'$ and a linear location model

(1) $\boldsymbol{\nu}(\boldsymbol{\beta}) = X\boldsymbol{\beta}$,

where X is an $n \times r$ matrix consisting of r linearly independent vectors $\mathbf{x}_1, \ldots, \mathbf{x}_r$ (so that X has rank r) and the parameter $\boldsymbol{\beta}$ takes values in R^r.

The set of possible response locations is given by

$$\mathcal{L}(X) = \{b_1\mathbf{x}_1 + \cdots + b_r\mathbf{x}_r : b_1, \ldots, b_r \in R\};$$

it is the linear subspace generated by the column vectors of X; see Figure 9.4.

The least-squares estimate of $\boldsymbol{\beta}$ is the value $\mathbf{b} = \mathbf{b}(\mathbf{y})$ that minimizes

$$(2) \qquad S^2 = \sum_1^n (y_i - b_1x_{i1} - \cdots - b_rx_{ir})^2 = (\mathbf{y} - X\mathbf{b})'(\mathbf{y} - X\mathbf{b}).$$

As pictured in Figure 9.4, S^2 is the squared distance from $\mathbf{y}$ to $X\mathbf{b}$; the method is to choose the value for $\mathbf{b}$ that gets $X\mathbf{b}$ closest to the observation $\mathbf{y}$.

The least-squares estimate can be obtained by differentiating with respect to $b_1, \ldots, b_r$ as in formula (9.1.2):

$$(3) \qquad \begin{aligned} -\frac{1}{2}\frac{\partial S^2}{\partial b_1} &= \sum_1^n x_{i1}(y_i - b_1x_{i1} - \cdots - b_rx_{ir}) = 0, \\ &\vdots \\ -\frac{1}{2}\frac{\partial S^2}{\partial b_r} &= \sum_1^n x_{ir}(y_i - b_1x_{i1} - \cdots - b_rx_{ir}) = 0. \end{aligned}$$

These equations can be written in vector form as

$$(4) \qquad \begin{aligned} \mathbf{x}_1'(\mathbf{y} - X\mathbf{b}) &= 0 \\ &\vdots \\ \mathbf{x}_r'(\mathbf{y} - X\mathbf{b}) &= 0 \end{aligned} \qquad \text{or} \qquad X'(\mathbf{y} - X\mathbf{b}) = 0;$$

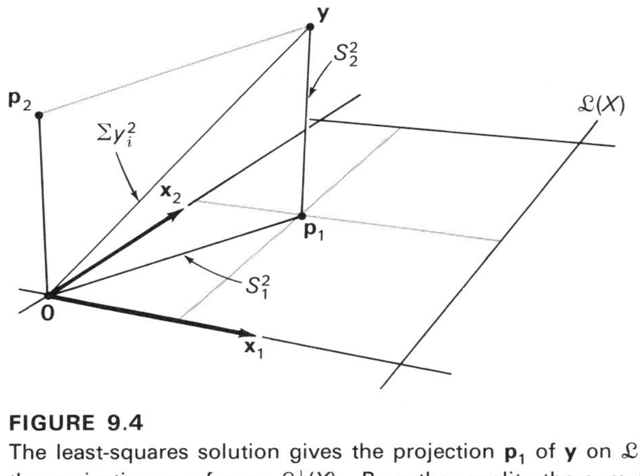

FIGURE 9.4
The least-squares solution gives the projection $\mathbf{p}_1$ of $\mathbf{y}$ on $\mathcal{L}(X)$; the residual is the projection $\mathbf{p}_2$ of $\mathbf{y}$ on $\mathcal{L}^{\perp}(X)$. By orthogonality the sum $S_1^2 + S_2^2$ of the squared lengths of the projections is the squared length $\Sigma\, y_i^2$ of the original vector.

they state that the residual vector $\mathbf{y} - X\mathbf{b}$ is orthogonal to $\mathbf{x}_1, \ldots, \mathbf{x}_r$; that is, orthogonal to $\mathcal{L}(X)$. Thus the least-squares solution gives the projection of $\mathbf{y}$ onto $\mathcal{L}(X)$: it has an orthogonal residual $\mathbf{y} - X\mathbf{b}$, which lies in $\mathcal{L}^{\perp}(X)$.

The equations (4) can be written as

$$(5) \qquad X'\mathbf{y} - X'X\mathbf{b} = 0$$

and solved to give the least-squares solution,

$$(6) \qquad \mathbf{b} = \mathbf{b}(\mathbf{y}) = (X'X)^{-1}X'\mathbf{y};$$

the coordinates of $\mathbf{b}$ are called *regression coefficients,* the regression coefficients of $\mathbf{y}$ on $\mathbf{x}_1, \ldots, \mathbf{x}_r$. Note that we assumed earlier that X has rank r and thus have that $X'X$ is a nonsingular symmetric $r \times r$ matrix with an inverse.

The least-squares fitted location is then given by

$$(7) \qquad \mathbf{p}_1 = X\mathbf{b}(y) = X(X'X)^{-1}X'y = P_1\mathbf{y};$$

this is the projection of $\mathbf{y}$ on $\mathcal{L}(X)$; see Figure 9.4. Note that $\mathbf{p}_1$ is obtained by applying the projection matrix

$$(8) \qquad P_1 = X(X'X)^{-1}X'$$

to the vector $\mathbf{y}$. The squared length of the projection $\mathbf{p}_1$ is easily obtained using equations (5) and (6):

$$(9) \qquad S_1^2 = \mathbf{b}'X'X\mathbf{b} = \mathbf{b}'X'\mathbf{y} = \mathbf{y}'X\mathbf{b}$$

$$(10) \qquad = \mathbf{y}'X(X'X)^{-1}X'\mathbf{y}.$$

The residual vector is given by

$$\mathbf{p}_2 = \mathbf{y} - X\mathbf{b}(\mathbf{y}) = (I - X(X'X)^{-1}X')\mathbf{y} = P_2\mathbf{y},$$

where I is the identity matrix. This is the projection of $\mathbf{y}$ on $\mathcal{L}^{\perp}(X)$, for clearly its "residual" $X\mathbf{b}$ in $\mathcal{L}(X)$ is orthogonal to $\mathcal{L}^{\perp}(X)$. Note that $\mathbf{p}_2$ is obtained by applying the projection matrix

$$(11) \qquad P_2 = I - X(X'X)^{-1}X'$$

to the vector $\mathbf{y}$. The squared length of the projection $\mathbf{p}_2$ is easily calculated. The orthogonality of $\mathbf{p}_1$ and $\mathbf{p}_2$ gives the decomposition

$$(12) \qquad \mathbf{y}'\mathbf{y} = (\mathbf{p}_1 + \mathbf{p}_2)'(\mathbf{p}_1 + \mathbf{p}_2) = \mathbf{p}_1'\mathbf{p}_1 + \mathbf{p}_2'\mathbf{p}_2;$$

thus the squared length of $\mathbf{p}_2$ is

$$(13) \qquad S_2^2 = \mathbf{y}'\mathbf{y} - S_1^2 = \mathbf{y}'\mathbf{y} - \mathbf{y}'X\mathbf{b}$$

$$(14) \qquad = \mathbf{y}'(I - X(X'X)^{-1}X')\mathbf{y}.$$

This decomposition into orthogonal components is presented in Table 9.4.

The least-squares method has been organized as a method of obtaining orthogonal components of a given vector. The geometry and algebra will allow us to examine some nontrivial and useful examples and will provide us with some basic results for the analysis of variance in Chapters 11 and 12.

TABLE 9.4
Components of vector **y**; corresponding squared lengths

Subspace	Dimension	Component	Squared Length
$\mathcal{L}(X)$	r	$\mathbf{p}_1 = X\mathbf{b}$	$S_1^2 = \mathbf{y}'X\mathbf{b}$
$\mathcal{L}^{\perp}(X)$	$n - r$	$\mathbf{p}_2 = \mathbf{y} - X\mathbf{b}$	$S_2^2 = \mathbf{y}'\mathbf{y} - \mathbf{y}'X\mathbf{b}$
	n	$\mathbf{y}$	$\mathbf{y}'\mathbf{y} = \sum y_i^2$

C NUMERICAL EXAMPLE

The general level of yield for a chemical process is approximately affine in terms of pressure measured in deviation x from some reference pressure. The yield y and input pressure x for five performances are as follows; also see Figure 9.5:

x	-2	-1	0	1	2
y	2	3	3	7	10

The location model is linear in terms of **1** and **x**; thus it has the form

$$\boldsymbol{\nu}(\beta_1, \beta_2) = \beta_1 \begin{pmatrix} 1 \\ 1 \\ 1 \\ 1 \\ 1 \end{pmatrix} + \beta_2 \begin{pmatrix} -2 \\ -1 \\ 0 \\ 1 \\ 2 \end{pmatrix} = \begin{pmatrix} 1 & -2 \\ 1 & -1 \\ 1 & 0 \\ 1 & 1 \\ 1 & 2 \end{pmatrix} \begin{pmatrix} \beta_1 \\ \beta_2 \end{pmatrix}$$

with two vectors **1** and **x** recording the values of 1 and x. We now fit this model using the least-squares formulas of Section B.

The vectors **1** and **x** have the following inner product matrix and inverse matrix:

$$X'X = \begin{pmatrix} 5 & 0 \\ 0 & 10 \end{pmatrix}, \qquad (X'X)^{-1} = \begin{pmatrix} 1/5 & 0 \\ 0 & 1/10 \end{pmatrix}.$$

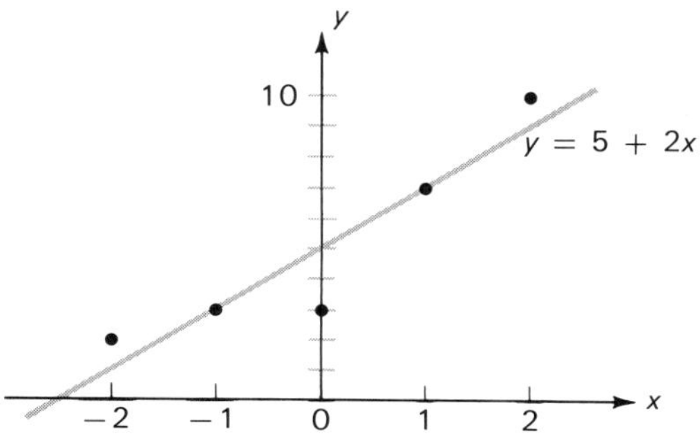

FIGURE 9.5
Fitting the line $y = b_1 + b_2 x$ to the five points as marked.

The vectors **1** and **x** have the following inner product matrix with **y**:

$$X'y = \begin{pmatrix} 25 \\ 20 \end{pmatrix}.$$

The least-squares estimate of $\boldsymbol{\beta}$ is

$$\mathbf{b} = (X'X)^{-1}X'y = \begin{pmatrix} 1/5 & 0 \\ 0 & 1/10 \end{pmatrix}\begin{pmatrix} 25 \\ 20 \end{pmatrix} = \begin{pmatrix} 5 \\ 2 \end{pmatrix}.$$

The fitted vector is the projection

$$\mathbf{p}_1 = X\mathbf{b} = \begin{pmatrix} 1 & -2 \\ 1 & -1 \\ 1 & 0 \\ 1 & 1 \\ 1 & 2 \end{pmatrix}\begin{pmatrix} 5 \\ 2 \end{pmatrix} = \begin{pmatrix} 1 \\ 3 \\ 5 \\ 7 \\ 9 \end{pmatrix}$$

onto $\mathcal{L}(X)$, and the fitted line is $y = 5 + 2x$. The squared length of the fitted vector is

$$S_1^2 = \mathbf{y}'X\mathbf{b} = (25, 20)\begin{pmatrix} 5 \\ 2 \end{pmatrix} = 165.$$

The residual vector is the projection

$$\mathbf{p}_2 = \mathbf{y} - X\mathbf{b} = \begin{pmatrix} 2 \\ 3 \\ 3 \\ 7 \\ 10 \end{pmatrix} - \begin{pmatrix} 1 \\ 3 \\ 5 \\ 7 \\ 9 \end{pmatrix} = \begin{pmatrix} 1 \\ 0 \\ -2 \\ 0 \\ 1 \end{pmatrix}$$

onto $\mathcal{L}^{\perp}(X)$. The squared length of the residual is

$$S_2^2 = \mathbf{y}'\mathbf{y} - \mathbf{y}'X\mathbf{b} = 171 - 165 = 6.$$

The decomposition into orthogonal components is summarized in Table 9.5.

TABLE 9.5
Components of vector **y**; corresponding squared lengths

Subspace	Dimension	Component	Squared Length
$\mathcal{L}(\mathbf{1}, \mathbf{x})$	2	(1, 3, 5, 7, 9)	165
$\mathcal{L}^{\perp}(\mathbf{1}, \mathbf{x})$	3	(1, 0, −2, 0, 1)	6
	5	(2, 3, 3, 7, 10)	171

D EXERCISES

1 The response vector $\mathbf{y} = (3, 7, 6, 5, 9)'$ was obtained with pressure settings given by $\mathbf{x} = (-2, -1, 0, 1, 2)'$; see the example in Section C.
(a) Calculate the projection $\mathbf{p}_1$ and the squared length S_1^2; record the fitted line.
(b) Calculate the projection $\mathbf{p}_2$ and the squared length S_2^2.
(c) Record the orthogonal decomposition in a table.

2 The general level of a yield y is approximately affine in terms of an input variable x. The following data were obtained:

x	1	2	3	4	5
y	2	3	3	7	10

Fit the linear location model $\mathbf{m} = \boldsymbol{\beta}_1 \mathbf{1} + \boldsymbol{\beta}_2 \mathbf{x}$, where $\mathbf{1}$ and $\mathbf{x}$ record the values of 1 and x.

(a) Calculate the projection $\mathbf{p}_1$ and the squared length S_1^2; record the fitted line.

(b) Calculate the projection $\mathbf{p}_2$ and the squared length S_2^2.

(c) Record the orthogonal decomposition in a table.

3 The response vector $\mathbf{y} = (3, 7, 6, 5, 9)'$ was obtained with input settings $\mathbf{x} = (1, 2, 3, 4, 5)'$; see Exercise 2.

(a) Calculate the projection $\mathbf{p}_1$ and the squared length S_1^2; record the fitted line.

(b) Calculate the projection $\mathbf{p}_2$ and the squared length S_2^2.

(c) Record the orthogonal decomposition in a table.

4 For the example in Section C, calculate the projection matrices P_1 and P_2.

5 For the example in Exercise 2, calculate the projection matrices P_1 and P_2.

6 Let $(y_{11}, y_{12}, y_{21}, y_{22})$ have location model

$$\begin{pmatrix} m_{11} \\ m_{12} \\ m_{21} \\ m_{22} \end{pmatrix} = \beta_0 \begin{pmatrix} 1 \\ 1 \\ 1 \\ 1 \end{pmatrix} + \beta_1 \begin{pmatrix} 1 \\ 1 \\ -1 \\ -1 \end{pmatrix}.$$

Find the least-squares estimate of (β_0, β_1) and record the projection $\mathbf{p}_1$.

7 Let $(y_{11}, \ldots, y_{1m}, y_{21}, \ldots, y_{2n}, y_{31}, \ldots, y_{3p})$ have location model $(\theta_1, \ldots, \theta_1, \theta_2, \ldots, \theta_2, \theta_3, \ldots, \theta_3)$ with $(\theta_1, \theta_2, \theta_3)$ in R^3; see the last example in Section A. Find the least-squares estimate of $(\theta_1, \theta_2, \theta_3)$.

(a) Calculate the projection $\mathbf{p}_1$ and the squared length S_1^2.

(b) Calculate the projection $\mathbf{p}_2$ and the squared length S_2^2.

8 Suppose that we allow the possibility of quadratic dependence for the example in Section C. Consider the response vector $\mathbf{y} = (2, 3, 3, 7, 10)'$ and the linear location model

$$\mathbf{m}(\beta_1, \beta_2, \beta_3) = \beta_1 \begin{pmatrix} 1 \\ 1 \\ 1 \\ 1 \\ 1 \end{pmatrix} + \beta_2 \begin{pmatrix} -2 \\ -1 \\ 0 \\ 1 \\ 2 \end{pmatrix} + \beta_3 \begin{pmatrix} 2 \\ -1 \\ -2 \\ -1 \\ 2 \end{pmatrix} = \begin{pmatrix} 1 & -2 & 2 \\ 1 & -1 & -1 \\ 1 & 0 & -2 \\ 1 & 1 & -1 \\ 1 & 2 & 2 \end{pmatrix} \begin{pmatrix} \beta_1 \\ \beta_2 \\ \beta_3 \end{pmatrix}$$

with three vectors recording the values of $1, x, x^2 - 2$. Find the least-squares estimate of $(\beta_1, \beta_2, \beta_3)$.

(a) Calculate the projection $\mathbf{p}_1$ and the squared length S_1^2; record the fitted parabola $y = b_1 + b_2 x + b_3(x^2 - 2)$.

(b) Calculate the projection $\mathbf{p}_2$ and the squared length S_2^2.

(c) Record the orthogonal decomposition in a table.

9 Let $X\mathbf{b}_1$ and $X\mathbf{b}_2$ be two points in the location subspace $\mathcal{L}(X)$. Show that the squared distance between them is given by the quadratic form

$$(15) \qquad (\mathbf{b}_2 - \mathbf{b}_1)' X' X (\mathbf{b}_2 - \mathbf{b}_1).$$

E PROBLEMS

10 Consider the response $\mathbf{y} = (y_1, \ldots, y_n)'$ and the linear location model

$$\mathbf{m} = \beta_1 \begin{pmatrix} 1 \\ \vdots \\ 1 \end{pmatrix} + \beta_2 \begin{pmatrix} x_1 - \bar{x} \\ \vdots \\ x_n - \bar{x} \end{pmatrix}.$$

(a) Show that the least-squares estimate of (β_1, β_2) is

(16) $$(b_1, b_2) = \left(\bar{y}, \left(\sum (x_i - \bar{x})y_i\right) \Big/ \left(\sum (x_i - \bar{x})^2\right)\right)$$

and that the fitted line is $y = \bar{y} + b_2(x - \bar{x})$.

(b) Show that the squared length of the residual is

(17) $$S_2^2 = \sum (y_i - b_1 - b_2(x_i - \bar{x}))^2 = \sum y_i^2 - n\bar{y}^2 - \frac{\left[\sum (x_i - \bar{x})y_i\right]^2}{\sum (x_i - \bar{x})^2}.$$

11 Consider the response $\mathbf{y} = (y_1, \ldots, y_n)'$ and the linear location model

$$\mathbf{m} = \alpha \begin{pmatrix} 1 \\ \vdots \\ 1 \end{pmatrix} + \beta \begin{pmatrix} x_1 \\ \vdots \\ x_n \end{pmatrix}.$$

(a) Show that the least-squares estimate of (α, β) is

(18) $$(a, b) = (\bar{y} - b\bar{x}, b),$$

where $b = (\Sigma (x_i - \bar{x})y_i)/(\Sigma (x_i - \bar{x})^2)$, and that the fitted line is $y = \bar{y} + b(x - \bar{x})$.

(b) Show that the squared length of the residual is

(19) $$S_2^2 = \sum (y_i - a - bx_i)^2 = \sum y_i^2 - n\bar{y}^2 - \frac{\left(\sum (x_i - \bar{x})y_i\right)^2}{\sum (x_i - \bar{x})^2}.$$

12 The projection $\mathbf{p}_1$ of $\mathbf{y}$ into $\mathcal{L}(X)$ is obtained by applying the projection matrix $P_1 = X(X'X)^{-1}X'$ to the response $\mathbf{y}$; the projection $\mathbf{p}_2$ of $\mathbf{y}$ into $\mathcal{L}^{\perp}(X)$ is obtained by applying the projection matrix $P_2 = I - X(X'X)^{-1}X'$ to the response $\mathbf{y}$.

(a) A matrix P is idempotent if $PP = P$. Show that P_1 and P_2 are idempotent.

(b) An idempotent matrix is an orthogonal projection matrix if it is symmetric: $P = P'$. Show that P_1 and P_2 are symmetric.

13 The least-squares solution for a linear model can be obtained by algebra rather than calculus. The sum of squares in formula (2) can be written as

(20) $$S^2 = (\mathbf{y} - X\mathbf{b})'(\mathbf{y} - X\mathbf{b})$$

$$S^2 = \mathbf{b}'X'X\mathbf{b} - \mathbf{b}'X'\mathbf{y} - \mathbf{y}'X\mathbf{b} + \mathbf{y}'\mathbf{y}$$

$$= (\mathbf{b} - \cdots)'X'X(\mathbf{b} - \cdots) + \cdots.$$

Choose an entry for the blank locations so that the third line multiplied out produces the second line. Thus deduce (a) the least-squares solution for $\mathbf{b}$; (b) the squared length of the residual vector.

14 Consider a response $\mathbf{y} = (y_1, \ldots, y_n)$ and the linear location model $X\boldsymbol{\beta}$ with $\boldsymbol{\beta}$ in R^r. If the various coordinates are expected to differ from the true location by differing amounts, then a weighted-least-squares method is sometimes used and is based on minimizing

$$S^2 = (\mathbf{y} - X\mathbf{b})'M^{-1}(\mathbf{y} - X\mathbf{b}),$$

where M is some appropriate positive definite symmetric matrix. Use the method in Problem 13 to show that the weighted-least-squares estimate is

$$\mathbf{b} = (X'M^{-1}X)^{-1}X'M^{-1}\mathbf{y}$$

and the squared length of the residual is

$$S_2^2 = \mathbf{y}'M^{-1}\mathbf{y} - \mathbf{y}'M^{-1}X\mathbf{b}.$$

Examples of this weighted least squares will be given later.

9.3

COMBINING UNBIASED ESTIMATORS

The theory of estimation is concerned with finding a function on the sample space whose values will cluster in a reasonable manner about the value of a particular parameter of interest. In an application there would be substantial hope, then, that the calculated value of such a function was a good approximation to the true value of the parameter.

The great wealth of functions on a sample space leads to a major problem in choosing one that is in some sense best. The concept that has been most fruitful both theoretically and in applications is that of *unbiasedness,* the central concept in this and the next two sections. There are other approaches, such as the decision theoretic, which provide extensions and alternatives, but none with the immediate relevance and usefulness of the unbiasedness approach. For example, with 10 measurements on the gravitational constant, should we use the average, $\bar{y}$, or the median, $\tilde{y}$, or some other sample function to obtain our combined estimate of the gravitational constant?

A UNBIASEDNESS

For the statistical model we let y be the response with values in a sample space $\mathcal{S}$, and θ be the parameter with values in a parameter space Ω. The parameter θ indexes the various distributions for the response y; in an application one of the parameter values is the true value which designates the distribution that is a reasonable approximation to the actual distribution of the response. We can now be somewhat more general than in Chapter 8: rather than record a density function for each parameter value, we merely assume that each parameter value designates a distribution on the sample space, a distribution that can be specified in some way—by density function, or by distribution function, or by probability measure.

Now suppose that we are interested in a particular real-valued parameter $\beta(\theta)$ and that $t(y)$ is some real-valued function that we are considering as an estimator for $\beta(\theta)$. Note that t is a function from $\mathcal{S}$ into R and β is a function from Ω into R. Then we define:

DEFINITION 1

A function t is an **unbiased estimator** for a parameter β if

(1) $\quad E(t(y)\,|\,\theta) = \beta(\theta)$

for each θ in Ω.

As an example, consider a sample of n from some distribution on R that has a finite mean μ. The mean μ can be viewed as a function of the distribution from which it is calculated. From formula (5.4.5) or (7.2.1) we have

$$E(\bar{y}) = \mu$$

for each distribution with finite mean. Thus in this general context $\bar{y}$ is an unbiased estimator of μ.

As a second example, consider a sample of n from some distribution on R that has a finite variance σ^2. The variance σ^2 can be viewed as a function of the distribution from which it is calculated. From formula (5.4.9) or (7.2.13) we have

$$E(s_y^2) = \sigma^2$$

for each distribution with finite variance. Thus in this general context s_y^2 is an unbiased estimator of σ^2.

Now consider a much more specific model involving a sample from the normal(μ, σ). Of course, $\bar{y}$ remains an unbiased estimator of μ, and s_y^2 remains an unbiased estimator of σ^2. In fact, it is true generally: if t is an unbiased estimate of β for a model with θ in Ω, then t is unbiased for the more specific model with θ in Ω^*, where $\Omega^* \subset \Omega$.

In Section 7.2 we gave a general justification for the use of unbiasedness: essentially that the values of t are sometimes larger than the true β and sometimes smaller than the true β but that the mean value is equal to β; that is, the distribution of t is located in a certain sense at the parameter value corresponding to the underlying response distribution.

The definition of unbiasedness extends immediately to cover a vector estimator $\mathbf{t}(y) = (t_1(y), \ldots, t_r(y))'$ for a vector parameter $\boldsymbol{\beta}(\theta) = (\beta_1(\theta), \ldots, \beta_r(\theta))'$:

DEFINITION 2

A function $\mathbf{t}$ is an **unbiased estimator** for a parameter $\boldsymbol{\beta}$ if

$$(2) \qquad E(\mathbf{t}(y) \mid \theta) = \boldsymbol{\beta}(\theta)$$

for each θ in Ω; that is, if

$$E(t_u(y) \mid \theta) = \beta_u(\theta), \qquad u = 1, 2, \ldots, r,$$

for each θ in Ω.

As an example consider a sample of n from a distribution on R that has a finite mean μ and finite variance σ^2. Then from the early discussions we have that $(\bar{y}, s_y^2)$ is an unbiased estimator for (μ, σ^2).

B COMPARISON OF UNBIASED ESTIMATORS

The property of unbiasedness says that the distribution of an estimator is located in a certain sense at the value of the parameter being estimated. We now consider the comparison of unbiased estimators. For this we would think immediately of the scale of the distribution and examine, say, the variance or perhaps the reciprocal variance or precision.

Sometimes it is convenient to have a numerical comparison of estimators. For this we give a definition of the comparative efficiency of two unbiased estimators of a parameter.

DEFINITION 3

The **efficiency** of an unbiased estimator t_1 of $\beta(\theta)$ with respect to an unbiased estimator t_2 of $\beta(\theta)$ is

$$(3) \qquad \text{Eff}\,(t_1, t_2) = \frac{1/\text{var}\,(t_1|\theta)}{1/\text{var}\,(t_2|\theta)} = \frac{\text{precision of } t_1}{\text{precision of } t_2} = \frac{\text{var}\,(t_2|\theta)}{\text{var}\,(t_1|\theta)}.$$

The efficiency is a function of θ—but for a lot of common examples it turns out to be independent of θ. Some other aspects of efficiency are examined in Section 9.4.

EXAMPLE 1

The normal(μ, σ): Let y be a sample from the normal(μ, σ) with (μ, σ) in $R \times R^+$. We have examined the sample average $\bar{y}$ as a natural estimator of μ; we have

$$E(\bar{y}|\mu, \sigma) = \mu, \qquad \text{var}\,(\bar{y}|\mu, \sigma) = \sigma^2/n.$$

The sample median $\tilde{y}$ is also a very simple function on the sample space, and we would naturally associate it with the center of the distribution (for an even sample let $\tilde{y}$ be the average of the 2 central values); we have

$$E(\tilde{y}|\mu, \sigma) = \mu, \qquad \text{var}\,(\tilde{y}|\mu, \sigma) \approx \pi\sigma^2/2n;$$

the mean follows from the symmetry of the distribution of $\tilde{y} - \mu$; the approximate variance is indicated by Problem 6.1.13. Thus the efficiency of $\tilde{y}$ with respect to $\bar{y}$ is

$$\text{Eff}\,(\tilde{y}, \bar{y}) = \frac{2n/\pi\sigma^2}{n/\sigma^2} = \frac{2}{\pi} \approx 0.64.$$

The variance of a reasonable estimator of a parameter is often proportional to the reciprocal of the sample size; this happens commonly with models that have a fixed region of positive density. We then have a simple interpretation for efficiency. Suppose we compare the use of the median $\tilde{y}$ for a sample of size 100 with the use of the mean $\bar{y}$ for a sample of size 64. We then have

$$E(\tilde{y}) = \mu, \qquad \text{var}\,(\tilde{y}) = \pi\sigma^2/200 = \sigma^2/64,$$

$$E(\bar{y}) = \mu, \qquad \text{var}\,(\bar{y}) = \sigma^2/64.$$

Thus the two estimators have the same location and the same scale, and thus in a sense are equivalent. And we can then, in this context, say that $\tilde{y}$ uses 64 percent of a sample by comparison with $\bar{y}$.

The comparison of vector estimates is somewhat more complicated. Each coordinate of an estimator has a variance, and in addition each pair of coordinates has a covariance. Consider an unbiased estimator $\mathbf{t}$ of the parameter $\boldsymbol{\beta}(\theta)$, and let $\Sigma(\theta)$ be the variance matrix:

$$(4) \qquad \Sigma(\theta) = \begin{pmatrix} \sigma_{11} & \cdots & \sigma_{1r} \\ \vdots & & \\ \sigma_{r1} & \cdots & \sigma_{rr} \end{pmatrix};$$

then σ_{11} is the variance of the first coordinate $t_1, \ldots, \sigma_{rr}$ is the variance of the rth coordinate t_r, and, for example, σ_{r1} is the covariance of t_r and t_1.

Now suppose that we consider a real-valued parameter $l_1\beta_1(\theta) + \cdots + l_r\beta_r(\theta) = \boldsymbol{l}'\boldsymbol{\beta}(\theta)$, where $l_1, \ldots, l_r$ are given real numbers. Clearly the estimator $l_1t_1 + \cdots + l_rt_r = \boldsymbol{l}'\mathbf{t}$ is an unbiased estimator of $\boldsymbol{l}'\boldsymbol{\beta}(\theta)$; its variance can be obtained from (5.3.18),

$$\text{var}\,(\boldsymbol{l}'\mathbf{t}) = \boldsymbol{l}'\Sigma(\theta)\boldsymbol{l} \tag{5}$$

which is a quadratic form with matrix $\Sigma(\theta)$.

Consider the comparison of two unbiased estimators $\mathbf{t}_1$ and $\mathbf{t}_2$ for the parameter $\boldsymbol{\beta}(\theta)$. We will say that $\mathbf{t}_1$ is *better* than $\mathbf{t}_2$ if

$$\Sigma_{22}(\theta) - \Sigma_{11}(\theta) \tag{6}$$

is positive semidefinite. If (6) holds, we will say that the matrix $\Sigma_{11}(\theta)$ is smaller than the matrix $\Sigma_{22}(\theta)$. For some background detail, see Supplement Section 2.5C.

If $\mathbf{t}_1$ is better than $\mathbf{t}_2$ then the relation (6) says that the first coordinate of $\mathbf{t}_1$ has smaller ($\leq$) variance than the first coordinate of $\mathbf{t}_2, \ldots$, and that the rth coordinate of $\mathbf{t}_1$ has smaller ($\leq$) variance than the rth coordinate of $\mathbf{t}_2$. In addition, the relation (6) together with (5) shows that the variance of $\boldsymbol{l}'\mathbf{t}_1$ is smaller ($\leq$) than the variance of $\boldsymbol{l}'\mathbf{t}_2$ for any given vector $\boldsymbol{l}$: note that these are unbiased estimates of $\boldsymbol{l}'\boldsymbol{\beta}(\theta)$ and use details from Section 2.5C.

C COMBINING UNBIASED ESTIMATORS

Suppose that we have two unbiased estimators t_1 and t_2 for the real-valued parameter $\beta(\theta)$ and wish to examine linear combinations as possible estimators for $\beta(\theta)$.

THEOREM 4

If t_1 and t_2 are unbiased estimators of $\beta(\theta)$, then

(a) *A linear combination of t_1 and t_2 that is unbiased for $\beta(\theta)$ has the form*

$$at_1 + (1 - a)t_2, \tag{7}$$

where the sum of the weights is 1.

(b) *Among such linear unbiased estimators the variance at the parameter value θ_0 is minimized by having*

$$\frac{a}{1-a} = \frac{1/(\text{var}\,(t_1|\theta_0) - \text{cov}\,(t_1, t_2|\theta_0))}{1/(\text{var}\,(t_2|\theta_0) - \text{cov}\,(t_1, t_2|\theta_0))}; \tag{8}$$

the weights are proportional to the reciprocal of excess of variance over covariance.

(c) *If the estimators are independent or uncorrelated, then the variance at the parameter value θ_0 is minimized by having*

$$\frac{a}{1-a} = \frac{1/\text{var}\,(t_1|\theta_0)}{1/\text{var}\,(t_2|\theta_0)}; \tag{9}$$

the weights are proportional to the precisions.

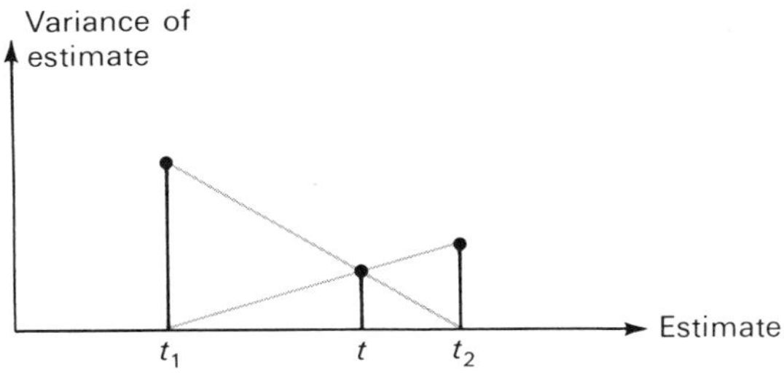

FIGURE 9.6
Independent estimates t_1 and t_2; the minimum variance linear combination t.

In the uncorrelated case the precision (at θ_0) of this best linear unbiased estimator is

$$\frac{1}{\text{var }(t_1|\theta_0)} + \frac{1}{\text{var }(t_2|\theta_0)}; \tag{10}$$

see Figure 9.6. We consider two examples before recording the proof.

EXAMPLE 2

Two instruments are available for measuring chlorophyll content. Five measurements with the first instrument gave an estimate $t_1 = 3.17$; the estimating procedure is unbiased and has var $(t_1) = (0.11)^2$; five measurements with the second instrument gave an estimate $t_2 = 3.11$; the estimating procedure is unbiased and has var $(t_2) = (0.07)^2$. By formula (9) the best linearly combined estimate is

$$t = \frac{\dfrac{1}{0.0121}\,3.17 + \dfrac{1}{0.0049}\,3.11}{\dfrac{1}{0.0121} + \dfrac{1}{0.0049}} = 3.127;$$

and by formula (10) it has variance

$$\text{var }(t) = \left(\frac{1}{0.0121} + \frac{1}{0.0049}\right)^{-1} = 0.0035 \approx (0.06)^2.$$

EXAMPLE 3

An incomplete block design in an agricultural investigation gives two unbiased estimates t_1 and t_2 of a yield comparison. Some theoretical calculations give

$$\text{var }(t_1) = 0.126\sigma^2, \qquad \text{var }(t_2) = 0.276\sigma^2, \qquad \text{cov }(t_1, t_2) = -0.100\sigma^2.$$

By formula (8) the best linear unbiased estimator is

$$t = \left(\frac{1}{0.226} + \frac{1}{0.376}\right)^{-1}\left(\frac{t_1}{0.226} + \frac{t_2}{0.376}\right).$$

Proof of Theorem 4 (a) Consider a linear combination $at_1 + bt_2$ that is unbiased for $\beta(\theta)$; then

$$E(at_1(y) + bt_2(y)) = a\beta(\theta) + b\beta(\theta) = \beta(\theta),$$

and hence $b = 1 - a$, provided, of course, that we exclude the ultratrivial $\beta(\theta) \equiv 0$.

(b) The variance of $at_1 + (1 - a)t_2$ at θ_0 is

$$a^2 \operatorname{var}(t_1|\theta_0) + 2a(1 - a)\operatorname{cov}(t_1, t_2|\theta_0) + (1 - a)^2 \operatorname{var}(t_2|\theta_0).$$

Setting the derivative with respect to a equal to zero gives

$$2a(\operatorname{var}(t_1|\theta_0) - \operatorname{cov}(t_1, t_2|\theta_0)) = 2(1 - a)(\operatorname{var}(t_2|\theta_0) - \operatorname{cov}(t_1, t_2|\theta_0)),$$

which is equivalent to the quoted condition. Alternatively, we can complete the square as suggested in Problem 9.2.13.

There is an analogous version of Theorem 4 which covers vector estimators. For simplicity we give here the special version for uncorrelated vector estimators:

THEOREM 5

If $\mathbf{t}_1$ and $\mathbf{t}_2$ are uncorrelated unbiased estimators for $\boldsymbol{\beta}(\theta)$ with variance matrices $\Sigma_{11}(\theta)$ and $\Sigma_{22}(\theta)$, then

(a) *Any linear combination of $\mathbf{t}_1$ and $\mathbf{t}_2$ that is unbiased for $\boldsymbol{\beta}$ has the form*

(11) $$A\mathbf{t}_1 + (I - A)\mathbf{t}_2,$$

where A is an $r \times r$ matrix and I is the $r \times r$ identity matrix [provided that $\beta_1(\theta), \ldots, \beta_r(\theta)$ are linearly independent].

(b) *Among such linear unbiased estimators the variance matrix at the parameter value θ_0 is minimized by the estimator*

(12) $$(\Sigma_{11}^{-1}(\theta_0) + \Sigma_{22}^{-1}(\theta_0))^{-1}(\Sigma_{11}^{-1}(\theta_0)\mathbf{t}_1 + \Sigma_{22}^{-1}(\theta_0)\mathbf{t}_2);$$

the weighting matrices are "proportional" to the inverse variance matrices or the precision matrices.

(c) *The precision matrix (at θ_0) of this best linear unbiased estimator is*

(13) $$\Sigma_{11}^{-1}(\theta_0) + \Sigma_{22}^{-1}(\theta_0),$$

the sum of the component precisions.

The proof follows as a special case of the generalized Gauss–Markov Theorem in Problem 11.

D THE GAUSS–MARKOV THEOREM FOR LINEAR LOCATION

The Gauss–Markov theorem is concerned with unbiased estimation for models that have the linear location structure discussed in Section 9.2. For many applications it provides theoretical support for the use of the least-squares and weighted-least-squares methods.

Consider a response vector $\mathbf{y} = (y_1, \ldots, y_n)'$ with a linear location model $X\boldsymbol{\beta}$ as discussed in Section 9.2A. Specifically, consider

$$\begin{pmatrix} y_1 \\ \vdots \\ y_n \end{pmatrix} = \begin{pmatrix} x_{11} & \cdots & x_{1r} \\ \vdots & & \vdots \\ x_{n1} & \cdots & x_{nr} \end{pmatrix} \begin{pmatrix} \beta_1 \\ \vdots \\ \beta_r \end{pmatrix} + \begin{pmatrix} v_1 \\ \vdots \\ v_n \end{pmatrix},$$

(14) $$\mathbf{y} = X\boldsymbol{\beta} + \mathbf{v},$$

where X is an $n \times r$ matrix of full rank r and where the variation $\mathbf{v} = (v_1, \ldots, v_n)'$ has mean value $\mathbf{0}$. For the present we consider the case where the coordinates have common variance σ^2 and zero covariance

(15) $$\text{VAR}\,(\mathbf{v}) = \begin{pmatrix} \sigma^2 & & 0 \\ & \ddots & \\ 0 & & \sigma^2 \end{pmatrix} = \sigma^2 I.$$

As an example, consider a response that has linear location in terms of an input variable x and has common variance for all values of x. Then for n observations, we have

$$\begin{aligned} y_1 &= \beta_1 + \beta_2 x_1 + v_1, \\ &\vdots \\ y_n &= \beta_1 + \beta_2 x_n + v_n, \end{aligned}$$

where the v's have common variance σ^2 and zero covariance.

There is some interest in thinking of the y's as providing indirect estimators for the parameters. We have that

y_1	is unbiased estimator for	$\beta_1 + \beta_2 x_1$
y_2	is unbiased estimator for	$\beta_1 + \beta_2 x_2$
$\dfrac{y_2 - y_1}{x_2 - x_1}$	is unbiased estimator for	β_2
$\dfrac{x_2 y_1 - x_1 y_2}{x_2 - x_1}$	is unbiased estimator for	β_1

and similarly for any pair with $x_j - x_i \neq 0$. Thus in looking for best estimates of β_1 and β_2, we can think in terms of combining or taking linear combinations of the indirect estimators $y_1, \ldots, y_n$.

THEOREM 6

(a) *A linear estimator* $A\mathbf{y}$ *based on an* $r \times n$ *matrix* A *is an unbiased estimator of* $\boldsymbol{\beta}$ *if* $AX = I$.

(b) *Among such linear unbiased estimators the variance matrix is smallest for the least-squares estimator*

(16) $$\mathbf{b} = (X'X)^{-1}X'\mathbf{y}.$$

(c) *The variance matrix of this best linear unbiased estimator is*

(17) $$(X'X)^{-1}\sigma^2.$$

The proof is recorded in Section E. For the definition of smaller matrices see Supplement Section 2.5C.

EXAMPLE 4

Consider the example concerning the yield from a chemical process as discussed in Section 9.2C. We now suppose that the variation has independent coordinates with common variance σ^2. The observed response is $\mathbf{y} = (2, 3, 3, 7, 10)'$ and the model is

$$\begin{pmatrix} y_1 \\ y_2 \\ y_3 \\ y_4 \\ y_5 \end{pmatrix} = \beta_1 \begin{pmatrix} 1 \\ 1 \\ 1 \\ 1 \\ 1 \end{pmatrix} + \beta_2 \begin{pmatrix} -2 \\ -1 \\ 0 \\ 1 \\ 2 \end{pmatrix} + \begin{pmatrix} v_1 \\ v_2 \\ v_3 \\ v_4 \\ v_5 \end{pmatrix}.$$

Theorem 5 is concerned with estimators for (β_1, β_2) that are linearly constructed from the response vector; the theorem says that the least-squares estimator $\mathbf{b} = (b_1, b_2)'$ has the minimum variance matrix. From Section 9.2C we have the estimate

$$\mathbf{b} = \begin{pmatrix} 5 \\ 2 \end{pmatrix}.$$

The variance matrix of this estimator is

$$(X'X)^{-1}\sigma^2 = \begin{pmatrix} 1/5 & 0 \\ 0 & 1/10 \end{pmatrix} \sigma^2.$$

Thus 5 is the calculated value of the unbiased estimator b_1 for β_1; this estimator has variance $\sigma^2/5$, which is the minimum among linear unbiased estimators of β_1. And 2 is the calculated value of the unbiased estimator b_2 for β_2; this estimator has variance $\sigma^2/10$, which is the minimum among linear unbiased estimators of β_2. We will see later that an unbiased estimator for σ^2 is given by

$$s^2 = \frac{S_2^2}{n - r},$$

where S_2 is the squared length of the residual; the value here of $s^2 = 6/3 = 2$.

E PROOF OF THE GAUSS–MARKOV THEOREM

First suppose that $A\mathbf{y}$ is unbiased estimator for $\boldsymbol{\beta}$. Then

$$E(A\mathbf{y}) = AE(\mathbf{y}) = AX\boldsymbol{\beta} = \boldsymbol{\beta};$$

and it follows that $AX = I$, as asserted in part (a) of the theorem.

Now consider the variance matrix for the estimators $A\mathbf{y}$. From formula (5.3.17) we have

$$\text{(18)} \qquad \text{VAR}\,(A\mathbf{y}) = A\sigma^2 IA' = \sigma^2 AA'.$$

Thus our problem is to examine matrices A that satisfy $AX = I$ in search of one that minimizes the matrix (18).

We follow the method of completing the square as suggested for a simpler context in Problem 9.2.13. The matrix AA' has already the form of a "completed square"; we rearrange it, however, as a "completed square" in a way that makes essential use of the constraint $AX = I$ satisfied by A,

$$(19) \qquad AA' = (A - \cdots)(A - \cdots)' + DD',$$

where DD' is an inner product matrix and the entry in the parentheses needs to use $AX = I$. We try the entry CX', which produces the products AX and $X'A'$ and has the yet undetermined matrix C:

$$(20) \qquad \begin{aligned} AA' &= (A - CX')(A - CX')' + DD' \\ &= AA' - AXC' - CX'A' + CX'XC' + DD' \\ &= AA' - C' - C + CX'XC' + DD'. \end{aligned}$$

The middle three terms collapse to one term if we choose $C = (X'X)^{-1}$:

$$AA' = AA' - (X'X)^{-1} - (X'X)^{-1} + (X'X)^{-1} + DD'.$$

This determines DD' and we obtain

$$(21) \qquad AA' = (A - (X'X)^{-1}X')(A - (X'X)^{-1}X')' + (X'X)^{-1}.$$

Note that $(X'X)^{-1}$ is positive definite and that

$$(22) \qquad (A - (X'X)^{-1}X')(A - (X'X)^{-1}X')'$$

is an inner product matrix and is thus positive semidefinite; for some background details see Supplement Section 2.5C. From formula (21) we then see that the matrix AA' subject to $AX = I$ can be made smallest by taking the inner product matrix (22) to the zero matrix; this choice for $A = (X'X)^{-1}X'$ gives

$$A\mathbf{y} = (X'X)^{-1}X'\mathbf{y},$$

which is the least-squares estimate as quoted in part (b) of the theorem.

The least-squares estimate makes the first term on the right of (21) equal to the zero matrix and thus $A'A$ equal to $(X'X)^{-1}$; it follows that the variance matrix for the least-squares estimate is

$$\sigma^2 AA' = \sigma^2 (X'X)^{-1},$$

as quoted in part (c) of the theorem.

F EXERCISES

1 A first determination of a distance by a surveyor has standard deviation 0.021; the observed value is 257.341; an independent determination has standard deviation 0.037; the observed value 257.361. Determine the best linearly combined estimate.

2 Let $\bar{y}_1 = \Sigma\, y_{1i}/m$ be an estimate of μ based on a first sample of m and $\bar{y}_2 = \Sigma\, y_{2j}/n$ be an estimate of μ based on a second sample of n from a distribution with finite variance. Derive the linear combination of $\bar{y}_1$ and $\bar{y}_2$ that has minimum variance.

3 Let $\mathbf{t}$ and $\mathbf{u}$ be independent estimates of $\boldsymbol{\beta} = (\beta_1, \beta_2)'$.

(a) If the variance matrices are

$$\begin{pmatrix} 1 & 0 \\ 0 & 1 \end{pmatrix}\sigma^2 \quad \text{and} \quad \begin{pmatrix} 1 & 0 \\ 0 & 1 \end{pmatrix}\sigma^2,$$

respectively, determine the best linearly combined estimate.

(b) If the variance matrices are

$$\begin{pmatrix} 1 & 0 \\ 0 & 1 \end{pmatrix}\sigma^2 \quad \text{and} \quad \begin{pmatrix} 1 & 0.95 \\ 0.95 & 1 \end{pmatrix}\sigma^2,$$

respectively, determine the best linearly combined estimate.

4 Consider the linear model $\mathbf{y} = \beta\mathbf{x} + \mathbf{v}$, where the v's are independent and each has zero mean and variance σ^2. Record the best linear unbiased estimate of β; determine its variance from Theorem 6. See Example 9.1.1.

5 Consider the linear model $\mathbf{y} = (y_{11}, \ldots, y_{1n}; y_{21}, \ldots, y_{2n})' = (\theta_1, \ldots, \theta_1; \theta_2, \ldots, \theta_2)' + \mathbf{v}$, where the v's are independent and each has zero mean and variance σ^2. Show that the best linear unbiased estimate of (θ_1, θ_2) is $(\bar{y}_1, \bar{y}_2)$; determine its variance matrix from Theorem 6. See Problem 9.1.11.

6 Consider the linear model

$$\mathbf{y} = \alpha\begin{pmatrix} 1 \\ \vdots \\ 1 \end{pmatrix} + \beta\begin{pmatrix} x_1 \\ \vdots \\ x_n \end{pmatrix} + \mathbf{v},$$

where the v's are independent and each has zero mean and variance σ^2. Determine the variance matrix of the best linearly unbiased estimate of (α, β); use Theorem 6 and recall Problem 9.2.11.

G PROBLEMS

7 Consider the linear model

$$\mathbf{y} = X\boldsymbol{\beta} + \mathbf{v},$$

where $\mathbf{v}$ is a sample from a normal$(0, \sigma)$ with $(\boldsymbol{\beta}, \sigma^2)$ in $R^r \times R^+$; assume that X has full rank r. Use Theorem 9.1.4 and the results from Section 9.2 to show that the maximum-likelihood estimate of $\boldsymbol{\beta}$ is

$$\widehat{\boldsymbol{\beta}} = (X'X)^{-1}X'\mathbf{y}$$

and the maximum-likelihood estimate of σ^2 is

$$\widehat{\sigma}^2 = \frac{1}{n}(\mathbf{y}'\mathbf{y} - \mathbf{y}'X\widehat{\boldsymbol{\beta}}).$$

Use Theorem 6 to show that $\widehat{\boldsymbol{\beta}}$ has mean $\boldsymbol{\beta}$ and variance matrix $(X'X)^{-1}\sigma^2$.

8 Let $t_1, \ldots, t_k$ be independent unbiased estimators for θ with known variances $\sigma_1^2, \ldots, \sigma_k^2$, respectively. Use Theorem 6 to show that

$$(\sigma_1^{-2} + \cdots + \sigma_k^{-2})^{-1}(\sigma_1^{-2}t_1 + \cdots + \sigma_k^{-2}t_k)$$

is the linear unbiased estimate with minimum variance

$$(\sigma_1^{-2} + \cdots + \sigma_k^{-2})^{-1}.$$

9 A large-sample property closely related to unbiasedness is consistency. Consider an estimator t_n for $\beta(\theta)$ based on a sample of n. Then $\{t_n\}$ is said to be a consistent estimator if

$$\lim_{n\to\infty} P(\beta(\theta) - \epsilon \le t_n(\mathbf{y}) \le \beta(\theta) + \epsilon) = 1$$

for all $\epsilon > 0$, that is, if $t_n(\mathbf{y}) \xrightarrow{P} \beta(\theta)$. Show that if t_n is unbiased for each n and $\lim_{n\to\infty}$ var $(t_n|\theta) = 0$ for each θ, then $\{t_n\}$ is a consistent estimator.

10 *A proof of Theorem 5:* Let $\mathbf{d} = \mathbf{t}_1 - \mathbf{t}_2$.

(a) Show that $E(\mathbf{d}) = 0$.

(b) Show that the covariance matrix of $\mathbf{u} = (\Sigma_{11}^{-1}\mathbf{t}_1 + \Sigma_{22}^{-1}\mathbf{t}_2)$ with $\mathbf{d}$ is the zero matrix.
(c) Show that $A\mathbf{u} + B\mathbf{d}$ generates all linear combinations of $\mathbf{t}_1$ and $\mathbf{t}_2$.
(d) Determine conditions so that $A\mathbf{u} + B\mathbf{d}$ is an unbiased estimator of $\boldsymbol{\beta}$.
(e) Find the unbiased estimator with minimum variance matrix.

11 *Generalized Gauss–Markov theorem:* Consider a response $\mathbf{y} = X\boldsymbol{\beta} + \mathbf{v}$, where X is an $n \times r$ matrix of full rank r and the variation $\mathbf{v}$ has mean $\mathbf{0}$ and variance matrix $\sigma^2 M$ with $(\boldsymbol{\beta}, \sigma^2)$ in $R^r \times R^+$; the matrix M is given. Prove:

THEOREM 7

(a) *A linear estimator $A\mathbf{y}$ based on an $r \times n$ matrix A is an unbiased estimator of $\boldsymbol{\beta}$ if $AX = I$.*

(b) *Among such linear unbiased estimators the variance matrix is smallest for the weighted-least-squares estimator*

$$\mathbf{b} = (X'M^{-1}X)^{-1}X'M^{-1}\mathbf{y}$$

as given in Problem 9.2.14.

(c) *The variance matrix of this best linear unbiased estimator is*

$$(X'M^{-1}X)^{-1}\sigma^2.$$

Method: Modify the pattern of proof in Section E.

12 Use Theorem 7 to prove Theorem 5.

9.4

THE INFORMATION INEQUALITY FOR UNBIASED ESTIMATORS

Least squares and the linear combination of estimators are methods primarily concerned with location parameters for response variables. In this section we examine more general parameters, including those that refer to the shape or form of the distribution, and we obtain a lower bound for the variance of unbiased estimates. The inequality that expresses this lower bound was developed by Frechét, Fisher, Cramér, and Rao; it is sometimes called the *information inequality,* as it involves the Fisher information function defined in the large-sample Section 8.6.

The information inequality can sometimes be used in reverse to construct good unbiased estimates; but the success of this procedure is limited to the special exponential models discussed in Section 8.5.

The information inequality is essentially a local result on the parameter space. Accordingly, to develop the inequality, we suppose that we are in a situation where we *know* that the true parameter value is in some small interval $(\theta_0 \pm \delta)$ and we *require* that the estimate be unbiased only for this small interval. The application of the inequality, however, is not restricted to such special situations.

A PROPERTIES OF THE SCORE FUNCTION

In Chapter 8 we saw how the likelihood function was a basic ingredient to the process of statistical inference. In particular, the slope of the log-likelihood function, the *score function,* had several important properties. The score function is the basis for the information inequality; we first collect some properties of the score function.

For notation let **y** be the full response under investigation with sample space $\mathcal{S}$, and let θ be the parameter with real values in the parameter space $\Omega = R$. We assume that the statistical model

$$\{f(\cdot\,|\,\theta) : \theta \in \Omega\}$$

is given by density functions and satisfies the regularity conditions as discussed in Section 8.1.

The score function $S(\mathbf{y}\,|\,\cdot)$ was defined in Section 8.1D:

$$S(\mathbf{y}\,|\,\theta) = \frac{\partial}{\partial\theta} \ln f(\mathbf{y}\,|\,\theta) = \frac{\partial}{\partial\theta} l(\mathbf{y}\,|\,\theta).$$

The following properties of the score function were developed in Section 8.6B on the basis of Assumption 8.6.1:

$$(1) \qquad E(S(\mathbf{y}\,|\,\theta)\,|\,\theta) = 0,$$

$$(2) \qquad \text{var}\,(S(\mathbf{y}\,|\,\theta)\,|\,\theta) = J(\theta) = -E(S'(\mathbf{y}\,|\,\theta)\,|\,\theta).$$

The first equation says that the mean slope of the log-likelihood function is zero at the true parameter value. The second says that the variance of the slope at the true parameter value is related to the mean curvature at the true parameter value. Assumption 8.6.1 requires the density function to depend on θ in a very respectable way. The function $J(\theta)$ is the Fisher information function for the full response **y** and is defined by the left side of (2).

We now derive a third result closely connected with those just described. Let $t(\mathbf{y})$ be a function with mean value $\beta(\theta)$. Again we assume that differentiation with respect to θ can be carried through the sample-space summation or integration sign; we record the analysis for the integration case; the summation case is parallel. In the pattern of Section 8.6B we differentiate the integral

$$\int t(\mathbf{y}) f(\mathbf{y}\,|\,\theta)\, d\mathbf{y} = \beta(\theta),$$

with respect to θ:

$$\int t(\mathbf{y}) \frac{\partial}{\partial\theta} f(\mathbf{y}\,|\,\theta)\, d\mathbf{y} = \int t(\mathbf{y}) S(\mathbf{y}\,|\,\theta) f(\mathbf{y}\,|\,\theta)\, d\mathbf{y} = \beta'(\theta).$$

Thus we obtain

$$(3) \qquad E(t(\mathbf{y}) S(\mathbf{y}\,|\,\theta)\,|\,\theta) = \frac{d}{d\theta} E(t(\mathbf{y})\,|\,\theta).$$

Note that (3) reduces to (1) by setting $t(\mathbf{y}) \equiv 1$. In summary: *derivative outside E gives score inside E.*

B THE INFORMATION INEQUALITY

For the moment let us picture ourselves in an application where we know that the true parameter value θ is very close to the value θ_0. And let us suppose that we are prepared to use unbiased estimation even though an unbiased estimator may well give values far outside a small interval $(\theta_0 \pm \delta)$.

An estimator $t(\mathbf{y})$ that is unbiased generally has the properties

$$(4) \qquad E(t(\mathbf{y}) \mid \theta) = \theta = \theta_0 + (\theta - \theta_0),$$

where we have reexpressed the right-hand side relative to the value θ_0. From this we are led to the following definition of unbiasedness for θ close to θ_0:

DEFINITION 1

t is **locally unbiased(θ_0)** for the parameter θ if

$$(5) \qquad E(t(\mathbf{y}) \mid \theta) = \theta_0 + (\theta - \theta_0) + o(\theta - \theta_0)$$

or, equivalently, if

$$(6) \qquad E(t(\mathbf{y}) \mid \theta_0) = \theta_0,$$

$$(7) \qquad \frac{\partial}{\partial \theta} E(t(\mathbf{y}) \mid \theta) \bigg|_{\theta_0} = 1.$$

This definition says that formula (4) holds to a first derivative approximation for θ near θ_0; see Figure 9.7. Note that $o(\theta - \theta_0)$ represents a remainder that goes to zero as a proportion of $\theta - \theta_0$.

We have, of course, that if t is unbiased generally, then it is locally unbiased(θ_0) for each θ_0 and that if t is locally unbiased(θ_0) for *each* θ_0, then it is unbiased generally.

Now let t be a locally unbiased(θ_0) estimator for θ. Then formulas (3) and (7) give

$$E(t(\mathbf{y}) S(\mathbf{y} \mid \theta_0) \mid \theta_0) = 1.$$

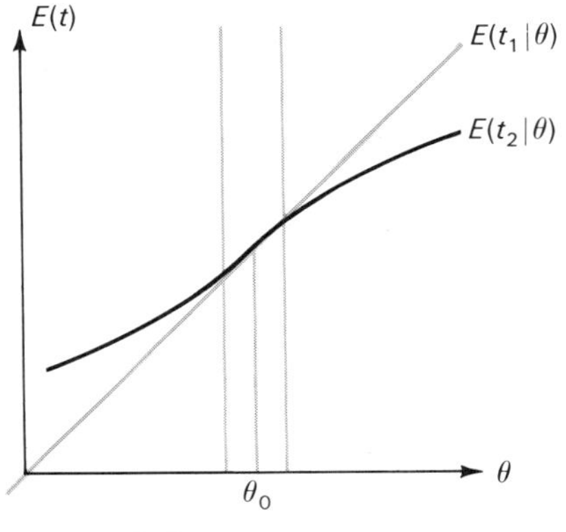

FIGURE 9.7
t_1 is unbiased for θ; t_2 is locally unbiased(θ_0) for θ.

We recall that covariance is mean product minus product means. Thus using formula (1) we obtain

(8) $\quad \operatorname{cov}(t(\mathbf{y}), S(\mathbf{y}|\theta_0)|\theta_0) = 1.$

We now record the information inequality:

THEOREM 2

Under the regularity assumptions in Section 8.6B, if t is locally unbiased(θ_0) for θ, then

(9) $\quad \operatorname{var}(t(\mathbf{y})|\theta_0) \geq J^{-1}(\theta_0)$

with equality if and only if $t(\mathbf{y})$ and $S(\mathbf{y}|\theta_0)$ are affinely related; that is, $S(\mathbf{y}|\theta_0) = a + ct(\mathbf{y})$ for some a, c.

Proof We apply the covariance inequality (5.3.24) to formula (8):

$\operatorname{var}(t(\mathbf{y})|\theta_0) \operatorname{var}(S(\mathbf{y}|\theta_0)|\theta_0) \geq 1^2;$

rearrangement gives formula (9).

The results in the theorem apply immediately, of course, to an estimate that is unbiased generally.

EXAMPLE 1

The location normal: Let $(y_1, \ldots, y_n)$ be a sample from the normal(θ, σ_0) with θ in R. Then we have

$$S(\mathbf{y}|\theta_0) = \frac{\bar{y} - \theta_0}{\sigma_0^2/n}, \qquad J(\theta) = \frac{n}{\sigma_0^2}$$

as derived in Section 8.6B. Consider an unbiased estimator t of θ, or, more generally, a locally unbiased(θ_0) estimator of θ. Then from Theorem 2 we have

$$\operatorname{var}(t(\mathbf{y})|\theta_0) \geq \frac{\sigma_0^2}{n}$$

with equality if and only if $t(\mathbf{y})$ and $S(\mathbf{y}|\theta_0)$ are affinely related.

Our usual estimator of θ is the sample average $\bar{y}$. It is unbiased and has variance σ_0^2/n, which is the lower bound in the inequality. The inequality then says that we cannot find a better estimate, better in the sense of smaller variance.

Thus in summary we have that $\bar{y}$ is an unbiased estimator of θ and it has smaller ($\leq$) variance than any other unbiased estimator of θ; accordingly, we call $\bar{y}$ a uniformly minimum variance (UMV) unbiased estimator of θ.

The theorem also says that $\bar{y}$ is the unique UMV unbiased estimator; for it is the only *affine* function of $S(\mathbf{y}|\theta_0)$ that is unbiased for θ.

The inequality (9) is sometimes used with Definition 9.3.1 to give an absolute measure of the efficiency of an estimator:

(10) $$\text{Eff}\,(t\,|\,\theta) = \frac{1/\text{var}\,(t\,|\,\theta)}{J(\theta)}.$$

We will see cases where the UMV unbiased estimator has variance larger than the information lower bound (in Section 9.5). In such cases the above efficiency would not be 100 percent. This is to suggest caution in interpreting formula (10). Its usefulness is primarily in a large-sample context as discussed in Section D, which follows.

C THE BEST LOCALLY UNBIASED ESTIMATOR

We now use the information inequality in reverse to construct a locally unbiased(θ_0) estimator that has minimum variance at the information lower bound. Theorem 2 says that a locally unbiased(θ_0) estimator with variance at the lower bound must be an affine function of the score $S(\mathbf{y}\,|\,\theta_0)$. Thus we consider affine functions

$$t(\mathbf{y}) = a + cS(\mathbf{y}\,|\,\theta_0)$$

of the score and see if a, c can be determined so that $t(\mathbf{y})$ is locally unbiased(θ_0).

Condition (6) for local unbiasedness gives

$$\begin{aligned} E(a + cS(\mathbf{y}\,|\,\theta_0)\,|\,\theta_0) &= \theta_0 \\ &= a + c \cdot 0; \end{aligned}$$

thus $a = \theta_0$. Condition (7) gives

$$\begin{aligned} E\big((a + cS(\mathbf{y}\,|\,\theta_0))S(\mathbf{y}\,|\,\theta_0)\,|\,\theta_0\big) &= 1 \\ &= c\,\text{var}\,(S(\mathbf{y}\,|\,\theta_0)\,|\,\theta_0); \end{aligned}$$

thus $c = J^{-1}(\theta_0)$. It follows that the estimator

(11) $$t(\mathbf{y}) = \theta_0 + J^{-1}(\theta_0)S(\mathbf{y}\,|\,\theta_0)$$

is the unique locally unbiased(θ_0) estimator that has minimum variance at the information lower bound.

EXAMPLE 2

Poisson(θ): Consider a response y that is Poisson(θ) with θ in R^+. We have immediately that

$$S(y\,|\,\theta_0) = \frac{y}{\theta_0} - 1, \qquad J(\theta_0) = \frac{1}{\theta_0}.$$

We now calculate the best unbiased estimator for θ near θ_0:

$$t(y) = \theta_0 + \theta_0\left(\frac{y}{\theta_0} - 1\right) = y.$$

This estimate is independent of θ_0. Thus it is unbiased generally and it has variance at the information lower bound; it is UMV unbiased.

D LARGE-SAMPLE ESTIMATION

We have examined some aspects of inference for large samples in Section 8.6. Now consider estimation for large samples in terms of the results in this section.

For the statistical model

$$\{f(\cdot\,|\theta) : \theta \in \Omega\},$$

let $S(y|\theta)$ be the score function and $j(\theta)$ be the information. And let $(y_1, \ldots, y_n)$ be a sample from the model.

We know from Section 8.6 that a large sample determines almost conclusively a small interval in which the true parameter value must lie. If the parameter is believed to be in the neighborhood of θ_0, then the best local estimate (11) for θ is

$$\theta_0 + \frac{\sum_1^n S(y_i|\theta_0)}{nj(\theta_0)}. \tag{12}$$

If this value, say θ_1, is rather far from θ_0, then an appropriate improved estimate based on θ_1 would be

$$\theta_1 + \frac{\sum_1^n S(y_i|\theta_1)}{nj(\theta_1)}.$$

Typically, this procedure converges very rapidly (two or three iterations) to the maximum-likelihood estimate $\hat{\theta}$. In particular, we can examine the best local estimate as calculated at $\hat{\theta}$:

$$\hat{\theta} + \frac{\sum_1^n S(y_i|\hat{\theta})}{nj(\hat{\theta})} = \hat{\theta},$$

which is simplified by means of the likelihood equation (9.1.8). Thus the argument in terms of local estimates is tied closely to the maximum-likelihood method. The large-sample properties of the maximum-likelihood estimate were discussed in Section 8.6C. Note that the large-sample variance of the maximum-likelihood estimate is $J^{-1}(\theta) = 1/nj(\theta)$, which is at the information lower bound.

It is of interest that the Newton–Raphson iterative procedure for the maximum-likelihood estimate has the form

$$\theta_0 + \frac{\sum_1^n S(y_i|\theta_0)}{-\sum_1^n S'(y_i|\theta_0)}. \tag{13}$$

The local estimate (12) has the mean value

$$nj(\theta) = E\left(-\sum_1^n S'(y_i|\theta)\right)$$

in place of the denominator as given in (13). Very often the procedure using local estimates converges much faster than the Newton–Raphson procedure.

E WHEN DO THE BEST LOCAL ESTIMATORS GIVE THE BEST ESTIMATORS?

Consider an unbiased estimator t that has variance at the information lower bound for all values of θ. Then from formula (11) we have

$$t(\mathbf{y}) = \theta + J^{-1}(\theta)S(\mathbf{y}\,|\,\theta)$$

for all values of θ. We rearrange and then integrate with respect to θ:

$$S(\mathbf{y}\,|\,\theta) = -\theta J(\theta) + J(\theta)t(\mathbf{y}),$$

$$\ln f(\mathbf{y}\,|\,\theta) = \phi(\theta) + \psi(\theta)t(\mathbf{y}) + a(\mathbf{y}),$$

where $a(\mathbf{y})$ is the constant of integration and $\phi(\theta)$ and $\psi(\theta)$ are the obvious indefinite integrals. Thus we obtain

$$f(\mathbf{y}\,|\,\theta) = \gamma(\theta)\exp\{\psi(\theta)t(\mathbf{y})\}h(\mathbf{y}), \tag{14}$$

which has exponential form with one ψ-function.

We thus see an example of a theoretical method whose effectiveness is limited to models having exponential form. Recall the remarks in Section 8.5D. Also, we have seen that the effectiveness of the method is limited to a model having exponential form with one ψ-function. But does it always work for such models?

Suppose that we have the score and information as calculated with respect to θ and yet we are interested in unbiased estimation of the real parameter $\beta(\theta)$. Then it is straightforward (Problem 6) to show that if t is locally unbiased $(\beta(\theta_0))$ for $\beta(\theta)$, then

$$\operatorname{var}(t(\mathbf{y})\,|\,\theta_0) \geq (\beta'(\theta_0))^2 J^{-1}(\theta_0). \tag{15}$$

Now consider a statistical model having the exponential form (14). The parameter θ itself may not have an unbiased estimator with variance at the information lower bound. Problem 7 is to show that the only parameter having such an unbiased estimator is

$$\frac{\phi'(\theta)}{\psi'(\theta)} \tag{16}$$

or a simple affine adjustment thereof. Thus not only is the effectiveness of the method limited to a model having exponential form with one ψ-function, but it is limited to essentially one parametrization of such a model.

F THE ROUTE TO UMV UNBIASED ESTIMATORS

In conclusion, we note that if an estimation problem has a UMV unbiased estimator *with variance at the information lower bound,* then the procedure of finding the best local estimate produces that UMV estimator.

Some problems, however, have UMV unbiased estimators with variance *above the information lower bound*. We will see that the methods in Section 9.5 can sometimes be successful in producing these estimators. We will also see that some estimation problems, in fact, have no UMV unbiased estimator.

G SUPPLEMENT: FOR VECTOR PARAMETERS

Consider briefly the case of a vector parameter $\boldsymbol{\theta} = (\theta_1, \ldots, \theta_r)'$ with values in R^r or some connected open set of R^r. And suppose that we have the vector analogs of the regularity conditions in Section 8.6.

From Section 8.6G we have

(17) $$E(\mathbf{S}(\mathbf{y} \mid \boldsymbol{\theta}) \mid \boldsymbol{\theta}) = \mathbf{0},$$

(18) $$\text{VAR}\ (\mathbf{S}(\mathbf{y} \mid \boldsymbol{\theta}) \mid \boldsymbol{\theta}) = J(\boldsymbol{\theta}) = -E\left(\frac{\partial \mathbf{S}(\mathbf{y} \mid \boldsymbol{\theta})}{\partial \boldsymbol{\theta}'} \mid \boldsymbol{\theta}\right).$$

And by the procedure in Section A we have

(19) $$E(t_u(\mathbf{y}) S_v(\mathbf{y} \mid \boldsymbol{\theta}) \mid \boldsymbol{\theta}) = \frac{\partial}{\partial \theta_v} E(t_u(\mathbf{y}) \mid \boldsymbol{\theta}),$$

or in matrix form,

(20) $$E(\mathbf{t}(\mathbf{y}) \mathbf{S}'(\mathbf{y} \mid \boldsymbol{\theta}) \mid \boldsymbol{\theta}) = \frac{d}{d\boldsymbol{\theta}'} E(\mathbf{t}(\mathbf{y}) \mid \boldsymbol{\theta}),$$

where the prime on the **S** here denotes the transpose of the vector.

An estimator **t** is locally unbiased($\boldsymbol{\theta}_0$) if

(21) $$E(\mathbf{t}(\mathbf{y}) \mid \boldsymbol{\theta}_0) = \boldsymbol{\theta}_0,$$

(22) $$\frac{\partial}{\partial \theta_u} E(t_v(\mathbf{y}) \mid \boldsymbol{\theta}) \Big|_{\boldsymbol{\theta}=\boldsymbol{\theta}_0} = \delta_{uv},$$

where $\delta_{uv} = 1$ if $u = v$ and 0 otherwise. And by formulas (17), (20), and (22) a locally unbiased($\boldsymbol{\theta}_0$) estimator satisfies

(23) $$\text{COV}\ (\mathbf{t}(\mathbf{y}), \mathbf{S}(\mathbf{y} \mid \boldsymbol{\theta}_0) \mid \boldsymbol{\theta}_0) = I,$$

the $r \times r$ identity matrix.

The information inequality is then obtained from Problem 5.5.10. The matrix

(24) $$\text{VAR}\ (\mathbf{t}(\mathbf{y}) \mid \boldsymbol{\theta}_0) - J^{-1}(\boldsymbol{\theta}_0)$$

is positive semidefinite. The best locally unbiased($\boldsymbol{\theta}_0$) estimator is

(25) $$\boldsymbol{\theta}_0 + \mathbf{J}^{-1}(\boldsymbol{\theta}_0)\mathbf{S}(\mathbf{y} \mid \boldsymbol{\theta}_0).$$

An unbiased estimator with variance matrix at the information lower bound has the form

$$\mathbf{t}(\mathbf{y}) = \boldsymbol{\theta} + \mathbf{J}^{-1}(\boldsymbol{\theta})\mathbf{S}(\mathbf{y} \mid \boldsymbol{\theta}),$$

and this occurs only if the model has exponential form with r ψ-functions.

H EXERCISES

1 Let $(x_1, \ldots, x_n)$ be a sample from the Bernoulli distribution with p in $\Omega = (0, 1)$. Derive the minimum variance locally unbiased(p_0) estimate of p. Is a UMV unbiased estimate available?

2 Let $(y_1, \ldots, y_n)$ be a sample from the exponential distribution $f(y \mid \theta) = \theta^{-1} \exp\{-y/\theta\}$ on R^+ with θ in $\Omega = R^+$. Derive the minimum variance locally unbiased(θ_0) estimate of θ. Is a UMV unbiased estimate available?

3 Let $(y_1, \ldots, y_n)$ be a sample from the normal(μ_0, σ^2) with σ^2 in R^+. Derive the minimum variance locally unbiased(σ_0^2) estimate of σ^2. Is a UMV unbiased estimate available?

4 Let $(y_1, \ldots, y_n)$ be a sample from $f(y|\theta) = \theta y^{\theta-1}$ on $(0, 1)$ and zero otherwise with θ in $\Omega = R^+$. Derive the minimum variance locally unbiased(θ_0) estimate of θ. Is a UMV unbiased estimate for θ with variance at the information bound available?

5 Consider independent models: $f_1(y|\theta)$ with $S_1(y|\theta)$ and $J_1(\theta)$; and $f_2(y|\theta)$ with $S_2(y|\theta)$ and $J_2(\theta)$. For local unbiased(θ_0) estimation we have $\theta_0 + J_1^{-1}(\theta_0)S_1(y|\theta_0)$ from the first model and $\theta_0 + J_2^{-1}(\theta_0)S_2(y|\theta_0)$ from the second model. Combine these local estimates by the methods in Section 9.3C and show that the combined estimate is the best locally unbiased(θ_0) estimate for the combined model.

I PROBLEMS

6 Let $\beta(\theta)$ be a continuously differentiable monotone increasing function mapping R into R with $\beta'(\theta) = d\beta(\theta)/d\theta$. Let $S(\mathbf{y}|\theta)$ and $J(\theta)$ be the score and information relative to θ. Show that the score, information, and best locally unbiased$(\beta(\theta_0))$ estimate relative to $\beta(\theta)$ are

$$S(y|\theta)/\beta'(\theta), \qquad J(\theta)/(\beta'(\theta))^2, \qquad \beta(\theta_0) + \beta'(\theta_0)\frac{S(y|\theta_0)}{J(\theta_0)}.$$

7 For the exponential model (14) show that the only parameter with an unbiased estimator at the information lower bound is $\phi'(\theta)/\psi'(\theta)$ or an affine function thereof; note that $\phi(\theta) = \ln \gamma(\theta)$. The formulas in Problem 6 are not needed.

8 For Exercise 4 determine the parameter that has an unbiased estimator with variance at the information lower bound.

9 Let y have a distribution with density

$$f(y|\theta) = \theta(\theta + 1)y(1 - y)^{\theta-1} \qquad \text{on } 0 < y < 1$$

and zero otherwise. Determine the parameter that has an unbiased estimator at the information lower bound.

10 Consider a distribution on R^{m+n} for

$$\begin{pmatrix}\mathbf{x}\\ \mathbf{y}\end{pmatrix} = (x_1, \ldots, x_m, y_1, \ldots, y_n)'$$

with mean $\mathbf{0}$ and variance matrix

$$\Sigma = \begin{pmatrix}\Sigma_{11} & \Sigma_{12}\\ \Sigma_{21} & \Sigma_{22}\end{pmatrix}$$

(as correspondingly partitioned). Consider the variance matrix of $\mathbf{y} - \Sigma_{21}\Sigma_{11}^{-1}\mathbf{x}$ and show that $\Sigma_{22} - \Sigma_{21}\Sigma_{11}^{-1}\Sigma_{12}$ is positive semidefinite and is zero if and only if $\mathbf{y} = C\mathbf{x}$, where C is $m \times n$. Compare with Problem 5.5.10.

J PROBLEMS FOR VECTOR PARAMETERS

11 Consider independent models: $f_1(\mathbf{y}_1|\boldsymbol{\theta})$ with $\mathbf{S}_1(\mathbf{y}_1|\boldsymbol{\theta})$ and $\mathbf{J}_1(\boldsymbol{\theta})$; and $f_2(\mathbf{y}_2|\boldsymbol{\theta})$ with $\mathbf{S}_2(\mathbf{y}_2|\boldsymbol{\theta})$ and information matrix $\mathbf{J}_2(\boldsymbol{\theta})$. Combine the best locally unbiased$(\boldsymbol{\theta}_0)$ estimator by the methods in Section 9.3C and show that the combined estimator is the best locally unbiased$(\boldsymbol{\theta}_0)$ estimator from the combined model.

12 Let $(y_1, \ldots, y_n)$ be a sample from a normal(μ, σ) with $\boldsymbol{\theta} = (\mu, \sigma^2)$ in $R \times R^+$. Calculate $\mathbf{S}(\mathbf{y}|\boldsymbol{\theta})$ and $\mathbf{J}(\boldsymbol{\theta})$. Compare with Problem 8.6.12.

13 (*continuation*) Derive the minimum variance-matrix locally unbiased(θ_0) estimator of θ. Is an unbiased estimator with variance matrix at the information lower bound available?

14 (*continuation*) Calculate the variance matrix of $(\bar{y}, s_y^2)$ and compare with the information lower bound.

15 Prove formula (20) for the derivative of the mean of a vector estimate.

16 Prove the information inequality (24) for vector parameters. Show that (24) is the zero matrix if and only if $\mathbf{t}(y)$ is an affine function of $\mathbf{S}(y|\theta_0)$. Use Problem 10.

17 (*continuation*) Show that the best locally unbiased($\boldsymbol{\theta}_0$) estimator of $\boldsymbol{\theta}$ is $\mathbf{t} = \boldsymbol{\theta}_0 + \mathbf{J}^{-1}(\boldsymbol{\theta}_0)\mathbf{S}(y|\boldsymbol{\theta}_0)$.

18 (*continuation*) Show that the information lower bound is attained for the unbiased estimator of the vector parameter $(\theta_1, \ldots, \theta_r)$ only if the model is exponential with r ψ-functions.

19 (*continuation*) Consider the reduced exponential model $f(y|\boldsymbol{\theta}) = \exp\{\phi(\boldsymbol{\theta}) + \boldsymbol{\psi}'(\boldsymbol{\theta})\mathbf{a}(y)\}h(y)$ with r ψ-functions and with $\boldsymbol{\theta}$ in R^r. Show that the only vector parameter that has an unbiased estimator with variance matrix at the information lower bound is $(\partial\boldsymbol{\psi}'(\boldsymbol{\theta})/\partial\boldsymbol{\theta})^{-1}\partial\phi(\boldsymbol{\theta})/\partial\boldsymbol{\theta}$ or an affine equivalent thereof.

20 For the normal model in Problem 12 show that the only parameter with an unbiased estimator at the information lower bound is $(\mu, \mu^2 + \sigma^2)$ or an affine equivalent; use Problem 19.

9.5

LIKELIHOOD AND COMPLETENESS FOR UNBIASED ESTIMATORS

When the likelihood function was introduced in Section 8.1 we noted that a statistical model says the same thing about each sample point with a particular likelihood function. This led us in Section 8.4 to define the likelihood statistic, a simple statistic that indexes the possible likelihood functions.

In this section we will see that the choice of an unbiased estimator should be restricted to those that are based on the likelihood statistic; of course this agrees with the general views in Chapter 8 but is based here on minimum variance. And we will see that if the likelihood statistic has a property of completeness, then such an estimate based on the likelihood statistic is unique.

A ESTIMATORS BASED ON THE LIKELIHOOD STATISTIC

For notation let y be the full response under investigation with sample space $\mathcal{S}$, and let θ be the parameter with values in a parameter space Ω. We assume that the statistical model,

$$\{f(\cdot|\theta) : \theta \in \Omega\},$$

is given in terms of density functions as discussed in Section 7.1. This makes available the likelihood statistic as defined in Section 8.4B; let s designate the likelihood statistic.

From the probability Section 5.5 we know that variance is reduced by taking a conditional mean. This is how we will improve an initial estimator to obtain an estimator based on the likelihood statistic. For a real-valued parameter $\beta(\theta)$ we have the following Rao–Blackwell theorem:

THEOREM 1

If $t(y)$ is an unbiased estimator of $\beta(\theta)$ and $s(y)$ is the likelihood statistic, then
(a) *The regression of $t(y)$ on $s(y)$, which is the function*

(1) $$E(t(y) : s(y) = s) = r(s)$$

is an unbiased estimator of $\beta(\theta)$ and is based on the likelihood statistic $s(y)$.
(b) *The variance is reduced in going from the estimator t to the estimator r,*

(2) $$\text{var}\,(t(y)\,|\,\theta) \geq \text{var}\,(r(s(y))\,|\,\theta)$$

with equality if and only if the estimator $t(y) = r(s(y))$ is already a function of the likelihood statistic $s(y)$ (with probability 1).

The theorem shows that an unbiased estimator not based on the likelihood statistic can be improved by taking its mean value given the likelihood statistic. The theorem remains true if $s(y)$ is any sufficient statistic; the likelihood statistic provides the largest reduction and would thus be the preferred sufficient statistic. The proof is recorded in Section B.

Consider some simple examples. The results are rather elementary and indeed familiar, but they do illustrate the ideas involved.

EXAMPLE 1

Sample from the Bernoulli: Let $(x_1, \ldots, x_n)$ be a sample from the Bernoulli(p) distribution with p in $[0, 1]$. From Example 8.4.2 we know that $y = \Sigma_1^n x_i$ is the likelihood statistic.

We examine unbiased estimation of the parameter p. A very simple and elementary estimator is given by the first coordinate x_1:

$$E(x_1\,|\,p) = 1 \cdot p + 0 \cdot q = p.$$

Of course, this estimator is rather ridiculous in that it just takes two values, 0 and 1, but technically it is unbiased. We now use Theorem 1 to derive an improved estimator $r(y)$:

$$r(y) = E(x_1 : y) = 1\,\frac{y}{n} + 0\,\frac{n-y}{n} = \frac{y}{n};$$

for this note from symmetry that the conditional distribution of $(x_1, \ldots, x_n)$ given the sum $y = \Sigma x_i$ has equal probability $1\Big/\binom{n}{y}$ for each sequence consisting of y 1's and $n - y$ 0's, and thus that x_1 is equal to 0 and 1 with probabilities $(n - y)/n$ and y/n, respectively. Thus we see that the improved estimator is just the proportion $\hat{p} = y/n$, which we have used frequently before. Also we see that

$$\text{var}\,(x_1\,|\,p) = pq, \qquad \text{var}\,(y/n\,|\,p) = pq/n.$$

EXAMPLE 2

Sample from the location normal: Let $(y_1, \ldots, y_n)$ be a sample from the normal(θ, σ_0) distribution with θ in R. From Section 8.4A we know that $\bar{y}$ or $u_1 = \sqrt{n}\,\bar{y}$ is the likelihood statistic.

We examine unbiased estimation of the parameter θ. A very simple and immediate estimator is given by the first coordinate y_1:

$$E(y_1 | \theta) = \theta.$$

Again we are starting with a rather ridiculous initial estimator: the first coordinate—ignoring all other coordinates. We now use Theorem 1 to derive an improved estimator. For this we recall the distribution results obtained from the transformation in formula (8.2.10). This transformation can be solved, giving

$$y_1 = \frac{1}{\sqrt{n}} u_1 + \frac{1}{\sqrt{2}} u_2 + \cdots + \frac{1}{\sqrt{n(n-1)}} u_n,$$

where the u's are statistically independent with common variance σ^2 and with means all equal to 0 except for u_1. Thus

$$\begin{aligned} E(y_1 : u_1) &= \frac{1}{\sqrt{n}} \cdot u_1 + \frac{1}{\sqrt{2}} \cdot 0 + \cdots + \frac{1}{\sqrt{n(n-1)}} \cdot 0 \\ &= \bar{y}. \end{aligned}$$

Note that the improved estimator is the familiar sample average. Also note that

$$\operatorname{var}(y_1 | \theta) = \sigma_0^2, \qquad \operatorname{var}(\bar{y} | \theta) = \sigma_0^2/n.$$

B THE RAO–BLACKWELL THEOREM: ITS PROOF AND A VECTOR EXTENSION

Consider the proof of Theorem 1. We know that the conditional distribution given the likelihood statistic is independent of θ; see Theorem 8.4.3. Thus the conditional mean

$$E(t(y) : s(y) = s) = r(s)$$

does not depend on θ, and correspondingly $r(s(y))$ is a function on the sample space that does not involve θ. We now use formula (5.5.3) for taking a mean in two stages:

$$\beta(\theta) = E(t(y) | \theta) = E\big(E(t(y) : s(y) = s) | \theta\big) = E\big(r(s) | \theta\big). \tag{3}$$

Thus $r(s(y))$ is an unbiased estimate of $\beta(\theta)$, proving part (a) of the theorem.

We now consider the variance reduction formula (2). For this we use formula (5.5.4), expressing variance in terms of variance of regression and variance about regression:

$$\operatorname{var}(t | \theta) = E(\operatorname{var}(t : s) | \theta) + \operatorname{var}(r | \theta). \tag{4}$$

Thus the inequality (2) holds. Also we have equality if and only if $E(\operatorname{var}(t : s) | \theta) = 0$, or, equivalently, $\operatorname{var}(t : s) = 0$ (with probability 1), or, equivalently, t is a constant given s (with probability 1). This proves part (b) of the theorem.

For the case of a vector parameter $\boldsymbol{\beta}(\theta)$, we record the generalized Rao–Blackwell theorem for improving estimators:

THEOREM 2

If $\mathbf{t}(y)$ *is an unbiased estimator of* $\boldsymbol{\beta}(\theta)$ *and* $s(y)$ *is the likelihood statistic, then*

(a) *The regression of* $\mathbf{t}(y)$ *on* $s(y)$,

(5) $$E(\mathbf{t}(y) : s(y) = s) = \mathbf{r}(s),$$

is an unbiased estimator of $\boldsymbol{\beta}(\theta)$ *based on the likelihood statistic* $s(y)$.

(b) *The variance matrix is reduced in going from the estimator* $\mathbf{t}$ *to the estimator* $\mathbf{r}$; *thus*

(6) $$\text{VAR}\ (\mathbf{t}\,|\,\theta) - \text{VAR}\ (\mathbf{r}\,|\,\theta)$$

is positive semidefinite and is the zero matrix if and only if the estimator $\mathbf{t}(y) = \mathbf{r}(s(y))$ *is already a function of the likelihood statistic* $s(y)$ *(with probability 1).*

The proof follows by using (5.5.9) and (5.5.10) in place of (5.5.3) and (5.5.4).

Note that the theorem says to improve each coordinate by taking a mean as described in Theorem 1. This reduces the variance matrix; it also reduces the variance of any coordinate that is not already a function of the likelihood statistic; and it reduces the variance of any linear combination of coordinates that is not already a function of the likelihood statistic.

C COMPLETENESS, AND UNIQUENESS OF UNBIASED ESTIMATORS

Theorem 1 says that in looking for an unbiased estimator of a parameter we need examine only those based on the likelihood statistic. How many such estimators are there? We will see that for some statistical models there is only one; it must then be the UMV-unbiased estimator.

First a convenient definition:

DEFINITION 3

A parameter $\beta(\theta)$ *is estimable if there exists an unbiased estimator for it.*

Parameters can be found that do not have unbiased estimates. For example, with a sample of 2 from the Bernoulli(p) *only* the polynomials $a + bp + cp^2$ up to order 2 have unbiased estimates. And, for example, with a sample from an arbitrary continuous distribution on R, the median does *not* have an unbiased estimate.

We use the notation in Section A and we let $g(s\,|\,\theta)$ be the density for the likelihood statistic; we record the proof for the integration case. Suppose that $\beta(\theta)$ is estimable. Then by Theorem 1 there is *an* unbiased estimator based on the likelihood statistic.

Suppose there are two, $r_1(s(y))$ and $r_2(s(y))$. Then

$$\beta(\theta) = \int r_1(s)g(s\,|\,\theta)\,ds$$
$$\beta(\theta) = \int r_2(s)g(s\,|\,\theta)\,ds,$$

and hence

$$0 = \int r(s)g(s\,|\,\theta)\,ds, \tag{7}$$

where $r(s) = r_1(s) - r_2(s)$ is the difference between the two unbiased estimators. We call r an unbiased estimator of zero.

DEFINITION 4

A statistical model $\{g(\cdot\,|\,\theta) : \theta \in \Omega\}$ is **complete** if

$$\int r(s)g(s\,|\,\theta)\,ds = 0 \qquad \text{for } \theta \in \Omega \tag{8}$$

implies that $r(s) = 0$ (with probability 1), that is, if there are no nontrivial unbiased estimators of zero.

If the statistical model for the likelihood statistic is complete, then equation (7) implies that $r(s) = 0$ and any two unbiased estimators based on the likelihood statistic must be equal. This proves the following Lehmann–Scheffé theorem.

THEOREM 5

If the statistical model for the likelihood statistic is complete, then an estimable parameter has exactly one unbiased estimator based on the likelihood statistic.

Rather than say that a statistical model for the likelihood statistic is complete, we will, for convenience, say that the *likelihood statistic is complete,* provided that the context makes clear what statistical model is being used.

By Theorems 1 and 2 we know that variance (or VARiance matrix) is reduced by going to estimators based on the likelihood statistic. If the likelihood statistic is complete, then we know there is only one improved estimator and, accordingly, it is UMV unbiased. This gives the following:

COROLLARY 6

If the likelihood statistic is complete, then an unbiased estimator based on the likelihood statistic is the UMV unbiased estimator.

D EXAMPLES OF COMPLETENESS

Consider two examples of completeness, one discrete and one absolutely continuous.

EXAMPLE 3

Poisson(λ): Let s have a Poisson(λ) distribution with λ in R^+. We show that s is complete. For this let $r(s)$ be an unbiased estimator of zero:

$$\sum_0^\infty r(s)\lambda^s e^{-\lambda}/s! = 0 \qquad \text{for } \lambda \in R^+,$$

$$\sum_0^\infty \frac{r(s)}{s!}\lambda^s = 0 \qquad \lambda \in R^+.$$

This is a power series that is convergent to zero for all values of λ in R^+. The coefficients of a power-series expansion are unique; hence $r(s)/s! = 0$, or, equivalently, $r(s) = 0$. This establishes completeness.

EXAMPLE 4

The location normal: Let s have a normal(θ, τ_0) distribution with θ in R. We show that s is complete. For this, let $r(s)$ be an unbiased estimator of zero:

$$\int_{-\infty}^{\infty} r(s)\frac{1}{\sqrt{2\pi}\,\tau_0}\exp\left\{-\frac{1}{2\tau_0^2}(s-\theta)^2\right\} ds = 0,$$

$$\int_{-\infty}^{\infty}\left[r(s)\exp\left\{-\frac{s^2}{2\tau_0^2}\right\}\right]\exp\left\{s\frac{\theta}{\tau_0^2}\right\} ds = 0,$$

$$(9) \qquad \int_{-\infty}^{\infty} r^+(s)\exp\left\{-\frac{s^2}{2\tau_0^2}\right\}\exp\{st\}\,ds = \int_{-\infty}^{\infty} r^-(s)\exp\left\{-\frac{s^2}{2\tau_0^2}\right\}\exp\{st\}\,ds$$

for all θ in R or all $t = \theta/\tau_0^2$ in R, where r^+ and r^- are the positive and negative components of r: $r^+(s) = \max(0, r(s))$, $r^-(s) = \max(0, -r(s))$.

Let D be the value of the two expressions in (9) with $t = 0$. If $D \neq 0$, then we can write

$$(10) \qquad \int_{-\infty}^{\infty} p_1(s)\exp\{st\}\,ds = \int_{-\infty}^{\infty} p_2(s)\exp\{st\}\,ds$$

for all t in R where

$$p_1(s) = \frac{r^+(s)}{D}\exp\left\{-\frac{s}{2\tau_0^2}\right\}, \qquad p_2(s) = \frac{r^-(s)}{D}\exp\left\{-\frac{s}{2\tau_0^2}\right\}$$

are probability densities. Equation (9) says that p_1 and p_2 have the same moment-generating function and thus are identical. But from their definition p_1 and p_2 are nonzero on different sets, thus providing a contradiction. It follows that $D = 0$ and thus that $r^+(s) = 0$ and $r^-(s) = 0$, and thus $r(s) = 0$ with probability 1. This establishes completeness.

Statistical models can be *not* complete; see, for example, Exercise 5. Nevertheless, a UMV unbiased estimate may still exist.

E SOME EXAMPLES OF UMV UNBIASED ESTIMATORS

Consider a statistical model with a complete likelihood statistic. To construct the UMV unbiased estimator for a parameter $\beta(\theta)$ we have essentially two approaches. We can look for some unbiased estimator $t(y)$ and then take the regression on $s(y)$, obtaining the UMV unbiased estimator $r(s(y))$. This is based on Theorems 1 and 2 and as illustrated by Examples 1 and 2 in Section A. Alternatively, we can try to construct directly an unbiased estimator based on the likelihood statistic; if we succeed, then it *is* the UMV unbiased estimator. The first approach is amply illustrated in Section A; we discuss the second approach now.

Let $g(s\,|\,\theta)$ be the density function for the likelihood statistic. An estimable parameter $\beta(\theta)$ has an estimator $r(s)$ such that

$$(11) \qquad \beta(\theta) = \int r(s)g(s\,|\,\theta)\;ds.$$

From Corollary 8.4.4 we know that the likelihood function from $s = s(y)$ with the marginal model $g(s\,|\,\theta)$ is the same as the likelihood function from the antecedent y using the full model $f(y\,|\,\theta)$. Accordingly, we can write $g(s\,|\,\theta) = k(s)L(s\,|\,\theta)$, where $k(s)$ selects the appropriate representative from the likelihood function; hence

$$(12) \qquad \beta(\theta) = \int r(s)k(s)L(s\,|\,\theta)\;ds.$$

Thus we can think of an estimable parameter $\beta(\cdot)$ as a weighted combination (11) of functions $g(s\,|\,\cdot)$ or as a weighted combination (12) of the possible likelihood functions $L(s\,|\,\cdot)$. More positively, we try to *expand* the parameter (11) in terms of densities $g(s\,|\,\cdot)$ and thus obtain the coefficients $r(s)$ that give the UMV unbiased estimator.

EXAMPLE 5

Sample from the Poisson(θ): Let $(y_1, \ldots, y_n)$ be a sample from a Poisson(θ) distribution with θ in R^+. The likelihood statistic $s = \Sigma_1^n y_i$ is complete: Example 3 with $\lambda = n\theta$. The function $\bar{y}$ is an unbiased estimator of θ based on s; thus it is UMV unbiased.

Now suppose that we are interested in the parameter $e^{-\theta}$; this is the probability that a response y is zero. We could start with an initial estimator based on the indicator function h from the first coordinate,

$$h(y_1) = 1, 0 \qquad \text{according as } y_1 = 0, > 0;$$

and we could improve it using Theorem 1. Alternatively, we can try to expand $e^{-\theta}$ in terms of the likelihood functions:

$$e^{-\theta} = \sum_{s=0}^{\infty} r(s)\,\frac{(n\theta)^s e^{-n\theta}}{s!},$$

$$e^{(n-1)\theta} = \sum_{s=0}^{\infty} r(s)n^s\,\frac{\theta^s}{s!}.$$

Then identifying the coefficient of $\theta^s/s!$ on each side, we obtain

$(n - 1)^s = r(s)n^s,$

$$r(s) = \left(1 - \frac{1}{n}\right)^s.$$

Thus $(1 - 1/n)^s$ is the UMV unbiased estimator of $e^{-\theta}$.

EXAMPLE 6

Sample from the location normal: Let $(y_1, \ldots, y_n)$ be a sample from a normal(θ, σ_0) distribution with θ in R. The likelihood statistic $s = \bar{y}$ is complete: Example 4 with $\tau_0^2 = \sigma_0^2/n$.

The function $\bar{y}$ is an unbiased estimator of θ based on s; thus it is UMV unbiased.

In a similar way we see that $\bar{y}^2$ is the UMV unbiased estimator for $E(\bar{y}^2) = \theta^2 + \sigma_0^2/n$, and then that $\bar{y}^2 + (n - 1)\sigma_0^2/n$ is the UMV unbiased estimator for $\mu_2 = \theta^2 + \sigma_0^2$. In fact, any function of $\bar{y}$ with finite variance is the UMV unbiased estimator of its mean.

For each of these examples there is really only one parameter—however one chooses to express it. Thus, if we take s as the natural estimate of θ, we would use $\beta(s)$ as the appropriate estimate of $\beta(\theta)$. The property of unbiasedness is typically spoiled by a nonlinear function β. The fact that we would need a new estimator on the basis of UMV unbiasedness is really an indication of the arbitrariness of the unbiased property.

F WHEN DO THE METHODS WORK?

The procedures in this section give a UMV unbiased estimator for a model that has a complete likelihood statistic. The exercises and problems will illustrate the kinds of models that have this property: the exponential model with the number of ψ-functions equal to the dimension of the parameter; the variable carrier model with the number of boundary types equal to the dimension of the parameter; and fairly straightforward mixtures of these. For the models with a fixed region of positive density we are restricted here, as with the methods in Section 9.4, to models with simple exponential form. We have noted earlier in Section 8.5D that the common models having exponential form are in the continuous case essentially just the location normal, the scale normal, and the location-scale normal. And these indeed are the examples we are using here.

G EXERCISES

1 *The scale exponential:* Let $(y_1, \ldots, y_n)$ be a sample from the exponential distribution $f(y|\theta) = \theta^{-1} \exp\{-y/\theta\}$ on R^+ with θ in R^+. Show that the likelihood statistic $\Sigma\, y_i$ is complete; use Problem 17. Obtain the UMV unbiased estimator for θ.

2 *The scale normal:* Let $(y_1, \ldots, y_n)$ be a sample from the normal(μ_0, σ^2) with σ^2 in $\Omega = R^+$. Show that the likelihood statistic $\Sigma\,(y_i - \mu_0)^2$ is complete; use Problem 17. Obtain the UMV unbiased estimator for σ^2.

3 *The uniform*$(0, \theta)$: Let $(y_1, \ldots, y_n)$ be a sample from the uniform$(0, \theta)$ with θ in R^+. Show that the likelihood statistic $\max y_i$ is complete. Obtain the UMV unbiased estimator for θ.

Show that $2\bar{y}$ is an unbiased estimate of θ. How do the variances of these two unbiased estimators compare for n large?

4 *The Bernoulli(p)*: Let $(x_1, \ldots, x_n)$ be a sample from the Bernoulli(p) distribution with p in $[0, 1]$. Show that the likelihood statistic $y = \Sigma x_i$ is complete. Obtain the UMV unbiased estimator for p.

5 Let (y_1, y_2) be a sample from the uniform($\theta, \theta + 1$) with θ in R. Show that the likelihood statistic $(y_{(1)}, y_{(2)})$ is not complete; that is, find a nontrivial function of $(y_{(1)}, y_{(2)})$ that has mean value 0 for each θ value. In fact, there is no UMV unbiased estimate; Lehmann and Scheffé (1950).

H PROBLEMS

6 For a sample $(y_1, \ldots, y_n)$ from the Poisson(θ) in Example 5, find the UMV unbiased estimator of $e^{-\theta}$ by determining the regression of $h(y_1)$ on $s = \Sigma y_i$.

7 Let $(x_1, \ldots, x_n)$ be a sample from the Bernoulli(p) distribution with p in $[0, 1]$. The likelihood statistic $y = \Sigma x_i$ is complete; Exercise 4. Obtain the UMV unbiased estimator for p^2; for pq.

8 Let $(y_1, \ldots, y_k)$ be a sample from the binomial(n, p) distribution with p in $[0, 1]$. Determine the UMV unbiased estimate of q^n, the probability of a zero count; compare with Problem 6.

9 Let y have the hypergeometric(N, n, D) distribution with D in $\{0, 1, \ldots, N\}$. Show that y is complete. Obtain the UMV unbiased estimator for D/N.

10 Let $(y_1, \ldots, y_n)$ be a sample from the model $f(y|\theta) = k(\theta)\phi(y/\theta)h(y)$, where ϕ is the indicator function for the interval $(0, 1)$. Show that $\max y_i$ is complete.

I PROBLEMS FOR VECTOR PARAMETERS

11 *The location-scale normal:* Let $(y_1, \ldots, y_n)$ be a sample from the normal(μ, σ) with (μ, σ^2) in $R \times R^+$. Show that the likelihood statistic $(\Sigma y_i, \Sigma y_i^2)$ is complete; use Problem 17. Obtain the UMV-matrix unbiased estimator for (μ, σ^2).

12 Let $(y_1, \ldots, y_n)$ be a sample from the normal(μ, σ) with (μ, σ^2) in $R \times R^+$. Show that $s_y(n-1)^{1/2}\Gamma((n-1)/2)/2^{1/2}\Gamma(n/2)(\approx s_y(1 + 1/4(n-1)))$ is the UMV unbiased estimate of σ; use Problem 11.

13 Let $(y_1, \ldots, y_n)$ be a sample from the uniform(θ_1, θ_2) with $\theta_1 < \theta_2$ and θ_i in R. Show that the likelihood statistic $(\min y_i, \max y_i)$ is complete. Obtain the UMV-matrix unbiased estimator for (θ_1, θ_2).

14 *The location-scale exponential:* Let $(y_1, \ldots, y_n)$ be a sample from $f(y|\theta, \tau) = \tau^{-1}c(y - \theta) \exp\{-(y - \theta)/\tau\}$, where c is the indicator function for $(0, \infty)$. Show that the likelihood statistic is $(\min y_i, \Sigma y_i)$. Show that the likelihood statistic is complete; compare with Exercises 1 and 3.

15 Let $(y_1, \ldots, y_r)$ be multinomial $(n; p_1, \ldots, p_r)$ with $p_i \geq 0, \Sigma p_i = 1$. Show that $(y_1, \ldots, y_r)$ is complete. Obtain the UMV matrix unbiased estimator for $(p_1, \ldots, p_r)$.

16 Let $(y_i, \ldots, y_r)$ be general hypergeometric $(N, n, D_1/N, \ldots, D_r/N)$ with $D_i = 0, 1, \ldots$ and $\Sigma D_i = N$. Show that $(y_1, \ldots, y_r)$ is complete. Determine the UMV matrix unbiased estimator for $(D_1/N, \ldots, D_r/N)$.

17 *Exponential form:* Let y have density function

$$f(y|\boldsymbol{\theta}) = \gamma(\boldsymbol{\theta}) \exp\left\{\sum_1^r s_u(y)\theta_u\right\} h(y)$$

with respect to volume or a counting measure. Then by formula (8.4.14) $\mathbf{s} = (s_1(y), \ldots, s_r(y))'$ has density

$$g(\mathbf{s}|\boldsymbol{\theta}) = \gamma(\boldsymbol{\theta}) \exp\left\{\sum_1^r s_u\theta_u\right\} H(\mathbf{s}).$$

Then for the model with $\boldsymbol{\theta}$ in R^r (or in a subset of R^r containing an open rectangle), show that $\mathbf{s}$ is complete. *Method:* If the rectangle does not contain the origin, then set $\boldsymbol{\theta} = \boldsymbol{\phi} + \mathbf{a}$ and redistribute terms so that the new parameter has a rectangle containing the origin. Follow Example 4 and use the uniqueness theorem for multivariate moment-generating functions in Supplement Section 5.8E.

18 Consider the normal linear model $\mathbf{y} = X\boldsymbol{\beta} + \mathbf{v}$, where $\mathbf{v}$ is a sample from the normal$(0, \sigma_0)$ with $\boldsymbol{\beta}$ in R^r. Show that the likelihood statistic $X'\mathbf{y}$ is complete; use Problem 17. For future reference note that $\mathbf{b}(\mathbf{y}) = (X'X)^{-1}X'\mathbf{y}$ is an equivalent function.

19 Consider the normal linear model $\mathbf{y} = X\boldsymbol{\beta} + \sigma\mathbf{z}$, where $\mathbf{z}$ is a sample from the normal$(0, 1)$ with $\theta = (\boldsymbol{\beta}, \sigma^2)$ in $R^r \times R^+$. Show that the likelihood statistic $(X'\mathbf{y}, \mathbf{y}'\mathbf{y})$ is complete; use Problem 17 and recall Problem 8.5.7. For future reference note that $(\mathbf{b}(\mathbf{y}), s^2(\mathbf{y}))$ is an equivalent function, where $s^2(y)$ is the minimized sum of squares in Section 9.2B.

20 (*continuation*) Show that the least-squares estimates $b_1, \ldots, b_r$ are UMV unbiased; recall Theorem 9.1.4.

21 *Measuring wavelength:* Measurements are made to determine a period of oscillation β. Let y_0 be measured time at which the system is in a certain phase, and let $y_1, \ldots, y_n$ be the measured times of the next n recurrences to that phase. If the measurement error is normally distributed without bias and with common variance, then the linear model $\mathbf{y} = \alpha\mathbf{1} + \beta\mathbf{x} + \sigma\mathbf{z}$ is appropriate where $\mathbf{z}$ is a sample from the standard normal and $\mathbf{x} = (0, 1, 2, \ldots, n)'$. Find the UMV unbiased estimator b of β; use Problem 20 together with Problem 9.2.11.

22 (*continuation*) Consider the following estimators:

$$b_2 = \frac{y_n - y_0}{n} \qquad b_1 = \begin{cases} \dfrac{\sum_{n/2+1}^{n} y_i - \sum_{0}^{n/2-1} y_i}{(n^2 + 2n)/4} & (n \text{ even}), \\[2ex] \dfrac{\sum_{(n+1)/2}^{n} y_i - \sum_{0}^{(n-1)/2} y_i}{(n+1)^2/4} & (n \text{ odd}). \end{cases}$$

(a) Show that b_1 and b_2 are unbiased estimators of β. Calculate the variances of b, b_1, and b_2.

(b) Suppose that b_2 is available for $n = 300$. Find values of n for which b and b_1 would have the same precision.

23 Let $(y_1, \ldots, y_n)$ be a sample from some density function on R (Ω would be an index set for the class of density functions or the class of piecewise continuous density functions). The statistic $(y_{(1)}, \ldots, y_{(n)})$ is sufficient; see Problem 8.4.8. An equivalent form of the statistic is $(\Sigma y_i, \Sigma y_i^2, \ldots, \Sigma y_i^n)$; see Problem 3.1.9. The density functions as described above include those of the model $f(y) = \phi(\boldsymbol{\theta}) \exp\{-y^{2n} + \theta_1 y^1 + \cdots + \theta_n y^n\}$ having exponential form. Deduce that the statistic is complete; use Problem 17.

24 Consider a sample of $n > 1$ from the multivariate normal $(\boldsymbol{\mu}; \Sigma)$ with $\boldsymbol{\mu}$ in R^k and Σ in the space of positive definite symmetric matrices (an open set of $R^{k+k(k+1)/2}$); see Problem 8.5.8. Let

$$Y = \begin{pmatrix} y_{11} & \cdots & y_{1k} \\ \vdots & & \vdots \\ y_{n1} & \cdots & y_{nk} \end{pmatrix},$$

where the n rows record the n elements in the sample. Show that

$$(\bar{y}_1, \ldots, \bar{y}_k) = n^{-1}\mathbf{1}'Y,$$

$$\begin{pmatrix} s_{11} & \cdots & s_{1k} \\ \vdots & & \vdots \\ s_{k1} & \cdots & s_{kk} \end{pmatrix} = (Y - \mathbf{1}n^{-1}\mathbf{1}'Y)'(Y - \mathbf{1}n^{-1}\mathbf{1}'Y)/(n - 1),$$

where $\mathbf{1}$ is the 1-vector with n coordinates and the s_{ij} are sample covariances; Section 5.4D.

Show that the sample averages and covariances are UMV unbiased estimates of the means and covariances of the distribution.

25 Theorem 5 shows that a UMV unbiased estimate is unique under the assumption of a complete likelihood statistic. Show that the uniqueness holds without this assumption. For this, let $\sigma^2(\theta)$ be the variance of two UMV unbiased estimates t_1 and t_2 of a parameter $\beta(\theta)$, let $\sigma_{12}(\theta)$ be the covariance, and consider the variance of $at_1 + (1 - a)t_2$.

SUPPLEMENTARY MATERIAL

Supplement Section 9.6 is on pages 572–579.

Occasional cases arise in practice where the parameter value θ of interest has itself come from an earlier random system. If the probability characteristics of this earlier system are known, then ordinary probability theory prescribes that we examine the combined system: the earlier system together with the present system. It also prescribes that we obtain the conditional distribution of what we have not observed (θ) given what we know (the observed response y). Compare with the discussion in Sections 4.1F and 4.1G. The analysis of such combined systems, called *Bayes analysis,* is discussed in Supplement Section 9.6.

In other cases where there is no earlier system, just an unknown θ of a current system, some statisticians suggest that we use a probability distribution to present our feelings and prejudices about θ and then use the Bayes analysis just mentioned. Some discussion of this *subjective Bayes analysis* is included in the supplementary section.

NOTES AND REFERENCES

Following our pattern of approaching statistics from an open and general viewpoint, we have introduced this chapter on estimation by presenting some numerical procedures applicable to data without reference to a formal statistical model. We then introduced unbiased estimation and used the method of combining estimates to motivate the general Gauss–Markov theorem. The remainder of the chapter was then involved with the core material of traditional estimation theory: the Cramér–Rao information inequality developed in its essential, local form, and the Rao–Blackwell and Lehmann–Scheffé results on minimum variance unbiased estimation.

On least squares and the linear model:

Seal, H. L. (1967). Studies in the history of probability and statistics: XV. The historical development of the Gauss linear model. *Biometrika,* **54,** 1–24.

On the information inequality:

Rao, C. R. (1945). Information and the accuracy attainable in the estimation of statistical parameters. *Bull. Calcutta Math. Soc.,* **37,** 81–91.

Fraser, D. A. S. (1964). On local inference and information. *J. Roy. Statist. Soc.,* **B26,** 253–260.

On sufficiency and completeness:

Lehmann, E. L., and H. Scheffé (1950). Completeness, similar regions and unbiased estimation. *Sankhyā,* **10,** 305–340.

On Bayes estimation:

Bayes, T. (1763). Essay towards solving a problem in the doctrine of chances. *Biometrika,* **45,** 293–315 (1958).

Lindley, D. V. (1965). *Introduction to Probability and Statistics from a Bayesian Viewpoint:* Vol. 2, *Inference.* New York: Cambridge University Press.

In Chapters 7 and 8 we introduced estimation, tests of significance, and confidence intervals. Then in Chapter 9 we examined estimation in detail.

In Chapter 10 we shall examine in detail a formalized version of tests of significance. In Section 10.1 we investigate the formal framework for ***hypothesis testing*** *and derive a* ***fundamental lemma*** *for obtaining a best test for a very simple problem. In Section 10.2 we use the lemma to obtain best tests for more general problems involving* ***composite hypotheses*** *and we use the likelihood statistic (or sufficiency) to obtain improved tests. In Section 10.3 we examine some general methods for obtaining tests:* ***generalized likelihood ratio, unbiasedness, large sample,*** *and* ***sequential analysis.*** *Finally, in Section 10.4 we examine tests for the validity of a statistical model itself, the* ***chi-square*** *tests of fit.*

In Supplement Section 10.5 we examine ***confidence intervals*** *and* ***regions,*** *specifically the formal ties with hypothesis testing.*

10
TESTING STATISTICAL HYPOTHESES

10.1

THE FUNDAMENTAL LEMMA

Consider the biochemical example examined in Section 7.2B. Certain theoretical calculations had suggested the value $\mu = 14.8$ for the mean response, and we were interested in whether the observed data in Section 7.1E were in accord with this suggested value. We calculated the standardized deviation of the sample average $\bar{y}$ from the suggested mean 14.8, obtaining

$$z = \frac{16.8 - 14.8}{0.7} = 2.86.$$

This observed value was moderately extreme as assessed against the standard normal distribution. Accordingly, the value provided moderately strong evidence against the suggested or hypothesized value 14.8 for the mean μ.

We noted that a value of z in the central range of the standard normal would be in accord with the hypothesis; that a moderately extreme value would provide some evidence against the hypothesis; and that a value very far out on the tails would effectively contradict the hypothesis. Only in this last and very extreme case are we in the position of having grounds for a near-categorical statement or decision concerning the truth of the hypothesis.

In Chapters 7 and 8 we did not consider formally the question of how to choose a function for assessing the departure from a hypothesis. One way of obtaining criteria for such a choice is to introduce cutoff or critical values on the range of a function: ''accept'' the hypothesis on one side of a critical value; ''reject'' the hypothesis on the other side. We could then examine the probability of ''accepting'' the hypothesis and the probability of ''rejecting'' the hypothesis under the various possibilities for the true parameter. We would then have a basis for comparing various functions.

In Section 7.2G we mentioned some very serious aspects of making decisions in this formal way—when typically there is not the evidence to support a decision. We can, however, use the theory connected with choosing such a decision procedure to guide us in our choice of function for making the tests of significance we have discussed in preceding chapters. It is with this goal in mind that we now examine the very attractive mathematical theory of hypothesis testing.

A EXAMPLE

Consider a response y that has a normal(θ, σ_0) distribution with θ in R. And suppose that an investigator is concerned with the hypothesis $H_0: \theta \leq \theta_0$, where θ_0 is a value indicated by some earlier investigations. The alternative to this hypothesis is given by the alternative hypothesis, $H_1: \theta > \theta_0$; typically this alternative is something the investigator is hoping or searching for. Let $(y_1, \ldots, y_n)$ be a sample from the response distribution just described.

The test-of-significance approach is to see if the data provide evidence against the hypothesis. Our earlier discussions suggest the use of the sample average and comparing it with the normal(θ_0, $\sigma_0/\sqrt{n}$) distribution.

In this section we consider this example from the point of view of making decisions. As part of this we examine the following *test* or *decision procedure;*

$$\text{(1)} \qquad \begin{aligned} &\text{accept } H_0: \quad \text{if } \bar{y} < \theta_0 + 1.64\sigma_0/\sqrt{n}; \\ &\text{reject } H_0: \quad \text{if } \bar{y} \geq \theta_0 + 1.64\sigma_0/\sqrt{n}; \end{aligned}$$

the value 1.64 is the value exceeded with probability 5 percent by the standard normal.

We can easily examine the performance characteristics of the preceding test procedure. Let $\mathcal{P}(\theta)$ be the probability that the test procedure leads to the decision "reject H_0":

$$\text{(2)} \qquad \mathcal{P}(\theta) = P(\text{"reject } H_0\text{"}|\theta) = P(\bar{y} \geq \theta_0 + 1.64\sigma_0/\sqrt{n}\,|\,\theta).$$

We call $\mathcal{P}(\theta)$ the *power function* of the test procedure; see Figure 10.1.

The power function $\mathcal{P}(\theta)$ in formula (2) is easily calculated numerically. Note that the decision "reject H_0" has preimage

$$D = \{\bar{y} : \bar{y} \geq \theta_0 + 1.64\sigma_0/\sqrt{n}\}$$

on the space of values for the function $\bar{y}$, and has preimage

$$C = \left\{\mathbf{y} : \sum_1^n y_i/n \geq \theta_0 + 1.64\sigma_0/\sqrt{n}\right\}$$

on the space for the sample $(y_1, \ldots, y_n)$; see Figure 10.2. The power function $\mathcal{P}(\theta)$ can most easily be calculated from the distribution of the function $\bar{y}$:

$$\begin{aligned} \mathcal{P}(\theta) &= P\left(\frac{\bar{y} - \theta}{\sigma_0/\sqrt{n}} \geq 1.64 + \frac{\theta_0 - \theta}{\sigma_0/\sqrt{n}}\,|\theta\right) \\ &= P\left(z \geq 1.64 + \frac{\theta_0 - \theta}{\sigma_0/\sqrt{n}}\right) \\ &= 1 - G\left(1.64 + \frac{\theta_0 - \theta}{\sigma_0/\sqrt{n}}\right), \end{aligned}$$

where G is the distribution function for the standard normal. Note the normal-distribution-function shape in Figure 10.1.

If H_0 is true and $\theta \leq \theta_0$, then the decision "reject H_0" is incorrect; it is called a *Type I error*. For $\theta \leq \theta_0$ the function $\mathcal{P}(\theta)$ gives the probability for such a Type I error.

If H_1 is true and $\theta > \theta_0$, then the decision "accept H_0" is incorrect; it is called a *Type II error*. For $\theta > \theta_0$ the function $1 - \mathcal{P}(\theta)$ gives the probability for an incorrect decision. The probability of an incorrect decision is plotted in Figure 10.3.

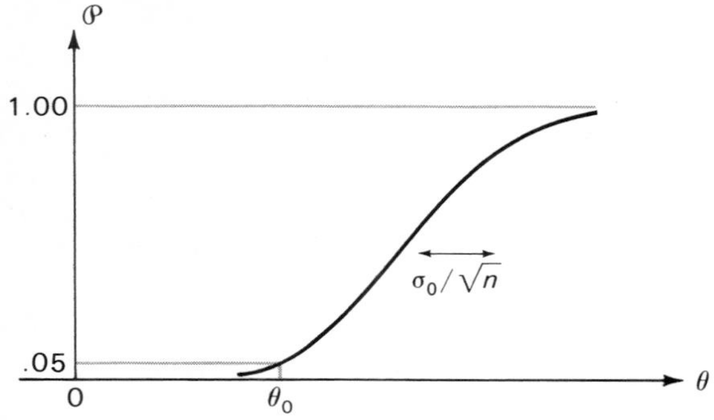

FIGURE 10.1
Power function of the test procedure (1).

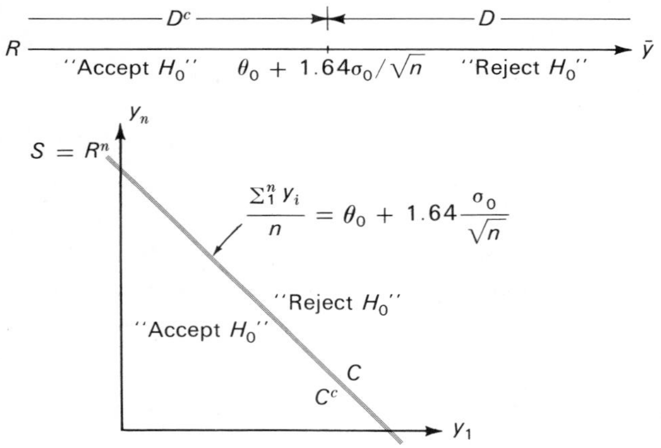

FIGURE 10.2
Set D of values for $\bar{y}$ that lead to "reject H_0"; and corresponding set C of values for $\mathbf{y}$ that lead to "reject H_0."

B THE GENERAL FORMULATION

Consider a random system with response $\mathbf{y}$. For the statistical model, let $\mathcal{S}$ be the sample space and $P(\cdot|\theta)$ be the probability measure for $\mathbf{y}$, with parameter θ taking values in the parameter space Ω. And suppose that the investigator has some hypothesis concerning the random system described by the preceding model. The investigator can consider each value θ and see whether the hypothesis holds for the corresponding description $P(\cdot|\theta)$. He can then assemble the values for which the hypothesis holds as the set H_0 on Ω and the values for which the hypothesis does not hold as the complementary set H_1. Thus the hypothesis induces a partition of Ω into two sets: H_0, called the *null hypothesis*, and H_1, called the *alternative hypothesis*.

The term "null hypothesis" derives from experimental applications concerned with cause–effect relationships. As part of determining whether there is evidence of a cause–effect relationship, an investigator considers a *null* hypothesis of no such

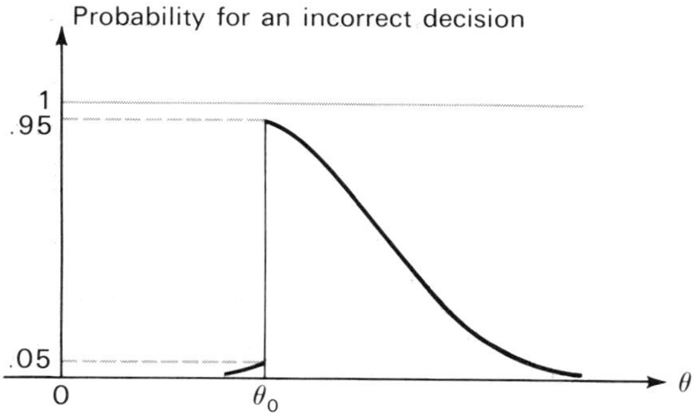

FIGURE 10.3
Probability of an incorrect decision using the test procedure (1).

relationship; he then assesses whether the model under the null hypothesis provides an explanation for any apparent effects in the data. As part of this assessment certain principles of experimentation are necessary. We return to this experimental situation in Chapter 12.

For notation let d_0 designate the decision "accept H_0" and d_1 designate the decision "reject H_0." The possible decisions then are given by the *decision space* $\{d_0, d_1\}$.

Now suppose that the investigator has selected a test procedure: then for each possible response value **y** he must have an associated decision, either d_0 or d_1. In effect, then, he has a *decision function* δ which maps the sample space $\mathcal{S}$ into the decision space $\{d_0, d_1\}$. It is convenient sometimes to think in terms of the preimage partition induced by δ on the sample space $\mathcal{S}$. For this let $C = \delta^{-1}(d_1)$ be the preimage of d_1, the set of sample points **y** that lead to the decision d_1. A test procedure δ can thus be presented in terms of a partition of the sample space into two sets: C, called the *critical region,* and C^c, called the *acceptance region*.

We can formalize this in terms of the following definitions:

DEFINITION 1

A **decision function** is a (measurable) function δ from the sample space $\mathcal{S}$ into the decision space $\{d_0, d_1\}$.

DEFINITION 2

A **critical region** is a (measurable) set $C = \delta^{-1}(d_1)$ on the sample space $\mathcal{S}$.

Now consider some theoretical aspects involved in choosing a decision function δ or equivalently in choosing a critical region C. For a test procedure, δ or C, the investigator will be interested in how it performs. This can be recorded in terms of the power function $\mathcal{P}_\delta$:

(3) $$\mathcal{P}_\delta(\theta) = P(\text{"reject } H_0\text{"} \mid \theta) = P(\delta(y) = d_1 \mid \theta) = P(C \mid \theta).$$

The investigator would want a test procedure for which $\mathcal{P}_\delta(\theta)$ is small for θ in H_0 and large for θ in H_1.

The performance characteristics of a test procedure can also be recorded in terms of the operating characteristic function OC_δ:

(4) $\quad OC_\delta(\theta) = P(\text{``accept } H_0\text{''}\,|\,\theta) = P(\delta(y) = d_0\,|\,\theta) = P(C^c\,|\,\theta).$

Clearly either function can be calculated from the other: as $C \cup C^c = \mathcal{S}$, we have

(5) $\quad P_\delta(\theta) + OC_\delta(\theta) = 1.$

C RELATED FUNCTIONS THAT DESCRIBE PERFORMANCE

If the hypothesis H_0 is true, then the decision d_1 is an incorrect decision; it is a Type I error or an error of the first kind. The probability of a Type I error with a decision function δ is

(6)
$$\begin{aligned} \alpha_\delta(\theta) &= \mathcal{P}_\delta(\theta) && \text{for } \theta \text{ in } H_0, \\ &= 0 && \theta \text{ in } H_1. \end{aligned}$$

If the hypothesis H_1 is true, then the decision d_0 is an incorrect decision; it is a Type II error or an error of the second kind. The probability of a Type II error with a decision function δ is

(7)
$$\begin{aligned} \beta_\delta(\theta) &= 0 && \text{for } \theta \text{ in } H_0, \\ &= 1 - \mathcal{P}_\delta(\theta) && \theta \text{ in } H_1. \end{aligned}$$

The probability for an incorrect decision is then given by

(8) $\quad \alpha_\delta(\theta) + \beta_\delta(\theta).$

This is plotted in Figure 10.3 for the test function (1).

In certain industrial applications it may be possible to determine the financial loss $l(d_i, \theta)$ entailed by the decision d_i when the parameter has the value θ. In such cases the investigator would be interested in the mean loss, or *risk,*

(9) $\quad \mathcal{R}_\delta(\theta) = E(l(\delta(\mathbf{y}), \theta)\,|\,\theta) = l(d_1, \theta)\mathcal{P}_\delta(\theta) + l(d_0, \theta)(1 - \mathcal{P}_\delta(\theta)).$

The loss function may have the reasonable property that there is no loss with a correct decision; then

$$\begin{aligned} l(d_0, \theta) &= 0 && \text{for } \theta \text{ in } H_0, \\ &= b(\theta) && \theta \text{ in } H_1, \\ l(d_1, \theta) &= a(\theta) && \text{for } \theta \text{ in } H_0, \\ &= 0 && \theta \text{ in } H_1. \end{aligned}$$

The risk function then has the form

$$\mathcal{R}_\delta(\theta) = a(\theta)\alpha_\delta(\theta) + b(\theta)\beta_\delta(\theta).$$

From a general point of view the investigator will want a test procedure for which the probability

(10) $\quad \alpha_\delta(\theta) + \beta_\delta(\theta)$

for incorrect decisions is small. Ideally he would want this to be small for each value of θ. In practice, however, he will usually find that the procedure that is best for certain θ values is not best for other θ values; he must then compromise in some way.

If a valid loss function is available, the investigator will want a test procedure for which the risk

(11) $\mathcal{R}_\delta(\theta)$

is small. Ideally he would want this small for each value of θ. In practice he will usually find that the procedure that is best for certain θ values is not best for other θ values; he must then compromise in some way.

D CONVENTIONAL CRITERIA FOR CHOOSING A TEST PROCEDURE

Consider the problem of choosing a good test procedure on the basis of performance characteristics. The usual approach involves restricting attention to those test procedures that have a certain bound α, say 5 percent or 1 percent, on the probability of a Type I error. The rationale is that deciding that a treatment effect exists when, in fact, it does not is serious and should be especially guarded against. Thus we restrict our attention to size α test procedures:

DEFINITION 3

A test procedure is a **size α test procedure** if

(12) $\mathcal{P}_\delta(\theta) = P(C|\theta) \le \alpha$ for θ in H_0;

it is an **exact size α test procedure** if

(13) $\sup_{\theta\in H_0} \mathcal{P}_\delta(\theta) = \sup_{\theta\in H_0} P(C|\theta) = \alpha.$

The test in Section A is an exact 5 percent test. It is also a 10 percent test, but as we shall see, a rather poor 10 percent test in comparison with other 10 percent tests.

The usual approach then is to search among the size α test procedures for one that maximizes the power,

(14) $\mathcal{P}_\delta(\theta) = P(C|\theta)$ for θ in H_1.

Of course, the test that maximizes the power $P(C|\theta)$ for one θ value in H_1 may not be the test that maximizes it for another θ value. Some sort of a compromise is then needed. The criteria (12) and (14) are illustrated in Figure 10.4.

E THE FUNDAMENTAL LEMMA

A hypothesis-testing problem is particularly simple if the hypothesis consists of a single parameter value and the alternative consists of a single parameter value. We might, for example, be testing the hypothesis that a response is normal(100, 5) against the alternative that it is normal(110, 5).

We say that a null or alternative hypothesis is *simple* if it has one parameter value, and is *composite* if it has more than one parameter value. Accordingly, we now consider the problem of testing a *simple hypothesis* against a *simple alternative*. Of course, most practical problems involve testing a composite hypothesis against a

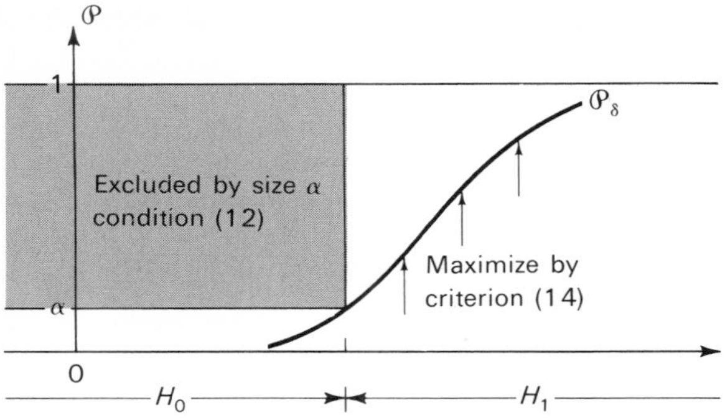

FIGURE 10.4
Size α condition (12); and maximum power criterion (14).

composite alternative. Fortunately, the solution of the special problem gives us the mechanics for obtaining solutions for many of the more general problems.

For the statistical model let $f(\cdot\,|\theta)$ be the density function with parameter θ in $\Omega = \{\theta_0, \theta_1\}$. We consider the problem of finding the best size α test of the hypothesis $H_0 = \{\theta_0\}$ against the alternative $H_1 = \{\theta_1\}$.

LEMMA 4

Fundamental Lemma: *A test with critical region C having*

(15) $$\frac{L(\mathbf{y}|\theta_1)}{L(\mathbf{y}|\theta_0)} = \frac{f(\mathbf{y}|\theta_1)}{f(\mathbf{y}|\theta_0)} \geq k \qquad \text{for } \mathbf{y} \text{ in } C,$$
$$\leq k \qquad\qquad C^c$$

is a most powerful test of $H_0 = \{\theta_0\}$ *against* $H_1 = \{\theta_1\}$ *at some size* α. *If the value k and the set C can be chosen to satisfy (exact size* α*)*

(16) $$\int_C f(\mathbf{y}|\theta_0)\, d\mathbf{y} = \alpha,$$

then C is most powerful among tests having size α.

The hypothesis-testing problem is one of finding a set C containing α of the $f(\mathbf{y}|\theta_0)$ probability and then as much as possible of the $f(\mathbf{y}|\theta_1)$ probability. Clearly points $\mathbf{y}$ with large values of $f(\mathbf{y}|\theta_1)/f(\mathbf{y}|\theta_0)$ should go in C first.

Also note that (15) does not define C for points having the likelihood ratio equal to k; for these points any assignment to C or C' is permissible subject to (16).

The proof is given after the following simple example.

EXAMPLE 1

The location normal: Consider a sample $\mathbf{y} = (y_1, \ldots, y_n)'$ from a normal(θ, σ_0) distribution with θ in R. To illustrate the lemma we take $\Omega = \{\theta_0, \theta_1\}$ and consider testing $H_0: \theta = \theta_0$ against $H_1: \theta = \theta_1$, where $\theta_0 < \theta_1$. By the lemma a most powerful test is to reject for large values of $L(\mathbf{y}|\theta_1)/L(\mathbf{y}|\theta_0)$ or, equivalently, to reject for large values of

(17) $$l(\mathbf{y}|\theta_1) - l(\mathbf{y}|\theta_0) = -\frac{1}{2\sigma_0^2}\sum (y_i - \theta_1)^2 + \frac{1}{2\sigma_0^2}\sum (y_i - \theta_0)^2$$
$$= \frac{\theta_1 - \theta_0}{\sigma_0^2}\sum \left(y_i - \frac{\theta_1 + \theta_0}{2}\right),$$

or equivalently ($\theta_1 > \theta_0$) to reject for large values of the likelihood statistic Σy_i or $\bar{y}$.

The most powerful test can then be obtained by calculating the critical value exceeded with probability α under the θ_0 distribution. The hypothesis distribution of $\bar{y}$ is normal(θ_0, $\sigma_0/\sqrt{n}$). Hence the size α test has critical region

$$C = \{\mathbf{y} : \bar{y} \geq \theta_0 + z_\alpha \sigma_0/\sqrt{n}\},$$

where z_α is the value exceeded with probability α under the standard normal distribution. Note that the test does not depend on θ_1 and hence is most powerful for each θ_1 value $> \theta_0$: it is thus *uniformly most powerful* UMP size α against the alternative $H_1 : \theta > \theta_0$.

Proof of Lemma 4 We present the proof for the absolutely continuous case; the discrete case is analogous. Consider a set C satisfying (15) and (16) for some α; and let D be any other size α test; then

$$\int_C f(\mathbf{y}|\theta_0)\, d\mathbf{y} = \alpha \geq \int_D f(\mathbf{y}|\theta_0)\, d\mathbf{y}.$$

We remove the set $C_0 = C \cap D$; let $C^* = C - C_0$ and $D^* = D - C_0$; hence

(18) $$\int_{C^*} f(\mathbf{y}|\theta_0)\, d\mathbf{y} \geq \int_{D^*} f(\mathbf{y}|\theta_0)\, d\mathbf{y}.$$

The points of C^* are inside C; thus by (15),

$$kf(\mathbf{y}|\theta_0) \leq f(\mathbf{y}|\theta_1).$$

The points of D^* are outside C; thus, by (15),

$$kf(\mathbf{y}|\theta_0) \geq f(\mathbf{y}|\theta_1).$$

We apply these inequalities to (18): we multiply through by k, then substitute in accord with the inequalities and obtain

$$\int_{C^*} f(\mathbf{y}|\theta_1)\, d\mathbf{y} \geq \int_{D^*} f(\mathbf{y}|\theta_1)\, d\mathbf{y}.$$

We now add back the integral over C_0, obtaining

$$\int_C f(\mathbf{y}|\theta_1)\, d\mathbf{y} \geq \int_D f(\mathbf{y}|\theta_1)\, d\mathbf{y}.$$

This says that the power of D is less than or equal to the power of C. Hence C has maximum power among tests of size α.

F A COMPLICATION WITH DISCRETE DISTRIBUTIONS

With a discrete distribution we can expect to have difficulty satisfying the exact size α condition (16). For suppose that we are to reject for large values of a discrete variable

y and find that under H_0 the discrete variable has, say, a Poisson(1) distribution. Then under H_0 we have

$$P(y \geq 3) = 0.080, \qquad P(y \geq 4) = 0.019.$$

Thus if we want a 5 percent test we must split the probability at the point $y = 3$. In effect, this can be accomplished by using a randomized test function.

DEFINITION 5

A **randomized test function** ϕ is a (measurable) real-valued function satisfying $0 \leq \phi(\mathbf{y}) \leq 1$.

A test function ϕ is used as follows. If $\phi(\mathbf{y}) = 1$, the decision is "reject H_0." If $\phi(\mathbf{y}) = 0$, the decision is "accept H_0." If $\phi(\mathbf{y}) = a$, then use some random system and with probability a choose the decision "reject H_0" and with probability $1 - a$ choose the decision "accept H_0." For a test with critical region C, note that the randomized test function ϕ is the indicator function of C.

The power function for a randomized test has the form

$$(19) \qquad \mathcal{P}_\phi(\theta) = E(\phi(\mathbf{y})\,|\,\theta) = \int \phi(\mathbf{y}) f(\mathbf{y}\,|\,\theta)\, d\mathbf{y}.$$

We now present the more general lemma in which the best size α test always exists.

LEMMA 6

For testing θ_0 against θ_1 the test

$$(20) \qquad \begin{aligned} \phi(\mathbf{y}) &= 1 \qquad \text{for } \frac{L(\mathbf{y}\,|\,\theta_1)}{L(\mathbf{y}\,|\,\theta_0)} > k, \\ &= a \qquad\qquad\qquad\quad = k, \\ &= 0 \qquad\qquad\qquad\quad < k, \end{aligned}$$

where a and k are chosen to satisfy (exact size α)

$$(21) \qquad \int \phi(\mathbf{y}) f(\mathbf{y}\,|\,\theta_0)\, d\mathbf{y} = \alpha,$$

is most powerful among tests having size α.

The proof is a straightforward extension of that for Lemma 4; see Problems 16 and 17.

The essential part of the lemma is that a and k can be chosen to satisfy the exact size α condition (21). For this, the Poisson example preceding the lemma suggests the method: define ϕ by

$$\begin{aligned} \phi(y) &= 1 \qquad\qquad\qquad\qquad \text{for } y \geq 4, \\ &= \frac{0.050 - 0.019}{0.080 - 0.019} \qquad\quad y = 3, \\ &= 0 \qquad\qquad\qquad\qquad\quad y \leq 2. \end{aligned}$$

This appropriately splits the probability at $y = 3$ so that

$$E(\phi(y)\,|\,\theta_0) = 1 \times 0.019 + \frac{0.050 - 0.019}{0.080 - 0.019}(0.080 - 0.019)$$
$$= 0.050;$$

thus the exact size 5 percent condition is satisfied.

G EXERCISES

1 The resistance was determined for five pieces of wire samples from production:

0.138, 0.140, 0.136, 0.142, 0.140

On the assumption that the variation is normal(0, 0.003), test at the 5 percent level the hypothesis $H_0{:}\theta \leq 0.135$, where θ is the mean response.

2 The yield of a certain variety of corn was determined on seven plots of land:

87.7, 87.2, 92.3, 94.2, 93.6, 85.6, 93.5

On the assumption that the variation is normal(0, 4.2), test at the 5 percent level the hypothesis $H_0{:}\theta \leq 85.0$, where θ is the mean response.

3 The breaking load was determined for a sample of six from the production run of a certain plastic:

18.3, 19.3, 21.2, 22.4, 17.1, 27.7

On the assumption that the variation is normal(0, 1.9), test at the 5 percent level the hypothesis $H_0{:}\theta \geq 21.0$, where θ is the mean response.

4 Sixteen measurements in a chemical response yield $\bar{y} = 285.3$. On the assumption that the variation is normal(0, 3.1), test at the 5 percent level the hypothesis $H_0{:}\theta \leq 280$, where θ is the mean response.

5 An investigator knows that a response is approximately normal with variance 1.44. Previously the mean had been 75, but a new treatment may have increased it. For testing the hypothesis that the mean θ is 75 against the alternative that it is, say, 76, determine the most powerful size 1 percent test for a sample size of 10. Plot the power function.

6 (*continuation*) For a 1 percent test, the power at $\theta = 76$ depends on the sample size n. Give an expression for the power and determine the minimum n so that there is a 95 percent chance of detecting a mean equal to 76.

H PROBLEMS

7 Let $(y_1, \ldots, y_n)$ be a sample from the exponential distribution $f(y\,|\,\theta) = \theta^{-1}\exp\{-y/\theta\}$ on R^+ (say, the interoccurrence times of Section 4.3A). Find the form of a most powerful test of $H_0{:}\theta = \theta_0$ against $H_1{:}\theta = \theta_1 > \theta_0$; the critical value can be presented in terms of a percentage point of a familiar distribution. Does the test depend on θ_1?

8 Let $(y_1, \ldots, y_n)$ be a sample from the normal(μ_0, σ) with σ in R^+.

(a) Find the most powerful size α test for $H_0{:}\sigma = \sigma_0$ against $H_1{:}\sigma = \sigma_1$ $(>\sigma_0)$. Does the test depend on σ_1?

(b) Give an expression for the power function of the test in terms of a chi-square distribution function.

9 (*continuation*)

(a) Find the most powerful size α test for $H_0{:}\sigma = \sigma_0$ against $H_1{:}\sigma = \sigma_1$ $(<\sigma_0)$. Does the test depend on σ_1?

(b) Give an expression for the power function of the test in terms of a chi-square distribution function.

10 Let $(y_1, \ldots, y_n)$ be a sample from the Poisson(θ) distribution. Find the form of the most powerful size α test of $H_0{:}\theta = \theta_0$ against $H_1{:}\theta = \theta_1$ $(>\theta_0)$. Does the test depend on θ_1?

11 (*continuation*) Find the form of the most powerful size α test of $H_0{:}\theta = \theta_0$ against $H_1{:}\theta = \theta_1$ $(<\theta_0)$. Does the test depend on θ_1?

12 Let $(x_1, \ldots, x_n)$ be a sample from the Bernoulli(p) distribution. Find the form of a most powerful size α test of $H_0{:}p = p_0$ against $H_1{:}p = p_1$ $(>p_0)$. Does the test depend on p_1?

13 (*continuation*) Find the form of the most powerful size α test of $H_0{:}p = p_0$ against $H_1{:}p = p_1$ $(<p_0)$. Does the test depend on p_1?

14 Let $(y_1, \ldots, y_n)$ be a sample from the uniform(0, θ) distribution. Find the form of the most powerful size α test of $H_0{:}\theta = \theta_0$ against $H_1{:}\theta = \theta_1$ $(>\theta_0)$. Does the test depend on θ_1?

15 (*continuation*) Find the form of the most powerful size α test of $H_0{:}\theta = \theta_0$ against $H_1{:}\theta = \theta_1$ $(<\theta_0)$. Does the test depend on θ_1?

I PROBLEMS: EXTENSIONS OF THE THEORY

16 *Part of Lemma 6:* Show that a and k can be chosen so that a test function ϕ given by (20) satisfies the exact size α condition (21). *Method:* Let G be the distribution function for

$$u = \frac{f(\mathbf{y}\,|\,\theta_1)}{f(\mathbf{y}\,|\,\theta_0)}$$

under the θ_0-distribution for $\mathbf{y}$. Note that the θ_0-probability for a zero denominator for u is zero; thus we need consider only the set where $f(\mathbf{y}\,|\,\theta_0) > 0$. Show that a and k can be determined in terms of characteristics of G.

17 *Part of Lemma 6:* Let ϕ satisfy (20) and (21) and let ψ be any other test of size α. Show that

$$(22) \qquad \mathcal{P}_\phi(\theta_1) - \mathcal{P}_\psi(\theta_1) = \int (\phi(\mathbf{y}) - \psi(\mathbf{y}))f(\mathbf{y}\,|\,\theta_1)\,d\mathbf{y}$$
$$\geq \int (\phi(\mathbf{y}) - \psi(\mathbf{y}))kf(\mathbf{y}\,|\,\theta_0)\,d\mathbf{y} = k(\mathcal{P}_\phi(\theta_0) - \mathcal{P}_\psi(\theta_0)).$$

Method: After dividing up the region of integration, use the likelihood ratio inequalities (20) together with the corresponding sign of $\phi(\mathbf{y}) - \psi(\mathbf{y})$.

18 *Extension of Lemma 6:* Let ψ be a most powerful size α test of $f(\mathbf{y}\,|\,\theta_0)$ against $f(\mathbf{y}\,|\,\theta_1)$. Show that ψ has the likelihood ratio form;

$$\psi(\mathbf{y}) = 1, \qquad f(\mathbf{y}\,|\,\theta_1)/f(\mathbf{y}\,|\,\theta_0) > k,$$
$$= 0, \qquad\qquad < k$$

and thus differs from the test ϕ in (20) only at points where the likelihood ratio equals k. *Method:* From size and power properties show that

$$\int (\phi(\mathbf{y}) - \psi(\mathbf{y}))(f(\mathbf{y}\,|\,\theta_1) - kf(\mathbf{y}\,|\,\theta_0))\,d\mathbf{y} \leq 0,$$

and from properties of ϕ show that the integrand is ≥ 0.

19 (*continuation*) If $E(\psi(\mathbf{y})\,|\,\theta_0) < \alpha$, then $k = 0$ and the maximum power at θ_1 is unity.

20 *Generalization of Lemma 6:* Let f_0 and f_1 be *real-valued* integrable functions relative to a measure μ. Among test functions ϕ for which

$$(23) \qquad \int \phi(\mathbf{y})f_0(\mathbf{y})\,d\mu(\mathbf{y}) = c$$

a necessary and sufficient condition that ϕ maximize

$$\int \phi(\mathbf{y})f_1(\mathbf{y})\,d\mu(\mathbf{y})$$

is that ϕ satisfy (23) and have the form

$$(24)\qquad \phi(\mathbf{y}) = 1, \qquad f_1(\mathbf{y}) > kf_0(\mathbf{y}),$$
$$= 0, \qquad < kf_0(\mathbf{y}),$$

for some k.

21 Consider responses $y_1, y_2, \ldots$ such that $(y_1, \ldots, y_k)$ has density function $f_k(y_1, \ldots, y_k | \theta)$. If

$$u_k = \frac{f_k(y_1, \ldots, y_k | \theta_1)}{f_k(y_1, \ldots, y_k | \theta_0)},$$

show that $E(u_{k+1} : y_1, \ldots, y_k | \theta_0) = u_k$. Thus the likelihood ratio is an example of a Martingale.

10.2

COMPOSITE HYPOTHESES

We are now in a position to examine some realistic hypothesis-testing problems. From Section 10.1 we have the solution for the problem of testing the hypothesis $\theta = \theta_0$ against the alternative $\theta = \theta_1$. As a problem in itself this is somewhat unrealistic, for it allows just two possible distributions for the response. We can, however, use the solution from the simple problem as a tool for solving more realistic problems, such as testing the following: the mean breaking strength θ for cable from a new production process does not exceed 12,500 lb (available with other processes) against the alternative that it is greater than 12,500 lb, that is, $H_0 : \theta \le 12{,}500$ versus $H_1 : \theta > 12{,}500$.

A THE LEAST FAVORABLE VALUE IN H_0

We first consider the problem of testing a composite hypothesis against a simple alternative:

$$(1)\qquad H_0 : \theta \text{ in } \omega \qquad \text{vs.} \quad H_1 : \theta = \theta_1.$$

We use ω to designate a subset of the initial parameter space Ω; the problem as just presented has an effective parameter space $\omega \cup \{\theta_1\}$.

The conventional approach to a hypothesis-testing problem was formulated in Section 10.1D. For the present problem we would search through the class of test functions ϕ that satisfy the size α condition,

$$(2)\qquad E(\phi(\mathbf{y}) | \theta) = \int_C f(\mathbf{y} | \theta)\, d\mathbf{y} \le \alpha \qquad \text{for } \theta \text{ in } \omega,$$

and try to find the particular test that maximizes the power

$$(3)\qquad E(\phi(\mathbf{y}) | \theta_1) = \int_C f(\mathbf{y} | \theta_1)\, d\mathbf{y}.$$

Our mathematical method with this problem is first to relax the condition (2), obtaining

$$(4)\qquad E(\phi(\mathbf{y}) | \theta_0) = \int_C f(\mathbf{y} | \theta_0)\, d\mathbf{y} \le \alpha,$$

where θ_0 is some particular value in ω deemed most difficult to distinguish from the alternative value θ_1, a test function ϕ satisfying condition (2) will satisfy the relaxed condition (4); thus the class of tests satisfying (4) is larger ($\supset$) than the original class satisfying (2). We look for the most powerful (θ_1) test in the larger class hoping that it will belong to the smaller class.

The mathematical method is to use the fundamental lemma 10.1.4 or 10.1.6 to determine the most powerful size α for the modified problem

(5) $\quad H_0: \theta = \theta_0 \quad \text{vs.} \quad H_1: \theta = \theta_1.$

If this most powerful test satisfies (2) rather than just (4) and is thus a size α test for the original hypothesis $\theta \in \omega$, then it is the most powerful size α test of the original problem (1).

In such a case, in which our choice of θ_0 value is successful, we call that θ_0 value the *least favorable value in* ω *against the alternative* θ_1. The success of the method depends essentially on making a suitable choice for θ_0. As a guiding principle we would choose the θ_0 value that gives a distribution $f(\cdot|\theta_0)$ "closest" to the alternative $f(\cdot|\theta_1)$, or most difficult to distinguish from the alternative $f(\cdot|\theta_1)$.

The hypothesis-testing problem (1) may, in fact, only be of secondary interest. Rather we may be interested in the more general problem

(6) $\quad H_0: \theta \text{ in } \omega \quad \text{vs.} \quad H_1: \theta \text{ in } \Omega - \omega,$

where the hypothesis and alternative are composite. If the most powerful (at θ_1) size α test for the problem (1) does not depend on θ_1 as it ranges over $\Omega - \omega$, then it is the uniformly most powerful UMP size α test for the more general problem (6).

EXAMPLE 1

The location normal: Consider a sample $\mathbf{y} = (y_1, \ldots, y_n)'$ from a normal(θ, σ_0) distribution with θ in R. We consider the hypothesis-testing problem

(7) $\quad H_0: \theta \le \theta_0 \quad \text{vs.} \quad H_1: \theta > \theta_0.$

We first determine the size α test that has maximum power at the parameter value θ_1 ($>\theta_0$). In effect, we are then considering the hypothesis-testing problem

(8) $\quad H_0: \theta \le \theta_0 \quad \text{vs.} \quad H_1: \theta = \theta_1.$

Our approach is to choose a value in H_0 that is most difficult to distinguish from the alternative θ_1; a natural first choice is the value θ_0. We thus temporarily examine

(9) $\quad H_0: \theta = \theta_0 \quad \text{vs.} \quad H_1: \theta = \theta_1.$

The most powerful test for this is available from Example 10.1.1:

(10) $\quad C = \{\mathbf{y}: \bar{y} \ge \theta_0 + z_\alpha \sigma_0/\sqrt{n}\}.$

We must now check whether this test has size α for the testing problem (8). The power function for $\theta \le \theta_0$ is

$$\begin{aligned}\mathcal{P}_C(\theta) &= P(\bar{y} \ge \theta_0 + z_\alpha \sigma_0/\sqrt{n}|\theta) \\ &= P\left(\frac{\bar{y} - \theta}{\sigma_0/\sqrt{n}} \ge \frac{\theta_0 - \theta}{\sigma_0/\sqrt{n}} + z_\alpha | \theta\right),\end{aligned}$$

which has maximum value α when $\theta_0 - \theta$ has its minimum value 0. Thus the test (10) has size α for the testing problem (8) and is thus the most powerful size α test for that problem (8).

Note that the test (10) does not depend on θ_1 and hence is most powerful for each θ_1 value $>\theta_0$; it is thus *uniformly most powerful* UMP size α for the original hypothesis-testing problem (7). See Figure 10.1 and note that we have found a test that has maximum power at each value of θ larger than θ_0. Thus we have accomplished what each of the arrows in Figure 10.4 indicates.

B THE LEAST FAVORABLE DISTRIBUTION

The method described in Section A may not work: there may be no θ_0 value for which the corresponding test satisfies the size α condition (2) for the full hypothesis: θ in ω. The method can, however, be extended in a simple way. Let λ be a probability measure defined over the hypothesis ω: such a measure can produce a mixture or average density

$$(11) \qquad f(\mathbf{y}|\lambda) = \int_{\omega} f(\mathbf{y}|\theta)\, d\lambda(\theta).$$

We would like to choose λ so that the mixture is most difficult to distinguish from the particular alternative.

Our extended method is first to relax condition (2) obtaining

$$(12) \qquad E(\phi(\mathbf{y})|\lambda) = \int_{C} f(\mathbf{y}|\lambda)\, d\mathbf{y} \le \alpha.$$

A test function ϕ satisfying (2) will satisfy the relaxed condition (12); use the Fubini Theorem 5.8.10. Thus the class of tests satisfying (12) is larger ($\supset$) than the original class satisfying (2). We look for the most powerful (θ_1) test in the larger class hoping that it will belong to the smaller class.

The mathematical method is to use the fundamental lemma 10.1.4 or 10.1.6 to determine the most powerful size α test of $f(\mathbf{y}|\lambda)$ against $f(\mathbf{y}|\theta_1)$. If this most powerful test satisfies (2) rather than just (12) and is thus of size α for the original hypothesis $\theta \in \omega$, then it is the most powerful size α test of the original problem (1). In such a case the distribution λ is called *least favorable* over ω against the alternative θ_1.

If the most powerful (θ_1) size α test does not depend on θ_1 in $\Omega - \omega$, then it is UMP size α for the more general problem (6).

Nontrivial illustrations can be rather involved and complicated; for an interesting illustration, see Problem 9. We present here a rather simple example.

EXAMPLE 2

For a sample $\mathbf{y} = (y_1, \ldots, y_n)'$ from the normal(θ, 1), consider the problem of testing the hypothesis $H_0 : \theta = \pm\delta_0$ against the alternative $H_1 : \theta = 0$. An intuitively reasonable choice for λ has probability 1/2 at each of $+\delta_0$, $-\delta_0$. The likelihood ratio test of $f(\mathbf{y}|\lambda)$ against $f(\mathbf{y}|0)$ is to reject for large values of

$$\frac{f(\mathbf{y}|0)}{f(\mathbf{y}|\lambda)} = \frac{\exp\left\{-(1/2)\sum y_i^2\right\}}{(1/2)\exp\left\{-(1/2)\sum (y_i - \delta_0)^2\right\} + (1/2)\exp\left\{-(1/2)\sum (y_i + \delta_0)^2\right\}}$$

$$= \frac{1}{\left[(1/2)\exp\left\{\delta_0 \sum y_i\right\} + (1/2)\exp\left\{-\delta_0 \sum y_i\right\}\right]\exp\left\{-n\delta_0^2/2\right\}}$$

or, equivalently, for small values of $\exp\{n\delta_0\bar{y}\} + \exp\{-n\delta_0\bar{y}\}$ or, equivalently, for small values of $|\bar{y}|$ (seen by plotting the sum of a positive and negative exponential). The size α test of the mixture λ thus has critical region $C = \{\mathbf{y}: |\sqrt{n}\,\bar{y}| < k\}$, where $(\sqrt{n}\,\delta_0 - k, \sqrt{n}\,\delta_0 + k)$ has probability α for the standard normal; note that $\sqrt{n}\,\bar{y}$ has a distribution that is a 50 : 50 mixture of the normal $(\pm\sqrt{n}\,\delta_0, 1)$ distributions, and note that the distribution of $|\bar{y}|$ is the same from the normal$(\delta_0, 1)$ as from the normal$(-\delta_0, 1)$ distribution (and hence the same as from the 50 : 50 mixture). It follows that the test has size α under the hypothesis H_0, and hence that the test is most powerful size α for H_0.

C MONOTONE LIKELIHOOD RATIO FORM

Consider the case of a real-valued parameter θ and the hypothesis-testing problem

(13) $\quad H_0: \theta \le \theta_0 \quad \text{vs.} \quad H_1: \theta > \theta_0.$

The method in Section A involves taking a value in the hypothesis, say θ_0, and a value in the alternative, say θ_1, and using the fundamental lemma to determine the most powerful test of θ_0 against θ_1. The resulting test is to reject for large values of the likelihood ratio,

(14) $$\frac{f(\mathbf{y}|\theta_1)}{f(\mathbf{y}|\theta_0)}.$$

Now suppose that the method produces a uniformly most powerful test for (13) at each size α. Then it follows generally that large values of the likelihood ratio (14) will correspond to large values of a function $t(\mathbf{y})$ for each value of $\theta_1 > \theta_0$. For this we present the following definition.

DEFINITION 1

The statistical model $\{f(\cdot|\theta): \theta \in R\}$ has **monotone likelihood ratio form** if there is a real-valued function t such that

(15) $$\frac{f(\mathbf{y}|\theta_1)}{f(\mathbf{y}|\theta_0)}$$

is a nondecreasing function of $t(\mathbf{y})$ for each choice of θ_0 and θ_1 with $\theta_1 > \theta_0$.

Note that Definition 1 implies that the likelihood statistic is a function of $t(\mathbf{y})$ and thus that $t(\mathbf{y})$ is a sufficient statistic. We recall from Section 8.5D that a real-valued likelihood statistic is intimately tied to models having exponential form with one

ψ-function. Thus a monotone likelihood ratio model itself is closely tied to models having exponential form with one ψ-function. As the prime example, consider the model

$$f(\mathbf{y}|\theta) = \gamma(\theta) \exp\{t(\mathbf{y})\psi(\theta)\}h(\mathbf{y}),$$

where $\psi(\theta)$ is a nondecreasing function of θ; this model has exponential form with one ψ-function. As specializations of this we have the usual location normal, scale normal, binomial, and Poisson examples.

The monotone likelihood ratio property gives immediately the following lemma.

LEMMA 2

For a model with monotone likelihood ratio form, a most powerful test of θ_0 against θ_1 has the form

$$(16) \qquad \begin{aligned} \phi(\mathbf{y}) &= 1 \qquad \text{for } t(\mathbf{y}) > t, \\ &= 0 \qquad\qquad\quad < t; \end{aligned}$$

that is, reject for large values of $t(\mathbf{y})$ with possible randomization at the critical value t.

Proof The fundamental lemma says that a most powerful test is obtained by rejecting for large values of the likelihood ratio (15) or by Definition 1, for large values of the function $t(\mathbf{y})$; Problem 10.1.18 says a most powerful test necessarily has this form.

A monotone likelihood ratio model has monotone power for a likelihood ratio test:

LEMMA 3

For a model with monotone likelihood ratio form, the probability

$$(17) \qquad P(t(\mathbf{y}) > t|\theta)$$

is a nondecreasing function of θ.

Proof From the monotone property in Definition 1, let k be the value such that

$$\begin{aligned} t(\mathbf{y}) &> t \quad \text{implies that} \quad \frac{f(\mathbf{y}|\theta_2)}{f(\mathbf{y}|\theta_1)} \geq k, \\ &\leq t \qquad\qquad\qquad\qquad\quad \leq k, \end{aligned}$$

where $\theta_1 < \theta_2$. Consider the following sum:

$$(18) \qquad P(t(y) \leq t|\theta_1) + P(t(y) > t|\theta_1) = 1,$$

$$\int_{t(y)\leq t} kf(y|\theta_1)\, dy + \int_{t(y)>t} kf(y|\theta_1)\, dy = k.$$

The integrand on the left is larger ($\geq$) than $f(y|\theta_2)$; and the integrand on the right is smaller ($\leq$) than $f(y|\theta_2)$. If we make these substitutions, we decrease the left integral and increase the right integral and obtain

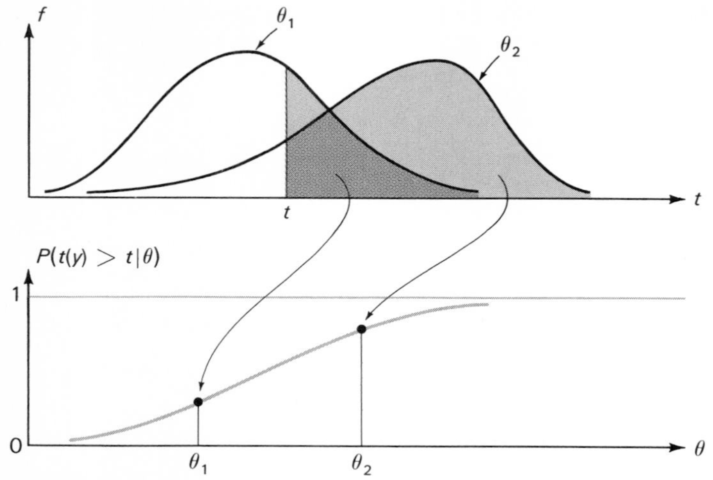

FIGURE 10.5
As the distribution shifts to the right under the monotone likelihood ratio property, the probability to the right of t increases.

$$(19) \qquad \int_{t(y)\le t} f(y|\theta_2)\,dy + \int_{t(y)>t} f(y|\theta_2)\,dy = 1.$$

The ratio of right integral to left integral in (18) is altered by the substitution; the ratio is increased ($\le$). A comparison of (18) and (19) then shows that

$$\int_{t(y)>t} f(y|\theta_1)\,dy \le \int_{t(y)>t} f(y|\theta_2)\,dy$$

and establishes the lemma. See Figure 10.5.

We now have the basic result for a monotone likelihood ratio model.

THEOREM 4

For a model with monotone likelihood ratio form the test

$$(20) \qquad \begin{aligned} \phi(y) &= 1 && \text{if } t(y) > k, \\ &= a && \quad = k, \\ &= 0 && \quad < k, \end{aligned}$$

where k and a satisfy

$$(21) \qquad E(\phi(y)|\theta_0) = \alpha,$$

is a uniformly most powerful size α test of $H_0: \theta \le \theta_0$ against $H_1: \theta > \theta_0$.

Proof By Lemma 2 the most powerful size α test of θ_0 against θ_1 is given by (20) with (21). Lemma 3 shows that the test has size α for the full hypothesis $\theta \le \theta_0$; it is thus the most powerful size α test for $\theta \le \theta_0$ against θ_1. The test does not depend on θ_1; it is thus UMP size α for $\theta \le \theta_0$ against $\theta > \theta_0$.

EXAMPLE 3

The location normal: Let $\mathbf{y} = (y_1, \ldots, y_n)'$ be normal(θ, σ_0) with θ in R. We show that the model has monotone likelihood ratio form:

$$\frac{f(\mathbf{y}|\theta_1)}{f(\mathbf{y}|\theta_0)} = \frac{\exp\left\{-\frac{1}{2\sigma_0^2}\sum (y_i - \theta_1)^2\right\}}{\exp\left\{-\frac{1}{2\sigma_0^2}\sum (y_i - \theta_0)^2\right\}}$$

$$= \exp\left\{\sum y_i \frac{\theta_1 - \theta_0}{\sigma_0^2}\right\} \exp\left\{-n\frac{\theta_1^2 - \theta_0^2}{2\sigma_0^2}\right\},$$

and this is a monotone increasing function of $\Sigma\, y_i$ or of $\bar{y}$. Then by Theorem 4 the UMP size α test of $\theta \le \theta_0$ against $\theta > \theta_0$ is to reject for large values of $\bar{y}$ with critical value determined by the θ_0 distribution. But this is just the test as determined in Example 1.

D ALTERNATIVES TO UMP

Our general approach has led us to look for UMP tests for hypothesis-testing problems. In Section C we have seen that UMP tests are closely tied to models having monotone likelihood ratio form. And these in turn are closely tied to models having exponential form with one ψ-function. Thus from Section 8.5D we see that the success of our approach is essentially limited to the location normal, the scale normal, the common binomial, and Poisson discrete models. What alternatives are available to us in other cases?

Consider the hypothesis-testing problem $H_0: \theta \le \theta_0$ against $H_1: \theta > \theta_0$ or perhaps the simpler problem $H_0: \theta = \theta_0$ against $H_1: \theta > \theta_0$. If a UMP size α test does not exist, then there are several directions in which we can proceed.

One possibility is to seek an important or distinctive value in the alternative and choose the most powerful test of θ_0 against θ_1:

$$(22) \qquad \phi(\mathbf{y}) = \begin{cases} 1 & \text{if } \dfrac{f(\mathbf{y}|\theta_1)}{f(\mathbf{y}|\theta_0)} > k, \\ a & \qquad\qquad = k, \\ 0 & \qquad\qquad < k, \end{cases}$$

where k and a are chosen to give exact size α under θ_0. With a strategically chosen value θ_1 the hope would be that the test had reasonable power for other θ values in H_1.

Another possibility is to determine the size α test satisfying

$$(23) \qquad \int \phi(\mathbf{y}) f(\mathbf{y}|\theta_0)\, d\mathbf{y} \le \alpha$$

and maximizing the power for the θ values just larger than θ_0, that is, maximizing the slope of the power function at θ_0. Such a test is called *locally most powerful*. The slope of the power function is available from (9.4.3) with appropriate regularity:

$$\frac{d}{d\theta}\int \phi(\mathbf{y}) f(\mathbf{y}|\theta)\, d\mathbf{y} = \int \phi(\mathbf{y}) S(\mathbf{y}|\theta) f(\mathbf{y}|\theta)\, d\mathbf{y}.$$

The possibility then is to determine the size α test satisfying (23) and maximizing

$$(24) \qquad \int \phi(\mathbf{y})S(\mathbf{y}\,|\,\theta_0)f(\mathbf{y}\,|\,\theta_0)\, d\mathbf{y}.$$

An obvious extension of the fundamental lemma is recorded in Problem 10.1.20; this extension says that the size α test of $H_0:\theta = \theta_0$ that maximizes the power slope is given by

$$\begin{aligned} \phi(\mathbf{y}) &= 1 \qquad \text{if } S(\mathbf{y}\,|\,\theta_0) > k, \\ &= a \qquad\qquad\qquad\quad = k, \\ &= 0 \qquad\qquad\qquad\quad < k, \end{aligned}$$

where k and a are chosen to satisfy $E(\phi(\mathbf{y})\,|\,\theta_0) = \alpha$. With such a test the hope would be that good power for θ close to θ_0 would mean reasonable power for other presumably more easily detected alternative values.

EXAMPLE 4

The location normal: Consider a sample $\mathbf{y} = (y_1, \ldots, y_n)'$ from a normal(θ, σ_0) with θ in R. And suppose that we want the size α test of $\theta = \theta_0$ which has maximum power slope at θ_0. The locally most powerful test is to reject for large values of

$$S(\mathbf{y}\,|\,\theta_0) = \frac{\bar{y} - \theta_0}{\sigma_0^2/n}$$

or equivalently for large values of $\bar{y}$. Thus we obtain, as we should expect, the test described in Examples 1 and 2.

E TESTS BASED ON THE LIKELIHOOD STATISTIC

The most powerful test of a simple hypothesis against a simple alternative is based on the likelihood ratio and accordingly is based on the likelihood statistic. Indeed, from Chapter 8 we know quite generally that a statistical procedure can reasonably be based completely on the likelihood statistic. We now see how this holds for general hypothesis-testing problems; we show that any test function ϕ has a corresponding test function ψ based on the likelihood statistic that has the same power function.

Consider the power function of a test ϕ and let s be the likelihood statistic; then by (5.5.3) we have

$$\mathcal{P}_\phi(\theta) = E(\phi(y)\,|\,\theta) = E\big(E(\phi(y):s)\,|\,\theta\big) = \mathcal{P}_\psi(\theta),$$

where we let

$$\psi(s) = E(\phi(y):s).$$

Thus the tests ϕ and ψ have the same power function for all θ in Ω. And, accordingly, we see that averaging over the conditional distribution given the likelihood statistic produces a test based on the likelihood statistic and with the same power function.

EXAMPLE 5

The location-scale normal: Consider a sample $\mathbf{y} = (y_1, \ldots, y_n)'$ from the normal(μ, σ) with $\theta = (\mu, \sigma)$ in $R \times R^+$. The preceding says that any test ϕ based on $(y_1, \ldots, y_n)$ has an equivalent test ψ based on the likelihood statistic $(\bar{y}, s_y)$, equivalent in the sense of having the same power function. Thus we can note that it suffices to consider only the test based on the likelihood statistic $(\bar{y}, s_y)$.

F EXERCISES

1 *Location normal:* Let $(y_1, \ldots, y_n)$ be a sample from the normal(θ, σ_0) with θ in R. Determine the UMP size α test of $H_0{:}\theta \geq \theta_0$ against $H_1{:}\theta < \theta_0$.

2 *The scale exponential:* Let $(y_1, \ldots, y_n)$ be a sample from the exponential(θ) distribution; Problem 10.1.7. Find the UMP size α test of $H_0{:}\theta \leq \theta_0$ against $H_1{:}\theta > \theta_0$.

3 *The scale normal:* Let $(y_1, \ldots, y_n)$ be a sample from the normal(μ_0, σ) with σ in R^+; Problem 10.1.8. Find the UMP size α test of $H_0{:}\sigma \leq \sigma_0$ against $H_1{:}\sigma > \sigma_0$.

4 (*continuation*) Find the UMP size α test of $H_0{:}\sigma \geq \sigma_0$ against $H_1{:}\sigma < \sigma_0$.

5 *The Poisson(θ) distribution:* Let $(y_1, \ldots, y_n)$ be a sample from the Poisson(θ) with θ in R^+; Problems 10.1.10 and 10.1.11. Find the UMP size α test of
(a) $H_0{:}\theta \leq \theta_0$ against $H_1{:}\theta > \theta_0$.
(b) $H_0{:}\theta \geq \theta_0$ against $H_1{:}\theta < \theta_0$.

6 *Binomial(n, p)*: Let $(x_1, \ldots, x_n)'$ be a sample from the Bernoulli(p) with p in [0, 1]; Problem 10.1.12 and 10.1.13. Find the form of the UMP size α test of
(a) $H_0{:}p \leq p_0$ against $H_1{:}p > p_0$.
(b) $H_0{:}p \geq p_0$ against $H_1{:}p < p_0$.

G PROBLEMS

7 *Uniform*(0, θ): Let $(y_1, \ldots, y_n)$ be a sample from the uniform(0, θ); Problems 10.1.14 and 10.1.15. Find the UMP size α test for $H_0{:}\theta = \theta_0$ against $H_1{:}\theta \neq \theta_0$; there is such an α test in contrast to Exercises 1–6.

8 *The general normal; scale:* Let $(y_1, \ldots, y_n)$ be a sample from a normal(μ, σ) with (μ, σ) in $R \times R^+$. Consider the composite hypothesis $H_0{:}\sigma \geq \sigma_0$, $\mu \in R$ against the simple alternative $H_1{:}\sigma = \sigma_1$ ($<\sigma_0$), $\mu = \mu_1$. As hypothesis distribution closest to the alternative, consider $\sigma = \sigma_0$, $\mu = \mu_1$.
(a) Show that the most powerful size α test of (μ_1, σ_0) against (μ_1, σ_1) is to reject if $\Sigma\,(y_i - \mu_1)^2 \leq \sigma_0^2\chi_{1-\alpha}^2$, where $\chi_{1-\alpha}^2$ is the point exceeded with probability $1 - \alpha$ for the chi-square(n) distribution. Some straightforward analysis shows that the preceding test has correct size α for the full hypothesis $H_0{:}\sigma \geq \sigma_0$, $\mu \in R$.
(b) Deduce that a UMP size α test does not exist for $H_0{:}\sigma \geq \sigma_0$ against $H_1{:}\sigma < \sigma_0$.

9 (*continuation*) Consider the composite hypothesis $H{:}\sigma \leq \sigma_0$, $\mu \in R$ against the simple alternative $H_1{:}\sigma = \sigma_1$ ($>\sigma_0$), $\mu = \mu_1$. By Section E it suffices to examine tests based on $(\bar{y}, \Sigma\,(y_i - \bar{y})^2)$. As a hypothesis mixture closest to the alternative, let $\sigma^2 = \sigma_0^2$ and let μ be normal$(\mu_1, ((\sigma_1^2 - \sigma_0^2)/n)^{1/2})$; this describes a probability measure λ.
(a) For $n \geq 2$, show that the most powerful test of $f(\mathbf{y}\,|\,\lambda)$ against $f(\mathbf{y}\,|\,\mu_1, \sigma_1^2)$ is to reject if $\Sigma\,(y_i - \bar{y})^2 \geq \sigma_0^2\chi_\alpha^2$, where χ_α^2 is the α point of the chi-square(n) distribution.
(b) Argue that the test is UMP size α for $H_0{:}\sigma \leq \sigma_0$ against $H_1{:}\sigma > \sigma_0$.

H FURTHER PROBLEMS

10 *The general normal; location:* Let $(y_1, \ldots, y_n)$ be a sample from the normal(μ, σ) with (μ, σ) in $R \times R^+$, and consider the problem $H_0: \mu = 0, \sigma \in R^+$ against $H_1: \mu > 0, \sigma \in R^+$. Some detailed analysis shows that for $\alpha < 1/2$, there does not exist a UMP test. Nevertheless, the test

$$\phi(y) = 1(0) \qquad \text{according as} \quad \sqrt{n}\,\bar{y}/s_y \geq (<)\, t_\alpha$$

is a very appropriate test; t_α is the value exceeded with probability α by the Student$(n - 1)$ distribution. Use Problem 3.6.2 to obtain an expression for the power function of this test.

11 Let $(y_{11}, \ldots, y_{1m})$ be a sample from the normal(μ_1, σ_0) and $(y_{21}, \ldots, y_{2n})$ be a sample from the normal(μ_2, σ_0) with (μ_1, μ_2) in R^2.

(a) Find the most powerful size α test of $H_0: \mu_1 = \mu_2$ against $H_1: \mu_1 = \mu_a, \mu_2 = \mu_b$ with $\mu_a > \mu_b$. *Method:* As parameter value closest to H_1, try $\mu_0 = (m\mu_a + n\mu_b)/(m + n)$; the test is to reject for large values of $\bar{y}_1 - \bar{y}_2$.

(b) Find the UMP size α test of $H_0: \mu_1 = \mu_2$ against $H_1: \mu_1 > \mu_2$.

12 Let $(y_{11}, \ldots, y_{1m})$ be a sample from the normal$(0, \sigma_1)$ and $(y_{21}, \ldots, y_{2n})$ be a sample from the normal$(0, \sigma_2)$ with (σ_1, σ_2) in $(0, \infty)^2$. Find the form of the most powerful test of $H_0: \sigma_1 = \sigma_2 = \sigma_0$ against $H_1: \sigma_1 = \sigma_a, \sigma_2 = \sigma_b$, where $\sigma_a > \sigma_b$. Show that σ_0 can be chosen so that the test is to reject for large values of

$$F = \frac{\sum y_{1j}^2/m}{\sum y_{2j}^2/n},$$

that is, for values of F that exceed the α point of the F distribution(m, n). Conclude that the test: reject if $F > F_\alpha$ is a UMP size α test for $H_0: \sigma_1 = \sigma_2$ against $H_1: \sigma_1 > \sigma_2$.

13 (*continuation*) Show that the power function of the preceding test is given by

$$1 - H\left(\frac{\sigma_2^2}{\sigma_1^2} F_\alpha\right),$$

where H designates the distribution function for the F distribution(m, n).

14 Let $(y_1, \ldots, y_n)$ be a sample from a distribution having a density function on R. Let ζ designate a median of the density f.

(a) Find the most powerful test of $H_0: \zeta \leq \zeta_0$ against the alternative f_1. *Method:* Write $f_1 = p_1 f^- + q_1 f^+$, where f^- and f^+ are the conditional densities given $y < \zeta_0$ and given $y > \zeta_0$; consider the hypothesis density $f_0 = (1/2)f^- + (1/2)f^+$ as a density "closest" to f_1.

(b) Show that the sign test is UMP size α for the problem $H_0: \zeta \leq \zeta_0$ against $H_1: \zeta > \zeta_0$; recall Section 7.4B.

10.3

SOME OTHER METHODS IN HYPOTHESIS TESTING

For testing a statistical hypothesis we would naturally be interested in having a uniformly most powerful (UMP) test. But we have seen that such tests are rather rare. Indeed, for models where the region of positive density does not vary with θ, the occurrence is essentially limited to models having exponential form with one ψ-function. In practice, this almost limits such tests to the normal, binomial, and Poisson.

For testing a statistical hypothesis with the normal, binomial, and Poisson, we have obtained UMP tests for *one-sided* problems. In this section, we see that such tests do not exist for *two-sided* problems and we examine reasonable alternatives. We also examine several convenient methods for obtaining suitable tests for more general hypothesis-testing problems.

A TWO-SIDED TESTS FOR THE NORMAL

EXAMPLE 1

The normal with known variance: Let $(y_1, \ldots, y_n)$ be a sample from the normal(θ, σ_0) with θ in R, and consider the hypothesis-testing problem $H_0: \theta = \theta_0$ against $H_1: \theta \neq \theta_0$. From Example 10.1.1 we know that the size α test having maximum power for θ values larger than θ_0 has critical region

$$C_1 = \{\mathbf{y} : \bar{y} \geq \theta_0 + z_\alpha \sigma_0 / \sqrt{n}\}.$$

Correspondingly, from symmetry we know that the size α test having maximum power for θ values less than θ_0 has critical region

$$C_2 = \{\mathbf{y} : \bar{y} \leq \theta_0 - z_\alpha \sigma_0 / \sqrt{n}\}.$$

The power functions of these two tests are plotted in Figure 10.6. And from the essential uniqueness of C_1 and C_2 (Problem 10.1.18) we can conclude that there is no UMP size α test for the original problem.

We note that the test C_1 rejects the hypothesis with probability $<\alpha$ for θ values to the left of θ_0. And correspondingly, C_2 has the same inadequacy for θ values larger than θ_0. One reasonable requirement that we can make is that the test reject with probability at least α for θ values in the alternative; such a test is called *unbiased*. This would seem to restrict us to tests with a horizontal tangent at θ_0. The results recorded in Section D below then establish the very reasonable test

$$(1) \qquad C = \{\mathbf{y} : |\bar{y} - \theta_0| \geq z_{\alpha/2} \sigma_0 / \sqrt{n}\}$$

as the UMP unbiased size α test; see Figure 10.6. The corresponding test of significance was examined in Section 8.3A.

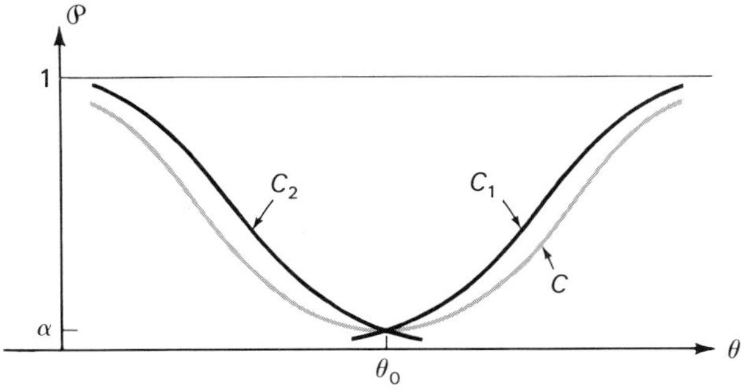

FIGURE 10.6
C_1 has maximum power for θ values $>\theta_0$; C_2 has maximum power for θ values $<\theta_2$; C has maximum power for $\theta \neq \theta_0$ among tests with a horizontal tangent to the power function at θ_0.

EXAMPLE 2

The normal with unknown variance: Let $(y_1, \ldots, y_n)$ be a sample from the normal(μ, σ) with (μ, σ) in $R \times R^+$, and consider the hypothesis-testing problem $H_0: \mu = \mu_0$, $\sigma \in R^+$ against $H_1: \mu \neq \mu_0$, $\sigma \in R^+$. We have noted in Problem 10.2.10 that there does not exist in general a UMP size α test for either of the one-sided problems. Thus there certainly does not exist one for the two-sided problem just presented. Again a reasonable requirement is that the test reject with probability at least α for θ values in the alternative. The continuity of the power function then implies that such a test reject the hypothesis with probability exactly α for θ values in the hypothesis $H_0: \mu = \mu_0$, $\sigma \in R^+$; such a test is called a *similar test*. The results quoted in Section E below then establish the very reasonable test

$$(2) \qquad C = \{\mathbf{y} : |\bar{y} - \mu_0| \geq t_{\alpha/2} s_y / \sqrt{n}\}$$

as the UMP similar size α test and as the UMP unbiased size α test. The corresponding test of significance was examined in Section 8.3C.

B THE GENERALIZED LIKELIHOOD RATIO

The most powerful test of a simple hypothesis against a simple alternative is to reject for large values of the likelihood ratio

$$\frac{L(\mathbf{y} \mid \theta_1)}{L(\mathbf{y} \mid \theta_0)} = \frac{f(\mathbf{y} \mid \theta_1)}{f(\mathbf{y} \mid \theta_0)}.$$

A generalized version of this ratio has been proposed for general hypothesis-testing problems.

Consider a statistical model with density function $f(\cdot \mid \theta)$ with θ in Ω and suppose that we are concerned with testing the hypothesis H_0 against the alternative $H_1 = \Omega - H_0$. The generalized likelihood ratio is defined by

$$(3) \qquad L(\mathbf{y}) = \frac{\sup_{\theta \text{ in } \Omega} f(\mathbf{y} \mid \theta)}{\sup_{\theta \text{ in } H_0} f(\mathbf{y} \mid \theta)} = \frac{L(\mathbf{y} \mid \hat{\hat{\theta}})}{L(\mathbf{y} \mid \hat{\theta})},$$

where $\hat{\hat{\theta}}(\mathbf{y})$ is the maximum-likelihood estimator in the full model with parameter space Ω and $\hat{\theta}(\mathbf{y})$ is the maximum-likelihood estimator in the special model with parameter space H_0. The generalized likelihood ratio test is to reject the hypothesis for large values of $L(\mathbf{y})$, where the critical value is determined so that the test has exact size α for the hypothesis H_0. Formula (3) used here parallels the ordinary likelihood ratio; other authors often use the reciprocal of (3).

This generalized likelihood ratio method has been remarkably successful in producing reasonable test procedures. It does, for example, give the compromise tests proposed in Examples 1 and 2. We now verify this for the more general Example 2.

EXAMPLE 3

The normal with unknown variance: Let $(y_1, \ldots, y_n)$ be a sample from the normal(μ, σ) with (μ, σ) in $R \times R^+$, and consider the hypothesis-testing problem $H_0: \mu = \mu_0$, $\sigma \in R^+$ against $H_1: \mu \neq \mu_0$, $\sigma \in R^+$. The density function is

$$f(\mathbf{y}|\mu, \sigma) = (2\pi\sigma^2)^{-n/2} \exp\left\{-\frac{1}{2\sigma^2}\sum (y_i - \mu)^2\right\}.$$

For the full model there are two parameters μ and σ and the maximum-likelihood estimators are

$$\widehat{\widehat{\mu}} = \bar{y}, \qquad \widehat{\widehat{\sigma}}^2 = n^{-1}\sum (y_i - \bar{y})^2;$$

thus

$$f(\mathbf{y}|\widehat{\widehat{\mu}}, \widehat{\widehat{\sigma}}) = \left(2\pi\frac{\sum (y_i - \bar{y})^2}{n}\right)^{-n/2} e^{-n/2}.$$

And for the special model of the hypothesis there is effectively one parameter σ (as $\mu = \mu_0$) and the maximum-likelihood estimators are

$$\widehat{\mu} = \mu_0, \qquad \widehat{\sigma}^2 = n^{-1}\sum (y_i - \mu_0)^2;$$

thus

$$f(\mathbf{y}|\widehat{\mu}, \widehat{\sigma}) = \left(2\pi\frac{\sum (y_i - \mu_0)^2}{n}\right)^{-n/2} e^{-n/2}.$$

The generalized likelihood ratio is then given by

$$L(\mathbf{y}) = \left(\frac{\sum (y_i - \mu_0)^2}{\sum (y_i - \bar{y})^2}\right)^{n/2}$$

$$= (1 + t^2/(n-1))^{n/2};$$

and the test is to reject for large values of $L(\mathbf{y})$ or for large values of $|t|$, where

$$t = \frac{\bar{y} - \mu_0}{s_y/\sqrt{n}}$$

is the usual t function. This gives the test discussed in Example 2.

One difficulty with the use of the generalized likelihood ratio method is that of determining the critical value. Fortunately, the large-sample distribution is often available using results from Sections 8.6D and 8.6G. For this we assume the regularity conditions used in Section 8.6. And, in addition, we assume that Ω is an open set in R^r and that H_0 is an $(r - s)$-dimensional region or surface in Ω; thus s is the number of effective parameters being tested or constrained by the hypothesis. It then follows from Section 8.6 and from properties of the multivariate normal that

(4) $\quad 2 \ln L(\mathbf{y})$

has a limiting chi-square(s) distribution. Note that the alternative definition for $L(\mathbf{y})$ would introduce a negative sign in (4).

For the generalized likelihood ratio in Example 3 note that $2 \ln L(\mathbf{y})$ is approximately t^2 for large n and that t^2 has a limiting chi-square(1) distribution.

C LARGE-SAMPLE TESTS

The test function based on the likelihood ratio or generalized likelihood ratio is usually easy to calculate but the critical value may be difficult to derive. Fortunately, as indicated in the preceding section, the large-sample distributions are usually available.

Let $(y_1, \ldots, y_n)$ be a sample from a distribution $f(\cdot|\theta)$ with θ in R.

The likelihood ratio for testing θ_0 versus θ_1 has the following logarithmic form for a single response y:

$$u = \ln f(y|\theta_1) - \ln f(y|\theta_0) = \ln \frac{f(y|\theta_1)}{f(y|\theta_0)}.$$

Suppose that μ_0 and σ_0^2 are the mean and variance of u under the hypothesis $\theta = \theta_0$. Then for a sample of n, the hypothesis (θ_0) distribution of the log-likelihood ratio

$$\sum_1^n u_i = \ln \frac{f(\mathbf{y}|\theta_1)}{f(\mathbf{y}|\theta_0)}$$

has limiting normal$(n\mu_0, \sqrt{n}\,\sigma_0)$ form; and the size α test has the approximate form

$$(5) \qquad \begin{aligned} \phi(\mathbf{y}) &= 1 \qquad \text{if } \frac{f(\mathbf{y}|\theta_1)}{f(\mathbf{y}|\theta_0)} \geq n\mu_0 + z_\alpha \sqrt{n}\,\sigma_0, \\ &= 0 \qquad\qquad\qquad < n\mu_0 + z_\alpha \sqrt{n}\,\sigma_0. \end{aligned}$$

The locally most powerful test of θ_0 against larger values of θ is based on the score function. The score function at θ_0 for a single response y is given by

$$S(y|\theta_0) = \frac{\partial}{\partial\theta} \ln f(y|\theta)\Big|_{\theta_0}.$$

Under the regularity assumptions in Section 8.6 we have

$$E(S(y|\theta_0)|\theta_0) = 0 \qquad \text{var}\,(S(y|\theta_0)|\theta_0) = j(\theta_0).$$

Then for a sample of n the hypothesis (θ_0) distribution of the score function

$$S(\mathbf{y}|\theta_0) = \sum_1^n S(y_i|\theta_0)$$

has limiting normal$(0, \sqrt{nj(\theta_0)})$ form and the approximate size α test has the form

$$(6) \qquad \begin{aligned} \phi(\mathbf{y}) &= 1 \qquad \text{if } S(\mathbf{y}|\theta_0) \geq z_\alpha \sqrt{nj(\theta_0)}, \\ &= 0 \qquad\qquad\qquad < z_\alpha \sqrt{nj(\theta_0)}. \end{aligned}$$

For more general testing problems we might turn to the generalized likelihood ratio. The large-sample hypothesis distribution of the generalized likelihood ratio was recorded in Section B and the approximate size α test thus has the form

$$\begin{aligned} \phi(\mathbf{y}) &= 1 \qquad \text{if } 2 \ln L(\mathbf{y}) \geq \chi_\alpha^2(1), \\ &= 0 \qquad\qquad\qquad < \chi_\alpha^2(1) \end{aligned}$$

where $\chi_\alpha^2(1)$ is the α-point on the right tail of the chi-square(1) distribution. See Example 3 and the concluding remarks in Section B.

D SUPPLEMENT: UNBIASED TESTS

In Example 1 we were looking for a suitable test for the two-sided problem. We noted that a reasonable requirement for a test was that it reject with probability at least α for θ values in the alternative. Accordingly, we define unbiased tests:

DEFINITION 1

A test with critical region C or test function ϕ is **unbiased of size α** for H_0 against H_1 if

(7) $$P(C\,|\,\theta) = E(\phi(\mathbf{y})\,|\,\theta) \le \alpha \qquad \text{for } \theta \text{ in } H_0,$$
$$\ge \alpha \qquad \theta \text{ in } H_1.$$

Now consider a real-valued parameter θ and the hypothesis-testing problem $H_0 : \theta = \theta_0$ against $H_1 : \theta \neq \theta_0$. In many problems the power function is continuously differentiable with respect to θ. It then follows easily that an unbiased test has a power function with a horizontal tangent at θ_0, and is thus a locally unbiased size α test:

DEFINITION 2

A test with critical region C or test function ϕ is a **locally unbiased size α** test of $H_0 : \theta = \theta_0$ against $H_1 : \theta \neq \theta_0$ if

(8) $$P(C\,|\,\theta_0) = E(\phi(\mathbf{y})\,|\,\theta_0) = \alpha,$$

(9) $$\frac{d}{d\theta} P(C\,|\,\theta)\bigg|_{\theta_0} = \frac{d}{d\theta} E(\phi(\mathbf{y})\,|\,\theta)\bigg|_{\theta_0} = 0.$$

With some regularity the tangent condition (9) can be given the alternative form

(10) $$E(\phi(\mathbf{y})S(\mathbf{y}\,|\,\theta_0)\,|\,\theta_0) = 0$$

on the basis of equation (9.4.3). Local unbiasedness provides a mathematically flexible way of finding best unbiased tests.

The generalized fundamental lemma in Problem 10.1.20 can be used to obtain UMP unbiased size α tests for the model

(11) $$f(\mathbf{y}\,|\,\theta) = \gamma(\theta) \exp\{t(\mathbf{y})\psi(\theta)\}h(\mathbf{y})$$

where $\psi(\theta)$ is continuously differentiable. We quote without proof the following theorem; for a proof, see Lehmann (1959).

THEOREM 3

For a model having exponential form (11), the UMP-unbiased size α test for $H_0 : \theta = \theta_0$ against $H_0 : \theta \neq \theta_0$ has the form

(12) $$\phi(\mathbf{y}) = 1 \qquad \text{if } t(\mathbf{y}) < t_1 \text{ or } > t_2,$$
$$= 0 \qquad t_1 < t(\mathbf{y}) < t_2$$

with possible randomization at t_1 and t_2 in the discrete case; the constants of the test are determined so that

$$(13)\qquad E(\phi(\mathbf{y})\,|\,\theta_0) = \alpha, \qquad \frac{d}{d\theta} E(\phi(\mathbf{y})\,|\,\theta)\bigg|_{\theta_0} = 0.$$

EXAMPLE 4

The normal with known variance: Let $(y_1, \ldots, y_n)$ be a sample from the normal(θ, σ_0) with θ in R, and consider testing $\theta = \theta_0$ against $\theta \neq \theta_0$. The density function can be written as

$$f(\mathbf{y}\,|\,\theta) = (2\pi\sigma_0^2)^{-n/2} \exp\left\{-\frac{n\theta^2}{2\sigma_0^2}\right\} \exp\left\{\sum y_i \frac{\theta}{\sigma_0^2}\right\} \exp\left\{-\frac{1}{2\sigma_0^2}\sum y_i^2\right\}$$

and thus has the required exponential form. We seek a test based on $\Sigma\, y_i$ or on $\bar{y}$ having the form (12) and satisfying (13). Clearly the test with critical region

$$C = \{\mathbf{y} : |\bar{y} - \theta_0| \geq z_{\alpha/2}\sigma_0/\sqrt{n}\}$$

given in Example 1 has the required size α. Symmetry easily gives the unbiasedness; see Figure 10.6.

E SUPPLEMENT: UNBIASED TESTS AND SIMILARITY

Consider testing the hypothesis H_0 against the alternative H_1, and suppose that the power function for any test is a continuous function of θ. An unbiased size α test satisfies conditions (8) and (9): the power is $\leq\alpha$ on H_0 and $\geq\alpha$ on H_1. From continuity it then follows that it is equal to α for points on the boundary between H_0 and H_1. Let ω be the boundary set of points on Ω; see Figure 10.7. Then we can note that an unbiased size α test for H_0 against H_1 is a similar size α test for ω:

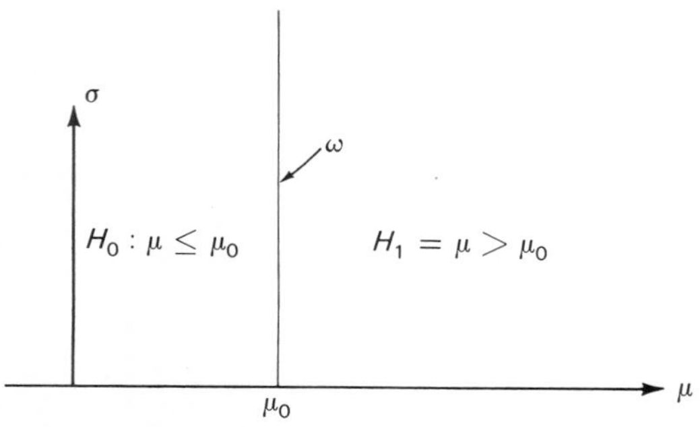

FIGURE 10.7
Hypothesis $H_0 : \mu \leq \mu_0$; alternative $H_1 : \mu > \mu_0$. The boundary set $\omega = \{(\mu_0, \sigma) : \sigma \in R^+\}$.

DEFINITION 4

A test with critical region C or test function ϕ is a **similar size α test** for ω if

(14) $$P(C \mid \theta) = E(\phi(\mathbf{y}) \mid \theta) = \alpha \qquad \text{for } \theta \text{ in } \omega.$$

The concepts of sufficiency and completeness can then be used to obtain UMP unbiased size α tests for a model

(15) $$f(\mathbf{y} \mid \theta, \boldsymbol{\psi}) = \gamma(\theta, \boldsymbol{\psi}) \exp\left\{t(\mathbf{y})\theta + \sum_1^r s_i(\mathbf{y})\psi_i\right\} h(\mathbf{y})$$

with $(\theta, \boldsymbol{\psi}')$ in an open set in R^{r+1}. We quote the following theorems:

THEOREM 5

For a model having exponential form (15), *the UMP unbiased size α test for $H_0 : \theta \le \theta_0$ against $H_1 : \theta > \theta_0$ has the form*

(16) $$\begin{aligned} \phi(\mathbf{y}) &= 1 \qquad \text{if } t(\mathbf{y}) > k(\mathbf{s}), \\ &= 0 \qquad\qquad\quad < k(\mathbf{s}) \end{aligned}$$

with possible randomization at $t(\mathbf{y}) = k(\mathbf{s})$ in the discrete case; the critical value $k(\mathbf{s})$ is chosen so that the test has conditional size α,

$$E(\phi(\mathbf{y}) : \mathbf{s} \mid \theta_0) = \alpha,$$

for each value of $\mathbf{s}$.

THEOREM 6

For a model having exponential form (15), *the UMP unbiased size α test for $H_0 : \theta = \theta_0$ against $H_1 : \theta \ne \theta_0$ has the form*

(17) $$\begin{aligned} \phi(\mathbf{y}) &= 1 \qquad \text{if } t(\mathbf{y}) < k_1(\mathbf{s}) \text{ or } > k_2(\mathbf{s}), \\ &= 0 \qquad\quad k_1(\mathbf{s}) < t(\mathbf{y}) < k_2(\mathbf{s}) \end{aligned}$$

with possible randomization at $t(\mathbf{y}) = k_1(\mathbf{s})$ and $t(\mathbf{y}) = k_2(\mathbf{s})$ in the discrete case; the critical values $k_1(\mathbf{s})$ and $k_2(\mathbf{s})$ are chosen so that the test is locally unbiased size α,

(18) $$E(\phi(\mathbf{y}) : \mathbf{s} \mid \theta_0) = \alpha, \qquad \frac{d}{d\theta} E(\phi(\mathbf{y}) : \mathbf{s} \mid \theta)\Big|_{\theta_0} = 0$$

conditionally given $\mathbf{s}$, *for each value of* $\mathbf{s}$.

EXAMPLE 5

The normal with unknown variance: Let $(y_1, \ldots, y_n)$ be a sample from the normal(μ, σ) with (μ, σ) in $R \times R^+$, and consider testing $\mu \le \mu_0$ against $\mu > \mu_0$ or testing $\mu = \mu_0$ against $\mu \ne \mu_0$. For simplicity we relocate and in effect examine the case $\mu_0 = 0$. The density can be written as

$$f(\mathbf{y}|\mu, \sigma) = (2\pi\sigma^2)^{-n/2} \exp\left\{-\frac{n\mu^2}{2\sigma^2}\right\} \exp\left\{\sum y_i \frac{\mu}{\sigma^2} + \sum y_i^2 \frac{-1}{2\sigma^2}\right\},$$

and this has the exponential form with $t = \Sigma y_i$ and $s = \Sigma y_i^2$. Under the hypothesis the conditional distribution given s is uniform on the sphere $s = \Sigma y_i^2$ with s given. The density generally given s has the exponential form(11). The ordinary t-function $\sqrt{n}\,\bar{y}/s_y$ can be shown to be a one-to-one function of Σy_i given the value of s. The conditional distribution of $\sqrt{n}\,\bar{y}/s_y$ under the hypothesis can fairly easily be seen to be the same as the marginal, that is, the Student$(n-1)$ distribution. It follows, then, fairly easily from Theorems 5 and 6 that $C_1 = \{\mathbf{y}: \bar{y} \geq t_\alpha s_y/\sqrt{n}\}$ gives the UMP unbiased size α test for the first problem and $C = \{\mathbf{y}: |\bar{y}| \geq t_{\alpha/2} s_y/\sqrt{n}\}$ gives the UMP unbiased size α test for the second problem.

These methods can be applied to the *normal* linear model (Section 9.2), giving the tests as examined in Section 11.4.

F SUPPLEMENT: SEQUENTIAL ANALYSIS

In Exercises 10.1.5 and 10.1.6 we considered a normal response with known variance and we were concerned with testing $H_0: \theta = 75$ against $H_1: \theta > 75$ at the 1 percent level. The problem was to determine the sample size so that the power was 0.95 at the alternative $\theta = 76$.

In this section we consider a sequential test of θ_0 against θ_1 such that observations are taken one by one until the θ_1 to θ_0 likelihood ratio becomes extreme—either above an upper bound, suggesting "reject," or below a lower bound, suggesting "accept." The mathematics is very attractive and there are approximations that make the method surprisingly easy to use. For applications the method seems particularly appropriate to industrial acceptance sampling of incoming manufactured items and to the developmental screening of new materials, drugs, and procedures.

Consider a response y and density function $f(\cdot\,|\theta)$ with θ in R. The sequential test is concerned nominally with a simple hypothesis $H_0: \theta = \theta_0$ against a simple alternative $H_1: \theta = \theta_1$, but effectively with the more reasonable $H_0: \theta \leq \theta_*$ against $H_1: \theta > \theta_*$, where θ_* is some intermediate value. After n observations the test is based on the likelihood ratio

$$L_n(y_1, \ldots, y_n) = \frac{f(y_1|\theta_1)}{f(y_1|\theta_0)} \cdots \frac{f(y_n|\theta_1)}{f(y_n|\theta_0)} \tag{19}$$

and is to

$$\begin{array}{lll} \text{reject } H_0 & \text{if } L_n(\mathbf{y}) \geq B, & \\ \text{continue sampling} & A < L_n(\mathbf{y}) < B, & (20) \\ \text{accept } H_0 & L_n(\mathbf{y}) \leq A & \end{array}$$

where "continue" means to take another sample value and repeat the procedure; see Figure 10.8. The values A and B are chosen to give the test the desired properties in terms of size and power.

Alternatively, the test can be expressed in logarithmic form with the advantages

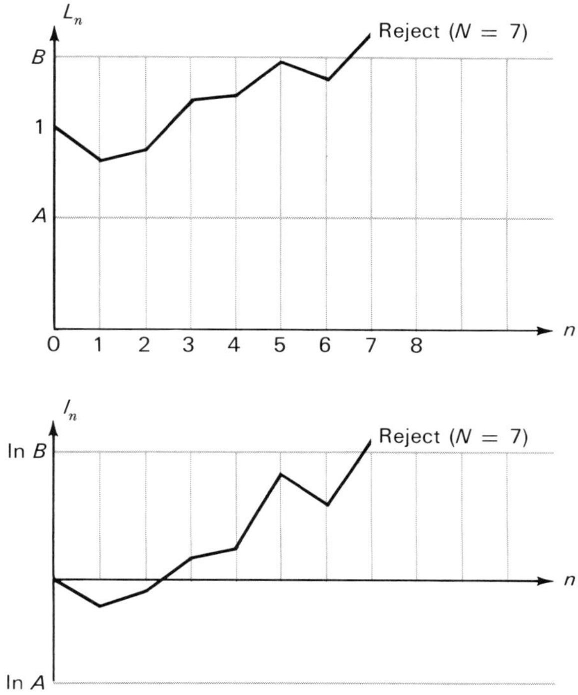

FIGURE 10.8
Sequential test based on the likelihood ratio $L_n(\mathbf{y})$ or the log-likelihood ratio $l_n(\mathbf{y})$.

that changes in log-likelihood are additive; then after n observations the test is based on the log-likelihood

$$(21) \qquad l_n(y_1, \ldots, y_n) = \ln \frac{f(y_1|\theta_1)}{f(y_1|\theta_0)} + \cdots + \ln \frac{f(y_n|\theta_1)}{f(y_n|\theta_0)}$$
$$= z_1 + \cdots + z_n,$$

where

$$z = \ln f(y|\theta_1) - \ln f(y|\theta_0)$$

and is to

$$(22) \qquad \begin{array}{lll} \text{reject } H_0 & \text{if } l_n(\mathbf{y}) \geq \ln B, \\ \text{continue} & \ln A < l_n(\mathbf{y}) < \ln B, \\ \text{accept } H_0 & l_n(\mathbf{y}) \leq \ln A. \end{array}$$

Let N be the sample size at which a terminal decision "reject H_0" or "accept H_0" is made. N can vary in value and is a function on the countable product sample space for the sequence $(y_1, y_2, \ldots)$.

The sequential test has the following optimum property established by Wald and Wolfowitz.

THEOREM 7

Among fixed and sequential decision procedures having error probabilities

$$P(\text{reject } H_0 | \theta_0) \le \alpha, \qquad P(\text{accept } H_0 | \theta_1) \le \beta$$

and finite $E(N|\theta_0)$ and $E(N|\theta_1)$, the sequential test (20) with error probabilities exactly α and β minimizes both $E(N|\theta_0)$ and $E(N|\theta_1)$.

Consider an example.

EXAMPLE 6

Let y be a response that is normal(θ, σ_0) and consider the sequential test of $H_0 : \theta = \theta_0$ against $H_1 : \theta = \theta_1$.

$$z = \ln \frac{f(y|\theta_1)}{f(y|\theta_0)} = \frac{1}{2\sigma_0^2}((y - \theta_0)^2 - (y - \theta_1)^2) = \frac{\theta_1 - \theta_0}{\sigma_0^2}\left(y - \frac{\theta_0 + \theta_1}{2}\right).$$

Thus the test after n observations is to

reject if $\displaystyle\sum_1^n \left(y_i - \frac{\theta_0 + \theta_1}{2}\right) \ge \frac{\sigma_0^2 \ln B}{\theta_1 - \theta_0}$,

continue $\displaystyle\frac{\sigma_0^2 \ln A}{\theta_1 - \theta_0} < \sum_1^n \left(y_i - \frac{\theta_0 + \theta_1}{2}\right) < \frac{\sigma_0^2 \ln B}{\theta_1 - \theta_0}$,

accept $\displaystyle\sum_1^n \left(y_i - \frac{\theta_0 + \theta_1}{2}\right) \le \frac{\sigma_0^2 \ln A}{\theta_1 - \theta_0}$.

Approximate values for A and B are available from the following lemma.

The critical steps in the proof of the fundamental lemma 10.1.4 can be used to prove the following lemma:

LEMMA 8

The Wald sequential test with size α and power $1 - \beta$ has

$$A \approx \frac{\beta}{1 - \alpha}, \qquad B \approx \frac{1 - \beta}{\alpha};$$

the exact values A and B give a tighter interval,

$$\frac{\beta}{1 - \alpha} \le A, \qquad B \le \frac{1 - \beta}{\alpha}.$$

G PROBLEMS

1 Let $y = (y_1, \ldots, y_n)'$ be a sample from a normal(θ, σ_0) with θ in R, and consider the hypothesis-testing problem $H_0 : \theta = \theta_0$ against $H_1 : \theta \ne \theta_0$. Determine the generalized likeli-

hood ratio $L(\mathbf{y})$ in Section B. How does the exact distribution of $L(\mathbf{y})$ compare with the limiting distribution recorded in Section B?

2 Let $(y_{11}, \ldots, y_{1m})$ be a sample of m from the normal(μ_1, σ) and $(y_{21}, \ldots, y_{2n})$ be a sample of n from the normal(μ_2, σ) with (μ_1, μ_2, σ) in $R^2 \times R^+$. Consider the problem of testing the hypothesis $H_0 : \mu_1 = \mu_2, \sigma \in R^+$ against the alternative $H_1 : \mu_1 \neq \mu_2, \sigma \in R^+$. Show that the generalized likelihood ratio test is to reject for large values of $|t|$, where

$$t = \frac{\bar{y}_2 - \bar{y}_1}{s(1/m + 1/n)^{1/2}}, \qquad s^2 = \frac{\sum_1^m (y_{1i} - \bar{y}_1)^2 + \sum_1^n (y_{2i} - \bar{y}_2)^2}{m + n - 2};$$

recall Section 7.2I and Problem 8.3.9. How does the large-sample distribution (Section B) of $2 \ln L(y)$ compare with the exact distribution recorded in Problem 8.3.9?

3 Let $(y_{11}, \ldots, y_{1m})$ be a sample of m from the normal(μ_1, σ_1) and $(y_{21}, \ldots, y_{2n})$ be a sample of n from the normal(μ_2, σ_2) with $(\mu_1, \mu_2, \sigma_1, \sigma_2)$ in $R^2 \times (R^+)^2$. Consider the problem of testing $H_0 : \sigma_1 = \sigma_2$ against $H_1 : \sigma_1 \neq \sigma_2$. Show that the generalized likelihood ratio is a function of

$$F = \frac{\sum (y_{1i} - \bar{y}_1)^2/(m - 1)}{\sum (y_{2i} - \bar{y}_2)^2/(n - 1)};$$

the usual test is to reject if $F \leq F_{1-\alpha/2}$ or $F \geq F_{\alpha/2}$, where F_α is the α point on the right tail of the $F(m - 1, n - 1)$ distribution. See Problems 10.2.12 and 10.2.13.

4 Consider the large-sample test given by formula (5). Suppose that u has mean μ and variance σ^2 under the alternative θ. Show that the power for a sample of n has the limiting form

$$1 - G\left(z_\alpha \frac{\sigma_0}{\sigma} - \frac{\mu - \mu_0}{\sigma/\sqrt{n}}\right),$$

where G is the standard normal distribution function; let $\mu = \mu_0 + \delta/\sqrt{n}$ with δ fixed.

5 Use Theorem 3 to show that C in Example 1 is UMP unbiased size α for the two-sided normal with known variance.

6 For the model (11) having exponential form with one ψ-function, use the score properties in Section 9.4 to show that the local unbiasedness conditions (8) and (9) can be replaced by

$$E(\phi(\mathbf{y})\,|\,\theta_0) = \alpha, \qquad E(\phi(\mathbf{y})t(\mathbf{y})\,|\,\theta_0) = \alpha E(t(\mathbf{y})\,|\,\theta_0).$$

7 Let y be a response with the normal(μ_0, σ) distribution with σ in R^+. Describe the sequential test of σ_0 against σ_1 $(\sigma_1 > \sigma_0)$ using error probabilities $\alpha = 0.05$ and $\beta = 0.05$; use the approximate formulas for A and B.

8 Let y be a response with the exponential (θ) distribution with θ in R^+. Describe the sequential test of θ_0 against θ_1 $(\theta_1 > \theta_0)$ with error probabilities $\alpha = 0.05$ and $\beta = 0.01$; use the approximate formulas for A and B.

9 Let y have a Poisson(θ) distribution. Describe the sequential test of θ_0 against θ_1 $(\theta_1 > \theta_0)$ with error probabilities $\alpha = 0.01$ and $\beta = 0.01$; use the approximate formulas for A and B.

10 Let $(y_1, \ldots, y_m)$ be a sample from the normal(μ, σ) and let $(w_1, \ldots, w_n)$ be an additional sample from the same normal, where n depends on the first sample variance s_y^2. An estimator for μ is required with variance δ^2 regardless of σ^2. Let us aim for δ_0^2, a value to be chosen later just less than δ^2. Given s_y^2, the estimator $a\bar{y} + b\bar{w}$ with $a + b = 1$ has variance $(a^2/m + b^2/n)$. Argue that a, b, and n can be chosen so that the estimated variance $(a^2/m + b^2/n)s_y^2 = \delta_0^2$. [For example, choose $n \geq 1$ as small as possible so that $s_y^2/(m + n)$ is just less than δ_0^2; then argue that values of a and b can be found to give equality.]

11 (*continuation*) Prove that

$$z = \frac{a\bar{y} + b\bar{w} - \mu}{(a^2/m + b^2/n)^{1/2}\sigma}$$

has a standard normal distribution given s_y^2 and hence that

$$t = \frac{a\bar{y} + b\bar{w} - \mu}{\delta_0}$$

has a Student($m - 1$) distribution, which has $E(t) = 0$ and var $(t) = (m - 1)/(m - 3)$. Thus the estimator $a\bar{y} + b\bar{w}$ has a distribution with Student form, and mean μ and variance δ^2, where we now choose $\delta_0^2 = (m - 3)\delta^2/(m - 1)$.

12 (*continuation*) For $m = 10$ give an expression for the minimum second sample size n to attain a 95 percent confidence interval for μ of prescribed length $2l$.

10.4

TESTING A STATISTICAL MODEL

We have been examining a lot of estimators and tests for the parameters of a statistical model. But we have not yet stopped to question the statistical model itself. The underlying assumption for the use of these estimators and tests is that the model is a valid model for the particular application. Sometimes, however, we will have applications where we want to check the validity of the model itself.

In Chapter 7 we examined a variety of inference procedures based on symmetry and large samples. The underlying assumption for those procedures was that the system being investigated was random, that the data formed a sample from a distribution. Supplement Section 7.5 was concerned with testing randomness, with testing the validity of the assumption.

Now, in a similar way, for the estimators and tests based on parametric models we are concerned with testing a statistical model, with testing the validity of the assumptions we have been using. The tests we examine are the chi-square tests of fit. There are many other tests of fit but most are concerned with particular applications or are designed to detect particular departures from the model. The chi-square tests cover a broad range of models and they are simple and accessible.

A TWO EXAMPLES OF TESTS OF FIT

In Section 2.1E we examined the Rutherford and Geiger data involving 2608 counts on the number of α-particles in a 7.5-second interval. The Poisson model is a fairly standard model for such frequency counts; indeed the frequency counts for radioactive disintegration form the classic application for the Poisson model.

We were concerned with whether the observed frequencies f_i are in reasonable agreement with the mean frequencies for some Poisson model. We determined a fitted Poisson model, we calculated the mean frequencies for the fitted model, and we then examined the deviations

$$d_i = 2(\sqrt{f_i} - \sqrt{e_i})$$

and then the sum of squares of these deviations,

$$\chi^2 = \sum d_i^2 = 4(\sqrt{f_i} - \sqrt{e_i})^2 = 14.60.$$

We noted that the distribution for χ^2 under the hypothesis of some Poisson distribution is approximately chi-square(10). The observed value 14.60 is a reasonable value for this chi-square distribution, and in this way we saw that the data are reasonable for the Poisson model.

In Section 2.2D we examined the Grummel and Dunningham data involving 250 observations on a continuous response. Some background information suggested that a normal distribution might be appropriate. To apply a chi-square type of test to continuous data it is necessary to group the observations and form cells, and thus obtain frequencies for various cells. The Grummel and Dunningham data were already grouped; some further grouping was done to avoid too-small frequencies.

We were concerned with whether the observed frequencies f_i were in reasonable agreement with the mean frequencies for a normal model. We determined a fitted normal model, we calculated the mean frequencies for the fitted model, and we then examined the deviations,

$$d_i = 2(\sqrt{f_i} - \sqrt{e_i}),$$

and then the sum of squares of these deviations,

$$\chi^2 = \sum d_i^2 = \sum 4(\sqrt{f_i} - \sqrt{e_i})^2 = 3.89.$$

We noted that the distribution for χ^2 under the hypothesis of some normal distribution is approximately chi-square(7). The observed value 3.89 is a reasonable value for this chi-square distribution, and in this way we saw that the data are reasonable for the normal model.

In this section we present the theory for these chi-square tests of fit and we examine a range of different applications. As part of this we obtain some tests for the parameters of a parametric model.

B THE CHI-SQUARE TESTS OF FIT

Consider the problem of testing a statistical model. In the discrete case, a model with parameter θ will record probabilities $p_i(\theta)$ for the various discrete sample points x_i. In the absolutely continuous case a model will record a density function on the sample space. As part of the chi-square approach we suppose that intervals or cells have been formed on the sample space; let $p_i(\theta)$ be the probability for the ith cell.

For a sample of n let f_i designate the frequency count of values at the ith sample point or in the ith cell. By Section 3.5B the distribution of the frequencies $(f_1, \ldots, f_k)$ is multinomial$(n, p_1(\theta), \ldots, p_k(\theta))$.

The multinomial distribution for large n can be rather unwieldy to calculate. Fortunately, we have a variety of limiting distributions, which we now record for convenient reference. Applications continue with Section C.

LEMMA 1

As $n \to \infty$ the limiting distribution of $(t_1, \ldots, t_k)$,

$$(1) \qquad t_1 = \frac{f_1 - np_1(\theta)}{(np_1(\theta))^{1/2}}, \ldots, t_k = \frac{f_k - np_k(\theta)}{(np_k(\theta))^{1/2}},$$

is the same as the conditional distribution of a sample of k from the standard normal given the condition $\Sigma_1^k t_i \sqrt{p_i(\theta)} = 0$.

LEMMA 2

As $n \to \infty$ the limiting distribution of $(z_1, \ldots, z_k)$,

$$z_1 = 2(\sqrt{f_1} - \sqrt{np_1(\theta)}), \ldots, z_k = 2(\sqrt{f_k} - \sqrt{np_k(\theta)}), \tag{2}$$

is the same as the conditional distribution of a sample of k from the standard normal given the condition $\Sigma_1^k z_i \sqrt{p_i(\theta)} = 0$.

Some details of the proof are discussed in Section I.

The limiting normal distributions just described lead immediately to limiting chi-square distributions.

THEOREM 3

As $n \to \infty$, the limiting distribution of

$$\chi_1^2 = \sum_1^k t_i^2 = \sum_1^k \frac{(f_i - np_i(\theta))^2}{np_i(\theta)} = \sum_1^k \frac{(f_i - e_i)^2}{e_i} \tag{3}$$

and the limiting distribution of

$$\chi_2^2 = \sum_1^k z_i^2 = \sum_1^k 4(\sqrt{f_i} - \sqrt{np_i(\theta)})^2 = \sum_1^k 4(\sqrt{f_i} - \sqrt{e_i})^2 \tag{4}$$

is chi-square$(k - 1)$.

The actual distribution of the χ_i^2 is discrete; but the limiting distribution is continuous. The proof follows by using Lemma 6.3.11 and Theorem 6.1.3 together with the normal-to-chi-square results from Section 8.2C.

Now consider the case in which some parameters of the model are estimated or fitted to the data. For this suppose that the true cell probability vector $\mathbf{p}(\boldsymbol{\theta}) = (p_1(\boldsymbol{\theta}), \ldots, p_k(\boldsymbol{\theta}))'$ is in a small neighborhood of the point $\mathbf{p}(\boldsymbol{\theta}_0) = (p_1(\boldsymbol{\theta}_0), \ldots, p_k(\boldsymbol{\theta}_0))'$ and that in this neighborhood the model $\mathbf{p}(\boldsymbol{\theta}) = (p_1(\boldsymbol{\theta}), \ldots, p_k(\boldsymbol{\theta}))'$ is a continuously differentiable surface that is approximately linear with dimension given by the number s of coordinates of the effective parameter $\boldsymbol{\theta} = (\theta_1, \ldots, \theta_s)'$.

THEOREM 4

As $n \to \infty$ the limiting distribution of $(z_1, \ldots, z_k)$,

$$z_1 = 2(\sqrt{f_1} - \sqrt{np_1(\widehat{\theta})}), \ldots, z_k = 2(\sqrt{f_k} - \sqrt{np_k(\widehat{\theta})}), \tag{5}$$

where $\widehat{\theta}$ is obtained by minimizing χ_2^2 in formula (4), is the same as the conditional distribution of a sample of k from the standard normal given the conditions

$$\sum_1^k z_i \sqrt{p_i(\theta)} = 0, \qquad \sum_1^k z_i \frac{1}{\sqrt{p_i(\theta)}} \frac{\partial p_i(\boldsymbol{\theta})}{\partial \theta_\alpha} = 0, \qquad \alpha = 1, \ldots, s. \tag{6}$$

And the limiting distribution of the minimum

(7) $$\chi_2^2 = \sum_1^k z_i^2 = \sum_1^k 4(\sqrt{f_i} - \sqrt{np_i(\hat{\theta})})^2$$

is chi-square$(k - 1 - s)$.

Some details of the proof are discussed in Section I. Maximum likelihood (using the multinomial) can also be used to determine the fitted parameter $\hat{\theta}$.

The t's and the z's provide two approximately equivalent ways of expressing the limiting normal form of the multinomial distribution. The use of the z's, however, provides a single reexpression $2\sqrt{f_i}$ of a frequency f_i rather than a reexpression $f_i/\sqrt{e_i}$ that varies with the parameter value being examined. We will see that this permits the intercomparison of a succession of hypotheses and provides a corresponding decomposition of the basic chi-square; see Section H below. Results analogous to (7) are, of course, available for χ_1^2 based on the t's; but the estimation should be done by maximum likelihood, and the chi-square decomposition just mentioned is not generally available.

C TEST OF A PROBABILITY MODEL

Consider observed frequencies $f_1, \ldots, f_k$ and a model that specifies the corresponding probabilities $p_1(\theta_0), \ldots, p_k(\theta_0)$. This is a probability model or a special statistical model in which there are no free parameters.

This probability model can be tested by calculating

$$\chi_1^2 = \sum_1^k \frac{(f_i - np_i(\theta_0))^2}{np_i(\theta_0)} \quad \text{or} \quad \chi_2^2 = \sum_1^k 4(\sqrt{f_i} - \sqrt{np_i(\theta_0)})^2$$

and comparing the value with the chi square$(k - 1)$ distribution. Large values of the χ^2 would indicate unusual deviations of the frequencies from the mean frequencies specified by the model, and in turn would provide evidence against the model.

D TEST OF A BINOMIAL MODEL

Perhaps the simplest situation for the chi-square method of Section C is the binomial. We examine briefly how the chi-square method works in this simple situation.

Consider the case with $k = 2$ and let y be the frequency for the first cell and $n - y$ the frequency for the second cell. And let p and $q = 1 - p$ be the corresponding probabilities with values specified by some theory; then the theory says that y has the binomial(n, p) distribution.

For the first chi-square we have

$$\chi_1^2 = \frac{(y - np)^2}{np} + \frac{(n - y - nq)^2}{nq} = \frac{(y - np)^2}{npq}$$

and by Theorem 3 this has a limiting chi-square(1) distribution under the theory. This corresponds of course to using $(y - np)/(npq)^{1/2}$ as a standard normal variable as indicated by the results in Example 6.2.3, and is in agreement with the large-sample methods suggested in Section 8.6E.

For the second chi-square we have

$$\chi_2^2 = 4(\sqrt{y} - \sqrt{np})^2 + 4(\sqrt{n-y} - \sqrt{nq})^2,$$

and by Theorem 3 this has a limiting chi-square(1) distribution under the theory. This chi-square can be pictured conveniently on binomial probability paper which has axes that are marked in square-root units. Thus the points $(\sqrt{np}, \sqrt{nq})$ and $(\sqrt{y}, \sqrt{n-y})$ can be plotted directly using the values np and nq for the first point, and the values y and $n - y$ for the second point; see Figure 10.9. The difference vector is

$$(z_1/2, z_2/2) = (\sqrt{y} - \sqrt{np}, \sqrt{n-y} - \sqrt{nq}).$$

By Lemma 2 the difference vector has a limiting normal(0, 1/2) distribution on the tangent to the circle at the point $(\sqrt{np}, \sqrt{nq})$. Thus the observed difference vector is easily assessed in units of length 1/2.

The second chi-square based on root frequencies can also be examined in terms of an *angular transformation:*

$$a = \sin^{-1}\sqrt{\frac{y}{n}} - \sin^{-1}\sqrt{p},$$

which is the angle a as pictured in Figure 10.9. Under the theory the angle a has a limiting normal distribution $(0, 1/2\sqrt{n})$. The angular transformation is available in most collections of statistical tables, for example, Fisher and Yates (1949). It provides a convenient analytic equivalent of the graphical use of binomial paper.

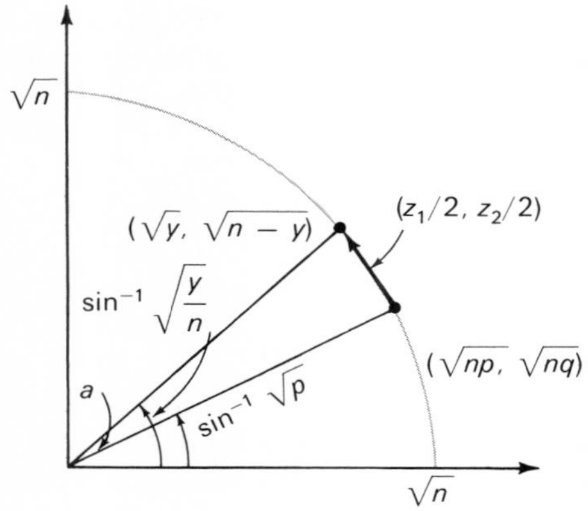

FIGURE 10.9

The difference vector$(z_1/2, z_2/2)$ from $(\sqrt{np}, \sqrt{nq})$ to $(\sqrt{y}, \sqrt{n-y})$ can be assessed in units of length 1/2. The difference angle from $(\sqrt{np}, \sqrt{nq})$ to $(\sqrt{y}, \sqrt{n-y})$ can be assessed in units of length $1/2\sqrt{n}$.

E TEST OF A PROBABILITY MODEL: AN EXAMPLE

The following example from genetics will be used repeatedly in this section.

EXAMPLE 1

The breeding of a certain type of flower can produce offspring that can have a magenta *M* or a red *R* flower, and can have a green *g* or a red *r* stigma; thus the possibilities are *Mg*, *Mr*, *Rg*, and *Rr*. For 220 offspring the data ($\mathfrak{D}$) are recorded in the following contingency table:

Observed frequencies

		g	*r*	*Total*
$\mathfrak{D}$	*M*	117	31	148
	R	55	17	72
	Total	172	48	220

A strong theory $\mathfrak{J}_0$ specifies the probabilities for the four kinds of offspring as 9/16, 3/16, 3/16, and 1/16. In the first display we present the corresponding mean frequencies e_i and root mean frequencies $\sqrt{e_i}$; in the last display we present the observed frequencies f_i and root frequencies $\sqrt{f_i}$; the intermediate display records the differences $\sqrt{f_i} - \sqrt{e_i}$:

Frequencies

		g	*r*	*Total*
$\mathfrak{J}_0$	*M*	123.75	41.25	165
	R	41.25	13.75	55
	Total	165	55	220

Root frequencies

	g	*r*
M	11.12	6.42
R	6.42	3.71

	g	*r*
M	−0.30	−0.85
R	1.00	0.41

		g	*r*	*Total*
$\mathfrak{D}$	*M*	117	31	148
	R	55	17	72
	Total	172	48	220

	g	*r*
M	10.82	5.57
R	7.42	4.12

The observed value of chi-square is

$$\chi^2 = 4((-0.30)^2 + (-0.85)^2 + (1.00)^2 + (0.41)^2) = 7.92.$$

The distribution for χ^2 under the strong theory $\mathfrak{J}_0$ is chi-square(3): the 5 percent point is 7.81; the 1 percent point is 11.3. Thus the observed χ^2 gives some moderate evidence against the hypothesis and the strong theory $\mathfrak{J}_0$ may not be applicable.

F TEST OF A STATISTICAL MODEL WITH FITTED PROBABILITIES

Consider observed frequencies $f_1, \ldots, f_k$ and a statistical model that specifies the corresponding probabilities $p_1(\boldsymbol{\theta}), \ldots, p_k(\boldsymbol{\theta})$, where the parameter θ has s-dimensional freedom as described in Section B; assume that $s < k - 1$.

This statistical model can be tested by calculating

$$\chi_2^2 = \sum_1^k 4(\sqrt{f_i} - \sqrt{np_i(\widehat{\theta})})^2$$

and comparing the value with the chi-square($k - 1 - s$) distribution; the fitted parameter can be obtained by minimizing chi-square as described in Theorem 4 (or by maximum likelihood using the multinomial). Large values of χ_2^2 would indicate unusual deviations from the statistical model and in turn would provide evidence against the model.

Consider the case where one set of marginal probabilities in a contingency table is estimated. We reexamine the data in Example 1.

EXAMPLE 2

A somewhat weaker theory, $\mathfrak{J}_1$, allows probabilities p_1 and q_1 for the colors *M* and *R* corresponding to rows but keeps the independence between flower and stigma color and keeps the probabilities 3/4 and 1/4 for the colors *g* and *r*. Thus the cells *Mg, Mr, Rg,* and *Rr* have probabilities $(3/4)p_1$, $(1/4)p_1$, $(3/4)q_1$, and $(1/4)q_1$, respectively, with p_1 in [0, 1].

The observed proportions for *M* and *R* are 148/220 and 72/220 and they give estimates (MLE) for p_1 and q_1. In the first display we present the corresponding mean frequencies e_i and root-mean frequencies $\sqrt{e_i}$; in the last display we present the observed frequencies f_i and root frequencies $\sqrt{f_i}$; the intermediate display records the differences $\sqrt{f_i} - \sqrt{e_i}$; note that our fitted frequencies have come closer to the data:

Frequencies

$\mathfrak{J}_1$		*g*	*r*	*Total*
	M	111	37	148
	R	54	18	72
	Total	165	55	220

Root frequencies

	g	*r*
M	10.54	6.08
R	7.35	4.24

	g	*r*
M	0.28	−0.51
R	0.07	−0.12

Frequencies

		g	r	*Total*
$\mathfrak{D}$	M	117	31	148
	R	55	17	72
	Total	172	48	220

Root frequencies

	g	r
M	10.82	5.57
R	7.42	4.12

The observed value of chi-square is

$$\chi^2 = 4((0.28)^2 + (-0.51)^2 + (0.07)^2 + (-0.12)^2) = 1.43.$$

The distribution for χ^2 under the theory $\mathfrak{I}_1$ is chi-square(2): the 5 percent point is 6.0; the 1 percent point is 9.2. Thus the observed χ^2 is a very reasonable value for the distribution and there is no evidence against the theory $\mathfrak{I}_1$.

G TEST OF INDEPENDENCE WITH A CONTINGENCY TABLE

The chi-square methods in this section can provide a test for the statistical independence of two response variables. For this suppose that the two variables are discrete or that they have been made discrete by grouping into cells.

For the first variable let $A_1, \ldots, A_k$ designate the k distinct possibilities and for the second variable let $B_1, \ldots, B_l$ designate the l distinct possibilities; thus any observation on the pair of variables corresponds to a combination A_iB_j. Then for n observations we can record the frequencies f_{ij} as in the following array:

	B_1	$\cdots$	B_l	*Total*
A_1	f_{11}	$\cdots$	f_{1l}	m_1
$\vdots$	$\vdots$		$\vdots$	
A_k	f_{k1}	$\cdots$	f_{kl}	m_k
Total	n_1	$\cdots$	n_l	n

Let $p_1, \ldots, p_k$ designate probabilities for the first variable ($\Sigma p_i = 1$) and $p'_1, \ldots, p'_l$ designate probabilities for the second variable ($\Sigma p'_j = 1$). Then the hypothesis of independence specifies a model with cell probabilities

$$p_{ij} = p_i p'_j;$$

this model has $k - 1$ free parameters for the rows and $l - 1$ free parameters for the columns.

The row probabilities can be estimated (MLE) by row proportions

$$\hat{p}_i = \frac{m_i}{n};$$

and the column probabilities by the column proportions

$$\hat{p}_j = \frac{n_j}{n}.$$

Thus the fitted model has mean frequencies $n\hat{p}_i\hat{p}_j = m_i n_j/n$, as recorded in the following array:

	B_1	$\cdots$	B_l	*Total*
A_1	$\frac{m_1 n_1}{n}$	$\cdots$	$\frac{m_1 n_l}{n}$	m_1
$\vdots$	$\vdots$		$\vdots$	$\vdots$
A_k	$\frac{m_k n_1}{n}$	$\cdots$	$\frac{m_k n_l}{n}$	m_k
Total	n_1	$\cdots$	n_l	n

This statistical model specifying independence between rows and columns can be tested by calculating

$$\chi_1^2 = \sum_{ij} \frac{(f_{ij} - m_i n_j/n)^2}{m_i n_j/n}$$

or

$$\chi_2^2 = \sum_{ij} 4(\sqrt{f_{ij}} - \sqrt{m_i n_j/n})^2,$$

and comparing the value with the chi-square$((k-1)(l-1))$ distribution; note that $kl - 1 - (k-1) - (l-1) = (k-1)(l-1)$. Large values of χ^2 would indicate unusual deviations from the independence model. The limiting distribution is sometimes slightly better approximated by using χ_1^2 than χ_2^2, particularly with the possibility of zero frequencies.

We now reexamine the data in Examples 1 and 2.

EXAMPLE 3

A still weaker theory, $\mathfrak{I}_2$, allows probabilities p_1 and q_1 for the colors *M* and *R* and probabilities p_2 and q_2 for the colors *g* and *r*, but independence between the flower color and stigma color.

The observed proportions for *M* and *R* are 148/220 and 72/220 and the observed proportions for *g* and *r* are 172/220 and 48/220; these provide estimates (MLE) for the row and column probabilities. In the first display we present the corresponding mean frequencies e_i and root-mean frequencies $\sqrt{e_i}$; in the last display the observed frequencies f_i and root frequencies $\sqrt{f_i}$; the intermediate display records the difference $\sqrt{f_i} - \sqrt{e_i}$; note that our fitted frequencies have come still closer to the data.

Frequencies

		g	*r*	*Total*
$\mathfrak{I}_2$	*M*	115.71	32.29	148
	R	56.29	15.71	72
	Total	172	48	220

Root frequencies

	g	*r*
M	10.76	5.68
R	7.50	3.96

Frequencies

$\mathfrak{D}$

	g	*r*	*Total*
M	117	31	148
R	55	17	72
Total	172	48	220

Root frequencies

	g	*r*
M	0.06	−0.11
R	−0.08	0.16

	g	*r*
M	10.82	5.57
R	7.42	4.12

The observed value of chi-square is

$$\chi^2 = 4((0.06)^2 + (-0.11)^2 + (-0.08)^2 + (0.16)^2) = 0.19.$$

The distribution for χ^2 under the theory $\mathfrak{I}_2$ is chi-square(1): the 5 percent point is 3.84; the 1 percent point is 6.63. Thus the observed χ^2 is a very reasonable value for the distribution, and there is no evidence against the theory $\mathfrak{I}_2$ (if anything, the observed value is rather small).

H A SUCCESSION OF HYPOTHESES OR THEORIES

The use of the second chi-square based on root frequencies allows intercomparisons within a succession of statistical models and provides a decomposition of the basic chi-square. We illustrate this by considering Examples 1, 2, and 3 in greater detail.

In the left column we record the root-frequency matrices for the fitted $\mathfrak{I}_0$, the fitted $\mathfrak{I}_1$, the fitted $\mathfrak{I}_2$, and the data $\mathfrak{D}$. On the right we record the successive difference matrices

Root frequencies

$\mathfrak{I}_0$

	g	*r*
M	11.12	6.42
R	6.42	3.71

$$\mathbf{p}_1 = \begin{bmatrix} -0.58 & -0.34 \\ 0.93 & 0.53 \end{bmatrix} \qquad SS_1 = 1.5978$$

$\mathfrak{I}_1$

	g	*r*
M	10.54	6.08
R	7.35	4.24

$$\mathbf{p}_2 = \begin{bmatrix} 0.22 & -0.40 \\ 0.15 & -0.28 \end{bmatrix} \qquad SS_2 = 0.3093$$

Root frequencies

		g	r
$\mathfrak{I}_2$	M	10.76	5.68
	R	7.50	3.96

$$\mathbf{p}_3 = \begin{array}{|cc|} 0.06 & -0.11 \\ -0.08 & 0.16 \end{array} \qquad SS_3 = 0.0477$$

		g	r
$\mathfrak{D}$	M	10.82	5.57
	R	7.42	4.12

The components (as vectors in R^4) and 4 times the squared lengths can be tabulated as in Sections 8.2B, 8.2C, and 9.2B; see Table 10.1.

The various components are approximately orthogonal, as is easily checked. They differ slightly from orthogonality because: (a) the fitted points lie on a sphere $\Sigma(\sqrt{f_i})^2 = 220$ with squared radius equal to 220; (b) the component models are represented by a curve and a surface on that sphere. Note, however, that the chi-square in Example 2,

$$\chi^2 = 1.43 \approx 1.24 + 0.19 = 1.43,$$

and the chi-square in Example 1,

$$\chi^2 = 7.92 \approx 6.39 + 1.24 + 0.19 = 7.82.$$

Thus we see that the original chi-squares have been separated into component chi-squares that are approximately orthogonal and refer to particular characteristics of the statistical model.

We can test for independence by examining 0.19 and comparing it with chi-square(1); the 5 percent point is 3.84; the 1 percent point is 6.63.

Assuming independence, we can test for the column probability $p_2 = 3/4$ by examining 1.24 and comparing it with chi-square(1): the 5 percent point is 3.84; the 1 percent point is 6.63.

And assuming the independence and the column probability we can test the row probability $p_1 = 3/4$ by examining 6.39 and comparing with chi-square(1): the 5 percent point is 3.84; the 1 percent point is 6.63. This test for the row probability is much more sensitive than the blanket test for the model in Example 1.

TABLE 10.1
Components and lengths of $\mathfrak{I}$

Source	Dimension	Component	4 · (squared length)
$p_1 = 3/4$	1	$\mathbf{p}_1$	6.39
$p_2 = 3/4$	1	$\mathbf{p}_2$	1.24
Independence	1	$\mathbf{p}_3$	0.19

I SUPPLEMENT: SOME DETAILS ON THE LIMITING DISTRIBUTIONS

Consider y with a Poisson(θ) distribution. Then it follows from Section 6.1D and Problem 6.1.10 that the density function for

$$t = \frac{y - \theta}{\theta^{1/2}} \quad \text{or} \quad z = 2(\sqrt{y} - \sqrt{\theta})$$

approaches the standard normal density function.

Now consider independent $y_1, \ldots, y_k$, where y_i has the Poisson(np_i) distribution. Then the preceding shows that the density function for $(t_1, \ldots, t_k)$ or $(z_1, \ldots, z_k)$ as given by formula (1) or (2) approaches the density function for a sample of k from the standard normal.

The multinomial $(n, p_1, \ldots, p_k)$ is obtained by conditioning $(\Sigma\, y_i = n)$ the preceding joint Poisson distribution. The sample lattice points are distributed uniformly along the plane $\Sigma\, y_i = n$. It follows that the conditional density along the surface $\Sigma\, y_i = n$ has the limiting normal form as described in Section B.

The limiting distribution is derived from the preceding by applying the normal sampling results from Section 8.2B.

J PROBLEMS

1 A die was tossed 1600 times:

Event	1	2	3	4	5	6
Frequency	301	308	340	214	196	241

Test the fit of the symmetric model having equal probabilities.

2 The progeny from a certain type of corn can be starchy S or sugary s and can be green G or white g. The data for $n = 3839$ progeny are

	G	g
S	1997	906
s	904	32

Test the hypothesis that the probabilities for SG, Sg, sG, and sg are 9/16, 3/16, 3/16, and 1/16, respectively.

3 (*continuation*) Test the hypothesis that the two attributes are statistically independent.

4 (*continuation*) Separate the chi-squares so that the hypothesis in Problem 2 can be tested, assuming the hypothesis in Problem 3; comment on the validity of such a test with the given data.

5 The following data (Freund) on blood types were obtained with random samples from three racial groups:

	O	A	B	AB
Race 1	176	148	96	72
Race 2	78	50	45	12
Race 3	15	19	8	7

Test the hypothesis that blood-type distribution is independent of racial group.

6 The following frequencies f_i were obtained (Morris) for d_i, the death month in months after the birth month for publicly prominent people:

d_i	−6	−5	−4	−3	−2	−1	0	1	2	3	4	5
f_i	24	31	20	23	34	16	26	36	37	41	26	34

Test the hypothesis that the probabilities are equal.

7 (*continuation*) Test the hypothesis that the months −1 and 0 have a probability p_1 and the remaining months have a probability p_2 (note that $2p_1 + 10p_2 = 1$).

8 (*continuation*) Separate the chi-squares and test the hypothesis in Problem 6 assuming the hypothesis in Problem 7.

9 One hundred plants were classified according to two attributes: large L or small l; white W or colored w. The observed frequencies are

	W	w	*Total*
L	40	20	60
l	15	25	40
Total	55	45	100

(a) Test the model that specifies equal probabilities in the four cells.

(b) Test the model that specifies equal $W - w$ probabilities and independence between the attributes.

(c) Test the model that specifies independence allowing arbitrary $W - w$ and arbitrary $L - l$ probabilities.

(d) Record the three-way decomposition of the chi-square; what tests are appropriate?

SUPPLEMENTARY MATERIAL

Supplement Section 10.5 is on pages 579–584.

Confidence intervals have been an integral part of the development in Chapters 7, 8, and 9; some theoretical aspects of confidence intervals are examined in the supplementary section in terms of the hypothesis-testing theory of this chapter. Some additional discussion on confidence intervals for the probability p in sampling from the Bernoulli(p) is included also.

NOTES AND REFERENCES

Some aspects of tests of significance have been formalized in the theory of hypothesis testing developed in this chapter. This theory originated in the work of Neyman and Pearson (1933), and has been thoroughly exposited in Lehmann (1959). Tests of models using the ordinary chi-square test have been surveyed in Lancaster (1969).

Fisher, R. A., and F. Yates (1949). *Statistical Tables for Biological, Agricultural and Medical Research Workers*. New York: Macmillan Publishing Co.

Lancaster, H. O. (1969). *The Chi-Squared Distribution*. New York: John Wiley & Sons, Inc.

Lehmann, E. L. (1959). *Testing Statistical Hypotheses*. New York: John Wiley & Sons, Inc.

Neyman, J., and E. S. Pearson (1933). On the problem of the most efficient tests of statistical hypotheses. *Phil. Trans. Roy. Soc. London,* **A231,** 289–337.

In Chapters 9 and 10 we examined in detail the basic methods of statistical inference.

In Chapter 11 we shall examine statistical inference for the very common and widely used statistical models that involve a linear structure for location, the ***linear models.*** *Much of the development can be done not just for normal error form but for arbitrary error form. In Section 11.1 we investigate the simple* ***location-scale model*** *and obtain inference methods for normal error and inference methods for arbitrary error. (The computer program for this is available.) In Section 11.2 some details of least squares concerning projections and residuals are assembled. Statistical inference for the* ***regression*** *or linear model is examined in Sections 11.3 and 11.4.*

11

LINEAR MODELS

11.1

THE LOCATION-SCALE MODELS

We now examine a large and very important class of statistical models, the linear models. In preceding chapters we have repeatedly examined the location and the location-scale models using a normal distribution for the variation or error. These models are special cases of the linear model discussed in this chapter. In this first section we examine the location-scale model using an arbitrary distribution for the variation.

In Chapter 9 we examined some estimation results for the linear model for a vector response. In Sections 11.2 to 11.4 we investigate this linear model using an arbitrary distribution for the variation: in Section 11.2, some vector results; in Section 11.3, the linear model itself; and in Section 11.4, an example.

The location-scale model covers applications where the input variables to the system are kept constant. The unknowns then are the general level of the response and the scaling of the variation. It is also possible to include a parameter that allows a range of distributions for the variation.

The linear model more generally covers applications with a random system under varied conditions; in particular, it includes the standard models of experimental design where the input variables of the system are changed by design.

A THE LOCATION-SCALE MODEL

Consider a real-valued response y. Suppose that earlier experience with the system under investigation has identified the distribution that describes the variation in the response. Let z designate a variable presenting this variation, and f be the known density function for z. And suppose that under present conditions the response scaling of the variation is unknown and the general level of the response is unknown. Let σ designate the response scaling and μ designate the general level of the response; this gives $y = \mu + \sigma z$.

Thus for the variation we have a probability model as given by the known density function f. And for the response we have the statistical model

$$\{y = \mu + \sigma z : (\mu, \sigma) \in R \times R^{+}\},$$

giving the set of possible functions for the response y in terms of the variation z.

Consider the question of a standardized presentation for the variation. If f can be standardized so that $E(z) = 0$, $\operatorname{var}(z) = 1$, then the general level μ is the mean response and the scaling σ is the response standard deviation. Perhaps more realistically, if f is standardized so that the median of z is zero and the interval $(-1, +1)$ contains 68.26 percent of the probability (as with the standard normal), then the general level μ is the median response and the interval $(\mu - \sigma, \mu + \sigma)$ contains 68.26 percent of the response values; call σ in this case the response standard error. This second standardization seems particularly appropriate with a distribution f that allows substantially more probability in the tails than the normal; especially so with a distribution such as the Cauchy, for which the mean and standard deviation do not exist.

Now consider a sequence of independent repetitions of the system under investigation. Let $\mathbf{z} = (z_1, \ldots, z_n)'$ designate the sequence for the variation, and $\mathbf{y} = (y_1, \ldots, y_n)'$ designate the corresponding sequence for the response. Then for the variation $\mathbf{z}$ we have the probability model as given by the density function

$$(1) \qquad f(z_1) \cdots f(z_n)$$

on R^n. And for the response we have the statistical model

$$(2) \qquad \{\mathbf{y} = \mu\mathbf{1} + \sigma\mathbf{z} : (\mu, \sigma) \in R \times R^+\},$$

giving the set of possible functions for the response $\mathbf{y}$ in terms of the variation $\mathbf{z}$.

B DATA FOR *n* INDEPENDENT REPETITIONS

Now consider an *observed* sequence $\mathbf{y} = (y_1, \ldots, y_n)'$ from n independent repetitions of the system under investigation. The observed response sequence is a function

$$(3) \qquad \mathbf{y} = \mu\mathbf{1} + \sigma\mathbf{z}$$

of a *realized* sequence $\mathbf{z}$ from the distribution (1) for variation.

What information does the observed $\mathbf{y}$ give concerning the realized $\mathbf{z}$ from the distribution? With no information concerning the values of μ and σ, it follows that

$$\mathbf{z} = -\sigma^{-1}\mu\mathbf{1} + \sigma^{-1}\mathbf{y}$$

can be any point on the half-plane

$$(4) \qquad \mathcal{L}^+(\mathbf{1}; \mathbf{y}) = \{a\mathbf{1} + c\mathbf{y} : a \in R, c \in R^+\}$$

hinged on the line $\mathcal{L}(\mathbf{1})$ and passing through the observed response $\mathbf{y}$; see Figure 11.1. Thus $\mathcal{L}^+(\mathbf{1}; \mathbf{z}) = \mathcal{L}^+(\mathbf{1}; \mathbf{y})$ and all that we effectively observe concerning the realized $\mathbf{z}$ is the half-plane passing through it; we have no differential information concerning the position of $\mathbf{z}$ on that half-plane.

From Sections 4.1 and 4.2 we know that the proper description of the realized $\mathbf{z}$ is given by the conditional distribution given the observed information that $\mathbf{z}$ lies on the half-plane (4). We derive this in Section D, but first we look for convenient coordinates to describe a half-plane and convenient coordinates for a point on such a half-plane.

C SIMPLE COORDINATES

Consider simple coordinates for a half-plane $\mathcal{L}^+(\mathbf{1}; \mathbf{z})$ and simple coordinates for a point $\mathbf{z}$ on such a half-plane; see Figure 11.2.

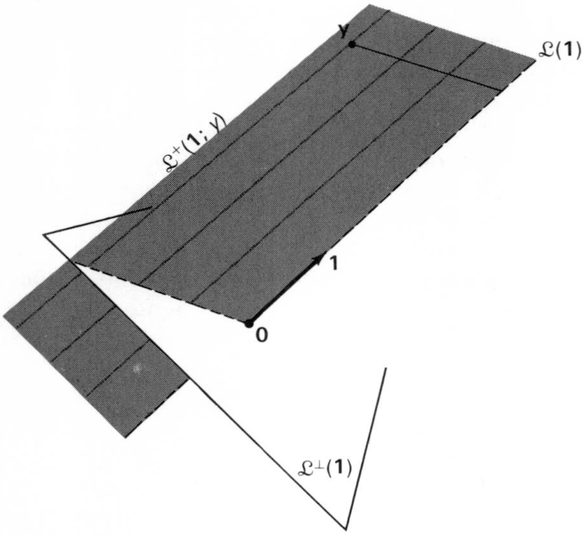

FIGURE 11.1
The possible values for **z** form the half-plane $\mathcal{L}^+(\mathbf{1}; \mathbf{y}) = \{a\mathbf{1} + c\mathbf{y} : a \in R, c \in R^+\}$ hinged on the line $\mathcal{L}(\mathbf{1})$ and passing through the observed **y**.

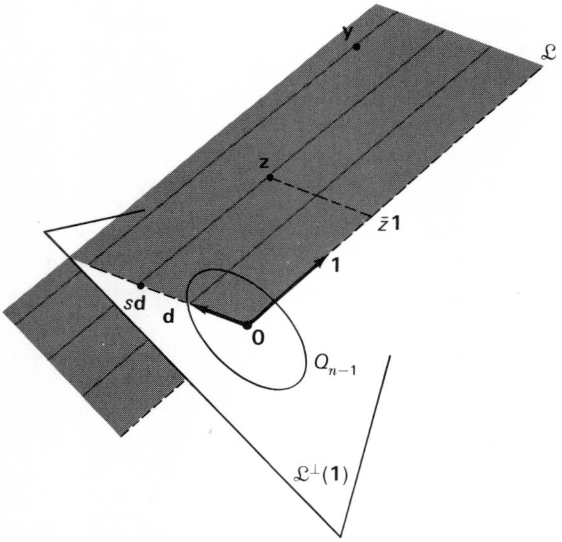

FIGURE 11.2
Unit orthogonal residual **d** identifies the half-plane $\mathcal{L}^+(\mathbf{1}; \mathbf{y})$. The point **z** has coordinates $\bar{z}$ and s relative to the basis vectors **1** and **d**.

The half-plane is hinged on the line $\mathcal{L}(\mathbf{1})$; accordingly, it is natural to use $\mathbf{1}$ as a basis vector. The vector $\mathbf{z}$ has projection $\bar{z}\mathbf{1}$ on the line $\mathcal{L}(\mathbf{1})$ and projection $\mathbf{z} - \bar{z}\mathbf{1}$ on the orthogonal complement $\mathcal{L}^{\perp}(\mathbf{1})$; thus

$$(5) \qquad \mathbf{z} = \bar{z}\mathbf{1} + (\mathbf{z} - \bar{z}\mathbf{1}) = \bar{z}\mathbf{1} + s(\mathbf{z})\mathbf{d}(\mathbf{z}),$$

where the residual $\mathbf{z} - \bar{z}\mathbf{1}$ produces the unit residual $\mathbf{d}(\mathbf{z})$ and the residual length $s(\mathbf{z})$,

$$(6) \qquad s^2(\mathbf{z}) = |\mathbf{z} - \bar{z}\mathbf{1}|^2 = \sum (z_i - \bar{z})^2, \qquad \mathbf{d}(\mathbf{z}) = \left(\frac{z_1 - \bar{z}}{s(\mathbf{z})}, \ldots, \frac{z_n - \bar{z}}{s(\mathbf{z})}\right)'.$$

This suggests using the vectors $\mathbf{1}$ and $\mathbf{d}(\mathbf{z})$ as orthogonal basis vectors for a point on a half-plane: thus

$$(7) \qquad \mathbf{z} = \bar{z}\mathbf{1} + s(\mathbf{z})\mathbf{d}(\mathbf{z})$$

with coordinates $(\bar{z}, s(\mathbf{z}))$ relative to the orthogonal vectors $\mathbf{1}$ and $\mathbf{d}(\mathbf{z})$.

The points $\mathbf{d}(\mathbf{z})$ are at unit distance from the origin and lie in the $(n - 1)$-dimensional subspace $\mathcal{L}^{\perp}(\mathbf{1})$. Thus the points $\mathbf{d}(\mathbf{z})$ form the unit sphere Q_{n-1} in $\mathcal{L}^{\perp}(\mathbf{1})$; see Figure 11.2.

Now consider the possible points $\mathbf{z}$ given the information provided by the observed response $\mathbf{y}$. From Section B we see that $\mathbf{z}$ and $\mathbf{y}$ must lie on the same half-plane hinged on the line $\mathcal{L}(\mathbf{1})$. Thus $\mathbf{d}(\mathbf{z}) = \mathbf{d}(\mathbf{y})$. For convenience we abbreviate $\mathbf{d}(\mathbf{y})$ as $\mathbf{d}$. We then see that the information provided by the observed response $\mathbf{y}$ is that $\mathbf{d}(\mathbf{z}) = \mathbf{d}$; thus we have the "observed" value $\mathbf{d}$ for the function $\mathbf{d}(\mathbf{z})$.

Now consider the relation (3) between the observed response $\mathbf{y}$ and the realized $\mathbf{z}$:

$$\begin{aligned} \mathbf{y} &= \bar{y}\mathbf{1} + s(\mathbf{y})\mathbf{d} = \mu\mathbf{1} + \sigma\mathbf{z} = \mu\mathbf{1} + \sigma(\bar{z}\mathbf{1} + s(\mathbf{z})\mathbf{d}) \\ &= (\mu + \sigma\bar{z})\mathbf{1} + \sigma s(\mathbf{z})\mathbf{d}. \end{aligned}$$

By equating coefficients relative to the basis vectors $\mathbf{1}$ and $\mathbf{d}$, we obtain

$$(8) \qquad \begin{aligned} \bar{y} &= \mu + \sigma\bar{z}, \\ s(\mathbf{y}) &= \sigma s, \end{aligned}$$

where we abbreviate $(\bar{z}, s(\mathbf{z})) = (\bar{z}, s)$.

D THE CONDITIONAL DISTRIBUTION GIVEN THE DATA

We now determine the marginal probability for what has been observed, the half-plane through $\mathbf{z}$ as given by $\mathbf{d}(\mathbf{z}) = \mathbf{d}$, and we determine the conditional probability for what cannot be observed, the position $(\bar{z}, s)$ of $\mathbf{z}$ on the observed half-plane.

The probability differential for the distribution on R^n is given by

$$(9) \qquad \prod_1^n f(z_i) \prod_1^n dz_i.$$

To obtain the marginal and conditional distributions we make the change of variable (7) to the new coordinates $\bar{z}$, s, and $\mathbf{d}$. For the density function the substitution is trivial. For the differential we argue as follows. At the point $\mathbf{z}$ we can measure distance in the direction $\mathbf{1}$ by $\sqrt{n}\, d\bar{z}$ (note that the vector $\mathbf{1}$ has length $\sqrt{n}$). At $\mathbf{z}$ we can measure distance in the orthogonal direction **d** by ds. Then in directions orthogonal to the half-plane we can measure area or volume on the sphere formed by the points $s\mathbf{d}(\mathbf{z})$ with s fixed. Let da be area or volume on the unit sphere formed by the points $\mathbf{d}(\mathbf{z})$; then $s^{n-2}\, da$ is area or volume on the sphere of radius s. We thus have length or volume measures in three orthogonal directions and obtain

$$\prod dz_i = \sqrt{n}\; d\bar{z} \cdot ds \cdot s^{n-2}\, da.$$

Accordingly, the probability differential for the distribution on R^n can be reexpressed as

$$\prod f(\bar{z} + sd_i)\sqrt{n}\; d\bar{z} \cdot ds \cdot s^{n-2}\, da.$$

By integrating out $\bar{z}$ and s we obtain the marginal density for **d**:

$$k(\mathbf{d}) = \int_0^{\infty} \int_{\infty}^{\infty} \prod f(\bar{z} + sd_i) s^{n-2}\, ds \sqrt{n}\; d\bar{z}. \tag{10}$$

This then gives the following factorization of the probability differential (9):

$$k(\mathbf{d})\, da \cdot k^{-1}(\mathbf{d}) \prod f(\bar{z} + sd_i) s^{n-2}\, ds \sqrt{n}\; d\bar{z}.$$

The probability differential is factored into the marginal probability for **d** times the conditional probability for $(\bar{z}, s)$ given **d**. Note that the norming constant $k(\mathbf{d})$ is obtained by the integration (10) over the half-plane $R \times R^+$.

We are in a position to be much more general than has been indicated so far. Suppose that we allow a parameter λ with values in Λ for the distribution $f = f_\lambda$ describing the variation. This then permits some imprecision in the information available concerning the distribution for variation.

The probability differential for the distribution on R^n is then given by

$$\prod f_\lambda(z_i) \prod dz_i.$$

This can be factored as before:

$$k_\lambda(\mathbf{d})\, da \cdot k_\lambda^{-1}(\mathbf{d}) \prod f_\lambda(\bar{z} + sd_i) s^{n-2}\, ds \sqrt{n}\; d\bar{z}, \tag{11}$$

into the marginal probability for the observed **d** times the conditional probability for $(\bar{z}, s)$ given **d**.

E INFERENCE CONCERNING THE PARAMETERS

Now consider statistical inference from an observed response **y**. As just described, we suppose that the distribution for the variation is given by one of the densities in the class

$$\{f_\lambda(z_1) \cdots f_\lambda(z_n) : \lambda \in \Lambda\}, \tag{12}$$

and that the observed response **y** is obtained from the realized **z** by one of the functions in the class

$$\{\mathbf{y} = \mu\mathbf{1} + \sigma\mathbf{z} : (\mu, \sigma) \in R \times R^+\}. \tag{13}$$

The only characteristic of the realized **z** that is observed is the value of $\mathbf{d}(\mathbf{z}) = \mathbf{d} = \mathbf{d}(\mathbf{y})$. The probability for this observed event is

$$k_\lambda(\mathbf{d})\, da. \tag{14}$$

Thus we have the observed likelihood function $L(\mathbf{d}|\cdot)$ for the parameter λ, where

(15) $$L(\mathbf{d}|\lambda) = ck_\lambda(\mathbf{d})$$

and the various values of c give the similarly shaped functions of λ. The likelihood function alone can sometimes provide sharp discriminations. In certain cases the likelihood statistic for the model $k_\lambda(\cdot)\,da$ may be accessible and the methods from Chapters 9 and 10 may be available for inferences concerning λ.

Now suppose that a value for λ is given—either from background information, or from sharp inferences based on (15), or as a tentative value in a range of values based on (15).

With λ given we have a distribution

(16) $$k_\lambda^{-1}(\mathbf{d}) \prod f(\bar{z} + sd_i)s^{n-2}\,ds\sqrt{n}\,d\bar{z}$$

describing the unknown position $(\bar{z}, s)$ of the realized variation $\mathbf{z}$; the relation to the observed position of the response is

(17) $$\bar{y} = \mu + \sigma\bar{z}, \qquad s(\mathbf{y}) = \sigma s,$$

where μ and σ are the general level and scaling for the response.

F INFERENCE CONCERNING μ

For inferences concerning μ we can appropriately simplify the equations (17):

(18) $$\frac{\bar{y} - \mu}{s(\mathbf{y})} = \frac{\bar{z}}{s} = T \quad \text{or} \quad \frac{\sqrt{n}\,(\bar{y} - \mu)}{s_y} = \frac{\sqrt{n}\,\bar{z}}{s/\sqrt{n-1}} = t = \sqrt{n(n-1)}\,T.$$

The marginal distribution of the ordinary t-function can be obtained from (16) by expressing $\bar{z}$ in terms of s and t and then integrating out s:

(19) $$g_\lambda(t:\mathbf{d})\,dt = k_\lambda^{-1}(\mathbf{d}) \int_0^\infty \prod f_\lambda(s(t/\sqrt{n^2 - n} + d_i))s^{n-1}\,ds \cdot (n-1)^{-1/2}\,dt.$$

A hypothesis $\mu = \mu_0$ can be tested by calculating

(20) $$t = \frac{\sqrt{n}\,(\bar{y} - \mu_0)}{s_y}$$

and comparing the calculated value with the distribution (19). This is an ordinary test of significance or hypothesis test as discussed in Chapters 7, 8, and 10, but in addition it is a *conditional* test given the value $\mathbf{d}(\mathbf{y})$ from the response. Thus appropriately here it uses more information, relevant information from the data.

A $1 - \alpha$ confidence interval for μ can be obtained from a $1 - \alpha$ interval for t:

$$\int_{t_1}^{t_2} g_\lambda(t:\mathbf{d})\,dt = 1 - \alpha;$$

the $1 - \alpha$ confidence interval for μ is

(21) $$(\bar{y} - t_2 s_y/\sqrt{n},\ \bar{y} - t_1 s_y/\sqrt{n}).$$

This is an ordinary $1 - \alpha$ confidence interval as discussed in Chapters 7 and 8, but in addition it is a *conditional* confidence interval given the value $\mathbf{d}(\mathbf{y})$ from the response; note that $E(1 - \alpha : \mathbf{d}) = 1 - \alpha$. Thus it uses more of the relevant information from the data.

G INFERENCE CONCERNING σ

For inferences concerning σ we can rewrite the second equation in (17):

$$(22) \qquad \frac{s(\mathbf{y})}{\sigma} = s \quad \text{or} \quad \frac{s_y}{\sigma} = \frac{s}{\sqrt{n-1}} = s_z.$$

The marginal distribution of this standard deviation s_z can be obtained from (16) by integrating out $\bar{z}$:

$$(23) \qquad h_\lambda(s_z : \mathbf{d})\, ds_z = k^{-1}(\mathbf{d}) \int_{-\infty}^{\infty} \prod f_\lambda(\bar{z} + \sqrt{n-1}\, s_z d_i) \sqrt{n}\, d\bar{z} (n-1)^{(n-1)/2} s_z^{n-2}\, ds_z.$$

A hypothesis $\sigma = \sigma_0$ can be tested by calculating

$$(24) \qquad s_z = \frac{s_y}{\sigma_0}$$

and comparing the calculated value with the distribution (23). This is an ordinary test of significance, but in addition is *conditional* given $\mathbf{d}(\mathbf{y})$ and thus uses more information from the data.

A $1 - \alpha$ confidence interval for σ can be obtained from a $1 - \alpha$ interval for s_z:

$$(25) \qquad \int_{s_1}^{s_2} h_\lambda(s_z : \mathbf{d})\, ds_z = 1 - \alpha;$$

the $1 - \alpha$ confidence interval for σ is

$$(26) \qquad (s_y/s_2,\ s_y/s_1).$$

This is an ordinary $1 - \alpha$ confidence interval, but in addition is *conditional* and thus uses more information from the data.

H EXAMPLE

Serum potassium determinations have been found to have variation with substantially more probability in the tails than the normal; a Student(λ) distribution with λ in the range 2 to 6 is indicated as an appropriate approximation. For a particular serum the following potassium determinations were obtained:

4.6	4.9	5.1	5.2	5.2
5.2	5.2	5.3	5.3	5.4
5.4	5.4	5.4	5.4	5.4
5.4	5.4	5.4	5.5	5.5
5.5	5.5	5.5	5.6	5.6
5.6	5.7	5.7	5.8	5.8

and we have

$$\bar{y} = 5.3967, \qquad s(\mathbf{y}) = 1.3526, \qquad s_y = 0.25118.$$

Let μ designate the median response and σ designate the standard error such that $(\mu \pm \sigma)$ contains 68.26 percent of the probability. For this we standardize the distribution f_λ for variation so that the median is zero and the interval $(-1, +1)$ contains 68.26 percent of the probability. As a distribution for variation we use the Student(λ)

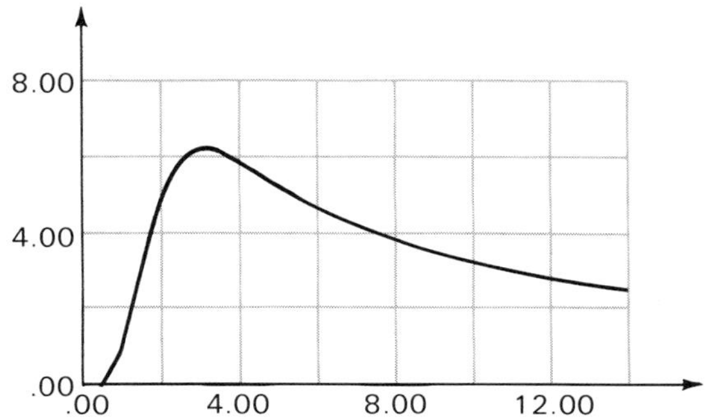

FIGURE 11.3
Likelihood function $L(\mathbf{d}|\lambda)$.

distribution with λ in the range $[1, \infty]$, which goes from the Cauchy to the normal. The Student distribution has median 0 but it has more probability in the interval $(-1, +1)$ than the 68.26 percent of the standard normal. Accordingly, let f_λ be the standardized Student(λ) distribution, that is, the Student(λ) distribution rescaled to have 68.26 percent probability in the interval $(-1, +1)$.

For the 30 determinations we have

$$5.3967 = \mu + \sigma\bar{z},$$

$$1.3526 = \sigma s,$$

where $\bar{z}$ and s give the position of the realized variation $\mathbf{z}$ on the half-plane $\mathcal{L}^+(\mathbf{1}; \mathbf{y})$ in R^{30}.

The likelihood function (15) for λ is recorded in Figure 11.3; it was obtained by the computer integration (10) over the upper half-plane. The values of the likelihood function are recorded relative to the value 1 for $\lambda = \infty$.

The likelihood function has maximum likelihood estimate near $\lambda = 3$ and is not sharply discriminating at that point. A related set of 35 determinations had a likelihood function very sharply peaked near $\lambda = 2$. The combined likelihood function has maximum likelihood between 2 and 3 and indicates a range from approximately 1 to 5.

The density (19) for the ordinary t-function is plotted in Figure 11.4 for the values of $\lambda = 1, 3, 5$, and ∞.

Consider now a 95 percent confidence interval for the median response μ. With $\lambda = 3$ the central 95 percent probability interval for t as obtained by computer integration is

$$(t_1, t_2) = (-2.209, 1.126).$$

Thus the 95 percent confidence interval for μ is

$$(\bar{y} - t_2 s_y/\sqrt{n}, \bar{y} - t_1 s_y/\sqrt{n}) = (5.345, 5.498).$$

Note that the asymmetry in the interval for t represents the expected bias for the particular spacing of the 30 observations (several extreme values on the left-hand tail); the calculated confidence interval has this bias removed from the raw estimate 5.40.

More generally, we record intervals for t and μ for $\lambda = 1, 3, 6, 9, \infty$, as shown in Table 11.1.

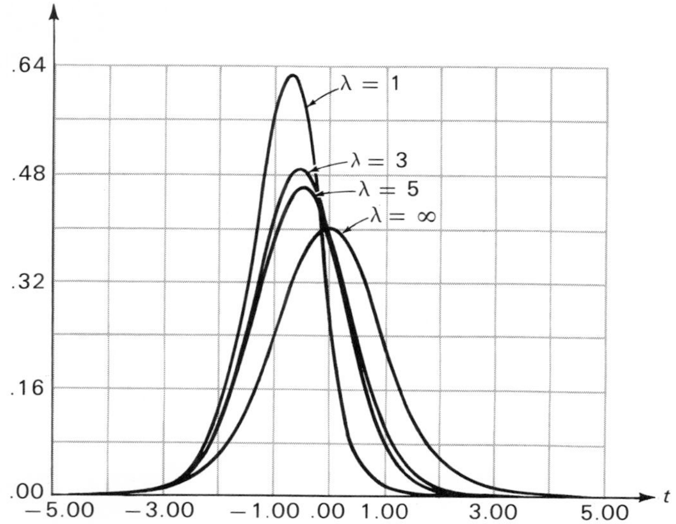

FIGURE 11.4
Density function (19) for the ordinary t-function is plotted for $\lambda = 1, 3, 5$, and ∞.

TABLE 11.1
Intervals for t and μ

λ	Percentage	t_1	t_2	μ_1	μ_2
1	90	−1.966	0.236	5.386	5.487
	95	−2.220	0.451	5.376	5.498
	99	−2.750	0.920	5.354	5.523
	99.9	−3.437	1.549	5.326	5.554
3	90	−1.930	0.837	5.358	5.485
	95	−2.209	1.126	5.345	5.498
	99	−2.780	1.722	5.318	5.524
	99.9	−3.499	2.482	5.283	5.557
6	90	−1.921	1.064	5.348	5.485
	95	−2.218	1.376	5.334	5.498
	99	−2.824	2.021	5.304	5.526
	99.9	−3.589	2.844	5.266	5.561
9	90	−1.893	1.194	5.342	5.483
	95	−2.199	1.517	5.327	5.498
	99	−2.827	2.184	5.297	5.526
	99.9	−3.620	3.033	5.258	5.563
∞	90	−1.699	1.699	5.319	5.475
	95	−2.045	2.045	5.303	5.490
	99	−2.756	2.756	5.270	5.523
	99.9	−3.659	3.659	5.229	5.564

The density (23) for s_z obtained by computer integration is plotted in Figure 11.5 for $\lambda = 1, 3, 5$, and ∞. With $\lambda = 3$ the central 95 percent confidence interval for s_z is (0.845, 1.768); the corresponding 95 percent confidence interval for σ is (0.142, 0.297).

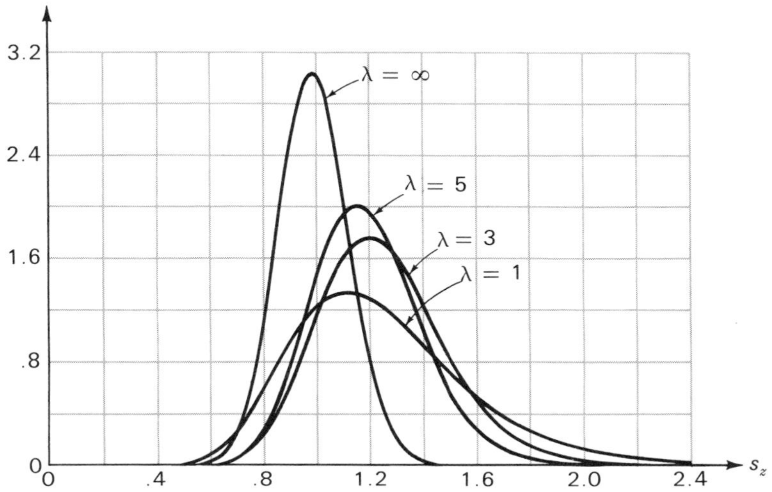

FIGURE 11.5
Density function (23) for the standard deviation s_z is plotted for $\lambda = 1, 3, 5$, and ∞.

ADDITIONAL EXAMPLES

The methods discussed in this section work for any distribution f that can be presented analytically or numerically to a computer for integration. Consider, however, two examples involving simple functions f.

EXAMPLE 1

Uniform variation: Consider a distribution for variation that is uniform(0, 1). Note then that $y = \mu + \sigma z$ presents μ as the lower limit for the response and σ as the range for the response. Let $\mathbf{y}$ be an observed response sequence. The conditional distribution of $(\bar{z}, s)$ given $\mathbf{d(s)} = \mathbf{d}$ is

$$
\begin{aligned}
k^{-1}(\mathbf{d}) \prod f(\bar{z} + sd_i)s^{n-2}\, ds \sqrt{n}\, d\bar{z} \\
&= cs^{n-2}\, ds\, d\bar{z} \qquad\qquad 0 < \bar{z} + sd_i < 1 \text{ for each } i \\
&= n(n-1)(d_{(n)} - d_{(1)})^{n-1} s^{n-2}\, d\bar{z}\, ds
\end{aligned}
$$

inside the triangle formed by $(\bar{z}, s) = (0, 0)$, $(1, 0)$, $(-d_{(1)}/(d_{(n)} - d_{(1)})$, $1/(d_{(n)} - d_{(1)}))$, and zero elsewhere. The marginal distributions for t and s_z are readily available.

EXAMPLE 2

Normal variation: Consider a distribution for underlying variation that is standard normal. Let $\mathbf{y}$ be a sample of n response values obtained by a transformation

$$\mathbf{y} = \mu\mathbf{1} + \sigma\mathbf{z}$$

of a realized sample $\mathbf{z}$ from the standard normal. The conditional distribution of $(\bar{z}, s)$ given the observed $\mathbf{d}$ is

$$k^{-1}(\mathbf{d})(2\pi)^{-n/2} \exp\left\{-(1/2) \sum (\bar{z} + sd_i)^2\right\} s^{n-2}\, ds \sqrt{n}\, d\bar{z}$$

$$= c \exp\left\{-\frac{n\bar{z}^2}{2}\right\} \exp\left\{-\frac{s^2}{2}\right\} s^{n-2}\, ds \sqrt{n}\; d\bar{z}$$

$$= \frac{\sqrt{n}}{(2\pi)^{1/2}} \exp\left\{-\frac{n\bar{z}^2}{2}\right\} d\bar{z} \cdot \frac{1}{\Gamma\left(\frac{n-1}{2}\right)} \exp\left\{-\frac{s^2}{2}\right\}\left(\frac{s^2}{2}\right)^{[(n-1)/2]-1} d\,\frac{s^2}{2},$$

where we note that $\Sigma\, d_i = 0$, $\Sigma\, d_i^2 = 1$, and for integration we have used just the norming constants for the normal and gamma density functions. From the preceding expression we see that $\bar{z}$ is normal$(0, 1/\sqrt{n})$, s^2 is chi-square$(n - 1)$, and $\bar{z}$ and s are statistically independent. Note that this conditional distribution does not depend on **d** and is thus equal to the marginal distribution. This is a substantially shorter and more direct derivation than the usual derivation; compare with Section 8.2C.

The distribution of the t-function, $t = \sqrt{n}\,\bar{z}/s_z$, follows from the normal and chi distributions just discussed; it is the Student$(n - 1)$ distribution. The distribution of the scaling function s is, of course, the chi$(n - 1)$ distribution.

Thus it follows that the tests and confidence intervals for μ and σ will agree with those developed in Chapters 8, 9, and 10 by the methods rather special to the normal distribution. Recall the discussion in Section 8.5D.

J PROBLEMS

1 Let (2.5, 2.8, 2.3) be an observed sequence for the model presented in Example 2. Determine a central 95 percent confidence interval (a) for μ; (b) for σ.

*2 Consider $\mathbf{y} = \mu\mathbf{1} + \sigma\mathbf{z}$, where $\mathbf{z}$ is a sample from the uniform(0, 1): see Example 1. Show that the distribution of $T = \bar{z}/s$ is given by

$$(n-1)\left(1 + \frac{T + d_{(1)}}{d_{(n)} - d_{(1)}}\right)^{-n} \frac{dT}{d_{(n)} - d_{(1)}}$$

on the interval $(-d_{(1)}, \infty)$.

*3 (*continuation*) Show that the marginal distribution of s is given by

$$n(n-1)[(d_{(n)} - d_{(1)})^{n-2}s^{n-2} - (d_{(n)} - d_{(1)})^{n-1}s^{n-1}][d_{(n)} - d_{(1)}]\, ds$$

on the interval $(0, 1/(d_{(n)} - d_{(1)}))$.

*4 Let (2.5, 2.8, 2.3) be an observed sequence for the model presented in Example 1. From Problem 2 determine the marginal distribution for T and derive a central 95 percent confidence interval for μ.

*5 (*continuation*) From Problem 3 determine the marginal distribution for s and derive a central 95 percent confidence interval for σ.

*6 Consider the location-scale model with the following distribution for variation: $f(z) = \exp\{-z\}$ on $(0, \infty)$ and zero elsewhere. Let $(y_1, \ldots, y_n)$ be an observed sequence.
(a) Determine the distribution describing $(\bar{z}, s)$.
(b) Determine the marginal distribution for $T = \bar{z}/s$.

7 Show that the response distribution consistent with an observed **d** has probability differential

$$k_\lambda^{-1}(\mathbf{d}) \prod f_\lambda\left(\frac{\bar{y} + s(\mathbf{y})d_i - \mu}{\sigma}\right) \frac{s^{n-2}(\mathbf{y})}{\sigma^n}\, ds(\mathbf{y}) \sqrt{n}\, d\bar{y},$$

where $(y, s(\mathbf{y}))$ refers to possible responses with given **d**. *Method:* Use (8) with (11).

K COMPUTER PROBLEMS

A computer program that handles the calculations discussed in this section is available; see the Notes and References at the end of this section.

The Student(λ) distribution with $0 < \lambda \leq \infty$ provides a convenient family of distributions for variation: the family ranges from long-tailed distributions like the Cauchy with $\lambda = 1$ through to the short-tailed normal with $\lambda \to \infty$. Each distribution has median zero. The scaling, however, is more complicated. The normal has 68.26 percent probability in $(-1, +1)$. The Student(λ) distribution has 68.26 percent probability in $(-l_\lambda, +l_\lambda)$; some values of l_λ are

λ	1	3	6	10	15	25	∞
l_λ	1.8367	1.1966	1.0903	1.0524	1.0343	1.0202	1.000

We shall speak of the standardized Student(λ) distribution if the Student(λ) distribution is rescaled so that $(-1, +1)$ contains 68.26 percent probability. Use the standardized Student family as the model describing variation.

For each of the following problems calculate the likelihood function for λ and the density functions for t and s_z. Form appropriate confidence intervals for the parameter μ.

8 The British statistician William S. Gosset, who published under the pseudonym "Student," examined in 1908 the following data, which compared the additional hours of sleep gained by using two soporific drugs. The following data represent the difference in additional hours gained from 10 patients given the two drugs at different times.

1.2, 2.4, 1.3, 1.3, 0.0, 1.0, 1.8, 0.8, 4.6, 1.4

Report 90 percent confidence intervals for μ.

9 In 1876, Charles Darwin examined the effects of cross- and self-fertilization in plants. The data from his experiment are given here as the differences (in eights of an inch) between cross- and self-fertilized plants of the same pair:

49	−67	8	16	6	23	28	41
14	29	56	24	75	60	−48	

Report 90 percent confidence intervals for μ.

10 The following 35 determinations on serum potassium were obtained:

5.20	5.20	5.20	5.20	5.20
5.20	5.20	5.10	5.10	5.10
5.00	5.00	5.00	5.00	5.00
4.90	6.60	6.10	6.10	5.30
5.30	5.30	5.30	5.20	4.90
4.90	4.80	4.80	4.80	4.80
4.80	4.70	4.70	4.60	4.50

Report 95 percent confidence intervals for μ.

11 The weight of coating, in hundredths of an ounce per square foot, is measured for 30 sheets.

160	153	158	155	160
159	148	153	138	150
160	138	173	160	148
137	170	154	147	164
160	159	162	146	160
149	152	151	155	177

Report 95 percent confidence intervals for μ. What do the results indicate about the choice of model for these data?

11.2

CALCULATING PROJECTIONS AND RESIDUALS

We now examine some convenient procedures for calculating projections and corresponding squared lengths. This is supplemental material for the least-squares method in Section 9.2. And it is basic material for the routine calculations used with the linear model in the present chapter.

In Section 8.2A we calculated the projection of a vector **y** onto each axis of a new orthogonal set of axes. We saw that the projections add to reconstruct the vector **y** and the squared lengths add to reconstruct the squared length of **y**.

In Section 9.2B we calculated the projection of a vector **y** onto a subspace generated by the r independent column vectors in a matrix X and we noted that the residual vector was the projection onto the orthogonal complement $\mathcal{L}^{\perp}(X)$. We saw that the projections add to reconstruct the vector **y** and the squared lengths add to reconstruct the squared length of **y**.

In this section we examine these calculations for a succession of orthogonal subspaces.

A PROJECTIONS AND SQUARED LENGTHS BY DIFFERENCING

Formulas are given in Section 9.2B for calculating the projection into a subspace of R^n and the squared length of the projection. First, consider an s-dimensional subspace $\mathcal{L}(X_1)$ generated by the s linearly independent column vectors of the matrix

$$X_1 = (\mathbf{x}_1 \cdots \mathbf{x}_s).$$

The projection of **y** and the squared length of the projection are

$$(1)\qquad \mathbf{p}_{(1)} = X_1\mathbf{b}_1, \qquad SS_{(1)} = \mathbf{y}'X_1\mathbf{b}_1,$$

where $\mathbf{b}_1 = (X_1'X_1)^{-1}X_1'\mathbf{y}$ gives the regression coefficients on the column vectors of X_1.

Now consider an r-dimensional subspace $\mathcal{L}(X)$ where $r > s$ and where $\mathcal{L}(X)$ contains the preceding subspace $\mathcal{L}(X_1)$. The column vectors of X could all be different from those of X_1; *or* they could include the vectors of X_1 and have $r - s$ additional vectors,

$$(2)\qquad X = (\mathbf{x}_1 \cdots \mathbf{x}_s\mathbf{x}_{s+1} \cdots \mathbf{x}_r) = (X_1\ X_2);$$

or in particular the vectors of X_2 could each be orthogonal to the vectors of X_1. The projection of **y** and the squared length of the projection are

$$(3)\qquad \mathbf{p}_{(2)} = X\mathbf{b}, \qquad SS_{(2)} = \mathbf{y}'X\mathbf{b},$$

where $\mathbf{b} = (X'X)^{-1}X'\mathbf{y}$ gives the regression coefficients on the column vectors of X.

Let $\mathcal{L}(X \perp X_1)$ be the orthogonal complement of $\mathcal{L}(X_1)$ in $\mathcal{L}(X)$; that is, it consists of all vectors **y** in $\mathcal{L}(X)$ that are orthogonal to X_1:

$$\mathcal{L}(X \perp X_1) = \{\mathbf{y} : \mathbf{y} \in \mathcal{L}(X);\ \mathbf{y} \perp \mathbf{x}_1, \ldots, \mathbf{x}_s\}.$$

In particular, if the vectors of X_2 in (2) are orthogonal to those of X_1, then $\mathcal{L}(X \perp X_1) = \mathcal{L}(X_2)$.

We now consider the $(r - s)$-dimensional subspace $\mathcal{L}(X \perp X_1)$ and show that the projection of $\mathbf{y}$ and the squared length of the projection are

(4) $$\mathbf{p}_2 = \mathbf{p}_{(2)} - \mathbf{p}_{(1)}, \qquad SS_2 = SS_{(2)} - SS_{(1)}$$

and can thus be obtained by differencing. We summarize this as shown in Table 11.2.

TABLE 11.2
Projections and corresponding squared lengths

Subspace	Dimension	Projection	Squared Length
$\mathcal{L}(X_1)$	s	$X_1\mathbf{b}_1$	$\mathbf{y}'X_1\mathbf{b}_1$
$\mathcal{L}(X \perp X_1)$	$r - s$	$X\mathbf{b} - X_1\mathbf{b}_1$	$\mathbf{y}'X\mathbf{b} - \mathbf{y}'X_1\mathbf{b}_1$
$\mathcal{L}^{\perp}(X)$	$n - r$	$\mathbf{y} - X\mathbf{b}$	$\mathbf{y}'\mathbf{y} - \mathbf{y}'X\mathbf{b}$
	n	$\mathbf{y}$	$\sum y_i^2$

The proof is straightforward. The projection and squared length of the projection are determined by a subspace and of course do not depend on the particular basis vectors for the subspace. For the proof, then, it is convenient to have the vectors of X_2 in (2) orthogonal to the vectors of X_1; that is, $X_2'X_1 = 0$, the zero matrix. Thus for the subspace $\mathcal{L}(X)$ we have

$$(X'X) = \begin{pmatrix} X_1'X_1 & 0 \\ 0 & X_2'X_2 \end{pmatrix}, \qquad (X'X)^{-1} = \begin{pmatrix} (X_1'X_1)^{-1} & 0 \\ 0 & (X_2'X_2)^{-1} \end{pmatrix},$$

$$\mathbf{b} = (X'X)^{-1}X'\mathbf{y} = \begin{pmatrix} (X_1'X_1)^{-1} & 0 \\ 0 & (X_2'X_2)^{-1} \end{pmatrix} \begin{pmatrix} X_1'\mathbf{y} \\ X_2'\mathbf{y} \end{pmatrix} = \begin{pmatrix} \mathbf{b}_1 \\ \mathbf{b}_2 \end{pmatrix},$$

where $\mathbf{b}_2 = (X_2'X_2)^{-1}X_2'\mathbf{y}$ gives the regression coefficients on the column vectors of X_2. Thus the projection and squared length for $\mathcal{L}(X)$ can be written

$$\mathbf{p}_{(2)} = X_1\mathbf{b}_1 + X_2\mathbf{b}_2, \qquad SS_{(2)} = \mathbf{y}'X_1\mathbf{b}_1 + \mathbf{y}'X_2\mathbf{b}_2,$$

and it follows that the projection $X_2\mathbf{b}_2$ and squared length $\mathbf{y}'X_2\mathbf{b}_2$ for the space $\mathcal{L}(X \perp X_1) = \mathcal{L}(X_2)$ are obtained by differencing. Of course, for differencing *we do not need* the special form for the vectors of X_2.

B INNER PRODUCTS OF RESIDUALS

Consider r linearly independent vectors $X = (\mathbf{x}_1 \cdots \mathbf{x}_r)$ and an additional set of s vectors $Y = (\mathbf{y}_1 \cdots \mathbf{y}_s)$. The inner product matrix for the $r + s$ vectors is

$$S = \begin{pmatrix} X' \\ Y' \end{pmatrix} (X\ Y) = \begin{pmatrix} X'X & X'Y \\ Y'X & Y'Y \end{pmatrix}.$$

Suppose that we operate on the first r rows of the matrix S: multiplying by real numbers and adding and subtracting rows until the first r rows and r columns become the identity matrix. We are then effectively performing the following premultiplication:

(5) $$\begin{pmatrix} (X'X)^{-1} & 0 \\ 0 & I \end{pmatrix} S = \begin{pmatrix} I & \mathbf{b}(\mathbf{y}_1) \cdots \mathbf{b}(\mathbf{y}_s) \\ Y'X & Y'Y \end{pmatrix} = \begin{pmatrix} I & B(Y) \\ Y'X & Y'Y \end{pmatrix},$$

where $\mathbf{b}(\mathbf{y}_j) = (X'X)^{-1}X'\mathbf{y}_j$ is the vector of regression coefficients of $\mathbf{y}_j$ on $\mathbf{x}_1, \ldots, \mathbf{x}_r$;

thus the upper right matrix

$$B(Y) = (X'X)^{-1}X'Y = (\mathbf{b}(\mathbf{y}_1) \cdots \mathbf{b}(\mathbf{y}_s))$$

contains the various **b** vectors obtained from $\mathbf{y}_1, \ldots, \mathbf{y}_s$ successively.

Now suppose that we take multiples of the first r rows and subtract them from the last s rows to produce zeros in the first r columns of those rows. We are then effectively performing the following matrix premultiplication:

$$(6) \qquad \begin{pmatrix} I & 0 \\ -Y'X & I \end{pmatrix} \begin{pmatrix} (X'X)^{-1} & 0 \\ 0 & I \end{pmatrix} S = \begin{pmatrix} I & B(Y) \\ 0 & Y'Y - Y'XB(Y) \end{pmatrix}.$$

We now show the interesting result that the bottom right $s \times s$ matrix is the inner product matrix of the residual vectors $\mathbf{y}_j - X\mathbf{b}(\mathbf{y}_j)$. We first assemble the residual vectors in an $n \times s$ matrix

$$(\mathbf{y}_1 - X\mathbf{b}(\mathbf{y}_1), \ldots, \mathbf{y}_s - X\mathbf{b}(\mathbf{y}_s)) = Y - XB(Y);$$

we then calculate the inner product matrix

$$(7) \qquad (Y - XB(Y))'(Y - XB(Y))$$
$$= Y'Y - Y'XB(Y) - B'(Y)X'Y + B'(Y)X'XB(Y)$$
$$= Y'Y - Y'XB(Y).$$

Note that the last three terms are simplified by substituting the expression for $B(Y)$.

Thus *the two left multiplications on the matrix S give the coefficients for the projection vectors and give the inner product matrix for the residual vectors.*

C EXAMPLE

Consider a biological system with an input variable x and a response variable y. The following five observations were obtained:

x	1	2	3	4	5
y	5	7	11	10	13

We first examine the trivial least-squares fit of the constant

$$(8) \qquad y = b \cdot 1.$$

We then examine the least-squares fit of an affine function of x:

$$(9) \qquad y = b_1 1 + b_2 x.$$

We then examine the least-squares fit of a quadratic function of x:

$$(10) \qquad y = b_1 1 + b_2 x + b_3 x^2.$$

For the calculations we use the methods and results discussed earlier in Sections A and B.

Let $\mathbf{x}_1$, $\mathbf{x}_2$, and $\mathbf{x}_3$ be the vectors recording the values of 1, x, and x^2, respectively. Some preliminary calculations are given in Table 11.3. The inner product matrix of $\mathbf{x}_1$,

TABLE 11.3
Calculations for some variables

1	x	x^2	x^3	x^4	y	xy	x^2y
1	1	1	1	1	5	5	5
1	2	4	8	16	7	14	28
1	3	9	27	81	11	33	99
1	4	16	64	256	10	40	160
1	5	25	125	625	13	65	325
5	15	55	225	979	46	157	617

$\mathbf{x}_2$, and $\mathbf{x}_3$ with $\mathbf{x}_1$, $\mathbf{x}_2$, $\mathbf{x}_3$, and $\mathbf{y}$ is recorded in the following array. An identity matrix is adjoined on the right; it will absorb and thus keep a record of any left matrix multiplication. The sum of squares of the y's is $\mathbf{y}'\mathbf{y} = \Sigma y_i^2 = 464$.

(11)

$\mathbf{x}_1$	$\mathbf{x}_2$	$\mathbf{x}_3$	$\mathbf{y}$	$\mathbf{x}_1$	$\mathbf{x}_2$	$\mathbf{x}_3$
5	15	55	46	1	0	0
15	55	225	157	0	1	0
55	225	979	617	0	0	1

D FITTING A CONSTANT

First consider the least-squares fit of the constant as represented by equation (8). This corresponds to finding the projection in the subspace generated by the 1-vector $\mathbf{x}_1$ ($= \mathbf{w}_1$ for later reference) or, equivalently, generated by the corresponding unit vector

$$\mathbf{v}_1 = \mathbf{x}_1/\sqrt{5} = (1, 1, 1, 1, 1)'/\sqrt{5};$$

the squared length 5 is the one–one element in the first array (11). In the pattern of Section B we perform row operations (left multiplication by a matrix) on (11) to reduce the first row–first column to 1 and then reduce the remainder of the first column to 0's:

(12)

$\mathbf{x}_1$	$\mathbf{x}_2$	$\mathbf{x}_3$	$\mathbf{y}$	$\mathbf{x}_1$	$\mathbf{x}_2$	$\mathbf{x}_3$
1	3	11	9.2	0.2	0	0
0	10	60	19	−3	1	0
0	60	374	111	−11	0	1

The second, third, and fourth columns contain regression coefficients on $\mathbf{x}_1$ above the dashed line and contain inner products of residual vectors below the dashed line. The right-hand array contains the matrix that has been left-multiplied into (11).

Consider the projection of $\mathbf{y}$ on the vector $\mathbf{x}_1$ and the squared length of the projection. From formulas (1) and (5) we have

$$\mathbf{p}_1 = 9.2\mathbf{x}_1, \qquad SS_1 = 46 \times 9.2 = 423.2.$$

We can summarize this in Table 11.4.

TABLE 11.4
Projections and corresponding squared lengths

Space	Basis	Dimension	Projection	Squared Length
$\mathcal{L}(\mathbf{x}_1)$	$\mathbf{v}_1$	1	$9.2\mathbf{x}_1$	423.2
$\mathcal{L}^{\perp}(\mathbf{x}_1)$		4	$\mathbf{y} - 9.2\mathbf{x}_1$	40.8
		5	$\mathbf{y}$	464.0

E FITTING AN AFFINE FUNCTION

Now consider the least-squares fit of the affine function represented by equation (9). This corresponds to finding the projection in the subspace generated by the vectors $\mathbf{x}_1$ and $\mathbf{x}_2$. We examine this in part as an extension from the preceding simpler case; as an extension we are concerned with $\mathcal{L}(\mathbf{x}_2 \perp \mathbf{x}_1)$.

The array (12) gives the coefficients for the projection of $\mathbf{x}_2$ on $\mathbf{x}_1$; we thus obtain the residual

$$\mathbf{w}_2 = \mathbf{x}_2 - 3\mathbf{x}_1 = (-2, -1, 0, 1, 2)'.$$

The array (12) also gives the squared length of this residual: the squared length is 10. The corresponding unit vector is

$$\mathbf{v}_2 = (\mathbf{x}_2 - 3\mathbf{x}_1)/\sqrt{10} = (-2, -1, 0, 1, 2)'/\sqrt{10}.$$

Note that $\mathbf{w}_2$ or $\mathbf{v}_2$ generates the orthogonal complement $\mathcal{L}(\mathbf{x}_2 \perp \mathbf{x}_1)$. Also note that the vector $\mathbf{w}_2$ records the values of the function $x - 3$, which is orthogonal to the function 1 at the points $\{1, 2, 3, 4, 5\}$. Now in the pattern of Section B we perform row operations (left multiplication by a matrix) on (11) to reduce the first two rows–first two columns to the identity matrix and the remainder of the first two columns to 0's. Note that this is equivalent to reducing the second row–second column element of (12) to 1 and the remainder of the second column to 0's:

(13)

$\mathbf{x}_1$	$\mathbf{x}_2$	$\mathbf{x}_3$	$\mathbf{y}$	$\mathbf{x}_1$	$\mathbf{x}_2$	$\mathbf{x}_3$
1	0	−7	3.5	1.1	−0.3	0
0	1	6	1.9	−0.3	0.1	0
0	0	14	−3	7	−6	1

The third and fourth columns contain regression coefficients on $\mathbf{x}_1$ and $\mathbf{x}_2$ above the dashed line and contain inner products of the residual vectors below the dashed line. The right-hand array contains the matrix that has been left-multiplied into (11).

Consider the projection of $\mathbf{y}$ on the vectors $\mathbf{x}_1$ and $\mathbf{x}_2$ and the squared length of the projection. From formulas (3) and (5) we have

$$\mathbf{p}_{(2)} = 3.5\mathbf{x}_1 + 1.9\mathbf{x}_2, \qquad SS_{(2)} = 46 \times 3.5 + 157 \times 1.9 = 459.3.$$

The projection of $\mathbf{y}$ on the orthogonal complement $\mathcal{L}(\mathbf{x}_2 \perp \mathbf{x}_1)$ is the projection of $\mathbf{y}$ on $\mathbf{w}_2$ or on $\mathbf{v}_2$; it is available using the step from (12) to (13),

$$\mathbf{p}_2 = 1.9\mathbf{w}_2, \qquad SS_2 = 19 \times 1.9 = 36.1.$$

We can then summarize in Table 11.5.

TABLE 11.5
Projections and corresponding squared lengths

Space	Basis	Dimension	Projection	Squared Length
$\mathcal{L}(\mathbf{x}_1)$	$\mathbf{v}_1$	1	$9.2\mathbf{w}_1$	423.2
$\mathcal{L}(\mathbf{x}_2 \perp \mathbf{x}_1)$	$\mathbf{v}_2$	1	$1.9\mathbf{w}_2$	36.1
$\mathcal{L}^{\perp}(\mathbf{x}_1, \mathbf{x}_2)$		3	$\mathbf{y} - 3.5\mathbf{x}_1 - 1.9\mathbf{x}_2$	4.7
		5	$\mathbf{y}$	464.0

F FITTING A QUADRATIC FUNCTION OF x

Now consider the least-squares fit of the quadratic function represented by equation (10). This corresponds to finding the projection in the subspace generated by $\mathbf{x}_1$, $\mathbf{x}_2$, and $\mathbf{x}_3$. We examine this in part as an extension from the preceding case; as an extension we are concerned with $\mathcal{L}(\mathbf{x}_3 \perp \mathbf{x}_1, \mathbf{x}_2)$.

The array (13) gives the coefficients for the projection of $\mathbf{x}_3$ on $\mathbf{x}_1$ and $\mathbf{x}_2$; we thus obtain the residual

$$\mathbf{w}_3 = \mathbf{x}_3 - (-7)\mathbf{x}_1 - 6\mathbf{x}_2 = (2, -1, -2, -1, 2)'.$$

The array (13) also gives the squared length of this residual: the squared length is 14. The corresponding vector is

$$\mathbf{v}_3 = (\mathbf{x}_3 + 7\mathbf{x}_1 - 6\mathbf{x}_2)/\sqrt{14} = (2, -1, -2, -1, 2)'/\sqrt{14}.$$

Note that $\mathbf{w}_3$ or $\mathbf{v}_3$ generates the orthogonal complement $\mathcal{L}(\mathbf{x}_3 \perp \mathbf{x}_1, \mathbf{x}_2)$. Also note that the vector $\mathbf{w}_3$ records the values of the function $x^2 + 7 - 6x$, which is orthogonal to the functions 1 and x at the points $\{1, 2, 3, 4, 5\}$.

Now in the pattern of Section B we perform row operations on (11) to reduce the first three rows–first three columns to the identity matrix and the remainder (here none) of the first three columns to 0's. Note that this is equivalent to reducing the third row–third column element of (13) to 1 and the remainder of the third column to 0's:

(14)

$\mathbf{x}_1$	$\mathbf{x}_2$	$\mathbf{x}_3$	$\mathbf{y}$	$\mathbf{x}_1$	$\mathbf{x}_2$	$\mathbf{x}_3$
1	0	0	2	4.6	-3.3	1/2
0	1	0	$3\frac{13}{70}$	-3.3	$2\frac{47}{70}$	$-3/7$
0	0	1	$-3/14$	1/2	$-3/7$	1/14

The fourth column contains regression coefficients of $\mathbf{y}$ on $\mathbf{x}_1$, $\mathbf{x}_2$, and $\mathbf{x}_3$. The right-hand array contains the matrix that has been left-multiplied into (11).

Consider the projection of $\mathbf{y}$ on the vectors $\mathbf{x}_1$, $\mathbf{x}_2$, and $\mathbf{x}_3$ and the squared length of the projection. From formulas (3) and (5) we have

$$\mathbf{p}_{(3)} = 2\mathbf{x}_1 + 3\tfrac{13}{70}\mathbf{x}_2 - \tfrac{3}{14}\mathbf{x}_3,$$

$$\begin{aligned} SS_{(3)} &= 46 \times 2 + 157 \times 3\tfrac{13}{70} + 617 \times (-\tfrac{3}{14}) \\ &= 459.96. \end{aligned}$$

The projection of $\mathbf{y}$ on the orthogonal complement $\mathcal{L}(\mathbf{x}_3 \perp \mathbf{x}_1, \mathbf{x}_2)$ is the projection of $\mathbf{y}$ on $\mathbf{w}_3$ or on $\mathbf{v}_3$; it is available using the step from (13) to (14):

$$\mathbf{p}_3 = -(3/14)\mathbf{w}_3, \qquad SS_3 = -3(-3/14) = 9/14 = 0.64.$$

We can then summarize in Table 11.6.

TABLE 11.6
Projections and corresponding squared lengths

Space	Basis	Dimension	Projection	Squared Length
$\mathcal{L}(\mathbf{x}_1)$	$\mathbf{v}_1$	1	$9.2\mathbf{w}_1$	423.2
$\mathcal{L}(\mathbf{x}_2 \perp \mathbf{x}_1)$	$\mathbf{v}_2$	1	$1.9\mathbf{w}_2$	36.1
$\mathcal{L}(\mathbf{x}_3 \perp \mathbf{x}_1, \mathbf{x}_2)$	$\mathbf{v}_3$	1	$-\frac{3}{14}\mathbf{w}_3$	0.64
$\mathcal{L}^{\perp}(\mathbf{x}_1, \mathbf{x}_2, \mathbf{x}_3)$		2	$\mathbf{y} - 2\mathbf{x}_1 - 3\frac{13}{70}\mathbf{x}_2 + \frac{3}{14}\mathbf{x}_3$	4.06
		5	$\mathbf{y}$	464.00

G SUPPLEMENT: AN ORTHONORMAL BASIS

We have been considering vectors $\mathbf{x}_1$, $\mathbf{x}_2$, and $\mathbf{x}_3$ which generated successively the subspaces $\mathcal{L}(\mathbf{x}_1)$, $\mathcal{L}(\mathbf{x}_1, \mathbf{x}_2)$, and $\mathcal{L}(\mathbf{x}_1, \mathbf{x}_2, \mathbf{x}_3)$. As part of this we derived orthogonal vectors $\mathbf{w}_1$, $\mathbf{w}_2$, and $\mathbf{w}_3$ which generated successively the same three subspaces. From the second column of (12) and the third column of (13), we have obtained the coefficients for expressing the $\mathbf{w}$'s in terms of the $\mathbf{x}$'s:

$$(15) \qquad W = (\mathbf{w}_1, \mathbf{w}_2, \mathbf{w}_3) = (\mathbf{x}_1, \mathbf{x}_2, \mathbf{x}_3)\begin{pmatrix} 1 & -3 & 7 \\ 0 & 1 & -6 \\ 0 & 0 & 1 \end{pmatrix} = X\begin{pmatrix} 1 & -3 & 7 \\ 0 & 1 & -6 \\ 0 & 0 & 1 \end{pmatrix}.$$

The corresponding orthonormal vectors $\mathbf{v}_1$, $\mathbf{v}_2$, and $\mathbf{v}_3$ were obtained from the $\mathbf{w}_1$, $\mathbf{w}_2$, and $\mathbf{w}_3$ using squared lengths from (11), (12), and (13):

$$(16) \qquad V = (\mathbf{w}_1, \mathbf{w}_2, \mathbf{w}_3)\begin{pmatrix} 1/\sqrt{5} & 0 & 0 \\ 0 & 1/\sqrt{10} & 0 \\ 0 & 0 & 1/\sqrt{14} \end{pmatrix}$$

$$= X\begin{pmatrix} 1/\sqrt{5} & -3/\sqrt{10} & 7/\sqrt{14} \\ 0 & 1/\sqrt{10} & -6/\sqrt{14} \\ 0 & 0 & 1/\sqrt{14} \end{pmatrix}.$$

We can also determine from the arrays (12) and (13) the coefficients for the projection of the $\mathbf{x}$'s on the orthogonal $\mathbf{w}$'s. From the first row of (12) and the second row of (13) we obtain the coefficients for expressing the $\mathbf{x}$'s in terms of the $\mathbf{w}$'s:

$$(17) \qquad X = (\mathbf{x}_1, \mathbf{x}_2, \mathbf{x}_3) = (\mathbf{w}_1, \mathbf{w}_2, \mathbf{w}_3)\begin{pmatrix} 1 & 3 & 11 \\ 0 & 1 & 6 \\ 0 & 0 & 1 \end{pmatrix} = W\begin{pmatrix} 1 & 3 & 11 \\ 0 & 1 & 6 \\ 0 & 0 & 1 \end{pmatrix}.$$

The vectors $\mathbf{x}_1$, $\mathbf{x}_2$, $\mathbf{x}_3$ can also be expressed directly in terms of the orthonormal vectors $\mathbf{v}_1$, $\mathbf{v}_2$, and $\mathbf{v}_3$:

$$(18)\qquad X = W\begin{pmatrix}1 & 3 & 11\\ 0 & 1 & 6\\ 0 & 0 & 1\end{pmatrix} = V\begin{pmatrix}\sqrt{5} & 0 & 0\\ 0 & \sqrt{10} & 0\\ 0 & 0 & \sqrt{14}\end{pmatrix}\begin{pmatrix}1 & 3 & 11\\ 0 & 1 & 6\\ 0 & 0 & 1\end{pmatrix}$$

$$= (\mathbf{v}_1, \mathbf{v}_2, \mathbf{v}_3)\begin{pmatrix}\sqrt{5} & 3\sqrt{5} & 11\sqrt{5}\\ 0 & \sqrt{10} & 6\sqrt{10}\\ 0 & 0 & \sqrt{14}\end{pmatrix} = VT,$$

where T is positive upper triangular (upper triangular with positive values on the diagonal).

The right-hand triangular matrices in (16) and (18) are inverses of each other.

H EXERCISES

A large number of significant figures can be needed to ensure accuracy for the regression coefficients and inner products. In general, fractions as illustrated in this section should be avoided; hand, desk, and large-scale computers make the necessary arithmetic accessible.

1 For an input variable x and response y the following data were obtained:

x	1.6	1.8	1.9	1.9	2.1	2.2
y	0.5	0.4	0.7	0.9	1.2	1.1

First fit the constant $y = b \cdot 1$ and then fit the affine function $y = b_1 \cdot 1 + b_2 x$. Record the projections and summary tables as in Sections D and E.

2 For an input variable x and response y the following data were obtained:

x	2.6	2.7	2.8	2.9	3.1
y	12.1	12.5	12.7	13.0	13.5

First fit the constant, then the affine function, then the quadratic function. Record the projections and summary tables as in Sections D, E, and F. Carry sufficient figures in calculations.

3 (*continuation*) Let $X = (\mathbf{x}_1, \mathbf{x}_2, \mathbf{x}_3)$, where $\mathbf{x}_1$, $\mathbf{x}_2$, and $\mathbf{x}_3$ record the values of 1, x, and x^2, respectively; and let $W = (\mathbf{w}_1, \mathbf{w}_2, \mathbf{w}_3)$ be the corresponding orthogonal vectors, and $V = (\mathbf{v}_1, \mathbf{v}_2, \mathbf{v}_3)$ be the corresponding orthonormal vectors. Determine the matrices T and V as in the decomposition (18).

4 The response SCI y was observed corresponding to input variables saturation x_1 and transisomers x_2:

x_1	1.0	1.1	1.2	1.3	1.0	1.1	1.2	1.3	1.0	1.1	1.2	1.3
x_2	0.04	0.04	0.04	0.04	0.20	0.20	0.20	0.20	0.38	0.38	0.38	0.38
y	67.1	64.0	44.3	45.1	69.8	58.5	46.3	44.1	74.5	60.7	49.1	47.6

Fit the constant $y = b1$; fit an affine function of the first input $y = b_1 1 + b_2 x_1$; fit an affine function of the two inputs $y = b_1 1 + b_2 x_1 + b_3 x_2$. Record the projections and summary tables as in Sections D, E, and F.

11.3

THE REGRESSION MODEL

We now examine the linear or *regression model,* in which the general level of the response vector is linear in terms of location parameters. For a variety of examples see Section 9.2. These examples indicate a broad range of applications in which input variables to a system cause a change in the general level of the responses. The regression model generalizes the location-scale model of Section 11.1.

A THE REGRESSION MODEL

Consider a real-valued response y. Suppose that earlier experience with the system under investigation has identified the distribution of the variation affecting the response. Let z designate a variable presenting this variation, and f be the known density function for z. And suppose that under present conditions the response scaling of the variation is unknown and the general level of the response varies in certain unknown ways relative to input variables for the system. Let σ designate the response scaling of the variation.

Consider the standardizations for f as presented in Section 11.1. If f can be standardized so that $E(z) = 0$, var $(z) = 1$, then the general level μ of the response is presented as the mean response and the scaling σ is the response standard deviation. Or if f is standardized so that the median of z is zero and the interval $(-1, +1)$ contains 68.26 percent of the probability, then the general level μ of the response is presented as the median response and the interval $(\mu - \sigma, \mu + \sigma)$ contains 68.26 percent of the response values; in this case σ is the response standard error.

Now consider a sequence of repetitions of the system under various settings for input variables. Let $\mathbf{z} = (z_1, \ldots, z_n)'$ designate the sequence for the variation and $\mathbf{y} = (y_1, \ldots, y_n)'$ designate the sequence for the response. And suppose that the response has a linear location model

(1) $$\mathbf{y} = X\boldsymbol{\beta} + \sigma\mathbf{z}$$

as discussed in Section 9.2A, where $X = (\mathbf{x}_1 \cdots \mathbf{x}_r)$ records r linearly independent vectors based on the input variables and $\boldsymbol{\beta}$ is the location parameter taking values in R^r.

Then for the variation $\mathbf{z}$ we have the probability model as given by the density function

(2) $$f(z_1) \cdots f(z_n)$$

on R^n. And for the response we have the statistical model

(3) $$\{\mathbf{y} = X\boldsymbol{\beta} + \sigma\mathbf{z} : (\beta, \sigma) \in R^r \times R^+\},$$

giving the set of possible functions for the response $\mathbf{y}$ in terms of the variation $\mathbf{z}$.

B DATA FOR n INDEPENDENT REPETITIONS

Now consider an *observed* response sequence $\mathbf{y} = (y_1, \ldots, y_n)'$ from n independent repetitions of the system under input conditions represented by X. The observed response sequence is a function

$$\mathbf{y} = X\boldsymbol{\beta} + \sigma\mathbf{z} \tag{4}$$

of a *realized* sequence $\mathbf{z}$ from the distribution (2).

Consider the information concerning the $\mathbf{z}$ realized from the distribution (2). With no information concerning the values of $\boldsymbol{\beta}$ and σ it follows that

$$\mathbf{z} = -\sigma^{-1}\beta_1\mathbf{x}_1 - \cdots - \sigma^{-1}\beta_r\mathbf{x}_r + \sigma^{-1}\mathbf{y}$$

can be any point in the half-$(r+1)$ space

$$\mathcal{L}^+(X; \mathbf{y}) = \{a_1\mathbf{x}_1 + \cdots + a_r\mathbf{x}_r + c\mathbf{y} : a_j \in R,\ c \in R^+\} \tag{5}$$

based on the subspace $\mathcal{L}(X)$ and passing through the observed response $\mathbf{y}$; see Figure 11.6. Thus all that we effectively observe concerning the realized $\mathbf{z}$ is the half-space passing through it; we have no differential information concerning the position of $\mathbf{z}$ in that half-space.

In Section C we obtain convenient coordinates for the half-space and convenient coordinates within the half-space. Then in Section D we derive the conditional distribution describing the realized $\mathbf{z}$ in the observed half-space.

C SIMPLE COORDINATES

We now use some least-squares results to obtain simple coordinates for a half-space $\mathcal{L}^+(X; \mathbf{z})$ and simple coordinates for a point $\mathbf{z}$ in such a half-space; see Figure 11.7. The half-space $\mathcal{L}^+(X; \mathbf{z})$ is hinged on the r-dimensional subspace $\mathcal{L}(X)$. Accordingly, it is reasonable to use the r vectors in X as basis vectors. The vector $\mathbf{z}$

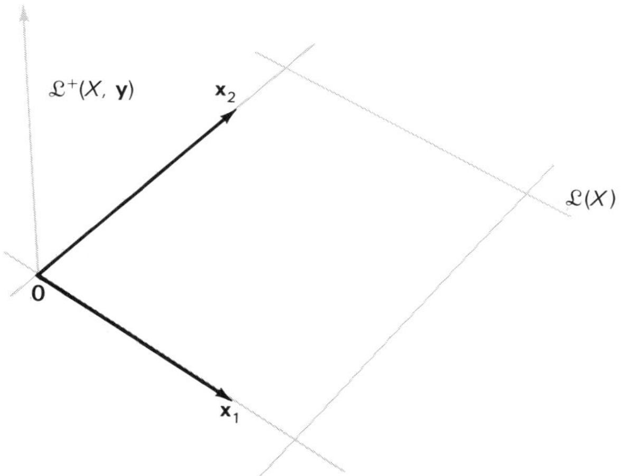

FIGURE 11.6
The r-dimensional base space $\mathcal{L}(X)$; an additional dimension by introducing the vector $\mathbf{y}$ gives the $(r+1)$-dimensional space $\mathcal{L}(X, \mathbf{y})$. $\mathcal{L}^+(X; \mathbf{y})$ is a positive half of $\mathcal{L}(X, \mathbf{y})$ obtained with a positive coordinate for $\mathbf{y}$.

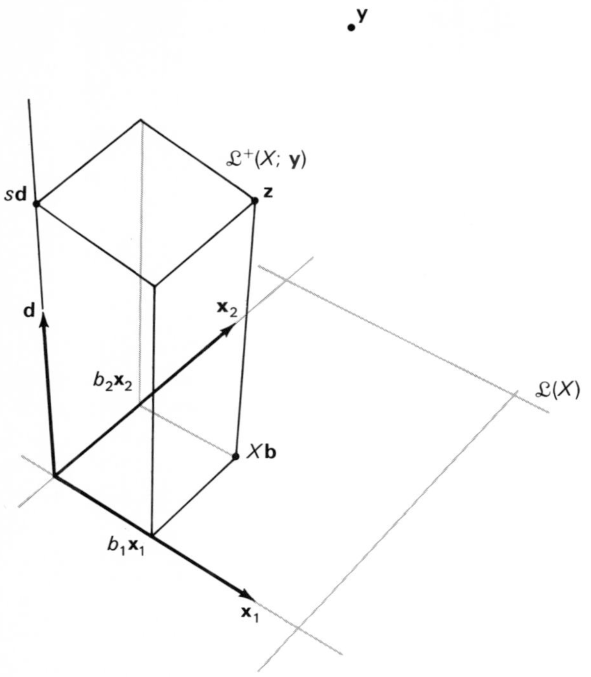

FIGURE 11.7
Coordinates $b_1, \ldots, b_r, s$ in terms of the basis $\mathbf{x}_1, \ldots, \mathbf{x}_r, \mathbf{d}$ in $\mathcal{L}^+(X; \mathbf{y})$.

has projection

$$(6) \qquad X\mathbf{b}(\mathbf{z}) = b_1(\mathbf{z})\mathbf{x}_1 + \cdots + b_r(\mathbf{z})\mathbf{x}_r$$

on the space $\mathcal{L}(X)$; the regression coefficients forming $\mathbf{b}(\mathbf{z})$ are given by formula (9.2.6). And the vector $\mathbf{z}$ has projection $\mathbf{z} - X\mathbf{b}(\mathbf{z})$ on the orthogonal complement $\mathcal{L}^\perp(X)$. Thus

$$(7) \qquad \mathbf{z} = X\mathbf{b}(\mathbf{z}) + (\mathbf{z} - X\mathbf{b}(\mathbf{z})) = X\mathbf{b}(\mathbf{z}) + s(\mathbf{z})d(\mathbf{z}),$$

where $s(\mathbf{z})$ is the length of the residual vector and $\mathbf{d}(\mathbf{z})$ is the unit residual vector. This suggests using the vectors X, $\mathbf{d}(\mathbf{z})$, that is, $\mathbf{x}_1, \ldots, \mathbf{x}_r, \mathbf{d}(\mathbf{z})$, as the $r + 1$ basis vectors for $\mathcal{L}^+(X, \mathbf{z})$; thus

$$(8) \qquad \mathbf{z} = X\mathbf{b}(\mathbf{z}) + s(\mathbf{z})\mathbf{d}(\mathbf{z}) = b_1(\mathbf{z})\mathbf{x}_1 + \cdots + b_r(\mathbf{z})\mathbf{x}_r + s(\mathbf{z})\mathbf{d}(\mathbf{z})$$

with coordinates $b_1(\mathbf{z}), \ldots, b_r(\mathbf{z}), s(\mathbf{z})$ relative to the vectors $\mathbf{x}_1, \ldots, \mathbf{x}_r, \mathbf{d}(\mathbf{z})$. Note that this is not an orthogonal basis but that the vector $\mathbf{d}(\mathbf{z})$ *is* orthogonal to the vectors of X.

The points $\mathbf{d}(\mathbf{z})$ are at unit distance from the origin and lie in the $(n - r)$-dimensional subspace $\mathcal{L}^\perp(X)$. Thus the points $\mathbf{d}(\mathbf{z})$ form the unit sphere Q_{n-r} in $\mathcal{L}^\perp(X)$.

Now consider the possible points $\mathbf{z}$ given the information provided by the observed response $\mathbf{y}$. From Section B we see that $\mathbf{z}$ and $\mathbf{y}$ must lie on the same half-space $\mathcal{L}^+(X; \mathbf{z}) = \mathcal{L}^+(X; \mathbf{y})$ hinged on $\mathcal{L}(X)$. Thus $\mathbf{d}(\mathbf{z}) = \mathbf{d}(\mathbf{y})$, which says they have the same $(r + 1)$st basis vector. For convenience we abbreviate $\mathbf{d}(\mathbf{y})$ as $\mathbf{d}$. We then see that the information provided by the observed response **y** is that $d(\mathbf{z}) = \mathbf{d}$: we have the "observed" value **d** for the function $\mathbf{d}(\mathbf{z})$.

Now consider the relation (4) between the observed response **y** and the realized **z**:

$$(9)\qquad \mathbf{y} = X\mathbf{b}(\mathbf{y}) + s(\mathbf{y})\mathbf{d} = X\boldsymbol{\beta} + \sigma\mathbf{z} = X\boldsymbol{\beta} + \sigma(X\mathbf{b}(\mathbf{z}) + s(\mathbf{z})\mathbf{d})$$
$$= X(\boldsymbol{\beta} + \sigma\mathbf{b}(\mathbf{z})) + \sigma s(\mathbf{z})\mathbf{d}.$$

By equating coefficients relative to the basis vectors $\mathbf{x}_1, \ldots, \mathbf{x}_r$, **d**, we obtain

$$(10)\qquad \begin{aligned} \mathbf{b}(\mathbf{y}) &= \boldsymbol{\beta} + \sigma\mathbf{b}, \\ s(y) &= \sigma s, \end{aligned}$$

where we abbreviate $(\mathbf{b}(\mathbf{z}), s(\mathbf{z})) = (\mathbf{b}, s)$. These can be compared with equations (11.1.8); we now have r location coordinates instead of 1.

D THE CONDITIONAL DISTRIBUTION GIVEN THE DATA

We now determine the marginal probability for what has been observed, the half-space through **z** as given by $\mathbf{d}(\mathbf{z}) = \mathbf{d}$, and we determine the conditional probability for what cannot be observed, the position $(\mathbf{b}, s)$ of **z** in the observed half-space. As in Section 11.1, we can be quite general and allow a parameter λ for the distribution $f = f_\lambda$ for the variation.

The probability differential for the distribution on R^n is given by

$$(11)\qquad \prod_1^n f_\lambda(z_i) \prod_1^n dz_i.$$

To obtain the marginal and conditional distributions we make the change of variable (8) to the new coordinates $\mathbf{b}(\mathbf{z})$, **d**. For the density function the substitution is trivial. For the differential we must argue in several steps. At the point **z** we first measure area or volume parallel to the subspace $\mathcal{L}(X)$. In the subspace we have coordinate $b_1(\mathbf{z}), \ldots, b_r(\mathbf{z})$ relative to the vectors X; suppose that we have alternative coordinates $a_1(\mathbf{z}), \ldots, a_r(\mathbf{z})$ relative to an orthonormal set of vectors V that also generate the subspace $\mathcal{L}(X)$. Let $X = VT$ express the X vectors as linear combinations of the V vectors; see (11.2.18) and note that T is $r \times r$. Then

$$(12)\qquad X = VT, \qquad X\mathbf{b} = VT\mathbf{b} = V\mathbf{a},$$

giving the relation $T\mathbf{b} = \mathbf{a}$ between the original coordinates **b** and the new coordinates **a**. Euclidean area or volume is given simply in terms of the **a** coordinates:

$$(13)\qquad da_1 \cdots da_r = d\mathbf{a} = |T|\,d\mathbf{b} = |V'V|^{1/2}|T|\,d\mathbf{b}$$
$$= |T'V'VT|^{1/2}\,d\mathbf{b} = |X'X|^{1/2}\,d\mathbf{b},$$

where of course $|V'V|$ is the determinant of an identity matrix. Thus $|X'X|^{1/2}\,d\mathbf{b}$ measures area or volume parallel to the subspace $\mathcal{L}(X)$. At **z** we can then measure distance in the orthogonal direction **d** by ds. Then in directions orthogonal to the half-space we can measure area or volume on the sphere formed by the points $s\mathbf{d}(\mathbf{z})$

with s fixed. Let da be area or volume on the unit sphere Q_{n-r} formed by the points $\mathbf{d}(\mathbf{z})$; then $s^{n-r-1}\,da$ is area or volume on the sphere of radius s. We then put together these three length or volume measures in the three orthogonal spaces and obtain

$$\prod dz_i = |X'X|^{1/2}\,d\mathbf{b}\,ds\,s^{n-r-1}\,da.$$

Thus the probability differential for the distribution on R^n can be reexpressed as

$$\prod f_\lambda\left(\sum b_u x_{ui} + sd_i\right)|X'X|^{1/2}\,d\mathbf{b}\,ds\,s^{n-r-1}\,da.$$

By integrating out the variables $\mathbf{b}$ and s over $R^r \times R^+$, we obtain the marginal density for $\mathbf{d}$:

$$(14)\qquad k_\lambda(\mathbf{d}) = \int_{-\infty}^{\infty}\int\int_0^{\infty} \prod f_\lambda\left(\sum b_u x_{ui} + sd_i\right)s^{n-r-1}\,ds|X'X|^{1/2}\,d\mathbf{b}.$$

This then gives the following factorization of the probability differential (11):

$$(15)\qquad k_\lambda(\mathbf{d})\,da \cdot k_\lambda^{-1}(\mathbf{d}) \prod f_\lambda\left(\sum b_u x_{ui} + sd_i\right) s^{n-r-1}\,ds|X'X|^{1/2}\,d\mathbf{b};$$

the probability differential is factored into the marginal probability for the observed $\mathbf{d}$ and the conditional probability for the unobserved $(\mathbf{b}, s)$. Note that the norming constant $k_\lambda(\mathbf{d})$ is obtained by integrating over the half-space $R^r \times R^+$; see formula (14).

E INFERENCE CONCERNING THE PARAMETERS

For statistical inference we follow the pattern in Section 11.1E. We have an observed $\mathbf{d}$ from a distribution involving λ. We can examine the likelihood function $L(\mathbf{d}\,|\,\cdot)$ for λ, where

$$(16)\qquad L(\mathbf{d}\,|\,\lambda) = ck_\lambda(\mathbf{d})$$

and perhaps obtain sharp discriminations concerning possible λ values. Perhaps the model $k_\lambda(\cdot)\,da$ may be accessible and of use.

Then for a given value for λ we can use the distribution

$$(17)\qquad k_\lambda^{-1}(\mathbf{d}) \prod f_\lambda\left(\sum b_u x_{ui} + sd_i\right)s^{n-r-1}\,ds|X'X|\,d\mathbf{b}$$

together with the equations (10) to make tests concerning $\beta_1, \ldots, \beta_r, \sigma$ or to calculate confidence intervals. In the remainder of this chapter we examine inference for the case of normal variation.

F DISTRIBUTIONS IN THE NORMAL CASE

Consider the case of a standard normal distribution for the variation z. Let $\mathbf{y}$ be a realized response and let $\mathbf{d} = \mathbf{d}(\mathbf{y})$ be the identified characteristic $\mathbf{d}(\mathbf{z})$ of the variation $\mathbf{z}$. Then from (17) we have the following distribution describing $b_1, \ldots, b_r, s$:

$$k^{-1}(\mathbf{d})(2\pi)^{-n/2}\exp\{-(1/2)(X\mathbf{b} + s\mathbf{d})'(X\mathbf{b} + s\mathbf{d})\}s^{n-r-1}\,ds|X'X|^{1/2}\,d\mathbf{b}$$

$$= \frac{|X'X|^{1/2}}{(2\pi)^{r/2}}\exp\{-(1/2)\mathbf{b}'(X'X)\mathbf{b}\}\,d\mathbf{b} \cdot \frac{1}{\Gamma\left(\dfrac{n-r}{2}\right)}\exp\left\{-\frac{s^2}{2}\right\}\left(\frac{s^2}{2}\right)^{[(n-r)/2]-1} d\frac{s^2}{2},$$

where we have used the orthogonality $X'\mathbf{d} = \mathbf{0}$ of the vectors $\mathbf{x}_1, \ldots, \mathbf{x}_r$ with $\mathbf{d}$, and for integration we have used just the norming constants for the normal and gamma densities. Thus we obtain the following:

LEMMA 1

Under the preceding assumption, $\mathbf{b}$ *is multivariate normal*$(\mathbf{0}; (X'X)^{-1})$; s^2 *is chi-square*$(n - r)$; $\mathbf{b}$ *is statistically independent of* s.

Note that the distribution does not depend on $\mathbf{d}$ and is thus equal to the marginal distribution.

The response distribution consistent with the observed value $\mathbf{d}$ can be obtained immediately from equation (10):

LEMMA 2

Under the preceding assumption: $\mathbf{b}(\mathbf{y})$ *is multivariate normal*$(\boldsymbol{\beta}; \sigma^2(X'X)^{-1})$; $s^2(\mathbf{y})$ *is* σ^2 *times a chi-square*$(n - r)$ *variable;* $\mathbf{b}(\mathbf{y})$ *is statistically independent of* $s(\mathbf{y})$.

It is of interest to record the analogous results for the use of coordinates $a_1, \ldots, a_r$ with respect to an orthonormal basis $\mathbf{v}_1, \ldots, \mathbf{v}_r$. From Section 11.2G we have $X = VT$; thus

(18) $$X\mathbf{b} = VT\mathbf{b} = V\mathbf{a},$$

where $\mathbf{a} = T\mathbf{b}$ and the particular matrix T is upper triangular. The inner product matrix $V'V$ is the $r \times r$ identity matrix.

LEMMA 3

Under the preceding assumptions: $a_1, \ldots, a_r, s$ *are statistically independent;* a_u *is standard normal;* s^2 *is chi-square*$(n - r)$.

For the response distribution relative to the orthonormal basis we have, from equation (10),

$$\mathbf{a}(\mathbf{y}) = \boldsymbol{\alpha} + \sigma\mathbf{a},$$

$$s(\mathbf{y}) = \sigma s,$$

where the new parameters $\alpha_1, \ldots, \alpha_r$ are given by $\boldsymbol{\alpha} = T\boldsymbol{\beta}$, as is seen from

(19) $$X = VT, \qquad X\boldsymbol{\beta} = VT\boldsymbol{\beta} = V\boldsymbol{\alpha}.$$

LEMMA 4

Under the preceding assumptions: $a_1(\mathbf{y}), \ldots, a_r(\mathbf{y}), s(\mathbf{y})$ *are statistically independent;* $a_u(\mathbf{y})$ *is normal* (α_u, σ); $s^2(\mathbf{y})$ *is* σ^2 *times a chi-square*$(n - r)$ *variable.*

TABLE 11.7
Projections and corresponding squared lengths

Source	Dimension	Projection	Squared Length	Distribution
$\mathbf{v}_1$	1	$a_1(\mathbf{y})\mathbf{v}_1$	$a_1^2(\mathbf{y})$	$\sigma^2\chi_1^2$ if $\alpha_1 = 0$
$\mathbf{v}_2$	1	$a_2(\mathbf{y})\mathbf{v}_2$	$a_2^2(\mathbf{y})$	$\sigma^2\chi_1^2$ if $\alpha_2 = 0$
$\vdots$				
$\mathbf{v}_r$	1	$a_r(\mathbf{y})\mathbf{v}_r$	$a_r^2(\mathbf{y})$	$\sigma^2\chi_1^2$ if $\alpha_r = 0$
$\mathcal{L}^\perp(X)$	$n - r$	$s(\mathbf{y})\mathbf{d}$	$s^2(\mathbf{y})$	$\sigma^2\chi_{n-r}^2$
	n	$\mathbf{y}$	$\sum y_i^2$	

These results are recorded partially in Table 11.7 following the pattern discussed in Section 11.2. The different projections are statistically independent, as are the different squared lengths. The residual squared length $s^2(\mathbf{y})$ has the distribution of σ^2 times a chi-square($n - r$) variable; if divided by its degrees of freedom, it provides an estimate of σ^2. The squared length for a component based on $\mathbf{v}_u$ has the distribution of σ^2 times a chi-square(1) distribution provided that the corresponding location parameter $\alpha_u = 0$; otherwise, it has a noncentral chi-square-type distribution (see Problem 3.6.3) and tends to be larger in value.

In Section A we noted that the vectors $\mathbf{x}_1, \ldots, \mathbf{x}_r$ are often obtained as values of r input variables. In such cases it can be of interest to estimate the response level at settings, say $l_1, \ldots, l_r$, for the input variables. This response level is given by

$$(20) \qquad y = \beta_1 l_1 + \cdots + \beta_r l_r = \boldsymbol{\beta}'\mathbf{l},$$

and the corresponding estimate is

$$(21) \qquad \hat{y} = b_1(\mathbf{y})l_1 + \cdots + b_r(\mathbf{y})\mathbf{l}_r = \mathbf{b}'(\mathbf{y})\mathbf{l}.$$

The distribution for this estimate is then obtained from, say, Example 5.7.5:

LEMMA 5

Under the model (4) with standard normal variation, the distribution of the estimate $\hat{y}$ is normal with mean $\boldsymbol{\beta}'\mathbf{l}$ and variance $\mathbf{l}'(X'X)^{-1}\mathbf{l}\sigma^2$.

11.4

INFERENCE WITH THE REGRESSION MODEL

We now examine statistical inference for the regression model of the preceding section. We do this in terms of the numerical example introduced in Section 11.2C.

The numerical example involved a random system with input variable x and response y. The data are as follows:

x	1	2	3	4	5
y	5	7	11	10	13

These are plotted in Figure 11.8.

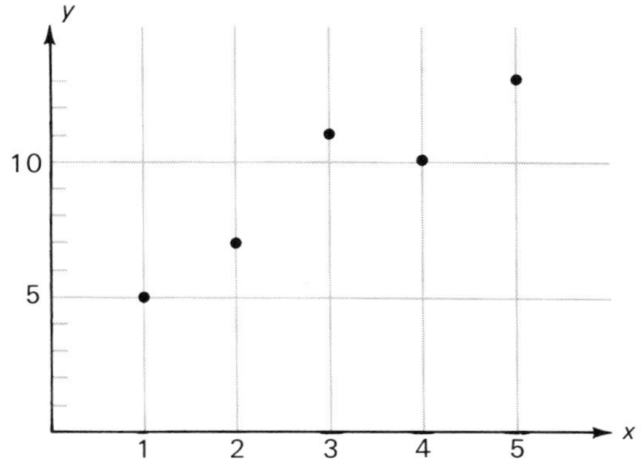

FIGURE 11.8
Data presented in Sections 11.2C and 11.4.

A THE QUADRATIC MODEL

We analyze the example on the assumption that the quadratic model

(1) $$y = \beta_1 + \beta_2 x + \beta_3 x^2 + \sigma z$$

adequately describes the response and that the underlying variation z is standard normal.

The *analysis-of-variance* table for fitting constant, linear, and then quadratic dependence on x was obtained in Section 11.2F and is presented here as Table 11.8. Dimension has been relabeled DF, for *degree of freedom,* and squared length has been relabeled SS, for *sums of squares*.

TABLE 11.8
Analysis-of-variance table

Source	Basis	DF	Projection	SS	MS
Constant	$\mathbf{v}_1$	1	$9.2\mathbf{x}_1 = 9.2\mathbf{w}_1$ $= (9.2, 9.2, 9.2, 9.2, 9.2)$	423.20	
Linear	$\mathbf{v}_2$	1	$-5.7\mathbf{x}_1 + 1.9\mathbf{x}_2 = 1.9\mathbf{w}_2$ $= (-3.8, -1.9, 0, 1.9, 3.8)$	36.10	
Quadratic	$\mathbf{v}_3$	1	$-1.5\mathbf{x}_1 + 1\frac{2}{7}\mathbf{x}_2 - \frac{3}{14}\mathbf{x}_3 = -\frac{3}{14}\mathbf{w}_3$ $= (-\frac{6}{14}, \frac{3}{14}, \frac{6}{14}, \frac{3}{14}, -\frac{6}{14})$	0.64	0.64
Residual		2	$\mathbf{y} - 2\mathbf{x}_1 - 3\frac{13}{70}\mathbf{x}_2 + \frac{3}{14}\mathbf{x}_3$ $= (\frac{2}{70}, -\frac{36}{70}, 1\frac{26}{70}, -1\frac{22}{70}, \frac{30}{70})$	4.06	2.03
		5	$(5, 7, 11, 10, 13)$	464.00	

B DISTRIBUTION THEORY AND OBSERVED VALUES, QUADRATIC MODEL

Consider the quadratic model derived from formula (1) together with (11.3.19):

$$(2)\qquad \mathbf{y} = \beta_1\mathbf{x}_1 + \beta_2\mathbf{x}_2 + \beta_3\mathbf{x}_3 + \sigma\mathbf{z} = \alpha_1\mathbf{v}_1 + \alpha_2\mathbf{v}_2 + \alpha_3\mathbf{v}_3 + \sigma\mathbf{z},$$

where $\mathbf{z}$ is a sample of five from the standard normal.

Under the preceding model we have from Section 11.3F that

$$(b_1(\mathbf{y}), b_2(\mathbf{y}), b_3(\mathbf{y}))$$

has a multivariate normal distribution with mean

$$(\beta_1, \beta_2, \beta_3)$$

and variance matrix

$$(3)\qquad \sigma^2(X'X)^{-1} = \sigma^2\begin{pmatrix} 4.6 & -3.3 & \frac{1}{2} \\ -3.3 & 2\frac{47}{70} & -\frac{3}{7} \\ \frac{1}{2} & -\frac{3}{7} & \frac{1}{14} \end{pmatrix},$$

with values from the array (11.2.14). The observed value of $(b_1(\mathbf{y}), b_2(\mathbf{y}), b_3(\mathbf{y}))$ is

$$(4)\qquad (2, 3\tfrac{17}{70}, -\tfrac{3}{14}),$$

and the fitted quadratic function is represented by the vector

$$(5)\qquad 2\mathbf{x}_1 + 3\tfrac{13}{70}\mathbf{x}_2 - \tfrac{3}{14}\mathbf{x}_3 = 9.2\mathbf{w}_1 + 1.9\mathbf{w}_2 - \tfrac{3}{14}\mathbf{w}_3,$$

as calculated in Section 11.2F.

And also we have that $s^2(\mathbf{y})$ has the distribution of $\sigma^2\chi^2$, where χ^2 has the chi-square$(n - r)$ distribution; here $n - r = 2$. The observed value of $s^2(y) = 4.06$.

C INFERENCE CONCERNING THE VARIANCE

The bottom element $s^2(\mathbf{y}) = 4.06$ in the SS column of the analysis-of-variance table is an observed value of $\sigma^2\chi^2$, where χ^2 is chi-square$(n - r = 2)$; the degrees of freedom $n - r$ is given in the corresponding DF column, $n - r = 2$. We take the ratio of sum of squares (SS) to degrees of freedom (DF), obtaining the observed *mean square* (MS):

$$(6)\qquad \text{MS} = \frac{\text{SS}}{\text{DF}} = \frac{4.06}{2} = 2.03;$$

this mean square is an unbiased estimate s_E^2 of the variance σ^2. A confidence interval for the variance σ^2 can be obtained from the chi-square$(n - r)$ distribution for

$$\chi^2 = \frac{(n - r)s_E^2}{\sigma^2},$$

following the pattern in Section 8.3B.

D INFERENCE CONCERNING QUADRATIC EFFECT

For the present example it is natural to investigate whether the quadratic term in (2) is needed for an adequate presentation of the response level. More generally it is natural to investigate whether the last introduced source for response variation is needed for an adequate response presentation.

We can test for the presence of a nonzero value for β_3 by calculating

$$(7) \qquad t_3 = \frac{b_3(\mathbf{y})}{s_E\sqrt{1/14}}$$

and comparing it with the Student($n - r = 2$) distribution. Note that $1/14$ is the reciprocal of the squared length of $\mathbf{w}_3$, and it is available from the right-hand portion of (3). Or more directly we can test by calculating the ratio of mean squares given in the analysis-of-variance table,

$$(8) \qquad F = \frac{\text{MS(quadratic)}}{\text{MS(residual)}}$$

and comparing it with the $F(1, 2)$ distribution. Note that if there is no quadratic effect, that is, $\alpha_3 = 0$ in the notation of Sections 11.2 and 11.3, then F is a ratio of mean squares based on ordinary chi-square variables and accordingly has the ordinary F distribution. On the other hand, if there is a quadratic effect, that is, $\alpha_3 \neq 0$, then F has the noncentral F distribution and tends to be larger in value. The t and F calculations are easily seen to be equivalent; they are just different ways of comparing the projection in $\mathcal{L}(\mathbf{v}_3)$ with the projection in $\mathcal{L}^{\perp}(X)$.

The observed value of $F = 0.64/2.03 = 0.32$. This is not an extremely large value; in fact, it is to the left of the center of the F distribution (which is approximately unity). Thus we have no evidence of a quadratic dependence. It is natural, then, to consider withdrawing to the linear model

$$y = \beta_1 + \beta_2 x + \sigma z$$

and analyzing accordingly. Note that our calculation sequence allows for this: examining quadratic after linear and constant.

On the assumption of the quadratic model we can, of course, calculate confidence intervals for the regression coefficients β_1, β_2, and β_3. For β_3 we have a Student($n - r = 2$) distribution for

$$(9) \qquad t_3 = \frac{b_3(\mathbf{y}) - \beta_3}{s_E\sqrt{1/14}}.$$

This gives the following 95 percent confidence interval for β_3:

$$(10) \qquad (-(3/14) \pm 4.30\sqrt{2.03}\sqrt{1/14}) = (-1.852, 1.423).$$

Recall the note following (7). Confidence intervals for β_2 and β_1 can be calculated analogously; for example, for β_2 we have

$$(3\tfrac{13}{70} \pm 4.30\sqrt{2.03}\sqrt{2\tfrac{47}{70}}) = (-6.83, 13.20).$$

E THE AFFINE MODEL

We now consider the analysis on the assumption that the affine model

(11) $y = \beta_1 + \beta_2 x + \sigma z$

adequately describes the response and the underlying variation z is standard normal.

The analysis-of-variance table for fitting constant, and then linear dependence on x, was obtained in Section 11.2E; now see Table 11.9.

TABLE 11.9
Analysis-of-variance table

Source	Basis	DF	Projection	SS	MS
Constant	$\mathbf{v}_1$	1	$9.2\mathbf{x}_1 = 9.2\mathbf{w}_1$ $= (9.2, 9.2, 9.2, 9.2, 9.2)$	423.20	
Linear	$\mathbf{v}_2$	1	$-5.7\mathbf{x}_1 + 1.9\mathbf{x}_2 = 1.9\mathbf{w}_2$ $= (-3.8, -1.9, 0, 1.9, 3.8)$	36.10	36.10
Residual		3	$\mathbf{y} - 3.5\mathbf{x}_1 - 1.9\mathbf{x}_2$ $= (-0.4, -0.3, 1.8, -1.1, 0)$	4.70	1.57
		5	(5, 7, 11, 10, 13)	464.00	

F DISTRIBUTION THEORY AND OBSERVED VALUES: AFFINE MODEL

Consider the quadratic model derived from formula (11) together with (11.3.19):

(12) $$\mathbf{y} = \beta_1\mathbf{x}_1 + \beta_2\mathbf{x}_2 + \sigma\mathbf{z}$$
$$= \alpha_1\mathbf{v}_1 + \alpha_2\mathbf{v}_2 + \sigma\mathbf{z},$$

where $\mathbf{z}$ is a sample of five from the standard normal.

Under the preceding model we have from Section 11.3 that

$$(b_1(\mathbf{y}), b_2(\mathbf{y}))$$

has a bivariate normal distribution with mean

$$(\beta_1, \beta_2)$$

and variance matrix

(13) $$\sigma^2(X'X)^{-1} = \sigma^2\begin{pmatrix} 1.1 & -0.3 \\ -0.3 & 0.1 \end{pmatrix}$$

with values from the array (11.2.13). Note that $(b_1(\mathbf{y}), b_2(\mathbf{y}))$ refers to the calculations for the affine model as examined in Section 11.2E; thus these b's are different from those considered earlier in Section B. The a's for the model here would, however, be the same as those for the model in Section B; an advantage from orthogonality. The observed value of $(b_1(\mathbf{y}), b_2(\mathbf{y}))$ is

(14) $\quad (3.5, 1.9)$

and the fitted affine function is represented by the vector

(15) $\quad 3.5\mathbf{x}_1 + 1.9\mathbf{x}_2 = 9.2\mathbf{w}_1 + 1.9\mathbf{w}_2,$

as calculated in Section 11.2E.

And also we have that $s^2(\mathbf{y})$ has the distribution of $\sigma^2\chi^2$, where χ^2 has the chi-square$(n - r)$ distribution; here $n - r = 3$. The observed value of $s^2(y) = 4.70$.

G INFERENCE CONCERNING LINEAR EFFECT

We now assume there is negligible quadratic effect and consider the analysis based essentially on the simpler model

(16) $\quad \mathbf{y} = \beta_1\mathbf{x}_1 + \beta_2\mathbf{x}_2 + \sigma\mathbf{z}.$

Tests and confidence intervals are available following the pattern in Section D.

This procedure based on negligible quadratic effect requires some caution. In Section D we tested for quadratic effect and we have now withdrawn to the simpler model. As part of withdrawing we pooled the quadratic SS and the residual SS of the table in Section A to obtain the residual SS of the table in Section E. This pooling procedure has some rather serious potentials for bias. For the present we can allow that the quadratic SS may contain minor quadratic effects and use the original table in Section A to obtain our error variance estimate and follow the pattern in Section C for inference concerning the variance.

Thus on the assumption of the affine model together with the precaution just mentioned for error variance, we can test for the presence of a nonzero value for β_2 by calculating

(17) $$t_2 = \frac{b_2(\mathbf{y})}{s_E\sqrt{0.1}}$$

with s_E from Section A and comparing it with the Student$(n - r)$ distribution; here $n - r = 2$, the DF for the error variance. Note that 0.1 is the reciprocal of the squared length of $\mathbf{w}_2$, and it is available from the right-hand portion of (13). Or more directly we can test by calculating

(18) $$F = \frac{\text{MS(linear)}}{\text{MS(residual)}}$$

and comparing it with the $F(1, 2)$ distribution; note that these DF come from the table in Section A.

The observed value of $F = 36.10/2.03 = 17.8$. The 5 percent point is 18.51 and the 1 percent point is 98.49. Thus there is some suggestion of a nonzero value for β_2. Note that if we had used the pooled estimate for the variance we would have had an additional DF for the denominator, and the F test would, in fact, give significance.

On the assumption of the linear model together with the precaution we can calculate confidence intervals for the regression coefficients β_1 and β_2. For β_2 we have a Student($m = 2$) distribution for

$$(19) \qquad t_2 = \frac{b_2(\mathbf{y}) - \beta_2}{s_E\sqrt{0.1}}$$

using the unpooled estimate. This gives the following 95 percent confidence interval for β_2:

$$(20) \qquad (1.9 \pm 4.30\sqrt{2.03}\sqrt{0.1}) = (-0.04, 3.84).$$

H PROBLEMS

1 Calculate formula (5) in terms of the orthonormal vectors $\mathbf{v}_1$, $\mathbf{v}_2$, and $\mathbf{v}_3$. Calculate formula (15) in terms of the orthonormal vectors $\mathbf{v}_1$ and $\mathbf{v}_2$.

2 For the model in Sections A and B, record numerical values for the entries in the matrix equation $X\mathbf{b} = V\mathbf{a}$ from (11.3.18); here $r = 3$. For the model in Sections E and B, record numerical values for the entries in the equation $X\mathbf{b} = V\mathbf{a}$; here $r = 2$. Recall Problem 11.2.3; compare with the preceding problem.

3 (*continuation of Problem 11.2.1*) Assume the affine model $y = \beta_1 + \beta_2 x + \sigma z$, where z is standard normal. Test the hypothesis: $\beta_2 = 0$. Form a 95 percent confidence interval for β_2.

4 (*continuation of Problem 11.2.2*) Assume the quadratic model $y = \beta_1 + \beta_2 x + \beta_3 x^2 + \sigma z$, where z is standard normal. Test the hypothesis $\beta_3 = 0$. If no significance, assume an affine model and test the hypothesis $\beta_2 = 0$; form a 95 percent confidence interval for β_2; recall the precaution in Section G.

5 (*continuation of Problem 11.2.4*) Assume the affine model $y = \beta_1 + \beta_2 x_1 + \beta_3 x_2$. Test the hypothesis $\beta_3 = 0$. If no significance, assume an affine model and test the hypothesis $\beta_2 = 0$; form a 95 percent confidence interval for β_2; recall the precaution in Section G.

NOTES AND REFERENCES

The linear model is the basic model for large areas of statistics, including the fundamental areas of experimental design. For material on linear models primarily based on normal variation, see the references cited below. For some discussion of linear models with other distribution forms for error, see Fraser (1976).

A computer program for analyzing the location-scale model is available to instructors on request; write to: Statistics Section, Attn.: Gordon Fick, Mathematics Department, University of Toronto, Canada.

Draper, N. R., and H. Smith (1967). *Applied Regression Analysis*. New York: John Wiley & Sons, Inc.

Fraser, D. A. S. (1976). Necessary analysis and adaptive inference. *J. Amer. Statist. Assoc.*, **71**.

Fraser, D. A. S., and G. H. Fick (1975). Necessary analysis and its implementation. *Proc. Symp. Statistics and Related Topics* (Carleton Mathematical Lectures, **12**, 5.01–5.30). Ottawa: Carleton University.

Plackett, R. L. (1960). *Principles of Regression Analysis*. Oxford: Clarendon Press.

Searle, S. R. (1971). *Linear Models*. New York: John Wiley & Sons, Inc.

Williams, E. J. (1959). *Regression Analysis*. New York: John Wiley & Sons, Inc.

In preceding chapters we have examined the basic methods of statistical inference and applied the methods to the very widely used linear models.

In Chapter 12 we shall survey ***experimental design,*** *the design of experimental investigations to determine cause–effect relations; in particular, we discuss the basic principles of experimentation,* ***control, randomization,*** *and* ***replication.*** *In Section 12.1 the* ***factorial design*** *is introduced and the* ***single-factor design*** *is examined in detail. In Section 12.2 the* ***two-factor design*** *is examined in detail. Then in Supplement Sections 12.3, 12.4, and 12.5 we examine* ***three-factor designs, randomized blocks,*** *and* ***analysis of covariance.***

12
THE DESIGN OF EXPERIMENTS

12.1

ONE-FACTOR DESIGN

The design of experiments, the most important part of statistics, has been mentioned briefly in Sections 1.1 and 7.1. In this chapter we discuss some central material of the design of experiments.

The investigation of a random system becomes an experiment if the investigator makes designed changes in the input variables or factors with the express purpose of determining cause–effect relations. In Sections 12.1 and 12.2 and Supplement Section 12.3 we examine the basic factorial design for one, two, and three factors, respectively. Then in Supplement Sections 12.4 and 12.5 we acknowledge the inevitable variation that occurs in the initial conditions and the input materials to a system and we examine two methods for minimizing the influence of such variation. If the conditions or materials can be grouped into blocks with less variation within a block, the randomized block design in Supplement Section 12.4 can be used with the factorial design. And if a concomitant variable that describes the variation is available, the analysis of covariance in Supplement Section 12.5 can be used with the factorial design. The development in this chapter emphasizes basic principles: the basic principles necessary to the design of experiments, and the basic principles underlying the algebraic results.

On an individual basis the algebraic results in this chapter can each be verified fairly directly. But the *why* is important—not just for the developments in this chapter but for the flexibility to handle new methods and new designs in a direct way. Thus at several points we may have somewhat more vector notation than is usual, but only so we can use the basic principles in a direct way to obtain the needed algebraic results.

A FACTORIAL DESIGN

We first discuss briefly some aspects of experimental design—and we do this in the context of a random system with a variety of input variables that may influence the response of the system. A *factor* is an input or input variable that may affect the response of the system: for example, the pressure at which a chemical process is operated, the kind of soil preparation for an agricultural experiment, the type of diet in a growth investigation, the variety of a grain in an agricultural experiment. The *level* of a factor is the particular value of the input or input variable: for example, a particular pressure, a particular diet, a particular variety of grain.

The *design factors* are the factors under study, the factors that the investigator changes in his designed investigation of the system. A particular combination of levels for the design factors is called a *treatment*.

The investigator when possible holds constant or *controls* other factors. This is done to ensure that such factors cannot change and thus cause changes in the response that might be attributable to the design factors.

This leaves the factors or influences that are not directly changed or controlled. These cover differences between *experimental units,* the basic units such as animals or plots of land that receive the treatments being investigated. They also cover things such as variation in input material, and show trends with respect to location and time. By extending the definition of experimental unit appropriately to include the basic unit at a given location and given time, these various factors or influences can all be embraced under "experimental units."

In practice there rarely is reliable information as to the actual variation from experimental unit to unit. Thus with treatments applied to units, an apparent difference in the effects of the treatments can be due, in fact, to differences between the units. The only reliable way of dealing with the actual variation from unit to unit is to randomly assign the treatments to the units. This *randomization* changes the variation from unit to unit into a random variation affecting the response; and this response variation or error can be directly assessed by the probability methods we have been developing.

We indicate design factors by capital letters: $A, B, C, \ldots$. And we add subscripts to denote the chosen levels for the factors:

$$\begin{aligned} &A: A_1, A_2, \ldots, A_a, \\ &B: B_1, B_2, \ldots, \quad B_b, \\ &C: C_1, C_2, \ldots, \quad\quad C_c. \end{aligned}$$

An experimental design is called a *factorial design* if each combination of levels is examined the same number of times, say n. For example, with three factors as indicated above, each treatment $A_iB_jC_k$ is examined n times, giving a total of $N = abcn$ observations on the system. The design is called *fully randomized* if the N observations, n for each treatment, are randomly assigned to N experimental units or to the N times or locations at which the system is performed. In this section we examine the fully randomized factorial design with one design factor, and then in Section 12.2 and Supplement Section 12.3 we examine the factorial design with two and three design factors, respectively.

In Supplement Section 12.4 we consider the case where experimental units can be grouped to form *blocks* with less variation within blocks than between blocks. The treatments can then be randomly assigned *within* blocks, producing the *randomized block design*.

In Supplement Section 12.5 we consider how some related or *concomitant* variable can give information on the variation attributable to the experimental units and how the *analysis of covariance* with such a concomitant variable can be used to assess more precisely the dependence of the response on the treatments.

B ONE-FACTOR DESIGN

Consider a single design factor A with chosen levels $A_1, \ldots, A_a$. And suppose that n_1 observations are planned for level $A_1, \ldots,$ and n_a observations for level A_a, making a total of $N = \Sigma n_j$ observations. For the one-factor design we can allow differing numbers at the different factor levels without serious complication. However, for several factors and with randomized blocks this departure from the formal definition of a factorial design can lead to severe difficulties. We assume that the observations are fully randomized against the experimental units, time, or any uncontrolled factor.

The response vector for the N observations can be presented in an array such as the following:

$$\begin{array}{c|c|llll} x_1 & A_1 & y_{11} & \cdots & y_{1n_1} & \\ \vdots & \vdots & & & & \\ x_a & A_a & y_{a1} & \cdots & & y_{an_a}. \end{array}$$

In certain cases the factor A may be represented by a controllable variable x; the values for such a variable are indicated in the left column.

An investigation is concerned with whether the factor A affects the response and, if so, in what manner. We assume that the response is presented in such a form that the variation has the same scaling at the various possible response levels. And we consider the analysis for the case of approximately normal variation. The analysis for other distribution forms with a parameter λ can be investigated using the theory in Section 11.3; this does, however, require integration over a space with dimension equal to the number of primary parameters.

If the factor A does not affect the response, then the model for a single observation and the model for the vector of N observations are

$$(1) \qquad y_{is} = \mu + \sigma z_{is}, \qquad \mathbf{y} = \mu \mathbf{1} + \sigma \mathbf{z},$$

where $\mathbf{z}$ is a sample of N from the standard normal. The fitted model for the general level is then the projection $\mathbf{p}_1(\mathbf{y})$ into $\mathfrak{L}(\mathbf{1})$; the projection and squared length of the projection are

$$(2) \qquad \begin{array}{c|llll} A_1 & \bar{y} & \cdots & \bar{y} & \\ \vdots & & & & \\ A_a & \bar{y} & \cdots & & \bar{y}, \end{array}$$

$$\mathrm{SS}_1 = \mathrm{SS}_{(1)} = N\bar{y}^2 = T_{..}^2/N.$$

Note that $T_{..} = \Sigma_{is} y_{is}$ is the overall total and $\bar{y} = T_{..}/N$ is the overall average.

If the factor A does affect the response and if we allow for this in full generality with a level μ_i at the factor level A_i, then the model is

$$(3) \qquad y_{is} = \mu_i + \sigma z_{is}, \qquad \mathbf{y} = \sum_{i=1}^{a} \mu_i \mathbf{1}_i + \sigma \mathbf{z},$$

where $\mathbf{z}$ is a sample of N from the standard normal and $\mathbf{1}_i$ is the vector with 1's in positions corresponding to the level A_i and 0's elsewhere, as given by the following array:

$$\begin{array}{c|lllll} A_1 & 0 & \cdots & 0 & & \\ \vdots & & & & & \\ A_i & 1 & \cdots & & 1 & \\ \vdots & & & & & \\ A_a & 0 & \cdots & & & 0. \end{array}$$

The fitted model is then the projection $\mathbf{p}_{(2)}(\mathbf{y})$ into $\mathfrak{L}(\mathbf{1}_1, \ldots, \mathbf{1}_a)$; it gives a separate average for each factor level. The projection and squared length are

$$(4) \quad \begin{array}{c|ccc} A_1 & \bar{y}_1 & \cdots & \bar{y}_1 \\ \vdots & & & \\ A_a & \bar{y}_a & \cdots & \bar{y}_a, \end{array}$$

$$SS_{(2)} = \sum_i n_i \bar{y}_i^2 = \sum_i T_{i\cdot}^2 / n_i.$$

Note that $T_{i\cdot} = \Sigma_s y_{is}$ is the total for the level A_i and $\bar{y}_{i\cdot} = T_{i\cdot}/n_i$ is the average for that level; for the justification note that the vectors $\mathbf{1}_1, \ldots, \mathbf{1}_a$ are orthogonal.

In going from the first model to the second model we are allowing the general level to change with the factor A. The fitted vector for such changes is the projection $\mathbf{p}_2 = \mathbf{p}_{(2)} - \mathbf{p}_1$ into $\mathcal{L}(\mathbf{1}_1, \ldots, \mathbf{1}_a \perp \mathbf{1})$; recall the discussion in Section 11.2A. The projection and squared length are

$$(5) \quad \begin{array}{c|ccc} A_1 & \bar{y}_1 - \bar{y} & \cdots & \bar{y}_1 - \bar{y} \\ \vdots & & & \\ A_a & \bar{y}_a - \bar{y} & \cdots & \bar{y}_a - \bar{y}, \end{array}$$

$$SS_2 = \sum_i n_i(\bar{y}_i - \bar{y})^2 = SS_{(2)} - SS_{(1)} = \sum_i T_{i\cdot}^2/n_i - T_{\cdot\cdot}^2/N.$$

Note that the projection and the squared length are obtained easily by differencing. And note also the interesting identity for the squared length, as calculated directly from the projection and by differencing.

The response vector itself and the square length of the response are

$$\begin{array}{c|ccc} A_1 & y_{11} & \cdots & y_{1n_1} \\ \vdots & & & \\ A_a & y_{a1} & \cdots & y_{an_a}, \end{array}$$

$$SS_{(3)} = \sum y_{is}^2.$$

The step from the second fitted model to the response vector is concerned with the deviations from the average at each factor level. The fitted vector is the projection $\mathbf{p}_3 = \mathbf{y} - \mathbf{p}_{(2)}$ into $\mathcal{L}^{\perp}(\mathbf{1}_1, \ldots, \mathbf{1}_a)$; the projection and squared length are

$$(6) \quad \begin{array}{c|ccc} A_1 & y_{11} - \bar{y}_1 & \cdots & y_{1n_1} - \bar{y}_1 \\ \vdots & & & \\ A_a & y_{a1} - \bar{y}_a & \cdots & y_{an_a} - \bar{y}_a, \end{array}$$

$$SS_3 = \sum (y_{is} - \bar{y}_i)^2 = SS_{(3)} - SS_{(2)} = \sum y_{is}^2 - \sum_i T_{i\cdot}^2/n_i.$$

TABLE 12.1
Cumulative table

	Source	Subspace	DF	Projection	SS
(1)	Mean	$\mathcal{L}(\mathbf{1})$	1	$\bar{y}\mathbf{1}$	$N\bar{y}^2 = T_{\cdot\cdot}^2/N$
(2)	A	$\mathcal{L}(\mathbf{1}_1, \ldots, \mathbf{1}_a)$	a	$\sum \bar{y}_i \mathbf{1}_i$	$\sum n_i \bar{y}_i^2 = \sum_i T_{i\cdot}^2/n_i$
(3)	Response	R^N	N	$\mathbf{y}$	$\sum y_{is}^2$

The cumulative table recording projections and squared lengths is given in Table 12.1. The difference table recording projections and squared lengths for orthogonal subspaces is then given in Table 12.2. This is an analysis-of-variance table, analyzing the variation among response values in accord with patterns based on the various models.

TABLE 12.2
Analysis-of-variance table

	Source	Space	DF	Projection	SS
1	Mean	$\mathcal{L}(\mathbf{1})$	1	$\bar{y}\mathbf{1}$	$N\bar{y}^2 = \mathrm{SS}_{(1)}$
2	A	$\mathcal{L}(\mathbf{1}_1, \ldots, \mathbf{1}_a \perp \mathbf{1})$	$a - 1$	$\sum (\bar{y}_i - \bar{y})\mathbf{1}_i$	$\sum n_i(\bar{y}_i - \bar{y})^2 = \mathrm{SS}_{(2)} - \mathrm{SS}_{(1)}$
3	Residual	$\mathcal{L}^{\perp}(\mathbf{1}_1, \ldots, \mathbf{1}_a)$	$N - a$	$\mathbf{y} - \sum \bar{y}_i\mathbf{1}_i$	$\sum (y_{is} - \bar{y}_i)^2 = \mathrm{SS}_{(3)} - \mathrm{SS}_{(2)}$
		R^N	N	$\mathbf{y}$	$\sum y_{is}^2$

C THE ANALYSIS

Now consider the problem of testing the hypothesis that changes in the factor A do not affect the response level.

For this we note some distribution results available from Section 11.3. Under the general model (3) we have that

$$\mathrm{SS}_3 = \sum (y_{is} - \bar{y})^2 \tag{7}$$

has the distribution of $\sigma^2\chi_3^2$, where χ_3^2 has the chi-square($N - a$) distribution. Under the special model (1) we have that

$$\mathrm{SS}_2 = \sum_i n_i(\bar{y}_i - \bar{y})^2 \tag{8}$$

has the distribution of $\sigma^2\chi_2^2$, where χ_2^2 has the chi-square($a - 1$) distribution; however, under the general model with the μ_i not all equal it tends to be larger in value and has the noncentral chi-square distribution in Problem 3.6.3.

If the hypothesis is true, then both (7) and (8) lead (division by appropriate DF) to mean squares (MS) that are estimates of σ^2. If the hypothesis is not true, then (8) tends to be larger. The hypothesis can be tested by calculating

$$F = \frac{\mathrm{SS}_2/(a - 1)}{\mathrm{SS}_3/(N - a)} = \frac{\mathrm{MS}_2}{\mathrm{MS}_3}$$

and comparing the value with the F distribution ($a - 1$, $N - a$); unusually large values are indicative of effects due to the factor A.

EXAMPLE 1

Consider a catalyst A at two levels A_1 and A_2 with three observations at level A_1 and two observations at level A_2. The five observations were randomized against time, giving the response data shown in Table 12.3 (rounded to the nearest integer for easy arithmetic). Some calculations based on the formulas in Section B are recorded on the right. The cumulative table is shown in Table 12.4. The analysis-of-variance table is obtained by differencing (Table 12.5).

TABLE 12.3
Response data

	Data			Totals	Calculations
A_1	4	3	5	12	$\sum y_{is}^2 = 55$
A_2	2	1		3	$12^2/3 + 3^2/2 = 52\frac{1}{2}$
				15	$15^2/5 = 45$

TABLE 12.4
Cumulative table

	Source	DF	Projection	SS
(1)	Mean	1	$\begin{pmatrix} 3 & 3 & 3 \\ 3 & 3 & \end{pmatrix}$	45
(2)	A	2	$\begin{pmatrix} 4 & 4 & 4 \\ 1\frac{1}{2} & 1\frac{1}{2} & \end{pmatrix}$	$52\frac{1}{2}$
(3)	Response	5	$\begin{pmatrix} 4 & 3 & 5 \\ 2 & 1 & \end{pmatrix}$	55

TABLE 12.5
Analysis-of-variance table

	Source	DF	Projection	SS	MS	F
1	Mean	1	$\begin{pmatrix} 3 & 3 & 3 \\ 3 & 3 & \end{pmatrix}$	45		
2	A	1	$\begin{pmatrix} 1 & 1 & 1 \\ -1\frac{1}{2} & -1\frac{1}{2} & \end{pmatrix}$	$7\frac{1}{2}$	$7\frac{1}{2}$	$7\frac{1}{2}/\frac{5}{6} = 9$
3	Residual	3	$\begin{pmatrix} 0 & -1 & 1 \\ \frac{1}{2} & -\frac{1}{2} & \end{pmatrix}$	$2\frac{1}{2}$	$\frac{5}{6}$	
	Total	5	$\begin{pmatrix} 4 & 3 & 5 \\ 2 & 1 & \end{pmatrix}$	55		

Now suppose that the general model (3) holds and we want to test the hypothesis that the factor A does not affect the general level of the response. We can test for the factor A by calculating

$$F = \frac{MS_2}{MS_3} = \frac{7\frac{1}{2}}{\frac{5}{6}} = 9$$

and comparing it with the $F(1, 3)$ distribution. The value 9 is slightly less than the 5 percent point 10.1, giving some moderate suggestion that the factor A affects the response at the levels tested.

D WITH A QUANTITATIVE FACTOR

Now suppose that the factor A represents an input variable x. If there *is* dependence on the factor A, the experimenter will want to know the form of the dependence, whether it depends, say, linearly, quadratically, or perhaps more generally on x. He can then examine some models that fall between the no-dependence case

(1) $$y_{is} = \mu + \sigma z_{is}, \qquad \mathbf{y} = \mu\mathbf{1} + \sigma\mathbf{z},$$

and the general dependence case

(3) $$\mu_{ij} = \mu_i + \sigma z_{is}, \qquad \mathbf{y} = \sum \mu_i \mathbf{1}_i + \sigma\mathbf{z}.$$

Consider the case of linear and quadratic dependence on x. The model presenting dependence that is at most linear is

(9) $$y_{is} = \beta_1 + \beta_2 x_i + \sigma z_{is}, \qquad \mathbf{y} = \beta_1\mathbf{1} + \beta_2 \sum x_i \mathbf{1}_i + \sigma\mathbf{z}.$$

And the model presenting dependence that is at most quadratic is

(10) $$y_{is} = \beta_1 + \beta_2 x_i + \beta_3 x_i^2 + \sigma z_{is},$$

$$\mathbf{y} = \beta_1\mathbf{1} + \beta_2 \sum x_i \mathbf{1}_i + \beta_3 \sum x_i^2 \mathbf{1}_i + \sigma\mathbf{z}.$$

The projections and squared lengths for the succession of models (1), (9), (10), and (3) can be obtained following the methods of Section 11.2. Alternatively some of them can be obtained directly by finding basis vectors for the successive orthogonal subspaces. For this, consider the sequence of vectors corresponding to the constant, linear, quadratic dependence on x:

$$\mathbf{1} = (1, \ldots, 1; \ldots; 1, \ldots, 1)',$$
$$\sum x_i \mathbf{1}_i = (x_1, \ldots, x_1; \ldots; x_a, \ldots, x_a)',$$
$$\sum x_i^2 \mathbf{1}_i = (x_1^2, \ldots, x_1^2; \ldots; x_a^2, \ldots, x_a^2)'.$$

The second vector can be orthogonalized to the first vector, giving

$$\sum (x_i - \bar{x})\mathbf{1}_i = (x_1 - \bar{x}, \ldots, x_1 - \bar{x}; \ldots; x_a - \bar{x}, \ldots, x_a - \bar{x})',$$

where $\bar{x} = \Sigma n_i x_i / N$ is the average for the N values of x. The third vector as orthogonalized to the first two vectors is

$$\sum (x_i^2 - c - dx_i)\mathbf{1}_i = (x_1^2 - c - dx_1, \ldots, x_1^2 - c - dx_1; \ldots;$$
$$x_a^2 - c - dx_a, \ldots, x_a^2 - c - dx_a)',$$

where the coefficients c and d are obtained from the usual least-squares equations,

$$Nc + \sum n_i x_i\, d = \sum n_i x_i^2,$$

$$\sum n_i x_i c + \sum n_i x_i^2 d = \sum n_i x_i^3.$$

Note, of course, that the solution for c and d can be obtained as part of the matrix procedures in Section 11.2 applied to

$$\begin{matrix} N & \sum n_i x_i & \sum n_i x_i^2 & \sum y_{is} \\ \sum n_i x_i & \sum n_i x_i^2 & \sum n_i x_i^3 & \sum x_i y_{is} \\ \sum n_i x_i^2 & \sum n_i x_i^3 & \sum n_i x_i^4 & \sum x_i^2 y_{is}. \end{matrix}$$

The analysis-of-variance table can be presented in terms of the orthogonalized vectors (Table 12.6). The projections are simply related to the expressions given in the table; for example, the quadratic projection and squared length are

$$\frac{\sum (x_i^2 - c - dx_i) y_{is}}{\sum n_i (x_i^2 - c - dx_i)^2} \sum (x_i^2 - c - dx_i) \mathbf{1}_i, \qquad \frac{\left(\sum (x_i^2 - c - dx_i) y_{is}\right)^2}{\sum n_i (x_i^2 - c - dx_i)^2}.$$

These are special cases of the formulas for the projection of a vector $\mathbf{y}$ on a vector $\mathbf{w}$ and the corresponding squared length.

$$\frac{\mathbf{w}'\mathbf{y}}{\mathbf{w}'\mathbf{w}} \cdot \mathbf{w}, \qquad \frac{\mathbf{w}'\mathbf{y}}{\mathbf{w}'\mathbf{w}} \cdot \mathbf{w}'\mathbf{y}.$$

To test for dependence beyond quadratic we calculate

$$F = \frac{\text{MS}_4}{\text{MS}_5}$$

TABLE 12.6
Analysis-of-variance table

	Source	Subspace	DF	SS
1	Mean	$\mathcal{L}(\mathbf{1})$	1	$\dfrac{\left(\sum y_{is}\right)^2}{N}$
2	A linear	$\mathcal{L}\left(\sum (x_i - \bar{x}) \mathbf{1}_i\right)$	1	$\dfrac{\left(\sum (x_i - \bar{x}) y_{is}\right)^2}{\sum n_i (x_i - \bar{x})^2}$
3	quadratic	$\mathcal{L}\left(\sum (x_i^2 - c - dx_i) \mathbf{1}_i\right)$	1	$\dfrac{\left(\sum (x_i^2 - c - dx_i) y_{is}\right)^2}{\sum n_i (x_i^2 - c - dx_i)^2}$
4	other		$a - 3$	Difference
	Subtotal		$a - 1$	$\sum n_i (\bar{y}_i - \bar{y})^2$
5	Residual	$\mathcal{L}^{\perp}(\mathbf{1}_1, \ldots, \mathbf{1}_a)$	$N - a$	$\sum (y_{is} - \bar{y}_i)^2$
	Total	R^n	N	$\sum y_{is}^2$

and compare it with the $F(a - 3, N - a)$ distribution.

Now suppose that we are prepared to assume that dependence is at most quadratic; then we can test for quadratic dependence by calculating

$$F = \frac{MS_3}{MS_5}$$

and comparing it with the $F(1, N - a)$ distribution.

And suppose we are prepared to assume that dependence is at most linear; then we can test for dependence by calculating

$$F = \frac{MS_2}{MS_5}$$

and comparing it with the F distribution $(1, N - a)$.

E EXAMPLE WITH A QUANTITATIVE FACTOR

Consider a biological system with a factor A at five levels corresponding to $x = 1, 2, 3, 4, 5$. Two observations were obtained at each factor level, giving the data of Table 12.7 (rounded to the nearest integer for easy arithmetic). The analysis-of-variance table without separation of linear and quadratic components is shown as Table 12.8.

TABLE 12.7
Response data

x	Data		Totals	Calculations
1	2	3	5	$y_{is}^2 = 236$
2	4	3	7	
3	5	6	11	
4	6	4	10	
5	7	6	13	$(5^2 + \cdots + 13^2)/2 = 232$
			46	$46^2/10 = 211.6$

TABLE 12.8
Analysis-of-variance table

Source	DF	SS
Mean	1	211.6
A	4	$20.4 = 232 - 211.6$
Residual	5	$4 = 236 - 232$
	10	236

The vector for linear dependence orthogonal to the mean is based on the input variable $x - 3$; recall Section 11.2E. The corresponding squared projection is

$$\frac{[-2(5) - 1(7) + 0(11) + 1(10) + 2(13)]^2}{8 + 2 + 0 + 2 + 8} = \frac{19^2}{20} = 18.05;$$

note that the value -2 for $x - 3$ occurs twice, -1 occurs twice, and so on.

The vector for quadratic dependence, orthogonal to the mean and linear vectors, is based on the input variable $x^2 - 6x + 7$; recall Section 11.2F. The corresponding squared projection is

$$\frac{[2(5) - 1(7) - 2(11) - 1(10) + 2(13)]^2}{8 + 2 + 8 + 2 + 8} = \frac{(-3)^2}{28} = 0.32.$$

The coefficients for these orthogonalized vectors can be obtained from the matrix procedures in Section 11.2:

$$\begin{array}{rrr} 10 & 30 & 110 \\ 30 & 110 & 450 \\ \hline 1 & 3 & 11 \\ 0 & 20 & 120 \\ \hline 1 & 0 & -7 \\ 0 & 1 & 6. \end{array}$$

The analysis-of-variance table can now be rewritten as shown in Table 12.9.

TABLE 12.9
Analysis-of-variance table

Source	DF	SS	MS
Mean	1	211.6	
A linear	1	18.05	18.05
A quadratic	1	0.32	0.32
Other	2	2.03	1.02
Residual	5	4	0.8
Total	10	236	

The more-than-quadratic and quadratic components show no significance. Then on the assumption that dependence is at most linear, the test for dependence is obtained by calculating

$$F = \frac{18.05}{0.8} = 22.5$$

and comparing it with $F(1, 5)$ distribution; the 5 percent value is 6.6 and the 1 percent value is 16.3. The observed value is extreme, thus providing strong evidence against the hypothesis of no linear dependence.

The example in Sections 11.2 and 11.4 is based on the row totals of the present example. Correspondingly, the sum of squares in the first four rows of the preceding table are 1/2 the corresponding values in the first table in Section 11.4A. The present analysis has two observations at each factor level and thus produces an estimate of the variance independent of the form of dependence on the input variable x. This permits a test for dependence beyond quadratic.

Comparison of the two analyses also illustrates two different calculations (by differencing and direct) for some elements in the analysis-of-variance table.

F EXERCISES

Recall the cautionary remarks at the beginning of Section 11.2H.

1 Fourteen pigs were randomly assigned to three diets for a two-month period. The following data record weight increase:

T_1	23	13	28	23	35
T_2	33	31	22	40	31
T_3	40	33	38	44	

Test for dependence on the diets.

2 Three methods of catalyzing a chemical process produced the following data:

C_1	47.2	49.8	48.4	48.7	
C_2	50.1	49.3	51.5	50.9	
C_3	49.1	53.2	51.2	52.8	52.3

The variation is known to be approximately normal. Construct the analysis-of-variance table and test for the equivalence of the catalytic methods.

3 A fertilizer A was used at five levels x on 10 plots of land, giving two yields for each level; the variation is known to be approximately normal.

x	2.6	2.7	2.8	2.9	3.1
y	12.2	12.5	12.5	13.1	13.5
	12.0	12.5	12.9	12.9	13.5

Construct the analysis-of-variance table with entries for Mean, Factor A, and Residual. Test for dependence on the factor A.

4 (*continuation*) Separate out components for linear, quadratic, and other dependence on the factor A; use the analysis from Problems 11.2.2 and 11.4.4. Determine a 95 percent confidence interval for the coefficient, giving the linear dependence on x.

5 Three procedures, T_1, T_2, and T_3, were used to determine the melting point of a hydrocarbon; the variation is approximately normal.

T_1	T_2	T_3
214.2	213.2	211.5
213.1	212.1	211.3
213.8		213.2

Test the equivalence of the procedures.

6 The compressive load at breakdown of three alloys, A_1, A_2, and A_3, were determined for samples of three from each; the variation is approximately normal.

A_1	A_2	A_3
38.5	40.1	40.1
39.8	41.5	43.2
37.2	39.3	42.2

Test the equivalence of the alloys for compressive strength.

12.2

TWO-FACTOR DESIGN

We now examine the fully randomized two-factor design. The analogous design with randomized blocks is examined briefly in Supplement Section 12.4.

A TWO-FACTOR DESIGN

Consider a random system with two design factors, A and B; let $A_1, \ldots, A_a$ be the chosen levels for the first factor and $B_1, \ldots, B_b$ be the chosen levels for the second factor. And suppose that n response observations are planned for each treatment A_iB_j, making a total of $N = abn$ observations. We assume that the observations are fully randomized against experimental units, time, or any uncontrolled factor.

The response vector for the N observations can be presented in a matrix array as follows:

			x_{21}				x_{2b}	
			B_1				B_b	
x_{11}	A_1	y_{111}	$\cdots$	y_{11n}	$\cdots$	y_{1b1}	$\cdots$	y_{1bn}
	$\vdots$		$\vdots$				$\vdots$	
x_{1a}	A_a	y_{a11}	$\cdots$	y_{a1n}	$\cdots$	y_{ab1}	$\cdots$	y_{abn}

In certain cases the factors may correspond to input variables x_1 and x_2; the values for such variables are indicated beside the factor levels.

An investigator is concerned with whether the factors A and B affect the response—and, if so, whether it is a simple dependence on one of the factors alone; or whether it is an *additive dependence* on both factors, so that a change in one factor shifts the response level by the same amount whatever the level of the other factor; or whether it is an *interactive dependence* on both factors with a response level special to each treatment A_iB_j. As before, we assume that the response is presented in such a form that the variation has the same scaling at the various possible response levels. And we consider the analysis for the case of approximately normal variation. The analysis for other distribution forms or with a parameter λ can be investigated using the theory in Section 11.3; and such analysis will typically need actual projections rather than just the squared lengths as recorded in the ordinary analysis-of-variance table.

If the factors do not affect the response, then the model for the response is

$$\mathbf{y} = \mu\mathbf{1} + \sigma\mathbf{z}, \tag{1}$$

where $\mathbf{z}$ is a sample of N from the standard normal. The fitted general-level vector and the corresponding squared length are

$$\begin{aligned} \mathbf{p}_1 &= \bar{y}\mathbf{1}, \\ SS_1 &= SS_{(1)} = N\bar{y}^2 = T^2_{\cdots}/N, \end{aligned} \tag{2}$$

where $T_{\cdots} = \Sigma_{ijs}\, y_{ijs}$ is the overall total and $\bar{y}$ is the overall average.

If only the factor A affects the response, the model is

$$\mathbf{y} = \sum \mu_i \mathbf{1}_{i\cdot} + \sigma\mathbf{z}, \tag{3}$$

where $\mathbf{1}_{i\cdot}$ is a vector with 1's in positions corresponding to the level A_i and 0's elsewhere; note that $\Sigma\, \mathbf{1}_{i\cdot} = \mathbf{1}$. The fitted general-level vector and the corresponding squared length are

$$\begin{aligned} \mathbf{p}_{(2)} &= \sum \bar{y}_{i\cdot}\mathbf{1}_{i\cdot}, \\ SS_{(2)} &= bn \sum \bar{y}^2_{i\cdot} = T^2_{i\cdot\cdot}/bn, \end{aligned} \tag{4}$$

where $T_{i\cdot\cdot} = \Sigma_{js} y_{ijs}$ is the total for the level A_i and $\bar{y}_{i\cdot} = T_{i\cdot\cdot}/bn$.

The projection into the space describing differences between the levels of the factor A, and the corresponding squared length are obtained by differencing

$$\begin{aligned} \mathbf{p}_2 &= \sum (\bar{y}_{i\cdot} - \bar{y})\mathbf{1}_{i}, \\ SS_2 &= bn \sum (\bar{y}_{i\cdot} - \bar{y})^2 = SS_{(2)} - SS_{(1)} = \sum_i T^2_{i\cdot\cdot}/bn - T^2_{\cdots}/N. \end{aligned} \tag{5}$$

Similarly, if only the factor B affects the response, the model is

$$\mathbf{y} = \sum \nu_j \mathbf{1}_{\cdot j} + \sigma\mathbf{z}, \tag{6}$$

where $\mathbf{1}_{\cdot j}$ is a vector with 1's in positions corresponding to the level B_j and 0's elsewhere; note that $\Sigma\, \mathbf{1}_{\cdot j} = \mathbf{1}$. The fitted general-level vector and the corresponding squared length are

$$\begin{aligned} p_{(2')} &= \sum \bar{y}_{\cdot j}\mathbf{1}_{\cdot j}, \\ SS_{(2')} &= an \sum \bar{y}^2_{\cdot j} = \sum T^2_{\cdot j\cdot}/an, \end{aligned} \tag{7}$$

where $T_{\cdot j\cdot} = \Sigma_{is} y_{ijs}$ is the total for the level B_j and $\bar{y}_{\cdot j} = T_{\cdot j\cdot}/an$ is the average for that level.

The projection into the space describing differences between the levels of the factor B and the corresponding squared length are obtained by differencing,

(8)
$$\mathbf{p}_{2'} = \sum (\bar{y}_{\cdot j} - \bar{y})\mathbf{1}_{\cdot j},$$
$$SS_{2'} = an \sum (\bar{y}_{\cdot j} - \bar{y})^2 = SS_{(2')} - SS_{(1)} = \sum_j T_{\cdot j \cdot}^2/an - T_{\cdots}^2/N.$$

The component $\Sigma (\bar{y}_{i\cdot} - \bar{y})\mathbf{1}_{i\cdot}$ describing A-differences is the projection of $\mathbf{y}$ into the subspace $\mathcal{L}(\mathbf{1}_{1\cdot}, \ldots, \mathbf{1}_{a\cdot} \perp \mathbf{1})$. The component describing B-differences is the projection of $\mathbf{y}$ into the subspace $\mathcal{L}(\mathbf{1}_{\cdot 1}, \ldots, \mathbf{1}_{\cdot b} \perp \mathbf{1})$. These subspaces are orthogonal and thus the A-difference component is orthogonal to the B-difference component. This is easily seen from an example:

$$\begin{pmatrix} 1 & 1 & 1 \\ -1 & -1 & -1 \end{pmatrix} \perp \begin{pmatrix} 2 & -1 & -1 \\ 2 & -1 & -1 \end{pmatrix}.$$

And it is easily verified generally: note that the inner product of $\mathbf{1}_{i\cdot}$ and $\mathbf{1}_{\cdot j}$ is n; thus the inner product of $\Sigma\, l_i \mathbf{1}_{i\cdot}$ and $\Sigma\, m_j \mathbf{1}_{\cdot j}$, where $\Sigma\, l_i = 0 = \Sigma\, m_j$, is $\Sigma\, l_i m_j n = 0 \cdot 0 \cdot n = 0$.

If the factors A and B can interact in their effect on the response, then the model is

(9) $$\mathbf{y} = \sum \mu_{ij}\mathbf{1}_{ij} + \sigma \mathbf{z},$$

where $\mathbf{1}_{ij}$ is a vector with 1's in positions corresponding to $A_i B_j$ and 0's elsewhere. The vectors $\mathbf{1}_{ij}$ are orthogonal; hence the vectors $\Sigma\, \mu_{ij}\mathbf{1}_{ij}$ form an ab-dimensional subspace. The fitted general-level vector and the corresponding squared length

(10)
$$\mathbf{p}_{(3)} = \sum \bar{y}_{ij}\mathbf{1}_{ij},$$
$$SS_{(3)} = n \sum \bar{y}_{ij}^2 = \sum_{ij} T_{ij\cdot}^2/n,$$

where $T_{ij\cdot} = \Sigma_s\, y_{ijs}$ is the total for the treatment $A_i B_j$ and $\bar{y}_{ij} = T_{ij\cdot}/n$ is the average value for that treatment.

Note the following vector relations:

$$\sum_i \mathbf{1}_{ij} = \mathbf{1}_{\cdot j}, \qquad \sum_j \mathbf{1}_{ij} = \mathbf{1}_{i\cdot}, \qquad \sum_{ij} \mathbf{1}_{ij} = \mathbf{1}.$$

Hence the subspaces for the mean, for the A effects, and for the B effects are component spaces of the ab-dimensional space $\mathcal{L}(\mathbf{1}_{ij}: \text{all } i, j)$. Now consider the AB *interaction,* the AB effects beyond that for the mean, for A levels, and for B levels. The projection into the space describing the AB interaction and the corresponding squared length can be obtained by differencing,

(11)
$$\mathbf{p}_3 = \mathbf{p}_{(3)} - \mathbf{p}_2 - \mathbf{p}_{2'} - \mathbf{p}_1 = \sum (\bar{y}_{ij} - \bar{y}_{i\cdot} - \bar{y}_{\cdot j} + \bar{y})\mathbf{1}_{ij},$$
$$SS_3 = n \sum (\bar{y}_{ij} - \bar{y}_{i\cdot} - \bar{y}_{\cdot j} + \bar{y})^2$$

$$= \sum_{ij} T_{ij\cdot}^2/n - \sum_i T_{i\cdot\cdot}^2/bn - \sum_j T_{\cdot j\cdot}^2/an + T_{\cdot\cdot\cdot}^2/N.$$

Note the algebraic identity obtained from the two ways of deriving the sum of squares.

The analysis-of-variance table (Table 12.10) summarizes some of the preceding results. Note that the SS entries are best computed by differencing, as in the second expressions in (5), (8), and (11). MS entries are calculated as SS/DF.

TABLE 12.10
Analysis-of-variance table

	Source	DF	Projection	SS
1	Mean	1	$\bar{y}\mathbf{1}$	$N\bar{y}^2$
2	A	$a - 1$	$\sum (\bar{y}_{i\cdot} - \bar{y})\mathbf{1}_{i\cdot}$	$bn \sum (\bar{y}_{i\cdot} - \bar{y})^2$
2′	B	$b - 1$	$\sum (\bar{y}_{\cdot j} - \bar{y})\mathbf{1}_{\cdot j}$	$an \sum (\bar{y}_{\cdot j} - \bar{y})^2$
3	$A \times B$	$(a - 1)(b - 1)$	$\sum (\bar{y}_{ij} - \bar{y}_{i\cdot} - \bar{y}_{\cdot j} + \bar{y})\mathbf{1}_{ij}$	$n \sum (\bar{y}_{ij} - \bar{y}_{i\cdot} - \bar{y}_{\cdot j} + \bar{y})^2$
4	Residual	$ab(n - 1)$	$\mathbf{y} - \sum \bar{y}_{ij}\mathbf{1}_{ij}$	$\sum (y_{ijs} - \bar{y}_{ij})^2$
		$abn = N$	$\mathbf{y}$	$\sum y_{ijs}^2$

B THE ANALYSIS

We consider the analysis for the case of normal variation. The analysis for other distribution forms is based on the theory in Section 11.3. We assume that the response y has variation σz, where z is standard normal.

The distribution theory is available from Section 11.3F and follows the pattern discussed in Section 12.1C. The distribution of an SS entry is that of $\sigma^2\chi^2$, where χ^2 has the "chi-square distribution" with degrees of freedom given by the corresponding DF entry; the chi-square distribution is ordinary chi-square if the true mean vector has zero projection in the particular subspace, and it is noncentral chi-square with values that tend to be larger if the mean vector has a nonzero projection.

Now consider the problem of testing for dependence on the factors A and B. The first test for dependence is made by calculating $F = MS_3/MS_4$ and comparing the value with the appropriate F distribution. Significance indicates dependence on the two factors; further analysis would be concerned with the form of the dependence on the two factors.

If the preceding test gives no significance, then an assumption that there is no interaction could be entertained; more detailed investigations would, of course, be needed to support such an assumption. Under such an assumption we can test separately for A effects and for B effects by calculating the F ratios

$$F = \frac{MS_2}{MS_4}, \qquad F = \frac{MS_{2'}}{MS_4}$$

and comparing the values with the appropriate F distribution.

Some of the arithmetic can be illustrated by a very simple numerical example. The example is so simple, however, that we have no degrees of freedom left for estimating the error variance and making appropriate tests; but earlier examples illustrate the procedure for making the tests.

EXAMPLE 1

Consider a response observed at three levels for a factor A and two levels for a factor B. The six observations are

	B_1	B_2	*Total*	
A_1	5	1	6	
A_2	1	1	2	
A_3	3	1	4	28
Total	9	3	12	
		30		24

together with some calculated numbers: $12^2/6 = 24$; $(6^2 + 2^2 + 4^2)/2 = 28$; and $(9^2 + 3^2)/3 = 30$. The sum of squares of the observations is 38.

The projections for the various models can be calculated from the array, row, column, and cell averages; the sums of squares are available from the numbers just recorded. It is convenient to arrange these in a partial ordering corresponding to relative generality:

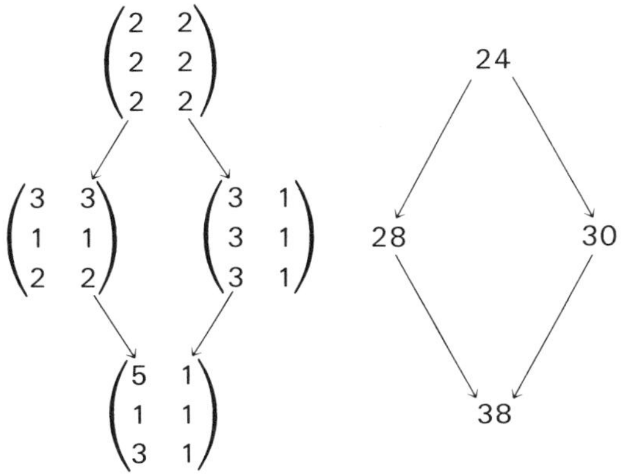

By differencing we then obtain the projection and squared lengths corresponding to orthogonal subspaces:

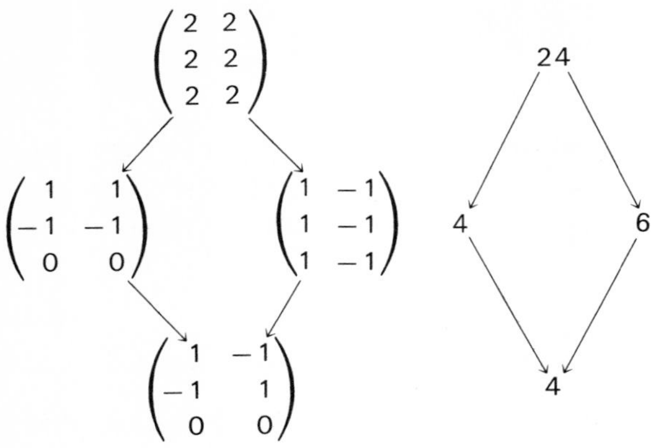

The analysis-of-variance table is available from the preliminary calculations (Table 12.11).

TABLE 12.11
Analysis-of-variance table

	Source	DF	SS
1	Mean	1	24
2	A	2	$4 = 28 - 24$
2′	B	1	$6 = 30 - 24$
3	$A \times B$	2	$4 = 38 -$ preceding
4	Residual	—	
			38

C FORM OF THE DEPENDENCE ON FACTORS A AND B

Dependence on the factors A and B can be investigated in part by separating out individual dimensions in the analysis-of-variance table.

Differences between various averages can be used as a basis for individual dimensions. For an example, suppose that the factor A has three levels. There are then two dimensions for A-differences. One dimension could be used say to measure how the average for A_2 and A_3 exceeds the value at A_1 and a second dimension (orthogonal) could examine the difference between A_2 and A_3. Such a separation is based on the orthogonal vectors $(-1, 1/2, 1/2)$ and $(0, -1, 1)$; of course, for the full design each coordinate as given would appear bn times. The resulting components of the sum of squares for the factor A would be

$$\frac{\left(-\sum_{js} y_{1js} + (1/2)\sum_{js} y_{2js} + (1/2)\sum_{js} y_{3js}\right)^2}{bn + (1/4)bn + (1/4)bn},$$

$$\frac{\left(-\sum_{js} y_{2js} + \sum_{js} y_{3js}\right)^2}{bn + bn}.$$

And suppose that the factor B has two levels. There is then one dimension for B differences, and the vector $(-1, +1)$ for the two levels generates the subspace; of course, for the full design each coordinate as given would appear an times. The resulting component of the sum of squares for the factor B would be

$$\frac{\left(-\sum_{is} y_{i1s} + \sum_{is} y_{i2s}\right)^2}{an + an}.$$

The vectors for the factor A and the vectors for the factor B can be combined to give multipliers for the interaction:

$$(12) \qquad \begin{pmatrix} -1 \\ \frac{1}{2} \\ \frac{1}{2} \end{pmatrix} (-1, 1) = \begin{pmatrix} 1 & -1 \\ -\frac{1}{2} & \frac{1}{2} \\ -\frac{1}{2} & \frac{1}{2} \end{pmatrix} = (a_{ij}),$$

$$(13) \qquad \begin{pmatrix} 0 \\ -1 \\ 1 \end{pmatrix} (-1, 1) = \begin{pmatrix} 0 & 0 \\ 1 & -1 \\ -1 & 1 \end{pmatrix} = (b_{ij});$$

of course, for the full design each coordinate in the arrays at right would appear n times. The resulting components of the sum of squares for the AB interaction would be

$$\frac{\left(\sum_{ijs} a_{ij} y_{ijs}\right)^2}{n \sum a_{ij}^2},$$

$$\frac{\left(\sum_{ijs} b_{ij} y_{ijs}\right)^2}{n \sum b_{ij}^2}.$$

If an input variable is available for the factor A, then it and various functions of it can be used to separate out individual dimensions following the pattern in Section 12.1D. Similarly, if an input variable is available for the factor B, then in a similar way various individual dimensions can be separated out again in the pattern of Section 12.1D.

Now consider the case of possible interaction and suppose that input variables x_1 and x_2 are available for the factors A and B. Various functions of the x's could provide components for structuring the response level. We illustrate the methods here by continuing with the linear and quadratic dependencies examined in Section 12.1. A quadratic expression in x_1 and x_2 has the form

$$a + a_1x_1 + a_{11}x_1^2 + a_2x_2 + a_{22}x_2^2 + a_{12}x_1x_2.$$

For x_1 we can follow the pattern in Section 12.1 and use the orthogonal functions

$$x_1 - \bar{x}_1, \qquad x_1^2 - c_1 - d_1x_1,$$

where

$$\sum (x_{1i}^2 - c_1 - d_1 x_{1i}) = 0,$$

$$\sum (x_{1i}^2 - c_1 - d_1 x_{1i}) x_{1i} = 0,$$

and $\bar{x}_1 = \Sigma\, x_{1i}/a$. These generate two dimensions corresponding to A-differences. And similarly for x_2, we can use the orthogonal functions

$$x_2 - \bar{x}_2, \qquad x_2^2 - c_2 - d_2 x_2,$$

where

$$\sum (x_{2j}^2 - c_2 - d_2 x_{2j}) = 0,$$
$$\sum (x_{2j}^2 - c_2 - d_2 x_{2j}) x_{2j} = 0,$$

TABLE 12.12
Analysis-of-variance table

Source		DF	SS
Mean		1	$\dfrac{\left(\sum y_{ijs}\right)^2}{N}$
A	Linear	1	$\dfrac{\left(\sum (x_{1i} - \bar{x}_1) y_{ijs}\right)^2}{bn \sum (x_{1i} - \bar{x}_1)^2}$
	Quadratic	1	$\dfrac{\left(\sum (x_{1i}^2 - c_1 - d_1 x_{1i}) y_{ijs}\right)^2}{bn \sum (x_{1i}^2 - c_1 - d_1 x_{1i})^2}$
	Other	$a - 3$	By differencing
B	Linear	1	$\dfrac{\left(\sum (x_{2j} - \bar{x}_2) y_{ijs}\right)^2}{an \sum (x_{2j} - \bar{x}_2)^2}$
	Quadratic	1	$\dfrac{\left(\sum (x_{2j}^2 - c_2 - d_2 x_{2j}) y_{ijs}\right)^2}{an \sum (x_{2j}^2 - c_2 - d_2 x_{2j})^2}$
	Other	$b - 3$	By differencing
$A \times B$	Quadratic	1	$\dfrac{\left(\sum (x_{1i} - \bar{x}_1)(x_{2j} - \bar{x}_2) y_{ijs}\right)^2}{n \sum (x_{1i} - \bar{x}_1)^2 \sum (x_{2j} - \bar{x}_2)^2}$
	Other	$(a - 1)(b - 1) - 1$	By differencing
Residual		$ab(n - 1)$	$\sum (y_{ijs} - \bar{y}_{ij})^2$
		N	$\sum y_{ijs}^2$

and $\bar{x}_2 = \Sigma\, x_{2j}/b$. These generate two dimensions corresponding to the B-differences. An orthogonal component for A-differences can be combined with an orthogonal component for B-differences to give an orthogonal component for the AB interaction; thus in the at-most-quadratic case we obtain the function

$$(x_1 - \bar{x}_1)(x_2 - \bar{x}_2)$$

as with the formulas (12) and (13).

The analysis-of-variance table then takes the form shown in Table 12.12. Some of the entries are available by differencing with respect to the earlier table. The various projections can be given explicitly by analogy from the sum-of-squares entries. For example, the projection onto $\mathfrak{L}(\Sigma\,(x_{1i} - \bar{x}_1)\mathbf{1}_{i\cdot})$ is

$$\frac{\sum (x_{1i} - \bar{x}_1) y_{ijs}}{bn \sum (x_{1i} - \bar{x}_1)^2} \cdot \sum (x_{1i} - \bar{x}_1)\mathbf{1}_{i\cdot\cdot}$$

Tests for dependence beyond the quadratic can be made by comparing the entries marked ''other'' against the residual.

On the assumption that dependence is at most quadratic, tests can be made by comparing the various quadratic and linear entries against the residual following the general pattern described in Section 12.1.

D EXERCISES

1 Twelve response observations were made randomly in time, two at each combination of three levels of a first factor temperature and two levels of a second factor pressure; the variation is known to be approximately normal. The data are as follows:

Temperature, x_1	Pressure, x_2: -1	Pressure, x_2: $+1$
73	14.2	18.3
	14.3	18.0
74	17.4	21.4
	17.6	21.0
75	20.7	23.8
	20.4	23.9

Construct the analysis-of-variance table separating Mean, T, P, $T \times P$, and Residual. Test for dependence on the factors.

2 (*continuation*) Separate the dependence on factor T into Linear and Other and make appropriate tests.

3 The effect of plate temperature T and filament lighting L on conductance was investigated for a certain kind of tube. Three observations were made at each combination of two levels for temperature and four levels for filament lighting:

	L_1	L_2	L_3	L_4
T_1	3774	4710	4176	4540
	4364	4180	4140	4530
	4374	4514	4398	3964
T_2	4216	3828	4122	4484
	4524	4170	4280	4332
	4136	4180	4226	4390

Make appropriate tests for dependence on the factors T and L.

4 The effects of fertilizer F and chemical seed treatment C on growth of wheat were investigated on 48 plots of land; three determinations were made at each combination of four levels for F and four levels for C:

	C_1	C_2	C_3	C_4
F_1	21.4	20.9	19.6	17.6
	21.2	20.3	18.8	16.6
	20.1	19.8	16.6	17.5
F_2	12.0	13.6	13.0	13.3
	14.2	13.3	13.7	14.0
	12.1	11.6	12.0	13.9
F_3	13.5	14.0	12.7	12.4
	11.9	15.6	12.9	13.7
	13.4	13.8	13.1	13.0
F_4	12.8	14.1	14.2	12.0
	13.8	13.2	13.6	14.6
	13.7	15.3	13.3	14.0

Make appropriate tests for dependence on the factors F and C.

SUPPLEMENTARY MATERIAL

Supplement Sections 12.3, 12.4, and 12.5 are on pages 584–598.

The one- and two-factor designs in preceding sections are the basic designs for experimentation. The supplementary material examines the extension to the *three-factor design* and by implication the general multifactor design. It also examines how information concerning variability from experimental unit to experimental unit can be used to increase the sensitivity of the experimental design: specifically, *randomized blocks* and the *analysis of covariance*.

NOTES AND REFERENCES

The emphasis in this chapter has been on principles of experimentation and on certain principles in the related mathematical analyses. The author believes that an introductory course on statistics should include some contact with these principles.

A full appreciation of the importance and substance of experimental design, however, requires close contact with the design process in various applications. For some additional readings, see the following:

Cox, D. R. (1958). *Planning of Experiments*. New York: John Wiley & Sons, Inc.

Fisher, R. A. (1971). *The Design of Experiments,* 9th ed. New York: Macmillan Publishing Co., Inc.

Snedecor, G. W., and W. G. Cochran (1967). *Statistical Methods,* 6th ed. Ames, Iowa: Iowa State University Press.

APPENDIX I
SUPPLEMENTARY MATERIAL

1.5

PROBABILITY

We have developed the basic mathematics for handling probabilities. A small amount of additional development gives some very useful results.

A SIMPLE FUNCTIONS AND PROBABILITY FORMULAS

Indicator functions can be very useful. In Section 1.1 we noted that the long-run average of an indicator function was the corresponding probability. In Section 1.2 we proved the relation

$$I_{A \cup B} = I_A + I_B - I_{AB},$$

which corresponds closely to the relation

$$P(A \cup B) = P(A) + P(B) - P(AB)$$

in Section 1.4. In this section we generalize the notion of an indicator function and establish more generally the correspondence just noted.

Consider a sample space $\mathcal{S}$ and a class $\mathcal{A}$ of subsets that form a σ-algebra.

DEFINITION 1

A real-valued function Y on $\mathcal{S}$ is a **simple function** if it takes only a finite number of values.

For an example, see Figure S.1.

Let $a_1, \ldots, a_m$ be the different values for the function Y, and let $A_1, \ldots, A_m$ be the corresponding sets on which these values are taken:

$$A_j = \{s : Y(s) = a_j\} = Y^{-1}(a_j).$$

We assume, of course, that the sets A_i belong to $\mathcal{A}$. The different sets $A_1, \ldots, A_m$ are clearly disjoint and they cover $\mathcal{S}$; they thus form a partition of $\mathcal{S}$. The simple function Y can then be written in the canonical form,

$$(1) \qquad Y = a_1 I_{A_1} + \cdots + a_m I_{A_m};$$

note that at any point s at most one term on the right is different from zero and the a-value of that term is the value of the function.

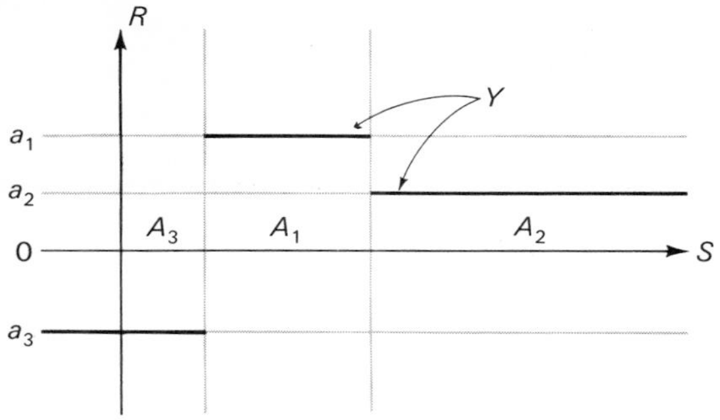

FIGURE S.1
Simple function Y takes the values a_1, a_2, and a_3, respectively, on A_1, A_2, and A_3.

EXAMPLE 1

Consider the rolling of a symmetric die and suppose that various characteristics are observed including the number of points. Let $\mathcal{S}$ be the sample space and $\mathcal{A}$ be the class of events; and let Y be the function on $\mathcal{S}$ that gives the number of points.

The function Y is a simple function that takes only the values 1, 2, . . ., 6; it can be put in the canonical form

$$Y = 1I_{E_1} + 2I_{E_2} + 3I_{E_3} + 4I_{E_4} + 5I_{E_5} + 6I_{E_6},$$

where E_j is the event "the number of points is j." Of course, Y can be expressed in other ways, for example,

$$Y = I_{E_{[1]}} + \cdots + I_{E_{[6]}},$$

where $E_{[j]}$ is the event "the number of points is j or more." Note that if the sample space $\mathcal{S}$ is $\{1, 2, \ldots, 6\}$ recording just the number of points, then Y is the identity function on $\mathcal{S}$.

Now suppose that we have found someone who will give us \$1 for every dot that appears when the die is rolled. If this game is played a large number of times, say N, then $Y = 1$ will occur approximately $N/6$ times, . . ., and $Y = 6$ will occur approximately $N/6$ times. The total amount received will be approximately

$$1 \cdot (1/6)N + 2 \cdot (1/6)N + \cdots + 6 \cdot (1/6)N$$

dollars and the average received per play will be

$$1 \cdot (1/6) + 2 \cdot (1/6) + \cdots + 6 \cdot (1/6) = 3.50$$

dollars. A fair price to play this game would then be \$3.50. Note that the preceding average is obtained by multiplying values by corresponding probabilities and summing. The same rule also works with the second expression for Y:

$$1 \cdot (6/6) + 1 \cdot (5/6) + \cdots + 1 \cdot (1/6) = 3.50.$$

Now consider in general this average value, or mean value, or "expected" value. Let P be a probability measure over the sample space $\mathcal{S}$; then

DEFINITION 2

The **mean value** of a simple function Y having canonical form

$$Y = a_1 I_{A_1} + \cdots + a_m I_{A_m}$$

is given by

$$(2) \qquad E(Y) = a_1P(A_1) + \cdots + a_mP(A_m).$$

For a random system with variation described by P, the mean $E(Y)$ represents the long-run average of the values of the function Y; compare with the example. This capital E is often called *expectation*.

Formula (2) for $E(Y)$ works whether or not Y is expressed in canonical form:

PROPOSITION 3

If a simple function Y is expressed as

$$Y = \sum_1^k b_j I_{B_j},$$

then

$$(3) \qquad E(Y) = \sum b_j P(B_j).$$

The proof is deferred until Section I.

Consider the constant function taking the value 1; then trivially from (2) we have that $E(1) = 1$. And, similarly, for a constant k we have $E(k) = k$.

The usefulness of E is largely based on the following linearity property:

THEOREM 4

If Y_1 and Y_2 are simple functions, then the mean of a linear combination is the linear combination of the means:

$$(4) \qquad E(k_1Y_1 + k_2Y_2) = k_1E(Y_1) + k_2E(Y_2).$$

Proof The proof is trivial using Proposition 3. Let

$$Y_1 = \sum a_i I_{A_i}, \qquad Y_2 = \sum b_j I_{B_j}$$

be representations for Y_1 and Y_2 in terms of simple functions. Then

$$\begin{aligned} E(k_1Y_1 + k_2Y_2) &= E\left(\sum k_1 a_i I_{A_i} + \sum k_2 b_j I_{B_j}\right) \\ &= k_1 \sum a_i P(A_i) + k_2 \sum b_j P(B_j) \\ &= k_1E(Y_1) + k_2E(Y_2). \end{aligned}$$

As a simple illustration, consider the relation

$$I_{A \cup B} = I_A + I_B - I_{AB}$$

from Section 1.2E. Applying E to both sides gives immediately the relation (1.4.3),

$$P(A \cup B) = P(A) + P(B) - P(AB)$$

in Section 1.4.

More generally, we have

$$\begin{aligned} I_{A \cup B \cup C} &= 1 - (1 - I_A)(1 - I_B)(1 - I_C) \\ &= I_A + I_B + I_C - I_{AB} - I_{BC} - I_{AC} + I_{ABC}. \end{aligned}$$

Applying E to both sides gives the relation (1.4.4)

$$P(A_1 \cup A_2 \cup A_3) = P(A_1) + P(A_2) + P(A_3) - P(A_1A_2) - P(A_1A_3) - P(A_2A_3) + P(A_1A_2A_3)$$

which was recorded in Section 1.4.

B PROBLEMS: PROBABILITY FORMULAS

The problems here use properties of the combinatorial function

$$\binom{n}{r} = \frac{n!}{r!(n-r)!} = \frac{n(n-1)\cdots(n-r+1)}{r!},$$

called *n choose r*. These will not be used in the main body of the text until Section 3.4.

1 Let $A_1, \ldots, A_m$ be subsets of the sample space $\mathcal{S}$ and P be a probability measure over $\mathcal{S}$. Use Problem 1.2.22 to prove that

$$(5) \qquad P(\cup_1^m A_i) = S_1 - S_2 + S_3 - + \cdots (-1)^{m-1}S_m,$$

where $S_r = \Sigma_{i_1<\cdots<i_r} P(A_{i_1} \cdots A_{i_r})$.

2 (*continuation*) Verify that the indicator function for E_r "exactly r of the events $A_1, \ldots, A_m$" can be expressed as

$$I_{E_r} = \sum_{\substack{i_1<\cdots<i_r \\ i_{r+1}<\cdots<i_m}} I_{i_1} \cdots I_{i_r}(1 - I_{i_{r+1}}) \cdots (1 - I_{i_m}),$$

where the summation is over those permutations $(i_1, \ldots, i_m)$ of $(1, \ldots, m)$ that satisfy the constraints indicated. Hence prove that

$$I_{E_r} = T_r - \binom{r+1}{r} T_{r+1} + \binom{r+2}{r} T_{r+2} - \cdots + (-1)^{m-r}\binom{m}{r} T_m,$$

where

$$T_r = \sum_{i_1<\cdots<i_r} I_{i_1} \cdots I_{i_r}.$$

3 (*continuation*) Prove the following generalization of formula (5):

$$(6) \qquad P(E_r) = S_r - \binom{r+1}{r} S_{r+1} + \binom{r+2}{r} S_{r+2} - \cdots (-1)^{m-r}\binom{m}{r} S_m$$

$$= S_r - (r+1)S_{r+1} + \frac{(r+2)(r+1)}{2} S_{r+2} - + \cdots;$$

see Problem 1.4.14 and note that $S_0 = 1$.

4 A deck contains N cards numbered $1, 2, \ldots, N$. The cards are thoroughly shuffled and dealt successively as the dealer calls $1, 2, \ldots, N$. Let E_r be the event "exactly r called correctly." Use Problem 3 to show that

$$P(E_r) = \frac{1}{r!}\left(1 - 1 + \frac{1}{2!} - \frac{1}{3!} \cdots \pm \frac{1}{(N-r)!}\right)$$

$$\approx \frac{1}{r!} e^{-1};$$

see Problem 1.4.16. Calculate an approximate value for the probability of no cards called correctly. Note that for medium to large N the number of cards called correctly is approximately Poisson(λ) with $\lambda = 1$ (see Example 1.3.6).

5 (*continuation of Problem 3*) Let $E_{[r]}$ be the event "at least r of $A_1, \ldots, A_m$." Recall Problem 1.4.15 and prove that

$$(7) \quad P(E_{[r]}) = S_r - \binom{r}{r-1} S_{r+1} + \binom{r+1}{r-1} S_{r+2} - \cdots (-1)^{m-r} \binom{m-1}{r-1} S_m$$

$$= S_r - rS_{r+1} + \frac{(r+1)r}{2} S_{r+2} - + \cdots;$$

use the combinatorial result in Problem 3.4.9.

C PROBLEMS: SIMPLE CLASSES OF SETS

As we have noted before, it is a long step from the intervals on R to the Borel sets on R. Something slightly more general than intervals is useful for studying probability measures.

Let $\mathcal{R}$ be a nonempty class of subsets of a sample space $\mathcal{S}$. Then we define the following properties (note that $A_1 - A_2 = A_1 A_2^c$):

Closure under difference: If $A_1, A_2 \in \mathcal{R}$, then $A_1 - A_2 \in \mathcal{R}$.
Closure under union: If $A_1, A_2 \in \mathcal{R}$, then $A_1 \cup A_2 \in \mathcal{R}$.
Closure under σ-union: If $A_i \in \mathcal{R}$, then $\cup_{i=1}^{\infty} A_i \in \mathcal{R}$.

6 $\mathcal{R}$ is a *ring* (set-theoretic) if it is closed under differences and union. Show that an equivalent definition is closure under symmetric difference and union; the symmetric difference $A_1 \triangle A_2$ is defined to be $(A_1 - A_2) \cup (A_2 - A_1)$.

7 $\mathcal{R}$ is an *algebra* if it is closed under union and complementation. Show that an equivalent definition is a *ring containing* $\mathcal{S}$.

8 $\mathcal{R}$ is a σ-ring if it is closed under difference and σ-union. Show that a σ-algebra can be defined as a σ-ring containing $\mathcal{S}$.

9 Let $\mathcal{R}$ be the class of finite unions of disjoint left open–right closed intervals $(a, b]$ on R; that is,

$$\mathcal{R} = \{\cup_{i=1}^{n} (a_i, b_i] : a_1 < b_1 \le a_2 < \cdots \le a_n < b_n\}.$$

Show that $\mathcal{R}$ is a ring. Some tedious but routine extensions of the proof establish the following: If $\mathcal{R}$ is the class of finite unions of disjoint rectangles

$$(a_1, b_1] \times \cdots \times (a_k, b_k]$$

on R^k, then $\mathcal{R}$ is a ring.

10 Let $\mathcal{A}_0 = \{A\}$ be a class of subsets A on a sample space $\mathcal{S}$. Show that there exists a σ-ring on $\mathcal{S}$ that contains $\mathcal{A}_0$.

11 (*continuation*) Let $K = \{\mathcal{A}_\alpha\}$ be the set of all σ-rings $\mathcal{A}_\alpha$ such that $\mathcal{A}_0 \subset \mathcal{A}_\alpha$ and α ranges over an index set. Show that $\mathcal{A} = \cap \mathcal{A}_\alpha$ is a σ-ring on $\mathcal{S}$.

12 (*continuation*) If $\mathcal{B}_\alpha \subset \mathcal{B}_\beta$, we say that $\mathcal{B}_\alpha$ is smaller than $\mathcal{B}_\beta$. Show that it is meaningful to speak of the smallest σ-ring containing a class $\mathcal{A}_0$ of subsets of $\mathcal{S}$; let $S(\mathcal{A}_0)$ designate this σ-ring.

D THE EXTENSION THEOREM

It is easy to define the length or some other measure of an interval. In this section we quote an important extension theorem that then says a measure is uniquely determined for the logical consequences of intervals—the Borel sets.

Consider a sample space $\mathcal{S}$ and a ring $\mathcal{R}$ of subsets of $\mathcal{S}$; a ring is defined in Problem 6. And suppose that we have a measure μ defined on the ring $\mathcal{R}$; a measure on a ring is defined as before: σ-additive (for countable unions that *belong* to $\mathcal{R}$) and nontrivial. A measure μ on $\mathcal{R}$ is called σ-finite if each A in $\mathcal{R}$ can be expressed as $\cup_1^{\infty} A_i$, where each $\mu(A_i) < \infty$.

THEOREM 5

If μ is a measure on a ring $\mathcal{R}$ and if μ is σ-finite, then there is a unique extension of μ to the smallest σ-ring $S(\mathcal{R})$ containing $\mathcal{R}$.

The proof is not simple and may be found in advanced books on measure theory.

E MEASURING LENGTH, AREA, AND VOLUME

In Section 1.3A we considered the real line R as a uniform rod having unit weight for unit length. The weight measure of a set A as a piece of the rod then gives the length measure of the set A. Consider this more formally.

On the real line R we define the length of an interval $(a, b]$ by

$$\mu((a, b]) = b - a.$$

We could consider open intervals or closed intervals, but the left open–right closed intervals have some advantages; see Problem 9. In the same way we define the length of $\cup_1^k (a_i, b_i]$, where the components are disjoint, by

$$\mu(\cup_1^k (a_i, b_i]) = \sum_1^k (b_i - a_i).$$

The class of finite unions of left closed–right open intervals forms a ring $\mathcal{R}$; Problem 9. The Extension Theorem in Section D then says there is a unique extension of μ to the smallest σ-ring $S(\mathcal{R})$ containing $\mathcal{R}$. This σ-ring is, however, the σ-algebra of Borel sets. Thus length as defined for intervals extends uniquely to a length measure for all Borel sets on R. This is called *Lebesgue measure* on the line R.

Now consider the plane R^2, or, more generally, the Cartesian product R^k. We define the volume of a rectangle $(a_1, b_1] \times \cdots \times (a_k, b_k]$ by

$$\mu\left((a_1, b_1] \times \cdots \times (a_k, b_k]\right) = \prod_1^k (b_i - a_i).$$

And for a finite union of disjoint rectangles we define it to be the sum of the component volumes. The class of finite unions of left open–right closed rectangles forms a ring $\mathcal{R}$; Problem 9. The Extension Theorem then gives a unique extension of μ to the smallest σ-ring containing $\mathcal{R}$; this σ-ring is the Borel class $\mathcal{B}^k$ on R^k. Thus volume as defined for rectangles extends uniquely to a volume measure for all Borel sets on R^k. This extension is called Lebesgue measure on R^k.

F PROBLEMS: LENGTH MEASURE

13 Let μ be Lebesgue measure on R.
   (a) Show that $\mu(\{x\}) = 0$, the measure of a single point is zero.
   (b) Show that $\mu(A) = 0$ if A is finite or countable.

14 Consider the real line R and let $A_0, A_1, \ldots$ be a sequence of sets defined as follows: $A_0 = [0, 1]$, $A_1 = [0, 1/3] \cup [2/3, 1]$, $A_2 = [0, 1/9] \cup [2/9, 1/3] \cup [2/3, 7/9] \cup [8/9, 1], \ldots$, where A_{n+1} is obtained from A_n by deleting the open middle third of each component interval. The intersection set $A = \cap_1^\infty A_i$ is called the *Cantor ternary set*. Let μ be Lebesgue measure.
   (a) Calculate $\mu(A_0)$, $\mu(A_1)$, and $\mu(A_2)$.
   (b) Calculate $\mu(A_n)$ and deduce $\mu(A)$.

15 (*continuation*) Let x belong to A, and hence to $A_0, A_1, A_2, \ldots$; let $i_1 = 0(1)$ if x is in the lower (upper) interval formed in going from A_0 to A_1; let $i_2 = 0(1)$ if x is in the lower (upper) interval in going from A_1 to $A_2, \ldots$. Thus x produces $(i_1, i_2, \ldots)$, or $.i_1 i_2 i_3 \cdots$ as a binary-decimal sequence. Note that all binary-decimal sequences can be obtained. Show that A has the same cardinality as the interval $[0, 1]$ itself, that is, the same as the continuum.

G PROBLEMS: LIMIT SETS

16 Let $A_1, A_2, \ldots$ be a sequence of subsets of a sample space $\mathcal{S}$.

(a) Let U be the set of points that belong to an infinity of the sets $A_1, A_2, \ldots$; we write $U = \limsup_{i \to \infty} A_i$. Prove that

$$U = \cap_{n=1}^{\infty} \cup_{i=n}^{\infty} A_i = \cap_{n=1}^{\infty} B_n;$$

note that $B_1, B_2, \ldots$ is a decreasing sequence where $B_n = \cup_{i=n}^{\infty} A_i$.

(b) Let I_i be the indicator function for A_i. Prove that

$$I_U = \limsup_{i \to \infty} I_i.$$

This is a generalization of part of Problem 1.4.19. The definition of limsup for functions is recorded in Section 3.6G.

17 Let $A_1, A_2, \ldots$ be a sequence of subsets of a sample space $\mathcal{S}$.

(a) Let L be the set of points that belong to all but a finite number of the sets $A_1, A_2, \ldots$; we write $L = \liminf_{i \to \infty} A_i$. Prove that

$$L = \cup_{n=1}^{\infty} \cap_{i=n}^{\infty} A_i = \cup_{n=1}^{\infty} C_n;$$

note that $C_1, C_2, \ldots$ is an increasing sequence, where $C_n = \cap_{i=n}^{\infty} A_i$.

(b) Let I_i be the indicator function for A_i. Prove that

$$I_1 = \liminf_{i \to \infty} I_i.$$

This is a generalization of part of Problem 1.4.19.

18 ***Borel–Cantelli Lemma:*** *Let $A_1, A_2, \ldots$ be a sequence of elements of a σ-algebra $\mathcal{A}$. If $\Sigma_1^{\infty} P(A_i) < \infty$, then the probability that an infinity of the A_i occur is zero.* See Problem 16 and note that the event just mentioned is $\limsup A_i$. Also note that $\cup_{i=n}^{\infty} A_i$ is monotone decreasing, and that $P(\cup_n^{\infty} A_i) \le \Sigma_n^{\infty} P(A_i)$. Prove the lemma.

H PROBLEMS: LIMIT FUNCTIONS

The following problem examines in depth what it means to have a sequence of functions converge to a limit function.

19 Let $Y, Y_1, Y_2, \ldots$ be functions mapping $\mathcal{S}$ into R; that is, $Y_i: \mathcal{S} \to R$. Let $C_i(\epsilon)$ be the set of points for which Y_i is at least ϵ close to Y:

$$C_i(\epsilon) = \{s: |Y_i(s) - Y(s)| \le \epsilon\}.$$

(a) Let $C_{[j]}(\epsilon)$ be the set of points for which each of $Y_j, Y_{j+1}, \ldots$ is at least ϵ close to Y. Show that

$$C_{[j]}(\epsilon) = \cap_{i=j}^{\infty} C_i(\epsilon) = \left\{s: \sup_{i=j}^{\infty} |Y_i(s) - Y(s)| \le \epsilon\right\}.$$

Note that $C_{[j]}(\epsilon)$ is monotone increasing in j.

(b) Let $C(\epsilon)$ be the set of points for which the sequence $Y_1, Y_2, \ldots$ eventually is ϵ close to Y. Show that

$C(\epsilon) = \cup_{j=1}^{\infty} C_{[j]}(\epsilon)$.

Note that $C(\epsilon)$ is monotone decreasing as $\epsilon \to 0$.

(c) Let C be the set of points for which the sequence $Y_1, Y_2, \ldots$ converges to Y. Show that

$$C = \left\{s : \lim_{i\to\infty} Y_i(s) = Y(s)\right\} = \cap_{n=1}^{\infty} C\left(\frac{1}{n}\right).$$

(d) If all the sets $\{s : |Y_i(s) - Y(s)| \leq a\}$ belong to a σ-algebra $\mathcal{A}$, prove that $C \in \mathcal{A}$. For a probability application of this problem, see Section 6.3E.

I MEAN VALUE OF SIMPLE FUNCTIONS

We now prove Proposition 3, which states that the mean of a simple function can be calculated directly from any expression for the simple function. The proposition asserts that the mean of a simple function

$$Y = \sum_{1}^{k} b_j I_{B_j}$$

can be calculated as

$$F = \sum_{1}^{k} b_j P(B_j)$$

without reduction to canonical form. The method of proof is to make a succession of changes on Y and the corresponding changes on F until Y has been put in canonical form; we then check if the expression for F is as given in Definition 2.

The sets $B_1, \ldots, B_k$ may not be a partition of $\mathcal{S}$. Form all the intersection sets

$$B_1^{i_1} \cap \cdots \cap B_k^{i_k},$$

where $i_k = 0, 1$, and B_1^1 designates B_1 and B_1^0 designates B_1^c. Let $C_1, \ldots, C_N$ be the nonempty intersection sets. Then $\{C_1, \ldots, C_N\}$ is a partition of $\mathcal{S}$ and each B_i is a union of C's. If B_1 is replaced by $C_1 \cup \cdots \cup C_t$, say, then

$$I_{B_1} = \sum_{1}^{t} I_{C_i}, \qquad P(B_1) = \sum_{1}^{t} P(C_i).$$

It follows then that Y and F can be put in the form

$$Y = b_1 \sum_{1}^{t} I_{C_i} + \cdots, \qquad F = b_1 \sum_{1}^{t} P(C_i) + \cdots.$$

If several terms, say $b_1 I_{C_1}, \ldots, b_u I_{C_1}$ involve the same set C_1, then they can be collected. Let $c_1 = b_1 + \cdots + b_u$; then

$$b_1 I_{C_1} + \cdots + b_u I_{C_1} = c_1 I_{C_1}, \qquad b_1 P(C_1) + \cdots + b_u P(C_1) = c_1 P(C_1).$$

It follows then that Y and F can be put in the form

$$Y = \sum c_j I_{C_j}, \qquad F = \sum c_j P(C_j).$$

If several terms have the same coefficient, say $c_1 = \cdots = c_v$, then they can be combined. Let $a_1 = c_1 = \cdots = c_v$ and $A_1 = C_1 \cup \cdots \cup C_v$; then

$$c_1 I_{C_1} + \cdots + c_v I_{C_v} = a_1 I_{A_1}, \qquad c_1 P(C_1) + \cdots + c_v P(C_v) = a_1 P(A_1).$$

It follows then that Y and F can be put in the form

$$Y = \sum a_i I_{A_i}, \qquad F = \sum a_i P(A_i);$$

but Y is now in canonical form and F is the defining expression for $E(Y)$.

J MEAN OF DISCRETE FUNCTIONS

We defined an indicator function I_A as an alternative way of representing an event; and we associated a value to the indicator function, the probability $P(A)$. We then defined a simple function, in effect, a linear combination of indicator functions; and we associated a value to it, the linear combination of the values associated with the indicator functions. We now extend further and define a *discrete function* Y as a real-valued function that takes a countable number of values, say $a_1, a_2, \ldots$; thus

$$Y = a_1 I_{A_1} + a_2 I_{A_2} + \cdots,$$

where I_{A_i} is the indicator function for the set A_i of points at which the function takes the value a_i; it is assumed that the A_i are in the class $\mathcal{A}$. And we associate a value to the discrete function, the mean value

$$E(Y) = a_1 P(A_1) + a_2 P(A_2) + \cdots$$

provided the series converges absolutely; otherwise, we say that $E(Y)$ does not exist.

We can then prove analogs of Proposition 3 and Theorem 4. The analog of Theorem 4 is: *if Y_1 and Y_2 are discrete functions and if* $E(Y_1)$ and $E(Y_2)$ *exist, then* $E(k_1Y_1 + k_2Y_2) = k_1E(Y_1) + k_2E(Y_2)$.

K PROBLEMS: MEAN VALUE; SIMPLE AND DISCRETE FUNCTIONS

20 If the distribution of a discrete function Y is given by the Poisson(λ) distribution, then show that $E(Y) = \lambda$.

21 The mean value of a simple function Y has been defined as the summation of values weighted by corresponding probabilities; thus $Y = \Sigma_1^k a_i I_{A_i}$ with distinct values $a_1, \ldots, a_k$ has mean

$$E(Y) = \sum_1^k a_i P(A_i) = \int Y(s)\, dP(s),$$

and can be called the integral of Y with respect to P. Now let Y be a simple function with nonnegative values $a_1, \ldots, a_k$. Then the integral of Y over the set E with respect to the measure μ is defined by

$$\int_E Y(s)\, d\mu(s) = \sum_1^k a_i \mu(A_i \cap E),$$

where $0 \cdot \infty$ if it occurs is taken to be zero.

(a) If $\nu(E) = \int_E Y(s)\, d\mu(s)$, then prove that ν is a measure. The function Y modulates the measure μ, giving an adjusted measure ν.

(b) Let Y_1 and Y_2 be simple functions taking nonnegative values, and $k_1, k_2 \geq 0$, then prove

$$\int_E (k_1 Y_1(s) + k_2 Y_2(s))\, d\mu(s) = k_1 \int_E Y_1(s)\, d\mu(s) + k_2 \int_E Y_2(s)\, d\mu(s).$$

The integral is linear in the integrand.

2.5

PROBABILITY ON THE LINE AND PLANE

In applications we frequently obtain a sequence of determinations of a response or a sequence of values from different responses, and then want to do similar numerical calculations on the different elements in the sequence. For such calculations, vector and matrix methods are essential. In this section we briefly survey vector and matrix methods; some have been touched on in the preceding section and others will become important as we progress through later chapters. We conclude the section with some material on components of a distribution: discrete, absolutely continuous, and singular.

A VECTOR ALGEBRA

Consider points $\mathbf{z} = (z_1, \ldots, z_n)'$ in R^n. We can add points $\mathbf{z}_1 + \mathbf{z}_2 = (z_{11} + z_{12}, \ldots, z_{n1} + z_{n2})'$ by adding coordinate by coordinate; and we can multiply by a scalar $c\mathbf{z} = (cz_1, \ldots, cz_n)'$, where c is real. With these operations R^n is a vector space. Call the points *vectors*.

Let $\mathbf{a}_1, \ldots, \mathbf{a}_r$ be vectors in R^n. Then the set

(1) $\mathcal{L}(\mathbf{a}_1, \ldots, \mathbf{a}_r) = \{x_1\mathbf{a}_1 + \cdots + x_r\mathbf{a}_r : x_i \in R\}$

of all linear combinations of the $\mathbf{a}_i$'s is closed under addition and scalar multiplication: it is a *linear subspace* of R^n, called the linear subspace *generated by* $\mathbf{a}_1, \ldots, \mathbf{a}_r$. It is convenient sometimes to present the vectors in the form of the matrix

$$A = (\mathbf{a}_1 \cdots \mathbf{a}_r) = \begin{pmatrix} a_{11} & \cdots & a_{1r} \\ \vdots & & \vdots \\ a_{n1} & \cdots & a_{nr} \end{pmatrix}$$

and write $\mathcal{L}(\mathbf{a}_1, \ldots, \mathbf{a}_r)$ or $\mathcal{L}(A)$; see Figure S.2.

The vectors $\mathbf{a}_1, \ldots, \mathbf{a}_r$ are called *linearly dependent* if there is a linear combination $x_1\mathbf{a}_1 + \cdots + x_r\mathbf{a}_r = \mathbf{0}$ which gives the zero vector and has at least one coefficient different from zero. Otherwise they are called *linearly independent*. For example, the vectors (1, 1, 1, 1) and (1, 2, 3, 4) are linearly independent and the vectors (1, 1, 1, 1), (1, 2, 3, 4), and (2, 3, 4, 5) are linearly dependent.

Now consider $\mathcal{L}(\mathbf{a}_1, \ldots, \mathbf{a}_r)$. If the vectors $\mathbf{a}_1, \ldots, \mathbf{a}_r$ are linearly dependent, then the linear combination can be solved for one vector in terms of the others and $\mathcal{L}(\mathbf{a}_1, \ldots, \mathbf{a}_r)$ can be generated by one less vector. This can be continued until a linearly independent set of vectors remains. Let s be the number of vectors remaining. Then s is uniquely determined; it is called the *dimension* of the subspace $\mathcal{L}(\mathbf{a}_1, \ldots, \mathbf{a}_r)$ and the corresponding set of vectors is called a *basis* for the subspace; s is also called the *rank* of A and cannot be larger than n or r.

Now consider n vectors $\mathbf{a}_1, \ldots, \mathbf{a}_n$ with matrix A and examine the equation

(2) $\mathbf{z} = A\mathbf{x}$ or $\mathbf{z} = x_1\mathbf{a}_1 + \cdots + x_n\mathbf{a}_n$,

where $\mathbf{x}$ is also a vector with n coordinates. We can view this as a linear transformation from a space R^n for $\mathbf{x}$ to a space R^n for $\mathbf{z}$. Or we can view $x_1, \ldots, x_n$ as a set of new coordinates for a $\mathbf{z}$ in the subspace $\mathcal{L}(\mathbf{a}_1, \ldots, \mathbf{a}_n)$. The n vectors are linearly independent according as $\mathcal{L}(\mathbf{a}_1, \ldots, \mathbf{a}_n)$ is the whole of R^n, or as the transformation just described is nonsingular, or as the determinant of the matrix A is nonzero, or as the rank of A is n.

B INNER PRODUCTS

Now consider the length of a vector and the inner product of two vectors. The squared length of a vector $\mathbf{z}$ in R^n is given by

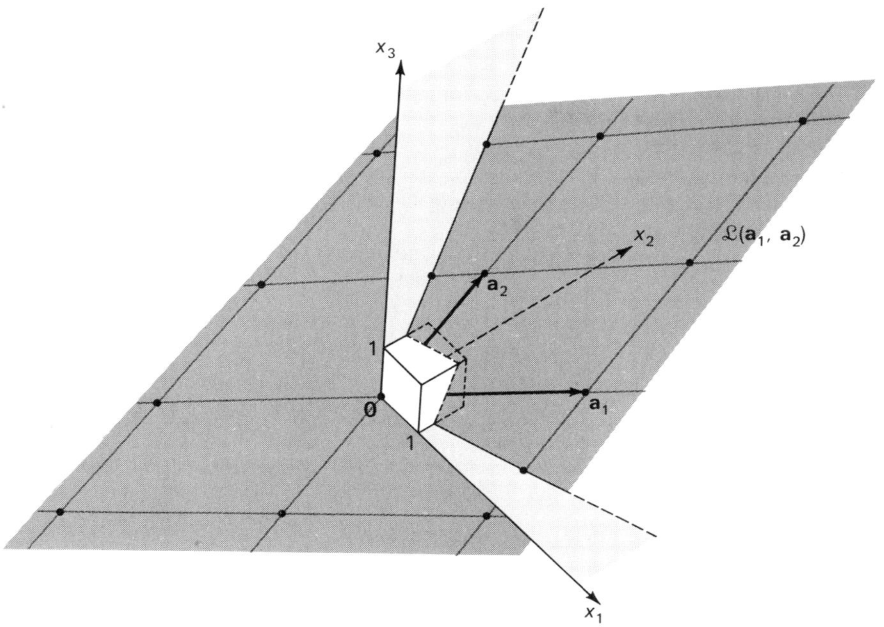

FIGURE S.2
Plane $\mathcal{L}(\mathbf{a}_1, \mathbf{a}_2)$ generated by $\mathbf{a}_1$ and $\mathbf{a}_2$ in R^3; the unit cube at the origin suggests the spatial position of the axes.

(3) $\quad |\mathbf{z}|^2 = z_1^2 + \cdots + z_n^2 = \mathbf{z}'\mathbf{z},$

where $\mathbf{z}'$ as a row vector is multiplied into $\mathbf{z}$ as a column vector by matrix multiplication; the inner product of vectors $\mathbf{z}_1$ and $\mathbf{z}_2$ is given by

(4) $\quad (\mathbf{z}_1, \mathbf{z}_2) = z_{11}z_{12} + \cdots + z_{n1}z_{n2} = \mathbf{z}_1'\mathbf{z}_2.$

Now suppose that we have several vectors $\mathbf{a}_1, \ldots, \mathbf{a}_r$ and want the squared lengths and inner products. Then we can calculate the inner product matrix,

(5) $$M = A'A = \begin{pmatrix} \mathbf{a}_1' \\ \vdots \\ \mathbf{a}_r' \end{pmatrix} (\mathbf{a}_1 \quad \cdots \quad \mathbf{a}_r) = \begin{pmatrix} \mathbf{a}_1'\mathbf{a}_1 & \cdots & \mathbf{a}_1'\mathbf{a}_r \\ \vdots & & \vdots \\ \mathbf{a}_r'\mathbf{a}_1 & \cdots & \mathbf{a}_r'\mathbf{a}_r \end{pmatrix} = \begin{pmatrix} m_{11} & \cdots & m_{1r} \\ \vdots & & \vdots \\ m_{r1} & \cdots & m_{rr} \end{pmatrix}$$

by matrix multiplication. Note that m_{11} is the squared length of $\mathbf{a}_1$, and $m_{12} = m_{21}$ is the inner product of $\mathbf{a}_1$ and $\mathbf{a}_2$. Also note that any square submatrix on the diagonal is also an inner product matrix—for a subset of the vectors $\mathbf{a}_1, \ldots, \mathbf{a}_r$.

As an example, consider the vectors $\mathbf{a}_1 = (1, 1, 1, 1, 1)'$ and $\mathbf{a}_2 = (1, 2, 3, 4, 5)'$ in R^5. The inner product matrix is

$$M = \begin{pmatrix} 1 & 1 & 1 & 1 & 1 \\ 1 & 2 & 3 & 4 & 5 \end{pmatrix} \begin{pmatrix} 1 & 1 \\ 1 & 2 \\ 1 & 3 \\ 1 & 4 \\ 1 & 5 \end{pmatrix} = \begin{pmatrix} 5 & 15 \\ 15 & 55 \end{pmatrix}.$$

Thus the length of $\mathbf{a}_1$ is $\sqrt{5}$ and of $\mathbf{a}_2$ is $\sqrt{55}$; the inner product of $\mathbf{a}_1$ and $\mathbf{a}_2$ is 15.

The angle θ between two vectors $\mathbf{z}_1$ and $\mathbf{z}_2$ can be defined by

(6) $\quad (\mathbf{z}_1, \mathbf{z}_2) = |\mathbf{z}_1||\mathbf{z}_2| \cos\theta$

provided that $|\mathbf{z}_1| \neq 0$ and $|\mathbf{z}_2| \neq 0$; that is, provided neither $\mathbf{z}_1$ nor $\mathbf{z}_2$ is the zero vector $\mathbf{0}$. Two vectors $\mathbf{z}_1$ and $\mathbf{z}_2$ are called *orthogonal* if $(\mathbf{z}_1, \mathbf{z}_2) = \mathbf{z}_1'\mathbf{z}_2 = 0$. In the preceding example the cosine of the angle between $\mathbf{a}_1$ and $\mathbf{a}_2$ is

$$\cos\theta = \frac{15}{\sqrt{5}\cdot\sqrt{55}} = 0.9045$$

and the angle $\theta = 0.4406$.

As a second example, consider the vectors $\mathbf{a}_1 = (1, 1, 1, 1, 1)'$ and $\mathbf{a}_{2\cdot 1} = (-2, -1, 0, 1, 2)'$. The inner product matrix is

$$N = \begin{pmatrix} 1 & 1 & 1 & 1 & 1 \\ -2 & -1 & 0 & 1 & 2 \end{pmatrix} \begin{pmatrix} 1 & -2 \\ 1 & -1 \\ 1 & 0 \\ 1 & 1 \\ 1 & 2 \end{pmatrix} = \begin{pmatrix} 5 & 0 \\ 0 & 10 \end{pmatrix}.$$

Thus the vectors $\mathbf{a}_1$ and $\mathbf{a}_{2\cdot 1}$ are orthogonal.

Consider two vectors $\mathbf{a}_1$ and $\mathbf{a}_2$. We can always find the *component* of $\mathbf{a}_2$ that is orthogonal to $\mathbf{a}_1$; that is, choose b so that $\mathbf{a}_2 - b\mathbf{a}_1$ is orthogonal to $\mathbf{a}_1$:

$$(\mathbf{a}_2 - b\mathbf{a}_1, \mathbf{a}_1) = \mathbf{a}_2'\mathbf{a}_1 - b\mathbf{a}_1'\mathbf{a}_1 = 0.$$

For the initial example we have

$$15 - b\cdot 5 = 0 \quad \text{or} \quad b = 3;$$

thus

$$\mathbf{a}_2 - 3\mathbf{a}_1 = (1, 2, 3, 4, 5)' - 3(1, 1, 1, 1, 1)' = (-2, -1, 0, 1, 2)'$$

is orthogonal to $\mathbf{a}_1$. Note that this orthogonal component is the vector $\mathbf{a}_{2\cdot 1}$ used in the second example. The part removed $3\mathbf{a}_1$ is the *projection* of $\mathbf{a}_2$ on $\mathbf{a}_1$.

C POSITIVE DEFINITE QUADRATIC FORMS

Consider the equation (2) that expresses certain points $\mathbf{z}$ in R^n in terms of vectors $\mathbf{a}_1, \ldots, \mathbf{a}_n$:

$$\mathbf{z} = A\mathbf{x} \quad \text{or} \quad \mathbf{z} = x_1\mathbf{a}_1 + \cdots + x_n\mathbf{a}_n.$$

The squared length of $\mathbf{z}$ can be expressed in terms of $\mathbf{x}$:

$$\text{(7)} \qquad \mathbf{z}'\mathbf{z} = \mathbf{x}'A'A\mathbf{x} = \mathbf{x}'M\mathbf{x} = \sum_{i,j=1}^{n} m_{ij}x_ix_j.$$

An expression of the form (with M symmetric)

$$\text{(8)} \qquad \mathbf{x}'M\mathbf{x} = \sum_{i,j=1}^{n} m_{ij}x_ix_j$$

is called a *quadratic form* and M is called the *matrix of the quadratic form*. A matrix M is called *positive definite* if $\mathbf{x}'M\mathbf{x} > 0$ for all $\mathbf{x} \neq 0$ and *positive semidefinite* if $\mathbf{x}'M\mathbf{x} \geq 0$ for all $\mathbf{x}$. Correspondingly, the quadratic form is called positive definite and positive semidefinite.

Note from equation (7) that an inner product matrix is positive semidefinite; it is also true conversely that a positive semidefinite matrix can be written as an inner product matrix. Also note that if A has full rank n, then no nontrivial linear combination of $\mathbf{a}_1, \ldots, \mathbf{a}_n$ can be $\mathbf{0}$; thus an inner product matrix of linearly independent vectors is positive definite; it is also true conversely that a positive definite matrix can be written as the inner product matrix for linearly independent vectors.

Now consider the transformation $\mathbf{z} = A\mathbf{x}$ and assume that A is of full rank n. The transformation from $\mathbf{x}$ to $\mathbf{z}$ is then a one-to-one transformation of R^n into R^n. The transformation dilates, contracts, rotates, and shears. The sphere $\mathbf{z}'\mathbf{z} = \sum z_i^2 = c^2$ for $\mathbf{z}$ corresponds to the ellipsoid $\mathbf{x}'M\mathbf{x} = \sum m_{ij}x_ix_j = c^2$ for $\mathbf{x}$.

A positive semidefinite matrix M_2 is called *greater than* or *equal to* a positive semidefinite matrix M_1 if $\mathbf{x}'M_2\mathbf{x} - \mathbf{x}'M_1\mathbf{x} = \mathbf{x}'(M_2 - M_1)\mathbf{x}$ is positive semidefinite. If M_2 is greater than or equal to M_1, then clearly a diagonal element of M_2 is greater than or equal to the corresponding diagonal element of M_1. For the case of positive definite matrices, M_2 greater than M_1 can be interpreted as the ellipsoid $\mathbf{x}'M_2\mathbf{x} = c^2$ is inside the ellipsoid $\mathbf{x}'M_1\mathbf{x} = c^2$.

For positive definite matrices it can be shown that M_2 greater than or equal to M_1 implies M_1^{-1} greater than or equal to M_2^{-1}; see Problem 12.

D THE SINGULAR COMPONENT ON *R*

Consider a distribution on R with measure P and distribution function F. In Section 2.1 we defined and separated out the discrete component:

$$\mathcal{S}_d = \{y : p(y) > 0\}, \qquad F_d(y) = P((-\infty, y] \cap \mathcal{S}_d) = \sum_{t \le y} p(t).$$

And in Section 2.2 we defined and separated out the absolutely continuous component:

$$\mathcal{S}_a = \{y : F'(y) \text{ exists}\}, \qquad F_a(y) = P((-\infty, y] \cap \mathcal{S}_a) = \int_{-\infty}^{y} f(t)\, dt.$$

Is there any probability left? The remaining sample space and "distribution function" are

$$\mathcal{S}_s = R - \mathcal{S}_d - \mathcal{S}_a, \qquad F_s(y) = F(y) - F_d(y) - F_a(y)$$
$$= P((-\infty, y] \cap \mathcal{S}_s).$$

Is $\mathcal{S}_s = \varnothing$ and $F_s(y) = 0$?

The function F_d records cumulative discrete probability. The difference function $F - F_d$ has the discrete probability removed; it is thus a continuous function. The function F_a records cumulative probability at points where the derivative exists. The difference function $F - F_d - F_a$ thus has the derivative removed where it exists; it is thus a continuous function that has derivative 0 throughout $\mathcal{S}_a$. But in Section 2.2 we noted that the length measure of $R - \mathcal{S}_a$ is zero. Thus F_s is continuous and has derivative 0 except on a set of length measure zero. Is there a nontrivial function F_s?

Surprisingly, an example can be constructed fairly easily. Initially: consider a uniform distribution on $A_0 = [0, 1]$; let G_0 be the distribution function. First: delete the middle third of the preceding interval and consider a uniform distribution on the remainder $A_1 = [0, 1/3] \cup [2/3, 1]$; let G_1 be the distribution function. Second: delete the middle thirds of the preceding intervals and consider a uniform distribution on the remainder $A_2 = [0, 1/9] \cup [2/9, 3/9] \cup [6/9, 7/9] \cup [8/9, 1]$; let G_2 be the distribution function. Continue in this manner and let $G(y) = \lim G_n(y)$. See Figure S.3 and compare with Problems 1.5.14 and 1.5.15. In a sense G describes the uniform distribution on the Cantor ternary set.

From the construction it is easy to see intuitively that G is continuous with $G(0) = 0$, $G(1) = 1$; and also that $G'(y) = 0$ on all the removed intervals and thus is zero except on the Cantor ternary set $A = \cap_1^\infty A_n$; Problem 1.5.14 is to show that the length measure of the Cantor set is zero.

An explicit expression can, in fact, be given for $G(y)$. Let y in $[0, 1]$ be expressed in a ternary decimal expansion:

$$y = .j_1 j_2 j_3 \cdots \left(= \sum_{\alpha=1}^{\infty} j_\alpha 3^{-\alpha} \right), \qquad j_\alpha = 0, 1, 2.$$

This is unique except for instances involving repeating 2's: round up if preceded by a 1 (thus change $0.1012\dot{2}$ to $0.1020\dot{0}$); retain if preceded by a 0 (thus use $0.102\dot{2}$ in preference to $0.110\dot{0}$).

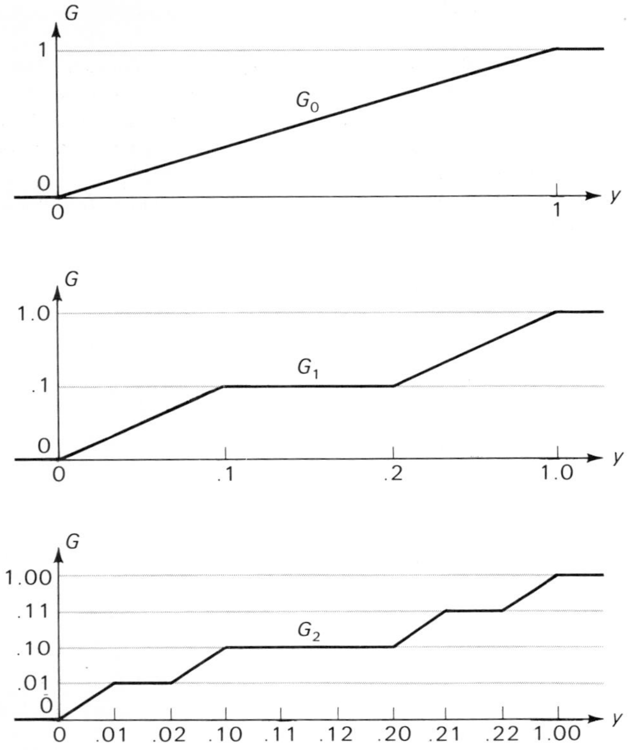

FIGURE S.3
Functions G_0, G_1, G_2, . . ., y in ternary, G in binary.

A point y gets removed by the construction procedure if a 1 occurs in the expansion; let $n(y)$ be the subscript of the first j that is equal to 1 with $n(y) = \infty$ if there is no such j. The value of $G(y)$ can then be expressed in binary form as

$$G(y) = .i_1 i_2 i_3 \cdots = \sum_{\alpha=1}^{n(y)-1} i_\alpha 2^{-\alpha} + 2^{-n(y)},$$

where $i_\alpha = j_\alpha/2$. Thus $G(0.20220102) = 0.101101$ using ternary and binary decimals.

E THE SINGULAR COMPONENT ON R^k

A distribution on R^k can be decomposed into a discrete component, an absolutely continuous component, and a singular component, just as on R. The possibilities for the singular component are, however, much wider with $k > 1$.

For an example, consider (y_1, y_2) on R^2 and suppose that we put a normal distribution along certain line segments. Specifically, let z designate a standard normal variable and define

$$\begin{aligned}(y_1, y_2) &= (z, z) && \text{if } |z| < c,\\ &= (-z, -z) && |z| \ge c.\end{aligned}$$

This is a singular distribution but a rather special one with its split components; this distribution can be useful for other purposes later. Sketch this distribution as a one-dimensional density function sitting on the three line segments.

F EXERCISES AND PROBLEMS

1 Calculate the angle between the two vectors for each of the following pairs:

$$\begin{pmatrix}1\\1\\1\\1\end{pmatrix}, \begin{pmatrix}1\\0\\0\\0\end{pmatrix}; \quad \begin{pmatrix}1\\1\\1\\1\end{pmatrix}, \begin{pmatrix}1\\2\\3\\4\end{pmatrix}; \quad \begin{pmatrix}1\\1\\1\\1\end{pmatrix}, \begin{pmatrix}1\\4\\9\\16\end{pmatrix}; \quad \begin{pmatrix}1\\2\\3\\4\end{pmatrix}, \begin{pmatrix}1\\4\\9\\16\end{pmatrix}.$$

2 Calculate the inner product matrix for the column vectors in

$$A = \begin{pmatrix}1 & 1 & 1 & 1\\ 1 & 1 & -1 & -1\\ 1 & -1 & 1 & -1\\ 1 & -1 & -1 & 1\end{pmatrix}.$$

3 Calculate the inner product matrix for the column vectors in

$$A = \begin{pmatrix}1 & 1 & 1\\ 1 & 2 & 4\\ 1 & 3 & 9\\ 1 & 4 & 16\end{pmatrix}.$$

4 Let $\mathbf{a}_1 = (1, 1, 1, 1)'$ and $\mathbf{a}_2 = (1, 2, 3, 4)'$. Find b so that $\mathbf{a}_{2\cdot 1} = \mathbf{a}_2 - b\mathbf{a}_1$ is orthogonal to $\mathbf{a}_1$; record $\mathbf{a}_{2\cdot 1}$.

5 Let $\mathcal{L}^{\perp}(\mathbf{1})$ consist of all the vectors $\mathbf{z}$ which are orthogonal to the 1-vector $\mathbf{1} = (1, 1, \ldots, 1)'$. Show that $\mathcal{L}^{\perp}(\mathbf{1})$ is the plane $\Sigma_1^n z_i = 0$.

6 Show that the quadratic form $\Sigma\, m_{ij}x_ix_j$ with

$$M = \begin{pmatrix}1/n & \cdots & 1/n\\ \vdots & & \vdots\\ 1/n & \cdots & 1/n\end{pmatrix} = n^{-1}\mathbf{1}\mathbf{1}'$$

is equal to $n\bar{x}^2$, where $\bar{x} = \Sigma\, x_i/n$.

7 Show that the quadratic form $\Sigma\, n_{ij}x_ix_j$ with

$$N = \begin{pmatrix}1 - 1/n & -1/n & \cdots & -1/n\\ -1/n & 1 - 1/n & \ddots & \vdots\\ & & \ddots & \\ \vdots & \ddots & 1 - 1/n & -1/n\\ -1/n & \cdots & -1/n & 1 - 1/n\end{pmatrix} = I - n^{-1}\mathbf{1}\mathbf{1}'$$

is equal to $\Sigma\, x_i^2 - n\bar{x}^2 = \Sigma\,(x_i - \bar{x})^2$.

8 A matrix M is idempotent if $MM = M$. Show that M and N in Problems 6 and 7 are idempotent and symmetric.

9 (*continuation of Problem 6*) Show that $M\mathbf{x} = \bar{x}\mathbf{1} = (\bar{x}, \ldots, \bar{x})'$, a point in $\mathcal{L}(\mathbf{1})$; it is the projection of $\mathbf{x}$ on $\mathcal{L}(\mathbf{1})$.

10 (*continuation of Problem 7*) Show that $N\mathbf{x} = \mathbf{x} - \bar{x}\mathbf{1} = (x_1 - \bar{x}, \ldots, x_n - \bar{x})'$, a point in $\mathcal{L}^{\perp}(\mathbf{1})$; it is the projection of $\mathbf{x}$ on $\mathcal{L}^{\perp}(\mathbf{1})$.

11 The distribution function G in Section D is called the Cantor function. Show that the Cantor function is continuous.

12 For positive definite matrices, show that if M_1 is greater than M_2, then M_2^{-1} is greater than M_1^{-1} (simultaneously diagonalize M_1 and M_2).

3.6

MARGINAL PROBABILITY

In this section we develop further several of the topics in the preceding sections. This is optional material presented here to round out the development.

A FACTORIZATION OF THE DISTRIBUTION FUNCTION

Proposition 6 in Section 3.3D records the factorization

$$F_{XY}(x, y) = F_X(x)F_Y(y)$$

as a necessary and sufficient condition for the statistical independence of X and Y. In that section we proved the necessity—that independence implies the factorization. We now prove the sufficiency—that the factorization $F_{XY}(x, y) = F_X(x)F_Y(y)$ implies independence. From Section 2.3B we know that the distribution function on R^2 uniquely determines the probability measure P_{XY} for (X, Y) on the plane. What can we say about this measure on the basis of the factored expression F_XF_Y for the distribution function? From Section 3.3E we know that the measure P_X for the first coordinate and P_Y for the second coordinate uniquely combine on the basis of independence to give a measure on R^2. This product measure, of course, agrees with F_XF_Y on the sets $(-\infty, x] \times (-\infty, y]$; thus it must be the unique probability measure P_{XY} determined from the distribution function. Hence X and Y are statistically independent.

B SOME NORMAL DISTRIBUTION THEORY

The normal distribution provides an approximation for the variation found with many measurement and response variables. And it happens to be essentially the only distribution for which certain calculations of statistical interest are simple and direct. These calculations concern certain marginal distributions of common functions. We shall now derive some of these distributions.

Consider a standard normal distribution for z. From Example 3.2.2 we have that z^2 has a chi-square(1) distribution. If we rescale by a factor σ and have a normal(0, σ) distribution for $v = \sigma z$, then $v^2/\sigma^2 = z^2$ has a chi-square(1) distribution, or equivalently, v^2 has the distribution of $\sigma^2\chi^2$, where χ^2 has the chi-square(1) distribution.

Now consider independent χ_1^2 and χ_2^2, where χ_i^2 is chi-square(n_i). The joint probability element is

$$\frac{1}{\Gamma(n_1/2)}\frac{1}{\Gamma(n_2/2)} e^{-\chi_1^2/2}\left(\frac{\chi_1^2}{2}\right)^{(n_1/2)-1} e^{-\chi_2^2/2}\left(\frac{\chi_2^2}{2}\right)^{(n_2/2)-1}\frac{d\chi_1^2}{2}\frac{d\chi_2^2}{2}.$$

We derive the marginal distribution of $\chi^2 = \chi_1^2 + \chi_2^2$. For this, we first change the variable from (χ_1^2, χ_2^2) to (χ^2, χ_2^2) as in the convolution example 3.2.8: the Jacobian is 1; hence, by substitution, we obtain the joint probability element

$$\frac{1}{\Gamma(n_1/2)}\frac{1}{\Gamma(n_2/2)} e^{-\chi^2/2}\left(\frac{\chi^2-\chi_2^2}{2}\right)^{(n_1/2)-1}\left(\frac{\chi_2^2}{2}\right)^{(n_2/2)-1}\frac{d\chi_2^2}{2}\frac{d\chi^2}{2}$$

$$= \frac{1}{\Gamma(n_1/2)}\frac{1}{\Gamma(n_2/2)} e^{-\chi^2/2}\left(\frac{\chi^2}{2}\right)^{[(n_1+n_2)/2]-2}\left(\frac{\chi^2-\chi_2^2}{\chi^2}\right)^{(n_1/2)-1}\left(\frac{\chi_2^2}{\chi^2}\right)^{(n_2/2)-1}\frac{d\chi_2^2}{2}\frac{d\chi^2}{2}$$

on the range $0 < \chi^2 < \infty$, $0 < \chi_2^2 < \chi^2$. We then integrate out the unwanted χ_2^2: the essential part of the integration is as follows:

$$\int_0^{\chi^2} \left(\frac{\chi^2 - \chi_2^2}{\chi^2}\right)^{(n_1/2)-1} \left(\frac{\chi_2^2}{\chi^2}\right)^{(n_2/2)-1} (1/2)\, d\chi_2^2 = \int_0^1 (1 - u)^{(n_1/2)-1} u^{(n_2/2)-1}\, du \cdot \frac{\chi^2}{2},$$

$$= \frac{\Gamma(n_1/2)\Gamma(n_2/2)}{\Gamma\left(\dfrac{n_1 + n_2}{2}\right)} \frac{\chi^2}{2}$$

where we have substituted $u = \chi_2^2/\chi^2$ and have used Problem 2.2.22. The result of the integration on the full expression is

$$(1) \qquad \frac{1}{\Gamma\left(\dfrac{n_1 + n_2}{2}\right)} e^{-\chi^2/2} \left(\frac{\chi^2}{2}\right)^{[(n_1+n_2)/2]-1} \frac{d\chi^2}{2}.$$

Thus $\chi^2 = \chi_1^2 + \chi_2^2$ is chi-square($n_1 + n_2$).

We now combine the results in the preceding paragraphs. Consider a sample $(z_1, \ldots, z_n)$ from the standard normal. It follows first that $(z_1^2, \ldots, z_n^2)$ is a sample from the chi-square(1); and then that $\chi^2 = z_1^2 + \cdots + z_n^2$ is chi-square(n). If we rescale by a factor σ and have a sample $(v_1, \ldots, v_n)$ from the normal(0, σ), then $\Sigma_1^n v_i^2$ has the distribution of $\sigma^2\chi^2$, where χ^2 is chi-square(n).

Now consider independent z, χ^2 where z is standard normal and χ^2 is chi-square(n). The joint probability element is

$$\frac{1}{\sqrt{2\pi}} e^{-z^2/2} \frac{1}{\Gamma(n/2)} e^{-\chi^2/2} \left(\frac{\chi^2}{2}\right)^{(n/2)-1} dz \frac{d\chi^2}{2}.$$

We derive the marginal distribution of $T = z/\chi$. For this we first change the variable from (z, χ^2) to (T, χ^2) parallel to the convolution problem 3.2.30: the Jacobian change is given by $dz = \chi\, dt$; hence by substitution we obtain

$$\frac{1}{\sqrt{\pi}} \frac{1}{\Gamma(n/2)} e^{-\chi^2T^2/2} e^{-\chi^2/2} \left(\frac{\chi^2}{2}\right)^{[(n+1)/2]-1} \frac{d\chi^2}{2}\, dT$$

on the range $0 < \chi^2 < \infty$, $-\infty < T < \infty$. We then integrate out the unwanted χ^2: the essential part of the integration is as follows:

$$\int_0^\infty e^{-\chi^2(1+T^2)/2} \left(\frac{\chi^2}{2}\right)^{[(n+1)/2]-1} \frac{d\chi^2}{2} = \int_0^\infty e^{-x} x^{[(n+1)/2]-1}\, dx (1 + T^2)^{-(n+1)/2},$$

$$= \Gamma\left(\frac{n+1}{2}\right)(1 + T^2)^{-(n+1)/2}$$

where $x = \chi^2(1 + T^2)/2$ and Problem 2.2.21 is used; the result of the integration on the full expression is

$$(2) \qquad \frac{\Gamma\left(\dfrac{n+1}{2}\right)}{\Gamma(1/2)\Gamma(n/2)} (1 + T^2)^{-[(n+1)/2]}\, dT,$$

using $\sqrt{\pi} = \Gamma(1/2)$. Thus $T = z/\chi$ is canonical Student(n) and $t = z/(\chi/\sqrt{n})$ is ordinary Student(n); see Problem 2.2.19.

We now combine the results in the preceding paragraphs. If we have a sample $(v, v_1, \ldots, v_n)$ from the normal(0, σ) distribution, then

$$t = \frac{v}{\left(\sum_{1}^{n} v_i^2/n\right)^{1/2}}$$

has the Student(n) distribution.

Now consider independent χ_1^2 and χ_2^2, where χ_i^2 is chi-square(n_i). The joint probability element is

$$\frac{1}{\Gamma(n_1/2)}\frac{1}{\Gamma(n_2/2)} e^{-\chi_1^2/2}\left(\frac{\chi_1^2}{2}\right)^{(n_1/2)-1} e^{-\chi_2^2/2}\left(\frac{\chi_2^2}{2}\right)^{(n_2/2)-1}\frac{d\chi_1^2}{2}\frac{d\chi_2^2}{2}.$$

We derive the marginal distribution of $G = \chi_1^2/\chi_2^2$. For this we first change the variable from (χ_1^2, χ_2^2) to (G, χ_2^2) as in the convolution problem 3.2.30: the Jacobian change is given by $d\chi_1^2 = \chi_2^2\, dG$; hence by substitution we obtain

$$\frac{1}{\Gamma(n_1/2)}\frac{1}{\Gamma(n_2/2)} e^{-\chi_2^2(1+G)/2}\left(\frac{\chi_2^2}{2}\right)^{[(n_1+n_2)/2]-1} G^{(n_1/2)-1}\, dG\, \frac{d\chi_2^2}{2}$$

on the range $0 < G < \infty$, $0 < \chi_2^2 < \infty$. We then integrate out the unwanted χ_2^2: the essential part of the integration is as follows:

$$\int_0^\infty e^{-\chi_2^2(1+G)}\left(\frac{\chi_2^2}{2}\right)^{[(n_1+n_2)/2]-1}\frac{d\chi_2^2}{2} = \int_0^\infty e^{-x}x^{[(n_1+n_2)/2]-1}\, dx(1+G)^{-(n_1+n_2)/2},$$

$$= \Gamma\left(\frac{n_1+n}{2}\right)(1+G)^{-(n_1+n_2)/2},$$

where we have substituted $x = \chi_2^2(1+G)$ and have used Problem 2.2.21; the result of the integration on the full expression is

$$\text{(3)} \qquad \frac{\Gamma\left(\frac{n_1+n_2}{2}\right)}{\Gamma(n_1/2)\Gamma(n_2/2)}\frac{G^{(n_1/2)-1}}{(1+G)^{[(n_1+n_2)/2]}}\, dG.$$

Thus $G = \chi_1^2/\chi_2^2$ is canonical $F(n_1, n_2)$; and $F = (\chi_1^2/n_1)/(\chi_2^2/n_2)$ is ordinary $F(n_1, n_2)$; see Problem 2.2.18.

We now combine results in preceding paragraphs. If we have a sample $(v_1, \ldots, v_{n_1}, v_{n_1+1}, \ldots, v_{n_1+n_2})$ from the normal(o, σ) distribution, then

$$F = \frac{\sum_{1}^{n_1} v_i^2/n_1}{\sum_{n_1+1}^{n_1+n_2} v_j^2/n_2}$$

has the $F(n_1, n_2)$ distribution.

C FURTHER NORMAL DISTRIBUTION RESULTS: THEORY AND PROBLEMS

Consider the normal(δ, 1) distribution for x. We derive the marginal distribution for $\chi^2 = x^2$. The probability element for x is

$$g(x)\, dx = \frac{1}{\sqrt{2\pi}} e^{-(x-\delta)^2/2}\, dx = \frac{1}{\sqrt{2\pi}} e^{-\delta^2/2} e^{x\delta} e^{-x^2/2}\, dx.$$

The probability element for $\chi^2 = x^2$ is then obtained from Example 3.2.7:

$$(4) \qquad h(\chi^2)\,d\chi^2 = (1/2)(g(\chi) + g(-\chi))\chi^{-1}\,d\chi^2$$

$$= \frac{1}{\sqrt{2\pi}} e^{-\delta^2/2}\left(1 + \chi^2\frac{\delta^2}{2!} + \chi^4\frac{\delta^4}{4!} + \cdots\right) e^{-\chi^2/2}\chi^{-1}\,d\chi^2$$

$$= e^{-\delta^2/2}\sum_{r=0}^{\infty}\frac{(\delta^2/2)^2}{r!}\cdot\frac{1}{\Gamma\left(\frac{2r+1}{2}\right)}\left(\frac{\chi^2}{2}\right)^{[(2r+1)/2]-1} e^{-\chi^2/2}\frac{d\chi^2}{2},$$

where the constants have been rearranged according to

$$\sqrt{2\pi}\;2r(2r-1)\cdots 2.1 = 2^r r!\,\frac{2r-1}{2}\,\frac{2r-3}{2}\cdots\frac{1}{2}\,\Gamma(1/2)2^{r+(1/2)}$$

to give chi-square density expressions. The distribution (4) is called the noncentral chi-square(1) distribution with noncentrality δ^2; note the Poisson($\delta^2/2$) mixture of chi-square($1 + 2r$)'s.

1 *Noncentral normal*(δ, 1) *distribution:* Consider the normal(δ, 1) distribution for x. Show that the density function for x can be expressed as

$$(5) \qquad f(x) = e^{-\delta^2/2}\frac{1}{\sqrt{2\pi}}e^{-x^2/2}\sum_{r=0}^{\infty}\frac{\delta^r}{r!}x^r.$$

2 *Noncentral canonical Student*(n) *distribution with noncentrality* δ*:* Consider the noncentral normal(δ, 1) distribution for x and an independent chi(n) distribution for χ. Use the method in Section B to show that the marginal density for $T = x/\chi$ has density function

$$(6) \qquad g(T) = e^{-\delta^2/2}\sum_{r=0}^{\infty}\frac{\delta^r}{r!}2^{r/2}\frac{\Gamma\left(\frac{n+r+1}{2}\right)}{\Gamma(1/2)\Gamma(n/2)}\frac{T^r}{(1+T^2)^{(n+r+1)/2}}$$

called the noncentral canonical Student(n) distribution. The rescaled distribution for $t = x/(\chi^2/n)^{1/2}$ is called the ordinary Student(n) distribution with noncentrality δ; see Figure S.4.

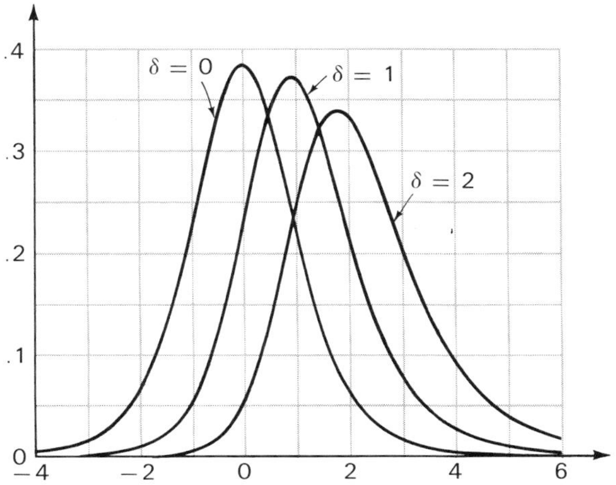

FIGURE S.4
Noncentral Student(6) distribution with noncentrality δ^2.

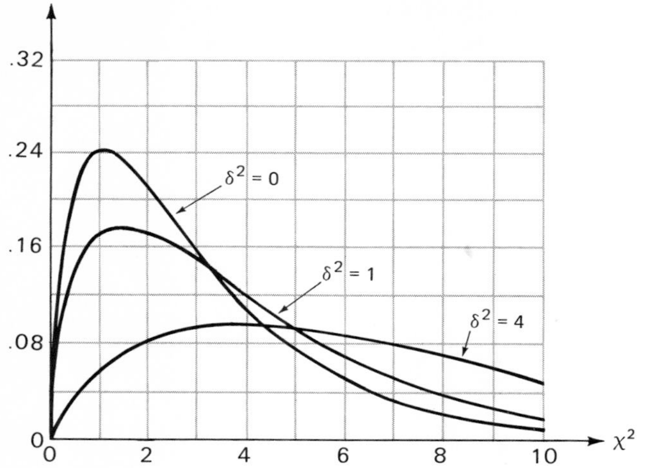

FIGURE S.5
Noncentral chi-square(3) distribution with noncentrality δ^2.

3 *Noncentral chi-square(m) distribution with noncentrality δ^2:* Consider a normal(δ, 1) distribution for x and an independent standard normal sample $(z_1, \ldots, z_{m-1})$ [or an independent chi-square($m - 1$) distribution for $\chi_2^2 = z_1^2 + \cdots + z_{m-1}^2$]. Show that the marginal distribution for χ^2,

$$\chi^2 = x^2 + \chi_2^2 = x^2 + z_1^2 + \cdots + z_{m-1}^2$$

has density

$$(7) \qquad e^{-\delta^2/2} \sum_{r=0}^{\infty} \frac{(\delta^2/2)^r}{r!} f(\chi^2 : m + 2r),$$

where $f(\chi^2 : m)$ is the chi-square(m) density; apply the convolution formula(3.2.25) term by term to the results in formula (3.6.4). Note the preceding form: a Poisson combination of chi squares; see Figure S.5.

4 *Noncentral canonical F(m, n) distribution with noncentrality δ^2:* Consider a noncentral chi-square(m) distribution for χ_1^2 with noncentrality δ^2 and an independent chi-square(n) distribution for χ_2^2. Show that the marginal distribution for $G = \chi_1^2/\chi_2^2$ has density

$$(8) \qquad e^{-\delta^2/2} \sum_{r=0}^{\infty} \frac{(\delta^2/2)^r}{r!} f(G : m + 2r, n),$$

where $f(G : m, n)$ is the canonical $F(m, n)$ distribution; apply the convolution procedure in Section B term by term to the results of Problem 3; see Figure S.6.

5 Let $\chi_1^2, \ldots, \chi_k^2$ be statistically independent and χ_i^2 have a chi-square(m_i) distribution. Let $w = \chi_1^2 + \cdots + \chi_k^2$ and $u_1 = \chi_1^2/w, \ldots, u_{k-1} = \chi_{k-1}^2/w$ as in Problem 3.3.9.
(a) Show that w is statistically independent of $(u_1, \ldots, u_{k-1})$: make the change of variable from $(\chi_1^2, \ldots, \chi_k^2)$ to $(\chi_1^2, \ldots, \chi_{k-1}^2, w)$ and then to $(u_1, \ldots, u_{k-1}, w)$.
(b) Then deduce that w is chi-square($\Sigma\, m_i$) and $(u_1, \ldots, u_{k-1})$ is Dirichlet($m_1/2, \ldots, m_k/2$).

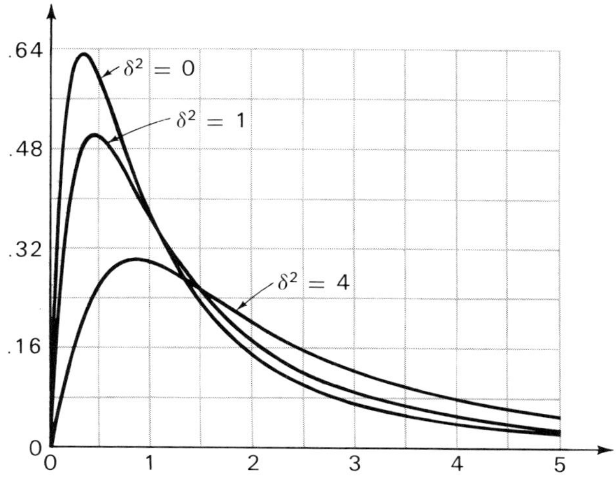

FIGURE S.6
Noncentral F(4, 4) distribution with noncentrality δ^2.

D INDEPENDENCE FOR NORMAL DISTRIBUTIONS: PROBLEMS

6 Show that $ay_1^2 + by_1y_2 + cy_2^2 + dy_1 + ey_2 + f$ can be expressed as $g(y_1) + h(y_2)$ on R^2 if and only if $b = 0$. Show that $f(y_1, y_2) = \exp\{ay_1^2 + by_1y_2 + cy_2^2 + dy_1 + ey_2 + f\}$ can be expressed as $g(y_1)h(y_2)$ on R^2 if and only if $b = 0$.

7 The bivariate normal$(\mu_1, \mu_2, \sigma_1, \sigma_2, \rho)$ distribution is recorded in Problem 2.4.24. Show that y_1 and y_2 are statistically independent if and only if $\rho = 0$; use Problem 6.

8 Show that $\Sigma\, a_i x_i^2 + 2\,\Sigma\, b_{iu} x_i y_u + \Sigma\, c_u y_u^2 + \Sigma\, d_i x_i + \Sigma\, e_u y_u + f$ can be expressed as $g(\mathbf{x}) + h(\mathbf{y})$ on R^{k+l} if and only if $b_{iu} = 0$ for all i and u.

9 *The multivariate normal distribution on R^{k+l}:* Let $\mathbf{y}_1 = (y_1, \ldots, y_k)'$, $\mathbf{y}_2 = (y_{k+1}, \ldots, y_{k+l})'$, $\boldsymbol{\mu}_1 = (\mu_1, \ldots, \mu_k)'$, $\boldsymbol{\mu}_2 = (\mu_{k+1}, \ldots, \mu_{k+l})'$,

$$\Sigma^{11} = \begin{pmatrix} \sigma^{11} & \cdots & \sigma^{1k} \\ \vdots & & \vdots \\ \sigma^{k1} & \cdots & \sigma^{kk} \end{pmatrix}, \quad \Sigma^{22} = \begin{pmatrix} \sigma^{k+1\,k+1} & \cdots & \sigma^{k+1\,k+l} \\ \vdots & & \vdots \\ \sigma^{k+l\,k+1} & \cdots & \sigma^{k+l\,k+l} \end{pmatrix}, \quad \Sigma^{12} = \begin{pmatrix} \sigma^{1\,k+1} & \cdots & \sigma^{1\,k+l} \\ \vdots & & \vdots \\ \sigma^{k\,k+1} & \cdots & \sigma^{k\,k+l} \end{pmatrix},$$

$$\Sigma^{-1} = \begin{pmatrix} \Sigma^{11} & \Sigma^{12} \\ \Sigma^{21} & \Sigma^{22} \end{pmatrix}, \quad \Sigma = \begin{pmatrix} \Sigma_{11} & \Sigma_{12} \\ \Sigma_{21} & \Sigma_{22} \end{pmatrix},$$

where Σ and Σ^{-1} are defined in Section 2.4G. Suppose that $\mathbf{y} = (\mathbf{y}_1', \mathbf{y}_2')'$ has the multivariate normal$(\boldsymbol{\mu} = (\boldsymbol{\mu}_1', \boldsymbol{\mu}_2')'; \Sigma)$ distribution with density

$$f(\mathbf{y}) = (2\pi)^{-(k+l)/2}|\Sigma|^{-1/2} \exp\left\{-\frac{1}{2}\begin{pmatrix} \mathbf{y}_1 - \boldsymbol{\mu}_1 \\ \mathbf{y}_2 - \boldsymbol{\mu}_2 \end{pmatrix}' \Sigma^{-1} \begin{pmatrix} \mathbf{y}_1 - \boldsymbol{\mu}_1 \\ \mathbf{y}_2 - \boldsymbol{\mu}_2 \end{pmatrix}\right\}.$$

Use partitioned-matrix multiplication and show that $\mathbf{y}_1$ and $\mathbf{y}_2$ are statistically independent if and only if $\Sigma^{12} = 0$ (the zero matrix); if and only if $\Sigma_{12} = 0$.

E ORDER STATISTIC DISTRIBUTIONS: THEORY AND PROBLEMS

We have noted before that the normal distribution is often used to describe the variation in many measurement and response variables. We will see later that it suffices with the normal to consider the sample average $\bar{y} = \Sigma y_i/n$ and the sample variance $s_y^2 = \Sigma (y_i - \bar{y})^2/(n-1)$. Some of the distribution theory in Sections B and C will be used later for statistical analysis based on $\bar{y}$ and s_y^2.

For other continuous distribution forms it is often convenient to examine the distribution of the order statistics. Let $\mathbf{y} = (y_1, \ldots, y_n)'$ designate a point in R^n; and let $y_{(1)}$ be the smallest of the coordinates y_i, $y_{(2)}$ the second smallest, . . ., $y_{(n)}$ the largest; thus $y_{(1)} < \cdots < y_{(n)}$. The vector $\mathbf{y}_0 = (y_{(1)}, \ldots, y_{(n)})'$ is called the *order statistic* and a coordinate $y_{(i)}$ is called the ith order statistic.

More formally we can view $\mathbf{y}_0$ as a function carrying $\mathbf{y}$ in R^n to $\mathbf{y}_0$ in R^n, actually the subset of R^n having $y_{(1)} < \cdots < y_{(n)}$.

Consider the distribution of a sample $\mathbf{y} = (y_1, \ldots, y_n)'$ from a distribution with density function f. The density for $\mathbf{y}$ is

$$f(y_1) \cdots f(y_n).$$

We first derive the marginal density for the order statistic $\mathbf{y}_0$. For this we follow the pattern in Section 3.2D: the mapping $\mathbf{y} \to \mathbf{y}_0$ is an $n!$-to-1 mapping except at points having two or more equal coordinates; the volume change for any component of the mapping is $+1$, just a reordering of coordinates; the density function at any component point is $f(y_{(1)}) \cdots f(y_{(n)})$ with factors rearranged in a convenient order; it follows that the joint density for $\mathbf{y}_0$ is

$$(9) \qquad n!\, f(y_{(1)}) \cdots f(y_{(n)})$$

for $y_{(1)} < \cdots < y_{(n)}$ and zero elsewhere.

We now derive the marginal distribution for, say, the largest value $y_{(n)}$ in the sample. For this we integrate first with respect to $y_{(1)}$ over the effective range $(-\infty, y_{(2)})$,

$$\int_{-\infty}^{y_{(2)}} n!\, f(y_{(1)}) \cdots f(y_{(n)})\, dy_{(1)} = n!\, F(y_{(2)}) f(y_{(2)}) \cdots f(y_{(n)}).$$

Next we integrate with respect to $y_{(2)}$ over the effective range $(-\infty, y_{(3)})$,

$$\int_{-\infty}^{y_{(3)}} n!\, F(y_{(2)}) f(y_{(2)}) \cdots f(y_{(n)})\, dy_{(2)} = \frac{n!}{2} F^2(y_{(3)}) f(y_{(3)}) \cdots f(y_{(n)}),$$

where $\int_{-\infty}^{a} F(u) f(u)\, du = \int_{-\infty}^{a} F(u)\, dF(u) = (1/2)F^2(a)$; continuing in this way we obtain the marginal density

$$(10) \qquad nF^{n-1}(y_{(n)}) f(y_{(n)}).$$

Note that this result can also be obtained from Exercise 3.2.8.

For further marginal distributions we take advantage of the inverse probability integral transformation in Section 2.4D. Consider the distribution of a sample $(u_1, \ldots, u_n)$ from the uniform(0, 1) distribution; let $(u_{(1)}, \ldots, u_{(n)})$ be the order statistic. Now consider a continuous distribution function F with inverse function $y = Y(u) = F^{-1}(u)$. Then it follows from Section 2.4D and Proposition 3.3.5 that $(y_1, \ldots, y_n) = (Y(u_1), \ldots, Y(u_n))$ has the distribution of a sample from F; and in addition by the monotonicity of Y we have $y_{(1)} = Y(u_{(1)}), \ldots, y_{(n)} = Y(u_{(n)})$. Thus we can derive order statistic distributions for the uniform and then transform to the general case.

To illustrate the preceding we determine the marginal distribution of $y_{(r)} = Y(u_{(r)})$, the rth smallest order statistic for a sample from a continuous distribution function F; suppose that the distribution is absolutely continuous with density function f. The probability differential for $(u_{(1)}, \ldots, u_{(n)})$ on $(0, 1)^n$ is

$$(11) \qquad n!\, du_{(1)} \cdots du_{(n)}.$$

We first integrate $u_{(1)}$ on $(0, u_{(2)})$, $u_{(2)}$ on $(0, u_{(3)})$, . . ., $u_{(r-1)}$ on $(0, u_{(r)})$, and obtain, in the pattern of formula (10),

$$\frac{n!}{(r-1)!} u_{(r)}^{r-1}\, du_{(r)} \cdots du_{(n)}.$$

We then integrate $u_{(n)}$ on $(u_{(n-1)}, 1)$, $u_{(n-1)}$ on $(u_{(n-2)}, 1)$, ..., $u_{(r+1)}$ on $(u_{(r)}, 1)$, and obtain

$$(12) \qquad \frac{n!}{(r-1)!(n-r)!} u_{(r)}^{r-1}(1 - u_{(r)})^{n-r}\, du_{(r)}$$

on the range (0, 1). Finally, we apply the inverse probability integral transform: we have $u_{(r)} = F(y_{(r)})$ and $du = f(y)\,dy$; thus the probability density for $y_{(r)}$ is

$$(13) \qquad \frac{n!}{(r-1)!(n-r)!} F^{r-1}(y_{(r)})(1 - F(y_{(r)}))^{n-r} f(y_{(r)})$$

$$= \frac{\Gamma(n+1)}{\Gamma(r)\Gamma(n-r+1)} F^{r-1}(y_{(r)})[(1 - F(y_{(r)})]^{n-r} f(y_{(r)}).$$

10 Consider the distribution of the order statistic $(u_{(1)}, \ldots, u_{(n)})$ for a sample from the uniform(0, 1). The amount of probability in the intervals between adjacent order statistics is given by the coverages $c_1 = u_{(1)}$, $c_2 = u_{(2)} - u_{(1)}$, ..., $c_n = u_{(n)} - u_{(n-1)}$, $c_{n+1} = 1 - u_{(n)}$. Show that $\mathbf{c} = (c_1, \ldots, c_n)'$ has density equal to $n!$ in the region $\{\mathbf{c}: 0 < c_i;\ \Sigma_1^n c_i < 1\}$ and equal to 0 elsewhere; the same distribution applies for any n of the $n + 1$ original c's. Verify that this is Dirichlet(1, ..., 1) with $k = n$.

11 *(continuation)* Let $C_1 = c_1 + \cdots + c_{r_1}$, $C_2 = c_{r_1+1} + \cdots + c_{r_1+r_2}$, ..., $C_k = c_{r_1+\cdots+r_{k-1}+1} + \cdots + c_{r_1+\cdots+r_k}$, where $\Sigma_1^k r_j < n + 1$. Show that $(C_1, \ldots, C_k)$ has the Dirichlet$(r_1, \ldots, r_{k+1})$ distribution, where $r_{k+1} = n + 1 - \Sigma_1^k r_j$; recall the convolution result in Problem 3.2.26.

12 *(continuation)* Let $U_1 = u_{(r_1)} = C_1$, $U_2 = u_{(r_1+r_2)} = C_1 + C_2$, $U_3 = u_{(r_1+r_2+r_3)} = C_1 + C_2 + C_3$, ... Show that $(U_1, \ldots, U_k)$ has the density function

$$(14) \qquad \frac{\Gamma(n+1)}{\Gamma(r_1)\cdots\Gamma(r_{k+1})} U_1^{r_1-1}(U_2 - U_1)^{r_2-1} \cdots (U_k - U_{k-1})^{r_k-1}(1 - U_k)^{r_{k+1}-1}$$

for $0 < U_1 < \cdots < U_k < 1$ and zero elsewhere.

13 Consider the distribution of the order statistic $(y_{(1)}, \ldots, y_{(n)})$ for a sample from the distribution F with density function f. Use Problem 12 to show that the marginal density for $(x_1, \ldots, x_k) = (y_{(r_1)}, y_{(r_1+r_2)}, \ldots, y_{(r_1+\cdots+r_k)})$ is

$$(15) \qquad \frac{\Gamma(n+1)}{\Gamma(r_1)\cdots\Gamma(r_{k+1})} F^{r_1-1}(x_1)(F(x_2) - F(x_1))^{r_2-1} \cdots$$

$$(F(x_k) - F(x_{k-1}))^{r_k-1}(1 - F(x_k))^{r_{k+1}-1} f(x_1) \cdots f(x_k)$$

on $-\infty < x_1 < \cdots < x_k < \infty$ and zero elsewhere; note that $r_{k+1} = n + 1 - r_1 - \cdots - r_k$.

14 Let $(y_1, \ldots, y_n)$ be a sample from the distribution F with density f. Show that the order statistic $(y_{(1)}, \ldots, y_{(n)})$ and rank statistic $(r_1, \ldots, r_n)$ are statistically independent; see Exercise 3.2.15.

F PRODUCT SPACES

In Section 3.3E we considered two independent random systems; and we constructed the probability model for the combined system from models for the component systems using independence and the extension theorem. We also indicated how we could continue and combine additional models one by one; as an illustration we constructed the models for a sample from the Bernoulli, the Poisson, and the normal.

Now suppose that we have probability models $(\mathcal{S}_t, \mathcal{A}_t, P_t)$ for each value of t in an index set T; the index set T can be finite, countable, or even larger; let s_t designate a point in $\mathcal{S}_t$. For example, s_t could be voltage at time t and we would be interested in the voltage function s_t on the time interval $(0, \infty)$. Certainly, we can consider s_t as being random: we would, however, have some difficulty in considering duplicate observations at time t; for we would need to consider duplicates of the whole system.

Consider a combined model based on independence. The combined response s has a coordinate s_t for each value of the index t; thus s is a function from T into $\cup_T \mathcal{S}_t$, where the tth coordinate is restricted to the subset $\mathcal{S}_t$ of $\cup_T \mathcal{S}_t$. The space of such responses s is called the corresponding *product space* and is designated $\mathcal{S} = \Pi_T \mathcal{S}_t$.

We would certainly be interested in an event that involves assertions concerning a finite number of coordinates; consider the rectangles $\Pi_T A_t$, where only a finite number of the A_t are not equal to the corresponding $\mathcal{S}_t$'s. We then define the product algebra $\mathcal{A} = \Pi_t \mathcal{A}_t$ to be the smallest σ-algebra that includes these rectangles.

On the basis of independence we define P for any rectangle by $P(\Pi_T A_t) = \Pi_T P_t(A_t)$, where of course only a finite number of the $P_t(A_t)$ are different from 1. Then by additivity this produces P for any finite union of disjoint rectangles. Additivity for this function P is immediate; continuity at $\emptyset$ requires some details of proof. It follows then by the Extension Theorem in Section 1.5D that P is uniquely defined on the σ-algebra $\mathcal{A}$. We write $P = \Pi_T P_t$.

Thus we have obtained the arbitrary-product space

$$(\mathcal{S}, \mathcal{A}, P) = \left(\prod_T \mathcal{S}_t, \prod_T \mathcal{A}_t, \prod_T P_t\right).$$

We can very easily, in fact, be much more general. As an alternative to independence, suppose that for each finite subset $K = \{t_1, \ldots, t_k\}$ of T we have a probability measure P_K on the corresponding class $\Pi_K \mathcal{A}_t$ of events; we assume that the probability measures are consistent in the sense that if $K' \subset K$, then $P_{K'}$ is the appropriate marginal measure from P_K.

We then proceed as before. We form the space $\mathcal{S} = \Pi_T \mathcal{S}_t$, and the algebra $\mathcal{A} = \Pi_T \mathcal{A}_t$. Then for the measure P on a set $\Pi_T A_t$ we use the measure P_K for the set K of coordinates having A_t not equal to $\mathcal{S}_t$. The consistency of the P_K, then, assures that P is a measure and the Extension Theorem uniquely determines P on the σ-algebra $\mathcal{A}$; for this general case some topological properties are needed for the spaces $\mathcal{S}_t$.

Thus we have obtained a general product space $(\Pi_T \mathcal{S}_t, \Pi_T \mathcal{A}_t, P)$, where P is derived from mutually consistent probability measures P_K defined on finite products.

G MEASURABLE FUNCTIONS: THEORY AND PROBLEMS

We now return to the measurable functions of Section 3.1. For this discussion we shall assume that any given space has a σ-algebra associated with it. And, in particular, if the space is R, then the σ-algebra is the Borel class $\mathcal{B}^1$; and if the space is R^k, then the σ-algebra is the Borel class $\mathcal{B}^k$; and if the space is a product space $\mathcal{S}_1 \times \mathcal{S}_2$, then the σ-algebra is the smallest σ-algebra containing the product sets $A_1 \times A_2$, where A_i is in the σ-algebra $\mathcal{A}_i$ for the space $\mathcal{S}_i$.

We first record some propositions; the proofs are assembled at the end of the section.

PROPOSITION 1

Any simple function is measurable; any discrete function is measurable.

PROPOSITION 2

Any continuous function (real- or vector-valued) *is measurable.*

PROPOSITION 3

If Y_1 and Y_2 are measurable functions on the same space, then (Y_1, Y_2) into the product space is measurable.

PROPOSITION 4

If X is measurable into a space on which a measurable Y is defined, then the composite function YX [taking s into $Y(X(s))$] *is measurable.*

PROPOSITION 5

If $Y_1, Y_2, \ldots$ are measurable functions (real- or vector-valued) *and if*

$$\lim_{i\to\infty} Y_i(s) = Y(s)$$

for all s, then $Y = \lim Y_i$ is measurable.

By applying these in various combinations, we can obtain a great wealth of measurable functions. For example, if X and Y are real-valued measurable functions, then $X + Y$, $X \times Y$, cX, max (X, Y), min (X, Y), $p(X, Y)$ are measurable, where c is real and p is a polynomial. Also, the positive and negative parts of Y, the functions $Y^+ = \max(Y, 0)$ are $Y^- = \max(-Y, 0)$, are measurable.

Before recording the proofs it seems reasonable to consider just how remote and strange a measurable function can be. Consider the case of a bounded measurable function Y with $-M \leq Y(s) < M$, where M is an integer. Let $A_n(i)$ be the preimage of the interval $[(i - 1)/n, i/n)$ for $i = -Mn + 1, -Mn + 2, \ldots, Mn$:

$$A_n(i) = \left\{s : \frac{i-1}{n} \leq Y(s) < \frac{i}{n}\right\}.$$

And let X_n and Y_n be simple functions approximating Y from below and above:

$$X_n = \sum_{-Mn+1}^{Mn} \frac{i-1}{n} I_{A_n(i)}, \qquad Y_n = \sum_{-Mn+1}^{Mn} \frac{i}{n} I_{A_n(i)}.$$

Then clearly $X_n \leq Y < Y_n$ and $Y_n - X_n = 1/n$. And it follows that

$$\lim_{n\to\infty} X_n = Y, \qquad \lim_{n\to\infty} Y_n = Y$$

and the convergence is uniform. Thus: *a bounded measurable function is the uniform limit of a sequence of simple functions.*

Proof of Proposition 1 A simple function is a discrete function. A discrete function has the form

$$Y = \sum_1^\infty a_i I_{A_i},$$

where A_i is the preimage of the value a_i for the function and the A_i are in the class $\mathfrak{A}$. The preimage, then, of any set B is a finite or countable union of A_i's and is thus in the class $\mathfrak{A}$.

Proof of Proposition 4 Let C be a set in the σ-algebra in the image space of Y; then $Y^{-1}(C)$ is a set in the σ-algebra on the image space of X; then $X^{-1}Y^{-1}(C)$ is a set in the σ-algebra on the given space for X. But

$$X^{-1}Y^{-1}(C) = \{s : Y(X(s)) \in C\} = (YX)^{-1}(C);$$

thus YX is measurable.

For the remaining proofs we need the following preliminary lemma:

LEMMA 6

Let Y be a function from $\mathcal{S}$ with σ-algebra $\mathcal{A}$ into $\mathcal{J}$ with σ-algebra $\mathcal{B}$. If each B in $\mathcal{B}_0$ has $Y^{-1}(B)$ in $\mathcal{A}$ and if *$\mathcal{B}$ is the smallest σ-algebra containing $\mathcal{B}_0$, then Y is measurable.*

Proof Let $\mathcal{B}^*$ consist of sets B whose preimages are in $\mathcal{A}$:

$$\mathcal{B}^* = \{B : Y^{-1}(B) \in \mathcal{A}\},$$

then $\mathcal{B}_0 \subset \mathcal{B}^*$. From the properties (3.1.9) of the inverse function we have: if B is in $\mathcal{B}^*$, then B^c is in $\mathcal{B}^*$; if B_i is in $\mathcal{B}^*$, then $\cup_1^\infty B_i$ is in $\mathcal{B}^*$. Thus $\mathcal{B}^*$ is a σ-algebra and it must contain the smallest σ-algebra $\mathcal{B}$. It follows that B in $\mathcal{B}$ implies $Y^{-1}(B)$ in $\mathcal{A}$ and hence that Y is measurable.

Proof of Proposition 3 Let Y_1 map into $\mathcal{J}_1$ with σ-algebra $\mathcal{B}_1$ and Y_2 map into $\mathcal{J}_2$ with σ-algebra $\mathcal{B}_2$. Then for B_1 in $\mathcal{B}_1$, $Y_1^{-1}(B_1)$ is in $\mathcal{A}$; and for B_2 in $\mathcal{B}_2$, $Y_2^{-1}(B_2)$ is in $\mathcal{A}$. Thus consider

$$\begin{aligned}(Y_1, Y_2)^{-1}(B_1 \times B_2) &= \{s : (Y_1(s), Y_2(s)) \in B_1 \times B_2\}\\ &= \{s : Y_1(s) \in B_1, Y_2(s) \in B_2\} = Y_1^{-1}(B_1) \cap Y_2^{-1}(B_2),\end{aligned}$$

which is a set in $\mathcal{A}$. Thus the preimage of any product set is in $\mathcal{A}$ and hence, by Lemma 6, (Y_1, Y_2) is measurable.

Proof of Proposition 2 Let Y be a continuous function from R into R. And consider the preimage $Y^{-1}((a, b))$ of an open interval (a, b). Let s be a point in the preimage. Then continuity gives: for any interval $(Y(s) - \epsilon, Y(s) + \epsilon)$, there is a δ such that $(s - \delta, s + \delta)$ is mapped into $(Y(s) - \epsilon, Y(s) + \epsilon)$. With ϵ sufficiently small, this says that $Y^{-1}((a, b))$ contains an open interval about the point s. Thus $Y^{-1}((a, b))$ is open. The Borel class is the smallest σ-algebra containing the open intervals, or containing the open sets; hence, by Lemma 6, Y is measurable. Similarly for R^k to R^l.

We now consider the proof of Proposition 5 for real-valued functions; the proof for vector-valued functions then follows from Proposition 3. Actually, it is convenient to be somewhat more general and let Y and even the Y_i's be extended real-valued functions taking real values or values $+\infty$ and $-\infty$. The Borel sets extend also in an obvious manner, and for measurability it suffices (Lemma 6) to verify that $Y^{-1}((c, \infty])$ is in $\mathcal{A}$ for all real c. The proof needs two preliminary results:

LEMMA 7

If $Y_1, Y_2, \ldots$ are measurable, then $W = \sup Y_i$ is measurable; also $\inf Y_i$ is measurable.

Proof We have

$$\begin{aligned}s \in W^{-1}((c, \infty]) \;&\leftrightarrow\; W(s) \in (c, \infty] \;\leftrightarrow\; Y_i(s) \in (c, \infty] \text{ for some } i\\ &\leftrightarrow\; s \in Y_i^{-1}((c, \infty]) \text{ for some } i \;\leftrightarrow\; s \in \cup_1^\infty Y_i^{-1}((c, \infty]).\end{aligned}$$

But each $Y_i^{-1}((c, \infty])$ is in $\mathcal{A}$; hence $\cup_1^\infty Y_i^{-1}((c, \infty])$ is in $\mathcal{A}$; thus W is measurable.

LEMMA 8

If $Y_1, Y_2, \ldots$ are measurable, then $X = \limsup Y_i$ is measurable; also, $\liminf Y_i$ is measurable.

Proof The lim sup or upper limit point of a sequence of real numbers $a_1, a_2, \ldots$ is a value a such that any $a + \delta$ $(\delta > 0)$ is exceeded by a finite number of the a_i, and any $a - \delta$ $(\delta > 0)$ is exceeded by a countable number of the a_i, provided that such an a exists; otherwise, it is $+\infty$. Formally,

$$\limsup_{i\to\infty} Y_i(s) = \inf_{j=1}^{\infty} \sup \{Y_j(s), Y_{j+1}(s), \ldots\}.$$

The lim sup for a sequence of functions is then given by

$$\limsup_{i\to\infty} Y_i = \inf_{j=1}^{\infty} \sup \{Y_j, Y_{j+1}, \ldots\}.$$

By Lemma 7 we have that sup $\{Y_j, Y_{j+1}, \ldots\}$ is measurable. And then using the analogous result for inf, we have that lim sup Y_i is measurable.

Proof of Proposition 5 If the limit lim Y_i exists, then lim Y_i = lim inf Y_i = lim sup Y_i and these are measurable by Lemma 8.

---

15 Let Y be a real-valued measurable function. Prove:
(a) Y is the uniform limit of a sequence of discrete functions.
(b) Y is the limit of a sequence of simple functions.

16 Let Y_1 and Y_2 be measurable functions from $\mathcal{S}$ into R. Use Propositions 1, . . ., 5 to prove that $Y_1 + Y_2$ is measurable.

17 Let μ be a measure on the σ-algebra $\mathcal{A}$ of subsets of $\mathcal{S}$; let Y be a measurable function from $\mathcal{S}$ into $\mathcal{T}$ with class $\mathcal{B}$ of subsets; and let μ_Y be defined by

$$\mu_Y(B) = \mu(Y^{-1}(B)).$$

Now let $\mathcal{S}_0 = \{1, 2, 3, 4, 5, 6\} \times \{1, 2, 3, 4, 5, 6\}$; let μ be the modified cardinality measure:

$$\mu(A) = c(A \cap \mathcal{S}_0)$$

for $A \subset R^2$; and let Y be the sum function given by $y = Y(x_1, x_2) = x_1 + x_2$. Calculate μ_Y and compare with P_Y in Exercise 3.1.2.

4.3

CONDITIONAL PROBABILITY

In science, engineering, and business, there are situations where we want to record a response at a sequence of time points or even record it continuously in time. For this, then, we need not a finite number of coordinates yielding a point in R^k but a countable number or a continuum of coordinates. The space is more complicated, but the probability model is easily constructed; for some formalities, see Supplement Section 3.6F. We now examine briefly some probability models having a countable number or a continuum of coordinates; conditional probability is used directly or implicitly in the construction of the models. These are examples of *stochastic processes* and they give an indication of the breadth and importance of this area of probability theory.

For notation we let $s = \{s_t\}$ designate the combined response. And let s_t designate the response at index t or at time t; and $\mathcal{S}_t$ designate the corresponding sample space. We call the index *time* even though in some applications it may refer to space and even take values in R^2, or R^3, or more generally. In accord with this, we call s_t the *state* of the system at time t and call $\mathcal{S}_t$ the *state space* at time t. A realized response $s = \{s_t\}$ is then a *function s* which records the state of the system for various times t. From Supplement Section 3.6F we have available the probability theory to describe the distribution of such a function s.

A POISSON PROCESSES

Suppose that we start observing a radioactive disintegration and keep a cumulative record of the number of particles observed; let n_t be the number of particles in the interval $(0, t]$. We might then obtain a record as illustrated in the first half of Figure S.7; this is an example of a Poisson process. Or in a more general sense, suppose that we keep a record of the number n_t of people alive in a society at time t. As t increases, a function n_t would increase at the time of a birth and decrease at the time of a death, as indicated in the second half of Figure S.7; this is an example of a birth-and-death process. We examine some theory concerning Poisson processes.

Suppose that a certain kind of event can occur uniformly and at random in time. Let n_t designate the number of occurrences in the time interval $(0, t]$; then $\{n_t\}$ is a Poisson process. We present this formally:

DEFINITION 1

The process $\{n_t\}$ is a **Poisson process** if the following hold:

(a) Independent increments: for $0 < t_1 < t_2 < \cdots < t_k$, the increments $n_{t_2} - n_{t_1}$, $\ldots, n_{t_k} - n_{t_{k-1}}$ are statistically independent.

(b) Homogeneous increments: the distribution of the increment $n_{t+h} - n_t$ with $h > 0$ does not depend on t.

(c) No cascade: let $p_t(i)$ be the probability $P(n_t = i)$; then as $t \to 0$,

$$\frac{p_t(1)}{t} \to \lambda, \qquad \frac{1 - p_t(0)}{t} \to \lambda, \qquad \lambda > 0.$$

Note that condition (c) says that the probability of one event in an interval is proportional to the length t of the interval as $t \to 0$; and that the probability of one or more event in an interval is the same proportion of the length t as $t \to 0$; that is, the probability of more than one event, as a proportion of t, goes to zero as $t \to 0$.

The homogeneity and independence give

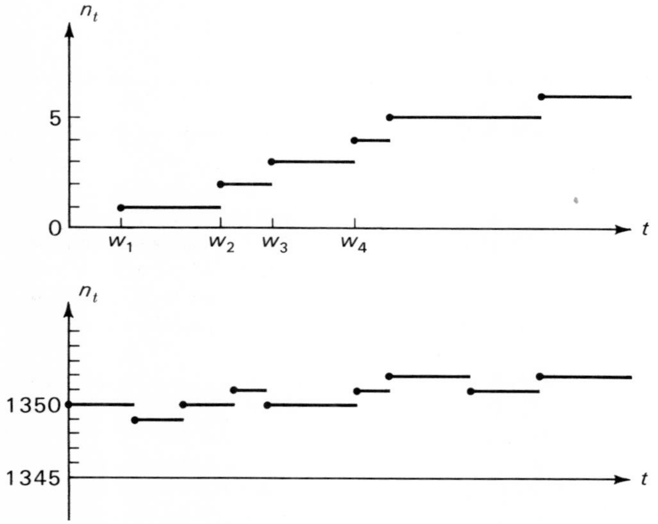

FIGURE S.7
Poisson process; and birth-and-death process.

$$p_{t+h}(0) = p_h(0)p_t(0).$$

Thus

$$\frac{p_{t+h}(0) - p_t(0)}{h} = \frac{p_h(0) - 1}{h} p_t(0),$$

$$\frac{dp_t(0)}{dt} = -\lambda p_t(0),$$

$$p_t(0) = e^{-\lambda t},$$

where an integration constant has been determined by the necessary $p_0(0) = 1$. In a similar way we obtain

$$p_{t+h}(i) = p_h(0)p_t(i) + p_h(1)p_t(i-1) + \cdots + p_h(i)p_t(0),$$

$$\frac{p_{t+h}(i) - p_t(i)}{h} = \frac{p_h(0) - 1}{h} p_t(i) + \frac{p_h(1)}{h} p_t(i-1) + \cdots + \frac{p_h(i)}{h} p_t(0),$$

$$\frac{dp_t(i)}{dt} = -\lambda p_t(i) + \lambda p_t(i-1),$$

$$\frac{d}{dt}(p_t(i)e^{\lambda t}) = \lambda(p_t(i-1)e^{\lambda t}).$$

We then obtain successively $p_t(1), p_t(2), \ldots$ using the necessary $p_0(i) = 0$ $(i > 0)$:

$$p_t(i) = \frac{(\lambda t)^i}{i!} e^{-\lambda t}.$$

Thus for any particular t the distribution of n_t is Poisson(λt).

Now let $w_1, w_2, \ldots$ be the time of the first, second, ... occurrence of the event. And let $t_1, t_2, \ldots$ be the interoccurrence times:

$$t_1 = w_1,\ t_2 = w_2 - w_1,\ t_3 = w_3 - w_2, \ldots.$$

The probability $P(t_1 > t)$ is the probability that the first occurrence time exceeds t, or equivalently that there are no occurrences in the interval $(0, t]$; thus

$$P(t_1 > t) = e^{-\lambda t}; \qquad P(t_1 \le t) = 1 - e^{-\lambda t},$$

and we see that t_1 has the exponential(λ) distribution or, equivalently, λt_1 has the standard exponential distribution.

The homogeneity and independence then show that $(\lambda t_1, \ldots, \lambda t_r)$ is a sample from the standard exponential; thus

$$\lambda w_r = \lambda t_1 + \cdots + \lambda t_r$$

has the gamma(r) distribution: see the reproductive property for the gamma distribution as presented in terms of the chi-square distribution in Exercise 3.3.5.

The probability $P(w_r > t)$ is the probability that the rth occurrence time exceeds t or, equivalently, that there are less than r occurrences in the interval $(0, t]$; thus

$$\int_t^\infty \frac{1}{\Gamma(r)} (\lambda u)^{r-1} e^{-\lambda u} \lambda \, du = P(w_r > t) = P(n_t < r) = \sum_{i=0}^{r-1} \frac{(\lambda t)^i}{i!} e^{-\lambda t},$$

which gives a direct probability verification of the gamma-Poisson relation in Problem 3.5.33.

B MARKOV PROCESSES

Consider a system, say an atom, that can be in one of three states $s = 1, 2, 3$; let s_t be the state at time t. And suppose the system is without memory, so that the probabilities for time $t + h$

($h > 0$) given the state at time t do not depend on the states at times earlier than t. We might then obtain a record as illustrated in the first half of Figure S.8; this is an example of a discrete Markov process.

Or consider the vertical elevation y_t of a particle in a fluid where changes in y_t are due to impacts by adjacent molecules. We might then obtain a record as illustrated in the second half of Figure S.8; this is an example of a Brownian process, a particular kind of continuous Markov process. We examine briefly some theory concerning Markov processes.

Consider a stochastic process $y = (y_t) = \{y_t : t \in J\}$, where the index t can range over the integers Z or the reals R.

DEFINITION 2

The process $\{y_t\}$ is a **Markov process** if the conditional distribution of y_{t_k} given $(y_{t_1}, \ldots, y_{t_{k-1}})$ with $t_1 < t_2 < \cdots < t_k$ is equal to the conditional distribution given $y_{t_{k-1}}$ only.

In other words, the probability for a future state given the present state is not altered by information concerning past states. A Markov process is called *homogeneous* if the distribution of y_{t+h} given y_t with $h > 0$ is independent of t. The Poisson process and certain birth-and-death processes are examples of Markov processes.

For some theory, consider the special case of a homogeneous Markov process with finite state space $\{1, 2, \ldots, s\}$ and with discrete time space $\{0, 1, 2, \ldots\}$. Let $p_1(i, j)$ be the probability of going to state j at time $t + 1$ from state i at time t; this can be presented as the one-step transition matrix

$$P = \begin{pmatrix} p_1(1, 1) & \cdots & p_1(1, s) \\ \vdots & & \vdots \\ p_1(s, 1) & \cdots & p_1(s, s) \end{pmatrix}.$$

And similarly, let $p_n(i, j)$ be the probability of going to state j at time $t + n$ from state i at time t; this can be presented as an n-step transition matrix

$$P_n = \begin{pmatrix} p_n(1, 1) & \cdots & p_n(1, s) \\ \vdots & & \vdots \\ p_n(s, 1) & \cdots & p_n(s, s) \end{pmatrix}.$$

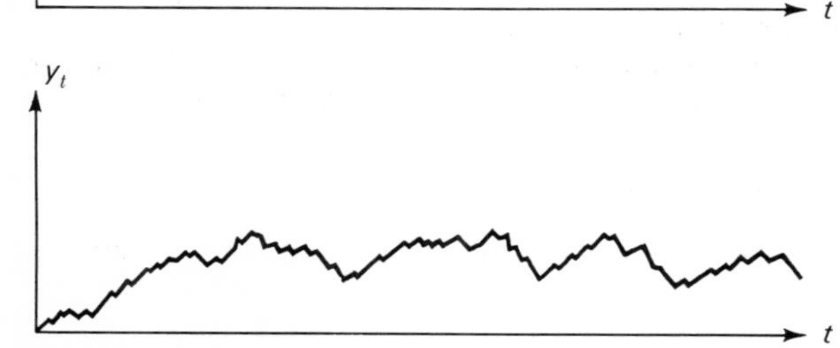

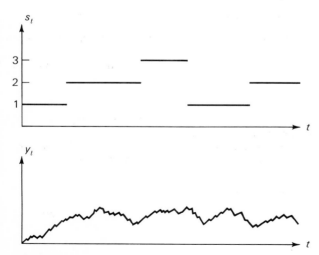

FIGURE S.8
Discrete Markov process; and Brownian process (a kind of Markov process).

Now consider going from state i at time t to state j at time $t + 2$. Let k designate the possible states $1, \ldots, s$ at time $t + 1$; then

$$p_2(i, j) = \sum_{k=1}^{s} p_1(i, k)p_1(k, j).$$

Note that this is the summation formula for matrix multiplication; thus

$$P_2 = P^2.$$

And by the same method

$$P_n = P^n;$$

thus the n-step transition matrix is the nth power of the one-step matrix.

In certain cases a transition matrix P can be factored $P = ADB$, where D is a diagonal matrix,

$$D = \begin{pmatrix} d_1 & & 0 \\ & \ddots & \\ 0 & & d_s \end{pmatrix},$$

and B and A record the left and right eigenvectors with $BA = I$. The n-step transition matrix is then given immediately,

$$\begin{aligned} P_n = P^n &= ADB \cdot ADB \cdot \cdots \cdot ADB \\ &= AD^nB \end{aligned}$$

by taking the nth power of the diagonal elements of D.

C NORMAL PROCESSES

For an electronic system, consider the voltage v_t at a certain point as a function of time t; this gives a function on the line R. Or consider the height h_{yt} of the ocean surface as a function of location y along a meridian at time t; this gives a function over the plane R^2.

When many small statistically independent components are added together, the resulting sum has a distribution that is approximately normal (this will be examined formally in Section 6.2). For a variety of systems the response has this property of being the resultant of many small component sources of variation, and accordingly the response is found to be approximately normal.

Consider the voltage v_t at a point in a small electronic system; without stimulation electrons can be discharged from the cathode of a vacuum tube at random in accord with a Poisson process; the passage of an electron to the anode can induce a transient voltage fluctuation; the superposition of many such voltage fluctuations can produce a voltage v_t such that $(v_{t_1}, \ldots, v_{t_n})$ has a multivariate normal distribution. This leads us to examine normal or Gaussian processes.

DEFINITION 3

The process $\{y_t : t \in R\}$ is a **normal process** if $(y_{t_1}, \ldots, y_{t_n})$ has a multivariate normal distribution for each $(t_1, \ldots, t_n)$.

A normal process is called *homogeneous* if the distribution of $(y_{t_1+h}, \ldots, y_{t_n+h})$ is independent of h. A normal process is called a *Brownian process* if it has

1. Independent increments: for $t_1 < t_2 < \cdots < t_n$, the increments $y_{t_2} - y_{t_1}, \ldots, y_{t_n} - y_{t_{n-1}}$ are statistically independent.
2. Homogeneous symmetrical increments: the distribution of $y_{t+h} - y_t$ with $h > 0$ is symmetrical about 0 and is independent of t.

Consider briefly the description for the Brownian process suggested by the second part of Figure S.8. Let $y_0 = 0$ and let σ be the scaling of the normal distribution of y_1. The distribution of y_n is then the sum of n independent normal$(0, \sigma)$ variables and thus has a normal$(0, \sqrt{n}\sigma)$ distribution; see Problem 3.3.11. It then follows from a continuity argument that y_t has a normal $(0, \sqrt{t}\sigma)$ distribution.

D THE POLYA URN

We conclude with a simple model involving colored balls in an urn, the Polya urn. The model has application in the study of occurrences of a contagious disease and in the distribution of plants and animals over area.

An urn contains b black balls and r red balls. The balls are thoroughly mixed, one ball is drawn, its color is observed, and it is replaced in the urn together with s balls of the same color and d balls of a different color. The procedure is repeated successively.

Let B_i and R_i be the occurrence of a black ball and a red ball, respectively, on the ith draw. Then,

$$P(B_1) = \frac{b}{b+r}, \qquad P(R_1) = \frac{r}{b+r}.$$

And, for example, for black on the second draw given the color on the first draw,

$$P(B_2 : B_1) = \frac{b+s}{(b+s)+(r+d)}, \qquad P(B_2 : R_1) = \frac{b+d}{(b+d)+(r+s)}.$$

Now let E_x be the event x black balls in n draws from the urn. The calculation of $P(E_x)$ can be very complicated.

Consider a simple case called the Polya urn: $s > 0$, $d = 0$. Then, for example,

$$\begin{aligned} P(B_1 \cdots B_x R_{x+1} \cdots R_n) &= \frac{b(b+s)\cdots(b+(x-1)s)\cdot r(r+s)\cdots(r+(n-x-1)s)}{(b+r)(b+r+s)\cdots(b+r+(n-1)s)} \\ &= \frac{\frac{b}{s}\left(\frac{b}{s}+1\right)\cdots\left(\frac{b}{s}+(x-1)\right)\cdot\frac{r}{s}\left(\frac{r}{s}+1\right)\cdots\left(\frac{r}{s}+n-x-1\right)}{\frac{b+r}{s}\left(\frac{b+r}{s}+1\right)\cdots\left(\frac{b+r}{s}+n-1\right)} \\ &= \left(\frac{b}{s}\right)^{[x]}\left(\frac{r}{s}\right)^{[n-x]} \Big/ \left(\frac{b+r}{s}\right)^{[n]}, \end{aligned}$$

where $N^{[n]} = N(N+1)\cdots(N+n-1)$. The probability for any other ordering of x B's and $n-x$ R's involves only a rearrangement of the factors above; thus $P(E_x)$ is $\binom{n}{x}$ times the preceding expression:

$$P(E_x) = \frac{\left(\frac{b}{s}\right)^{[x]}}{x!}\,\frac{\left(\frac{r}{s}\right)^{[n-x]}}{(n-x)!}\,\frac{n!}{\left(\frac{b+r}{s}\right)^{[n]}} = \frac{\binom{-b/s}{x}\binom{-r/s}{n-x}}{\binom{-(b+r)/s}{n}}.$$

These urn models have been used to describe the number of occurrences of a contagious disease: an incidence of the disease can infect others and alter the population.

E EXERCISE

1 A player throws a pair of symmetrical dice until he gets 3 points or 7 points; let n be the number of tosses. He wins if he gets 3 points; he loses if he gets 7 points.

(a) Determine the probability that n is even; that n is a multiple of 3.
(b) Show that {Win, Lose} is statistically independent of n.

F PROBLEMS

2 *The Poisson(λ) process:* Show that the distribution of $w_1, \ldots, w_n$ is given by

$$f(w_1, \ldots, w_n) = \lambda^n \exp\{-\lambda w_n\} \qquad \text{if } 0 < w_1 < \cdots < w_n$$

and zero otherwise.

3 *The Poisson(λ) process given the value n at a specified time t:* Show that the conditional distribution of $(w_1, \ldots, w_n)$ given $n_t = n$ is the distribution of the order statistic $(u_{(1)}, \ldots, u_{(n)})$ of a sample from the uniform(0, t) distribution.

4 (*continuation*) *The Poisson(λ) process given the time of the* $(n + 1)$st *occurrence:* Show that the conditional distribution of $(w_1, \ldots, w_n)$ given $w_{n+1} = t$ is the order statistic distribution described in Problem 3.

5 (*continuation*) For simplicity take $\lambda = 1$. Calculate expressions for

$$P(w_r > p : w_{n+1} = t) = P(n_p < r : w_{n+1} = t)$$

and obtain the beta-binomial relation in Problem 3.5.28. Use the order statistic theory in Supplement Section 3.6E.

6 Calculate the two-step transition matrix for the Markov process with one-step matrix

$$P = \begin{pmatrix} 0.8 & 0.1 & 0.1 \\ 0 & 0.7 & 0.3 \\ 0 & 0.6 & 0.4 \end{pmatrix}.$$

Let $p_t(i)$ be the probability of state i at time t. If $(p_0(1), p_0(2), p_0(3)) = (0.4, 0.3, 0.3)$, determine $(p_1(1), p_1(2), p_1(3))$ and $(p_2(1), p_2(2), p_2(3))$. What will the value of $p_n(1)$ be for large n?

7 A symmetric coin is tossed at times $t = 0, 1, 2, \ldots$. Let n_t be the number of heads at time t. Treat (n_t) as a Markov process and calculate the one-step transition probability $p_1(i, j)$ from state i at time t to state j at time $t + 1$. Present this as a matrix with rows $0, 1, 2, \ldots$ and columns $0, 1, 2, \ldots$. Calculate the two-step transition matrix.

8 Consider a Poisson(λ) process (n_t). Treat (n_t) as a Markov process and calculate the transition probability $p_h(i, i + k)$ from state i at time t to state $i + k$ at time $t + h$. Present this as a transition matrix P_h.

9 Consider a state space consisting of two points; the general homogeneous Markov process can be represented by the one-step transition matrix

$$P = \begin{pmatrix} 1-\beta & \beta \\ \alpha & 1-\alpha \end{pmatrix}.$$

(a) Show that for $\alpha + \beta \neq 0, 1$:

$$P = \begin{pmatrix} 1 & -\beta \\ 1 & \alpha \end{pmatrix} \begin{pmatrix} 1/(\alpha+\beta) & 0 \\ 0 & 1/(\alpha+\beta) - 1 \end{pmatrix} \begin{pmatrix} \alpha & \beta \\ -1 & 1 \end{pmatrix}$$

and calculate the n-step transition matrix P_n.

(b) Determine $\lim_{n\to\infty} P_n$.

(c) Examine the case $\alpha + \beta = 1$, also the case $\alpha + \beta = 0$.

10 Let t be an index with range $J = \{1, 1/2, 1/2^2, 1/2^3, \ldots\}$. Let $\mathcal{S}_t = (0, 1)$, $\mathcal{B}_t$ be the Borel class on (0, 1), and P_t be the uniform distribution on (0, 1). Consider $(\Pi\, \mathcal{S}_t, \Pi\, \mathcal{B}_t, \Pi\, P_t)$ and set $s = (s_t) = \{s_t : t \in J\}$ designate a typical outcome.

(a) Calculate the probability that the function s_t is less than the function e^{-t}; that is, $P(s_t < e^{-t}$ for all $t)$.

(b) Calculate the probability $P(s_t < e^{at}$ for all $t)$, where $a < 0$.
(c) Determine the distribution of $w = \sup (t^{-1} \ln s_t)$.

5.8

MEAN VALUE

We have noted that a moment-generating function uniquely determines a corresponding distribution. This uniqueness or unicity result is a very powerful tool in theoretical analyses. It is also useful for routine calculations with distributions provided the mathematics is tractable—that is, density functions involving powers, exponentials, logarithms, and trigonometric functions. In the first sections of this supplement we establish the uniqueness results for moment-generating functions, characteristic functions, and moment sequences.

The formal theory for mean values needs the concept of the general integral. In Section 5.1 we briefly described the general integral, then in subsequent sections used ordinary summations and integrations. In the later sections of this supplement we develop the basic concepts and properties for the general integral.

A THE UNIQUENESS THEOREM FOR DISCRETE DISTRIBUTIONS

Consider a discrete distribution for x on the real line R and suppose that all the probability is on the set Z of integers. Let p be the probability function and c be the corresponding characteristic function. Then

$$(1) \qquad c(t) = E(e^{ixt}) = \sum_{x=-\infty}^{\infty} e^{ixt}p(x).$$

The function $e^{iyt} = \cos yt + i \sin yt$ is a periodic function of t for y an integer; we have

$$(2) \qquad \int_{-\pi}^{\pi} e^{iyt}\, dt = 2\pi \qquad \text{if } y = 0$$

and zero if $y = \pm 1, \pm 2, \ldots$. Now multiply both sides of (1) by e^{-iyt} and integrate from $-\pi$ to π:

$$\int_{-\pi}^{\pi} e^{-iyt}c(t)\, dt = \int_{-\pi}^{\pi} \sum_{x=-\infty}^{\infty} e^{i(x-y)t}p(x)\, dt.$$

The integrand on the right has $|e^{i(x-y)t}| \leq 1$, and it is integrable with respect to x on Z and t on R. Thus by the Fubini Proposition 5.8.10 we can interchange the order of summation and integration:

$$(3) \qquad \int_{-\pi}^{\pi} e^{-iyt}c(t)\, dt = \sum_{x=-\infty}^{\infty} \int_{-\pi}^{\pi} e^{i(x-y)t}\, dt p(x) = 2\pi p(y)$$

for y equal to an integer. This gives us the inversion formula,

$$(4) \qquad p(y) = \frac{1}{2\pi} \int_{-\pi}^{\pi} e^{-iyt}c(t)\, dt$$

and from it we can note that a distribution on the integers is uniquely determined by its characteristic function.

B THE GENERAL UNIQUENESS THEOREM

Consider a distribution for x on the real line R. Let P be the probability measure and c be the characteristic function:

$$(5) \qquad c(t) = \int_{-\infty}^{\infty} e^{ixt}\, dP(x).$$

Let P^* be some probability measure on R and c^* be the corresponding characteristic function. We multiply the preceding equation by e^{-iyt} and integrate using P^* as the probability measure for t:

$$\int_{-\infty}^{\infty} e^{-iyt}c(t)\, dP^*(t) = \int_{-\infty}^{\infty} \left(\int_{-\infty}^{\infty} e^{i(x-y)t}\, dP(x)\right) dP^*(t).$$

The integrand on the right has $|e^{i(x-y)t}| \le 1$ and is integrable with respect to the product distribution for x and t. Thus by the Fubini Proposition 5.8.10, we can interchange the order of integration:

$$(6) \qquad \int_{-\infty}^{\infty} e^{-iyt}c(t)\, dP^*(t) = \int_{-\infty}^{\infty} \left(\int_{-\infty}^{\infty} e^{i(x-y)t}\, dP^*(t)\right) dP(x)$$

$$= \int_{-\infty}^{\infty} c^*(x-y)\, dP(x).$$

This is called the *Parseval relation;* a simpler and more symmetrical version is obtained with $y = 0$.

Now consider a fruitful choice for P^*. The normal distribution has the remarkable property that its density and its characteristic function have the same functional form. Thus for the central normal with variance σ^2 and the normal with variance $1/\sigma^2$, we have Table S.1. We multiply (6) by $1/\sqrt{2\pi}\sigma$ and use the second of these distributions for P^*:

$$(7) \qquad \frac{1}{2\pi}\int_{-\infty}^{\infty} e^{-iyt}c(t)e^{-\sigma^2t^2/2}\, dt = \int_{-\infty}^{\infty} \frac{1}{\sqrt{2\pi}\sigma} e^{-(x-y)^2/2\sigma^2}\, dP(x).$$

The remarkable property that we noted for the normal has produced a density function in the integrand on the right side. For interpretation think of x with distribution P and u with the normal(0, σ) distribution. The right side then is a slight generalization of the convolution formula (3.2.25) as applied to give the density function for $y = x + u$; note that y does have a density even if x does not. Let f_σ be the density for y, and F_σ be the corresponding distribution function; then

$$f_\sigma(y) = \frac{1}{2\pi}\int_{-\infty}^{\infty} e^{-iyt}c(t)e^{-\sigma^2t^2/2}\, dt,$$

$$F_\sigma(y_2) - F_\sigma(y_1) = \frac{1}{2\pi}\int_{-\infty}^{\infty} \frac{e^{-iy_2t} - e^{-iy_1t}}{-it}\, c(t)e^{-\sigma^2t^2/2}\, dt,$$

TABLE S.1
Functions for the normal

Density Function (t)	Characteristic Function (u)
$\frac{1}{\sqrt{2\pi}\sigma} e^{-t^2/2\sigma^2}$	$e^{-\sigma^2u^2/2}$
$\frac{\sigma}{\sqrt{2\pi}} e^{-\sigma^2t^2/2}$	$e^{-u^2/2\sigma^2}$

where we have used $\int_{y_1}^{y_2} e^{Ay}\, dy = (e^{Ay_2} - e^{Ay_1})/A$ with the Fubini Proposition 5.8.10. Our goal has been to obtain the distribution function for x. In fact, we have obtained the distribution function for $y = x + u$, where u is normal(0, σ). If σ is small, the effect of u should be small. Indeed, our study of limits in Section 6.1 will show that we can obtain the distribution function F for x by letting $\sigma \to 0$. This gives us the inversion formula,

$$(8) \qquad F(y_2) - F(y_1) = \lim_{\sigma \to 0} \frac{1}{2\pi} \int_{-\infty}^{\infty} \frac{e^{-iy_2t} - e^{-iy_1t}}{-it} c(t) e^{-\sigma^2 t^2/2}\, dt$$

for points y_1 and y_2 of continuity for the function F; and we can note from this that a distribution on the real line is uniquely determined by its characteristic function.

Now suppose that c is integrable with respect to length on the real line. Then by the Proposition 5.8.8 we can take the limit inside the integral sign on the right side of (8):

$$F(y_2) - F(y_1) = \frac{1}{2\pi} \int_{-\infty}^{\infty} \frac{e^{-iy_2t} - e^{-iy_1t}}{-it} c(t)\, dt.$$

The density function f is then given (use Proposition 5.8.10) by the inversion formula,

$$(9) \qquad f(y) = \frac{1}{2\pi} \int_{-\infty}^{\infty} e^{-iyt} c(t)\, dt.$$

Let $\mathcal{P}$ be the set of all probability measures P on the real line. And let $\mathcal{C}$ be the set of all characteristic functions of distributions on the real line. The defining equation (5) gives a mapping $\mathcal{P} \to \mathcal{C}$; the equation (8) shows that the mapping is one to one.

C THE UNIQUENESS THEOREM FOR MOMENT-GENERATING FUNCTIONS

Now consider a distribution for x on the real line R and suppose the moment-generating function exists. Let P be the probability measure and m be the moment-generating function,

$$(10) \qquad m(t) = \int_{-\infty}^{\infty} e^{xt}\, dP(x);$$

by Definition 5.7.1 the integral exists for an interval (t_1, t_2) about the origin. The analogous function of a complex variable $w = t + iu$,

$$(11) \qquad m(w) = \int_{-\infty}^{\infty} e^{x(t+iu)}\, dP(x)$$

exists if the real part of w lies in the interval (t_1, t_2); this follows by noting that

$$|e^{xw}| = |e^{xt} e^{ixu}| = e^{xt}$$

and using results from Section I. Complex function theory shows that the function $m(w)$ is determined by its values along a segment such as the interval (t_1, t_2) on the real axis. Thus the moment-generating function m determines the function (11), which in turn determines the characteristic function

$$c(u) = m(iu),$$

and hence determines the distribution P. Thus, a distribution on the real line is uniquely determined by the moment-generating function.

D THE UNIQUENESS THEOREM FOR MOMENT SEQUENCES

Consider a distribution for x on the real line R. Let P be the probability measure and $\mathbf{m}$ be the moment sequence:

$$\mathbf{m} = (E(x^0), E(x^1), E(x^2), \ldots)' = (\mu_0, \mu_1, \mu_2, \ldots)'.$$

If the series

$$(12) \qquad \sum_{r=0}^{\infty} \mu_r \frac{t^r}{r!}$$

converges for some interval containing the origin, then a detailed analysis shows that the characteristic function is determined and thus shows that the distribution is uniquely determined by its moment sequence. Note that the series (12) has the series properties of a moment-generating function; indeed, analysis shows that it *is* the moment-generating function.

It is possible to construct different distributions that have the same moment sequence. A necessary and sufficient condition for uniqueness of the distribution has been found—that the series

$$\sum_{r=0}^{\infty} \mu_{2r}^{-1/(2r)}$$

diverge.

E UNIQUENESS ON R^k

Consider a distribution for $\mathbf{x}$ on R^k. The characteristic function c given by

$$c(\mathbf{t}) = E(e^{i\mathbf{x}'\mathbf{t}})$$

always exists. A generalization of the methods in Section B show that the distribution is uniquely determined by its characteristic function.

The moment-generating function m given by

$$m(\mathbf{t}) = E(e^{\mathbf{x}'\mathbf{t}})$$

is said to exist if the mean exists for an open rectangle containing the origin. A generalization of the methods in Section C shows that the distribution is uniquely determined by its moment-generating function.

The moment array has the $(r_1, \ldots, r_k)$th moment

$$\mu_{r_1 \cdots r_k} = E(x_1^{r_1} \cdots x_k^{r_k}), \qquad r_i = 0, 1, \ldots.$$

If the moment-generating type of series (12) converges for some rectangle containing the origin, it can be shown that the distribution is uniquely determined by the moment array.

F THE INTEGRAL OF A NONNEGATIVE FUNCTION

The discussion in Section 5.1A lead us to think of the mean

$$E(Y) = \int Y(s)\, dP(s)$$

as the summation of values of the function Y weighted by corresponding probabilities. It is easy to be much more general and think of the summation of values of the function Y weighted by some measure μ.

Consider a sample space $\mathcal{S}$, a class $\mathcal{A}$ of events, and a measure μ. And let g be a nonnegative function mapping $\mathcal{S}$ into $[0, \infty]$. For the moment we are allowing the value $+\infty$; this, in fact, leads to simpler initial results.

If g is a simple function, then we define the integral of g with respect to μ by

$$(13) \qquad \int g(s)\, d\mu(s) = \sum a_j \mu(g^{-1}(a_j)) = \sum a_j \mu_g(a_j),$$

where the a_j are the values of the function g and where μ_g in the final expression is the marginal measure for g; see Problem 3.6.17. This integral is really just a summation; it was presented earlier in Problem 1.5.21 together with two basic properties.

Now consider some general function g mapping $\mathcal{S}$ into $[0, \infty]$. The integral of g with respect to μ is defined by

$$(14) \qquad \int g(s)\, d\mu(s) = \sup_{Y \leq g} \int Y(s)\, d\mu(s),$$

where the supremum is taken over all simple functions Y having $0 \leq Y(s) \leq g(s)$. Note that the integral has a value in $[0, \infty]$ and that we are, in effect, approximating g from below by simple functions. If we think of g plotted over the space $\mathcal{S}$, then the integral can be thought of as the area under g with respect to a length measure μ.

In the preceding definition the supremum is taken over all simple functions Y having $0 \leq Y(s) \leq g(s)$. Actually, it suffices to consider simple functions such as $X(g)$ having $0 \leq X(g(s)) \leq g(s)$. For if we have any simple function Y, then we can find a bigger function $X(g)$,

$$X(g(s)) = \max \{a_i : a_i \leq g(s)\},$$

where the a_i are the values of the function Y. It follows, then, that the integral

$$(15) \qquad \int g(s)\, d\mu(s) = \int_0^\infty y\, d\mu_g(y)$$

depends only on values of g and on the marginal measure for those values; this justifies the right-hand expression.

From the definition of the integral note that if $g_1 \leq g_2$, then

$$\int g_1(s)\, d\mu(s) \leq \int g_2(s)\, d\mu(s).$$

The integral of g over a set A is obtained by the use of indicator functions:

$$(16) \qquad \int_A g(s)\, d\mu(s) = \int I_A(s) g(s)\, d\mu(s).$$

We, of course, assume that the functions g are measurable (Definition 3.1.2) and that sets A are measurable (belong to the σ-algebra $\mathcal{A}$).

G PROPERTIES OF THE INTEGRAL FOR NONNEGATIVE FUNCTIONS

The integral of a nonnegative function has a very fundamental continuity property:

THEOREM 1

Monotone convergence theorem: *If $\{g_n\}$ is a monotone sequence of measurable functions $0 \leq g_1(s) \leq g_2(s) \leq \cdots$, then*

$$(17) \qquad \lim_{n\to\infty} \int g_n(s)\, d\mu(s) = \int \lim_{n\to\infty} g_n(s)\, d\mu(s).$$

Proof: Let $g(s) = \lim g_n(s)$: the limit exists because of the monotonicity and the inclusion of the value $+\infty$; g is measurable by Proposition 3.6.5. Consider the integrals

$$\int g_n(s)\, d\mu(s), \qquad \int g(s)\, d\mu(s).$$

Any simple function that approximates g_n from below also approximates g from below; thus the left integral is less than or equal to the right integral, and

$$\lim_{n\to\infty} \int g_n(s)\, d\mu(s) \leq \int g(s)\, d\mu(s).$$

Now consider a simple function Y that approximates g from below. The function g_n is above cY $(0 < c < 1)$ on the set

$$A_n = \{s : g_n(s) \geq cY(s)\},$$

and of course

$$\int g_n(s)\, d\mu(s) \geq \int_{A_n} g_n(s)\, d\mu(s) \geq c \int_{A_n} Y(s)\, d\mu(s).$$

We now note that $\lim A_n = \mathcal{S}$. For certainly, the sequence $A_1, A_2, \ldots$ is monotone increasing; and if $Y(s) = 0$, then $s \in A_1$ and hence $s \in A_n$; and if $Y(s) > 0$, then $g_n(s) \to g(s)$ implies that $g_n(s) \geq cY(s)$ for some n and hence $s \in A_n$.

By Problem 1.5.21, the right-hand integral can be viewed as a measure of A_n; the monotone property for measures then allows us to take the limit on both sides as $n \to \infty$:

$$\lim_{n\to\infty} \int g_n(s)\, d\mu(s) \geq c \int_{\mathcal{S}} Y(s)\, d\mu(s) = c \int Y(s)\, d\mu(s).$$

With arbitrary $c < 1$ and then with arbitrary $Y \leq g$, we obtain

$$\lim_{n\to\infty} \int g_n(s)\, d\mu(s) \geq \int g(s)\, d\mu(s).$$

This together with the first paragraph gives the equality stated in the theorem.

A function f mapping $\mathcal{S}$ into $[0, \infty]$ can be used as a density function for a measure just as such a function has been used as a density for probability.

PROPOSITION 2

A measurable function f from $\mathcal{S}$ into $[0, \infty]$ produces a new measure ν from a given measure μ:

$$(18) \qquad \nu(A) = \int_A f(s)\, d\mu(s);$$

for brevity we write $d\nu = f\, d\mu$.

For a simple function this was presented in Problem 1.5.21. For a general measurable function, the proof uses the Monotone Convergence Theorem; see Problem 6.

H THE INTEGRAL OF A REAL-VALUED FUNCTION

Consider a sample space $\mathcal{S}$, a class $\mathcal{A}$ of events, and a measure μ. And let g be a measurable function from $\mathcal{S}$ into R. The function g can be decomposed into positive and negative components

$$g^+(s) = \max\{0, g(s)\}, \qquad g^-(s) = \max\{0, -g(s)\},$$

and we have $g = g^+ - g^-$. We then define the integral of g with respect to μ by

$$(19) \qquad \int g(s)\, d\mu(s) = \int g^+(s)\, d\mu(s) - \int g^-(s)\, d\mu(s)$$

provided both terms on the right are finite, and we then say that g is *integrable;* otherwise we say the integral *does not exist* or *diverges*. Note that we no longer allow the value $+\infty$ for an integral. Also note that if g is integrable, then any function f having $|f| \leq |g|$ is also integrable; we say that f is dominated by g.

We can use the result (15) to see that the integral depends only on values of the function g and the marginal measure for those values:

$$(20) \qquad \int g(s)\, d\mu(s) = \int_{-\infty}^{\infty} y\, d\mu_g(y).$$

The integral of g over a set A is obtained by the use of indicator functions:

$$(21) \qquad \int_A g(s)\, d\mu(s) = \int I_A(s) g(s)\, d\mu(s).$$

If the measure μ gives the ordinary length measure on R, and if g is continuous, then the integral (19) agrees with the ordinary Riemann integral.

I PROPERTIES OF THE INTEGRAL FOR REAL-VALUED FUNCTIONS

Many of the ordinary properties of the Riemann integral also hold for the general integral. In addition, there are important properties involving limits as indicated in part by the monotone convergence theorem. We now record some of these properties, properties that are useful in probability and statistics; some proofs are included, others are organized as problems in Section J.

PROPOSITION 3

Linearity in the integrand: *If g_1 and g_2 are integrable over the set A, then*

$$(22) \qquad \int_A (ag_1(s) + bg_2(s))\, d\mu(s) = a \int_A g_1(s)\, d\mu(s) + b \int_A g_2(s)\, d\mu(s)$$

for real a and b.

PROPOSITION 4

σ-additive in the domain: *If g is integrable on $\mathcal{S}$, then*

$$(23) \qquad \int_{\cup_1^\infty A_i} g(s)\, d\mu(s) = \int_A g(s)\, d\mu(s),$$

where $\{A_1, A_2, \ldots\}$ is a partition of A; also,

$$\int_B g(s)\, d\mu(s) = 0$$

if $\mu(B) = 0$.

Thus the integral acts like a generalized measure that takes positive and negative values. The generalized measure depends essentially on g; for if

$$(24) \qquad \int_B g(s)\, d\mu(s) = 0$$

for all B in $\mathcal{A}$, then $g = 0$ except perhaps on a set having μ measure 0.

PROPOSITION 5

Linear in the measure: *If g is integrable with respect to μ_1 and μ_2, then*

(25) $$\int_A g(s)\, d(b\mu_1(s) + c\mu_2(s)) = b\int_A g(s)\, d\mu_1(s) + c\int_A g(s)\, d\mu_2(s),$$

where b, $c \geq 0$.

Proof: The linear combination of measures is clearly a measure. For $g \geq 0$ let Y be a simple function approximating g from below:

$$\sum a_i\,(b\mu_1(Y^{-1}(a_i)) + c\mu_2(Y^{-1}(a_i))) = b\sum a_i\mu_1(Y^{-1}(a_i)) + c\sum a_i\mu_2(Y^{-1}(a_i)).$$

It follows then that (25) holds for $g \geq 0$. And then by the decomposition, it holds for general g.

PROPOSITION 6

Absolute value: *For any integrable g*

(26) $$\left|\int_A g(s)\, d\mu(s)\right| \leq \int_A |g(s)|\, d\mu(s).$$

Proof: The left and right sides are given respectively by

$$\left|\int_A g^+(s)\, d\mu(s) - \int_A g^-(s)\, d\mu(s)\right|, \qquad \int_A g^+(s)\, d\mu(s) + \int_A g^-(s)\, d\mu(s).$$

The result (26) then follows from the absolute-value property for real numbers.

PROPOSITION 7

A measurable function f from $\mathcal{S}$ into $[0, \infty]$ produces a new measure ν,

$$\nu(A) = \int_A f(s)\, d\mu(s)$$

and

(27) $$\int_A g(s)\, d\nu(s) = \int_A g(s)f(s)\, d\mu(s)$$

for g integrable with respect to ν.

PROPOSITION 8

Dominated convergence theorem: *If $g_1, g_2, \ldots$ is a convergent sequence of measurable functions, and if $|g_n| \leq h$ where h is integrable with respect to μ, then*

(28) $$\lim_{n\to\infty}\int_A g_n(s)\, d\mu(s) = \int_A \lim_{n\to\infty} g_n(s)\, d\mu(s).$$

PROPOSITION 9

Dominated derivative theorem: *If $g(\cdot, t_0)$ is integrable with respect to μ and if*

$$\left|\frac{\partial}{\partial t} g(s, t)\right| \leq h(s)$$

for all t in the interval $(t_0 - \delta, t_0 + \delta)$ for some $\delta > 0$ and if h is integrable with respect to μ, then

(29) $$\frac{d}{dt}\int_A g(s, t)\, d\mu(s) = \int_A \frac{\partial}{\partial t} g(s, t)\, d\mu(s)$$

at $t = t_0$.

Proof: Apply Proposition 8 to the function

$$\frac{g(s, t_0 + u) - g(s, t_0)}{u} = \frac{\partial}{\partial t} g(s, t)\bigg|_{t_0+\theta u},$$

where θ is given by the mean value theorem.

PROPOSITION 10

Fubini theorem: *If $\mathcal{S}_1$ and $\mathcal{S}_2$ are sample spaces with classes $\mathcal{A}_1$ and $\mathcal{A}_2$ of events and with measures μ_1 and μ_2 and if g is integrable on $\mathcal{S}_1 \times \mathcal{S}_2$ with respect to $\mu = \mu_1 \times \mu_2$* (definitions as in Section 3.6F), *then*

$$(30) \qquad \int_{\mathcal{S}_1\times\mathcal{S}_2} g(s_1, s_2)\, d\mu(s_1, s_2) = \int_{\mathcal{S}_1} \left[\int_{\mathcal{S}_2} g(s_1, s_2)\, d\mu_2(s_2)\right] d\mu_1(s_1),$$

where the term in brackets may diverge but only for s_1 values in a set of μ_1-measure zero. In particular, if g factors so that the variables separate, then

$$(31) \qquad \int_{\mathcal{S}_1\times\mathcal{S}_2} g_1(s_1)g_2(s_2)\, d\mu(s_1, s_2) = \int_{\mathcal{S}_1} g_1(s_1)\, d\mu_1(s_1)\cdot\int_{\mathcal{S}_2} g_2(s_2)\, d\mu_2(s_2).$$

Proof: See textbooks on measure theory.

J PROBLEMS CONCERNING THE PROPOSITIONS IN SECTION I

1 If g_1 and g_2 are measurable functions from $\mathcal{S}$ into $[0, \infty]$, show that

$$\int_{\mathcal{S}} (g_1(s) + g_2(s))\, d\mu(s) = \int_{\mathcal{S}} g_1(s)\, d\mu(s) + \int_{\mathcal{S}} g_2(s)\, d\mu(s).$$

Method: Let $Y_{11}, Y_{12}, \ldots$ be a monotone increasing sequence of simple functions such that $\lim Y_{1i} = g_1$ (as in Section 5.1A); and let $Y_{21}, Y_{22}, \ldots$ be a monotone increasing sequence of simple functions such that $\lim Y_{2i} = g_2$; use Problem 1.5.21 and the monotone convergence Theorem 1.

2 (*continuation*) If $g(s) = \Sigma_1^N g_n(s)$ is a sum of measurable functions from $\mathcal{S}$ into $[0, \infty]$, show that

$$\int g(s)\, d\mu(s) = \sum_1^N \int g_n(s)\, d\mu(s).$$

3 (*continuation*) If $g(s) = \Sigma_1^\infty g_n(s)$ is a convergent series of measurable functions from $\mathcal{S}$ into $[0, \infty]$, show that

$$\int g(s)\, d\mu(s) = \sum_1^\infty \int g_n(s)\, d\mu(s).$$

Method: Monotone Convergence Theorem 1.

4 (*continuation*) If g_n is a sequence of measurable functions from $\mathcal{S}$ into $[0, \infty]$, show that

$$\int_{\mathcal{S}} \liminf_{n\to\infty} g_n(s)\, d\mu(s) = \lim_{n\to\infty} \int_{\mathcal{S}} \inf\,\{g_n(s), g_{n+1}(s), \ldots\}\, d\mu(s).$$

5 **Fatou lemma:** If g_n is a sequence of measurable functions from $\mathcal{S}$ into $[0, \infty]$, show that

$$\int_{\mathcal{S}} \liminf_{n\to\infty} g_n(s)\, d\mu(s) \le \liminf_{n\to\infty} \int_{\mathcal{S}} g_n(s)\, d\mu(s).$$

6 *Proof of Proposition 2:* Let f be a measurable function from $\mathcal{S}$ into $[0, \infty]$. If $A_1, A_2, \ldots$ is a sequence of disjoint subsets of $\mathcal{S}$, show that

$$\int_{\cup_1^\infty A_i} f(s)\, d\mu(s) = \sum_1^\infty \int_{A_i} f(s)\, d\mu(s).$$

Method: With integrals over $\mathcal{S}$, use $I_{\cup A_i} = \Sigma I_{A_i}$ and Problem 3.

7 If a measurable function $f \geq 0$ changes μ into ν as in Proposition 2, show that

$$\int_{\mathcal{S}} g(s)\, d\nu(s) = \int_{\mathcal{S}} g(s) f(s)\, d\mu(s)$$

for any measurable function $g \geq 0$. *Method:* With $g = I_A$, the equation follows from Proposition 2; prove for a simple function g; then use the Monotone Convergence Theorem to prove for a general g.

8 *Proof of part of Proposition 3:* If g is integrable over A, show that

$$\int_A a g(s)\, d\mu(s) = a \int_A g(s)\, d\mu(s)$$

for any real a. *Method:* Examine $a = 0, >0, <0$; use the decomposition $g = g^+ - g^-$.

9 *Proof of part of Proposition 3:* If g_1 and g_2 are integrable over A, show that

$$\int_A (g_1(s) + g_2(s))\, d\mu(s) = \int_A g_1(s)\, d\mu(s) + \int_A g_2(s)\, d\mu(s).$$

Method: Let $g = g_1 + g_2$, and note that $g^+ + g_1^- + g_2^- = g^- + g_1^+ + g_2^+$.

10 *Proof of Proposition 4:* If g is integrable on $\mathcal{S}$, then prove that

$$\int_{\cup_1^\infty A_i} g(s)\, d\mu(s) = \sum_1^\infty \int_{A_i} g(s)\, d\mu(s),$$

where the A_i are disjoint. *Method:* Use Problem 6 for g^+ and for g^-.

11 *Proof of the addendum to Proposition 4:* If

$$\int_A g(s)\, d\mu(s) = 0$$

for all A in $\mathcal{A}$, then prove that $g = 0$ almost everywhere with respect to μ. *Method:* Let $A_n = \{s : g(s) > 1/n\}$ and $A_+ = \{s : g(s) > 0\}$; show that $\mu(A_n) = 0$ and $\mu(A_+) = 0$. Similarly, $\mu(A_-) = 0$.

12 *Proof of Proposition 7:* If a measurable function $f \geq 0$ changes μ into ν as in Proposition 2, show that

$$\int_{\mathcal{S}} g(s)\, d\nu(s) = \int_{\mathcal{S}} g(s) f(s)\, d\mu(s)$$

for any measurable function g. *Method:* Use Problem 7 for g^+ and g^-.

13 *Proof of Proposition 8:* If $g_1, g_2, \ldots$ is a convergent sequence of measurable functions and if $|g_n| < h$, where h is integrable with respect to μ, show that

$$\lim_{n\to\infty} \int_{\mathcal{S}} g_n(s)\, d\mu(s) = \int_{\mathcal{S}} \lim_{n\to\infty} g_n(s)\, d\mu(s).$$

Method: (a) $|g_n^+|$ and $|g_n^-| \leq h$; hence g_n is integrable.
(b) Let $g = \lim g_n$ and apply Problem 5 to $2h - |g_n - g|$, obtaining

$$\limsup_{n\to\infty} \int_{\mathcal{S}} |g_n - g|\, d\mu \leq 0$$

and hence

$$\lim_{n\to\infty} \int_{\mathcal{S}} |g_n - g| \, d\mu = 0.$$

(c) Apply Proposition 6.

14 Use the properties of the general integral to prove the properties of the mean as cited in Section 5.1C.

K EXERCISES AND PROBLEMS

15 Let n be the counting measure, where $n(A)$ is the number of integers in A; see the measure μ_1 in Section 1.3B with $\mathcal{S}_0 = Z$, the set of integers.

(a) Show that the Poisson(λ) distribution is given by

$$P(A) = \int_A \frac{\lambda^x e^{-\lambda}}{\Gamma(x + 1)} \, dn(x).$$

(b) Show that the binomial(n, p) distribution is given by

$$P(A) = \int_A \binom{n}{x} p^x (1 - p)^{n-x} \, dn(x).$$

16 If $X_1, \ldots, X_k$ are integrable with respect to μ and if $|Y| \le \max\{|X_1|, \ldots, |X_k|\}$, show that Y is integrable with respect to μ.

17 *Schwartz inequality:* Let X and Y be real-valued measurable functions and suppose that X^2 and Y^2 are integrable with respect to μ.

(a) Show that $|XY| \le X^2 + Y^2$ and hence that XY is integrable.

(b) Show that

$$\left| \int X(s)Y(s) \, d\mu(s) \right|^2 \le \int X^2(s) \, d\mu(s) \cdot \int Y^2(s) \, d\mu(s)$$

with equality if and only if $c_1 X = c_2 Y$ (for some c_1 and c_2 not both zero) almost everywhere with respect to μ. *Method:* Consider the quadratic expression in $c_1, c_2 \in R$ given by

$$\int (c_1 X(s) + c_2 Y(s))^2 \, d\mu(s).$$

18 Consider a distribution for y on the real line with distribution function F. Show that

$$E(y) = \int_0^\infty (1 - F(y)) \, dy - \int_{-\infty}^0 F(y) \, dy.$$

Method: Apply the Fubini Proposition 5.8.10 to the following expression:

$$E(y) = \int_{(0,\infty)} \left(\int_{(0,y)} dt \right) dP(u) - \int_{(-\infty,0]} \left(\int_{[y,0]} dt \right) dP(u).$$

Recall the simpler version in Problem 5.2.11.

6.3

LIMITING FUNCTIONS

In preceding sections of Chapter 6 we have been considering a sequence of distributions on the real line and have examined how the shape of a distribution can converge to a limit. In this section we consider a sequence of functions and examine various definitions for the convergence of the sequence of functions. We then record the proofs for two theorems in Section 6.1. And in conclusion we examine the Strong Law of Large Numbers.

A LIMITING FUNCTIONS

In Section 6.1 we considered the Poisson(λ) distribution for y and showed that the distribution of $2(\sqrt{y} - \sqrt{\lambda})$ converges to the standard normal distribution. This example has a distribution on R for each value of λ and the shape of the distribution converges to the standard normal as $\lambda \to \infty$; this is convergence in the sense of Definition 6.1.

In Section 6.2 we considered a distribution on R^n for a sample from a distribution with mean μ and variance σ^2 and we showed that the shape of the distribution of $(\Sigma_1^n y_i - n\mu)/\sqrt{n}\sigma$ converges to the standard normal as $n \to \infty$.

This second example can be viewed somewhat differently. We could consider a distribution on R^∞ for $(y_1, y_2, \ldots)$, where the coordinates are independent and identically distributed with mean μ and variance σ^2. The function $(\Sigma_1^n y_i - n\mu)/\sqrt{n}\sigma$ is a function of the first n coordinates but can be viewed as a function on R^∞. As n varies we obtain a sequence of functions and the shape of the distribution converges to the standard normal as $n \to \infty$.

We now consider several types of convergence for a sequence $(Y_1, Y_2, \ldots)$ of real-valued functions. The functions Y_i are measurable functions defined on a space $\mathcal{S}$ with σ-algebra $\mathcal{A}$ and probability measure P.

B CONVERGENCE IN DISTRIBUTION

The first definition refers only to the marginal distributions of the functions.

DEFINITION 1

A sequence $Y_1, Y_2, \ldots$ **converges in distribution** to Y designated $Y_n \xrightarrow{\mathfrak{D}} Y$ if $F_n \to F$ as defined by Definition 6.1.1, where F_n and F are the marginal distribution functions for Y_n and Y, respectively.

This definition does not say that a value $Y_n(s)$ is close to the corresponding value $Y(s)$; only that the distribution for $Y_n(s)$ is close to that for $Y(s)$, as discussed in Section 6.1.

C CONVERGENCE IN PROBABILITY

We now consider a definition that does say that a value $Y_n(s)$ is close to $Y(s)$, at least for "most" of the values for s.

DEFINITION 2

A sequence $Y_1, Y_2, \ldots$ **converges in probability** to Y designated $Y_n \xrightarrow{P} Y$ or $\operatorname{plim} Y_n = Y$ if

$$(1) \qquad \lim_{n\to\infty} P(|Y_n(s) - Y(s)| \le \epsilon) = 1$$

for each $\epsilon > 0$.

An equivalent definition for $Y_n \xrightarrow{P} Y$ is that $Y_n - Y \xrightarrow{\mathfrak{D}} 0$; this follows from the alternative expression (6.1.10) for convergence to a constant.

Note that if $y_n \xrightarrow{\mathfrak{D}} y$, then we cannot conclude that $y_n - y \xrightarrow{\mathfrak{D}} 0$ and thus cannot conclude that $y_n \xrightarrow{P} y$.

Convergence in probability is stronger than convergence in distribution:

LEMMA 3

If $Y_n \xrightarrow{P} Y$, then $Y_n \xrightarrow{\mathfrak{D}} Y$.

Proof: Let $W_n = Y_n - Y$ give the deviation from the limit function; then from (1) we obtain that $W_n \xrightarrow{\mathfrak{D}} 0$. We can then write

$$Y_n = Y + W_n$$

and use Theorem 6.1.4 to show that Y_n and Y have the same limiting distribution; that is, $Y_n \xrightarrow{\mathfrak{D}} Y$.

D CONVERGENCE IN MEAN(*r*)

We now consider a definition that says the mean deviation goes to zero.

DEFINITION 4

A sequence $Y_1, Y_2, \ldots$ **converges in mean of order** r ($r \geq 1$) to Y designated $Y_n \xrightarrow{(r)} Y$ if

$$(2) \qquad E(|Y_n(s) - Y(s)|^r) \to 0$$

as $n \to \infty$.

Convergence in mean of order r is stronger than convergence in probability:

LEMMA 5

If $Y_n \xrightarrow{(r)} Y$, then $Y_n \xrightarrow{P} Y$.

Proof: The Chebyshev inequality in Problem 5.2.12 gives

$$(3) \qquad P(|Y_n(s) - Y(s)| > t) \leq \frac{E(|Y_n(s) - Y(s)|^r)}{t^r}$$

for $t > 0$. The right side goes to zero, hence the left side also, thus giving convergence in probability.

Also note that convergence in mean of order r is stronger than convergence in mean of order s with $1 \leq s \leq r$.

E CONVERGENCE ALMOST SURELY

As a final definition we consider convergence at all sample points except perhaps on a set having probability zero.

DEFINITION 6

A sequence $Y_1, Y_2, \ldots$ **converges almost surely** to Y, designated $Y_n \xrightarrow{\text{a.s.}} Y$ if

(4) $P(\lim Y_n(s) = Y(s)) = 1.$

The sequence $Y_1(s), Y_2(s), \ldots$ may not converge to $Y(s)$, but the points where it does not converge form a set having probability zero.

To appreciate the definition of almost-sure convergence we should take a very close look at the concept of a limit. Consider sample points that have ϵ closeness for the jth function:

(5) $C_j(\epsilon) = \{s : |Y_j(s) - Y(s)| \le \epsilon\}.$

Now consider sample points that have ϵ closeness from the nth function onward:

(6) $$C_{[n]}(\epsilon) = \left\{s : \sup_{j=n}^{\infty} |Y_j(s) - Y(s)| \le \epsilon\right\} = \cap_{j=n}^{\infty} C_j(\epsilon).$$

Note that $C_{[1]}(\epsilon), C_{[2]}(\epsilon), \ldots$ is a monotone increasing sequence of sets. Now consider sample points at which the sequence in the limit is ϵ close:

(7) $$C(\epsilon) = \cup_{n=1}^{\infty} C_{[n]}(\epsilon) = \lim_{n\to\infty} C_{[n]}(\epsilon).$$

Note that $C(1), C(1/2), C(1/3), \ldots$ is a monotone decreasing sequence of sets. Now consider sample points at which the sequence converges:

(8) $$C = \cap_{i=1}^{n} C(1/i) = \lim_{i\to\infty} C(1/i).$$

The preceding analysis shows that $Y_n \xrightarrow{\text{a.s.}} Y$ if and only if $P(C) = 1$. Note that C is defined by countable unions and intersections and thus is measurable; recall Problem 1.5.19.

F CRITERIA FOR ALMOST-SURE CONVERGENCE

We now examine several ways of testing for almost sure convergence.

LEMMA 7

$Y_n \xrightarrow{\text{a.s.}} Y$ if and only if

(9) $$P\left(\sup_{j=n}^{\infty} |Y_j(s) - Y(s)| > \epsilon\right) \to 0$$

as $n \to \infty$ for any $\epsilon > 0$.

Proof: From the analysis at the end of Section E we have that $Y_n \xrightarrow{\text{a.s.}} Y$ if and only if

$$P(C) = P\left(\lim_{i\to\infty} C(1/i)\right) = \lim_{i\to\infty} P(C(1/i)) = 1.$$

The sequence $C(1/1), C(1/2), C(1/3)$ is monotone decreasing; thus $P(C) = 1$ if and only if

$$P(C(1/i)) = 1$$

for each i. The sequence $C_{[1]}(\epsilon), C_{[2]}(\epsilon), C_{[3]}(\epsilon), \ldots$ is monotone increasing; thus $P(C(1/i)) = 1$ for each i if and only if

(10) $$\lim_{n\to\infty} P(C_{[n]}(1/i)) = 1$$

for each i; but this is equivalent to condition (9) in the lemma.

We can use the preceding lemma to show that almost-sure convergence implies convergence in probability.

COROLLARY 8

If $Y_n \xrightarrow{\text{a.s.}} Y$*, then* $Y_n \xrightarrow{P} Y$.

Proof: Note that

$$(11) \qquad C_{[n]}(\epsilon) = \left\{s: \sup_{j=n}^{\infty} |Y_j(s) - Y(s)| \leq \epsilon\right\} \subset \{s: |Y_n(s) - Y(s)| \leq \epsilon\}$$

$$= C_n(\epsilon).$$

Almost-sure convergence means that $P(C_{[n]}(\epsilon)) \to 1$, while convergence in probability means that $P(C_n(\epsilon)) \to 1$; the lemma follows from the inequality (11).

Almost-sure convergence is more difficult to work with than convergence in probability. The source of this difficulty can be found in the supremum operation in formula (9). One approach to this difficulty lies in the Kolmogorov inequality; see Problems 5.5.8 and 5.5.9 and the Strong Law later in this section. Another possibility lies in the following lemma:

LEMMA 9

If $\Sigma_1^{\infty} P(|Y_n(s) - Y(s)| > \epsilon)$ *is finite, for each* $\epsilon > 0$, then $Y_n \xrightarrow{\text{a.s.}} Y$.

Proof: We can use Lemma 7:

$$P\left(\sup_{j=n}^{\infty} |Y_j(s) - Y(s)| > \epsilon\right) \leq \sum_{k=n}^{\infty} P(|Y_k(s) - Y(s)| > \epsilon);$$

the right side is a tail sum of a convergent series and thus goes to zero as $n \to \infty$. Thus the criterion in Lemma 7 is satisfied.

COROLLARY 10

If $\Sigma_1^{\infty} E(|Y_n(s) - Y(s)|^r)$ *is finite for some* $r > 0$*, then* $Y_n \xrightarrow{\text{a.s.}} Y$.

Proof: Use the Chebyshev inequality in Problem 5.2.12.

G PROBLEMS

1 If $Y_n \xrightarrow{(r)} Y$, show that $Y_n \xrightarrow{(s)} Y$ for $1 \leq s \leq r$.

2 Let $Y_n(w) = w/n$, where w is standard Cauchy. Show that $Y_n \xrightarrow{\text{a.s.}} Y$ does not imply that $Y_n \xrightarrow{(r)} Y$.

3 Let $Y_{2^k+c}(u) = 1$ if $(c-1)/2^k < u \leq c/2^k$ and $= 0$ otherwise ($c = 1, \ldots, 2^k$), and let u be uniform(0, 1). Show that $Y_n \xrightarrow{(r)} Y$ does not imply that $Y_n \xrightarrow{\text{a.s.}} Y$.

H PROOF OF THE CONTINUITY THEOREM

The following lemma is needed for the Continuity Theorem 6.1.3.

LEMMA 11

If $F_n \to F$ in accordance with Definition 6.1.1, then

$$\int_{+\infty}^{\infty} g(y)\, dP_n(y) \to \int_{-\infty}^{\infty} g(y)\, dP(y) \tag{12}$$

as $n \to \infty$ for any bounded continuous function g (the integrations are with respect to the measures P_n and P corresponding to F_n and F).

Proof: By using a scale factor we see it is sufficient to prove the theorem for a function having $|g(y)| \le 1$.

Consider a continuity interval B for the limiting distribution such that there is at most ϵ probability in B^c. And suppose that we choose n sufficiently large that the distribution F_n assigns at most 2ϵ probability to B^c.

Now subdivide B into component continuity intervals B_j so that the variation of g is less than ϵ on each B_j. Let $h(y)$ be an approximating discrete function based on the subdivision, and let $h(y) = 0$ on B^c. Consider the following integrals:

$$\int_{-\infty}^{\infty} h(y)\, dP_n(y), \qquad \int_{-\infty}^{\infty} h(y)\, dP(y).$$

The left (right) integral is within $\epsilon + 2\epsilon$ $(\epsilon + \epsilon)$ of the left (right) integral in (12). But the left integral is a linear combination of the $P_n(B_j)$; and the right integral is the same linear combination of the $P(B_j)$; accordingly, the left integral converges to the right integral. With ϵ arbitrary this proves the lemma.

We can extend the lemma and allow F_n to converge to a function F without requiring F to be a distribution function; of course, F will be monotone nondecreasing with values in $[0, 1]$. With this extension, however, we must assume that $g(-\infty) = g(+\infty) = 0$.

We now record the proof for the Continuity Theorem 6.1.3.

Proof: First assume that $F_n \to F$. We have

$$c_n(t) = \int_{-\infty}^{\infty} e^{iyt}\, dF_n(y).$$

The integrand $e^{iyt} = \cos yt + i \sin yt$ is a bounded continuous function; hence by Lemma 11 we have

$$\lim_{n\to\infty} c_n(t) = \int_{-\infty}^{\infty} e^{iyt}\, dF(y) = c(t).$$

Now assume that $c_n(t) \to c(t)$ for real t. Given any bounded sequence of real numbers we can extract a subsequence that is convergent. Now consider the sequence $\{F_n\}$ of distribution functions. For a rational number r_1 we can extract a subsequence $\{F_{n_i}\}$ that is convergent at r_1; for a second rational number r_2 we can extract a further subsequence that is also convergent at r_2; and so on. We then form a special subsequence by taking the first function in the original sequence, the second in the second sequence, the third in the third, and so on following this diagonalization procedure. We then have a subsequence $\{F_{(n)}\}$ which is convergent at the rationals. Let F be the limit function.

The Parseval relation (5.8.7) gives

$$\int_{-\infty}^{\infty} e^{-iyt} c_{(n)}(t) \frac{\sigma}{\sqrt{2\pi}} e^{-\sigma^2 t^2/2}\, dt = \int_{-\infty}^{\infty} e^{-(x-y)^2/2\sigma^2}\, dF_{(n)}(x).$$

Let $n \to \infty$ and use the Dominated Convergence Theorem 5.8.8 for the left side and Lemma 11 as extended for the right side:

$$\int_{-\infty}^{\infty} e^{-iyt} c(t) \frac{\sigma}{\sqrt{2\pi}} e^{-\sigma^2 t^2/2}\, dt = \int_{-\infty}^{\infty} e^{-(x-y)^2/2\sigma^2}\, dF(x).$$

Now let $\sigma \to \infty$. We obtain

$$c(0) = F(+\infty) - F(-\infty).$$

But c is a characteristic function; hence $c(0) = 1$ and we obtain that $F(+\infty) = 1$, $F(-\infty) = 0$.

We now use the Uniqueness Theorem in Section 5.8B to show that there can be only one such function F. By the first part of the present theorem, any such limiting F has the characteristic function c. The Uniqueness Theorem, however, says there is only one such F. Thus all convergent subsequences have the same limit. But this means that the original sequence is convergent to F; for otherwise we could construct a subsequence that converged to something different from F.

I PROOF OF THEOREM 6.1.4

Consider the sample space R^2 for (x_n, y_n). Let c be a point of continuity of F; then for any $\epsilon > 0$, there are points of continuity $c - \delta$, $c + \delta$ such that

$$F(c) - F(c - \delta) < \epsilon, \qquad F(c + \delta) - F(c) < \epsilon.$$

Consider:

$$P(y_n \le c - \delta, |x_n| < \delta) \le P(x_n + y_n \le c) \le P(y_n \le c + \delta \quad \text{or} \quad |x_n| \ge \delta).$$

As $n \to \infty$, $P(|x_n| < \delta) \to 1$; it follow that the left side has limit $F(c - \delta)$ and the right side has limit $F(c + \delta)$. Thus the limiting values for $P(x_n + y_n \le c)$ are in the range $F(c) \pm \epsilon$. With arbitrary $\epsilon > 0$, this proves that $\lim P(x_n + y_n \le c) = F(c)$.

J THE STRONG LAW OF LARGE NUMBERS

For the Weak Law of Large Numbers we considered a distribution for $\mathbf{y} = (y_1, \ldots, y_n)'$ on R^n, where the coordinates y_i are statistically independent and identically distributed with mean μ. The law says that the marginal distribution of $\bar{y}_n = \Sigma_1^n y_i/n$ piles up at the mean μ as $n \to \infty$; that is, $\bar{y}_n$ converges in distribution to the constant μ ($\bar{y}_n \xrightarrow{\mathfrak{D}} \mu$).

In this section we examine a conceptually more incisive question. We envisage sampling from a distribution with mean μ obtaining $(y_1, y_2, \ldots)$; we contemplate the successive sample means $\bar{y}_1$, $\bar{y}_2, \ldots$, where $\bar{y}_n = \Sigma_1^n y_i/n$ and we question whether the sequence $\bar{y}_1, \bar{y}_2, \ldots$ has μ at its mathematical limit. The Strong Law of Large Numbers says that this mathematical convergence occurs with probability 1. This result was developed from Borel (1909) and Cantelli (1917) to Kolmogorov (1930), giving the Kolmogorov Theorem of the Strong Law of Large Numbers:

THEOREM 12

Strong law of large numbers: *If $(y_1, y_2, \ldots)$ is a sequence with coordinates that are statistically independent and identically distributed with mean $E(y) = \mu$, then $\bar{y}_n = \Sigma_1^n y_i/n$ converges to μ almost surely; that is, $\bar{y}_n \xrightarrow{\text{a.s.}} \mu$.*

The Strong Law is proved in three stages: Section K, the Borel–Cantelli lemma, which is a basic means of examining a condition on a countable number of coordinates; Section L, the Kolmogorov lemma, which is essentially a proof of the Strong Law with finite variance; and Section M, the main part of the proof of the Kolmogorov theorem.

K BOREL–CANTELLI LEMMA

Consider a sample space $\mathcal{S}$ with σ-algebra $\mathcal{A}$ and probability measure P.

LEMMA 13

Borel–Cantelli lemma: *If $\Sigma_1^\infty P(A_n)$ is finite for events $A_1, A_2, \ldots$ in $\mathcal{A}$, then the probability that an infinity of the A_n occur is zero, that is,*

(13) $\quad P(\limsup A_n) = 0.$

Proof: The definition of lim sup is given in Problem 1.5.16:

$$\limsup_{n\to\infty} A_n = \cap_{n=1}^\infty \cup_{j=n}^\infty A_j = \lim_{n\to\infty} \cup_{j=n}^\infty A_j.$$

Note that $\cup_{j=n}^\infty A_j$ is monotone decreasing as $n \to \infty$, thus justifying the use of the limit in the expression on the right side. We calculate:

$$P\left(\lim_{n\to\infty} \cup_{j=n}^\infty A_j\right) \le P(\cup_{j=n}^\infty A_j) \le \sum_{j=n}^\infty P(A_j);$$

the right side is a tail sum of a convergent series and thus goes to zero as $n \to \infty$.

L KOLMOGOROV LEMMA

Consider a distribution for $(y_1, y_2, \ldots)$ on R^∞ with statistically independent coordinates; we do not require that the coordinates have a common distribution.

LEMMA 14

Kolmogorov lemma: *If $(y_1, y_2, \ldots)$ is a sequence with statistically independent coordinates and with $E(y_n) = 0$, var $(y_n) = \sigma_n^2$, and if*

(14) $$\sum_{n=1}^\infty \frac{\sigma_n^2}{n^2} < \infty$$

is finite, then $\bar{y}_n = \Sigma_1^n y_i/n$ converges to 0 almost surely; that is, $\bar{y}_n \xrightarrow{\text{a.s.}} 0$.

Proof: To use the criterion for almost-sure convergence in Lemma 7, we must show that

(15) $$P\left(\sup_{j=n}^\infty |\bar{y}_n| > \epsilon\right) \to 0$$

for each $\epsilon > 0$ as $n \to \infty$. The procedure is to separate the sequence into batches as follows: $\bar{y}_1$; $\bar{y}_2, \bar{y}_3$; $\bar{y}_4, \bar{y}_5, \bar{y}_6, \bar{y}_7$; . . ., and use the Kolmogorov inequality (Problem 5.5.9) within batches, and the Borel–Cantelli lemma from batch to batch.

Let $S_n = \Sigma_1^n y_i$; thus $\bar{y}_n = S_n/n$. Let A_k be the event that an ϵ discrepancy occurs somewhere in the kth batch:

(16) $$A_k = \left\{\left|\frac{S_n}{n}\right| > \epsilon \text{ for some } n \text{ with } 2^{k-1} \le n < 2^k\right\}$$
$$\subset \{|S_n| > 2^{k-1}\epsilon \text{ for some } n \text{ with } 2^{k-1} \le n < 2^k\}.$$

We now apply the Kolmogorov inequality (Problem 5.5.9) to the S_n with $2^{k-1} \le n < 2^k$, and note the trivial $E(S_{2^{k}-1}^2) \le E(S_{2^k}^2)$:

$$(17)\qquad P(A_k) \le \frac{E(S_{2^k}^2)}{(2^{k-1}\epsilon)^2} = \frac{4}{\epsilon^2}\,\frac{\sum_{n=1}^{2^k} \sigma_n^2}{2^{2k}}.$$

We now use the Borel–Cantelli lemma to see if more than a finite number of the A_k occur, that is, to see if the event on the left side of (15) occurs for arbitrarily large values of n. For this we check whether the following series is convergent:

$$\sum_1^\infty P(A_k) \le \frac{4}{\epsilon^2}\sum_{k=1}^{\infty}\sum_{n=1}^{2^k}\frac{\sigma_n^2}{2^{2k}} = \frac{4}{\epsilon^2}\sum_{n=1}^{\infty}\sum_{2^k\ge n}\frac{\sigma_n^2}{2^{2k}}$$

$$\le \frac{4}{\epsilon^2}\cdot\sum_{n=1}^{\infty}\sigma_n^2\cdot\frac{4}{3}\frac{1}{n^2} = \frac{16}{3\epsilon^2}\sum_{n=1}^{\infty}\frac{\sigma_n^2}{n^2} < \infty;$$

the first step uses (17); the second step involves a change in the order of summation; the third step uses the sum of a geometric series with ratio 1/4; the fourth step involves a rearrangement; and the final step uses (14). Thus by the Borel–Cantelli lemma the condition (15) is satisfied and we have convergence almost surely.

M KOLMOGOROV THEOREM

We now prove the Kolmogorov Theorem 12 as presented in Section J. We must show that $\bar{y}_n \xrightarrow{\text{a.s.}} \mu$ or $\bar{y}_n - \mu = \Sigma_1^n (y_i - \mu)/n \xrightarrow{\text{a.s.}} 0$. It thus suffices to examine the case with $\mu = 0$.

The method of proof is to use the Kolmogorov Lemma 14. This requires finite variance, so we truncate the initial distributions in accord with position in the sequence; we define

$$(18)\qquad \begin{aligned} w_n &= y_n \qquad \text{if } -n < y_n < n \\ &= 0 \qquad \text{otherwise.} \end{aligned}$$

Let L_n be the event $|y_n| \ge n$ or equivalently the event $w_n \ne y_n$. Note the mean and variance of w_n:

$$(19)\qquad E(w_n) = \mu_n = \int_{-n}^{n} y\, dP(y) \xrightarrow[n\to\infty]{} \mu = 0.$$

$$(20)\qquad \operatorname{var}(w_n) \le \int_{|y|<n} y^2\, dP(y) \le \sum_{j=1}^{n} j \int_{j-1\le|y|<j} |y|\, dP(y)$$

$$= \sum_{j=1}^{n} ja_j,$$

where a_j is the integral in the second-to-last expression.

We check whether the truncated sequence $(w_1, w_2, \ldots)$ satisfies the criterion in the Kolmogorov Lemma 14:

$$\sum_{n=1}^{\infty}\frac{\operatorname{var}(w_n)}{n^2} \le \sum_{n=1}^{\infty}\sum_{j=1}^{n}\frac{ja_j}{n^2} = \sum_{j=1}^{\infty} ja_j \sum_{n=j}^{\infty}\frac{1}{n^2}$$

$$< \sum_{j=1}^{\infty} ja_j\frac{2}{j} = 2E(|y|) < \infty;$$

the first step uses (20); the second step involves a change in the order of summation; the third step uses a bound for $\Sigma_j^\infty n^{-2}$ obtained from $\int_j^\infty y^{-2}\, dy$; the fourth uses the definition of a_j; and the fifth step uses the existence of $E(y)$. Thus

$$P\left(\lim\left(\sum_1^n w_i/n - \sum_1^n \mu_i/n\right) = 0\right) = 1.$$

But $\mu_i \to 0$ and thus $\Sigma_1^n \mu_i/n \to 0$; it follows that

$$P\left(\lim \sum_1^n w_i/n = 0\right) = 1.$$

We now show that with probability 1 just a finite number of the w_i are not equal to the corresponding y_i; that is, just a finite number of the events L_i occur. For this we use the Borel–Cantelli lemma:

$$\sum_{n=1}^{\infty} P(L_n) = \sum_{n=1}^{\infty} \int_{|y|\geq n} dP(y) \leq \sum_{n=1}^{\infty} \sum_{k=n}^{\infty} \frac{a_{k+1}}{k}$$

$$= \sum_{k=1}^{\infty} \frac{a_{k+1}}{k} \sum_{n=1}^{k} 1 \leq E(|y|) < \infty.$$

If only a finite number of the w_i are not equal to the corresponding y_i, then $\lim \bar{y}_n = \lim \bar{w}_n$, provided the second limit exists. It follows that

$$P(\lim \bar{y}_n = 0) = 1,$$

which completes the proof.

N LIMITS OF PROPORTIONS AND AVERAGES

We have now come full circle. In Chapter 1 we interpreted the probability of an event as the empirical limit of the proportion as the number of repetitions becomes very large. And in Chapter 5 we interpreted a mean as the empirical limit of the average value as the number of repetitions becomes large.

We have now been able to build a probability model on a countable product space to describe a continuing sequence of repetitions. The model describes the sample proportion and shows that with probability 1 it converges to the probability. And more generally the model describes the sample average and shows that with probability 1 it converges to the mean. Thus in a consistent way the model describes and describes correctly the process on which, in fact, the model was based. This is a measure of the coherence of the model.

O PROBLEMS

4 (*continuation of Exercises 6.2.3 and 6.2.5*) Show that $m_s \xrightarrow{\text{a.s.}} \mu_s$; that $g(m_s) \xrightarrow{\text{a.s.}} g(\mu_s)$ with g continuous.

5 (*continuation of Problems 6.2.7 and 6.2.8*) Show that $s_x^2 \xrightarrow{\text{a.s.}} \sigma^2$; that $s_{12} \xrightarrow{\text{a.s.}} \sigma_{12}$.

6 (*continuation of Problem 6.2.9*) Show that $\bar{m}_s \xrightarrow{\text{a.s.}} \bar{\mu}_s$.

7.5

TESTING RANDOMNESS

We have been investigating statistical inference under the assumption that the system being examined is random. The alternative, of course, is that the system is not random—that conditions are changing, thus causing response fluctuations, or that various determining factors are not under control, thus causing corresponding response fluctuations.

In this section we examine several simple tests of randomness, of whether we have a sample from a distribution. With scientific and sample survey investigations the randomness is often formally introduced by random sampling methods and by randomization of treatments to experimental units. For some of these investigations a simple test for randomness can be of interest as a routine check that conditions are, in fact, being kept essentially constant. We present one test of this kind in Sections C and D; it illustrates some probability theory and some combinatorial methods and is easy to apply.

With industrial applications the question of randomness is of much more specific interest—is some factor not under control, thus causing corresponding fluctuations in the response? For such cases there is a need for procedures that help choose among various factors for an assignable cause for any unusual variation found in the response values. In Sections A and B we examine the quality-control chart. The use of such charts has been an important aid in controlling quality in industrial production.

Our approach to inference in this chapter uses simple results directly available from probability theory in preceding chapters. At the same time the approach is consistent with our view that a first approach to inference should be quite general and not immediately involve a highly structured framework concerning the form of the distribution or concerning the production of the response. In succeeding chapters we shall examine inference with more structured models.

A QUALITY-CONTROL CHARTS

Consider a manufacturing or industrial process and suppose that we have a variable or characteristic v that is of special importance. For example, consider the production of bearings or moldings: we might take a sample of 5 each hour, record the values $y_1, \ldots, y_5$ of a critical dimension, and obtain the characteristic $v = (y_1 + \cdots + y_5)/5$, representing average performance at that time; alternatively, we might calculate the sample variance and obtain the characteristic $v = s_y^2$ representing the variability at that time. Or, for example, we might observe an interval of time and obtain a Poisson variable v recording the number of occurrences of some simple event such as an error or an accident. Or, for example, we might take a sample of 10 each half-day and record the binomial variable v, the number of items that fail a go–no go test.

The quality-control method involves sampling routinely and systematically and recording the resulting values $v_1, v_2, \ldots, v_N$ on a chart plotted against time. Limits are drawn that embrace a reasonable range if production is random: a central value CL is obtained as the average

$$(1) \qquad \text{CL} = \sum_1^N v_i/N;$$

an upper control limit UCL and lower control limit LCL are then formed as

$$(2) \qquad \text{CL} \pm 3\widehat{\sigma}_v,$$

where $\widehat{\sigma}_v$ is an estimated standard deviation obtained in ways to be described in Section B. Values $v_{N+1}, v_{N+2}, \ldots$ are then plotted continually against time and the eye can check easily for values outside the control limits or for any unusual pattern in the sequence of values.

Consider an example involving the mass production of a plastic molding. A sample of 5 was taken each hour, and each unit in the sample was measured with respect to a critical dimension. The data are recorded in Table S.2 and the sample averages $v_i = \bar{y}_i$ are plotted against time in Figure S.9.

If the production is random or in statistical control, we can estimate a central value and then estimate upper and lower control limits based on a conventional 3 standard deviations. For the central value CL we use the average of the 27 sample averages as an overall estimate of the mean μ for the process:

TABLE S.2
Plastic molding data

i	y_{i1}	y_{i2}	y_{i3}	y_{i4}	y_{i5}	$v_i = \bar{y}_i$	R_i
1	140	143	137	134	135	137.8	9
2	138	143	143	145	146	143.0	8
3	139	133	147	148	139	141.2	15
4	143	141	137	138	140	139.8	6
5	142	142	145	135	136	140.0	10
6	136	144	143	136	137	139.2	8
7	142	147	137	142	138	141.2	10
8	143	137	145	137	138	140.0	8
9	141	142	147	140	140	142.0	7
10	142	137	145	140	132	139.2	13
11	137	147	142	137	135	139.6	12
12	137	146	142	142	140	141.4	9
13	142	142	139	141	142	141.2	3
14	137	145	144	137	140	140.6	8
15	144	142	143	135	144	141.6	9
16	140	132	144	145	141	140.4	13
17	137	137	142	143	141	140.0	6
18	137	142	142	145	143	141.8	8
19	142	142	143	140	135	140.4	8
20	136	142	140	139	137	138.8	6
21	142	144	140	138	143	141.4	6
22	139	146	143	140	139	141.4	7
23	140	145	142	139	137	140.6	8
24	134	147	143	141	142	141.4	13
25	138	145	141	137	141	140.4	8
26	140	145	143	144	138	142.0	7
27	145	145	137	138	140	141.0	8

(3) $\quad \text{CL} = \hat{\mu} = \bar{v} = 140.64.$

For the standard deviation we can proceed in a variety of ways that are discussed in Section B; for the present data we use the estimate $\hat{\sigma}_v = 1.657$. For the upper and lower control limits (UCL and LCL), we use 3 standard deviations on either side of the central value CL:

$$\text{(4)} \qquad \begin{aligned} \text{UCL} &= \hat{\mu} + 3\hat{\sigma}_{\bar{y}} = 140.64 + 3 \times 1.657 = 145.62, \\ \text{LCL} &= \hat{\mu} - 3\hat{\sigma}_{\bar{y}} = 140.64 - 3 \times 1.657 = 135.67. \end{aligned}$$

These values are plotted as horizontal lines in Figure S.9.

If the estimated control limits are reasonably close to the true limits, then values outside the limits will occur less than 1 percent of the time (normal case) and certainly less than 1/9 of the time (the extreme case under Chebyshev).

The usual practice is to continue plotting sample averages $v_i = \bar{y}_i$ based on the systematic sampling procedure and to watch for anomalous patterns or occurrences. If a value goes outside the limits, the engineer would be called to search for an assignable cause for the departure. It is frequently found that the variation in a manufactured item is gradually reduced as assignable causes of variation are discovered and corrected.

The chart can also provide early warnings for progressive departures, often long before the effects would be discovered in the routine use of the manufactured units. One type of departure that is easily observed is a succession of high values or a succession of low values; as a rule of

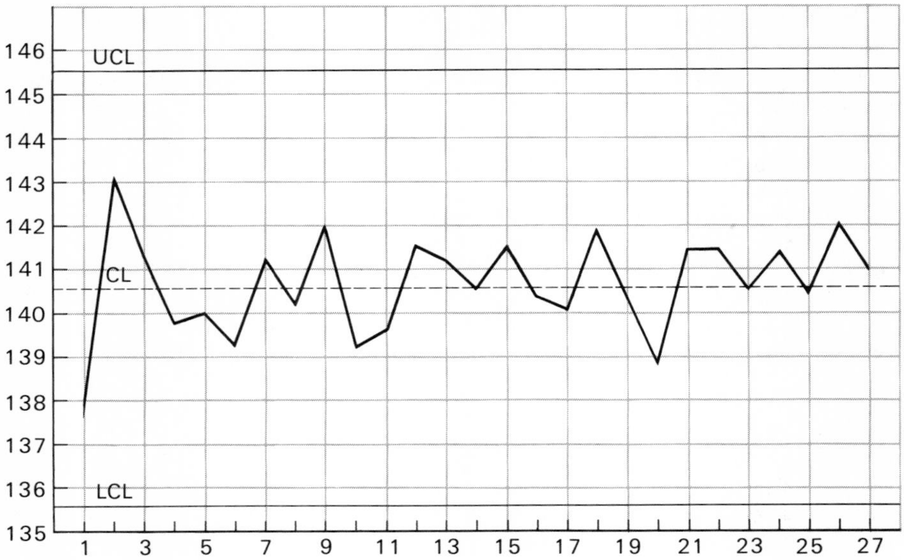

FIGURE S.9
Sample averages plotted against time: center line, CL; upper control limit, UCL; lower control limit, LCL.

thumb, a succession of 6 or 7 values above the median or a succession of 6 or 7 values below the median can reasonably be viewed as an anomalous result.

B ESTIMATING THE STANDARD DEVIATION

Consider an estimate $\widehat{\sigma}_v$ for the standard deviation for use in the quality-control chart. First, consider the case where $v_i = \bar{y}_i$ is based on samples of n. We can estimate the variance σ^2 for an individual response by calculating the individual sample variances,

$$(5) \qquad s_i^2 = \frac{1}{n-1} \sum_{j=1}^{n} (y_{ij} - \bar{y}_i)^2$$

and then forming an overall variance estimate

$$(6) \qquad s^2 = \frac{1}{N} \sum_{1}^{N} s_i^2 = \frac{1}{N(n-1)} \sum_{ij} (y_{ij} - \bar{y}_i)^2.$$

For the standard deviation $\sigma_v = \sigma_{\bar{y}}$ we would then have

$$(7) \qquad \widehat{\sigma}_v = \frac{s}{\sqrt{n}}.$$

An alternative and numerically easier method is to calculate the range for the individual samples

$$R_i = \max_{j=1}^{n} y_{ij} - \min_{j=1}^{n} y_{ij},$$

and then form the average range

$$\bar{R} = \frac{1}{N}\sum R_i.$$

Consider the range R for a sample of n from the standard normal; let f_n be its density (use Problem 3.6.13); then

$$E(R) = \int_0^\infty R f_n(R)\, dR = l_n$$

gives the factor by which the range misestimates the standard deviation for the case of the normal. Accordingly, $\bar{R}/l_n$ is an estimate of σ and

$$k_n\bar{R} = \frac{1}{\sqrt{n}\, l_n}\bar{R}$$

is an estimate of $\sigma_{\bar{y}}$. We record some values for k_n:

n	2	3	4	5	6	7	8	9	10
k_n	0.627	0.341	0.243	0.192	0.161	0.140	0.124	0.112	0.103

The standard deviation $\sigma_{\bar{y}}$ in Section A was estimated as

$$\hat{\sigma}_{\bar{y}} = 0.192\bar{R} = 0.192 \cdot \frac{233}{27} = 1.657.$$

Now consider the case where v is a Poisson(θ) variable. The central value CL $= \bar{v} = \Sigma_1^n v_i/N$ gives an estimate of θ. Accordingly, $\sqrt{\bar{v}}$ is an estimate of the standard deviation $\sqrt{\theta}$ and the control limits would be

$$\text{UCL} = \bar{v} + 3\sqrt{\bar{v}},$$

$$\text{LCL} = \bar{v} - 3\sqrt{\bar{v}}.$$

Now consider the case where v is a binomial(n, p) variable. The central value CL $= \bar{v} = \Sigma_1^n v_i/N$ gives an estimate of the mean number np of events. Accordingly, the standard deviation $\sigma = (npq)^{1/2}$ can then be estimated as $(\bar{v}(1 - \bar{v}/n))^{1/2}$, and the control limits would be

$$\text{UCL} = \bar{v} + 3(\bar{v}(1 - \bar{v}/n))^{1/2},$$

$$\text{LCL} = \bar{v} - 3(\bar{v}(1 - \bar{v}/n))^{1/2}.$$

C THE RUN TEST

Now consider a sequence of response observations on some system and suppose that we want to test whether the system is random or in statistical control. A simple test can be made by examining the sequence in terms of values above (A) or below (B) the median value of the sequence. If there were a trend or a change in level somewhere along the sequence then there would be a tendency for A's to be clumped together and for B's to be clumped together—with the result that the number of runs of A's and of B's would tend to be smaller than under randomness. The run test involves counting the number of runs and comparing the total with the distribution for such under the hypothesis of randomness.

Consider an example. A manufacturing process mass produces an electronic component. A sample is taken every 20 minutes and a critical property is measured. The resulting sequence of 15 measurements is

0.212	0.211	0.213	0.212	0.213
B	B	B	B	B

0.212	0.213	0.214	0.215	0.216
B	B	Med.	A	A
0.215	0.215	0.217	0.218	0.217
A	A	A	A	A

The median value is 0.214 and the values are denoted B for below the median and A for above the median. Let U be the number of runs of A's and B's in the sequence of A and B. There is one run of B's followed by one run of A's; thus $U = 2$.

The distribution of U under randomness will be derived in Section D below; some percentage points are recorded in Table S.3. The table is calculated so that $U_{99\%}$ and $U_{95\%}$ are the largest values satisfying

$$P(U \leq U_{99\%}) \leq 1\%, \qquad P(U \leq U_{95\%}) \leq 5\%;$$

and $U_{5\%}$ and $U_{1\%}$ are the smallest values satisfying

$$P(U \geq U_{5\%}) \leq 5\%, \qquad P(U \geq 1\%) \leq 1\%.$$

For the example there are $n_1 = 7$ values above the median and $n_2 = 7$ values below the median. The value $U = 2$ is the smallest value possible; it is below the 95 percentage point 4 and below the 99 percentage point 3. It provides moderately strong evidence against randomness. An extremely large value of U could also be of concern—perhaps suggesting an oscillation or periodic effect underlying the observations.

TABLE S.3
Distributions of U

$n_1 = n_2$	$U_{99\%}$	$U_{95\%}$	$U_{5\%}$	$U_{1\%}$
5	2	3	9	9
6	2	3	10	11
7	3	4	11	12
8	4	5	12	13
9	4	6	13	15
10	5	6	15	16
11	6	7	16	17
12	6	8	17	19
13	7	9	18	20
14	8	10	19	21
15	9	11	20	22
16	10	11	22	23
17	10	12	23	25
18	11	13	24	26
19	12	14	25	27
20	13	15	26	28
21	14	16	27	29
22	14	17	28	31
23	15	17	30	32
24	16	18	31	33
25	17	19	32	34
26	18	20	33	35
27	19	21	34	36
28	19	22	35	38
29	20	23	36	39
30	21	24	37	40

Many tests can be designed and developed with the object of obtaining sensitivity to various patterns such as trends up or down or cyclic effects. The run test, however, is a convenient and simple test, and it is useful for illustrating some probability methods and calculations.

D THE DISTRIBUTION OF RUNS UNDER RANDOMNESS

Consider the distribution of the number of runs U under randomness—that is, *the sequence is a* sample *from some distribution*. For this we can easily examine a generalization in which we have n_1 values above a certain percentile and n_2 values below the percentile. The symmetry under randomness gives equal probability

$$1\Big/\binom{n_1+n_2}{n_1}$$

to each possible sequence of n_1 A's and n_2 B's.

Let r_1 be the number of runs of A's and r_2 be the number of runs of B's. For example, we might have

$$\text{AAA}_{\text{BB}}\text{A}_{\text{BB}}\text{AAA}_{\text{BB}}, \qquad r_1 = 3, \quad r_2 = 3,$$

with $U = r_1 + r_2 = 6$. Note that $|r_1 - r_2| \le 1$; also note that if $|r_1 - r_2| = 1$, there is only one way in which an A sequence with r_1 runs can be interlaced with a B sequence with r_2 runs; and that if $|r_1 - r_2| = 0$, there are two ways in which an A sequence with r_1 runs can be interlaced with a B sequence with r_2 $(= r_1)$ runs. For this let

$$\delta(r_1, r_2) = 2,\ 1,\ 0 \quad \text{according as } |r_2 - r_1| = 0,\ 1,\ \text{other}.$$

Now consider the number of different ways in which a sequence of n_1 A's can be separated into r_1 runs of A's: the number of ways of choosing $r_1 - 1$ separation points from the $n_1 - 1$ spaces between A's is $\binom{n_1 - 1}{r_1 - 1}$. Similarly, the number of different ways in which r_2 B runs can be formed is $\binom{n_2 - 1}{r_2 - 1}$. It follows, then, that the probability function for (r_1, r_2) is

$$p(r_1, r_2) = \frac{\binom{n_1 - 1}{r_1 - 1}\binom{n_2 - 1}{r_2 - 1}}{\binom{n_1 + n_2}{n_1}}\delta(r_1, r_2).$$

The probabilities for U can then be calculated easily:

$$P(U = 2r + 1) = \frac{\binom{n_1 - 1}{r}\binom{n_2 - 1}{r - 1} + \binom{n_1 - 1}{r - 1}\binom{n_2 - 1}{r}}{\binom{n_1 + n_2}{n_1}},$$

$$P(U = 2r) = \frac{2\binom{n_1 - 1}{r - 1}\binom{n_2 - 1}{r - 1}}{\binom{n_1 + n_2}{n_1}}.$$

The mean of U is fairly easy to calculate; the variance is substantially more complicated. The mathematically direct calculation is concerned with the general factorial moment for (r_1, r_2).

For this we rewrite the hypergeometric identity [Problem 3.4.7 or implicit in formula (3.5.3)] as

$$\sum_i \binom{m_1}{d+i}\binom{m_2}{i} = \sum_i \binom{m_1}{m_1-d-i}\binom{m_2}{i} = \binom{m_1+m_2}{m_1-d} = \binom{m_1+m_2}{d+m_2}$$

and use it twice in the following calculation:

$$\sum_{r_1,r_2} \binom{n_1-1}{r_1+d-1}\binom{n_2-1}{r_2-1}\delta(r_1, r_2)$$

$$= \sum_r \binom{n_1-1}{r+d}\binom{n_2-1}{r-1} + 2\sum_r \binom{n_1-1}{r+d-1}\binom{n_2-1}{r-1} + \sum_r \binom{n_1-1}{r+d-2}\binom{n_2-1}{r-1}$$

$$= \binom{n_1+n_2-2}{d+n_2} + 2\binom{n_1+n_2-2}{d+n_2-1} + \binom{n_1+n_2-2}{d+n_2-2} = \binom{n_1+n_2}{d+n_2}.$$

Now consider the joint factorial-type moment:

$$E(r_1-1)^{(h_1)}(r_2-1)^{(h_2)} = \sum_{r_1,r_2} (r_1-1)^{(h_1)}(r_2-1)^{(h_2)}p(r_1, r_2)$$

$$= (n_1-1)^{(h_1)}(n_2-1)^{(h_2)} \sum_{r_1,r_2} \frac{\binom{n_1-h_1-1}{r_1-h_1-1}\binom{n_2-h_2-1}{r_2-h_2-1}}{\binom{n_1+n_2}{n_1}}\delta(r_1, r_2)$$

$$= (n_1-1)^{(h_1)}(n_2-1)^{(h_2)}\frac{\binom{n_1+n_2-h_1-h_2}{h_2-h_1+n_2-h_2}}{\binom{n_1+n_2}{n_1}}$$

$$= \frac{(n_1-1)^{(h_1)}(n_2-1)^{(h_2)}}{\binom{n_1+n_2}{n_1}}\binom{n_1+n_2-h_1-h_2}{n_2-h_1}.$$

The means, variances, and covariances of r_1 and r_2 can then be calculated:

$$E(U) = E(r_1) + E(r_2) = \frac{2n_1n_2}{n_1+n_2} + 1,$$

$$\text{var}(U) = \text{var}(r_1) + \text{var}(r_2) + 2\,\text{cov}(r_1, r_2) = \frac{2n_1n_2(2n_1n_2-n_1-n_2)}{(n_1+n_2)^2(n_1+n_2-1)}.$$

The calculations for the distribution of U with n_1 and n_2 large can be excessive. Fortunately, the distribution approaches normal form rapidly and the approximation is generally adequate for n_1 and $n_2 \geq 10$: treat

$$z = \frac{U - E(U)}{SD(U)}$$

as standard normal (see Problem 4).

E EXERCISES

1 Twenty-four samples of four dispensers were taken at regular intervals from a bottling process. The weight of the contents was determined for each dispenser and the average and range for each sample are as follows:

Number	Average $\bar{y}$	Range R	Number	Average $\bar{y}$	Range R
1	471.5	19	13	457.5	4
2	462.2	31	14	431.0	14
3	458.5	22	15	454.2	19
4	476.5	27	16	474.5	18
5	461.8	17	17	475.8	18
6	462.8	38	18	455.8	24
7	464.0	38	19	497.8	34
8	461.0	37	20	448.5	21
9	450.2	19	21	453.0	58
10	479.0	41	22	469.8	32
11	452.8	15	23	474.0	10
12	467.2	28	24	461.0	36

Calculate the upper and lower control limits for the sample averages. Plot the quality-control chart and comment on the results obtained. The specifications are that individual charges should have a mean of 454 and a standard deviation of 9 or less; comment on the production process.

2 (*continuation*) Control charts are sometimes used for the variation as given by say the sample range. The values of R are plotted as were the values of $\bar{y}$; the overall average $\bar{R} = \Sigma R_i/N$ is calculated and lower and upper control limits are obtained as LCL $= D_1\bar{R}$ and UCL $= D_2\bar{R}$. The values of D_1 and D_2 are obtained as 3 standard deviation limits based on the distribution of the range for normal sampling:

n	2	3	4	5	6	7	8	9	10
D_1	0	0	0	0	0	0.08	0.14	0.18	0.22
D_2	3.27	2.58	2.28	2.12	2.00	1.92	1.86	1.82	1.78

Calculate the upper and lower control limits for the sample ranges. Plot the quality-control chart and comment on the results.

3 Millikan in 1930 obtained the following sequence of measurements on the charge on an electron (column by column):

4.781	4.764	4.777	4.809	4.761	4.769
4.795	4.776	4.765	4.790	4.792	4.806
4.769	4.771	4.785	4.779	4.758	4.779
4.792	4.789	4.805	4.788	4.764	4.785
4.779	4.772	4.768	4.772	4.810	4.790
4.775	4.789	4.801	4.791	4.799	4.777
4.772	4.764	4.785	4.788	4.779	4.749
4.791	4.774	4.783	4.783	4.797	4.781
4.782	4.778	4.808	4.740	4.790	
4.769	4.791	4.771	4.775	4.747	

Make a run test for randomness. Use the tables as far as possible. Calculate the observed level of significance using the normal approximation.

F PROBLEMS

4 Consider the run test as $N = n_1 + n_2 \to \infty$ with $n_1 = \alpha_1 N$ and $n_2 = \alpha_2 N$. Let $p_N(U)$ be the probability function and consider

$$f_N(z) = p_N(E(U) + z\sigma_U)\sigma_U$$

as $N \to \infty$. Use Stirling's formula as described in Problem 7.4.24. Thus verify the limiting normality.

5 Consider the two-sample problem as described in Problem 7.4.14. The two samples are combined and then ordered from smallest to largest. The values in this sequence are then replaced by A's if they originated from the first sample and by B's if they originated from the second sample. Let U be the number of runs of A's and B's.

(a) Show that the distribution of U in this section is appropriate with $n_1 = m$ and $n_2 = n$.

(b) For the data in Exercise 7.2.5 and Problem 7.2.7, test the hypothesis that the location shift $\theta = 0$.

9.6

BAYES ESTIMATION

A biophysicist is investigating radiation effects on mice. A particular black mouse can be homozygous black *BB* or heterozygous Black–brown *Bb* with probabilities 1/3 and 2/3, respectively. The distribution of a certain characteristic being investigated is different for the two-gene combinations. How can the investigator use the prior probabilities 1/3 and 2/3 for the gene combinations *BB* and *Bb*?

In this section we examine some methods of inference for the case where the parameter value in whole or part has arisen from a performance of some antecedent random system with known probabilities.

A THE ANALYSIS OF THE BAYESIAN MODEL

For notation let y be the response under investigation with sample space $\mathcal{S}$, and let θ be the parameter with parameter space Ω. We assume that the statistical model

$$\{f(\cdot\,|\,\theta) : \theta \in \Omega\}$$

is given by density functions as discussed in Section 8.1.

We now suppose that the true value of the parameter θ is a realization from some separate random system with known probability properties. Often in such cases the true value of θ is a characteristic of some individual or experimental unit and the individual or experimental unit has been obtained from a finite population by random sampling (Section 7.3) or from some random system such as occurs with the combination of genes in mating. Typically, then, the response y is a measurement or experimental result on the sampled individual or unit. Let $p(\cdot)$ be the density function for the characteristic θ relative to some volume or counting measure on the parameter space Ω.

We thus have a *prior probability model,*

$$p(\cdot),$$

describing the source of the true parameter value, and we have a *statistical model,*

$$\{f(\cdot\,|\,\theta) : \theta \in \Omega\},$$

describing the production of the response y for any given value for θ.

The Bayesian approach is to enlarge the viewpoint and examine the combined model describing the generation of the θ value and the production of the response value y based on that θ value. The enlarged sample space is $\Omega \times \mathcal{S}$; the class of events is the appropriate product class as discussed in Section 3.3E; and the distribution is given by the joint density $p(\theta)f(y\,|\,\theta)$ relative to the volume or counting measure on $\Omega \times \mathcal{S}$. We thus obtain a probability model describing the combined outcome (θ, y).

The analysis of the combined model uses basic probability theory. Consider an observed response y. We then have a realization (θ, y) from the combined probability model and the value of the second coordinate is known. The principles of conditional probability in Chapter 4 then give the conditional distribution of θ given y as the appropriate description of the unknown θ.

The joint probability model has probability differential

(1) $\quad p(\theta)f(y \mid \theta)\, d\theta\, dy.$

The conditional distribution for θ given y is

(2) $\quad q(\theta \mid y)\, d\theta = k^{-1}(y)p(\theta)f(y \mid \theta)\, d\theta,$

where the norming constant $k(y)$ is the marginal density for y and is obtained by integration,

(3) $\quad k(y) = \int p(t)f(y \mid t)dt.$

This conditional distribution (2) is the appropriate description for θ based on the observed y and the combined model (1).

The distribution $p(\theta)\, d\theta$ is called the *prior distribution;* it describes the true θ before any measurement or experimental investigation of it. The conditional distribution $q(\theta \mid y)\, d\theta$ is called the *posterior distribution;* it describes the true θ after the measurement or experimental investigation has produced the value y. Note the simple form of the posterior density: *prior density times likelihood,*

$$q(\theta \mid y) \propto p(\theta)L(y \mid \theta),$$

with a norming constant obtained by integration; see Figure S.10.

The inference procedure is essentially solved with the production of the posterior distribution itself. Of course, certain characteristics of the posterior distribution may be of particular interest. As examples, consider the mean, the quartiles (the 25 percent point and the 75 percent point), or say a central 95 percent probability interval (a_1, a_2), where

$$\int_{a_1}^{a_2} q(\theta \mid y)\, d\theta = 95\%.$$

The particular choice of characteristic, however, would be a matter for the judgment of a particular user at a particular time.

B EXAMPLES

Consider examples: involving an individual sampled randomly from a population; and involving a characteristic obtained from random gene combination in mating.

EXAMPLE 1

Let θ be an intelligence characteristic and suppose a measuring procedure is available which gives a response y that is normally distributed about the true value θ with standard deviation 5. Also suppose that an individual is randomly sampled from a population in which the intelligence characteristic is known to be normally distributed with mean 100 and standard deviation 10, and that the measuring procedure applied to the individual gives $y = 120$.

The prior density, the likelihood function, and the posterior density are as follows:

$$p(\theta) = \frac{1}{\sqrt{2\pi}\cdot 10} \exp\left\{-\frac{1}{2\cdot 10^2}(\theta - 100)^2\right\},$$

$$L(y|\theta) = c \exp\left\{-\frac{1}{2 \cdot 5^2}(y-\theta)^2\right\},$$

(4) $$q(\theta \mid y) = k^{-1}(y)p(\theta)f(y \mid \theta)$$

$$= c(y) \exp\left\{-\frac{1}{2}\left(\frac{1}{10^2}+\frac{1}{5^2}\right)\theta^2 + \left(\frac{100}{10^2}+\frac{y}{5^2}\right)\theta + \cdots\right\}$$

$$= \frac{1}{\sqrt{2\pi}\tau} \exp\left\{-\frac{1}{2\tau^2}(\theta - \nu(y))^2\right\}.$$

Thus the posterior distribution for θ is normal with mean

$$\nu(y) = \left(\frac{100}{10^2}+\frac{y}{5^2}\right)\left(\frac{1}{10^2}+\frac{1}{5^2}\right)^{-1}$$

and precision (reciprocal variance)

$$\frac{1}{\tau^2} = \frac{1}{10^2} + \frac{1}{5^2} = \frac{1}{20}.$$

For the observed $y = 120$ the posterior distribution is normal (116, 4.47). See Figure S.10.

Note that the posterior mean is the weighted average of 100 and y with weights propor-

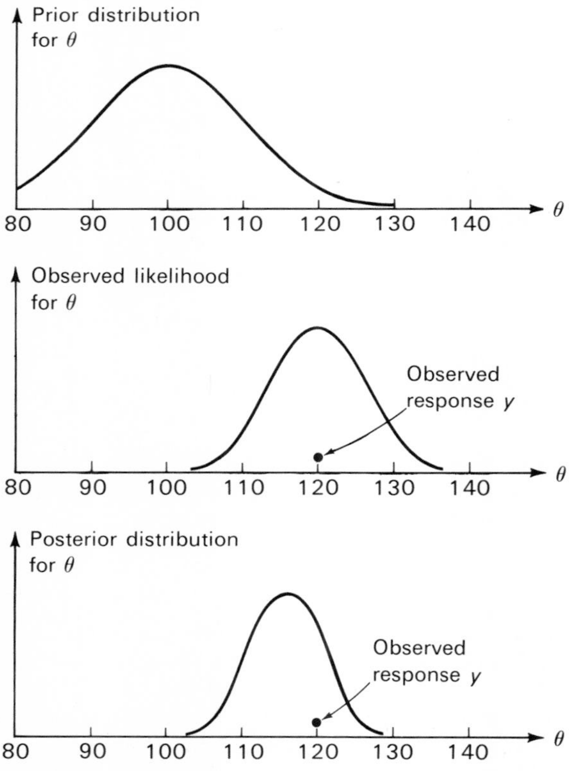

FIGURE S.10
Prior is normal(100, 10); likelihood is normal(120, 5); posterior is normal(116, 4.47).

tional to precision; see the procedure in Section 9.3C. Also note that the posterior variance is the variance of such an average; see Section 9.3C.

The measuring procedure alone would indicate the value 120 with a scaling unit of 5. The effect of enlarging the model and using the prior distribution is to displace the indicated value toward the population mean 100 and to reduce the scaling unit slightly.

EXAMPLE 2

A classical example of Bayesian analysis in genetics is concerned with a black mouse that is an offspring of a *Bb* mouse mated with a *Bb* mouse. Capital *B* denotes a dominant gene for the color black and small *b* denotes a recessive gene for the color brown. An offspring receives one gene from each parent, and it could be either of the parent's genes with equal probability. Thus the prior distribution for an offspring from *Bb* to *Bb* is

θ	*BB*	*Bb*	*bb*
Probability	1/4	1/2	1/4

An offspring is black if it receives a dominant gene *B*, and it is brown otherwise. Thus the information that the mouse is black identifies the mouse as being *BB* or *Bb*. It follows that the posterior distribution given the mouse color is

θ	*BB*	*Bb*	*bb*
Probability	1/3	2/3	0

Now suppose that the particular mouse is tested by mating it with a brown mouse *bb*. And suppose that the seven offspring are all black. The probability of a black offspring from $\theta = BB$ mated with *bb* is 1 (offspring are necessarily *Bb*); and the probability that 7 are black is 1. The probability of a black offspring from $\theta = Bb$ mated with *bb* is 1/2; and the probability that 7 are black is $(1/2)^7 = 1/128$. Thus

θ	*BB*	*Bb*
$f(y \mid \theta)$	1	1/128

Multiplying the two functions together (modulating one by the other) gives

θ	*BB*	*Bb*
Posterior $\propto$	1/3	1/192
Posterior	64/65	1/65

The likelihood function gives a ratio 128:1 in favor of homozygous *BB* to heterozygous *Bb*. Combining this with the probability information concerning the source of the mouse produces the modified ratio 64:1. Note that the result from the testing operation alone is the likelihood ratio 128:1; that the result from the prior process alone is the ratio 1:2; and that the result from the testing operation combined with the source information is the modified ratio 64:1.

C BAYES SOLUTION

Consider the combined model (1) as discussed in Section A. We now investigate some theoretical aspects of choosing an estimator $t(y)$ for the parameter θ. We do this for theoretical interest on the basis of the combined model (1). Of course from the principles of probability theory we do not, in fact, have this initial option: the conditional distribution (2) *is* the proper description of θ and choosing an estimate must be based directly on it.

Suppose that we have a loss function $l(t, \theta)$, which presents the loss to us if we present the estimate t when the parameter has, in fact, the value θ. Accordingly, if we use an estimator $t(y)$ our mean loss or risk is given by

$$\mathcal{R} = E\big(l(t(y), \theta)\big) = \int l(t(y), \theta)p(\theta)f(y|\theta))\, d\theta\, dy.$$

It is reasonable from some points of view to consider choosing the function t in such a way as to minimize the preceding risk $\mathcal{R}$. The risk $\mathcal{R}$ can be written

$$\mathcal{R} = \int \left[\int l(t(y), \theta)q(\theta|y)\, d\theta \right] k(y)\, dy; \tag{5}$$

with the *conditional* risk given y recorded in the square brackets.

Now suppose that for each value of y we choose $t(y)$ to minimize the conditional risk

$$\int l(t(y), \theta)q(\theta|y)\, d\theta. \tag{6}$$

If we choose the function t in this way, we obtain an integrand $[\cdots]$ for (5) that is smaller ($\leq$) than any other integrand; accordingly, such a function *t minimizes the risk* (5). In short: we minimize the risk generally if we minimize the conditional risk given y. This agrees with the general result we mentioned at the beginning.

Now for a real parameter θ, suppose that we use squared error

$$l(t, \theta) = (t - \theta)^2$$

as the loss in estimating t when the parameter is θ. The minimum-risk estimate $t(y)$ is obtained by minimizing

$$\int (t(y) - \theta)^2 q(\theta|y)\, d\theta;$$

thus the minimum mean-square-error estimate is given by the conditional mean,

$$t(y) = \int \theta q(\theta|y)\, d\theta,$$

of the posterior distribution; see Problem 5.3.29.

An estimator obtained by minimizing the risk with the combined model discussed in this section is called a *Bayes solution*.

D SOME OTHER USES OF BAYESIAN ANALYSIS

Consider a random system having a response y with sample space $\mathcal{S}$ and a parameter θ with parameter space Ω. We assume that the statistical model

$$\{f(\cdot|\theta):\theta \in \Omega\}$$

is given by density functions. For an application recall that a model presents the various possibilities for response performances as identified by earlier experience with the same or similar systems.

Now suppose that the true θ is some physical characteristic of the system and that there is no random system as an identifiable source for the true θ. Some statisticians recommend the following approaches to the analysis of an observed value with this model:

1. There is no reason to prefer one θ value to another; hence assign a uniform density ($p(\theta)$ = constant) to the parameter space and perform a Bayesian analysis as in Section A. The resultant is the likelihood function but normed and treated as a posterior density. If Ω is finite, then $p(\theta)$ is defined; if Ω is continuous, a uniform density relative to one metric in θ will typically not be a uniform density with respect to another metric; and if Ω is unbounded, a constant density must be truncated in order to be integrable. This method is called the *method of insufficient reason*.
2. The vague and general information concerning θ does not sharply differentiate among the possible values in a reasonable central range embraced by the likelihood function. Hence assign an approximately constant prior and perform a Bayesian analysis. The result is the likelihood function but treated as a posterior density. This is the *method of the diffuse prior*.
3. The investigator consults his feelings and impressions concerning the various θ values and assigns a density that represents the shadings up and down from one θ value to another. He then performs a Bayesian analysis as in Section A. This is the *method of the personal prior*.

Whatever the justifications, these extended Bayesian methods do produce solutions, as many solutions as there are priors. This is in rather sharp contrast to a somewhat limited range of results from the other theory in this chapter and in Chapter 10, and is certainly an attractive feature of the methods. A solution, however, is at most as good as the arguments that support it, and the prior distribution can be a very personal ingredient of the investigator. Certainly solutions can be assessed pragmatically—in terms of how they work in practice. But, of course, from this point of view it suffices to obtain solutions from any plausible source.

If the extended Bayesian methods are to be favored because they produce solutions, it should be noted that simpler methods produce as many solutions, in fact more solutions. We can start with the likelihood function; apply weight functions for any plausible reason; integrate it; section it; profile it; use any device associated with the Bayesian or other approaches. No probability arguments are needed, and there is perhaps less chance, then, that a result will be labeled and hence taken as a probability. These alternative methods produce as many solutions; in fact, they produce more because the likelihood function itself can be obtained and it provides a point evaluation rather than the density evaluation of the Bayesian approach.

An argument sometimes advanced for the extended Bayesian methods notes that subjective elements are involved in the selection of a model and in the choice of methods for analysis; and it thus recommends that subjective assessments also be made of one parameter value against another. This argument overlooks some important distinctions. The choice of a model properly is based on objective evidence concerning the system being investigated, and a model provides the set of possibilities for the system being investigated; the subjective prior, on the other hand, is placing shadings or preferences on the various possibilities. It is one thing to allow a set of possibilities for a system and it is a different thing to introduce preferences and shadings among the possibilities. The extended Bayesian approach provides adjusted preferences or shadings. It does not produce objective assessments from the data and model alone.

The principles of the experimental method in science have long been concerned with excluding any effects from an investigator's preferences or beliefs. The extended Bayesian approach treats these subjective elements as the primary input, and in doing this it runs counter to the basic principles of experimentation.

Another argument for the extended Bayesian approach comes from decision theory—that optimum decisions should be based on a prior distribution and a Bayesian analysis using a utility function. In any observational or experimental investigation a first objective can always be to assemble the information concerning the set of possibilities presented by the model. The Bayesian approach says that this information should be combined with a prior; the decision-theory argument proposes this as a necessity for optimum decisions. Of course, an investigation can have a multitude of different end users, and this argument may perhaps be appropriate for certain individual end users. The scientific approach, however, would say that the presentation of results from an investigation should be objective and not involve prior shadings and preferences on the set of possibilities.

The question remains concerning some end user and how he is to use the information from the observational or experimental investigation. The extended Bayesian use of a prior and a utility function provides one approach; the justification for this approach derives from the Bayesian acceptance of prior probabilities. Some recent theoretical results indicate, however, that probabilities are not available in general to describe unknowns; this contradicts the usual justification for extended Bayesian methods. Of course, an end user can examine separately the information from the investigation and his prior impressions concerning the unknown of the system, and can then make his judgments and decisions. This may be a safer route than the examination of a formal combination of the information from the investigation and the prior impressions concerning the unknown.

E EXERCISES

1 A certain capability θ is approximately normal(100, 20) in a relatively homogeneous population. A measuring procedure for θ is normal(θ, 10). An individual randomly sampled from the population has the measurement $y = 110$.
(a) Determine the posterior distribution for θ.
(b) Record the estimate of θ using the mean of the posterior distribution.

2 A certain tissue characteristic θ in a species of fish is normal(8, 1.5). A rapid method of measuring θ gives a normal(θ, 0.5) distribution for the measurement y. A fish randomly sampled from the population has the measurement $y = 10.5$; estimate the value of θ using the mean of the posterior distribution.

3 Suppose that the mouse in Example 2 was tested by mating it with a brown mouse *bb* and the offspring were 7 black and 1 brown. Determine the posterior distribution for the genetic type of the mouse.

F PROBLEMS

4 A particular population consists of two racial groups A_1 and A_2 in the proportion 1 to 5. The distribution of a capability θ in racial group A_1 is normal(85, 10) and in A_2 is normal(115, 10). A measuring procedure for θ is normal (θ, 10). An individual randomly sampled from the population obtains $y = 110$ without record of the racial group.
(a) Determine the posterior distribution for θ.
(b) Estimate of θ using the mean of the posterior distribution.
(c) Calculate the posterior probability for racial group A_1.
(d) If the individual is of racial group A_1, calculate the mean of the posterior distribution of θ.

Note that the posterior distribution here is a normal likelihood function located at 110 with scale 10 multiplied by a factor that decreases very rapidly from left to right; the factor is the right tail of the normal(85, 10). Thus the posterior is shifted sharply to the left in comparison with the likelihood. The validity, of course, depends heavily on how well the tail of the normal describes the population; ordinarily the normal is not viewed as a good approximation out on the tails.

5 Let $\mathbf{y} = (y_1, \ldots, y_n)'$ be a sample from a normal distribution (θ, σ_0) with θ in R. And suppose that the true θ is a realized value from a normal distribution (ν_0, τ_0). Determine the posterior distribution for θ given $\mathbf{y}$.

6 (*continuation*) (a) Determine the estimator $t(\mathbf{y})$ for θ that minimizes the mean-squared error $\mathcal{R} = E((t - \theta)^2)$. (b) Determine the estimator $t(\mathbf{y})$ for θ that minimizes the mean absolute error $\mathcal{R} = E(|t - \theta|)$.

7 Let $\mathbf{x} = (x_1, \ldots, x_n)'$ be a sample from the Bernoulli(p) distribution with p in [0, 1]. And suppose that the true p is a realized value from a beta distribution(α_0, β_0). Determine the posterior distribution for p.

8 Let $y = (y_1, \ldots, y_n)$ be a sample from the Poisson(θ) distribution with θ in R^+. And suppose that the true θ is a realized value from the distribution for $\beta_0 u$, where u is gamma(p_0). Determine the posterior distribution of θ.

9 Let $\mathbf{y}$ have a multivariate normal($\boldsymbol{\theta}$; Σ_0) distribution with $\boldsymbol{\theta}$ in R^k. And suppose that the true $\boldsymbol{\theta}$ is a realized value from the multivariate normal($\boldsymbol{\tau}_0$; M_0). Determine the posterior distribution of $\boldsymbol{\theta}$.

10.5

CONFIDENCE INTERVALS AND CONFIDENCE REGIONS

Confidence intervals are one of the basic methods of statistical inference. Consider some examples from preceding chapters. For this let $(y_1, \ldots, y_n)$ be a sample from a distribution on the real line.

1. If the distribution has finite mean θ and finite variance, then

   $$(\bar{y} \pm z_{\alpha/2} s_y/\sqrt{n})$$

   is an approximate (large-sample) $1 - \alpha$ confidence interval for θ.
2. If, in addition, the distribution form is normal, then

   $$(\bar{y} \pm t_{\alpha/2} s_y/\sqrt{n})$$

   is an exact $1 - \alpha$ confidence interval for θ [based on the Student($n - 1$) distribution].
3. If, in addition, the variance is known to be σ_0^2, then

   $$(\bar{y} \pm z_{\alpha/2} \sigma_0/\sqrt{n})$$

   is an exact $1 - \alpha$ confidence interval for θ.
4. If the distribution has a density function $f(\cdot|\theta)$ satisfying the regularity discussed in Section 8.6, then

   $$(\hat{\theta}(\mathbf{y}) \pm z_{\alpha/2}/\sqrt{nj(\hat{\theta}(\mathbf{y}))})$$

   is an approximate (large-sample) confidence interval for θ.

In this section we discuss some theory and some methods for confidence intervals and confidence regions.

A GENERAL THEORY OF CONFIDENCE REGIONS

Consider a statistical model with a response y taking values in a sample space $\mathcal{S}$ and a parameter θ taking values in a parameter space Ω. A $1 - \alpha$ confidence interval or region is an interval or set on Ω that depends on the response y: for each y we obtain a set in Ω. Thus formally a confidence region is a function from the sample space $\mathcal{S}$ into a class of subsets on Ω; the essential probability property is recorded in the following definition.

DEFINITION 1

A function $B(\cdot)$ carrying a point y in $\mathcal{S}$ to a set $B(y)$ on Ω is a **$1 - \alpha$ confidence region** for the parameter θ if

(1) $P(\theta \in B(y) \mid \theta) \geq 1 - \alpha$

for each θ in Ω.

Confidence regions are very closely connected with hypothesis-testing methods. To show this connection, let $C(\theta_*)$ be a size α critical region for testing the hypothesis $\theta = \theta_*$; that is,

$$P(C(\theta_*) \mid \theta_*) \leq \alpha.$$

And let $A(\theta_*) = \mathcal{S} - C(\theta_*)$ be the corresponding acceptance region:

(2) $P(A(\theta_*) \mid \theta_*) \geq 1 - \alpha.$

We then establish a correspondence between confidence regions and acceptance regions. Given the acceptance regions $A(\theta)$ for θ in Ω, we examine an observed response y and collect together the parameter values that are acceptable for that response:

(3) $B(y) = \{\theta : y \in A(\theta)\}.$

Clearly we have the following logical equivalence:

(4) $\theta \in B(y) \leftrightarrow y \in A(\theta).$

For we can view $A(\theta)$ and $B(y)$ as the first- and second-axis sections through a set D on $\mathcal{S} \times \Omega$; see Figure S.11. From formula (4) we then have

(5) $P(\theta \in B(y) \mid \theta) = P(y \in A(\theta) \mid \theta)$ for $\theta \in \Omega$,

which establishes the following lemma.

LEMMA 2

$B(\cdot)$ is a $1 - \alpha$ confidence region for θ if and only if, for each θ_*, $A(\theta_*)$ is a $1 - \alpha$ acceptance region for testing $\theta = \theta_*$.

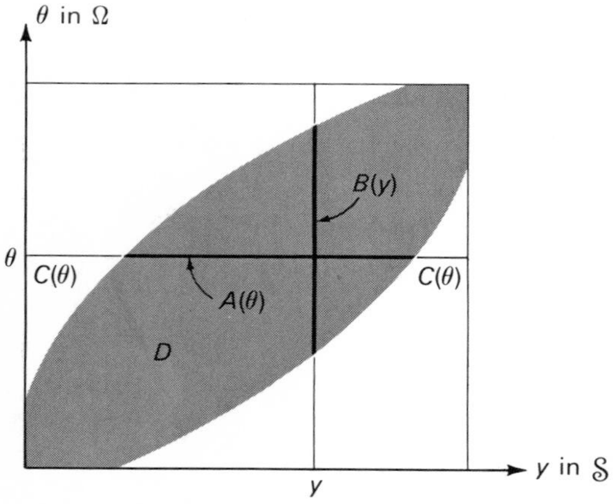

FIGURE S.11
The θ distribution can be viewed as a distribution along the θ-section of $\mathcal{S} \times \Omega$. $A(\theta)$ is a $1 - \alpha$ acceptance region for the value θ; $B(y)$ is the $1 - \alpha$ confidence region obtained from a value y.

We thus have a formal verification of the procedure we followed in Chapters 7 and 8: test each of the parameter values and collect together those that are acceptable when assessed against the observed response.

B CONFIDENCE INTERVAL FOR THE BINOMIAL PROBABILITY p

Consider a sample $(x_1, \ldots, x_n)$ from the Bernoulli(p) distribution with p in $[0, 1]$. We apply the general approach discussed in the preceding section. For each value p, a $1 - \alpha$ acceptance interval $A(p) = [a_1(p), a_2(p)]$ can be formed so that

(6) $$\sum_{a_1(p)}^{a_2(p)} \binom{n}{y} p^y (1 - p)^{n-y} \geq 1 - \alpha,$$

where the interval is an interval of values for the likelihood statistic $y = \Sigma x_i$. Clearly we would want to come as close to $1 - \alpha$ as possible, and we do have some freedom in determining an interval that is approximately central to the distribution.

For the confidence level $1 - \alpha = 95$ percent and for various sample sizes, the acceptance regions are plotted in Figure S.12 (using $\hat{p} = y/n$ rather than y itself). A 95 percent confidence region is then obtained from the appropriate vertical section: find the observed $\hat{p}$ on the horizontal scale and on the vertical section obtain the interval between the curves for the particular sample size.

C THE USE OF PIVOTAL QUANTITIES

A somewhat more direct approach to confidence regions is by means of pivotal quantities:

DEFINITION 3

A function

(7) $t = q(y, \theta)$

defined on $\mathcal{S} \times \Omega$ is a **pivotal quantity** if it has a fixed distribution (independent of θ) as derived from the θ-distribution for y.

For an illustration consider a sample $(y_1, \ldots, y_n)$ from a normal (μ, σ) with (μ, σ) in $R \times R^+$. Then the familiar t quantity

$$t = \frac{\bar{y} - \mu}{s_y/\sqrt{n}}$$

has a fixed distribution, the Student$(n - 1)$ distribution. Note that the pivotal quantity involves μ and not σ^2. Also recall that we used t in Section 8.3C to determine a confidence interval for μ.

Now, in general, let T be a set having probability $1 - \alpha$ according to the fixed distribution of the pivotal quantity $t = q(y, \theta)$ in formula (7):

(8) $A(\theta) = \{y : q(y, \theta) \in T\}$

is a $1 - \alpha$ acceptance region for testing the value θ. And

(9) $B(y) = \{\theta : q(y, \theta) \in T\}$

is a $1 - \alpha$ confidence region for the parameter θ. Note that this definition (9) for $B(y)$ is the general equivalent of "rearranging the inequalities" as, say, in Section 7.2C.

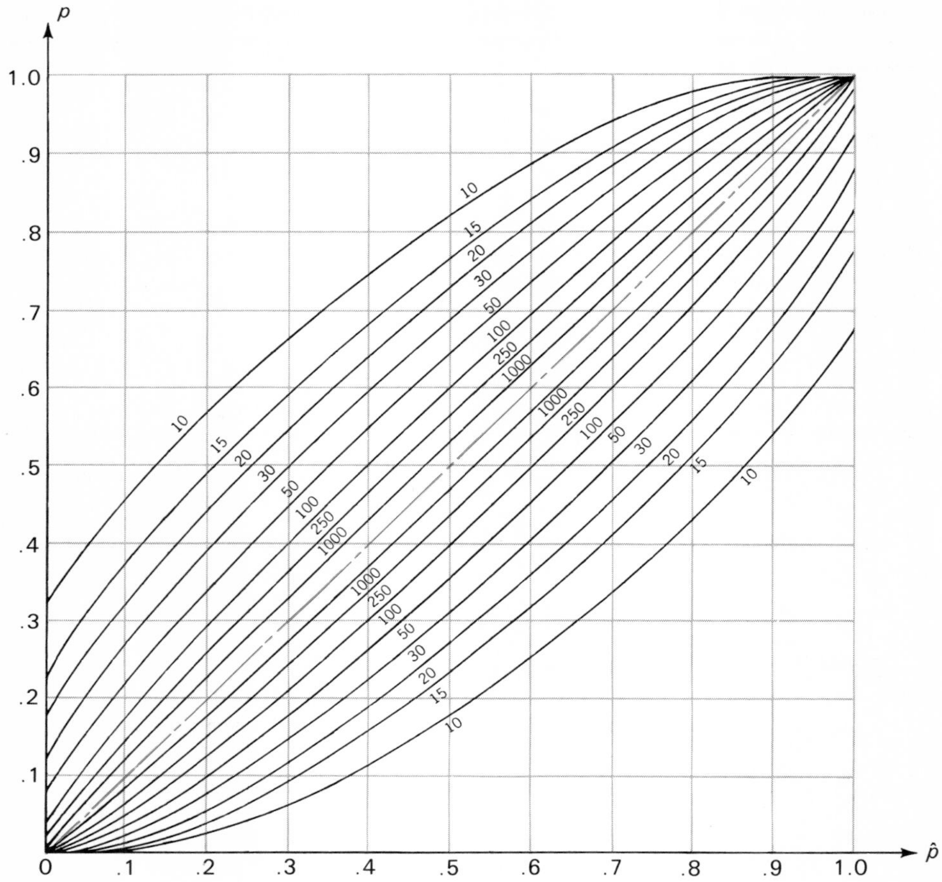

FIGURE S.12
Ninety-five percent confidence intervals for p when n = 10, 15, 20, 30, 50, 100, 250, 1000.

We would naturally choose a pivotal quantity that is a function of the likelihood statistic; this derives from properties of the likelihood statistic in Chapter 8.

D THE POWER OF A CONFIDENCE REGION

Let $A(\theta_*) = \mathcal{S} - C(\theta_*)$ be a $1 - \alpha$ acceptance region for testing the parameter value θ_*. And let $B(y)$ be the corresponding $1 - \alpha$ confidence region for the parameter θ.

The power function for testing the value θ_* is given by

$$P_{\theta_*}(\theta) = P(C(\theta_*)\,|\,\theta) = P(\mathcal{S} - A(\theta_*)\,|\,\theta)$$
$$= P(\theta^* \notin B(y)\,|\,\theta).$$

Thus the power function gives the probability that the confidence region does not cover the value θ^* when the true value is θ. For all θ values except θ_*, this is the probability of not covering a wrong parameter value. Thus in certain problems it is possible to obtain a uniformly most powerful confidence interval. See Problem 10.2.7 concerning the parameter θ of the uniform(0, θ) distribution.

And more generally it is possible to obtain uniformly most powerful unbiased confidence intervals; see examples in Section 10.3A concerning the parameter μ with the normal(μ, σ_0) and the normal(μ, σ).

E CONFIDENCE REGIONS FOR COMPONENT PARAMETERS

Our general discussion in Section A is concerned with the full parameter θ of a statistical model. Often, however, we are concerned with a coordinate of θ or some function, say $\beta(\theta)$ of θ. For example, with the normal(μ, σ) we may be concerned primarily with μ and obtain a confidence interval as mentioned in Section C.

We now discuss briefly the theory of confidence regions for component parameters such as $\beta(\theta)$. Let $\beta(\theta)$ take values in a parameter space Λ:

DEFINITION 4

A function $B(\cdot)$ carrying a point y in $\mathcal{S}$ into a set $B(y)$ on Λ is a **$1 - \alpha$ confidence region** for the parameter $\beta(\theta)$ if

$$P(\beta(\theta) \in B(y)\,|\,\theta) \geq 1 - \alpha$$

for each θ in Ω.

Let $A(\beta_*)$ be an acceptance region for testing the hypothesis: $\beta(\theta) = \beta_*$ at level α:

(10) $$P(A(\beta_*)\,|\,\theta) \geq 1 - \alpha$$

for all θ having $\beta(\theta) = \beta_*$.

Then the definition

(11) $$B(y) = \{\beta : y \in A(\beta)\},$$

gives the logical equivalence

(12) $$\beta \in B(y) \leftrightarrow y \in A(\beta).$$

From this formula (12) we then have

$$P(\beta(\theta) \in B(y)\,|\,\theta) = P(y \in A(\beta(\theta))\,|\,\theta) \qquad \text{for } \theta \in \Omega,$$

which established the following lemma.

LEMMA 5

$B(\cdot)$ is a $1 - \alpha$ confidence region for $\beta(\theta)$ if and only if, for each β_* in Λ, $A(\beta_*)$ is a $1 - \alpha$ acceptance region for testing $\beta(\theta) = \beta_*$.

F EXERCISES

Confidence intervals were derived for the normal location parameter in Section 8.3. The present Section D, together with the examples in Section 10.3A, show that those confidence intervals are uniformly most powerful unbiased; thus the location-parameter confidence intervals in the exercises in Section 8.3D are examples of UMP unbiased confidence intervals.

1 A sample of 5 was obtained on a nitrogen determination known to be normally distributed: 12.7, 13.3, 12.9, 13.0, and 13.1. On the assumption that $\sigma^2 = 0.06$, determine the UMP unbiased 95 percent confidence interval for the mean μ.

2 (*continuation*) With no information concerning σ, determine the UMP unbiased 95 percent confidence interval for the mean μ.

Consider samples from the normal (μ_1, σ) and normal (μ_2, σ) with common variance σ^2. A central $1 - \alpha$ confidence interval for $\mu_2 - \mu_1$ was proposed in Problem 8.3.9; Theorem 10.3.6 with Section D shows that the confidence interval is a UMP unbiased confidence interval. A $1 - \alpha$ confidence interval for σ^2 was proposed in Problem 8.3.7; the amount of probability on each tail of the chi-square distribution can, in fact, be chosen to produce an unbiased test and an unbiased confidence interval, which are then UMP unbiased by Theorem 10.3.6; usually, however, $\alpha/2$ is taken on each tail rather than bothering with this unbiasedness adjustment.

3 Strength determinations for two different materials are known to be normally distributed with means μ_1 and μ_2 and common variance σ^2. A sample of 5 on the first material gave $\bar{y}_1 = 0.275$, $s_1^2 = 0.00045$, and a sample of 6 on the second material gave $\bar{y}_2 = 0.295$, $s_2^2 = 0.00039$. Determine the UMP unbiased 95 percent confidence interval for $\mu_2 - \mu_1$.

4 (*continuation*) Determine a central 95 percent confidence interval for the common variance σ^2.

G PROBLEMS

5 Let $(y_1, \ldots, y_n)$ be a sample from the uniform$(0, \theta)$ with θ in R^+. Determine the UMP size $1 - \alpha$ confidence interval for θ; see Problem 10.2.7 and use the distribution results from Section 3.6E.

6 Let $(y_1, \ldots, y_n)$ be a sample from the uniform $(\theta - 1/2, \theta + 1/2)$ with θ in R. Give an expression for the confidence level using $(y_{(1)}, y_{(n)})$ as a confidence interval for θ; some distribution results are available in Section 3.6E and Problem 3.6.13.

7 (*continuation*) Suppose that an investigator receives a sample having $y_{(n)} - y_{(1)} = r > 1/2$. What can he assert concerning the interval covering or not covering θ? Use the distribution results in Problem 3.6.13 to determine the probability that $(y_{(1)}, y_{(n)})$ brackets θ given that $y_{(n)} - y_{(1)}$ has an observed value r.

8 Let s_1^2 be an unbiased estimate of σ_1^2 with a scaled chi-square(m) distribution, and s_2^2 be an independent estimate of σ_2^2 with a scaled chi-square(n) distribution. Then show that

$$\left(\frac{s_1^2}{s_2^2}F^{-1}_{2\frac{1}{2}\%}(m, n), \quad \frac{s_1^2}{s_2^2}F_{2\frac{1}{2}\%}(n, m)\right)$$

is a 95 percent confidence interval for σ_1^2/σ_2^2, where $F_\alpha(m, n)$ is the α point on the tail of the $F(m, n)$ distribution.

12.3

THREE-FACTOR DESIGN

We now examine the fully randomized three-factor design. The analogous design with randomized blocks can be obtained using the methods in Section 12.4.

A THE THREE-FACTOR DESIGN

Consider a random system with three design factors A, B, C; let the chosen levels for the factors be as follows:

A	$A_1, \ldots, A_a$
B	$B_1, \ldots, B_b$
C	$C_1, \ldots, C_c$

Suppose that n response observations are made at each treatment $A_iB_jC_k$ for a total of $N = abcn$ observations. And suppose that the N observations have been fully randomized against experimental units, time, or any uncontrolled factor.

Let y_{ijks} be the sth observation at the treatment $A_iB_jC_k$. A matrix array for AB as in Section 12.2 can be presented for each level of the factor C; this can be useful for some of the calculations. In a similar manner, a matrix array for AC can be presented for each level of the factor B, and a matrix array for BC can be presented for each level of the factor A; these also can be useful for some of the calculations.

Let $\mathbf{y}$ designate the vector in R^N. And let $\mathbf{1}_{i\cdot\cdot}$ be a vector with 1's in the A_i positions and 0's elsewhere; similarly for $\mathbf{1}_{\cdot j\cdot}$ and $\mathbf{1}_{\cdot\cdot k}$ with respect to B_j and C_k. And let $\mathbf{1}_{ij\cdot}$ be a vector with 1's in the A_iB_j positions and 0's elsewhere; similarly for $\mathbf{1}_{i\cdot k}$ and $\mathbf{1}_{\cdot jk}$. And let $\mathbf{1}_{ijk}$ be a vector with 1's in the n $A_iB_jC_k$ positions and 0's elsewhere. Note the identities such as $\Sigma_k \mathbf{1}_{ijk} = \mathbf{1}_{ij\cdot}$.

Various projections and squared lengths can be separated out using the two-factor design in A and B with cn observations, in A and C with bn observations, and in B and C with an observations. In particular, for the mean we obtain the projection and squared length:

$$\mathbf{p}_1 = \bar{y}\mathbf{1},$$

$$\text{SS}_1 = N\bar{y}^2 = T^2_{\cdots\cdot}/N.$$

For A effects, B effects, and C effects we obtain from Section 12.2 the following projections and squared lengths:

$$\mathbf{p}_2 = \sum \bar{y}_{i\cdot\cdot}\mathbf{1}_{i\cdot\cdot} - \mathbf{p}_1 = \sum (\bar{y}_{i\cdot\cdot} - \bar{y})\mathbf{1}_{i\cdot\cdot},$$

$$\text{SS}_2 = bcn \sum (\bar{y}_{i\cdot\cdot} - \bar{y})^2 = \sum_i T^2_{i\cdot\cdot\cdot}/bcn - \text{SS}_1,$$

$$\mathbf{p}_3 = \sum \bar{y}_{\cdot j\cdot}\mathbf{1}_{\cdot j\cdot} - \mathbf{p}_1 = \sum (\bar{y}_{\cdot j\cdot} - \bar{y})\mathbf{1}_{\cdot j\cdot},$$

$$\text{SS}_3 = acn \sum (\bar{y}_{\cdot j\cdot} - \bar{y})^2 = \sum_j T^2_{\cdot j\cdot\cdot}/acn - \text{SS}_1,$$

$$\mathbf{p}_4 = \sum \bar{y}_{\cdot\cdot k}\mathbf{1}_{\cdot\cdot k} - \mathbf{p}_1 = \sum (\bar{y}_{\cdot\cdot k} - \bar{y})\mathbf{1}_{\cdot\cdot k},$$

$$\text{SS}_4 = abn \sum (\bar{y}_{\cdot\cdot k} - \bar{y})^2 = \sum_k T^2_{\cdot\cdot k\cdot}/abn - \text{SS}_1.$$

These projection vectors are orthogonal vectors and they are obtained as projections of $\mathbf{y}$ into orthogonal subspaces of dimensions $a - 1$, $b - 1$, and $c - 1$; of course, each of these subspaces is orthogonal to $\mathcal{L}(\mathbf{1})$.

The interactions AB, BC, and AC can also be separated out using the analysis of the two-factor design. The projection and squared length for the AB interaction, then BC, then AC are:

$$\mathbf{p}_5 = \sum \bar{y}_{ij\cdot}\mathbf{1}_{ij\cdot} - \mathbf{p}_1 - \mathbf{p}_2 - \mathbf{p}_3 = \sum (\bar{y}_{ij\cdot} - \bar{y}_{i\cdot\cdot} - \bar{y}_{\cdot j\cdot} + \bar{y})\mathbf{1}_{ij\cdot},$$

$$SS_5 = cn \sum (\bar{y}_{ij\cdot} - \bar{y}_{i\cdot\cdot} - \bar{y}_{\cdot j\cdot} + \bar{y})^2 = \sum_{ij} T_{ij\cdot\cdot}^2/cn - SS_1 - SS_2 - SS_3,$$

$$\mathbf{p}_6 = \sum \bar{y}_{\cdot jk}\mathbf{1}_{\cdot jk} - \mathbf{p}_1 - \mathbf{p}_3 - \mathbf{p}_4 = \sum (\bar{y}_{\cdot jk} - \bar{y}_{\cdot j\cdot} - \bar{y}_{\cdot\cdot k} + \bar{y})\mathbf{1}_{\cdot jk},$$

$$SS_6 = an \sum (\bar{y}_{\cdot jk} - \bar{y}_{\cdot j\cdot} - \bar{y}_{\cdot\cdot k} + \bar{y})^2 = \sum_{jk} T_{\cdot jk\cdot}^2/an - SS_1 - SS_3 - SS_4,$$

$$\mathbf{p}_7 = \sum \bar{y}_{i\cdot k}\mathbf{1}_{i\cdot k} - \mathbf{p}_1 - \mathbf{p}_2 - \mathbf{p}_4 = \sum (\bar{y}_{i\cdot k} - \bar{y}_{i\cdot\cdot} - \bar{y}_{\cdot\cdot k} + \bar{y})\mathbf{1}_{i\cdot k},$$

$$SS_7 = bn \sum (\bar{y}_{i\cdot k} - \bar{y}_{i\cdot\cdot} - \bar{y}_{\cdot\cdot k} + \bar{y})^2 = \sum_{ik} T_{i\cdot k\cdot}^2/bn - SS_1 - SS_2 - SS_4.$$

Again we have orthogonality, a benefit of the factorial design. For consider a vector in the subspace for the AB interaction,

$$\sum c_{ij}\mathbf{1}_{ij\cdot}, \qquad \text{where } \sum_i c_{ij} = 0 = \sum_j c_{ij}$$

and a vector in the subspace for the AC interaction,

$$\sum b_{ik}\mathbf{1}_{i\cdot k}, \qquad \text{where } \sum_i b_{ik} = 0 = \sum_k b_{ik}.$$

The inner product of $\mathbf{1}_{ij\cdot}$ and $\mathbf{1}_{i\cdot k}$ is n; thus the inner product of $\Sigma\, c_{ij}\mathbf{1}_{ij\cdot}$ and $\Sigma\, b_{ik}\mathbf{1}_{i\cdot k}$ is

$$\sum_{ijk} nc_{ij}b_{ik} = n \sum_{ij} c_{ij} \sum_k b_{ik} = 0.$$

Also it is easily checked, for example, that an AB interaction vector is orthogonal to a C-effect vector. Thus the two-way interactions are obtained as projections of $\mathbf{y}$ into orthogonal subspaces of dimensions $(a-1)(b-1)$, $(b-1)(c-1)$, $(a-1)(c-1)$, and each of these subspaces is orthogonal to the subspaces for the mean and for the main effects of the factors A, B, and C.

Now consider the ABC interaction. A vector giving the level μ_{ijk} at the treatment $A_iB_jC_k$ is

$$\sum \mu_{ijk}\mathbf{1}_{ijk}.$$

The vectors $\mathbf{1}_{ijk}$ are orthogonal; hence the vectors $\Sigma\, \mu_{ijk}\mathbf{1}_{ijk}$ form an abc-dimensional subspace of R^N. The projection of $\mathbf{y}$ into this subspace and the squared lengths are

$$\sum \bar{y}_{ijk}\mathbf{1}_{ijk},$$

$$n \sum \bar{y}_{ijk}^2 = \sum_{ijk} T_{ijk\cdot}^2/n.$$

Note the following vector relations:

$$\sum_k \mathbf{1}_{ijk} = \mathbf{1}_{ij\cdot}, \qquad \sum_{jk} \mathbf{1}_{ijk} = \mathbf{1}_{i\cdot\cdot}, \qquad \sum_{ijk} \mathbf{1}_{ijk} = \mathbf{1}.$$

Hence the subspaces for the mean, for the main effects, and for the two-way interactions are component subspaces of $\mathcal{L}(\mathbf{1}_{ijk}\text{: all } i, j, k)$. The three-way interaction ABC is the projection of $\mathbf{y}$ into the orthogonal complement of these components within the space $\mathcal{L}(\mathbf{1}_{ijk}\text{: all } i, j, k)$; thus the projection and squared length are

$$\mathbf{p}_8 = \sum \bar{y}_{ijk}\mathbf{1}_{ijk} - \mathbf{p}_1 - \mathbf{p}_2 - \mathbf{p}_3 - \mathbf{p}_4 - \mathbf{p}_5 - \mathbf{p}_6 - \mathbf{p}_7$$

$$= \sum (\bar{y}_{ijk} - \bar{y}_{ij\cdot} - \bar{y}_{i\cdot k} - \bar{y}_{\cdot jk} + \bar{y}_{i\cdot\cdot} + \bar{y}_{\cdot j\cdot} + \bar{y}_{\cdot\cdot k} - \bar{y})\mathbf{1}_{ijk},$$

$$SS_8 = n \sum (\bar{y}_{ijk} - \bar{y}_{ij\cdot} - \bar{y}_{i\cdot k} - \bar{y}_{\cdot jk} + \bar{y}_{i\cdot\cdot} + \bar{y}_{\cdot j\cdot} + \bar{y}_{\cdot\cdot k} - \bar{y})^2$$

$$= \sum_{ijk} T^2_{ijk\cdot}/n - SS_1 - SS_2 - SS_3 - SS_4 - SS_5 - SS_6 - SS_7.$$

Note the substantial identity obtained from the two ways of deriving the sum of squares SS_8.

The residual is obtained as the projection into $\mathcal{L}^\perp(\mathbf{1}_{ijk}\text{: all } i, j, k)$; the projection and squared length are

$$\mathbf{p}_9 = \mathbf{y} - \sum \bar{y}_{ijk}\mathbf{1}_{ijk},$$

$$SS_9 = \sum_{ijks} (y_{ijks} - \bar{y}_{ijk})^2 = \sum_{ijks} y^2_{ijks} - \sum_{ijk} T^2_{ijk\cdot}/n.$$

The cumulative table recording projections and squared lengths is as shown in Table S.4; note that for the projections a typical *ijks* coordinate is recorded inside the parentheses indicating the vector. The omitted elements are available by direct analogy.

The ordinary table describing projections into orthogonal subspaces is then obtained by the differencing described earlier. Thus we obtain Table S.5, giving the separation of **y** and $\Sigma\, y^2_{ijks}$ into components corresponding to the various dependencies on the factors *A*, *B*, and *C*. The omitted elements are available by analogy. For computation the SS entries are best calculated by differencing using the second expression recorded in each case.

TABLE S.4
Cumulative table

Source	DF	Projection	SS
Mean	1	$(\bar{y})$	$N\bar{y}^2_{\cdots} = T^2_{\cdots}/N$
A	a	$(\bar{y}_{i\cdot\cdot})$	$bcn \sum_i \bar{y}^2_{i\cdot\cdot} = \sum_i T^2_{i\cdots}/bcn$
B	b	$(\bar{y}_{\cdot j\cdot})$	...
C	c	$(\bar{y}_{\cdot\cdot k})$	...
AB	ab	$(\bar{y}_{ij\cdot})$	$cn \sum_{ij} \bar{y}^2_{ij\cdot} = \sum_{ij} T^2_{ij\cdot\cdot}/cn$
BC	bc	$(\bar{y}_{\cdot jk})$	...
AC	ac	$(\bar{y}_{i\cdot k})$	...
ABC	abc	$(\bar{y}_{ijk})$	$n \sum_{ijk} \bar{y}^2_{ijk} = \sum T^2_{ijk\cdot}/n$
Response	$abcn$	(y_{ijks})	$\sum_{ijks} y^2_{ijks}$

TABLE S.5
Analysis-of-variance table

	Source	DF	SS
1	Mean	1	$N\bar{y}^2 = T^2_{\cdots}/N$
2	A	$a - 1$	$bcn \sum (\bar{y}_{i\cdot\cdot} - \bar{y})^2 = \sum T^2_{i\cdots}/bcn - SS_1$
3	B	$b - 1$	$\cdots$
4	C	$c - 1$	$\cdots$
5	AB	$(a - 1)(b - 1)$	$cn \sum (\bar{y}_{ij\cdot} - \bar{y}_{i\cdot\cdot} - \bar{y}_{\cdot j\cdot} + \bar{y})^2 = T^2_{ij\cdot\cdot}/cn - SS_1 - SS_2 - SS_3$
6	BC	$(b - 1)(c - 1)$	$\cdots$
7	AC	$(a - 1)(c - 1)$	$\cdots$
8	ABC	$(a - 1)(b - 1)(c - 1)$	$n \sum (\bar{y}_{ijk} - \bar{y}_{ij\cdot} - \bar{y}_{i\cdot k} + \bar{y}_{\cdot jk} + \bar{y}_{i\cdot\cdot} + \bar{y}_{\cdot j\cdot} + \bar{y}_{\cdot\cdot k} - \bar{y})^2$ $= \sum T^2_{ijk\cdot}/n - \sum_1^7 SS_\alpha$
9	Residual	$abc(n - 1)$	$\sum (y_{ijks} - \bar{y}_{ijk})^2 = \sum y^2_{ijks} - \sum_1^8 SS_\alpha$
		N	$\sum y^2_{ijks}$

B THE ANALYSIS

We consider the analysis for the case of approximately normal variation. For this we assume that the response y has variation σz, where z is standard normal. By Section 11.4F, in the pattern discussed in Section 12.1C, we note that an SS entry has the distribution of $\sigma^2\chi^2$, where χ^2 has the chi-square distribution with degrees of freedom given by the corresponding DF entry; the chi-square distribution is ordinary chi-square if the mean vector has zero component in the particular subspace, and it is noncentral chi-square with values that tend to be larger if the mean vector has a nonzero component in the subspace.

The first test for dependence on the factors A, B, and C is made by comparing the mean square for the ABC interaction with the mean square for the residuals. Significance would indicate that the response level depended on the factors A, B, and C; some more detailed analysis perhaps using regression on input variables could investigate the form of the dependence.

On the assumption that there is no ABC interaction, then the various two-factor interactions can be tested against the residual. A significant value would indicate dependence, an interactive dependence, on the corresponding pair of factors.

On the assumption that there are no two-way interactions, then the various main effects can be tested against the residual. A significant value would indicate dependence, a direct additive dependence, on the corresponding factor.

The form of the dependence on the factors A, B, and C can be investigated in part by separating out individual dimensions in the analysis of variance. This can be done by examining specific comparisons between groups of levels as in Section 12.2C. Or with input variables corresponding to the various factors, it can be done by separating out specific linear and quadratic dependencies following again the pattern in Section 12.2C.

C THE 2^3 DESIGN

The analysis of the three-factor design is particularly simple if each factor is examined at just two levels: the 2^3 design. The same simplicity carries over to the 2^r design with r factors each at two levels.

TABLE S.6
Analysis-of-variance table

Source	Dimension	SS
Mean	1	u^2
A	1	u_A^2
B	1	u_B^2
C	1	u_C^2
AB	1	u_{AB}^2
BC	1	u_{BC}^2
AC	1	u_{AC}^2
ABC	1	u_{ABC}^2
	8	$\sum y^2$

For simplicity consider the case with $n = 1$ and suppose that we know that there is no three-factor interaction so that the corresponding SS entry can be used to provide an estimate of the error variance. The more general case with $n > 1$ follows easily from the analysis here and earlier in this section.

For notation let y_{ab}, for example, be the response at the upper level of the first and second factors, A and B, and the lower level of the third factor C. And consider the following orthogonal transformation to a new coordinate vector $\mathbf{u}$:

$$\begin{pmatrix} u \\ u_A \\ u_B \\ u_C \\ u_{AB} \\ u_{BC} \\ u_{AC} \\ u_{ABC} \end{pmatrix} = \frac{1}{\sqrt{8}} \begin{pmatrix} 1 & 1 & 1 & 1 & 1 & 1 & 1 & 1 \\ 1 & 1 & 1 & 1 & -1 & -1 & -1 & -1 \\ 1 & 1 & -1 & -1 & 1 & 1 & -1 & -1 \\ 1 & -1 & 1 & -1 & 1 & -1 & 1 & -1 \\ 1 & 1 & -1 & -1 & -1 & -1 & 1 & 1 \\ 1 & -1 & -1 & 1 & 1 & -1 & -1 & 1 \\ 1 & -1 & 1 & -1 & -1 & 1 & -1 & 1 \\ 1 & -1 & -1 & 1 & -1 & 1 & 1 & -1 \end{pmatrix} \begin{pmatrix} y_{abc} \\ y_{ab} \\ y_{ac} \\ y_a \\ y_{bc} \\ y_b \\ y_c \\ y \end{pmatrix}$$

Note, for example, that $\sqrt{8}\, u_A$ has $+1$ for each observation at the upper level of A and -1 for each observation at the lower level of A. And note that $\sqrt{8}\, u_{AB}$ has coefficients that are the product of those for A and for B; it examines the A-factor difference as it changes from lower to upper level for the factor B. Thus each of the u's corresponds to a particular entry of the analysis-of-variance table. But we have only one dimension for each entry; hence the table is as shown in Table S.6. For precise assessments this design needs to be replicated so as to have an estimate for error separate from the main effects and interactions.

D EXERCISE

1 A three-factor design was used to examine a spinning band laboratory fractionating column. The factors involved were A, band clearance; B, boil-up rate; and C, band speed; the response was the number of equivalent theoretical plates based on a refractive-index calculation. The data are as follows:

A	B	C		Response	
0.05	I	750	1500	11.8	20.9
	II	750	1500	8.5	16.2
0.10	I	750	1500	9.9	18.3
	II	750	1500	8.1	16.0

Calculate the analysis-of-variance table and consider appropriate tests.

12.4

RANDOMIZED BLOCKS

At the beginning of Chapter 12 we briefly discussed the need to randomly assign treatments to experimental units. This randomization changes the inevitable variation among units into the dominant variation affecting response values. The ability, then, of a design to detect treatment effects depends fundamentally on the magnitude of treatment effects in comparison with the variation among units. Any reduction in this variation among units thus produces a corresponding increase in the ability of a design to detect treatment effects.

Now suppose that units or times or locations can be grouped into *blocks* on the basis of some accessible characteristic so that in general there is less variation between units from the same block than between units from different blocks. If the number of units in a block is equal to the number of treatments under consideration and if the treatments are randomly assigned within each block, then the design is called the *randomized block design*. The response variation that is appropriate to judging treatment comparisons is then the variation within blocks. Effective grouping to form blocks can thus produce a substantial increase in the sensitivity of the design. In this section we discuss some aspects of randomized blocks.

If an experimental unit has some numerical variable or characteristic that measures the effective variation between units, then such a *concomitant variable* can account for a substantial part of the response variation. In the next section we examine how the analysis of covariance can be used to remove the effects of such variation and thus increase the sensitivity of the design.

A UNIFORMITY TRIALS

In agricultural experimentation a response observation will typically require a full growing season from the time of implementation. Plots of land will often differ substantially in their effects on growing material. These and many other things place tremendous emphasis on the need for careful design and proper analysis. As part of the investigation of experimental procedures, plots of land are sometimes seeded and growths recorded without the application of any treatments. Such investigations are called *uniformity trials* and they can give some indications concerning the effectiveness of techniques such as blocking and covariance analysis.

Several researchers conducted uniformity trials on the root weights for marigolds. The data obtained for five blocks, each consisting of four plots, are shown in Table S.7. The total sum of squares for the 20 observations is 2,276,781.

If the data are analyzed as a fully randomized design, Table S.8 is obtained. The estimated

TABLE S.7
Uniformity-trial data

B_1	B_2	B_3	B_4	B_5	
376	316	326	317	321	
371	338	326	343	332	
355	336	335	330	317	
356	356	343	327	318	
1458	1346	1330	1317	1288	6739
				2,274,953.25	2,270,706.05

TABLE S.8
Analysis-of-variance table

Source	DF	SS	MS
Mean	1	2,270,706.05	
Residual	19	6,074.95	319.73
	20	2,276,781.00	

variance for the variation from plot to plot is 319.73 and the standard deviation is 17.88.

Now if the data are analyzed as a randomized block design that recognizes the five distinct blocks of land, Table S.9 is obtained. The estimated variance from plot to plot within blocks is 121.85 and the standard deviation is 11.04. Thus the recognition of the blocks has produced a substantial reduction in the variation that would affect any treatment comparisons that could have been under investigation.

TABLE S.9
Analysis-of-variance table; separation of block components

Source	DF	SS	MS
Mean	1	2,270,706.05	
Blocks	4	4,247.20	1061.8
Residual	15	1,827.75	121.85
	20	2,276,781.00	

B THE RANDOMIZED BLOCK DESIGN

Now consider treatments $T_1, \ldots, T_t$, and suppose that we have b blocks each consisting of t plots. The randomized block design is obtained by assigning treatments to units randomly within each block. The resulting response values can be presented in an array as follows:

	B_1		B_b
T_1	y_{11}	$\cdots$	y_{1b}
	$\vdots$		$\vdots$
T_t	y_{t1}	$\cdots$	y_{tb}

The analysis is based on separating out the mean accounting for one dimension, then the block differences accounting for $b - 1$ dimensions; the remaining $tb - b = (t - 1)b$ dimensions refer to differences within blocks. The analysis then separates out $t - 1$ dimensions for the treatment differences, leaving $(t - 1)(b - 1)$ dimensions referring to other differences within blocks. This separation is the standard separation for the two-factor design in Section 12.2 and we obtain Table S.10. The test for treatment effects based on normal variation uses the F for treatment MS divided by residual MS.

C EXAMPLE

Suppose that we have two types of fertilizer, F_1 and F_2, and two types of ground preparation, G_1 and G_2. The two-factor design then involves four treatments $T_1 = F_1G_1$, $T_2 = F_1G_2$, $T_3 = F_2G_1$,

TABLE S.10
Analysis-of-variance table

Source	DF	SS
Mean	1	$T_{..}^2/tb$
Blocks	$b - 1$	$\sum T_{.j}^2/t - T_{..}^2/tb$
Treatments	$t - 1$	$\sum T_{i.}^2/b - T_{..}^2/tb$
Residual	$(t - 1)(b - 1)$	Difference
	tb	$\sum y_{ij}^2$

and $T_4 = F_2G_2$. And suppose that these four treatments are randomly assigned to four units in each of five blocks.

The data in Section A can be used to construct an example here. Four "treatment effects" were chosen and then randomly assigned within each block to the numbers as given in Section A; the following data were obtained:

	B_1	B_2	B_3	B_4	B_5		
$T_1 =$	345	327	324	306	321	1623	
$T_2 =$	365	305	315	332	306	1623	
$T_3 =$	366	347	337	338	329	1717	
$T_4 =$	382	367	354	341	332	1776	2,274,104.6
	1458	1346	1330	1317	1288	6739	
					2,274,953.25		2,270,706.05

The total sum of squares is 2,279,795.

The analysis-of-variance table for this randomized block design is shown in Table S.11. The estimated variance from plot to plot within blocks is 120.27 and the standard deviation is 10.97.

If we ignore the rather special nature of the four treatments, we could test in general for treatment effects by calculating F for the treatment MS divided by the residual MS. The observed value is 9.42 and the 5 and 1 percent points for F on 3 over 12 degrees of freedom are 3.49 and 5.95; the observed value is extreme, giving strong evidence for treatments effects.

The analysis, however, must take account of the two-factor design for the treatments. The totals and some related two-factor calculations are

	G_1	G_2		
F_1	1623	1623	3246	
F_2	1717	1776	3493	2,273,756.5
	3340	3399	6739	
	2,270,880.1			2,270,706.05

The three dimensions for treatments can be separated into one dimension for F, one dimension for G, and one dimension for the $F \times G$ interaction (Table S.12).

The first test for dependence on the factors F and G is made by calculating F for the interactive effect $F \times G$; the observed value is 1.45, and the 5 and 1 percent values on 1 over 12 degrees of freedom are 4.75 and 9.33.

On the assumption that there is no interactive effect, tests for F effects and G effects can be

TABLE S.11
Analysis-of-variance table; separation of blocks and treatments

Source	DF	SS	MS	F
Mean	1	2,270,706.05		
Blocks	4	4,247.20		
Treatments	3	3,398.55	1,132.85	9.42
Residual	12	1,443.20	120.27	
	20	2,279,795.00		

TABLE S.12
Analysis-of-variance table; separation of individual treatment components

Source	DF	SS	MS	F
Mean	1	2,270,706.05		
Blocks	4	4,247.20		
F	1	3,050.45	3,050.45	25.36
G	1	174.05	174.05	1.45
$F \times G$	1	174.05	174.05	1.45
Residual	12	1,443.20	120.27	
	20	2,279,795.0		

made by examining the corresponding MS's divided by the residual MS: for the factor G there is no significance; for the factor F the observed value of the function F is 25.36, and the 5 and 1 percent values are 4.74 and 9.33. The test for dependence on the factor F is highly significant, providing very substantial evidence for effects due to fertilizers.

The data for this example were constructed from the uniformity trial data by using the effects $+11$ for F_2 observations and -11 for F_1 observations and no other effects; the treatment effects were, of course, randomized to plots within each block.

D EXERCISE

1 Consider an experiment to test the effect of temperature on the yield of a chemical process. Six samples of raw material are obtained from each of 3 batches of material; each is used as an input for the process thus giving 18 runs. The temperatures are assigned randomly within batches, each temperature is used for 2 runs within each batch.

Temperature	Blocks 1	2	3
500	27.3	29.4	28.8
	27.4	29.6	28.8
550	30.4	32.8	32.0
	30.4	32.7	31.9
600	29.5	31.5	31.5
	29.3	31.5	31.4

Construct the analysis-of-variance table separating Mean, Blocks, Temperature, $B \times T$, Determinations. Make appropriate tests; the experimental error for testing temperature is given by $B \times T$.

12.5

ANALYSIS OF COVARIANCE

In the preceding section we considered how experimental units can sometimes be organized into blocks with less variation within blocks than between blocks. The restricted randomization of treatments-to-units-within-blocks then gives a design that typically has greater precision and greater sensitivity.

In a variety of problems, however, there may be one or several variables that indicate or measure the fluctuations attributable to the units. The use of such concomitant variables in addition to randomization and perhaps blocking can frequently reduce the relevant variation from unit to unit and thus increase the precision and sensitivity. The analysis, however, involves the essential use of the regression methods from Sections 11.2 and 11.3. And it goes outside the basic experimental methods of manipulating factors, controlling factors, and randomizing against possible effects of uncontrolled factors, and it does this primarily because these variables cannot be manipulated, just recorded.

Consider some examples. The effect of several diets on the weight of a certain kind of animal: a possible concomitant variable is the initial weight of an animal. The effect of various treatments on the yield of a particular crop: a possible concomitant variable is the preceding year's yield if uniformity trials were performed. The effects of several treatments on the yield of rubber trees: a possible concomitant variable is some average yield from preceding years if the yields were obtained without treatments.

The computational methods for the factorial designs are organized to take advantage of orthogonality within sets of 1-vectors and between certain subspaces. By contrast the direct regression methods of Section 11.2 used a step-by-step procedure that calculates inner products of residual vectors. If we now introduce concomitant variables to a factorial design, we have an essential need for the step-by-step procedure. But—we can benefit computationally by using the orthogonality as far as possible and just terminally using the step-by-step procedure (which usually destroys orthogonality relations). We recall from Section 11.2 that the projection into an orthogonal complement and its squared length do not depend on the order in which components are removed.

A THE FULLY RANDOMIZED DESIGN

Consider a fully randomized design with response y and treatments $T_1, \ldots, T_t$. Let x be a concomitant variable that can be observed for each experimental unit, and suppose for simplicity here that the response level without treatments is at most linear in x, linearity providing in many cases a reasonable approximation over a moderate range. The data array can be presented as

$$\begin{array}{c|ccc} \multirow{2}{*}{T_1} & x_{11} & \cdots & x_{1n_1} \\ & y_{11} & \cdots & y_{1n_1} \\ \vdots & & & \\ \multirow{2}{*}{T_t} & x_{t1} & \cdots & y_{tn_t} \\ & y_{t1} & \cdots & y_{tn_t}. \end{array}$$

If the treatments are part of a factorial design, then $n_1 = \cdots = n_t$. The data can appear as in Figure S.13.

The investigation typically will involve the following succession of models: an overall general level, $\mathbf{y} = \beta\mathbf{1} + \sigma\mathbf{z}$; dependence on the concomitant variables, which here is affine dependence on x, $\mathbf{y} = \beta_1\mathbf{1} + \beta_2\mathbf{x} + \sigma\mathbf{z}$; dependence on the concomitant variables with allowance for different levels for different treatments, $\mathbf{y} = \Sigma\gamma_i\mathbf{1}_i + \delta\mathbf{x} + \sigma\mathbf{z}$, where $\mathbf{1}_i$ is the 1-vector for treatment T_i. This succession of models is represented by the sequence of solid arrows in the following display presenting fitted models:

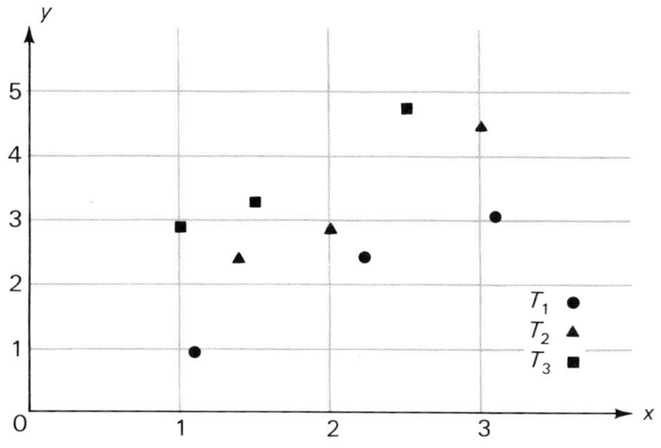

FIGURE S.13
Three observations at each of three treatments.

$$
\begin{array}{ccc}
\mathbf{y} = b\mathbf{1} + & \longrightarrow & \mathbf{y} = b_1\mathbf{1} + b_2\mathbf{x} + \\
\downarrow & & \downarrow \\
\mathbf{y} = \sum c_i \mathbf{1}_i + & \dashrightarrow & \mathbf{y} = \sum c_i \mathbf{1}_i + d\mathbf{x} +
\end{array} \tag{1}
$$

We have noted in the introductory remarks that there are computational advantages to using the simplicity and orthogonality of factorial designs as long as possible and only terminally using the regression procedures. In the preceding array this corresponds to following the dashed arrows involving the various fits appropriate to a factorial design and then using regression to obtain the fitted model in the right-hand column. This is the computation procedure we follow in this section.

First we consider the cumulative table appropriate to the particular factorial design. Ordinarily, this records the squared length of the projections of the response in the succession of subspaces. In the present case we record the inner product matrix for the projections of both **y** and the concomitant vectors in the succession of subspaces. For example, the projections of **x** and **y** on the vector **1** are $\bar{x}\mathbf{1}$ and $\bar{y}\mathbf{1}$, where $\bar{x} = T_{..}(x)/N$ and $\bar{y} = T_{..}(y)/N$, and these capital T's are used to denote totals as in earlier sections of this chapter; thus the inner product matrix is

$$
\begin{pmatrix} T_{..}^2(x)/N & T_{..}(x)T_{..}(y)/N \\ T_{..}(y)T_{..}(x)/N & T_{..}^2(y)/N \end{pmatrix}.
$$

In this way we obtain Table S.13.

The values in Table S.13 are now complemented with respect to the inner product matrix,

TABLE S.13
Cumulative table

	Source	DF	SPxx	SPxy	SPyy
(1)	Mean	1	$T_{..}^2(x)/N$	$T_{..}(x)T_{..}(y)/N$	$T_{..}^2(y)/N$
(2)	Treatments	t	$\sum T_{i\cdot}^2(x)/n_i$	$\sum T_{i\cdot}(x)T_{i\cdot}(y)/n_i$	$\sum T_{i\cdot}^2(y)/n_i$
(3)	Response	N	$\sum x_{ij}^2$	$\sum x_{ij}y_{ij}$	$\sum y_{ij}^2$

(2) $$\begin{pmatrix} \sum x^2 & \sum xy \\ \sum yx & \sum y^2 \end{pmatrix},$$

for the initial vectors. By doing this we obtain the successive inner product matrices for *residuals* (Table S.14).

TABLE S.14
Cumulative table; residuals

	Source	DF	SPxx	SPxy	SPyy
[0]	Null	N	$\sum x_{ij}^2$	$\sum x_{ij}y_{ij}$	$\sum y_{ij}^2$
[1]	Mean	$N-1$	$\sum x_{ij}^2 - \dfrac{T_{..}^2(x)}{N}$	$\sum x_{ij}y_{ij} - \dfrac{T_{..}(x)T_{..}(y)}{N}$	$\sum y_{ij}^2 - \dfrac{T_{..}^2(y)}{N}$
[2]	Treatments	$N-t$	$\sum x_{ij}^2 - \sum \dfrac{T_{i\cdot}^2(x)}{n_i}$	$\sum x_{ij}y_{ij} - \sum \dfrac{T_{i\cdot}(x)T_{i\cdot}(y)}{n_i}$	$\sum y_{ij}^2 - \sum \dfrac{T_{i\cdot}^2(y)}{n_i}$

Now recall the step-by-step procedure described in Section 11.2. For example, to introduce linear regression on x after fitting the mean involves a single step using the residuals of **y** and **x** with respect to the mean (note the penultimate remarks in the third paragraph of Section 12.3E); thus in the present notation the regression coefficient is $\mathrm{SP}_{[1]}xy/\mathrm{SP}_{[1]}xx$ using the values in row [1], and the reduction in the sum of squares for the response is given by $(\mathrm{SP}_{[1]}xy)^2/\mathrm{SP}_{[1]}xx$.

We now record in Table S.15 the sum of squares of residuals obtained following the solid arrows in the display (1) earlier in this section. Note that a prime indicates a shift from the left side of the display to the right side.

The sum of squares for the orthogonal components can be obtained by differencing Table S.15, giving the results shown in Table S.16. Thus with allowance made for linear dependence on x we can test for treatment effects by calculating the treatment MS divided by residual MS in Table S.16.

B EXAMPLE

Consider a response y whose level is known to adjust linearly with respect to a concomitant variable x. And suppose that three treatments T_1, T_2, and T_3 are applied treatments that could provide an additive adjustment to the response level. The data shown in Table S.17 were obtained. Some calculated results are given with the table:

$$35.64 = \sum T_{i\cdot}^2(x)/n_i, \qquad 84.5521 = \sum T_{i\cdot}^2(y)/n_i,$$
$$52.5413 = \sum T_{i\cdot}(x)T_{i\cdot}(y)/n_i;$$

the total sums of squares and products are

(3) $$\begin{matrix} 40.12 & & 90.9700 \\ & 57.815 & \end{matrix}$$

The cumulative table recording inner products for projections into the successive linear spaces is shown in Table S.18.

TABLE S.15
Cumulative table; residuals

	Source	DF	SS Reduction	SS
[0]	Null	N		$SP_{[0]}yy = \sum y_{ij}^2$
[1]	Mean	$N - 1$		$SP_{[1]}yy = \sum y_{ij}^2 - \frac{T_{..}^2(y)}{N}$
[1′]	Linear (x)	$N - 2$	$\frac{SP_{[1]}^2xy}{SP_{[1]}xx}$	$SP_{[1]}yy - \frac{SP_{[1]}^2xy}{SP_{[1]}xx}$
[2′]	Treatments	$N - t - 1$	$\frac{SP_{[2]}^2xy}{SP_{[2]}xx}$	$SP_{[2]}yy - \frac{SP_{[2]}^2xy}{SP_{[2]}xx}$

TABLE S.16
Analysis-of-covariance table

	Source	DF	SS
1	Mean	1	$\frac{T_{..}^2(y)}{N}$
1′	Linear (x)	1	$\frac{SP_{[1]}^2xy}{SP_{[1]}xx}$
2′	Treatments	$t - 1$	$SP_{[1]}yy - SP_{[2]}yy - \left(\frac{SP_{[1]}^2xy}{SP_{[1]}xx} - \frac{SP_{[2]}^2xy}{SP_{[2]}xx}\right)$
3	Residual	$N - t - 1$	$SP_{[2]}yy - \frac{SP_{[2]}^2xy}{SP_{[2]}xx}$

TABLE S.17
Treatment data

T_1		T_2		T_3		Total	
x	y	x	y	x	y	x	y
1.1	0.95	1.4	2.43	1.0	2.86		
2.2	2.42	2.0	2.88	1.5	3.27		
3.1	3.04	3.0	4.44	2.5	4.71		
6.4	6.41	6.4	9.75	5.0	10.84	17.8	27.00
				35.64	84.5521	35.2044	81.0000
				52.5413		53.4000	

TABLE S.18
Cumulative table

	Source	DF	SPxx	SPxy	SPyy
(1)	Mean	1	35.2044	53.4000	81.0000
(2)	Treatments	3	35.6400	52.5413	84.5521
(3)	Response	9	40.1200	57.8150	90.9700

TABLE S.19
Cumulative table; residuals

	Source	DF	SP*xx*	SP*xy*	SP*yy*
[0]	Null	9	40.1200	57.8150	90.9700
[1]	Treatments	8	4.9156	4.4150	9.9700
[2]	Residual	6	4.4800	5.2737	6.4179

TABLE S.20
Cumulative table; residuals

	Source	DF	SS Reduction	SS
[0]	Null	9		90.9700
[1]	Mean	8		9.9700
[1′]	Linear (x)	7	3.9654	6.0046
[2′]	Treatments	5	6.2080	0.2099

The inner products for residuals are obtained by complementing with respect to (3), Table S.19. The regression on x can now be removed at each stage after the mean, Table S.20.

The sums of squares for the orthogonal components relative to the succession of models (1) is now obtained by differencing; see Table S.21. The observed F for testing treatments is 69 and the 5 and 1 percent points are 5.79 and 13.27. Under the assumed model this provides overwhelming evidence against the hypothesis of no treatment effects. It is easy to check that a test neglecting covariance analysis would give an F value just larger than 1; and this would give no evidence against the hypothesis of no treatment effects.

TABLE S.21
Analysis-of-covariance table

	Source	DF	SS	MS	F
1	Mean	1	81.0000		
1′	Linear (x)	1	3.9654		
2′	Treatments	2	5.7947	2.897	$F = 69$
3	Residual	5	0.2099	0.042	
		9	90.9700		

C EXERCISE

1 Fourteen pigs were randomly assigned to 3 diets for a 2-month period; see Exercise 12.1.1. The following table records (initial weight, final weight) for each pig:

T_1	(69, 82)	(73, 96)	(87, 110)	(98, 126)	(105, 140)
T_2	(52, 74)	(59, 90)	(67, 98)	(76, 109)	(82, 122)
T_3	(53, 86)	(58, 98)	(66, 104)	(71, 115)	

Calculate the analysis-of-variance table and make appropriate tests for dependence on the diets.

APPENDIX II
TABLES

TABLE 1

Normal(0, 1) distribution: right-tail probabilities $\alpha = P(z \geq z_\alpha) = 1 - G(z_\alpha)$

z_α	0.00	0.01	0.02	0.03	0.04	0.05	0.06	0.07	0.08	0.09
0.0	0.5000	0.4960	0.4920	0.4880	0.4840	0.4801	0.4761	0.4721	0.4681	0.4641
0.1	0.4602	0.4562	0.4522	0.4483	0.4443	0.4404	0.4364	0.4325	0.4286	0.4247
0.2	0.4207	0.4168	0.4129	0.4090	0.4052	0.4013	0.3974	0.3936	0.3897	0.3859
0.3	0.3821	0.3783	0.3745	0.3707	0.3669	0.3632	0.3594	0.3557	0.3520	0.3483
0.4	0.3446	0.3409	0.3372	0.3336	0.3300	0.3264	0.3228	0.3192	0.3156	0.3121
0.5	0.3085	0.3050	0.3015	0.2981	0.2946	0.2912	0.2877	0.2843	0.2810	0.2776
0.6	0.2743	0.2709	0.2676	0.2643	0.2611	0.2578	0.2546	0.2514	0.2483	0.2451
0.7	0.2420	0.2389	0.2358	0.2327	0.2297	0.2266	0.2236	0.2206	0.2177	0.2148
0.8	0.2119	0.2090	0.2061	0.2033	0.2005	0.1977	0.1949	0.1922	0.1894	0.1867
0.9	0.1841	0.1814	0.1788	0.1762	0.1736	0.1711	0.1685	0.1660	0.1635	0.1611
1.0	0.1587	0.1562	0.1539	0.1515	0.1492	0.1469	0.1446	0.1423	0.1401	0.1379
1.1	0.1357	0.1335	0.1314	0.1292	0.1271	0.1251	0.1230	0.1210	0.1190	0.1170
1.2	0.1151	0.1131	0.1112	0.1093	0.1075	0.1056	0.1038	0.1020	0.1003	0.09853
1.3	0.09680	0.09510	0.09342	0.09176	0.09012	0.08851	0.08691	0.08534	0.08379	0.08226
1.4	0.08076	0.07927	0.07780	0.07636	0.07493	0.07353	0.07215	0.07078	0.06944	0.06811
1.5	0.06681	0.06552	0.06426	0.06301	0.06178	0.06057	0.05938	0.05821	0.05705	0.05592
1.6	0.05480	0.05370	0.05262	0.05155	0.05050	0.04947	0.04846	0.04746	0.04648	0.04551
1.7	0.04457	0.04363	0.04272	0.04182	0.04093	0.04006	0.03920	0.03836	0.03754	0.03673
1.8	0.03593	0.03515	0.03438	0.03362	0.03288	0.03216	0.03144	0.03074	0.03005	0.02938
1.9	0.02872	0.02807	0.02743	0.02680	0.02619	0.02559	0.02500	0.02442	0.02385	0.02330
2.0	0.02275	0.02222	0.02169	0.02119	0.02068	0.02018	0.01970	0.01923	0.01876	0.01831
2.1	0.01786	0.01743	0.01700	0.01659	0.01618	0.01578	0.01539	0.01500	0.01463	0.01426
2.2	0.01390	0.01355	0.01321	0.01287	0.01255	0.01222	0.01191	0.01160	0.01130	0.01101
2.3	0.01072	0.01044	0.01017	$0.0^{2}9903$	$0.0^{2}9642$	$0.0^{2}9387$	$0.0^{2}9137$	$0.0^{2}8894$	$0.0^{2}8656$	$0.0^{2}8424$
2.4	$0.0^{2}8198$	$0.0^{2}7976$	$0.0^{2}7760$	$0.0^{2}7549$	$0.0^{2}7344$	$0.0^{2}7143$	$0.0^{2}6947$	$0.0^{2}6756$	$0.0^{2}6569$	$0.0^{2}6387$
2.5	$0.0^{2}6210$	$0.0^{2}6037$	$0.0^{2}5868$	$0.0^{2}5703$	$0.0^{2}5543$	$0.0^{2}5386$	$0.0^{2}5234$	$0.0^{2}5085$	$0.0^{2}4940$	$0.0^{2}4799$
2.6	$0.0^{2}4661$	$0.0^{2}4527$	$0.0^{2}4396$	$0.0^{2}4269$	$0.0^{2}4145$	$0.0^{2}4025$	$0.0^{2}3907$	$0.0^{2}3793$	$0.0^{2}3681$	$0.0^{2}3573$
2.7	$0.0^{2}3467$	$0.0^{2}3364$	$0.0^{2}3264$	$0.0^{2}3167$	$0.0^{2}3072$	$0.0^{2}2980$	$0.0^{2}2890$	$0.0^{2}2803$	$0.0^{2}2718$	$0.0^{2}2635$
2.8	$0.0^{2}2555$	$0.0^{2}2477$	$0.0^{2}2401$	$0.0^{2}2327$	$0.0^{2}2256$	$0.0^{2}2186$	$0.0^{2}2118$	$0.0^{2}2052$	$0.0^{2}1988$	$0.0^{2}1926$
2.9	$0.0^{2}1866$	$0.0^{2}1807$	$0.0^{2}1750$	$0.0^{2}1695$	$0.0^{2}1641$	$0.0^{2}1589$	$0.0^{2}1538$	$0.0^{2}1489$	$0.0^{2}1441$	$0.0^{2}1395$
3.0	$0.0^{2}1350$	$0.0^{2}1306$	$0.0^{2}1264$	$0.0^{2}1223$	$0.0^{2}1183$	$0.0^{2}1144$	$0.0^{2}1107$	$0.0^{2}1070$	$0.0^{2}1035$	$0.0^{2}1001$
3.1	$0.0^{3}9676$	$0.0^{3}9354$	$0.0^{3}9043$	$0.0^{3}8740$	$0.0^{3}8447$	$0.0^{3}8164$	$0.0^{3}7888$	$0.0^{3}7622$	$0.0^{3}7364$	$0.0^{3}7114$
3.2	$0.0^{3}6871$	$0.0^{3}6637$	$0.0^{3}6410$	$0.0^{3}6190$	$0.0^{3}5976$	$0.0^{3}5770$	$0.0^{3}5571$	$0.0^{3}5377$	$0.0^{3}5190$	$0.0^{3}5009$
3.3	$0.0^{3}4834$	$0.0^{3}4665$	$0.0^{3}4501$	$0.0^{3}4342$	$0.0^{3}4189$	$0.0^{3}4041$	$0.0^{3}3897$	$0.0^{3}3758$	$0.0^{3}3624$	$0.0^{3}3495$
3.4	$0.0^{3}3369$	$0.0^{3}3248$	$0.0^{3}3131$	$0.0^{3}3018$	$0.0^{3}2909$	$0.0^{3}2803$	$0.0^{3}2701$	$0.0^{3}2602$	$0.0^{3}2507$	$0.0^{3}2415$
3.5	$0.0^{3}2326$	$0.0^{3}2241$	$0.0^{3}2158$	$0.0^{3}2078$	$0.0^{3}2001$	$0.0^{3}1926$	$0.0^{3}1854$	$0.0^{3}1785$	$0.0^{3}1718$	$0.0^{3}1653$
3.6	$0.0^{3}1591$	$0.0^{3}1531$	$0.0^{3}1473$	$0.0^{3}1417$	$0.0^{3}1363$	$0.0^{3}1311$	$0.0^{3}1261$	$0.0^{3}1213$	$0.0^{3}1166$	$0.0^{3}1121$
3.7	$0.0^{3}1078$	$0.0^{3}1036$	$0.0^{4}9961$	$0.0^{4}9574$	$0.0^{4}9201$	$0.0^{4}8842$	$0.0^{4}8496$	$0.0^{4}8162$	$0.0^{4}7841$	$0.0^{4}7532$
3.8	$0.0^{4}7235$	$0.0^{4}6948$	$0.0^{4}6673$	$0.0^{4}6407$	$0.0^{4}6152$	$0.0^{4}5906$	$0.0^{4}5669$	$0.0^{4}5442$	$0.0^{4}5223$	$0.0^{4}5012$
3.9	$0.0^{4}4810$	$0.0^{4}4615$	$0.0^{4}4427$	$0.0^{4}4247$	$0.0^{4}4074$	$0.0^{4}3908$	$0.0^{4}3747$	$0.0^{4}3594$	$0.0^{4}3446$	$0.0^{4}3304$
4.0	$0.0^{4}3167$	$0.0^{4}3036$	$0.0^{4}2910$	$0.0^{4}2789$	$0.0^{4}2673$	$0.0^{4}2561$	$0.0^{4}2454$	$0.0^{4}2351$	$0.0^{4}2252$	$0.0^{4}2157$

α	0.0	0.1	0.2	0.3	0.4	0.5	0.6	0.7	0.8	0.9
4	$0.0^{4}317$	$0.0^{4}207$	$0.0^{4}133$	$0.0^{5}854$	$0.0^{5}541$	$0.0^{5}340$	$0.0^{5}211$	$0.0^{5}130$	$0.0^{6}793$	$0.0^{6}479$
5	$0.0^{6}287$	$0.0^{6}170$	$0.0^{7}996$	$0.0^{7}579$	$0.0^{7}333$	$0.0^{7}190$	$0.0^{7}107$	$0.0^{8}599$	$0.0^{8}332$	$0.0^{8}182$
6	$0.0^{9}987$	$0.0^{9}530$	$0.0^{9}282$	$0.0^{9}149$	$0.0^{10}777$	$0.0^{10}402$	$0.0^{10}206$	$0.0^{10}104$	$0.0^{11}523$	$0.0^{11}260$

TABLE 2

Student(*f*)distribution: right-tail percentage points t_α; $\alpha = P(t \geq t_\alpha)$

f \ α	0.25	0.1	0.05	0.025	0.01	0.005
1	1.000	3.078	6.314	12.706	31.821	63.657
2	0.816	1.886	2.920	4.303	6.965	9.925
3	0.765	1.638	2.353	3.182	4.541	5.841
4	0.741	1.533	2.132	2.776	3.747	4.604
5	0.727	1.476	2.015	2.571	3.365	4.032
6	0.718	1.440	1.943	2.447	3.143	3.707
7	0.711	1.415	1.895	2.365	2.998	3.499
8	0.706	1.397	1.860	2.306	2.896	3.355
9	0.703	1.383	1.833	2.262	2.821	3.250
10	0.700	1.372	1.812	2.228	2.764	3.169
11	0.697	1.363	1.796	2.201	2.718	3.106
12	0.695	1.356	1.782	2.179	2.681	3.055
13	0.694	1.350	1.771	2.160	2.650	3.012
14	0.692	1.345	1.761	2.145	2.624	2.977
15	0.691	1.341	1.753	2.131	2.602	2.947
16	0.690	1.337	1.746	2.120	2.583	2.921
17	0.689	1.333	1.740	2.110	2.567	2.898
18	0.688	1.330	1.734	2.101	2.552	2.878
19	0.688	1.328	1.729	2.093	2.539	2.861
20	0.687	1.325	1.725	2.086	2.528	2.845
21	0.686	1.323	1.721	2.080	2.518	2.831
22	0.686	1.321	1.717	2.074	2.508	2.819
23	0.685	1.319	1.714	2.069	2.500	2.807
24	0.685	1.318	1.711	2.064	2.492	2.797
25	0.684	1.316	1.708	2.060	2.485	2.787
26	0.684	1.315	1.706	2.056	2.479	2.779
27	0.684	1.314	1.703	2.052	2.473	2.771
28	0.683	1.313	1.701	2.048	2.467	2.763
29	0.683	1.311	1.699	2.045	2.462	2.756
30	0.683	1.310	1.697	2.042	2.457	2.750
40	0.681	1.303	1.684	2.021	2.423	2.704
60	0.679	1.296	1.671	2.000	2.390	2.660
120	0.677	1.289	1.658	1.980	2.358	2.617
$N(0, 1)$ ∞	0.674	1.282	1.645	1.960	2.326	2.576

TABLE 3

Chi-Square(m) distribution: right-tail percentage points χ^2_α; $\alpha = P(\chi^2 \geq \chi^2_\alpha)$

m \ α	0.995	0.975	0.050	0.025	0.010	0.005
1	$0.0^4 3927$	$0.0^3 9821$	3.84146	5.02389	6.63490	7.87944
2	0.010025	0.050636	5.99147	7.37776	9.21034	10.5966
3	0.071721	0.215795	7.81473	9.34840	11.3449	12.8381
4	0.206990	0.484419	9.48773	11.1433	13.2767	14.8602
5	0.411740	0.831211	11.0705	12.8325	15.0863	16.7496
6	0.675727	1.237347	12.5916	14.4494	16.8119	18.5476
7	0.989265	1.68987	14.0671	16.0128	18.4753	20.2777
8	1.344419	2.17973	15.5073	17.5346	20.0902	21.9550
9	1.734926	2.70039	16.9190	19.0228	21.6660	23.5893
10	2.15585	3.24697	18.3070	20.4831	23.2093	25.1882
11	2.60321	3.81575	19.6751	21.9200	24.7250	26.7569
12	3.07382	4.40379	21.0261	23.3367	26.2170	28.2995
13	3.56503	5.00874	22.3621	24.7356	27.6883	29.8194
14	4.07468	5.62872	23.6848	26.1190	29.1413	31.3193
15	4.60094	6.26214	24.9958	27.4884	30.5779	32.8013
16	5.14224	6.90766	26.2962	28.8454	31.9999	34.2672
17	5.69724	7.56418	27.5871	30.1910	33.4087	35.7185
18	6.26481	8.23075	28.8693	31.5264	34.8053	37.1564
19	6.84398	8.90655	30.1435	32.8523	36.1908	38.5822
20	7.43386	9.59083	31.4104	34.1696	37.5662	39.9968
21	8.03366	10.28293	32.6705	35.4789	38.9321	41.4010
22	8.64272	10.9823	33.9244	36.7807	40.2894	42.7956
23	9.26042	11.6885	35.1725	38.0757	41.6384	44.1813
24	9.88623	12.4011	36.4151	39.3641	42.9798	45.5585
25	10.5197	13.1197	37.6525	40.6465	44.3141	46.9278
26	11.1603	13.8439	38.8852	41.9232	45.6417	48.2899
27	11.8076	14.5733	40.1133	43.1944	46.9630	49.6449
28	12.4613	15.3079	41.3372	44.4607	48.2782	50.9933
29	13.1211	16.0471	42.5569	45.7222	49.5879	52.3356
30	13.7867	16.7908	43.7729	46.9792	50.8922	53.6720
40	20.7065	24.4331	55.7585	59.3417	63.6907	66.7659
50	27.9907	32.3574	67.5048	71.4202	76.1539	79.4900
60	35.5346	40.4817	79.0819	83.2976	88.3794	91.9517
70	43.2752	48.7576	90.5312	95.0231	100.425	104.215
80	51.1720	57.1532	101.879	106.629	112.329	116.321
90	59.1963	65.6466	113.145	118.136	124.116	128.299
100	67.3276	74.2219	124.342	129.561	135.807	140.169

Source: "Tables of the Percentage Points of the χ^2-Distribution," *Biometrika,* Vol. 32 (1941), pp. 188–189, by Catherine M. Thompson; reproduced by permission of Professor E. S. Pearson and the Biometrika Trustees.

TABLE 4

The $F(m, n)$ distribution: right-tail percentage points F_α; $\alpha = P(F \geq F_\alpha)$

α	n \ m	1	2	3	4	5	6	7	8	9	10	12	15	20	30	60	120	∞
0.10	1	39.9	49.5	53.6	55.8	57.2	58.2	58.9	59.4	59.9	60.2	60.7	61.2	61.7	62.3	62.8	63.1	63.3
0.05		161	200	216	225	230	234	237	239	241	242	244	246	248	250	252	253	254
0.025		648	800	864	900	922	937	948	957	963	969	977	985	993	1000	1010	1010	1020
0.01		4,050	5,000	5,400	5,620	5,760	5,860	5,930	5,980	6,020	6,060	6,110	6,160	6,210	6,260	6,310	6,340	6,370
0.005		16,200	20,000	21,600	22,500	23,100	23,400	23,700	23,900	24,100	24,200	24,400	24,600	24,800	25,000	25,200	25,400	25,500
0.10	2	8.53	9.00	9.16	9.24	9.29	9.33	9.35	9.37	9.38	9.39	9.41	9.42	9.44	9.46	9.47	9.48	9.49
0.05		18.5	19.0	19.2	19.2	19.3	19.3	19.4	19.4	19.4	19.4	19.4	19.4	19.5	19.5	19.5	19.5	19.5
0.025		38.5	39.0	39.2	39.2	39.3	39.3	39.4	39.4	39.4	39.4	39.4	39.4	39.4	39.5	39.5	39.5	39.5
0.01		98.5	99.0	99.2	99.2	99.3	99.3	99.4	99.4	99.4	99.4	99.4	99.4	99.4	99.5	99.5	99.5	99.5
0.005		199	199	199	199	199	199	199	199	199	199	199	199	199	199	199	199	199
0.10	3	5.54	5.46	5.39	5.34	5.31	5.28	5.27	5.25	5.24	5.23	5.22	5.20	5.18	5.17	5.15	5.14	5.13
0.05		10.1	9.55	9.28	9.12	9.01	8.94	8.89	8.85	8.81	8.79	8.74	8.70	8.66	8.62	8.57	8.55	8.53
0.025		17.4	16.0	15.4	15.1	14.9	14.7	14.6	14.5	14.5	14.4	14.3	14.3	14.2	14.1	14.0	13.9	13.9
0.01		34.1	30.8	29.5	28.7	28.2	27.9	27.7	27.5	27.3	27.2	27.1	26.9	26.7	26.5	26.3	26.2	26.1
0.005		55.6	49.8	47.5	46.2	45.4	44.8	44.4	44.1	43.9	43.7	43.4	43.1	42.8	42.5	42.1	42.0	41.8
0.10	4	4.54	4.32	4.19	4.11	4.05	4.01	3.98	3.95	3.93	3.92	3.90	3.87	3.84	3.82	3.79	3.78	3.76
0.05		7.71	6.94	6.59	6.39	6.26	6.16	6.09	6.04	6.00	5.96	5.91	5.86	5.80	5.75	5.69	5.66	5.63
0.025		12.2	10.6	9.98	9.60	9.36	9.20	9.07	8.98	8.90	8.84	8.75	8.66	8.56	8.46	8.36	8.31	8.26
0.01		21.2	18.0	16.7	16.0	15.5	15.2	15.0	14.8	14.7	14.5	14.4	14.2	14.0	13.8	13.7	13.6	13.5
0.005		31.3	26.3	24.3	23.2	22.5	22.0	21.6	21.4	21.1	21.0	20.7	20.4	20.2	19.9	19.6	19.5	19.3
0.10	5	4.06	3.78	3.62	3.52	3.45	3.40	3.37	3.34	3.32	3.30	3.27	3.24	3.21	3.17	3.14	3.12	3.11
0.05		6.61	5.79	5.41	5.19	5.05	4.95	4.88	4.82	4.77	4.74	4.68	4.62	4.56	4.50	4.43	4.40	4.37
0.025		10.0	8.43	7.76	7.39	7.15	6.98	6.85	6.76	6.68	6.62	6.52	6.43	6.33	6.23	6.12	6.07	6.02
0.01		16.3	13.3	12.1	11.4	11.0	10.7	10.5	10.3	10.2	10.1	9.89	9.72	9.55	9.38	9.20	9.11	9.02
0.005		22.8	18.3	16.5	15.6	14.9	14.5	14.2	14.0	13.8	13.6	13.4	13.1	12.9	12.7	12.4	12.3	12.1
0.10	6	3.78	3.46	3.29	3.18	3.11	3.05	3.01	2.98	2.96	2.94	2.90	2.87	2.84	2.80	2.76	2.74	2.72
0.05		5.99	5.14	4.76	4.53	4.39	4.28	4.21	4.15	4.10	4.06	4.00	3.94	3.87	3.81	3.74	3.70	3.67
0.025		8.81	7.26	6.60	6.23	5.99	5.82	5.70	5.60	5.52	5.46	5.37	5.27	5.17	5.07	4.96	4.90	4.85
0.01		13.7	10.9	9.78	9.15	8.75	8.47	8.26	8.10	7.98	7.87	7.72	7.56	7.40	7.23	7.06	6.97	6.88
0.005		18.6	14.5	12.9	12.0	11.5	11.1	10.8	10.6	10.4	10.2	10.0	9.81	9.59	9.36	9.12	9.00	8.88

0.10	7	3.59	3.26	3.07	2.96	2.88	2.83	2.78	2.75	2.72	2.70	2.67	2.63	2.59	2.56	2.51	2.49	2.47
0.05		5.59	4.74	4.35	4.12	3.97	3.87	3.79	3.73	3.68	3.64	3.57	3.51	3.44	3.38	3.30	3.27	3.23
0.025		8.07	6.54	5.89	5.52	5.29	5.12	4.99	4.90	4.82	4.76	4.67	4.57	4.47	4.36	4.25	4.20	4.14
0.01		12.2	9.55	8.45	7.85	7.46	7.19	6.99	6.84	6.72	6.62	6.47	6.31	6.16	5.99	5.82	5.74	5.65
0.005		16.2	12.4	10.9	10.1	9.52	9.16	8.89	8.68	8.51	8.38	8.18	7.97	7.75	7.53	7.31	7.19	7.08
0.10	8	3.46	3.11	2.92	2.81	2.73	2.67	2.62	2.59	2.56	2.54	2.50	2.46	2.42	2.38	2.34	2.31	2.29
0.05		5.32	4.46	4.07	3.84	3.69	3.58	3.50	3.44	3.39	3.35	3.28	3.22	3.15	3.08	3.01	2.97	2.93
0.025		7.57	6.06	5.42	5.05	4.82	4.65	4.53	4.43	4.36	4.30	4.20	4.10	4.00	3.89	3.78	3.73	3.67
0.01		11.3	8.65	7.59	7.01	6.63	6.37	6.18	6.03	5.91	5.81	5.67	5.52	5.36	5.20	5.03	4.95	4.86
0.005		14.7	11.0	9.60	8.81	8.30	7.95	7.69	7.50	7.34	7.21	7.01	6.81	6.61	6.40	6.18	6.06	5.95
0.10	9	3.36	3.01	2.81	2.69	2.61	2.55	2.51	2.47	2.44	2.42	2.38	2.34	2.30	2.25	2.21	2.18	2.16
0.05		5.12	4.26	3.86	3.63	3.48	3.37	3.29	3.23	3.18	3.14	3.07	3.01	2.94	2.86	2.79	2.75	2.71
0.025		7.21	5.71	5.08	4.72	4.48	4.32	4.20	4.10	4.03	3.96	3.87	3.77	3.67	3.56	3.45	3.39	3.33
0.01		10.6	8.02	6.99	6.42	6.06	5.80	5.61	5.47	5.35	5.26	5.11	4.96	4.81	4.65	4.48	4.40	4.31
0.005		13.6	10.1	8.72	7.96	7.47	7.13	6.88	6.69	6.54	6.42	6.23	6.03	5.83	5.62	5.41	5.30	5.19
0.10	10	3.29	2.92	2.73	2.61	2.52	2.46	2.41	2.38	2.35	2.32	2.28	2.24	2.20	2.15	2.11	2.08	2.06
0.05		4.96	4.10	3.71	3.48	3.33	3.22	3.14	3.07	3.02	2.98	2.91	2.84	2.77	2.70	2.62	2.58	2.54
0.025		6.94	5.46	4.83	4.47	4.24	4.07	3.95	3.85	3.78	3.72	3.62	3.52	3.42	3.31	3.20	3.14	3.08
0.01		10.0	7.56	6.55	5.99	5.64	5.39	5.20	5.06	4.94	4.85	4.71	4.56	4.41	4.25	4.08	4.00	3.91
0.005		12.8	9.43	8.08	7.34	6.87	6.54	6.30	6.12	5.97	5.85	5.66	5.47	5.27	5.07	4.86	4.75	4.64
0.10	12	3.18	2.81	2.61	2.48	2.39	2.33	2.28	2.24	2.21	2.19	2.15	2.10	2.06	2.01	1.96	1.93	1.90
0.05		4.75	3.89	3.49	3.26	3.11	3.00	2.91	2.85	2.80	2.75	2.69	2.62	2.54	2.47	2.38	2.34	2.30
0.025		6.55	5.10	4.47	4.12	3.89	3.73	3.61	3.51	3.44	3.37	3.28	3.18	3.07	2.96	2.85	2.79	2.72
0.01		9.33	6.93	5.95	5.41	5.06	4.82	4.64	4.50	4.39	4.30	4.16	4.01	3.86	3.70	3.54	3.45	3.36
0.005		11.8	8.51	7.23	6.52	6.07	5.76	5.52	5.35	5.20	5.09	4.91	4.72	4.53	4.33	4.12	4.01	3.90
0.10	15	3.07	2.70	2.49	2.36	2.27	2.21	2.16	2.12	2.09	2.06	2.02	1.97	1.92	1.87	1.82	1.79	1.76
0.05		4.54	3.68	3.29	3.06	2.90	2.79	2.71	2.64	2.59	2.54	2.48	2.40	2.33	2.25	2.16	2.11	2.07
0.025		6.20	4.77	4.15	3.80	3.58	3.41	3.29	3.20	3.12	3.06	2.96	2.86	2.76	2.64	2.52	2.46	2.40
0.01		8.68	6.36	5.42	4.89	4.56	4.32	4.14	4.00	3.89	3.80	3.67	3.52	3.37	3.21	3.05	2.96	2.87
0.005		10.8	7.70	6.48	5.80	5.37	5.07	4.85	4.67	4.54	4.42	4.25	4.07	3.88	3.69	3.48	3.37	3.26

TABLE 4

The $F(m, n)$ distribution: right-tail percentage points F_α; $\alpha = P(F \geq F_\alpha)$ (*continued*)

α	n \ m	1	2	3	4	5	6	7	8	9	10	12	15	20	30	60	120	∞
0.10	20	2.97	2.59	2.38	2.25	2.16	2.09	2.04	2.00	1.96	1.94	1.89	1.84	1.79	1.74	1.68	1.64	1.61
0.05		4.35	3.49	3.10	2.87	2.71	2.60	2.51	2.45	2.39	2.35	2.28	2.20	2.12	2.04	1.95	1.90	1.84
0.025		5.87	4.46	3.86	3.51	3.29	3.13	3.01	2.91	2.84	2.77	2.68	2.57	2.46	2.35	2.22	2.16	2.09
0.01		8.10	5.85	4.94	4.43	4.10	3.87	3.70	3.56	3.46	3.37	3.23	3.09	2.94	2.78	2.61	2.52	2.42
0.005		9.94	6.99	5.82	5.17	4.76	4.47	4.26	4.09	3.96	3.85	3.68	3.50	3.32	3.12	2.92	2.81	2.69
0.10	30	2.88	2.49	2.28	2.14	2.05	1.98	1.93	1.88	1.85	1.82	1.77	1.72	1.67	1.61	1.54	1.50	1.46
0.05		4.17	3.32	2.92	2.69	2.53	2.42	2.33	2.27	2.21	2.16	2.09	2.01	1.93	1.84	1.74	1.68	1.62
0.025		5.57	4.18	3.59	3.25	3.03	2.87	2.75	2.65	2.57	2.51	2.41	2.31	2.20	2.07	1.94	1.87	1.79
0.01		7.56	5.39	4.51	4.02	3.70	3.47	3.30	3.17	3.07	2.98	2.84	2.70	2.55	2.39	2.21	2.11	2.01
0.005		9.18	6.35	5.24	4.62	4.23	3.95	3.74	3.58	3.45	3.34	3.18	3.01	2.82	2.63	2.42	2.30	2.18
0.10	60	2.79	2.39	2.18	2.04	1.95	1.87	1.82	1.77	1.74	1.71	1.66	1.60	1.54	1.48	1.40	1.35	1.29
0.05		4.00	3.15	2.76	2.53	2.37	2.25	2.17	2.10	2.04	1.99	1.92	1.84	1.75	1.65	1.53	1.47	1.39
0.025		5.29	3.93	3.34	3.01	2.79	2.63	2.51	2.41	2.33	2.27	2.17	2.06	1.94	1.82	1.67	1.58	1.48
0.01		7.08	4.98	4.13	3.65	3.34	3.12	2.95	2.82	2.72	2.63	2.50	2.35	2.20	2.03	1.84	1.73	1.60
0.005		8.49	5.80	4.73	4.14	3.76	3.49	3.29	3.13	3.01	2.90	2.74	2.57	2.39	2.19	1.96	1.83	1.69
0.10	120	2.75	2.35	2.13	1.99	1.90	1.82	1.77	1.72	1.68	1.65	1.60	1.54	1.48	1.41	1.32	1.26	1.19
0.05		3.92	3.07	2.68	2.45	2.29	2.18	2.09	2.02	1.96	1.91	1.83	1.75	1.66	1.55	1.43	1.35	1.25
0.025		5.15	3.80	3.23	2.89	2.67	2.52	2.39	2.30	2.22	2.16	2.05	1.94	1.82	1.69	1.53	1.43	1.31
0.01		6.85	4.79	3.95	3.48	3.17	2.96	2.79	2.66	2.56	2.47	2.34	2.19	2.03	1.86	1.66	1.53	1.38
0.005		8.18	5.54	4.50	3.92	3.55	3.28	3.09	2.93	2.81	2.71	2.54	2.37	2.19	1.98	1.75	1.61	1.43
0.10	∞	2.71	2.30	2.08	1.94	1.85	1.77	1.72	1.67	1.63	1.60	1.55	1.49	1.42	1.34	1.24	1.17	1.00
0.05		3.84	3.00	2.60	2.37	2.21	2.10	2.01	1.94	1.88	1.83	1.75	1.67	1.57	1.46	1.32	1.22	1.00
0.025		5.02	3.69	3.12	2.79	2.57	2.41	2.29	2.19	2.11	2.05	1.94	1.83	1.71	1.57	1.39	1.27	1.00
0.01		6.63	4.61	3.78	3.32	3.02	2.80	2.64	2.51	2.41	2.32	2.18	2.04	1.88	1.70	1.47	1.32	1.00
0.005		7.88	5.30	4.28	3.72	3.35	3.09	2.90	2.74	2.62	2.52	2.36	2.19	2.00	1.79	1.53	1.36	1.00

Source: "Tables of Percentage Points of the Inverted Beta Distribution," *Biometrika,* Vol. 33 (1943), by Maxine Merrington and Catherine M. Thompson; reproduced by permission of Professor E. S. Pearson and the Biometrika Trustees.

ANSWERS

CHAPTER 1: PROBABILITY

1.1 $\mathcal{B}' = \{\varnothing, H, T, E, H \vee T = E^c, H \vee E = T^c, T \vee E = H^c, \mathfrak{U}\}$; corresponding probabilities are 0, 0.53, 0.47, 0, 1, 0.53, 0.47, 1. **1.2** $\varnothing$ (null event); P, T, U, $P \vee T$, 0 to 10 coliform; $P \vee U$, 0 coliform or more than 10; $T \vee U$, at least one coliform; $\mathfrak{U}$. **1.5** In correspondence with Exercise 1.2: $\varnothing$, P, $S \wedge P^c$, U, S, $P \vee U$, P^c, $\mathfrak{U}$. **1.8** Necessary and sufficient condition for the cardinality of $\mathcal{B}$ to be 2^k is that k be the number of nonnull events among A_1A_2, $A_1A_2^c$, $A_1^cA_2$, $A_1^cA_2^c$.

2.1 $\mathcal{S} = \{1, 2, 3, 4, 5, 6\}$: $A = \{2, 4, 6\}$, $B = \{3, 6\}$, $C = \{1, 2, 3\}$, $A \wedge B = \{6\}$, $A \wedge C = \{2\}$, $A^c \vee B^c = \{1, 2, 3, 4, 5\}$. **2.3** $\mathcal{S} = \{1, 2, 3, 4, 5, 6\}^2$; $A = \{(1, 5), (2, 4), (3, 3), (4, 2), (5, 1)\}$, $B = \{(1, 6), (2, 5), (3, 4), (4, 3), (5, 2), (6, 1)\}$, $\{(1, 4), (2, 5), (3, 6), (4, 1), (5, 2), (6, 3)\}$, $\{(1, 5), (2, 6), (5, 1), (6, 2)\}$. **2.4** $\mathcal{S} = \{(1, 1), (1, 2), (1, 3), (1, 4), (1, 5), (1, 6), (2, 2), (2, 3), (2, 4), (2, 5), (2, 6), (3, 3), (3, 4), (3, 5), (3, 6), (4, 4), (4, 5), (4, 6), (5, 5), (5, 6), (6, 6)\}$; $A = \{(1, 5), (2, 4), (3, 3)\}$, $\{(1, 4), (2, 5), (3, 6)\}$, $\varnothing$. **2.6** $\mathcal{S} = \{(g, g), (g, b), (b, g), (b, b)\}$, $\{(g, g)\}$, $\{(g, b), (b, g)\}$, $\{(b, b)\}$. **2.8** $\mathcal{S} = \{(y_1, y_2): y_1 \neq y_2, y_i = AS, KS, \ldots, 2C\}$. $52 \times 51 = 2652$, $\{(y_1, y_2): y_1 \neq y_2, y_i = \text{AS, KS}, \ldots, 2\text{S}\}$. $13 \times 12 = 156$. **2.10** $\mathcal{S} = \{(1, 2, 3, 4), (1, 2, 4, 3), \ldots$, giving all 24 rearrangements of $(1, 2, 3, 4)\}$; $A_1 = \{(1, 2, 3, 4), (1, 2, 4, 3), (1, 3, 2, 4), (1, 3, 4, 2), (1, 4, 2, 3), (1, 4, 3, 2)\}$; $E_2 = \{(1, 2, 4, 3), (1, 4, 3, 2), (1, 3, 2, 4), (4, 2, 3, 1), (3, 2, 1, 4), (2, 1, 3, 4)\}$, $E_3 = \varnothing$, $E_4 = \{(1, 2, 3, 4)\}$. **2.13** $\{(y_1, y_2): y_1 + y_2 \leq 10\}$, $\{(y_1, y_2): y_1 \leq 5, y_2 \leq 5\}$, $\{(y_1, y_2)$: at least one $y_i \leq 3\}$. **2.15** $\{y_1: y_1 \leq 5\} \times R$, $R \times \{y_2: y_2 \leq 5\}$, $\{(y_1, y_2): y_1 + y_2 \leq 5\}$, $\{(y_1, y_2): -5 \leq y_2 - y_1 \leq 5\}$. **2.20** Given $A_1, A_2, \ldots \in \mathcal{A}$; then $A_1^c, A_2^c, \ldots \in \mathcal{A}$; then $\cup_1^\infty A_i^c \in \mathcal{A}$; then $(\cup_1^\infty A_i^c)^c = \cap_1^\infty A_i \in \mathcal{A}$ (using Problem 2.19); thus closed under σ-intersection. **2.23** $\mathcal{P}(\mathcal{S})$ is a σ-algebra; it contains $\mathcal{A}_0$.

3.1 $\mathcal{S}_0 = \{0, 1\}^3$; $\bar{P}(A) = c(A \cap \mathcal{S}_0)/8$; 1/8, 3/8, 3/8, 1/8. **3.3** $p(0, 0) = q^2$, $p(1, 0) = p(0, 1) = pq$, $p(1, 1) = p^2$; $P(\text{H}_1) = p^2 + pq = p$, $P(\text{T}_1) = qp + q^2 = q$. **3.4** $\bar{P}(A) = c(A \cap \mathcal{S})/12$; $\bar{P}(E_0) = 6/12$, $\bar{P}(E_1) = 6/12$, $\bar{P}(E_2) = 0$. **3.7** $\mathcal{S} = \{(y_1, y_2): y_1 \neq y_2, y_i = 1, 2, 3, 4\}$; same as for Exercise 3.4. **3.8** 0.6, 0.2; 0.5. **3.10** 0.2, 0.2; 0.04; $\pi/100$. **3.11** $A = A \cup \varnothing \cup \varnothing \cup \cdots$; then $\mu(A) = \mu(A) + \Sigma_1^\infty \mu(\varnothing)$; then $\Sigma_1^\infty \mu(\varnothing) = 0$; then $\mu(\varnothing) = 0$; note that A itself could be just $\varnothing$. **3.12** Let $A = A_1 \cup \cdots \cup A_k$, where the A_i are disjoint; then $A = A_1 \cup \cdots A_k \cup \varnothing \cup \varnothing \cdots$ with disjoint components; then $\mu(A) = \mu(A_1) + \cdots + \mu(A_k) + \Sigma\, \mu(\varnothing) = \Sigma_1^k \mu(A_i)$, using Problem 11.

4.1 $3/6 + 3/6 - 9/36 = 3/4$. **4.2** $1 - P(E_4) = 23/24$. **4.4** $E_0 = A_1^cA_2^c$, $E_1 = A_1^cA_2 \cup A_1A_2^c$, $E_2 = A_1A_2$; $48 \cdot 47/52 \cdot 51$, $2 \cdot 48 \cdot 4/52 \cdot 51$, $4 \cdot 3/52 \cdot 51$. **4.6** $1 - 1/52!$, 0. **4.8** $\Sigma_1^\infty 1/2^{3i} = 1/7$. **4.10** 6/24; $1 - 6/24$. **4.12** 2/10, 2/10, $4/10 - 4/100$. **4.13** $P(A_1 \cup A_2 \cup A_3) = P(A_1 \cup A_2) + P(A_3) - P(A_1A_3 \cup A_2A_3)$ and one further step. **4.17** $\cup_{k+1}^\infty A_j$ is monotone decreasing to ϕ, thus $\lim P(\cup_{k+1}^\infty A_j) = 0$; $P(\cup_1^\infty A_i) = P(A_1) + \cdots + P(A_k) + P(\cup_{k+1}^\infty A_j)$; take the limit; $P(\cup_1^\infty A_i) = \Sigma_1^\infty P(A_i)$.

5.7 $A_2 - A_1 = (A_1 \cup A_2)\, \Delta\, A_1$; $A_1\, \Delta\, A_2 = (A_1 - A_2) \cup (A_2 - A_1)$. **5.9** Let A and B be finite unions of disjoint left open–right closed intervals on R; then $A \cup B$ and $A - B$ are also finite unions of left open–right closed intervals. **5.20** $E(Y) = \Sigma_0^\infty yp(y) = \Sigma_1^\infty yp(y) = \Sigma_1^\infty \lambda^y e^{-\lambda}/(y-1)! = \lambda \Sigma_0^\infty \lambda^x e^{-\lambda}/x! = \lambda$.

CHAPTER 2: PROBABILITY ON THE LINE AND PLANE

1.1 $F(y) = 0, [y]/6, 1$ according as $y < 0$, $0 \le y \le 6$, $6 < y$; $5/6 - 3/6$, $1 - 4/6$. **1.3** 0.3679, 0.3679/1, 0.3679/2 = 0.1840, 0.1840/3 = 0.0613, 0.0613/4 = 0.0153, . . .; 0.3679, 0.3679, 1 − 0.3679 − 0.3679. **1.4** $e^{-2}(1+2)$, $2e^{-2}$, $1 - 5e^{-2}$. **1.8** $F(y) = 0, [y]/N, 1$ according as $y < 0$, $0 \le y \le N$, $N < y$; $[N/2]/N$, $([3N/4] - [N/4])/N$. **1.10** See Section 3.5A, 14/16, 174/256. **1.12** See Section 3.5C, 68/70, 55/70. **1.13** $G_d(+\infty) = F_d(+\infty)/F_d(+\infty) = 1$; other properties of a distribution function hold; $G_d(y) = \Sigma_{t \le y} p(t)/\Sigma p(u)$ and thus G_d is a discrete distribution function. **1.16** $y \in S_d \leftrightarrow p(y) > 0 \leftrightarrow y \in A_n$ for some n; thus $S_d = \cup A_n = \lim A_n$; $c(A_n) < n$; thus S_d is finite or countable.

2.1 Monotone with $G(-\infty) = 0$, $G(+\infty) = 1$; $g(y) = 2y$ on (0, 1) and zero otherwise; $1/\sqrt{2}$. **2.2** 0.866, 0.5, 0.159; $0.01e^{-1/2}/\sqrt{2\pi}$. **2.4** 0.325, 0.0116. **2.6** $g(z) \ge 0$, $\int g(z)\,dz = 1$; $G(z) = 1 - e^{-z}$ on $(0, \infty)$ and 0 otherwise; ln 2. **2.8** $g(z) \ge 0$, $\int g(z)\,dz = \tan^{-1} z|_{-\infty}^{\infty}/\pi = 1$; $G(z) = 1/2 + (\tan^{-1} z)/\pi$; $(1 + (y-\mu)^2/\sigma^2)^{-1}/\pi\sigma$. **2.10** $g(z) \ge 0$, $\int g(z)\,dz = 1$; $G(z) = 1 - (1+z)^{-\alpha}$ on $(0, \infty)$ and 0 otherwise; $2^{1/\alpha} - 1$. **2.14** $G(z) = \exp\{-e^{-z}\}$; $-\ln \ln 2$. **2.16** $\Gamma(m/2) \int g(\chi^2)\,d\chi^2 = \int_0^\infty e^{-\chi^2/2}(\chi^2/2)^{m/2-1}\,d\chi^2/2 = \Gamma(m/2)$ with $m > 0$; $P(\chi^2/2 \le t) = \int_{\chi^2/2 \le t} e^{-\chi^2/2}(\chi^2/2)^{m/2-1}\,d\chi^2/2 = \int_0^t h(u)\,du$ with $p = m/2$ and $t > 0$. **2.21** $\Gamma(1) = \int_0^\infty e^{-t}\,dt = 1$; $\Gamma(p+1) = -e^{-t}t^p|_0^\infty + p\int_0^\infty e^{-t}t^{p-1}\,dt = p\Gamma(p)$; substitute $t = u^2/2$ in $\Gamma(1/2)$. **2.22** $|\partial(x, y)/\partial(t, u)|_+ = t$, $\Gamma(p)\Gamma(q) = \int_0^\infty \int_0^1 e^{-t}t^{p+q-1}u^{p-1}(1-u)^{q-1}\,dt\,du = \Gamma(p+q)B(p, q)$.

3.1 $(a_1, b_1]_1 F_1 (a_2, b_2]_2 F_2$; $(\mathbf{a}, \mathbf{b}]F = (\ge 0)(\ge 0) \ge 0$, $F(\mathbf{y} + \mathbf{0}) = F_1(y_1 + 0)F_2(y_2 + 0) = F_1(y_1)F_2(y_2)$; $F(+\infty) = 1 \cdot 1 = 1$; $F(-\infty, y_2) = F(y_1, -\infty) = 0$. **3.4** c^2 for $c < 1$ and 1 for $c \ge 1$. **3.6** $f \ge 0$, $\int f(y_1, y_2)\,dy_1\,dy_2 = \int_0^\infty e^{-y_1/6}\,dy_1/6 \cdot \int_0^\infty e^{-y_2}\,dy_2 = 1$. **3.8** c; $\pi c^2/4$. **3.10** $f = 1$ on $(0, 1)^2$ and 0 otherwise; $2r$ if $r < 1/2$ and 1 if $r \ge 1/2$; $4r^2$ if $r < 1/2$ and 1 if $r \ge 1/2$; $1 - 4r + 4r^2$ if $r < 1/2$ and 0 if $r \ge 1/2$. **3.13** $\int_{-\infty}^{\infty} g(z_1, z_2)\,dz_2 = (\Gamma((n+1)/2)/\pi^{1/2}\Gamma(n/2)) \cdot [(\Gamma((n+2)/2)/\pi^{1/2}\Gamma((n+1)/2)) \int_{-\infty}^{\infty} (1 + t_2^2)^{-(n+2)/2}\,dt_2 \cdot (1 + z_1^2)^{-(n+1)/2}$, where $t_2 = z_2/(1 + z_1^2)^{1/2}$.

4.1 $F_W(w) = w$ on (0, 1). **4.4** $p_W(w) = 1/6$ on 1, 4, 9, 16, 25, 36 and 0 otherwise. **4.5** $\lambda \exp\{-\lambda A\}$ for $A > 0$ and 0 otherwise. **4.7** e^y for $y < 0$ and 1 otherwise. **4.9** $F_Y(y) = 1 - (1 + Z(y))^{-\alpha} = 1 - (1+y)^{-\alpha/p}$ for $y > 0$. **4.11** $F_Y(y) = 1 - G(-\ln y) = 1 - e^{-y}$ for $y > 0$. **4.24** $\sigma_{11} = \sigma_1^2$, $\sigma_{12} = \rho\sigma_1\sigma_2$, $\sigma_{22} = \sigma_2^2$; $\sigma^{11} = 1/\sigma_1^2(1-\rho^2)$, $\sigma^{12} = -\rho/\sigma_1\sigma_2(1-\rho^2)$, $\sigma^{22} = 1/\sigma_2^2(1-\rho^2)$.

5.1 $\cos^{-1}(1/2)$, $\cos^{-1}(\sqrt{5}/\sqrt{6})$, $\cos^{-1}(15/\sqrt{354})$, $\cos^{-1}(100/\sqrt{30}\sqrt{354})$. **5.2** $4I$, where I is 4×4 identity matrix. **5.4** (−3/2, −1/2, 1/2, 3/2). **5.6** $\mathbf{x}'M\mathbf{x} = n^{-1}\mathbf{x}'\mathbf{1}\mathbf{1}'\mathbf{x} = (\Sigma x_i)^2/n = n\bar{x}^2$.

CHAPTER 3: MARGINAL PROBABILITY

1.1 $p(s) = 1/8$ for $s = (i_1, i_2, i_3)$ with $i_j = 0, 1$. $P_Y(\{y\}) = 1/8, 3/8, 3/8, 1/8$ for $y = 0, 1, 2, 3$; e.g., note that $P_Y(\{1\}) = P(Y^{-1}(1)) = P(\{(1, 0, 0), (0, 1, 0), (0, 0, 1)\}) = 1/8 + 1/8 + 1/8$. **1.3** $F_Y(y) = y^2/2$, $1 - (2-y)^2/2$ for $0 < y < 1$, $1 < y < 2$; $f_Y(y) = y$, $2 - y$ for $0 < y < 1$, $1 < y < 2$ and 0 otherwise. **1.4** The line $y_1 + y_2 = c$; the circle $y_1^2 + y_2^2 = d$; preimage of (c, d) is $\{(c/2 \pm \sqrt{2d - c^2}/2,\ c/2 \mp \sqrt{2d - c^2}/2)\}$, which has two points provided that $2d > c^2$; preimage of (c', d') is $\{(c', d'), (d', c')\}$, which has two points provided that $c' < d'$. **1.6** $\bar{y} = c$ is a plane, $y_1 + y_2 + y_3 = 3c$ perpendicular to the vector $\mathbf{1} = (1, 1, 1)'$; $s_y^2 = d$ is a cylinder of radius $\sqrt{n-1}\sqrt{d}$ with axis the line $\mathfrak{L}(\mathbf{1})$. $(\bar{y}, s_y^2) = (c, d)$ is the intersection of the preceding, which gives a circle of radius $\sqrt{n-1}\sqrt{d}$ centered at (c, c, c) on the plane through (c, c, c) perpendicular to **1**. **1.8** $s \in Y^{-1}(\cup B_i) \leftrightarrow Y(s) \in \cup B_i \leftrightarrow Y(s) \in B_j$ for some $j \leftrightarrow s \in Y^{-1}(B_j)$ for some $j \leftrightarrow s \in \cup Y^{-1}(B_j)$. **1.10** Consider a rotation of axes through angle $\pi/4$; $F_W(w) = G(w/\sqrt{2})$, where G is the $N(0, 1)$ distribution function; $N(0, \sqrt{2})$.

2.1 H_1, H_2. **2.2** 0.10, 0.45, 0.80, 1.00; 0.20, 0.50, 0.80, 1.00. **2.3** $F_1(y_1) = 1-(1+y_1)^{-1}$, $F_2(y_2) = 1-(1+y_2)^{-1}$ on positive axes; $f_1(y_1) = (1+y_1)^{-2}$, $f_2(y_2) = (1+y_2)^{-2}$ on positive axes. **2.6** $F_1(y_1) = (G(y_1) + H(y_1))/2$, $F_2(y_2) = 0, 1/2, 1$ on $y_2 < 0$, $0 \le y_2 < 1$, $1 \le y_2 < \infty$. **2.9** $G(w) = 0, w, 1$ for $w < 0$, $0 \le w < 1$, $1 \le w$. $H(r) = 0, r^2, 1$ for $r < 0$, $0 \le r < 1$, $1 \le r$. **2.11** $F_1(y_1) = 1-(1+y_1)^{-\alpha}$, $F_2(y_2) = 1-(1+y_2)^{-\alpha}$ on positive axes. **2.12** $p(x_1, x_2) = 1/12$ at the 12 possible points. $p_Y(1) = p_Y(7) = 1/12$, $p_Y(y) = 1/6$ otherwise. **2.14** $\int_{y_1<y_2} f(y_1, y_2)\, dy_1\, dy_2 = \int_{y_1<y_2} f(y_2, y_1)\, dy_1\, dy_2 = \int_{y_1>y_2} f(y_1, y_2)\, dy_1\, dy_2$ and hence each is 1/2. **2.17** $f_1(y_1) = 2e^{-2y_1}$, $f_2(y_2) = 2(e^{-y_2} - e^{-2y_2})$ on positive axes. **2.18** $f_1(y_1) = 2\sqrt{(1-y_1^2)}/\pi$ on $(-1, +1)$ and zero otherwise, $f_2 = f_1$. **2.21** $f_i(y_i) = \int_0^\infty \alpha(\alpha+1)(1+y_i+y_j)^{-\alpha-2}\, dy_j = \alpha(1+y_i)^{-\alpha-1}$ on positive axis. **2.24** Integrate out y_k from 0 to ∞; last factor of constant cancels out; power increased by 1. **2.27** See Section 3.6B. **2.30** $|\partial(x, y)/\partial(v, y)| = y$.

3.1 Independence implies factorization (Proposition 6). Factorization gives $F_X(x) = g(x)h(+\infty)$, $F_Y(y) = g(+\infty)h(y)$; but $1 = g(+\infty)h(+\infty)$ so that $F_X(x)F_Y(y) = g(x)h(y)$, which is $F_{XY}(x, y)$. **3.2** Independence implies factorization (Proposition 7). Factorization gives $p_X(x) = g(x)\sum_y h(y)$, $p_Y(y) = \sum_x g(x)h(y)$; but $1 = \sum_x g(x) \sum_y h(y)$, so that $p_X(x)p_Y(y) = g(x)h(y)$, which is $p_{XY}(x, y)$. **3.5** Add chi-squares one at a time and correspondingly add degrees of freedom. **3.7** $216p_W(w) = 1, 3, 6, 10, 15, 21, 25, 27, 27, 25, 21, 15, 10, 6, 3, 1$ at $w = 3, 4, \ldots, 18$. **3.12** $|\partial(u_1, u_2)/\partial(z_1, z_2)| = \exp\{-(z_1^2 + z_2^2)/2\}/2\pi$; $f(z_1, z_2) = g(u_1, u_2)\,|\partial(u_1, u_2)/\partial(z_1, z_2)|$.

4.4 $(1+1)^n = 2^n$; $(1-1)^n = 0$. **4.12** $1/6^5$, $1/2^5$, $5^5/6^5$. **4.14** $1/2^5$, $10/2^5$. **4.16** $\binom{13}{5}4^5\Big/\binom{52}{5}$. **4.20** $\binom{n}{y_1 \cdots y_k}k^{-n}$.

5.1 1/8, 3/8, 3/8, 1/8 at 0, 1, 2, 3. **5.3** As for Exercise 5.1. **5.5** 1/27 at (3, 0, 0), (0, 3, 0), (0, 0, 3), 3/27 at (2, 1, 0), (1, 2, 0), (2, 0, 1), (1, 0, 2), (0, 2, 1), (0, 1, 2). 6/27 at (1, 1, 1). **5.7** As for Exercise 5.5. **5.11** $\binom{5}{y_1}\binom{5}{y_2}p_1^{y_1}(1-p_1)^{5-y_1}p_2^{y_2}(1-p_2)^{5-y_2}$ for $y_i = 0, 1, 2, 3, 4, 5$. **5.13** $20p(y) = 1, 9, 9, 1$ at 0, 1, 2, 3. **5.15** As for Exercise 5.13. **5.16** $\binom{4}{y}\binom{48}{13-y}\Big/\binom{52}{13}$ at $y = 0, 1, 2, 3, 4$. **5.18** 3/10, 3/10, 3/10, 1/10 at (1, 1, 1), (1, 0, 2), (0, 1, 2), (0, 0, 3). **5.20** $\binom{13}{x}\binom{13}{y}\binom{13}{z}\binom{13}{w}\Big/\binom{52}{13}$. **5.22** $\binom{D}{y}\binom{100-D}{5-y}\Big/\binom{100}{5}$ for $y = 0, 1, 2, 3, 4, 5$. $Q(D) = (100-D)^{(5)}/100^{(5)}$. **5.26** Note that $\binom{n}{y}\Big/\binom{n}{y-1} = (n-y+1)/y$. **5.29** $p_T(t) = p^t(1-p)^{n_1+n_2-t}\sum_y \binom{n_1}{t-y}\binom{n_2}{y}$. **5.34** $p_T(y) = e^{-\lambda_1-\lambda_2}(1/t!)\sum_y \binom{t}{y}\lambda_1^{t-y}\lambda_2^y$. **5.37** $p_1(y_1) = \binom{n}{y_1}p_1^{y_1}\sum_{y=0}^{n-y_1}\binom{n-y_1}{y}p_2^y p_3^{n-y_1-y}$. **5.49** $\binom{10}{y_0\, y_1 \cdots y_4}\prod p_i^{y_i}$ for $\sum y_i = 10$, where $p_i = \binom{4}{i}\binom{48}{13-i}\Big/\binom{52}{13}$.

6.10 $|\partial(u_{(1)}, \ldots, u_{(n)})/\partial(c_1, \ldots, c_n)| = 1$. **6.12** $|\partial(C_1, \ldots, C_k)/\partial(U_1, \ldots, U_k)| = 1$; $C_i = U_i - U_{i-1}$.

CHAPTER 4: CONDITIONAL PROBABILITY

1.1 1/3. **1.2** 0. **1.3** 1/2. **1.4** $\left[\binom{13}{x}\binom{39}{13-x}\binom{13-x}{y}\binom{26+x}{13-y}\Big/\binom{52}{13}\binom{39}{13}\right]\Big/\left[\binom{13}{4}\binom{39}{22}\Big/\binom{52}{26}\right]$. **1.7** Disjoint A_i, then $P_C(\cup A_i) = P(\cup A_iC)/P(C) = \sum P(A_iC)/P(C) =$

$\Sigma P_C(A_i)$; $P_C(\mathcal{S}) = P(\mathcal{S}C)/P(C) = 1$. **1.9** $365^{(n)}/365^n$; $(1 - \Sigma_0^{n-1} i/365) = (730 + n - n^2)/730$ or, better, $\exp\{-\Sigma_0^{n-1} i/365\} = \exp\{-n(n-1)/730\}$; 23 by direct calculation. **1.11** 244/495.

2.1 (405, 1080, 1080, 480, 80)/3125 at (0, 1, 2, 3, 4), respectively; (243, 810, 1080, 720, 240, 32)/3125 at (0, 1, 2, 3, 4, 5), respectively. **2.3** 3/10, 6/10, 1/10 at 0, 1, 2; 1/10, 6/10, 3/10 at 0, 1, 2. **2.5** Number of sequences of three defectives is $3^{(3)}$, of two good articles is $7^{(2)}$; the number of ways of interlacing so that the defective is last is $\binom{4}{2}$. **2.7** $\Pi(\lambda_i^{y_i} e^{-\lambda_i}/y_i!)/\lambda^y e^{-\lambda}/y! = \binom{y}{y_1 \cdots y_k} \Pi(\lambda_i/\lambda)^{y_i}$. **2.10** $6e^{-6y_2}$ on $(0, \infty)$. **2.12** $1/2\sqrt{(1 - y_1^2)}$ for y_2 in $(\pm\sqrt{(1 - y_1^2)})$ and zero otherwise. **2.15** $y_2/(1 + y_1)$ is Pareto($\alpha + 1$). **2.17** Uniform on parallelogram((0, 0), (1, 0), (2, 1), (1, 1)); uniform(0, t) for $0 < t < 1$ and uniform($t - 1$, 1) for $1 < t < 2$. **2.21** $(\Gamma(r_{p+1} + \cdots + r_{k+1})/\Gamma(r_{p+1}) \ldots \Gamma(r_{k+1}))(y_{p+1}/Y)^{r_{p+1}-1} \cdots (y_k/Y)^{r_k-1}((Y - \Sigma_{p+1}^k y_i)/Y)^{r_{k+1}-1}/Y^{k-p}$ on $0 < y_i$, $\Sigma_{p+1}^k y_i < Y$. **2.25** Number of sequence of r defectives $D^{(r)}$, of $n - r$ good is $(N - D)^{(n-r)}$, number of interlacings is $\binom{n}{r}$. **2.29** "$r - 1$ defectives in first $n - 1$" and "defective on nth."

3.1 $p(n) = q^{n-1}p$ with $p = 1 - q = 2/9$; see Example 1.4.4; 14/32, 98/386; P(win on nth) $= q^{n-1}(1/18)$, P(lose on nth) $= q^{n-1}(3/18)$. **3.3** $\Pi_1^n (\lambda e^{-\lambda(w_i - w_{i-1})}) e^{-\lambda(t - w_n)}/((\lambda t)^n e^{-\lambda t}/n!)$. **3.4** $\Pi_1^{n+1}(\lambda e^{-\lambda(w_i - w_{i-1})})/\Gamma^{-1}(n+1)(\lambda t)^n e^{-\lambda t}$. **3.6** $p_2(1, j) = 0.64, 0.21, 0.15$, $p_2(2, j) = 0, 0.67, 0.33$; $p_2(3, j) = 0, 0.66, 0.34$; $p_1(j) = 0.32, 0.43, 0.25$; $p_2(j) = 0.256, 0.483, 0.261$. **3.10** e^{-2}; e^{2a}; $F_W(w) = e^{2w}$ for $w < 0$ and 1 otherwise.

CHAPTER 5: THE MEAN VALUE FOR REAL AND VECTOR DISTRIBUTIONS

1.1 \$0.60, −\$0.40. **1.3** 3/4, 3/5. **1.5** $\Sigma_x 1 = \infty$.

2.1 2. **2.3** 2. **2.5** 2. **2.7** 2/5. **2.8** (1/2, 1, 3/2). **2.10** (1, 1). **2.11** $E(y) = \Sigma_0^\infty jp(j) = \Sigma_1^\infty p(j) + \Sigma_2^\infty p(j) + \cdots$. **2.13** $\int_A \phi(y)\, dP(y) \geq \int_A k\, dP(y)$. **2.15** $\Sigma |a_i - t|$ is continuous in t; $\Sigma |a_i - (t + \delta)| - \Sigma |a_i - (t - \delta)| = 2\delta$(number of $a_i < t$ less number of $a_i > t$) provided there is no a_i in interval $(t \pm \delta)$.

3.1 1. **3.3** 4/7. **3.6** 1/25. **3.7** (1/20)(5, −2, −3 / −2, 8, −6 / −3, −6, 9), 1/20(−2, −3).

3.29 $h(t) = E((x - t)^2)$, $h'(t) = 2t - 2\mu$, $h''(t) = 2$. **3.30** 50,000. **3.32** $\mathbf{t}'\Sigma\mathbf{t}$ is a variance, ≥ 0; hence Σ positive semidefinite. **3.36** $E(x_i x_j) = 0$ for $i \neq j$; $E(x_i^2) = E(r^2)/k = k/k$.

4.1 $\bar{y} = a + c\bar{x}$, $y_i - \bar{y} = c(x_i - \bar{x})$, $\Sigma(y_i - \bar{y})^2 = c^2 \Sigma(x_i - \bar{x})^2$. **4.3** 16.8, 2.6725, 1.6348. **4.5** (3, 9.2), (2.5, 4.75 / 4.75, 10.2), 0.9406. **4.7** $\Sigma \mathbf{y}_i = n\mathbf{a} + C\Sigma\mathbf{x}_i$, $\bar{\mathbf{y}} = \mathbf{a} + C\bar{\mathbf{x}}$; $\mathbf{y}_i - \bar{\mathbf{y}} = C(\mathbf{x}_i - \bar{\mathbf{x}})$, $(\mathbf{y}_i - \bar{\mathbf{y}})(\mathbf{y}_i - \bar{\mathbf{y}})' = C(\mathbf{x}_i - \bar{\mathbf{x}})(\mathbf{x}_i - \bar{\mathbf{x}})'C'$, $S_{\mathbf{yy}} = CS_{\mathbf{xx}}C'$. **4.10** $(1/(n+1), n(n+1))$, $(n, 1 / 1, n)/(n+1)^2(n+2)$. **4.12** $(1, 2, \ldots, n)/(n+1)$, $\operatorname{cov}(u_{(r)}, u_{(r+s)}) = r(n - r - s + 1)/(n+1)^2(n+2)$.

5.1 $2(1 - y)/3$, $(1 - y)^2/18$. **5.2** $1/4 = E((1 - x)/2)$, $3/80 = E((1 - x)^2/12) + \operatorname{var}((1 - x)/2)$. **5.5** $u_{(r)} + s(1 - u_{(r)})/(n - r + 1)$, $s(n - r - s + 1)(1 - u_{(r)})^2/(n - r + 1)^2(n - r + 2)$. **5.7** $I = \Sigma^{22}\Sigma_{22} + \Sigma^{21}\Sigma_{12} = \Sigma^{22}\Sigma_{22} + (\Sigma^{21}\Sigma_{11} = -\Sigma^{22}\Sigma_{21})\Sigma_{11}^{-1}\Sigma_{12}$.

6.1 $2/(r + 2)$. **6.3** $\mu_r = p$, $\bar{\mu}_r = q^r p + (-p)^r q$. **6.6** $(N + 1)^{(r+1)}/N(r + 1)$. **6.7** m_r is the sample average for $y = x^r$.

7.1 $(e^t + e^{2t} + e^{3t})/3$, $(1 + 2 + 3)/3$, $(1 + 4 + 9)/3$. **7.2** $(1 + 4e^t + 6e^{2t} + 4e^{3t} + e^{4t})/16 = (1 + e^t)^4/16$. **7.3** $\int_0^1 e^{ut}\, du = (e^t - 1)/t$, 1/2. **7.7** If independent, then "mean product = product mean" gives $c(t_1, t_2) = c_1(t_1)c_2(t_2)$. If $c(t_1, t_2) = g(t_1)h(t_2)$, then $c_1(t_1) = g(t_1)h(0)$, $c_2(t_2) = g(0)h(t_2)$, $1 = g(0)h(0)$; the joint distribution calculated from independence has characteristic function $g(t_1)h(0)g(0)h(t_2) = g(t_1)h(t_2)$, which equals $c(t_1, t_2)$; then use the uniqueness result cited in Section D. **7.29** $\Pi(q + pe^t)^{n_i} = (q + pe^t)^{\Sigma n_i}$. **7.31** $m(t) = \int_{-\infty}^{\infty} (1/\sqrt{2\pi}) e^{-x^2(1-2t)/2}\, dx = (1 - 2t)^{-1/2}$. **7.33** $\Pi(1 - 2t)^{-f_i/2} = (1 - 2t)^{-\Sigma f_i/2}$. **7.35** For $\Sigma_1^n x_i$, $m(t) = \Pi \exp\{\mu t + \sigma^2 t^2/2\}$; for $\bar{x}$, $m(t) = \exp\{n\mu t/n + n\sigma^2 t^2/2n^2\}$. **7.36** For $\Sigma_1^n \mathbf{x}_i$,

$m(\mathbf{t}) = \Pi \exp\{\boldsymbol{\mu}'\mathbf{t} + \mathbf{t}'\Sigma\mathbf{t}/2\}$; for $\bar{\mathbf{x}}$, $m(\mathbf{t}) = \exp\{n\boldsymbol{\mu}'\mathbf{t}/n + n\mathbf{t}'\Sigma\mathbf{t}/2n^2\}$. **7.38** Moment-generating function for $\mathbf{x}$ is $m_{\mathbf{x}}(\mathbf{a}) = m_{\mathbf{a}'\mathbf{x}}(1)$, which is determined by the distribution of $\mathbf{a}'\mathbf{x}$. **7.41** $\exp\{\mu_1 t_1 + \mu_2 t_2 + (\sigma_1^2 t_1^2 + 2\rho\sigma_1\sigma_2 t_1 t_2 + \sigma_2^2 t_2^2)/2\}$.

CHAPTER 6: LIMITING DISTRIBUTIONS AND LIMITING FUNCTIONS

1.1 0.706. **1.2** $\ln m(t) = -\sqrt{m/2}\,t - (m/2)\ln(1 - t/\sqrt{m/2}) = -\sqrt{m/2}\,t - (m/2)[-t/\sqrt{m/2} - (1/2)t^2/(m/2)\cdots] \to t^2/2$. **1.4** $P(|y_n - c| \le \epsilon) \to 1$ implies that $P(y_n < c - \epsilon) \to 0$, $P(y_n \le c + \epsilon) \to 1$. **1.7** Density for $u_{(1)}$ is $n(1 - u_{(1)})^{n-1}$, for $nu_{(1)} = y$ is $(1 - y/n)^{n-1} \to e^{-y}$ for y on $(0, \infty)$. **1.9** The uniform distribution for u spreads the discrete probability $\lambda^y e^{-y}/y!$ at y uniformly over $(y \pm 1/2)$ with density as given; $(w - \lambda)/\lambda^{1/2} = (y - \lambda)/\lambda^{1/2} + u/\lambda^{1/2}$, final term $\xrightarrow{\mathfrak{D}} 0$, and thus same limiting distribution by Theorem 4.

2.1 9604. **2.3** m_s is sample average for y^s. **2.15** $y_i - \mu$ has mean 0; Weak Law for $y_i - \mu$ says that $\bar{y}_n - \mu \xrightarrow{\mathfrak{D}} 0$, that is $\bar{y}_n \xrightarrow{\mathfrak{P}} \mu$.

CHAPTER 7: STATISTICAL INFERENCE

1.1 0.1392, (0.1392, 0.1392, 0.1392, 0.1392, 0.1392), (−0.0012, 0.0008, −0.0032, 0.0028, 0.0008), 0.0000052. **1.2** 21, (21, 21, 21, 21, 21, 21), (−2.7, −1.7, 0.2, 1.4, −3.9, 6.7), 14.456. **1.3** 90.5857, 90.5857 (1, 1, 1, 1, 1, 1, 1), (−2.8857, −3.3857, 1.7143, 3.6143, 3.0143, −4.9857, 2.9143), 13.0381.

2.1 $z = 5.3/0.775 = 6.839$, a very extreme value for the standard normal; subject to the assumptions, this is strong evidence against the hypothesis. (285.3 ± 1.52). (285.3 ± 2.00). **2.2** $z = 4.696$, an extreme value for the standard normal; subject to the assumptions, this is strong evidence against the hypothesis. **2.3** (21 ± 2.00). **2.4** $t = 0.5857/1.3648 = 0.429$, a reasonable value for the standard normal, no evidence against the hypothesis. (90.59 ± 2.73). **2.5** $t = -4.077$, (3.921 ± 0.333), (3.921 ± 0.500). **2.6** $z = 2.920$, a value beyond the 1% point of the standard normal; subject to the assumptions, this is moderately strong evidence against the hypothesis, (3.20 ± 2.15). **2.7** $t = 0.2573/0.2627 = 0.979$, a reasonable value for the standard normal, no evidence against the hypothesis. (0.257 ± 0.79). **2.8** As sample sizes are equal, test is same as in preceding answer. (0.257 ± 0.53). **2.9** $(\bar{y}_2 - \bar{y}_1 \pm 3(s_1^2/m + s_2^2/n)^{1/2})$. **2.10** $(\bar{y}_2 - \bar{y}_1 \pm 2(1/m + 1/n)^{1/2}((m - 1)s_1^2 + (n - 1)s_2^2)^{1/2}/(m + n - 2)^{1/2})$. **2.11** Without loss of generality, take $\mu = 0$; use $(n - 1)^2 s_y^4 = (\Sigma y_i^2 - n\bar{y}^2)^2$ and take the mean value.

3.1 (1452 ± 94.67). **3.3** 871.5, (871.5 ± 220.5). **3.5** (1325 ± 306).

4.1 $K = 5$, 6.25%, (0.136, 0.142). **4.2** $K = 1$, 21.9%, (17.1, 27.7). **4.3** $T = 0.021$, 6.25%, (0.136, 0.142). **4.4** $T = -36$, 6.25%, (17.1, 27.7). **4.7** $k = 3.5$, $p^7 + (1 - p)^7$. **4.8** $r_i =$ (number of $y_j < y_i$) $+ 1 = \Sigma_1^n c(y_j - y_i) + 1$. **4.10** 6; 2048(probability) = 1, 1, 1, 2, 2, 3, 4 at 0, 1, 2, 3, 4, 5, 6; observed level of significance is 28/2048 = 1.36%; some evidence against hypothesis. $(6 - 33)/11.25 = -2.4$; observed level of significance is 1.64%. **4.13** (3.385, 4.61). **4.16** $S = 4$. $(4 - 5.5)/1.20 = -1.25$; no evidence against hypothesis. **4.19** $U = 75$ or 76, $z = 0.95$; no evidence against hypothesis. **4.23** Observed $\Sigma y_{2j} = 86.0$; 20 possible values Σy_{2j} are 86.1, 86.0, . . ., 80.8, 80.7; observed level of significance is 20%; no evidence against hypothesis.

5.1 (463.5 ± 18.83). Sample number 19 above UCL, sample number 14 below LCL, process not in control. Nine samples outside control limits based on specifications; does not conform to specifications.

CHAPTER 8: THE LIKELIHOOD FUNCTION IN STATISTICAL INFERENCE

1.2 $c\exp\{-6.3/\theta\}/\theta^3$, $-6.3\theta^{-1} - 3\ln\theta + a$, $6.3\theta^{-2} - 3\theta^{-1}$; $\hat{\theta} = 2.1$. **1.4** $c[\exp\{-9.4 + 3\theta\}, \exp\{-5.2 + \theta\}, \exp\{1.6 - \theta\}, \exp\{9.4 - 3\theta\}$ on $(-\infty, 2.1)$,

(2.1, 3.4), (3.4, 3.9), (3.9, ∞)]; $\widehat{\theta} = 3.4$; (2.1, 45.0, 3.4), for example. **1.6** $c(\sigma^2)^{-n/2} \exp\{-\Sigma (y_i - \mu_0)^2/2\sigma^2\}$, $-n \ln(\sigma^2)/2 - \Sigma (y_i - \mu_0)^2/2\sigma^2 + a$, $-n/2\sigma^2 + \Sigma (y_i - \mu_0)^2/2\sigma^4$; $\Sigma (y_i - \mu_0)^2/n$; the likelihood function is seen to be a function of $\Sigma (y_i - \mu_0)^2$. **1.8** $c \exp\{-305/\theta\}/\theta^7$; 43.57.

2.1 $(\mathbf{u}_1, \mathbf{u}_2) = \mathbf{u}_1'\mathbf{u}_2 = \mathbf{y}_1'CC'\mathbf{y}_2 = \mathbf{y}_1'\mathbf{y}_2 = (\mathbf{y}_1, \mathbf{y}_2)$. **2.3** $2s_y^2 = (y_1 - y_2)^2/2 + (y_1 + y_2 - 2y_3)^2/6$; (9.5724, −0.1980, 0.2693). **2.5** $u_1 = \Sigma y_{ij}/\sqrt{5} = \sqrt{5}\,\bar{y}$, $u_2 = 2\,\Sigma y_{1j}/\sqrt{30} - 3\,\Sigma y_{2j}/\sqrt{30} = 6(\bar{y}_1 - \bar{y}_2)/\sqrt{30}$; for a sample of 2, $\Sigma (y_{1j} - \bar{y}_1)^2 = (y_{11} - y_{12})^2/2 = u_2^2$ (see Exercise 3); for a sample of 3, $\Sigma (y_{2j} - \bar{y}_2)^2 = u_3^2 + u_4^2$ (see Exercise 3). **2.6** First sample independent of second sample; hence $(\bar{y}_1, \Sigma (y_{1j} - \bar{y}_1)^2)$ independent of $(\bar{y}_2, \Sigma (y_{2j} - \bar{y}_2)^2)$; then use the result as quoted using the reproductive property of chi-square. **2.9** $\sigma_2^2F/\sigma_1^2 = (\sigma_2^2/\sigma_1^2)(\sigma_1^2\chi_1^2/(m-1))/(\sigma_2^2\chi_2^2/(n-1)) = (\chi_1^2/(n-1))/(\chi_2^2/(n-1))$, which is F, as quoted.

3.1 $z = (61.5 - 58.0)/(2.5/\sqrt{10}) = 4.43$; a very extreme value for the standard normal; strong evidence against the hypothesis; (61.5 ± 1.55), (61.5 ± 2.04). **3.2** $\chi^2 = 76/4 = 19$; beyond 1% point on the right-hand tail; moderately strong evidence against the hypothesis. $(\sqrt{76}/\sqrt{14.45}, \sqrt{76}/\sqrt{1.237}) = (2.29, 7.84)$. **3.3** $t = 4.461$; observed level is 0.5%; moderately strong evidence against the hypothesis. (686 ± 3.291), (686 ± 4.99). **3.7** Compare S^2/σ_0^2 with chi-square$(m + n - 2)$ distribution. $(S^2/\chi^2_{\alpha/2}, S^2/\chi^2_{1-\alpha/2})$, where χ^2_γ is γ point on the right tail of chi-square$(m + n - 2)$. **3.9** $t = [(\bar{y}_2 - \bar{y}_1 - (\mu_2 - \mu_1))/\sigma(1/m + 1/n)^{1/2}]/[s/\sigma]$, which has the form of [normal(0, 1)]/[chi$(m + n - 2)/\sqrt{(m + n - 2)}$] with independent square brackets; hence t, as cited. **3.12** Compare F with $F(m - 1, n - 1)$ distribution; $(s_1^2/s_2^2F_{\alpha/2}(m - 1, n - 1), s_1^2F_{\alpha/2}(n - 1, m - 1)/s_2^2)$.

4.1 $L(\mathbf{y}|\lambda) = c\lambda^s e^{-n\lambda}$ is determined by $s = \Sigma y_i$; different s values produce different functions of λ; thus s indexes the likelihood functions. **4.3** $L(\mathbf{y}|\sigma^2) = c(\sigma^2)^{-n/2} \exp\{-s/2\sigma^2\}$ is determined by $s = \Sigma (y_i - \mu_0)^2$; different s values produce different functions of σ^2, thus s is the likelihood statistic. **4.4** $L(\mathbf{y}|\theta) = c$ for $y_{(n)} - 1 < \theta < y_{(1)}$ and 0 for other θ values; thus $(y_{(1)}, y_{(n)})$ is the likelihood statistic. **4.8** For $(y_1, \ldots, y_n)$ with $y_i \neq y_j (i \neq j)$ the probability element for $(y_1, \ldots, y_n)$ is $\Pi f(y_i) \Pi dy_i$, and for $(y_{(1)}, \ldots, y_{(n)})$ is $n! \,\Pi f(y_{(i)}) \Pi dy_{(i)}$; but $\Pi f(y_i) = \Pi f(y_{(i)})$ and the Jacobian is 1 for one branch of mapping; thus the conditional distribution has probability $1/n!$ for each permutation of $(y_{(1)}, \ldots, y_{(n)})$, which clearly does not depend on f. **4.9** $L(\mathbf{x}|p) = cp^k(1 - p)^{s-k}$; these likelihood functions are indexed by $s = s(\mathbf{x}) = n$. **4.11** $L(\mathbf{y}_1, \ldots, \mathbf{y}_r|\mathbf{p}) = cp_1^{s_1} \cdots p_{k-1}^{s_{k-1}}(1 - p_1 - \cdots - p_{k-1})^{nr - s_1 - \cdots - s_{k-1}}$. These likelihood functions are indexed by $(s_1, \ldots, s_{k-1})$ or equivalently by $(s_1, \ldots, s_k)$ with $s_k = nr - s_1 - \cdots - s_{k-1}$, where $(s_1, \ldots, s_k)' = \mathbf{s} = \Sigma_1^r \mathbf{y}_j$. **4.14** $L(\mathbf{y}|\theta) = c(\theta_2 - \theta_1)^{-n}$ for $\theta_1 < y_{(1)}$, $y_{(n)} < \theta_2$ and zero otherwise; these likelihood functions are indexed by $(y_{(1)}, y_{(n)})$. **4.17** $L(\mathbf{y}|\theta) = c\theta^s(1 - \theta)^{n-s}$ with $s = y_1 + 2y_2 + 3y_3$; these likelihood functions are indexed by s. **4.21** The likelihood function (third form) from an observed $\mathbf{y}$ is given by the real number $L_3(\mathbf{y}) = \Pi f(y_i|\theta_2)/\Pi f(y_i|\theta_1)$. The likelihood function (third form) from $t = L_3(\mathbf{y})$ is (by the introductory remark in the problem) given by the real number $L_3(\mathbf{y})$, which is just t. Thus the likelihood function from an observed t is given by the identity function applied to t.

5.1 $f(x|p) = p^x q^{1-x} = q \cdot \exp\{x \ln(p/q)\}$; $\psi(p) = \ln(p/q)$; thus LS $= \Sigma x_i$. **5.3** $f(y|\theta) = \Gamma^{-1}(p)\theta^{-p} \exp\{y(-1/\theta)\}y^{p-1}$; $\psi(\theta) = -1/\theta$; thus Σy_i is LS. **5.7** $f(\mathbf{y}|\boldsymbol{\beta}, \sigma) = (2\pi\sigma^2)^{-n/2} \exp\{-\boldsymbol{\beta}X'X\boldsymbol{\beta}/2\sigma^2\} \cdot \exp\{\mathbf{y}'\mathbf{y}(-1/2\sigma^2) + \mathbf{y}'X \cdot \boldsymbol{\beta}/\sigma^2\}$; $(\psi_1, \ldots, \psi_{r+1}) = (-1/2\sigma^2, \beta_1/\sigma^2, \ldots, \beta_r/\sigma^2)$; thus $(\mathbf{y}'\mathbf{y}, \mathbf{y}'X)$ is LS.

6.1 $z = (0.5120 - 0.5000)/(1/25{,}600)^{1/2} = 1.92$; small indication against the hypothesis. (0.5120 ± 1.96(0.00625)) = (0.5120 ± 0.0122). **6.2** $y/\theta - 1$, $-y/\theta^2$, $1/\theta$. **6.3** $(113 \pm 1.96(1/113)^{-1/2}) = (113 \pm 20.8)$. **6.7** $y/\theta^2 - 1/\theta$, $-2y/\theta^3 + 1/\theta^2$, $1/\theta^2$. **6.12** $(S_1, S_2) = ((y - \mu)/\sigma^2, -1/2\sigma^2 + (y - \mu)^2/2\sigma^4)$, $\partial S_1/\partial\mu = -1/\sigma^2$, $\partial S_1/\partial\sigma^2 = -(y - \mu)/\sigma^4$, $\partial S_2/\partial\sigma^2 = 1/2\sigma^4 - (y - \mu)^2/\sigma^6$; $\mathbf{j} = (1/\sigma^2, 0 \,/\, 0, 1/2\sigma^4)$.

CHAPTER 9: ESTIMATION

1.1 $\widehat{\beta} = 2$; (2, 2, −2, −2), 16; (0, 2, 1, 1), 6. **1.2** (1, 1, −1, −1 / 1, 1, −1, −1 / −1, −1, 1, 1 / −1, −1, 1, 1)/4; (3, −1, 1, 1 / −1, 3, 1, 1 / 1, 1, 3, −1 / 1, 1, −1,

3)/4. **1.5** min $\Sigma (y_i - \lambda)^2$ gives $\bar{y}$; $\lambda = \bar{y}$ gives $\bar{y}$; $\Sigma y_i/\lambda - n = 0$ gives $\bar{y}$. **1.9** $(\hat{\beta}\mathbf{x})'(\hat{\beta}\mathbf{x}) = \hat{\beta}\mathbf{x}'\mathbf{x}\cdot\hat{\beta} = \Sigma y_i x_i \cdot \hat{\beta}$. **1.10** $\mathbf{x}$ is orthogonal to the residual $\mathbf{y} - \hat{\beta}\mathbf{x}$; $(\mathbf{y} - \hat{\beta}\mathbf{x})'(\mathbf{y} - \hat{\beta}\mathbf{x}) = \mathbf{y}'(\mathbf{y} - \hat{\beta}\mathbf{x}) = \mathbf{y}'\mathbf{y} - \mathbf{y}'\mathbf{x}\hat{\beta}$.

2.1 (4, 5, 6, 7, 8), 190; (−1, 2, 0, −2, 1), 10; 2, (4, 5, 6, 7, 8), 190 / 3, (−1, 2, 0, −2, 1), 10 / 5, (3, 7, 6, 5, 9), 200. **2.2** (1, 3, 5, 7, 9), 165; (1, 0, −2, 0, 1), 6; the same table as in Section C. **2.4**

$$P_1 = \frac{1}{10}\begin{pmatrix} 6 & 4 & 2 & 0 & -2 \\ 4 & 3 & 2 & 1 & 0 \\ 2 & 2 & 2 & 2 & 2 \\ 0 & 1 & 2 & 3 & 4 \\ -2 & 0 & 2 & 4 & 6 \end{pmatrix}, \qquad P_2 = \frac{1}{10}\begin{pmatrix} 4 & -4 & -2 & 0 & 2 \\ -4 & 7 & -2 & -1 & 0 \\ -2 & -2 & 8 & -2 & -2 \\ 0 & -1 & -2 & 7 & -4 \\ 2 & 0 & -2 & -4 & 4 \end{pmatrix}.$$

2.7 $(\bar{y}_1, \bar{y}_2, \bar{y}_3)$; $(\bar{y}_1, \ldots, \bar{y}_1, \bar{y}_2, \ldots, \bar{y}_2, \bar{y}_3, \ldots, \bar{y}_3)$, $m\bar{y}_1^2 + n\bar{y}_2^2 + p\bar{y}_3^2$; $(y_{11} - \bar{y}_1, \ldots, y_{1m} - \bar{y}_1, y_{21} - \bar{y}_2, \ldots, y_{2n} - \bar{y}_2; y_{31} - \bar{y}_3, \ldots, y_{3p} - \bar{y}_3)$, $\Sigma (y_{ij} - \bar{y}_i)^2$. **2.9** $(X\mathbf{b}_2 - X\mathbf{b}_1)'(X\mathbf{b}_2 - X\mathbf{b}_1) = (\mathbf{b}_2 - \mathbf{b}_1)'X'X(\mathbf{b}_2 - \mathbf{b}_1)$. **2.10** $(X'X) = (n, 0 \,/\, 0, \Sigma(x_i - \bar{x})^2)$, $X'\mathbf{y} = (\Sigma y_i, \Sigma y_i(x_i - \bar{x}))'$, $\mathbf{b} = (X'X)^{-1}X'\mathbf{y}$ = expression as given; $\mathbf{y}'\mathbf{y} - \mathbf{y}'X\mathbf{b} = \Sigma y_j^2 - \Sigma y_j(\bar{y} + (x_j - \bar{x})\,\Sigma y_i(x_i - \bar{x})/\Sigma (x_i - \bar{x})^2)$.

3.1 257.3459. **3.3** $(\mathbf{t} + \mathbf{u})/2$; $A\mathbf{t} + B\mathbf{u}$ with $A = (0.354318, 0.306698 \,/\, 0.306698, 0.354318)$, $B = (0.645682, -0.306698 \,/\, -0.306698, 0.645682)$. **3.4** $b = (\mathbf{x}'\mathbf{x})^{-1}\mathbf{x}'\mathbf{y}$, $(\mathbf{x}'\mathbf{x})^{-1}\sigma^2$. **3.12** Theorem 7 with the following notations:

$$E\begin{pmatrix}\mathbf{t}_1\\ \mathbf{t}_2\end{pmatrix} = \begin{pmatrix} I \\ I \end{pmatrix}\boldsymbol{\beta}, \qquad M = \begin{pmatrix} \Sigma_{11}(\theta_0) & 0 \\ 0 & \Sigma_{22}(\theta_0) \end{pmatrix}.$$

4.1 $t = p_0 + (n/p_0q_0)^{-1}(\Sigma x_i/p_0 - (n - \Sigma x_i)/q_0) = \bar{x}$; yes. **4.3** $\sigma_0^2 + (n/2\sigma_0^4)^{-1}(-n/2\sigma_0^2 + \Sigma (y_i - \mu_0)^2/2\sigma_0^4) = \Sigma (y_i - \mu_0)^2/n$; yes. **4.7** $S(y|\theta) = \phi'(\theta) + \psi'(\theta)t(y)$; $E(S) = 0$ gives $E(t) = -\phi'(\theta)/\psi'(\theta)$; an affine function of $S(y|\theta)$ that is independent of θ must be an affine function of t; thus the corresponding parameter is an affine function of $\phi'(\theta)/\psi'(\theta)$. **4.9** $(1/\theta + 1/(\theta + 1))$. **4.11** $(J_1(\boldsymbol{\theta}_0) + J_2(\boldsymbol{\theta}_0))^{-1}(J_1(\boldsymbol{\theta}_0)(\boldsymbol{\theta}_0 + J_1^{-1}(\boldsymbol{\theta}_0)\mathbf{S}_1(\mathbf{y}|\boldsymbol{\theta}_0)) + J_2(\boldsymbol{\theta}_0)(\boldsymbol{\theta}_0 + J_2^{-1}(\boldsymbol{\theta}_0)\mathbf{S}_2(\mathbf{y}|\boldsymbol{\theta}_0))) = \boldsymbol{\theta}_0 + (J_1(\boldsymbol{\theta}_0) + J_2(\boldsymbol{\theta}_0))^{-1}(\mathbf{S}_1(\mathbf{y}|\boldsymbol{\theta}_0)) + \mathbf{S}_2(\mathbf{y}|\boldsymbol{\theta}_0))$. **4.13**

$$\mathbf{t} = \begin{pmatrix}\mu_0 \\ \sigma_0^2\end{pmatrix} + \begin{pmatrix}\sigma_0^2/n & 0 \\ 0 & 2\sigma_0^4/n\end{pmatrix}\begin{pmatrix} n(\bar{y} - \mu_0)/\sigma_0^2 \\ -n/2\sigma_0^2 + \sum (y_i - \mu_0)^2/2\sigma_0^4\end{pmatrix} = \begin{pmatrix} \bar{y} \\ \sum (y_i - \mu_0)^2/n \end{pmatrix};$$

thus no UMV unbiased estimate at information lower bound; however, later we will determine a UMV unbiased estimate.

5.1 Distribution of $s = \Sigma y_i$ has density $\Gamma^{-1}(n)\theta^{-n}s^{n-1}\exp\{-s/\theta\}$; has exponential form; thus s is complete. $E(s/n) = E(\bar{y}) = \theta$; thus $\bar{y}$, a function of s, is UMV unbiased. **5.4** Distribution of y is binomial(n, p) with p in [0, 1]. $E(r(y)) = r(0)\binom{n}{0}q^n + r(1)\binom{n}{1}pq^{n-1} + \cdots \equiv 0$ for all p; with $p = 0$, obtain $r(0) = 0$; then divide out p and consider the limit as $p \to 0$, obtain $r(1) = 0$; continue. **5.5** $y_{(2)} - y_{(1)} - 1/3$. **5.9** $E(r(y)) = \Sigma\, r(y)\binom{D}{y}\binom{N-D}{n-y}\Big/\binom{N}{n}$ with y in $(\max(0, n - N + D), \min(n, D))$; take $D = 0$ and obtain $r(0) = 0$; then $D = 1$ and obtain $r(1) = 0$; continue. **5.13** Density for $(s_1, s_2) = (u_{(1)}, u_{(n)})$ is $n(n-1)(s_2 - s_1)^{n-2}/(\theta_2 - \theta_1)^n$ on (θ_1, θ_2) and zero otherwise; $E(r(s_1, s_2)) = 0$ implies that $\int_{\theta_1}^{\theta_2}[\int_{\theta_1}^{s_2} r(s_1, s_2)(s_2 - s_1)^{n-2}\,ds_1]\,ds_2 = 0$ $\theta_1 < \theta_2$; differentiate re θ_2, then re θ_1, and obtain $r(\theta_1, \theta_2)(\theta_2 - \theta_1)^{n-1} = 0$ for all $\theta_1 < \theta_2$. **5.17** Follow Example 4; $\int r^+(\mathbf{s})\exp\{\mathbf{s}'\boldsymbol{\theta}\}h(s)\,ds = \int r^-(\mathbf{s})\exp\{\mathbf{s}'\boldsymbol{\theta}\}h(s)\,ds$; then continue as Example 4, using uniqueness for multivariate moment-generating functions. **5.21** From Problem

9.2 we have $b = \Sigma (i - n/2)y_i/\Sigma (i - n/2)^2$. **5.23** Let $(s_1, \ldots, s_n) = (\Sigma y_i, \ldots, \Sigma y_i^n)$; then by Problem 17 $(s_1, \ldots, s_n)$ is complete. But the order statistic is equivalent to $(s_1, \ldots, s_n)$. **5.25** $\text{var}(at_1 + (1 - a)t_2) = a^2\sigma^2(\theta) + 2a(1 - a)\,\text{cov}(t_1, t_2) + (1 - a)^2\sigma^2(\theta) \geq \sigma^2(\theta)$, which implies that $\text{cov}(t_1, t_2) - \sigma^2(\theta) \geq 0$, which by covariance inequality implies that t_1 and t_2 are affinely related and hence equal.

6.1 $N(108, 4\sqrt{5})$; 108. **6.3** All probability for Bb. **6.5** $N((n\bar{y}/\sigma_0^2 + \nu_0/\tau_0^2)\omega^2, \omega)$, where $1/\omega^2 = n/\sigma_0^2 + 1/\tau_0^2$. **6.7** $\text{beta}(\alpha_0 + \Sigma x_i, \beta_0 + n - \Sigma x_i)$. **6.9** $N(\Omega(\Sigma_0^{-1}\mathbf{y} + M_0^{-1}\boldsymbol{\tau}_0); \Omega)$, where $\Omega^{-1} = \Sigma_0^{-1} + M_0^{-1}$.

CHAPTER 10: TESTING STATISTICAL HYPOTHESES

1.1 Critical values for $\bar{y} \geq 0.1372$; "reject at 5% level." **1.3** Critical values for $\bar{y} \leq 19.73$; "accept at 5% level." **1.5** Critical region is $\{\mathbf{y}: \bar{y} \geq 75 + 2.326 \times 0.3795 = 75.883\}$; $G((\theta - 75)/0.3795 - 2.326)$. **1.8** Critical region is $\{\mathbf{y}: \Sigma (y_i - \mu_0)^2 \geq \sigma_0^2\chi_\alpha^2(n)\}$. Test independent of $\sigma_1(>\sigma_0)$, $\mathcal{P}(\sigma) = 1 - H(\sigma_0^2\chi_\alpha^2(n)/\sigma^2)$, where H is the distribution function for chi-square(n). **1.10** Reject for large $t = \Sigma y_i$, that is, $\phi(y) = 1, a, 0$ according as $t = \Sigma y_i >, =, < k$, where $P(t > k) \leq \alpha < P(t \geq k)$ and $a = (\alpha - P(t > k))/[P(t \geq k) - P(t > k)]$ based on the Poisson($n\theta_0$). **1.14** Reject if $t = \max y_i > \theta_0$ and for t values in $(0, \theta_0)$ such that $\int_0^{\theta_0} \phi(t)nt^{n-1}\theta_0^{-n}\,dt = \alpha$. Test does not depend on $\theta_1(>\theta_0)$. **1.16** Let k be the α point on the right tail of the G distribution: $G(k) \geq 1 - \alpha \geq G(k - 0)$; choose $a = (G(k) - (1 - \alpha))/(G(k) - G(k - 0))$.

2.1 Critical region is $\{\mathbf{y}: \bar{y} \leq \theta - z_\alpha\sigma_0/\sqrt{n}\}$. **2.2** Critical region is $\{\mathbf{y}: 2\,\Sigma y_i \geq \theta_0\chi_\alpha^2(n)\}$. **2.5** (a) As in Problem 10.1.10; (b) Test function is $1 - \phi(\mathbf{y})$, where ϕ as in Problem 10.1.10 with size $1 - \alpha$. **2.7** $\phi(\mathbf{y}) = 1$ if $y_{(n)} > \theta_0$ or $<\theta_0\alpha^{1/n}$ and zero otherwise. **2.9** Note $\bar{y} = (\bar{y} - \mu) + \mu$; the first term is $N(0, \sigma_0^2/n)$; the second term under the mixture is $N(\mu_1, ((\sigma_1^2 - \sigma_0^2)/n)^{1/2})$; thus the "marginal" distribution of $\bar{y}$ is $N(\mu_1, \sigma_1/\sqrt{n})$, which is the same as under the alternative. The likelihood ratio then involves only the density ratio coming from $\Sigma (y_i - \bar{y})^2$. **2.13** $\mathcal{P}(\sigma_1, \sigma_2) = P((\Sigma y_{1j}^2/\sigma_1^2 m)/(\Sigma y_{2j}^2/\sigma_2^2 n) \geq \sigma_2^2 F_\alpha/\sigma_1^2) = 1 - H(\sigma_2^2 F_\alpha/\sigma_1^2)$.

3.1 $L(\mathbf{y}) = \exp\{n(\bar{y} - \theta_0)^2/2\sigma_0^2\}$; same distribution. **3.3** $L(\mathbf{y}_1, \mathbf{y}_2) = ((\Sigma (y_{1i} - \bar{y}_1)^2 + \Sigma (y_{2j} - \bar{y}_2)^2)/(m + n))^{(m+n)/2}(\Sigma (y_{1i} - \bar{y}_1)^2/m)^{-m/2}\,\Sigma (y_{2j} - \bar{y}_2)/n^{-n/2} = (F + (n - 1)/(m - 1))^{(m+n)/2}F^{-m/2}c(m, n)$. **3.6** $S(\mathbf{y}|\theta) = \gamma'(\theta)/\gamma(\theta) + \psi'(\theta)t(\mathbf{y})$, $E(S(\mathbf{y}|\theta)t(\mathbf{y})|\theta\} = 0 = E(\phi(\mathbf{y})\gamma'(\theta)/\gamma(\theta) + \phi(\mathbf{y})\psi'(\theta)t(\mathbf{y})|\theta\}$, then use $\gamma'(\theta)/\gamma(\theta)\psi'(\theta) = -E(t(\mathbf{y})|\theta)$ obtained from $E(S(\mathbf{y}|\theta)|\theta\} = 0$. **3.7** $z = \ln(\sigma_0/\sigma_1) - (1/2)(1/\sigma_0^2 - 1/\sigma_1^2)(y - \mu_0)^2$; reject if $\Sigma_1^n z_i \geq 2.944$, accept if $\Sigma_1^n z_i \leq -2.944$ and continue otherwise. **3.9** $z = y\ln(\theta_1/\theta_0) - (\theta_1 - \theta_0)$; reject if $\Sigma_1^n z_i \geq 4.595$, accept if $\Sigma_1^n z_i \leq -4.595$, and continue otherwise.

4.1 $\chi^2 = 4(15.94804) = 63.7922$ cf $\chi^2(5)$; $\chi_{0.005}^2(5) = 16.75$; observed value far out on the right tail; very strong evidence against the hypothesis. **4.5** $\chi^2 = 4(3.6616) = 14.646$ cf $\chi^2(6)$; $\chi_{0.025}^2(6) = 14.45$; moderate evidence against the hypothesis. **4.9** $\chi^2 = 4(3.3032) = 13.213$ cf $\chi^2(3)$; $\chi_{0.005}^2(3) = 12.84$; strong evidence against the hypothesis. $\chi^2 = 4(2.3658) = 9.463$; cf $\chi^2(2)$; $\chi_{0.01}^2(2) = 9.21$; strong evidence against the hypothesis. $\chi^2 = 4(2.1024) = 8.410$ cf $\chi^2(1)$; $\chi_{0.005}^2(1) = 7.88$; strong evidence against the hypothesis.

Source	Dimension	χ^2
Row probability	1	4.05
Column probability	1	1.00
Independence	1	8.41

Tests: Strong evidence against independence; no further tests available.

5.1 $(13 \pm 1.96\sqrt{.06}/\sqrt{5}) = (13 \pm 0.215)$. **5.2** $(13 \pm 2.78\sqrt{.05}/\sqrt{5}) =$

(13 ± 0.278). **5.6** $P(y_{(1)} < \theta < y_{(n)}) = 1 - 1/2^{n-1}$. **5.7** The interval $(y_{(1)}, y_{(n)})$ does cover θ. $P(y_{(1)} < \theta < y_{(n)} : y_{(n)} - y_{(1)} = r) = r/(1 - r)$ if $r < 1/2$ and 1 if $r \geq 1/2$.

CHAPTER 11: LINEAR MODELS

1.4 $\mathbf{d}' = (-0.033\dot{3}, 0.266\dot{6}, -0.233\dot{3})/0.355903 = (-0.09366, 0.74927, -0.65561)$; density for $u = (T - 0.65561)/1.40488$ is $2(1 + u)^{-3}$; central 95% interval for u is $(0.01274, 5.3246)$ and for $T = (0.67351, 8.13603)$; $(2.5333 - 8.13603 \times 0.355903, 2.5333 - 0.67351 \times 0.355903) = (-0.362, 2.294)$. **1.6** Density is $n(-nd_{(1)})^{n-1}\Gamma^{-1}(n-1) \exp\{-n\bar{z}\}s^{n-2}$ on $\bar{z} + sd_{(1)} > 0$ and zero elsewhere; density is $n(n-1)(-nd_{(1)})^{n-1}(nT)^{-n}$ on $(-d_{(1)}, \infty)$ and zero elsewhere. **1.8** The likelihood function is greater than 3 between $\lambda = 1$ and $\lambda = 5$ and is greater than 2 between $\lambda = 0.5$ and $\lambda = 9$. Ninety percent confidence intervals for μ are: $\lambda = 1$, $(1.029, 1.630)$; $\lambda = 3$, $(0.891, 1.813)$; $\lambda = 6$, $(0.857, 1.985)$; $\lambda = 9$, $(0.851, 2.073)$; and $\lambda = \infty$, $(0.817, 2.229)$.

2.1

$\mathcal{L}(\mathbf{x}_1)$	1	$0.8\mathbf{w}_1$	3.840
$\mathcal{L}(\mathbf{x}_2 \perp \mathbf{x}_1)$	1	$1.314\mathbf{w}_2$	0.394
$\mathcal{L}^{\perp}(\mathbf{x}_1, \mathbf{x}_2)$	4	$\mathbf{y} - 0.8\mathbf{w}_1 - 1.314\mathbf{w}_2$	0.126
	6	$\mathbf{y}$	4.360

where $\mathbf{w}_1 = (1, 1, 1, 1, 1, 1)'$, $\mathbf{w}_2 = \mathbf{x}_2 - 1.91667\mathbf{x}_1$.

2.2

$\mathcal{L}(\mathbf{x}_1)$	1	$12.760\mathbf{w}_1$	814.0880
$\mathcal{L}(\mathbf{x}_2 \perp \mathbf{x}_1)$	1	$2.730\mathbf{w}_2$	1.1028
$\mathcal{L}(\mathbf{x}_3 \perp \mathbf{x}_1, \mathbf{x}_2)$	1	$-1.002\mathbf{w}_3$	0.0037
$\mathcal{L}^{\perp}(\mathbf{x}_1, \mathbf{x}_2, \mathbf{x}_3)$	2	$\mathbf{y} - 12.760\mathbf{w}_1 - 2.730\mathbf{w}_2 + 1.002\mathbf{w}_3$	0.0055
	5	$\mathbf{y}$	815.2000

4.1 $20.5718\mathbf{v}_1 + 6.0083\mathbf{v}_2 - 0.80178\mathbf{v}_3$, $20.5718\mathbf{v}_1 + 6.0083\mathbf{v}_2$. **4.3** $F = 0.394/(0.126/4) = 12.51$ cf $F(1, 4)$; 5% and 1% points are 7.71, 21.20; significance at the 5% level (observed significance is 2.4%) $(1.314 \pm 2.776 \times .1775\sqrt{4.38}) = (1.314 \pm 1.031)$. **4.4** $F = 0.0037/(0.0055/2) = 1.34$ cf $F(1, 2)$; $F_{0.05}(1, 2) = 18.5$; no significance at the 5% level. $F = 1.1028/(0.0055/2) = 401$ cf $F(1, 2)$; $F_{0.005}(1, 2) = 199$; significance at the 1/2% level (observed significance is 0.25%) $(2.730 \pm 4.303(0.00275)^{1/2}(6.757)^{1/2}) = (2.730 \pm 0.587)$.

CHAPTER 12: THE DESIGN OF EXPERIMENTS

1.1

Mean	1	13,454			
T	2		458.85	229.425	$F = 5.18$
Residual	11		487.15	44.286	

$F_{0.05}(2, 11) = 3.98$, $F_{0.01}(2, 11) = 7.21$; significance at 5% level (observed significance is 2.6%). Evidence of dependence.

1.3

Mean	1	1628.176		
A	4	2.224	0.556	$F = 23.2$
Residual	5	0.120	0.024	

$F_{0.01}(4, 5) = 11.4$; significance at 1% level (observed significance is 0.2%). **1.4** Regression entries can be transferred from Problems 11.2.2 and 11.4.4 (multiply by 2 because MS's estimate variance for individuals, whereas former MS's estimate variance for an average of two individuals).

Mean	1	1628.1760	
A linear	1	2.2056	2.2056
quadratic	1	0.0074	0.0074
other	2	0.0110	0.0055
Residual	5	0.1200	0.024

Test for other *A* dependence $F = 0.0055/0.024 = 0.23$; no significance. Test for quadratic *A* dependence $F = 0.0074/0.024 = 0.31$, no significance. Test for linear *A* dependence $F = 2.2056/0.024 = 91.8$, $F_{0.01}(1, 5) = 16.3$; significance at 1% level (observed significance is 0.02%). $(2.73 \pm 2.57(0.024)^{1/2}(6.757/2)^{1/2}) = (2.73 \pm 0.732)$.

2.1

Mean	1	4446.7500		
T	2	72.0600	36.0300	$F = 1080$
P	1	39.6033	39.6033	$F = 1190$
$T \times P$	2	0.1867	0.0934	$F = 2.8$
Residual	6	0.2000	0.0333	

$F_{0.05}(2, 6) = 5.14$; no significance for $T \times P$ interaction (observed significance is 14%); $F_{0.01}(2, 6) = 10.9$, $F_{0.01}(1, 6) = 13.7$. Significance for *T* and for *P* at 1% level (observed significances are 0.0^52% and 0.0^54%).

2.2

T linear	1	72.0000	72.0000	$F = 2160$
other	1	0.0600	0.0600	$F = 1.8$

No significance for *T* other. Significance for *T* linear at 1% level (observed significance is 0.0^67%).

INDEX

absolutely continuous component, 52
absolutely continuous distribution, 53
  on R^k, 74
acceptance region, 413
acceptance sampling, 158, 175, 176
additive, 21
  finite, 28, 29
affine function, of normal variables, 234, 240
algebra
  of events, 9
  smallest, 19
  of subsets, 14, 515
almost surely, 34
alternative hypothesis, 412
analysis of covariance, 594
analysis of variance, 483
  analysis of covariance, 596
  one-factor design, 494
  randomized blocks, 591
  three-factor design, 588
  two-factor design, 504
angle, 522
angular transformation, 446
antecedent event, probability for, 157, 160
average, 18, 207
  distribution for normal, 323
  finite population
    mean, 289
    variance, 289
  limiting value, 563
  mean, 208
  variance, 208
average on R^k, 211
  mean, 211
  transformations, 216
  variance, 211

basis, 520
Bayes analysis, other uses, 576
Bayes formula, 161, 573
Bayes solution, 576
Bernoulli(p), 44
  probability-generating function, 237
  variance, 197
Bernoulli($p_1, \ldots, p_k$), 138
beta(p, q), 66
  generating functions, 239
  marginal from chi squares, 119
  mean, 205
  moments, 239
  variance, 205
beta function, 69
binary expansion, 517, 524
binary number, 33
  for events, 8
binomial(n, p), 25, 50
  characteristic function, 232
  completeness, 406
  confidence interval, 354
  convolution, 147, 240
  density function, 554
  derivation, 137
  distribution function, 28, 147
  factorial moments, 238
  generating functions, 238
  mean, 192
  mode, 147
  moment-generating function, 232
  normal limit, 253
  Poisson limit, 147, 245
  recursion relation, 147
  tests, 354
  variance, 202
binomial theorem, 132
birth-and-death process, 538
blocks, randomized, 590ff.
Boolean algebra, 9, 10
Borel–Cantelli lemma, 517, 561
Borel sets, 15, 16
Bose–Einstein, 135
Brownian process, 541

cardinality, 22
cardinality measure, 22
Cauchy
  characteristic function, 240
  convolution, 240
  IPIT, 95
Cauchy(μ, σ), 61
cause–effect, 274
central limit theorem, 259
  generalizations, 262, 263
  proof, 261
central moments, 226
change of variable
  density function, 83
    on R^k, 91
    on R^2, 87
  discrete, 81
  distribution function, 80
  probability differential, 84, 91
characteristic function, 231
characteristic function on R^k, 235
Chebyshev inequality, 189, 198, 223
chi square(1), 107
chi square(2), IPIT, 95
chi square(m), 66
  convolution, 118, 127, 240, 527
  generating functions, 239
  mean, 205
  moments, 239
  proportion, 127
  ratio, 127
  sum, 127
  tables, 601
  variance, 205
chi-square test, 443
  of normal, 59
  of Poisson, 48
combinatorial symbol, 130, 132
  generalized, 131, 133
combining models, 124
complement of event, 3
complement of unions, 18
complete statistic, 402
  example, 402
completeness, 402, 437
component, 522
composite hypothesis, 415
concomitant variable, 594
conditional distribution, for inference, 459
conditional mean, 217
conditional mean on R^k, 221
conditional probability, 155
  cautions, 161
  conditions for, 162
  independence, 159
  by probability density, 168
    general, 172
  by probability functions, 163
conditional sign symmetry, 296
conditional test, 437
conditional variance, 217
conditional variance on R^k, 221
confidence interval, 579
  acceptance region, 580
  binomial, 581
  component parameters, 583
  general formulation, 579
  large-sample, 353
  for location, 461
  for mean, 279, 461
    normal theory, 327, 330
    variance unknown, 283
  for mean response with regression model, 482
  for median, 295, 300, 461
  by pivotal quantity, 581
  power, 582
  for regression coefficient, 485
  for scale, 462
  for variance, 462
    normal theory, 328
  for variance ratio, normal theory, 584
confidence regions (*see* confidence interval)
continuity, point of, 247
continuity theorem, 248
  proof, 559
continuous component, 51
control, 491
control limits, 564
convergence
  almost surely, 556
    criteria, 557, 558
  in distribution, 246, 555
  of functions, 517
  in mean(r), 556
  in probability, 555
convergence to a constant, 251
convolution, of normal, 234
convolution formula
  absolutely continuous, 115
  difference, 118
  discrete, 110
  linear combination, 119
  proportion, 119
  ratio, 119
correlation, 203
  sample, 213
correlation inequality, sample, 214
count data, 136ff., 442ff.
counting measure, 554
counting sequences, 128, 129
counting sequences of sets, 131
counting sets, 130
covariance, 198
  linear combinations, 200
  sample, 212
    convergence, 264
covariance inequality, 203
  sample, 214
Cramér–Rao inequality, 389

critical region, 413
cumulants, 241

data reduction, 98
decision function, 413
decision space, 413
decision theory, 413ff., 577ff.
decisions, 284
degrees of freedom, 483
De Morgan's laws, 18
density function, of a measure, 549
dependence
  additive, 501
  interactive, 501
  linear, 520
descending factorial, 129, 132
design factors, 491
deviation vector, 364
difference of means, 285, 331, 332
  nonparametric, 303ff.
  normal
    confidence interval, 332
    test, 331
diffuse prior, 577
Dirichlet($r_1, \ldots, r_{k+1}$), 79
  conditional mean, 223
  conditional variance, 223
  conditionals, 175
  marginals, 118
  mean, 206
  variance, 206
discrete component, 43
discrete distribution, 24, 44
  on R^k, 73
discrete function, 519
disjoint events, 3
distribution, 43
distribution-free methods, 292ff.
distribution function, 40
  properties, 43
  on R^k, 70
    properties, 73
distribution generating function, 240
distributions
  absolutely continuous, 52
    on R^k, 74
  beta(p, q), 66
  binomial(n, p), 50
  Cauchy(μ, σ), 61
  chi-square, 66
  Dirichlet($r_1, \ldots, r_{k+1}$), 79
  discrete, 44
    on R^k, 73
  double exponential, 62
  exponential(θ), 60
  extreme value, 64
  $F(m, n)$, 66, 67
  gamma(p), 65
  geometric(p), 49
  hypergeometric(N, n, D/N), 51
  Laplace, 62
  logistic, 63
  lognormal, 63
  negative binomial(r, p), 50
  normal($\mu_1, \mu_2, \sigma_1, \sigma_2, \rho$), 95
  normal(μ, σ), 55
  normal($\boldsymbol{\mu}$; Σ), 93
  normal on R^k, 76
  Pareto(α), 62
    on R^k, 79
  Student(n), 67
    on R^k, 79
  triangular, 60
  uniform(a, b), 53
  uniform$\{1, \ldots, N\}$, 49
  Weibull(β), 63
dominated convergence theorem, 551
dominated derivative theorem, 551
double exponential, 62

efficiency, of unbiased estimators, 381
error distribution (*see* variation)
errors, 414
estimable parameter, 401
estimation, 362ff., 456ff.
  Bayes, 576
  conditional, 456ff.
  Gauss–Markov, 385, 389
  information inequality, 392
  large-sample, 394
  least-squares, 363, 370
  maximum-likelihood, 367
  for mean, 276
  method of moments, 366
  methods of combining, 379ff.
  sufficiency and completeness, 398ff.
  for variance, 281
event, 2
  for a function, 100, 104
  as a subset, 12
events, algebra of, 9
exact size α, 415, 418
expectation (*see* mean value)
experimental design, 273
experimental units, 271, 491
exponential form, 342
  completeness, 406
  likelihood statistic, 342
  location model, 345
  location scale model, 345
  reduced, 342
  sampling, 343
exponential(1)
  generating functions, 239
  mean, 205
  moments, 239
  sample property, 92
  variance, 205

exponential(θ), 60
  completeness, 405
  IPIT, 95
extension theorem, 515
extreme value, 64
  IPIT, 95
  transformation of, 94

$F(m, n)$
  canonical, 66
  generating functions, 239
  marginal from chi squares, 119
  mean, 205
  moments, 239
  from samples, 325
  standard, 67
  tables, 602, 603, 604
  variance, 205
F-statistic distribution
  normal theory, 528
    noncentral, 529
factorial design, 490
factorial function, 25
factorial moments, 228
factorization theorem, 337
factors, 273
Fatou lemma, 552
Fermi–Dirac, 135
finite populations, 286
Fisher information, 351, 356
Fisher test, 298
fitted response, 364
fitting affine function, 472
fitting quadratic function, 473
frequency counts (*see also* binomial; hypergeometric; multinomial)
  general, 143
    double counts, 148
Fubini theorem, 552
fundamental lemma, 416, 418, 420
  monotone likelihood, 425, 426
  unbiased tests, 435, 437

gamma function, 69
gamma(1), 85
gamma(p), 65
  convolution, 240
  generating functions, 239
  mean, 205
  moments, 239
  variance, 205
Gauss–Markov, 384, 385, 389
Gaussian process, 541
generating function cumulants, 241
geometric(p), 49
  factorial moments, 238
  generating functions, 238
  mean, 204
  variance, 204

hypergeometric(N, n, p), 51
  derivation, 140
  factorial moments, 238
  generating functions, 238
  mean, 204
  variance, 204
hypergeometric(N, n, p_1, . . ., p_k)
  completeness, 406
  conditionals, 174
  derivation, 142
  factorial moments, 239
  marginals, 147
  mean, 205
  variance, 205

IID, 125
independence
  conditional density function, 169
  by conditional probability, 155
  conditional probability function, 166
  by distribution function, 123, 127, 526
  normal(μ_1, μ_2, σ_1, σ_2, ρ), 531
  normal($\boldsymbol{\mu}$; Σ), 531
  by probability density, 124, 127
  by probability function, 123, 127
independent algebras, 121
independent events, 120
independent functions, 122
independent random variables, 122
indicator function, 4, 8, 16, 511
  positive axis, 302
  unit interval, 334
inference, 269ff. (*see also* confidence interval; estimation; tests)
  location-scale, 460
  mean, 276ff.
    finite population, 287
  median, 294ff.
  normal, 326
  regression coefficients, 480
information, 351, 390
  combining, 397
information inequality, 392, 396
  vector, 396
inner product, 520
inner products of residuals, 469
insufficient reason, 577
integral
  absolute value, 551
  linearity, 550
  nonnegative function, 547
  real-valued function, 549
interaction, 503
  contingency tables, 449
intersection of events, 3
inverse probability integral transformation, 87
inversion formula, 546
IPIT, 87

Jacobian, 88, 91
Jensen inequality, 286

k statistics, 241
Kolmogorov inequality, 223
Kolmogorov lemma, 561
Kolmogorov theorem, 562

Laplace, 62
  generating functions, 239
  mean, 205
  moments, 239
  variance, 205
large samples
  confidence intervals, 280, 284, 353, 357
  tests, 277, 282, 353, 357, 434
law of large numbers
  strong, 560
  weak, 256
least favorable distribution, 423
least favorable value, 421
least squares, 363
  fit of affine function, 378, 472
  fit of quadratic function, 473
  weighted, 378
least-squares equations, 373
least-squares estimate, 365
least-squares solution, 378
Lehmann–Scheffé, 402
likelihood
  independence, 336
  large samples, 348
likelihood function, 309, 312, 358
  assessment, 310
  Bernoulli, 315
  large-sample, distribution of, 352, 356
  normal, 314, 316
    distribution of, 349
  for shape parameter, 461, 480
  from sufficient statistic, 338
likelihood principle, 311
likelihood ratio, 313, 416, 418
  generalized, 432
    large samples, 433
likelihood statistic, 332, 334, 398
  Bernoulli, 334
  fixed dimension, 344
  hypothesis testing, 428
  normal, 334
  uniform, 334
liminf, 517, 537
limit, of functions, 517
limiting distribution, 246
  by characteristic function, 248
  by density function, 247
  by moment-generating function, 248
limiting function, 555
limsup, 517, 537
linear function, of normal variables, 234, 240
linear model (*see also* regression model)
  for location, 372
  maximum likelihood, 388
  normal
    completeness, 407
    likelihood statistic, 347
linear regression, 224
linear subspace, 520
linearly dependent, 520
linearly independent, 520
locally most powerful, 427
locally unbiased, 391, 396
locally unbiased test, 435
location
  confidence interval for, 461
  inference for, 461
  test of, 461
location-scale model, 456
log-likelihood function, 312
logistic, 63, 81
  generating functions, 239
  mean, 205
  moments, 239
  variance, 205
lognormal, 63
  mean, 205
  moments, 239
  variance, 205
loss function, 414

Mann–Whitney confidence interval, 304
Mann–Whitney test, 303
marginal distribution, 99
  by characteristic function, 236
  by conditional probability, 159
  by distribution function, 105, 106
  for frequencies, 136
  for general frequencies, 143
  by measure, 101
  by moment-generating function, 236
  by moment sequence, 236
  by probability density, 111, 112, 115
  by probability function, 109
marginal model, 99, 102
marginal probability, 101 (*see also* marginal distribution)
Markov process, 539
Martingale, 421
matching problem, 144, 514
maximum likelihood
  least absolute durations, 370
  least squares, 368
maximum likelihood estimate, 315, 367
maximum likelihood method, 367
Maxwell–Boltzmann, 135
mean, 186

mean (*continued*)
confidence interval for, 279, 283, 289, 295, 300 (*see also* large samples; location; mean of normal)
derivative of, 390, 396
on R^k, 190
symmetry, 187
test of, 277, 282, 289, 294, 298 (*see also* large samples; location; mean of normal)
mean of normal
confidence interval, 327, 330
test of, 326, 329
mean value
additivity, 513
bounded function, 182
discrete function, 519
general function, 183, 184
properties, 185
simple function, 181, 512
in terms of distribution function, 554
measurable function, 101, 104, 534
criterion for, 536
measurable space, 14
measure, 21
discrete, 22, 24
finite, 23
probability, 23
σ finite, 23
symmetric, 22
measurement
direct, 362
indirect, 362
wavelength, 407
measuring, 19
measuring area, 516
measuring length, 516
median
efficiency, 381
limiting distribution, 254
median test, 303
confidence interval, 303
minimal sufficient statistic, 338
minimum variance, 382, 384, 385, 389, 393, 396, 402, 404
MLE, 315, 367
moment-generating function, 231
properties of, 233
on R^k, 235
moments, 224
method of, 366
moments of sample, convergence, 264
monotone convergence theorem, 548
monotone likelihood ratio, 424
Mood-Westenberg test, 303
multinomial($n, p_1, \ldots, p_k$), 138
completeness, 406
conditional mean, 220
conditional variance, 220
conditionals, 166, 174
derivation, 138
factorial moments, 239
generating functions, 239
limiting distribution, 264
marginals, 147
mean, 205
Poisson limit, 147
variance, 205
multinomial theorem, 133
multiplication for probabilities, 158

negative binomial(k, p), 50
derivation, 147
factorial moments, 238
generating functions, 238
mean, 204
variance, 204
Newton–Raphson, 394
Neyman–Pearson lemma (*see* fundamental lemma)
noncentral chi square(m), 529
noncentral $F(m, n)$, 529
noncentral normal, 529
noncentral Student(n), 529
nonnormal variation, 457
nonparametric inference, 292ff.
normal
conditionals, 172
distribution theory, 319
orthogonal transformation, 321
tables, 599
normal(μ, σ), 55
central moments, 227
characteristic function, 234
completeness, 406
convolution, 127, 234, 240
mean, 188
moment-generating function, 234
variance, 197
normal(μ, σ_0), completeness, 403
normal($\boldsymbol{\mu}; \Sigma$), 93
completeness, 407
conditional, 177
conditional mean, 222
conditional variance, 222
convolution, 240
likelihood statistic, 347
marginals, 177
mean, 192
moment-generating function, 235
transformation, 240
variance matrix, 202
normal(μ_0, σ), completeness, 403
normal($\mu_1, \mu_2, \sigma_1, \sigma_2, \rho$), 95
conditional mean, 220
conditionals, 177
marginals, 177
mean, 206
variance, 206

normal(0, 1), 55
  square of, 107, 116, 240
normal data, 57
normal on R^k, 77
normal process, 541
null hypothesis, 412

one-factor design, 491
operating characteristic, 414
order statistic, 18
order-statistic distributions, 532, 533
orthogonal, 522
orthogonal components, 322, 365, 375, 376
orthogonal transformation, 320
orthonormal basis, 321, 474

parameter change, information inequality, 397
Pareto(α), 62
  generating functions, 239
  IPIT, 95
  mean, 205
  moments, 239
  on R^k, 79
    conditionals, 175
    marginals, 118
  transformation of, 94
  variance, 205
Parseval relation, 545
partition, 20
  of events, 3
  of a function, 103
percentile, limiting distribution, 254
personal prior, 577
PIT, 86
Pitman test, 304
pivotal quantity, 581
Poisson(λ), 2, 5, 45
  completeness, 403
  conditionals, 174
  convolution, 164, 240
  density function, 554
  distribution function, 28, 147
  factorial moments, 229, 238
  generating functions, 238
  limiting distribution, 453
  mean, 188, 204, 519
  mode, 147
  normal limit, 249, 250, 253
  recursion relation, 147
  square root, 250
  variance, 197, 204
Poisson data, 46
Poisson process, 538, 540
Polya urn, 542
positive definite, 522
  greater than, 523
posterior distribution, 573
power function, 413
power set, 14
precision, 574
preimage, 100
prior distribution, 572
probability
  a posteriori, 156
  a priori, 156
  at least one event, 36, 143, 514
  conditional, 155
  empirical example, 4, 6, 47, 58
  marginal, 101
  monotone property, 31
  r events, 143, 514
  r or more events, 515
probability density function, 52
  properties, 54
  on R^k, 74
    properties, 75
probability differential, 56
probability function, 43
  properties, 45
  on R^k, 73
    properties, 74
probability-generating function, 236
probability integral transformation, 86
probability measure
  alternative definition, 33
  definition, 23, 33
  symmetric, 26
probability model, 26
probability of a limit, 32
probability of complement, 29
probability of union, 30
probability space, 27
product algebra, 124
product measure, 124
product space, 124, 534
projection, 468, 522
  by differencing, 469
  one-factor design, 493
  into orthogonal complement, 469
  three-factor design, 585
  two-factor design, 503
projection matrix, 374
proportions, limiting value, 563

quadratic form, 522
quality control, 564
quantitative factor
  one-factor design, 495
  two-factor design, 507

random bits, 49
random digits, 49
random variable, 96, 102
randomization, 158, 274
randomized blocks, 590
randomized test function (*see* test function)

rank function, 302
rank statistic, 117
Rao–Blackwell, 399, 401
ratio of variances, normal
  confidence interval, 332
  tests of, 332
reciprocal variance, as weights, 574
reduction, 98, 99, 100
regression coefficients, 374, 470
  distribution theory, 480
regression model, 476
  conditional distribution, 479
  distribution theory, 479, 481
  inference, 482
regression of y on x, 218
replication, 274
residual vector, 374
ring, 515
risk, 414
run test, 567
  distribution, 569
Rutherford and Geiger, 46

σ algebra (*see* algebra)
sample: Bernoulli(p), 125
sample: normal(μ, σ), 126
sample: Poisson(λ), 126
sample from distribution, 125
sample point, 11
sample space, 11
scale
  confidence interval for, 462
  inference for, 462
  test of, 462
Schwartz inequality, 554
score equation, 367
score function, 312
  combining, 397
  local tests, 428
  normal, distribution of, 350
  properties, 351, 390, 396
selection, 153
sequential analysis, 438
sequential test, 439
  critical values, 440
  optimality, 440
serial system, 31
sign confidence interval, 295
sign function, 293
sign test, 294
signed deviation confidence interval, 299
signed deviation test, 298
signed rank confidence interval, 302
signed rank test, 302
similar test, 432, 437
simple function, 511
simple hypothesis, 415
simple random sampling, 287
singular component, 523, 524
size α, 415
SLLN, 560
smallest ring, 515
squared length
  by differencing, 469
  of projection, 374
  of residual, 374
standard deviation (*see also* variance)
  estimator, 406, 566
standard deviation of sample (*see* variance of sample)
state space, 537
statistical model, 271
  parametric, 308
statistically independent (*see* independence)
Stirling's formula, 133, 136
stochastic process, 537
stratified sample, 291
strong law of large numbers, 560
Student(n)
  canonical, 67
  limiting distribution, 247
  marginal from (n, χ), 119
  mean, 205
  moments, 240
  on R^k, 79
    conditionals, 175
    marginals, 118
  standard, 67
  tables, 600
  variance, 205
sufficient statistic, 337
  fixed dimension, 344
  hypothesis testing, 428
sums of squares, 483
symmetric distribution, 26 (*see also* distributions, uniform)
symmetry, use of, 4, 7, 26, 288ff., 292ff., 491, 501, 590ff.

t-statistic, 18
t-statistic distribution
  general theory, 461
  normal theory, 466, 527
    noncentral, 529
t-test, 438
ternary expansion, 524
test function, 418
tests, 412ff.
  of fit, 442
  of independence
    contingency tables, 449
  large sample, 353
  of location, 461
  of mean, 277
    finite population, 289
    variance unknown, 282
  of median, 294, 298
  of probability model, 445

of regression coefficients, 485
of scale, 462
of statistical model, fitted parameters, 448
of variance, 328, 484
of variance ratio, 332
three-factor design, 584
2^3 design, 588
transition matrix, 540
treatments, 491
triangular distribution, 60
two-factor design, 501
two-sample problems, 285
Type I error, 414
Type II error, 414

UMP, 417, 423, 426
UMP similar, 432
UMP unbiased, 431
UMV unbiased, 402
by completeness, 402, 405
examples, 404
by information, 393, 396
method, 395, 396
unbiased estimator, 276, 379, 380
better than, 382
combining, 382, 384, 388
comparison, 380, 382
efficiency, 381
exponential models, 395, 396
Gauss–Markov, 384, 385
general Gauss–Markov, 389
large sample, 394
likelihood statistic, 399, 401
local, 391, 393, 394, 396
minimum variance, 401
uniqueness, 402
sufficient statistic, 399, 401
variance reduction, 399, 401
unbiased test, 431, 435
uniform(a, b), 53
mean, 187
variance, 196
uniform$\{1, \ldots, N\}$, 49
factorial moments, 238
generating functions, 238
mean, 204
variance, 204
uniform(0, θ), completeness, 403
uniformity trials, 590
uniformly most powerful (*see* UMP)
union of complements, 18
union of events, 3
uniqueness theorem, 232, 544, 545, 547
characteristic function, 232, 546
moment-generating function, 232, 546, 547
moment sequence, 546, 547
unknowns of a random system, 270

variable
input, 1
response, 1
variance, 194
linear combination, 200
transformations, 195
zero, 198
variance about regression, 219
variance matrix, 199
formulas, 199, 200
variance matrix of sample, 213
transformations, 216
variance of normal
confidence interval, 328
test of, 328
variance of sample, 18, 58, 210, 273, 281
convergence, 264
distribution for normal, 323
finite population, mean, 289
mean, 210
unbiasedness, 380
variance, 210
variance ratio, 325
variance ratio test, 430
power function, 430
variation, 1, 456, 476
Student distribution for, 463
vector algebra, 520

Wald–Wolfowitz, 440
weak law of large numbers, 256
Weibull(β), 63
IPIT, 95
mean, 206
moments, 240
transformation of, 94
variance, 206
Wilcoxon test, 302, 304
WLLN, 256
alternative proof, 265
proof, 263